AF772252

MODERN MACHINING TECHNOLOGY

Advanced, Hybrid, Micro Machining and Super Finishing Technology

MODERN MACHINING TECHNOLOGY

Advanced, Hybrid, Micro Machining and Super Finishing Technology

BIJOY BHATTACHARYYA
BISWANATH DOLOI

ELSEVIER

ACADEMIC PRESS
An imprint of Elsevier

Academic Press is an imprint of Elsevier
125 London Wall, London EC2Y 5AS, United Kingdom
525 B Street, Suite 1650, San Diego, CA 92101, United States
50 Hampshire Street, 5th Floor, Cambridge, MA 02139, United States
The Boulevard, Langford Lane, Kidlington, Oxford OX5 1GB, United Kingdom

© 2020 Elsevier Inc. All rights reserved.

No part of this publication may be reproduced or transmitted in any form or by any means, electronic or mechanical, including photocopying, recording, or any information storage and retrieval system, without permission in writing from the publisher. Details on how to seek permission, further information about the Publisher's permissions policies and our arrangements with organizations such as the Copyright Clearance Center and the Copyright Licensing Agency, can be found at our website: www.elsevier.com/permissions.

This book and the individual contributions contained in it are protected under copyright by the Publisher (other than as may be noted herein).

Notices

Knowledge and best practice in this field are constantly changing. As new research and experience broaden our understanding, changes in research methods, professional practices, or medical treatment may become necessary.

Practitioners and researchers must always rely on their own experience and knowledge in evaluating and using any information, methods, compounds, or experiments described herein. In using such information or methods they should be mindful of their own safety and the safety of others, including parties for whom they have a professional responsibility.

To the fullest extent of the law, neither the Publisher nor the authors, contributors, or editors, assume any liability for any injury and/or damage to persons or property as a matter of products liability, negligence or otherwise, or from any use or operation of any methods, products, instructions, or ideas contained in the material herein.

Library of Congress Cataloging-in-Publication Data
A catalog record for this book is available from the Library of Congress

British Library Cataloguing-in-Publication Data
A catalogue record for this book is available from the British Library

ISBN 978-0-12-812894-7

For information on all Academic Press publications
visit our website at https://www.elsevier.com/books-and-journals

Publisher: Matthew Deans
Acquisition Editor: Brian Guerin
Editorial Project Manager: Isabella C. Silva
Production Project Manager: R. Vijay Bharath
Cover Designer: Miles Hitchen

Typeset by SPi Global, India

Working together to grow libraries in developing countries
ELSEVIER Book Aid International
www.elsevier.com • www.bookaid.org

Contents

Preface

With the advancement of the technology, complex shaped components with high-accuracy are increasingly in demand for various industries. Controlling of machining parameters for the specific industrial applications urgently need detail studies in the area of modern machining technologies. Some of materials are being machined neither by traditional nor by any non traditional machining processes solely, effectively and economically. Hence, hybrid machining processes (HMP) are being developed presently. Since miniaturization will continue as long as people require effective space, material and energy utilization with more efficient and better accuracy products, micromachining process will be more important. Recent development in the area of super finishing technologies should be known to achieve the desired level of finish.

There is dearth of books in the area of modern machining technologies, which includes all gamut of development in the study of modern machining and finishing processes. Therefore, authors have attempted to develop this book considering their vast research and teaching experience in these areas, as well as reviewing the knowledge based information from the several research findings reported in various reports and journal papers. The book on modern machining technology includes the introduction to advance machining as well as classifications. It highlights on the different advanced, hybrid and micro machining processes as well as super finishing techniques considering their material removal mechanisms, influence of different parameters as well as control strategies. Discussion is made in depth to each machining technology highlighting their present status, recent advancement, set-up details, advantages and limitations. The book contains eight chapters. Each chapter of the present book also includes important model questions for understanding of the fundamentals and related research issues on modern machining technologies.

This book highlights different needs of modern machining technology and finishing methods in the field of advanced manufacturing. The chronological development of machining starting from the prehistoric age up to the present days has been reported in Chapter 1. Detail classifications of modern machining processes based on different criteria which influence the processes and its outcomes are described in Chapter 2. It is essential to identify the correct nature and capabilities of various modern machining processes for efficient use of modern machining processes. Issues which must be analyzed in depth for the selection of modern machining processes have also been highlighted considering large number of quantitative as well as qualitative considerations.

Detail discussion on various advanced machining processes utilizing mechanical energy namely, Ultrasonic Machining (USM), Rotary Ultrasonic Machining (RUM), Abrasive Jet Machining (AJM), Water Jet Machining (WJM) and Ion Beam Machining (IBM) have been included in Chapter 3. The fundamentals, material removal mechanism and related modeling, setup details, effects of process parameters on different outcomes, advantages and limitations of these processes have been presented in this chapter. Recent advancements considering the latest research findings of all these advanced machining processes have also been highlighted.

Over the years different forms of thermal energy, e.g., electrical spark, plasma arc, laser beam and electron beam etc. are used for machining advanced materials. Some of the modern machining processes, e.g., Electro-Discharge Machining (EDM), Wire Electro Discharge Machining (WEDM), Plasma Arc Machining (PAM), Laser Beam Machining (LBM) and Electron Beam Machining (EBM) uses different forms of thermal energy and are become popular now-a-days due to their inherent advantages over conventional machining processes. Chapter 4 includes the material removal mechanism, setup details, various modeling, parametric influences, advantages and limitations, applications and recent advancement of these thermal energy based modern machining processes.

Chapter 5 describes the modern machining processes utilizing chemical and electrochemical energy namely Chemical Machining (CM) and Electrochemical Machining (ECM). Chemical Machining is discussed in detail where material removal is achieved by chemical reactions which are extensively employed to produce the precise components for different industrial applications such as semiconductor industries and MEMS. Electrochemical Chemical Machining (ECM) is an anodic dissolution process which has tremendous potential on account of versatility of applications and expected to be one of the most promising, successful and commercially utilized machining processes in the modern manufacturing industries. Different variants of ECM, e.g., EC Drilling, Deburring, EC Milling, Wire ECM, EC Sawing, EC Grinding, EC Honing and EC Turning as well as environmental impacts of ECM are also discussed in details in this chapter.

Chapter 6 includes the detailed discussion on compound hybrid machining processes such as Electrochemical Grinding (ECG), Electrochemical Arc Machining (ECAM), Electrochemical Discharge Machining (ECDM). The fundamental principles, applications and recent advancements of various Abrasive Assisted Advanced Machining (AAAM) processes such as abrasive water jet machining, abrasive assisted EDM, abrasive assisted ECM are also discussed. Ultrasonic Assisted Advanced Machining (UAMM) processes are also discussed. Laser assisted conventional machining as well as laser assisted nonconventional machining such as laser assisted EDM, laser assisted CM, laser assisted ECM, Laser assisted oxide film lithography have been elaborated in this chapter.

Chapter 7 deals with various modern micro machining methods which utilize energies, e.g., mechanical, thermal, chemical and electrochemical micromachining. Detail discussion on micro-USM, micro-EDM, micro LBM, micro ECM have been highlighted. It also covers different hybrid micromachining processes which combine two or more micro machining processes for achieving advantages of individual processes simultaneously. Latest development in the field of micro machining has also been included considering various sequential micro machining (SMM) processes.

For achieving damage free, nano-level or angstrom level surface finish on difficult to finish materials such as ceramics, glass, silicon wafer, diamond wafer, sapphire wafer etc., and also for finishing complex and free form surfaces with high accuracy, the advanced super finishing processes are very much needed. Various advanced finishing processes, namely Abrasive Flow Finishing (AFF), Chemical Mechanical Polishing (CMP), Elastic Emission Machining (EEM), Magnetic Abrasive Finishing (MAF), Magnetorheological Finishing (MRF) and Plasma Assisted Polishing (PAP) process have been discussed in Chapter 8.

No recent book is available in the area of modern machining technologies, which includes all topics of modern machining as well as finishing processes. Some websites as well as other sources are available in the internets which are not contained all the areas of machining and latest developments in the areas of advanced machining, hybrid machining, micromachining as well as super finishing processes. Hence, the present book will provide information and knowledge in the area of modern machining technology which are not available presently in a single book. This book will be most useful to undergraduate and postgraduate students, researchers, scientists, as well as application engineers for efficient use of knowledge in the various specialization, e.g., Mechanical, Manufacturing, Production and Industrial Engineering etc.

Finally significant suggestions from the readers are sought forward to improve quality of the book in future.

BIJOY BHATTACHARYYA AND BISWANATH DOLOI
Production Engineering Department, Jadavpur University, Kolkata, India

Acknowledgment

This book has become a reality due to the constant inspirations and encouragement received from the senior professors such as Late Professor S.K. Sorkhel of Jadavpur University, Prof. S.K. Basu of COE, Pune, Prof. V. Radhakrishnan of IIT Madras, Prof. V.K. Jain of IIT Kanpur, Prof. P.K. Mishra of IIT Kharagpur, and also from professors and scientists who have lent their support like Dr. R. Balasubramaniam of BARC, Mumbai, Prof. P.K. Jain of IIT Roorkee and Prof. U.S. Dixit of IIT Guwahati, etc.

Considerable amount of our research work presented in this book was accomplished by collaborative team works by colleagues, such as Dr. S. Chakraborty, Dr. S. Mitra, Dr. S. Sarkar, Dr. D. Banerjee, Dr. A.S. Kuar, and Dr. B.R. Sarkar of Production Engineering Department, JU as well as researchers, such as Dr. A. B. Puri, Dr. A. Manna, Dr. D. Dhupal, Dr. B.B. Pradhan, Dr. J. Munda, Dr. M. Malapati, Dr. S. Dhobe, Dr. Golam Kibria, Dr. Josephine Peter, Dr. H.C. Vela, Dr. Somanth Das, Dr. V. U Rathod, Dr. Bikash Ghosal, Dr. S.S. Anasane and others.

The book could not be completed without the active participation of research scholars and others who put their efforts at the best for carrying out research in this field as well as preparing the manuscript, figures, and other details. In this respect, Mr. Koushik Mishra, Mr. Santosh Kumar, Mr. Subhrajit Debnath, Mr. Subrata Mahata, Mr. Sandip Kunar, Mr. R.M. Tayade, Mr. Abhishek Sen, Mr. Mohit Pandey, Mr. S. Chaudhury and others of Non Traditional Machining Laboratory of the Production Engineering Department, JU are to be mentioned.

The authors also acknowledge the wholehearted support and encouragement of their family members especially Dr. Rita Ganguly, Ms. Jyotiprana Bhattacharya, Mrs. Nilima Doloi, Mr. Sayan Doloi during the writing and preparation of the manuscript of this book.

Financial support from the University Grants Commission (UGC), All India Council for Technical Education (AICTE), Department of Science and Technology (DST), Council of Scientific Industrial Research (CSIR), Bhabha Atomic Research Centre (BARC) and Aeronautics Research and Development Board (AR&DB) as well as some industries for carrying out research in this area have proved to be useful for utilizing research outcomes to enrich this book.

The team members of Elsevier like Publisher: Matthew Deans, Acquisition Editor: Brian Guerin, Editorial Project Manager: Isabella C. Silva, Production Project Manager: R. Vijay Bharath have put their constant effort in transforming this book into its final shape.

CHAPTER ONE

Need of advanced machining technology

Contents

1.1 Chronological development of machining

Human species is most intelligent among all other species and used to live in society by improving life trouble-free and pleasant by utilizing tools and expertise. In primitive age, initially wood, bone and stone were consecutively used as tools by human beings for shaping materials to accomplish their urgent needs of daily life. Subsequently, for shaping stronger metals to fabricated sophisticated items, iron tools were invented and evolution in machining technology started from those early days. It was in about 4000 BC, ancient Egypt used drilling and cutting tools for machining and since then its development continued. Tools were powered by physical labor, either by animal or human. Basic mechanically driven tools were widely used in the 17th century. More sophisticated tools powered by water, wind, steam and electricity were subsequently used in the 18th and 19th centuries.

It was only in the 20th century that machining was performed by different machine tools for removal of specified amount of metal from the workpiece utilizing wedge shaped cutting tools in order to produce a required shape with greater accuracy and faster machining rate. Machining accuracy as well as economy mainly depends on the type of tools, tool materials, energy source and kinematic structure of machine tools. With the advancement of civilization, needs of more sophisticated items were increased. Hence, to fulfill the societal needs newer challenges of machining were

Modern Machining Technology
https://doi.org/10.1016/B978-0-12-812894-7.00001-3
© 2020 Elsevier Inc.
All rights reserved.

encountered and resolved by utilizing newer tool materials with newer machining strategies. These advancements in machining process in terms of accuracy are exhibited through Fig. 1.1 [1]. To machine newly developed advanced materials with higher degree of accuracy and economy, manufacturing scientists developed newer strategies for machining through the development of various nontraditional machining processes. These machining are known as nontraditional or modern machining methods because traditional cutting tool are not utilized during machining where energy in its direct form is employed.

At the same time, development of control system for machine tools had also taken place to carry out much precision machining operations. In the 1950s, numerical control (NC) came into picture in the field of manufacturing technology and opened the way for computer numerical control (CNC) and direct numerical control (DNC) machining. Further enhancement in machining accuracy, repeatability, and flexibility in manufacturing process has been made possible with the advent of robotics in 1980s. Between the 1980s and the 1990s, revolution in the computer and electronics industries accelerated the improvement in machining accuracy. The demand for components with dimensions of the order of micrometers increased in the 1990s. Taniguchi and then McKeown with their co-researchers have considered equipment to achieve very fine shape and accuracy over the 20th century [2].

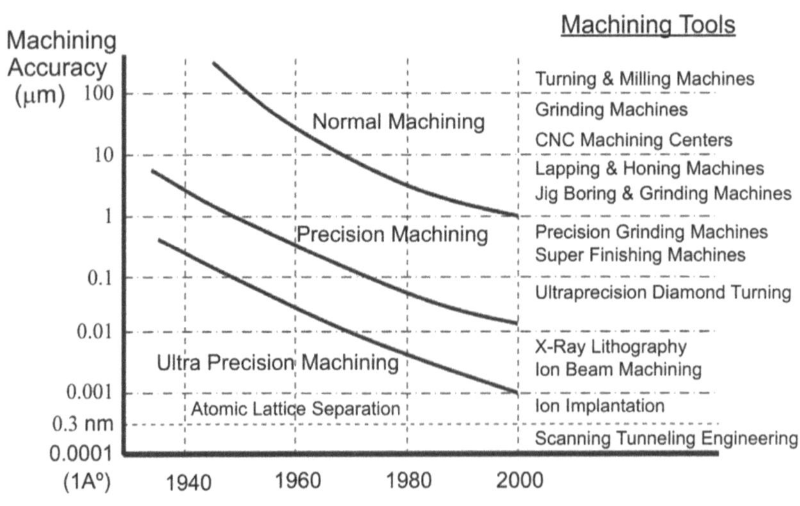

Fig. 1.1 Advancement of machining process in terms of accuracies [1].

To overcome the limitations of advanced machining processes as developed along the period, it was attempted to combine different machining mechanisms, thus several hybrid machining processes come into the modern machining scenario. Already established and recently developed methods of machining continue to be investigated for shaping of micro parts to specified small dimensions. Further development of various micromachining technologies took place to produce micro parts to enhance product functionality in limited space and save material and energy. For improvement of the machined surface condition, further improvement was necessitated in the area of finishing and subsequently several super finishing methods have been emerged.

1.2 Utility of advanced machining technology

With the advancement of material science and engineering different types of advanced engineering materials such as high strength and temperature resistant (HSTR) and exotic super alloys have been synthesized and manufactured which are commercially available and extensively used in the field of aerospace, aeronautics, defense, biomedical engineering, electronics industries and metal cutting industries. The greatly improved thermal, chemical and mechanical properties of material yielded enormous economic benefits to manufacturing industries through improved product performance. However complicated product design and closed tolerance are restricting traditional machining processes to machine them economically. This is because traditional machining is most often based on removing materials using tools harder than the workpieces. In addition to these, more and more complex shapes, low rigidity structure and mechanical components with tight tolerances and better surface finish were often needed for the above mentioned advanced materials. Traditional machining techniques are uneconomical and sometime impossible for those materials. Degree of accuracy and achievable surface finish of the products machined by traditional methods are not up to the mark.

Under these circumstances, to machine those advanced materials with higher degree of accuracy and economy, manufacturing scientists developed newer strategies for machining through the development of various advanced machining processes without employing traditional wedge shaped tools. The major reasons for the usefulness of the advanced machining process may be identified as follows [3]:

(i) Machinability factor of job material,
(ii) Job shape complexity,

(iii) Automation of data exchanges,
(iv) Precision requirements,
(v) Machined surface integrity, and
(vi) Requirements of miniaturization.

(i) Machinability factors of job material—Machinability of the materials indicate how easily it could be machined and characterized by three major outcomes of machining, e.g., cutting force, tool wear and surface roughness. Materials, e.g., high strength and temperature resistance alloy, super alloy with high strength to weight ratio, various ceramics and composites are very difficult to machine and encountered with high cutting forces, high tool wear and poor surface finish. As discussed above, materials with lower machinability are frequently used due to its better engineering properties and more difficult to machine by traditional tool based machining techniques. However, modern machining methods can be employed to machine those above mentioned materials easily and economically as these methods are independent of the machinability factors of the job materials. Rather these methods are influenced by job material properties, e.g., melting point, thermal conductivity, electrical conductivity, electrochemical equivalent, chemical inertness, etc. [4].

(ii) Job shape complexity—Machining of complex profiles having geometrical restrictions, critical design requirements, poor machining accessibility are difficult to carryout utilizing traditional tool based machining techniques. Hence, to overcome the shape complexity of the job, utility of modern machining methods where energy is utilized instead of cutting tool are unavoidable. Some typical examples of shape complexity are holes with square or any other shape, curved holes, free form surfaces, e.g., turbine blade; narrow channels, grooves with high aspect ratio, etc. as shown in Fig. 1.2, which are impossible to machine utilizing traditional machining methods.

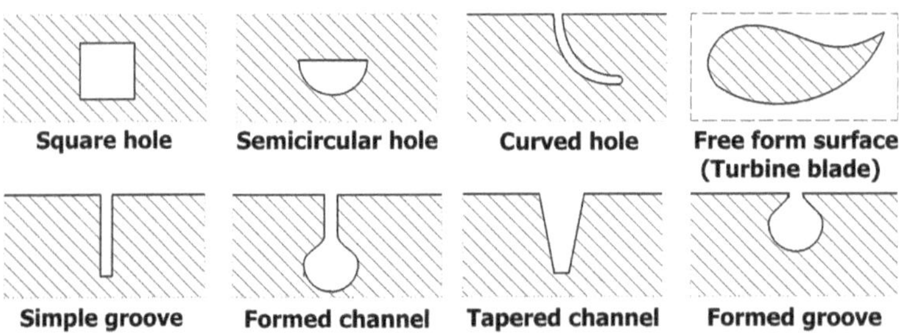

Fig. 1.2 Typical examples of shape complexity.

However, these complex shapes could be fabricated utilizing some of the modern machining practices [5].

(iii) Automation of data exchanges—In machining, accuracy and productivity mainly depends on how various activities are performed precisely. In this regard, automation plays a crucial role which depends on feedback control by data exchanges. Reduction of in process inventory, throughput time and cost of production can be achieved considerably if various data exchanges from different operation activities during machining are more automated. In this connection of automation of data exchange, introduction of CNC, CAD/CAM and CIM in manufacturing created considerable success. These techniques can be easily incorporated in case of modern machining methods where data exchanges are much simplier. Wire Electrodischarge Machining (WEDM), Laser Beam Machining (LBM), Electron Beam Machining (EBM), Ion Beam Machining (IBM), Water Jet Machining (WJM), etc. are some of the typical examples where CNC controls have been successfully implemented through automated data communications which further improved the modern machining performances.

(iv) Precision requirements—Machining can be classified into three major categories, e.g., normal machining, precision machining and ultra precision machining depending upon achievable machining accuracy as presented in Fig. 1.1. Hence, in machining precision requirements are becoming vital day by day. Precision of the machined components can be indicated by the limit of tolerance. However, tolerance (T) depends on the dimensions of machining of the parts (D). Precision products are identified by the following range of D/T ratio as follows:

$$D/T > 10^4 \quad (1.1)$$

To maintain this range of ratio, i.e., $>10^4$ for attaining a desired level of precision, if D reduces, T also has to be further scaled down [6]. This will be much more difficult to accomplish when scale of machining dimensions are becoming very low mainly in case of micro range machining or micro products. This level of precision may be achieved not by removing chips as in the case of traditional machining. However, it can be attained by removing molecules or atoms during machining as in the case of various modern machining techniques.

(v) Machined surface integrity—Quality of machined surface is important for different engineering applications. Surface integrity of the machined parts can be identified by top surface condition, e.g., surface finish, profile accuracy, etc. and subsurface condition, e.g., mechanical and thermal damages, micro surface cracks, change in micro structure, etc. Products machined by traditional machining techniques consist of various surface defects due to generation of high cutting forces as well as temperature in the machining zone as there is always physical contact in between tool and the workpiece. Moreover, achievable surface finish restricted by the nose radius of the cutting tool and profile accuracy depends on the robustness of the kinematic structures of the machine tools. However, in case of modern machining methods, there is no direct contact between the cutting element and the workpiece results in generation of better quality surfaces. With the modern machining methods, higher level of surface integrity, i.e., excellent surface finish, good subsurface condition can be achieved. Some of the thermal assisted modern machining processes cause subsurface defects, e.g., recast layer, heat affected zone (HAZ) and conversion zone which can be minimized by controlling major process parameters. Some of the advanced finishing processes may be useful for generation of surface with improved profile accuracy, high degree of surface finish as well as defect free sub surface.

(vi) Requirements of miniaturization—Very few components like wrist-watch parts were the basic industrial microproducts that were manufactured. However, there have been recent changes in societies' demand for small parts mainly due to the requirements of greater compactness in the utilization of space and portability. The mechanical and electrical devices that make up these items need to be produced in ever-decreasing size, with tightly specified dimensions and accuracies. Although these miniature devices may be manufactured by various procedures, their shaping through removal of material constitutes a major means of production. In this regard, utilization of traditional machining for micro machining applications is not successful due to several limitations, e.g., physical contact, tool size limitation and requirement of equipment precision.

Established modern machining methods can be successfully utilized and continue to be investigated for machining of such parts to specified small dimensions. Developed micromachining technologies utilizing modern machining methods are used successfully for producing microparts to enhance product functionality in limited space and save material and energy. These technologies are employed to produce MEMS, microsensors, and microactuators and also to machine micro-sized features. Another example is the fuel injection nozzles for automobiles where several factors of environmental safety have forced manufacturers to improve the nozzle. Miniaturization of medical tools is one of the effective application areas of micromachining. Micromachining technology plays an increasingly crucial role in the miniaturization of components [7, 8]. Hence, micromachining utilizing modern machining methods is the key technology to fulfill the miniaturization requirements for microproducts and related application fields, which are rapidly expanding.

These above mentioned requirements of manufacturing guide the production engineers to adopt all these modern machining methods. Several machining strategies have already been developed in this field. Each method is having their own advantages and limitations during machining applications. Depending on the application requirements, e.g., geometrical configurations, dimensional accuracy and quality of surface, etc., it is important to the manufacturing engineers to take decision about the suitability of the preferred machining methods. Sometimes other factors of these modern machining methods, e.g., input power requirements, economic consideration and process capabilities are also played important role during selection of modern machining methods.

References

[1] N. Taniguchi, Current status in, and future trends of, ultra-precision machining and ultrafine materials processing, Ann. CIRP 32 (2) (1983) 573–582.

[2] T. Masuzawa, State of the art of micromachining, Ann. CIRP 49 (2) (2000) 473–488.

[3] R. Snoeys, F. Staelens, W. Dekeyser, Current trends in non-conventional material removal processes, Ann. CIRP 35 (2) (1986) 467–480.

[4] G.F. Benedict, Non-Traditional Manufacturing Processes, Marcel Dekker, New York, 1987.

[5] J.A. McGeough, Advanced Methods of Machining, Chapman and Hall, London, 1988.

[6] V. Radhakrishnen, Challenges in meso-, micro-, and nanomanufacturing, in: V.K. Jain (Ed.), Micromanufacturing Processes, CRC Press, UK, 2013.

[7] J.A. McGeough (Ed.), Micromachining of Engineering Materials, Marcel Dekker, NY, USA, 2002.

[8] K.P. Rajurkar, et al., New developments in electrochemical machining, CIRP Ann. 48 (2) (1999) 567–579.

CHAPTER TWO

Classification of advanced machining technology

Contents

2.1 Introduction

With the need of emerging applications in the field of automobile, defense, power, aerospace, electronics and biomedical engineering, etc., metallurgical scientist and researchers are continuously occupied with the development of new advanced materials. With the progress of work materials of higher strength, greater hardness, light weight and high temperature resistance, it has become difficult to process them by conventional machining methods like turning, milling, drilling, etc. It has become well-established fact that the use of conventional machining is uneconomical for such high strength extra hard materials along with the reduction of machining performance. This elevated strength level would have great effect on machining economy if there was no parallel development in machining technology. This technological improvement maneuvered the urgent need of research in manufacturing technology and finally "Modern Machining Processes" had been grown up. Extensive research and development in this field has emerged number of modern machining processes. Electro-discharge machining, Ultrasonic machining, Laser machining, Chemical and Electrochemical machining are some of the few processes those can be considered under this category. These modern machining technologies utilize different types of energies for removal of material to generate desired shape, size and finish of the final products. Hence, it is important to

Modern Machining Technology
https://doi.org/10.1016/B978-0-12-812894-7.00002-5
© 2020 Elsevier Inc.
All rights reserved.

categorize those advanced machining processes to understand the basic mechanisms of metal removal. The applicability of modern machining processes is also differed from each other which need to be analyzed for better usefulness of those in the field of advanced manufacturing.

2.2 Classifications of modern machining processes

Modern machining methods can be classified based on different criteria which influence the processes and its outcomes. The basic classifications of the modern machining processes are based on the types of fundamental energy employed. Here, different types of input energy may be used to process the material and to convert the raw material into final finished product. Various forms of energy like mechanical, thermal, chemical, electrical, etc. are used to machine raw material into usable product. Table 2.1 shows basic classifications based on type of energy used in different modern machining process [1]. It also listed the mechanisms of metal removal, medium utilized, source of energy or tools, etc., with respect to different advanced machining processes. The process like Abrasive Jet Machining (AJM), Ultrasonic Machining (USM), Water Jet Machining (WJM) and Ion Beam Machining (IBM) are categorized under mechanical type of modern manufacturing process in which kinetic energy of abrasive, water jet as well as ions are utilized for the metal removal process. The process in which the kinetic energy of abrasive grains is responsible to remove the material from the workpiece surface is known as USM where, the abrasive particles in the form of slurry are driven into the work surface by a tool oscillating normal to work piece at higher frequency with very low amplitude. AJM and WJM are also utilized kinetic energy of abrasive and water respectively to machine the workpiece. Whereas, IBM utilizes kinetic energy of ions to remove atoms from the workpiece surface by the physical sputtering.

Chemical and electrochemical energies are utilized in case of Chemical Machining (CM) and Electrochemical Machining (ECM) respectively. Chemical Machining (CM) is the process where, desired amount of material is removed from the remaining portion of work surface by the application of reactive chemicals. However, ECM is a controlled anodic dissolution process that provides comparatively higher MRR than CM, which is independent of physical and mechanical properties of work materials due to which it has got very wide field of applications for conductive materials. In ECM

Table 2.1 Classification of modern machining processes

Type of energy	Mechanism of metal removal	Transfer media	Energy source	Processes
Mechanical	Plastic shear	Physical contact	Cutting tool	Tool based machining (TBM)
	Brittle fracture/ shear	High velocity particles	Pneumatic/ hydraulic pressure	Abrasive jet machining (AJM), ultrasonic machining (USM), water jet machining (WJM)
	Physical sputtering	Ion stream in vacuum	Kinetic energy of ion	Ion beam machining (IBM)
Chemical	Corrosive reaction	Reactive environment	Corrosive agent	Chemical machining (CM)
Electrochemical	Ion dissolution	Electrolyte	High current	Electrochemical machining (ECM), electrochemical grinding (ECG)
Thermoelectric	Fusion	Hot gases	Ionized material	Plasma arc machining (PAM)
	Melting/ ablation	Electrons	High voltage	Electro discharge machining (EDM)
	Melting/ ablation	Radiation	Amplified light	Laser beam machining (LBM)
	Melting/ ablation	Electron stream	Kinetic energy of electron	Electron beam machining (EBM)

there is no existence of tool wear, residual stresses, thermal damage and burrs.

Material can also be removed with the help of thermal energy utilizing spark discharges, focused beam of photons, focused beam of electrons as well as focused beam of ionized gas in case of Electro Discharge Machining (EDM), Laser beam machining (LBM), Electron Beam Machining (EBM) and Plasma beam Machining (PAM) respectively. The process like EDM is widely used for machining of hard and tough but electrically conductive materials. LBM, EBM and PAM are very fast as compare to the other advanced machining processes; PAM is less accurate machining process. However, their applications are limited where, lower thermal damage to the work surface is the primary requirements.

2.3 Selection of modern machining processes

To make efficient use of modern machining process, it is necessary to know the correct nature and capabilities of various modern machining processes. It is to be considered that, these methods are not a complete replacement of conventional machining techniques but their use becomes essential in demanding conditions. One machining process may effective under a typical machining condition which may not be equally suitable under the other machining conditions. Thus, it is essential to know suitability of these methods under particular situation. Hence, best possible selection of the modern machining processes is crucial for a particular machining problem [2]. Issues which must be analyzed in depth for the selection scheme for employing one of the modern machining processes are as following [3–5]:

(i) Depending on different machining parameters,
(ii) Based on work materials to be machined,
(iii) Depending on the various output criteria,
(iv) Based on cost effectiveness.

(i) *Depending on different machining parameters*—Major influencing parameters of different modern machining process are represented in Table 2.2. It shows dominant parameters, e.g., voltage, current, input power, nature of power (AC, DC, or pulsed DC), gap between tool/nozzle and job as well as other factors, e.g., type of medium, its velocity and pressure, etc. Selection of the best possible machining process under a particular condition will be governed by the range of parameters as listed in the table. However, it needs to study the suitability of the machining process for different material applications which may lead to the optimal selection of the most suitable process.

(ii) *Based on work materials to be machined*—Selection of the suitable process out of various modern machining processes also depends on type of work materials to be machined. Table 2.3 shows suitability of the various modern machining processes for different material applications. Some extensively used materials in modern engineering applications are listed in the table to exhibit the applicability of different machining processes. Table shows mainly the qualitative analysis of the suitability of different machining processes which help the selection process. In depth study is also necessary to know the impact of different types of work materials on various output criteria for best selection of the suitable process.

Table 2.2 Major parameters of different modern machining processes

Parameters	USM	AJM	WJM	IBM	ECM	CM	EDM	EBM	LBM	PAM
Voltage (kV)	0.22	0.44	0.44	7–30	0.01–0.03	NA	0.03–0.25	50–150	1.5	0.44
Current (amp)	12 (AC)	50–160 (AC)	50–100 (AC)	0.002–0.01 (DC)	1000–100,000 (DC)	NA	1–100 (pulsed DC)	0.025–1 (pulsed DC)	2 (average) 200 (peak)	50–1000 (DC)
Input power (kW)	2.64 (15–30 kHz) (20–100 μm)	22–70	22–40	0.014–0.3	10–3000	NA	0.03–25	1.5 (average) 150 (peak)	3	20–440
Gap (mm)	0.02–0.04	0.25–75	2.5–50	NA	0.1–0.5	NA	0.01–0.05	NA	NA	5–10
Medium	Abrasive in water (velocity: 150–300 m/s, pressure: 2–10 atm)	Abrasive in gas (velocity: 50–300 m/s, pressure: 2–10 atm)	Water (velocity: 150–300 m/s, pressure: 2–10 atm)	Vacuum	Electrolyte	Corrosive chemical	Liquid dielectric	Vacuum	Air	Argon

Table 2.3 Material applications

Process	Material							
	Aluminum	Steel	Super alloys	Ti	Refractories	Plastics	Ceramics	Glass
USM	Poor	Fair	Poor	Fair	Excellent	Poor	Excellent	Excellent
AJM	Fair	Fair	Good	Fair	Good	Fair	Good	Good
WJM	Fair	Good	poor	poor	Fair	Fair	Fair	Good
IBM	Good	Good	Good	Good	Poor	Poor	Poor	Poor
ECM	Fair	Excellent	Excellent	Fair	NA	NA	NA	NA
CM	Good	Good	Fair	Fair	NA	Poor	Fair	Good
EDM	Fair	Excellent	Excellent	Excellent	Poor	Poor	Poor	Poor
EBM	Fair	Excellent	Excellent	Excellent	Poor	Poor	Poor	Poor
LBM	Fair	Excellent	Excellent	Excellent	Good	Good	Good	Fair
PAM	Good	Good	Good	Fair	NA	NA	NA	NA

Table 2.4 Process capability

Process	Material removal rate (mm³/min)[a]	Cutting speed (mm/min)[a]	Power consumption (W/mm³/min)[a]	Surface damage depth (μm)[a]
USM	3–300	0.1–7	0.1–0.008	25
AJM	2.5–7.5	1000–20,000	5000–10,000	2.5
WJM	2–6	800–12,000	4000–8000	0.001
IBM	250e − 10 to 400e − 8	0.0003–0.01	10–100 keV	0.005
ECM	2000–16,000	2.5–12.7	5–200	5.0
CM	15	0.025	NA	50
EDM	4900	12.7	0.006–5	125
EBM	8.2	400–800	200–20,000	250
LBM	4.9	250–5000	600–2500	125
PAM	164,000	50–6000	0.1–2.5	500
Conventional turning	300,000	76,000	0.046	25

[a]Depends on type of workpieces and machining conditions.

(iii) *Depending on the various output criteria*—The process capabilities of various modern machining methods are listed in Table 2.4 which may guide the decision makers for the selection of best machining process under a particular machining condition [6]. In regard of the process capabilities; material removal rate, cutting speed, power consumptions and depth of surface damage are to be considered. During selection process, other two important output criteria must be taken into consideration during precision machining are surface finish and tolerance. In this connection, surface finish and accuracy in terms of tolerance of different machining processes have also been compared in Figs. 2.1 and 2.2. Moreover, the economic consideration also plays an important role during selection of best machining process.

(iv) *Based on cost effectiveness*—As identified above, process economy is one of the most crucial considerations during selection of best machining process under a particular condition. Table 2.5 shows a comparison on the basis of economic considerations which will help the selection process.

However, it has been found that, modern machining processes are sometimes incapable of providing solution to some of the tribulations for difficult

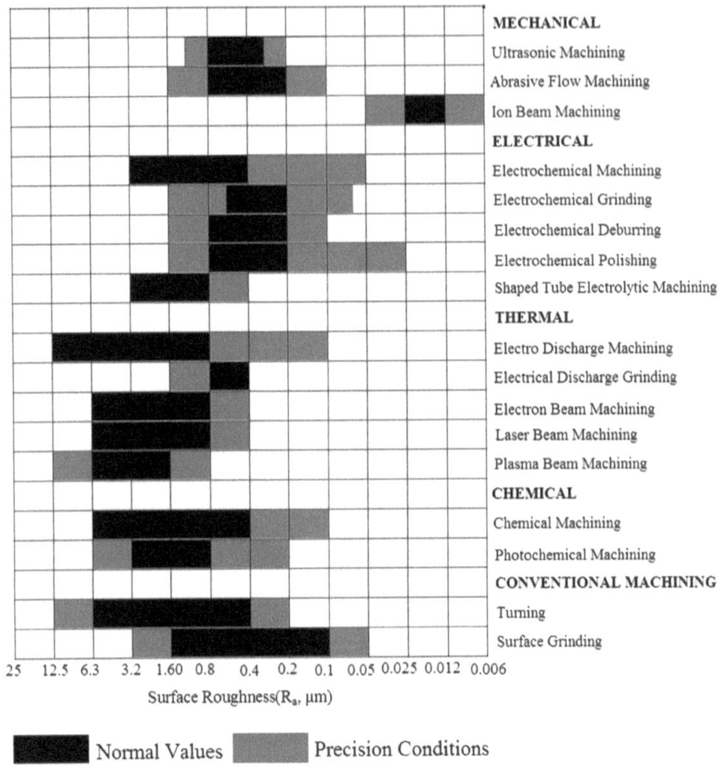

Fig. 2.1 Surface finish for different machining processes.

to machine materials for achieving desire shape complexity. These modern machining processes are also having some limitations with respect to type of materials, shape accuracy, surface quality as well as sub surface damage, etc. Hence, to overcome these limitations, manufacturing scientists have attempted to explore newer techniques by combining different modern machining methods. This concept of machining is known as hybrid machining processes. Hybrid modern machining processes are those processes where two or more than two machining methods are combined to enhance the competence of constituent machining processes. In this category, methods like Electric discharge grinding (EDG), Electrochemical grinding (ECG), and Electrochemical discharge machining (ECDM) and ultrasonic assisted electric discharge machining (UEDM) are some of the examples of hybrid processes, which have been developed to take worthiness of the individual modern machining processes [7].

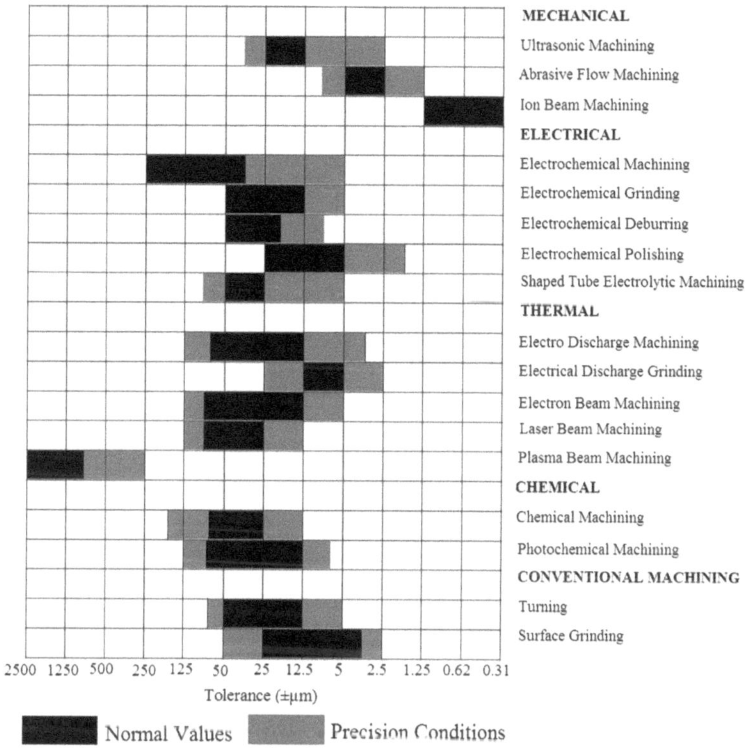

Fig. 2.2 Machining accuracy for different machining processes.

Table 2.5 Process economy

Process	Initial investment	Tooling and fixtures	Power requirement	Efficiency	Total consumption
USM	Medium	Low	Low	High	Medium
AJM	Medium	Low	Low	Low	Very low
WJM	Medium	Low	Low	Low	Very low
IBM	Very high	Medium	High	Low	Medium
ECM	High	Medium	High	Medium	High
CM	Low	High	NA	Medium	Medium
EDM	Medium	Medium	Low	High	Medium
EBM	Very high	Medium	High	Low	High
LBM	Medium	Low	Medium	Very low	Low
PAM	Very low	Low	Low	Medium	Low
Conventional Machining	Low	Medium	Low	High	Medium

To meet the needs of fabrication of micro components for various micro engineering applications, there has been an urgent necessity for developing newer machining methods or modifying the existing modern machining methods to achieve the desire capabilities by which micro features can be manufactured [8]. Hence, several micro machining methods have been developed which need to be investigated for further improvement. In this connection, it should be noted that not only the shape and size of the products are important but also the quality of machined surface should also be improved to a higher level. Several aspects of the machined surface, e.g., surface finish, profile accuracy; surface and subsurface damage play important roles during advanced engineering applications. Sometimes for some typical applications, requirements of surface finish, surface defects and flatness of the surface cannot be achieved by employing existing finishing process, e.g., grinding, honing, lapping, etc. In this connection, several ultra precision finishing techniques have been developed to encounter problems related to various factors of surface integrity mainly using loose abrasives with different techniques. By utilizing these techniques, typical surfaces with flatness of 0.1 μm, roughness of 0.01 μm and damaged layer depth of only a few nanometers can be achieved [2].

Figure 2.3 represents several modern machining and finishing processes which needs to be studied. Further development of these processes should be learned for solving several machining related problems as encountered in the field of present day manufacturing. Hence, authors have attempted to develop this book considering the needs of advance machining as well as its detail classifications. It highlights on different advanced, hybrid, and micro machining processes as well as super finishing techniques considering their material removal mechanisms, influence of different parameters as well as control strategies. Explanations are made in depth to each machining technology highlighting their present status, set-up details, advantages and limitations as well as recent advancement.

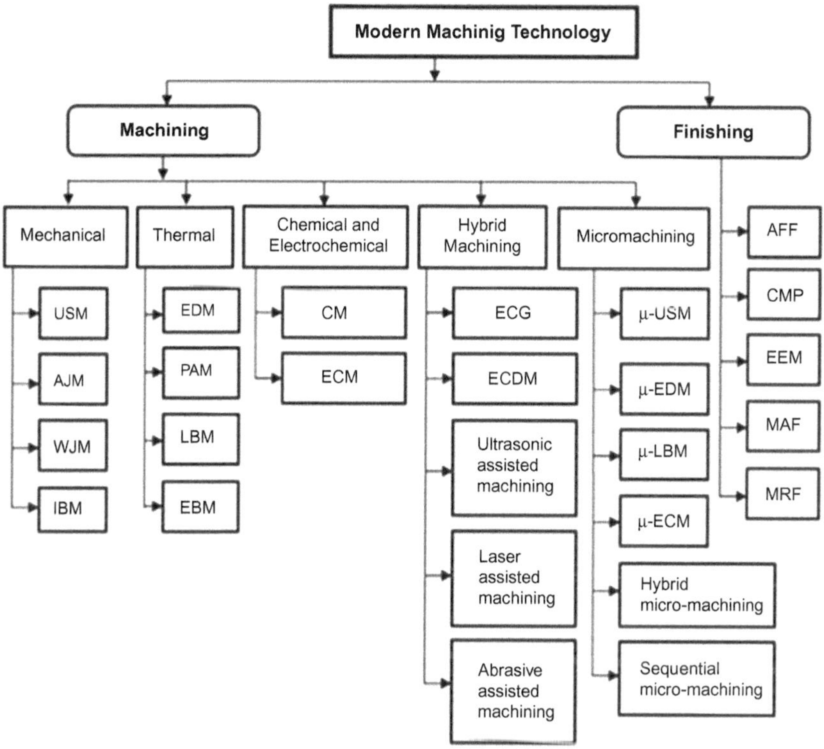

Fig. 2.3 Classification of modern machining Technologies.

References

[1] Machining Data Handbook, vol. 2, Machinability Data Center, Cincinnati, USA.
[2] R. Snoeys, F. Staelens, W. Dekeyser, Current trends in non-conventional material removal processes, Ann. CIRP 35 (2) (1986) 467–480.
[3] J. Carvill, Mechanical Engineer's Data Handbook, Butterworth-Heinemann, Oxford, 2003.
[4] ASM Handbook, Volume 16: Machining, ASM International, Bradley University, USA, 1989.
[5] Tool and Manufacturing Engineers Handbook, Vol. 1: Machining, fourth ed., Society of Manufacturing Engineers, 1983.
[6] J. Kozak, K.P. Rajurkar, Hybrid machining process evaluation and development, in: Keynote Paper, Second International Conference on Machining and Measurements of Sculptured Surfaces, Krakow, 2000, pp. 501–536.
[7] N. Taniguchi, Current status in, and future trends of, ultra-precision machining and ultrafine materials processing, Ann. CIRP 32 (2) (1983) 573–582.
[8] T. Masuzawa, State of the art of micromachining, Ann. CIRP 49 (2) (2000) 473–488.

CHAPTER THREE

Machining processes utilizing mechanical energy

Contents

Modern Machining Technology
https://doi.org/10.1016/B978-0-12-812894-7.00003-7

© 2020 Elsevier Inc.
All rights reserved.

SUBCHAPTER 3.1

Ultrasonic Machining (USM)

3.1.1 Introduction

Ultrasonic Machining (USM) is a mechanical type non-conventional machining method used for machining both electrically conductive and non-conductive materials, which have low ductility and a high hardness above 40 HRC such as ceramics, inorganic glasses, and quartz, etc. The ultrasonic frequency of the tool vibration is 20 kHz and above with amplitude of 25–75 μm. Water mixed with abrasive particles is supplied through the machining zone by a recirculation pump. The tool is fed toward the workpiece with static load and the workpiece material is hammered by numerous abrasive particles with high kinetic energy imparted by the vibrated tool with ultrasonic frequency. Flushing of abrasive slurry refreshes the machining area and also removes the materials from the workpiece.

The history of ultrasonic technology began with the development of piezoelectric effects in the year 1880. As a result of the experience in 1914–18 war, an interest in the subject developed and Langevin in France investigated the use of quartz transducers for transmitting and receiving ultrasonic waves of relatively low frequencies in water which provided promising means for the detection of submarines, and under water communication. The application of ultrasonic energy for machining was first reported through a paper authored by R.W. Wood and A.L. Loomis in 1927. However, USM process was traced back in 1945 when American engineer Lewis Balamuth invented the process and the patent was granted in 1945. Benefits of discovery of USM to industry were quickly realized, and in 1950 the production of USM-tools started. By 1953–54, the first ultrasonic machine tools mostly in the form of drilling and milling machines had been developed. By 1960s, ultrasonic machine tools of various types and sizes for a variety of purposes had been developed and some models had begun to come into regular production. A wide range of material specially hard materials, e.g., tungsten and titanium carbides, die and tool steels, etc., and brittle materials, e.g., germanium, silicon, ferrites, ceramics, glass, quartz, etc., could be effectively machined by this process. USM is used in wide range of industry including aerospace, electronics, optics, and automobile industries.

For finishing of the component processed by the electro spark machines, USM was used as a finishing process. For machining of electrically non-conducting, semi conductive and brittle material ultrasonic machining gained importance and prominence. Various types of ultrasonic machining systems such as stationary ultrasonic machining and rotary ultrasonic machining system have been developed through research and development on ultrasonic machining technology.

Through research and development on ultrasonic machining technology, rotary ultrasonic machining (RUM) is a hybrid ultrasonic machining process in which abrasive bonded tool vibrates at ultrasonic frequency as well as rotates. RUM works with the combined principle of ultrasonic machining and diamond grinding while diamond impregnated tool is used. RUM has many advantages in terms of high accuracy, better surface finish and high machining speed, high tool life over stationary USM.

RUM machine tool was first developed by Mr. Percy Legge in 1954 at United Kingdom Atomic Energy Authority (UKAEA) for hole drilling operation. RUM is also named as ultrasonic drilling, ultrasonic grinding, ultrasonic vibration grinding and ultrasonic twist grinding as RUM uses ultrasonic vibratory and rotary abrasive bonded tool for combined action of machining and grinding. Later on RUM process has been applied for various types of machining operations for milling, surface texturing, grinding and micromachining operations, etc.

3.1.1.1 Fundamentals of ultrasonic machining

Ultrasonic is the science of sound waves above the limits of human audibility. The term silent sound also has been used to denote ultrasonic waves. The frequency of a sound wave determines its tone or pitch. Low frequencies produce low or bass tones. High frequencies produce high or treble tones. Ultrasound is a sound with a pitch so high that it cannot be heard by the human ear. Frequencies above 18 kHz are usually considered to be ultrasonic. The range of ultrasonic frequency may vary from about 20 kHz to several hundred kHz as required in the areas of applications. There are two types of waves such as shear waves and longitudinal waves. Longitudinal waves are mostly used in ultrasonic applications since they are easily generated at the desired frequency and amplitude. The device for converting any type of energy into high frequency vibration is called ultrasonic transducer. Hence, in ultrasonic machining (USM) process, the ultrasonic transducer, also known as converter, transforms electrical energy from the 50/60 Hz power supply into high frequency mechanical vibrations. The horn connected with coupler transfers ultrasonic vibrations from the converter to the tool.

Depending upon the motion and type of tool used, there are two variant of ultrasonic machining such as stationary USM and rotary USM.

(a) Stationary Ultrasonic Machining (USM)

In stationary USM the tool made of soft material vibrates at ultrasonic frequencies in the order of 20 to 30 kHz with amplitude of vibration about 25 μm. It is pressed against the workpiece with a static load of a few kilograms and fed downwards continuously as the cavity is being cut and generated in the work. The rate of machining is not a function of load or imposed stresses on the tool. The excitement of the tool is mostly given by means of piezoelectric or magnetostrictive transducer. Figure 3.1.1 exhibits the fundamental principle of stationary ultrasonic machining. The abrasive slurry is made of abrasive particles suspended in liquid. It is supplied by circulation pump to the cutting zone between the tool and the work piece. The abrasive particles strike the work piece repeatedly and material is removed due to direct hammering of the abrasive particles against the work piece surface, micro chipping by impact of the free moving abrasive particles and cavitations effects from the abrasive slurry. For smooth machining operations the gap between the tool and the workpiece must be at least three times the average diameter of the abrasive grit used. The liquid medium generally water which carries abrasive particles performs many functions during machining operations such as acoustic bond between the workpiece and the tool, transfer of energy between the tool and the workpiece and it is also used as coolant at the tip of the tool and the surface of the workpiece. The

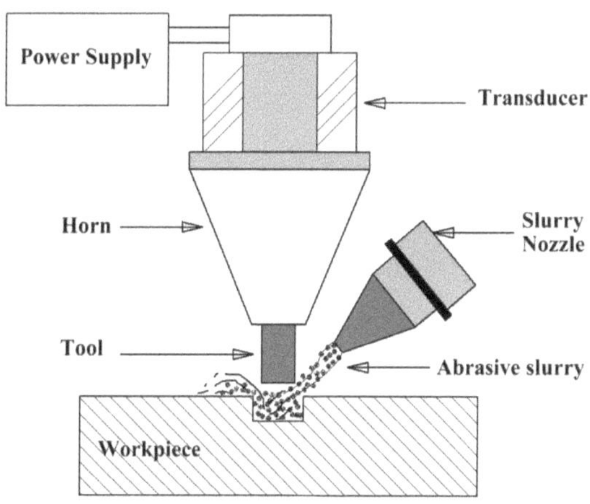

Fig. 3.1.1 Fundamental principle of stationary USM.

liquid medium carries away worn abrasive and swarf and also carries the abrasive to the cutting zone.

(b) Rotary Ultrasonic Machining (RUM)

Rotary Ultrasonic Machining (RUM) is a hybrid process which combines diamond grinding (DG) and ultrasonic machining (USM). In RUM process, generally abrasive (diamond) bonded rotary tool is used instead of supply of abrasive slurry surrounding the stationary tool. For both the cases stationary as well as rotary USM, tool vibrates at ultrasonic frequency. In RUM, the coolant (water) is either supplied through the hollow tool or through side nozzles for solid tool. The ultrasonic vibratory and rotary tool assembly consists of power supply unit, electric motor, transducer, coupler, horn, tool holder and abrasive bonded tool for rotary motion as well as ultrasonic vibration of abrasive particles. For RUM also piezoelectric or magnetostictive transducer can be used for generating ultrasonic vibration. The combined action of vibration and rotation of abrasive particles bonded on tool surface enhances the material removal rate from the workpiece. In RUM material is extracted from the workpiece due to grinding or abrasion action as well as repetitive hammering action of ultrasonically vibrated abrasive particles against the workpiece materials [1]. Figure 3.1.2 shows the basic principle of RUM.

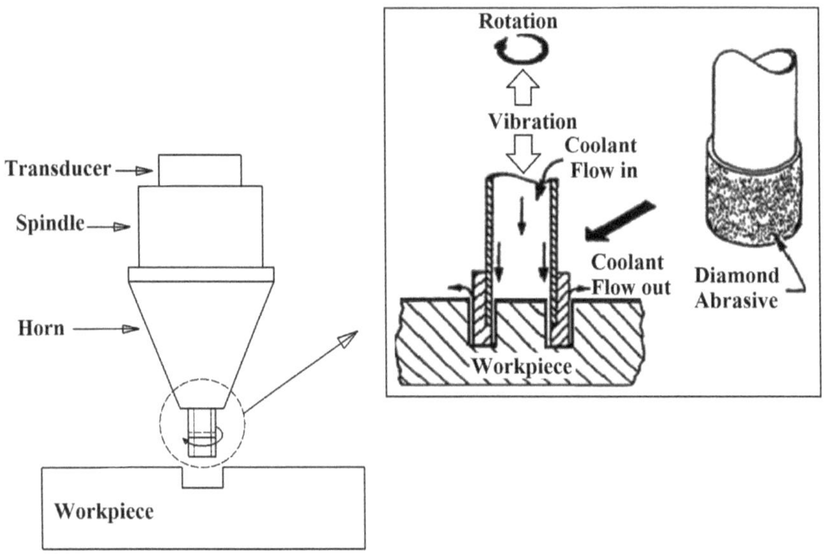

Fig. 3.1.2 Basic principle of RUM.

3.1.1.2 Need of ultrasonic machining

Modern machining works demand intricate shapes, tighter tolerances and finer and more precise dimensions with minimal surface damage and very specific surface characteristics in many end uses. Conventional forming and sintering techniques are often not able to meet these demands. Conventional machining is limited to the generation of complex shapes and sizes on hard and brittle materials. Non-Conventional Machining Processes such as Electrical discharge machining and Electro-Chemical Machining which can produce varieties of complex shapes, but it is only suitable for electrically conductive materials. Since laser beam machining is a thermal process, heat-affected damage does occur and it is an expensive process and not economical for thicker workpiece with large removal of material.

USM is a non-thermal, non-chemical and non-electrical machining process that does not change the chemical composition, material microstructure and physical properties of the workpiece after machining. USM process can be used to machine precision macro and micro-features, round and odd-shaped holes, blind cavities, and stepped holes, etc. Multiple features can be drilled simultaneously, often reducing the total machining time significantly. Ultrasonic machining generally uses of tooling material such as tungsten carbide, monel and stainless steel to provide efficient energy transmission to abrasive particles and minimize tool wear. As high dimensional accuracy, good surface finish and complex shape on the machined surface can be achieved with USM, the need of USM in modern industries is very high. For machining engineering ceramic materials that exhibits very attractive properties for modern industries. Properties of interest include high hardness, high thermal resistance, chemical inertness, tailored electrical conductivity, high strength-to-weight ratio and longer life expectancy, USM is a viable alternative machining process. Alumina based ceramics have been used in seal rings, medical prostheses, laser tubes, electronic substrates, ballistic armor, thermocouple tubes, electrical insulators, grinding media, thread guides and wear component, etc. USM is a mechanical material removal process that can be used for machining these types of hard and brittle materials. The ultrasonic machining process can be employed to machine precision micro-features, circular and non-circular micro-holes and blind micro-cavities. Ultrasonic machining is also suitable for machining of hard and brittle materials including Glass, Sapphire, Ferrite, PCD, Piezo-ceramics, Quartz, CVD Silicon Carbide and Bio-Ceramics, etc.

High accuracy and good surface finish, no heat generation during machining, capability of drilling circular and non-circular holes on very hard materials, no thermal effects on mechanical workpiece and ability to machine electrical conductive as well as non-conductive materials are some of the advantages of USM. Unlike other non-traditional processes such as laser beam machining and electro-discharge machining, etc., ultrasonic machining process generates the work surface with no thermal damage and low level of residual stress, which is important for the survival of brittle materials in service. The newest and most exciting class of advanced engineering materials are alumina, zirconia, silicon carbide, silicon nitride, boron carbide and boron nitride, etc., those can be machined by USM. There is tremendous need of ultrasonic machining and it has great potential application in the field of advanced manufacturing.

For further enhancing material removal rate and reducing hole inaccuracy such as conicity, overcut, circularity error, etc., rotary ultrasonic machining (RUM) is highly needed for *meso* (milimeters) range as well as micro range machining applications. For deeper hole drilling with high accuracy, abrasive bonded tool in RUM is very much effective instead of supply of abrasive slurry in stationary USM. To avoid the erosion of machined hole walls due to loose abrasive particles in the slurry and handling of abrasive slurry during flushing in stationary USM, RUM can be used to achieve close dimensional tolerance where water coolant is supplied for clean and sustainable machining. Sometimes to overcome the shortcoming of stationary USM, RUM is a viable option for various machining operations such as drilling, milling, grinding, texturing, etc., on varieties of hard and brittle, conducting and non-conducting materials for industrial applications.

3.1.2 Mechanism of material removal in USM

In ultrasonic machining, material removal from work piece mainly takes place due to mechanical abrasion caused by hammering and throwing action of abrasive particles. Cavitation erosion also takes place during ultrasonic machining. Material removal from workpiece during rotary ultrasonic machining is mainly due to hammering, extraction and abrasion actions of rotary and vibratory abrasive bonded tool. Material removal mechanisms of stationary and rotary USM are discussed subsequently through development of models of material removal rate.

3.1.2.1 Material removal mechanism in stationary USM

In stationary USM process the abrasive particles present in the abrasive slurry vibrates ultrasonically and due to repetitive impact of abrasive particles on the workpiece, the fatigue failure of workpiece materials occurs. As a result micro cracks are initiated mainly due to brittle fracture. The resulted micro cracks are propagated to fracture and remove the materials from the work-piece in the form of micro-chips. Figure 3.1.3 shows the scheme of material removal mechanism during stationary USM.

In the present model, material removal would be assumed to take place mainly due to impact of abrasives, between tool and workpiece, followed by indentation and brittle fracture of the workpiece [2]. The model does not consider the deformation of the tool. The following assumptions have been proposed for determining the mechanism of material removal during conventional ultrasonic machining:

(i) All abrasive particles are identical in shape and size
(ii) The work material removal rate is proportional to the volume of work material per impact
(iii) Local spherical bulges have identical shape and diameter
(iv) All impacts are identical

The abrasive particles are characterized by the mean grit diameter, d_g The relation between diameters of the local spherical bulges (d) and that of abrasive grit diameter is $d=\pi d_g^2$. During indentation by the abrasive grit onto the workpiece and the surface of the tool, the local spherical bulges contact the surfaces and the effect of impact on the workpiece is characterized by d.

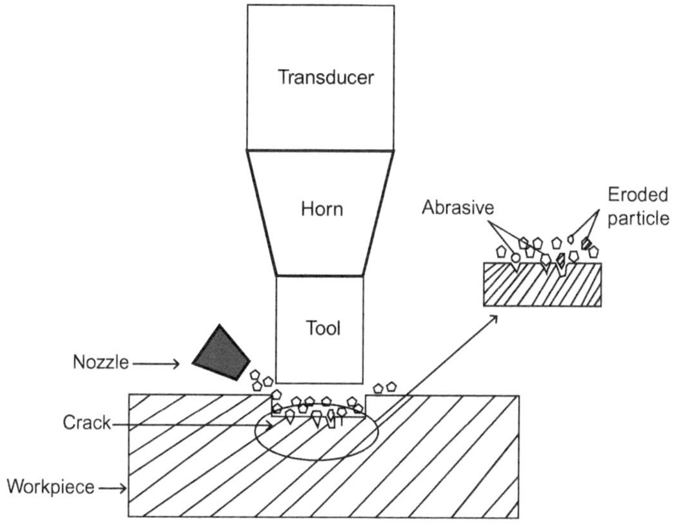

Fig. 3.1.3 Scheme of material removal mechanism during stationary USM.

The indentation due to direct impact of the abrasive particle on work material is exhibited in Fig. 3.1.4. As the indentation on the workpiece proceeds, the contact zone between the abrasive grit and workpiece is established and increases proportionately with duration of machining. The contact zone created by indention, is circular in nature and is characterized by its diameter "2 ×." At full indentation, the indentation depth in the workpiece is characterized by Δ x. Due to direct impact of abrasive particles on the brittle work material, brittle fracture takes place leading to hemi-spherical fracture of diameter "2 ×" at the point of contact between the grit and the workpiece [3]. Thus material removal per abrasive particle (P_a) is given as:

$$P_a = \frac{2}{3}\pi x^3 \tag{3.1.1}$$

Now considering the geometry of indentation as shown in Fig. 3.1.4:

$$AB^2 = AC^2 + BC^2$$

$$\left(\frac{d}{2}\right)^2 = \left(\frac{d}{2} - \Delta x\right)^2 + x^2$$

$$x^2 = d\Delta x \text{ Neglecting } \Delta x^2 \text{ as } \Delta x \ll d$$

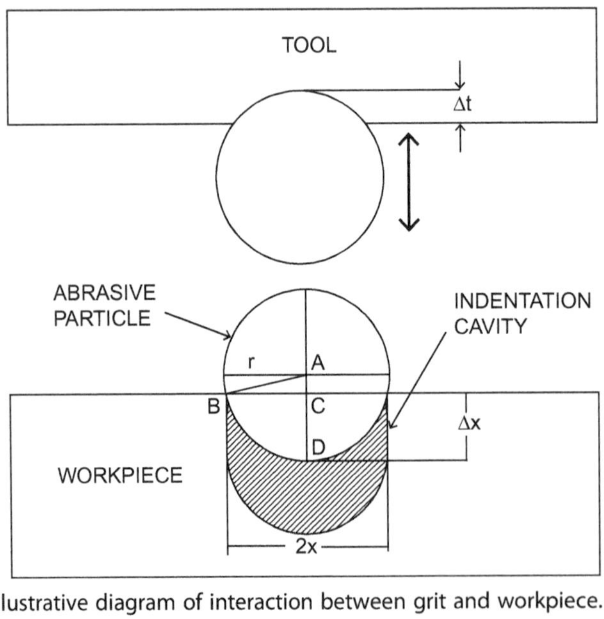

Fig. 3.1.4 Illustrative diagram of interaction between grit and workpiece.

Therefore, material removal per abrasive particle (Eq. 3.1.1) becomes

$$P_a = \frac{2}{3}\pi(d\Delta x)^{3/2} \tag{3.1.2}$$

If at any moment of time, there are an average "n" of grit particles and the tool is vibrating at a frequency "f," then material removal rate (MRR) can be expressed as:

$$\begin{aligned} MRR &= P_a.n.f \\ &= \frac{2}{3}\pi(d\Delta x)^{3/2}.n.f \end{aligned} \tag{3.1.3}$$

Now, considering the tool and the workpiece are pressed each other through contact points causing indentations on work material (Δ x) and tool material (Δ t), both of them would deform or wear out. And then the indentation process starts and finally completes with an indentation of Δ x and Δ t on the work and tool respectively. The tool vibrates in harmonic motion. The interaction of tool and workpiece with abrasive grains may occur during first quarter of each cycle. The cycle time is T. Out of this quarter; some part is used to engage the tool with abrasive grains. Thus the time of indentation τ can be roughly estimated as:

$$\frac{\Delta}{a_0} = \frac{\tau}{T/4}$$

where, a_0 is the amplitude of vibration of tool.

$$\tau = \frac{T\Delta}{4a_0} = \frac{T(\Delta x - \Delta t)}{4a_0} \tag{3.1.4}$$

During machining, the impulse of force on the tool and work would be balanced. Thus assuming F_{max} as the maximum indentation force per abrasive, the total impulse on the tool can be expressed as:

$$F_I = nf\frac{1}{2}F_{\max}\tau \tag{3.1.5}$$

In the USM, the tool is usually fed with an average force F. Thus

$$F = nf\frac{1}{2}F_{\max}\tau \tag{3.1.6}$$

$$F_{\max} = \sigma_w \pi x^2 \tag{3.1.7}$$

where, σ_w is the flow stress of the workpiece materials.

Therefore,

$$F=\frac{1}{2}\sigma_w\pi x^2\tau nf$$

$$F=\frac{1}{2}nf\sigma_w\pi x^2\frac{T(\Delta x+\Delta t)}{4a_0} \tag{3.1.8}$$

If "A" is total area of the tool facing the workpiece, then volume of abrasive slurry of one grain thickness is Ad_g.

If "n" is the number of grits then the total volume of n grits is $\frac{\pi d_g{}^3}{6}n$.

Thus, the concentration of abrasive grits in the slurry (C) is related as follows:

$$\frac{\pi d_g{}^3}{6}n=Ad_gC$$

where,

$$C=\frac{\pi d_g{}^2}{6A}n \tag{3.1.9}$$

Therefore, the number of grits in the abrasive slurry is estimated as:

$$n=\frac{6AC}{\pi d_g{}^2} \tag{3.1.10}$$

Now, assuming the indentation to be inversely proportional to the flow strength then,

$$\frac{\Delta t}{\Delta x}=\frac{\sigma_w}{\sigma_t}=\lambda \tag{3.1.11}$$

where, σ_t is the flow stress of tool material.

Thus, combining above equations, "F" can be written as:

$$F=\frac{16AC}{2\,\pi d_g{}^2}f\sigma_w\pi x^2\frac{T\Delta x(1+\lambda)}{4a_0}$$

or,

$$F=\frac{3AC}{d_g{}^2}(fT)\frac{\sigma_w}{4a_0}d\Delta_x{}^2(1+\lambda)$$

or,

$$F=\frac{3AC}{d_g{}^2}(fT)\frac{\sigma_w}{4a_0}\pi d_g{}^2d\Delta_x{}^2(1+\lambda) \tag{3.1.12}$$

From above equation, it can be written as:

$$\Delta_x^{\ 2} = \frac{4AF}{3\mu dAC\sigma_w(1+\lambda)}(fT)$$

Now, the material removal rate (MRR) can be expressed as,

$$\begin{aligned}
MRR &= P_a nf \\
&= \frac{2}{3}\pi x^3 nf \\
&= \frac{2}{3}\pi \frac{6CA}{\mu d_g^2} fx^3 \\
&= 4\frac{CA}{d_g^2} f(d\Delta x)^{3/2} \\
&= \frac{4CA}{d_g^2} f\left(\mu d_g^2 \Delta x\right)^{3/2} \\
&= 4CAd_g\mu^{3/2} f\left(\frac{4Fa_0}{3\mu AC\sigma_w(1+\lambda)}\right)^{3/4}
\end{aligned} \quad (3.1.13)$$

From above expressions with usual notations, it can be written as

$$\begin{aligned}
MRR &\propto \frac{(CA)^{1/4}(\mu a_0 F)^{3/4} d_g f}{\{\sigma_w(1+\lambda)\}^{3/4}} \\
&\propto (CA)^{1/4}\left\{\frac{\mu a_o F}{\sigma_w(1+\lambda)}\right\}^{3/4} d_g f \\
&\propto Ad_g f\frac{C^{1/4}(\rho a_0)^{3/4}\mu^{3/4}}{\sigma_w^{3/4}(1+\lambda)^{3/4}}
\end{aligned}$$

where σ_w is strength of work material, A is total surface area, λ is ratio of indentation of the tool and work material, a_0 is amplitude of the tool tip.

Based on above assumptions, the expression for total material removal rate with total active number of abrasive particles (N) on the contact surface can be written as:

$$MRR = 4ACNd_g\mu^{3/2} f\left[\frac{4Fa_0}{3\mu AC\sigma_w(1+\lambda)}\right]^{3/4} \quad (3.1.14)$$

In USM material removal primarily occurs due to direct hammering and indentation of the hard abrasive grits on the brittle work material.

Other than this brittle failure of the work material due to indentation some material removal such as micro-chipping by impact of the free moving abrasive particles, cavitation effects from the abrasive slurry and chemical action associated with the fluid employed may occur, but it is estimated to be rather insignificant. As the tool vibrates, it leads to indentation of the abrasive grits. During indentation, due to Hertzian contact stresses [3], cracks would develop just below the contact site, then as indentation progresses the cracks would propagate due to increase in stress and ultimately lead to brittle fracture of the work material under each individual interaction site between the abrasive grits and the workpiece. Moreover, a small amount of material removal might also be contributed by the mechanical abrasion of the hard micro abrasives in stationary USM.

The tool material should be such that indentation by the abrasive grits does not lead to brittle failure. Thus the tools are made of tough, strong and ductile materials like steel, stainless steel and other ductile metallic alloys.

Some researchers have developed several numbers of analytical models of material removal in ultrasonic machining process, considering different basic approaches to removal mechanism. Some developments in the relevant works have been discussed as follows:

Shaw [4] proposed the first theory to confirm the conception on ultrasonic machining process with assumptions for determining the characteristics of ultrasonic machining based on the model as follows:

(i) The work material removal rate is proportional to the volume of work material per impact

(ii) The work material removal rate is proportional to the number of particles making impact per cycle

(iii) The rate of work material removal is proportional to the frequency of tool vibration

(iv) All impacts are identical

(v) All abrasive grains are identical and spherical in shape, etc.

According to above assumptions, material removal rate due to direct hammering was given by [5]:

$$V_1 = K\left[\frac{FA}{H_w(1+\lambda)}\right]^{3/4} C^{1/4} d_m f \tag{3.1.15}$$

where F is mean static force over the time period T, A is amplitude of vibration; H is Brinell hardness (MPa or N/mm^2), λ is indentation ratio or ratio of tool and work material, C is concentration of abrasive in the slurry by mass,

d_m is mean diameter of abrasive grain (mm), f is cyclic frequency of vibration (cycles/s); K is a constant of proportionality and is given by

$$K=\frac{2}{3}(\pi K_1)^{1/4}K_2{}^{3/4} \tag{3.1.16}$$

Material removal rate due to particle impacting on work material is given by

$$V_2=8.6K_1\left[\frac{\rho_a}{H_w}\right]^{3/4}f^{5/2}A^{3/2}Cd_m \tag{3.1.17}$$

where, average diameter of abrasive grain is given by $d_1 = K_2 d_m^2$; and ρ_a is the density of work material (kg/mm^3). For a given area of tool face, the number of active abrasive grains.

(N) is given by

$$N=K_1\frac{C}{d_m^2} \tag{3.1.18}$$

G.E. Miller [6] also proposed analytical model for material removal with the following assumptions:

(i) Abrasive particles are of cubical size
(ii) Plastic deformation is directly proportional to the stress
(iii) Plastic flow stress equals Burger vector times Shear modulus
(iv) Cross sectional area of cut does not change during machining
(v) Viscosity effects in water slurry are negligible

According to above assumptions, the material removal rate is given by

$$V=K\left[\frac{Ad_m fFP_{atm}}{b_w G_w q_w R V_{wg}\rho_a}\right]\left(\frac{C}{C+1}\right) \tag{3.1.19}$$

But depth of cut per (D_c) unit time is given by

$$D_c=K\left[\frac{Ad_m fpP_{atm}}{b_w G_w q_w R V_{wg}\rho_a}\right]\left(\frac{C}{C+1}\right) \tag{3.1.20}$$

where, K is a constant of proportionality; G_w is shear modulus (MPa or N/mm^2); q_w is work hardening capacity (MPa or N/mm^2); R is tool radius (mm); b_w is Burger vector w (mm); V is volume of tool-work gap (mm^3) and P_{atm} is atmospheric pressure (MPa or N/mm^2).

Developments in application of ultrasonic machining processes have resulted several numbers of researchers to carry out investigations in the field of mechanism on conventional and hybrid of ultrasonic machining, called rotary ultrasonic machining. Some researchers have focused on development of mathematical and empirical relations to estimate material removal rate in conventional ultrasonic machining processes.

Wang and Rajurkar [7] carried out dynamic analysis of the ultrasonic machining on ceramic materials and observed the presence of two phenomena that contribute to material removal; the deformation at the point of impact, and the brittle structure below the impact zone. Based on dynamic impact test, it was observed that the material removal in the USM process appears to be a function of impact velocity, which is determined by the frequency and amplitude of the vibrating tool. Material at the impact zone was removed by fragmentation and by chipping micro-fracture due to the high compressive stresses developed in the machined area. According to the authors, based on the combined effects of impact indentation and fracture phenomenon by assuming work material is semi-infinite solid, axis of the moving grit is perpendicular to the free surface during the machining process and speed of abrasive grit is same as that of the vibrating cutting tool; Volumetric MRR in mm^3/s is given by

$$V = K_1(1000)^{-0.8} r_m \rho_m^{4/5} f^{8/5} A^{8/5} \tag{3.1.21}$$

where,

$$K_1 = \frac{c_3 A_1 K_5^2}{6} \text{ and } K_5 = \left[\frac{5\pi^3(1-\nu_w)}{2\mu_w}\cos^2(2\pi ft - \Phi)\right]^{2/5}$$

Here, c_3 is gap coefficient,

Φ = phase lag used in the equation of amplitude

φ = phase lag used in the equation of amplitude,

i.e., $A(t) = A\sin(2\pi f_t - \varphi)$

The above model could also be useful for predicting analytically an optimum grain size for a specified MRR.

Lee and Chan [8] proposed a model based on brittle fracture by assuming pre-existing flaws in the workpiece material for the initiation of median or lateral cracks. For the size of the median or radial crack C, relation with critical load P is given by $C^m = k.P$; with m being assumed equal to 1.5. Cutting tool was assumed as a slender column. The expression for determining volumetric MRR is given as:

$$MRR = \frac{2\pi k_1^2 f}{3N}(aW + bU)^2 \tag{3.1.22}$$

where,

$$a = \frac{T}{\Delta t};\ b = \frac{U_1 E_1}{L_1}\left[1 + \frac{T}{\pi \Delta t}\cos\left(\omega t_a + \Phi\right)\right]$$

Above equation also can be simplified to

$$MRR = \frac{k^1}{N}(aW + bU)^2 \tag{3.1.23}$$

where, φ = Phase used in the equation of amplitude, i.e., $U(t) = U\sin(\omega t + \varphi)$, t is shocking time, k_1 is a constant as mentioned in the assumptions, a and b are coefficients determined by the machining conditions, N is the number of the effective abrasive particles in the machining area; W is the static load; E is elastic modulus of tool and U is the amplitude of the tool tip. Any increase in tool tip vibration amplitude, static load and abrasive particle size causes increase in MRR and surface roughness on the workpiece.

Lee and Jianxin [9] reported the effects of whisker orientation on the material removal rates and mechanisms in ultrasonic machining of Al_2O_3/SiC ceramic composites. It was observed that material removal rate and machined surface roughness in ultrasonic machining of the ceramic composites varied according to the orientation of the hot pressing direction. The relation for estimating the volume of material removed is given by

$$MRR = K\sum N\pi E^{5/4} K_{IC}^{-1} H_v^{-2} P^{7/4} \tag{3.1.24}$$

where, E and Hv are Young's modulus and Vickers hardness of the material, and P is the mean vertical impact force acting from one particle to material during machining, N is number of impact particles, K_{IC} is fracture toughness of work material and K is a constant.

The above mentioned mathematical models can be utilized to determine material removal rate for specified material under certain conditions. These models identify the machining process parameters for stationary USM process which influences the material removal rate (MRR).

3.1.2.2 Material removal mechanism in rotary USM

Rotary Ultrasonic Machining (RUM) process is a combination of conventional grinding and static USM process, the material removal mechanism of RUM is also evolved combining the material removal mechanisms of conventional grinding and USM process. Three mechanisms such as hammering, extraction and abrasion actions are caused by rotational motion and vibrations of the tool as identified in RUM as follows [10].

(1) Indentation and crushing of work material beneath the impact of ultrasonic vibration causes the removal of material by "hammering action."

(2) Material removal by pure rotational motion of diamond tool considered same as in the case of conventional grinding and called "abrasion action."

(3) The simultaneous action of vibrations and rotational motion of the tool caused the removal of material by "extraction action."

Figure 3.1.5 shows schematic representation of mechanism of material removal in RUM.

For hard and brittle materials, brittle fracture of the work surface was observed as a foremost mode of material removal. A common phenomena occurred in RUM of brittle materials are formation of micro cracks and their propagation, material removal with larger chucks, etc. During hole drilling by RUM, the hammering action of diamond grits is responsible for removal of material toward the tool tip, whereas abrasion action of abrasives removes the materials from the walls of the drilled hole. Both actions result in removal of debris formed with the aid of pressurized coolant fluid at the walls of the drilled hole. In the RUM process, removal of material also took place in ductile mode during machining [12].

For rotary ultrasonic machining of ceramic materials, plastic deformation occurs and materials are removed in the form of ground chips due to diamond grinding actions in addition to brittle fracture rotary ultrasonic face milling (RUFM) of ceramic materials (such as magnesia stabilized zirconia, reaction-bonded and hot-pressed silicon nitride) as identified by Pei et al. [13].

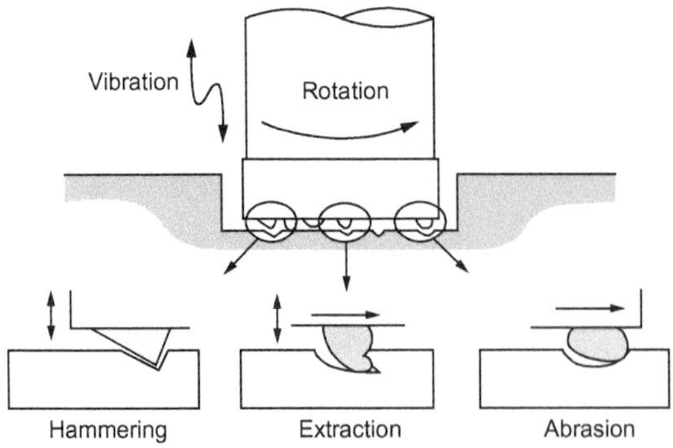

Fig. 3.1.5 Mechanism of material removal in RUM [11].

Brittle materials are plastically deformed when these are indented by a diamond particle. Plastic deformation occurred in many hard, ordinary brittle materials by scratching the surface with diamonds having a very small contact point and a light normal load. Ductile behavior in normally brittle materials had been produced in micro indentation, scratching and erosive cutting. In RUM, the interaction between individual diamond particles and work piece material can be described, to some extent, as that in erosive cutting, diamond turning and grinding. In RUM high stress is developed as the contact area of abrasive particle on the work material is small. As region of deformation is small enough, the stress may be large enough for plastic flow to occur before fracture. Pulverization is also identified as a new existing mode of material removal during the RUM of optical glass K9 by Lv et al. [14]. For machining tough and ductile materials like titanium alloys, stainless steel, etc., with RUM, the removal of material in mixed mode, i.e., plastic deformation and brittle fracture [15]. For rotary USM of ductile materials the plastic flow mechanism is more dominant over brittle failure and material is removed mainly due to with abrasion action. There is a significant amount of plastic deformation before failure takes place with the propagation and intersection of cracks observed for tough and ductile materials like RUM of stainless steel [16].

A mathematical model of material removal rate in rotary USM has been proposed based on theory of fracture mechanics considering the impact and grinding of abrasives in tool tip on the machined surface. The material removal rate (MRR) is expressed as follows:

$$\mathrm{MRR} = \mathrm{f}.\rho.\mathrm{V} \tag{3.1.25}$$

where f is the frequency of ultrasonic vibration; ρ is the density of workpiece materials and V is the total volume of materials removed in each vibration cycle.

The total volume of materials removed in each vibration cycle is determined as [17]:

$$V = \int_0^{R_o} \frac{2\pi r_1 n V_1}{S} dr_1 = \frac{\pi n c_1}{S}\left[R_o^{\,2}\left(1 + \frac{vf}{2fd}\right) + \frac{n_r \pi}{90df} R_o^{\,3}\right] \tag{3.1.26}$$

where,

$$c_1 = \pi\left(\frac{d}{2} - \frac{\delta}{3}\right)\delta^2 \tag{3.1.27}$$

d is the diameter of an abrasive particle; δ is the maximum impacting depth; n is the effective number of abrasive particles in the tool end face; n_r is the rotational speed of the tool; v_f is the feed rate of the workpiece; R_o is the radius of the tool end face; S is the area of the tool end face.

From the MRR model of rotary USM it is found that MRR depends on the static load, grit size and concentration of the abrasives, rotational speed of the tool, feed rate of the workpiece, mechanical properties of machined material and tool material.

3.1.3 Details of stationary ultrasonic machining system

The stationary ultrasonic machining system consists of power supply unit, tool vibration unit, workpiece holding unit, abrasive slurry supply unit and tool feed control unit. Figure 3.1.6 shows the schematic diagram of the total set up of ultrasonic machining system. The details of components of different subsystems of ultrasonic machine are given in subsequent subsections [2]. The photographic view of the total USM system is shown in Fig. 3.1.7.

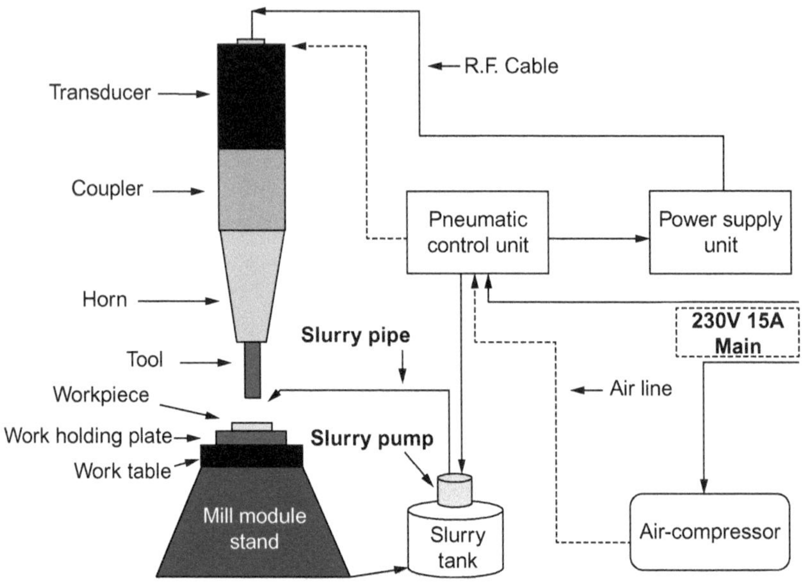

Fig. 3.1.6 Schematic diagram of the stationary ultrasonic machining system.

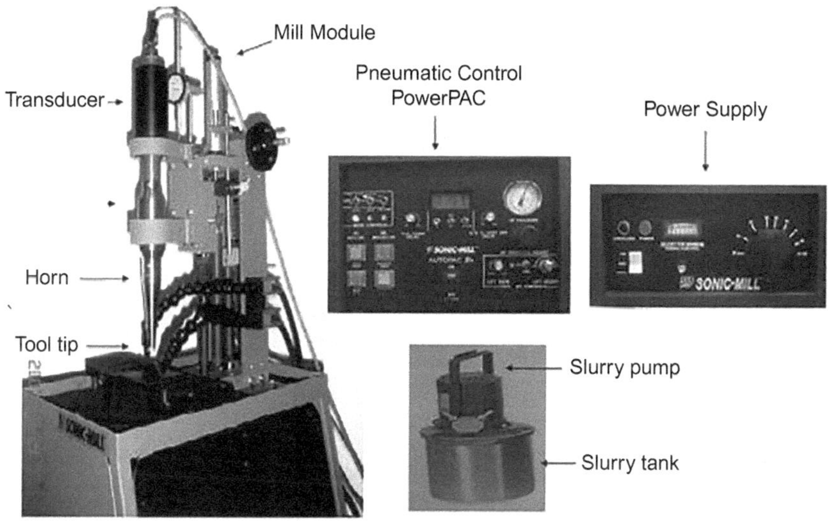

Fig. 3.1.7 Photographic view of stationary USM system. *(Courtesy Sonic Mill, USA.)*

3.1.3.1 Tool vibration unit

Tool vibration unit consists of various units such as transducer, coupler, horn, tool holder and tool. The tool vibration unit is also shown in Fig. 3.1.8.

(a) Transducer

Transducer is a device which generates ultrasonic vibration for ultrasonic machining. High power sine wave generator generates high frequency ($\geq$20 kHz) electrical power from low frequency electrical power. Transducer receives high frequency electrical energy and transforms the high frequency electrical oscillations into high frequency mechanical vibrations. In USM, generally two types of transducers such as piezoelectric and magnetostrictive transducers are used.

(i) Piezoelectric transducer

Piezoelectric transducer works with piezoelectric effect. When an electric current is passed through the piezoelectric material, then it expands. But when the current is removed the piezoelectric material reaches its original size. This phenomenon actually is called piezoelectric effect. Piezoelectric transducers use materials like quartz and lead zirconate titanate. Such piezoelectric transducers can be used up to a power of 1000 W. The heart of the transducer is piezoelectric element which expands and contracts when subjected to ultrasonic frequency

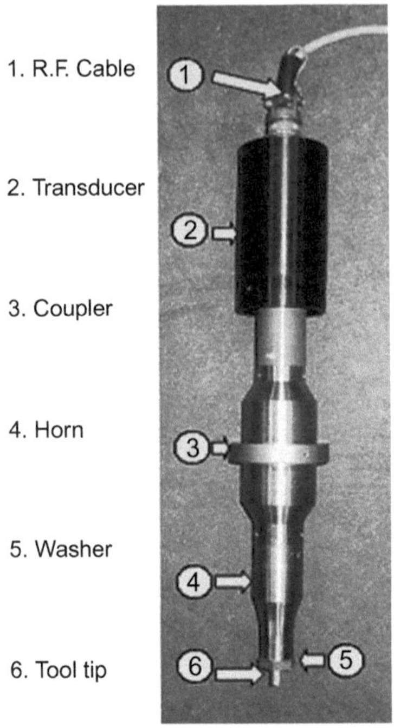

Fig. 3.1.8 Tool vibration unit. *(Courtesy Sonic Mill, USA.)*

of the power supply voltage. The piezoelectric converter is highly efficient and the degree of energy conversion is 96%. The amplitude of vibration of piezoelectric transducer is less which is upto 5 μm.

(ii) Magnetostrictive transducer

According to the magnetostriction effect, ferromagnetic metals and alloys are changed in their length when these are subjected to strong magnetic field. The ferromagnetic material is only responsible for positive or negative deformation. One coil is formed by wrapping around a stack made of magnetostrive material (iron-nickel alloy). The electric signal of ultrasonic frequency is supplied in to this coil. For minimizing eddy current and hysteresis losses, this stack is fully laminated and to dissipate the generated heat, it must be cooled. The stack expands and contracts at the same frequency with the help of the alternating magnetic field produced by the high frequency AC generator. If the frequency of the ac signal in the magnetic field is tuned by same value of the natural frequency of the transducer and the whole oscillating system, so it acts at mechanical resonance. This time oscillation amplitude becomes moderately large and the exciting power reaches its smallest value. Due to large waste heat of magnetostrictive transducer, the conversion efficiency is lower (20–35%). But high power upto 2.4 kW and amplitude of vibration upto 25 μm can be obtained for magnetostrictive transducers required for ultrasonic machining.

(b) Coupler

The coupler (also known as mechanical amplifier) is made of aluminum or titanium. It is mounted between the converter and the horn and modifies the amplitude of vibration applied to the horn. Coupler as shown in Fig. 3.1.8 is designed to resonate at the same frequency as the converter with which they are to be used and are usually mounted on at a nodal (minimum vibration) point of axial motion, minimizing the loss of energy and preventing the transmission of the sound into the support column. The coupler attaches between the converter and the horn.

(c) Horn

The amplitude of vibration of transducer is less than the required amplitude of vibration (25 to 100 μm) for USM. For efficient ultrasonic machining the amplitude of vibration is half of the average diameter of abrasive particle used for USM. The use of different size of horn can generate different amplitude of vibration for USM. The horn amplifies the amplitude of vibration of transducer upto the required range for USM. The horn as shown in Fig. 3.1.8 is a mechanical amplifier on concentrator that transfers and amplifies ultrasonic vibrations from the transducer on converter to the tool. It acts as a resonator to amplify the vibration. The shape of the horn alters the gain factor (ratio of output amplitude to input amplitude). The shapes of various types of horn such as exponential, conical and stepped cylindrical are used for USM. Titanium alloys are considered the best for horn fabrication, but horns are also made of brass, steel, and monel. These horns are generally used for tools that are over 1 in. (25.4 mm) in diameter. Tips are attached to the ½ inch (12.7 mm) and 1 in. (25.4 mm) horns available in a range of head thickness in order to maintain weight relationships with the horn for varying weights and lengths of tools. This allows the horn/tip/tool to operate at optimum frequency.

(d) Details of USM tools

The straight tubular cylindrical tools made of annealed stainless steel (S304) of 10.00 mm outer diameter, 1.3 mm wall thickness and 16.50 mm long were used for ultrasonic drilling operation [2]. Highly efficient tool material is generally tough, but not hard. Good examples of excellent tool materials according to Sonic-Mill Manual are 304 and 316 stainless steels. It yields good tool life and can be easily machined. Testing of tools is also possible in other metals that yield long life. The photographic view of the developed tubular cylindrical tool along with detail dimension is shown in Fig. 3.1.9. The straight tubular hexagonal tools made of the same annealed stainless steel (SS304) were fabricated to designed shape of 11.25 mm minor diameter, 12.72 mm major diameter, 8.70 mm hole diameter and 17.00 mm long [2]. The photographic view of the developed tubular hexagonal tool along with detail dimension is shown in Fig. 3.1.10. The cutting tools were fabricated from SS304 stainless steel because this type of material is an excellent tool material for USM and also locally available. The total weight of the

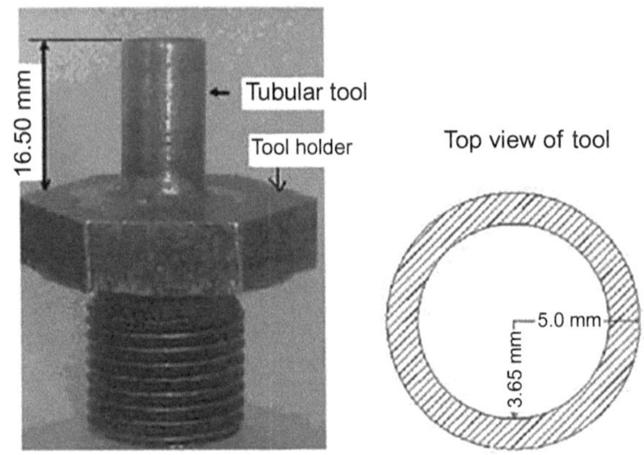

Fig. 3.1.9 Photographic and top view of tubular tool.

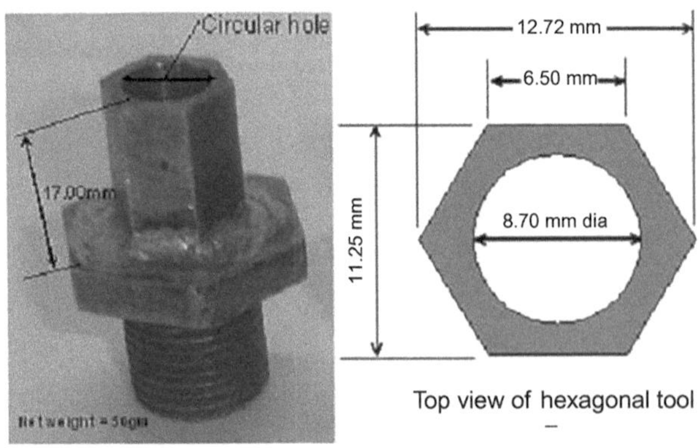

Fig. 3.1.10 Photographic and top view of hexagonal tool.

developed tools including tool holder were within standard weight of tool, i.e., 57 g. The tools were silver brazed to the replaceable threaded tool holder.

3.1.3.2 Workpiece holding unit

The workpiece holding unit consists of magnetic based worktable and workpiece holding plate for accommodating workpiece for machining operation. The workpiece holding arrangement with respect to position of the tool is shown in Fig. 3.1.11.

The magnetic base worktable consists of steel plate of 100 mm × 100 mm cross section and 8 mm thickness [2]. It is used to fix the workpiece holding plate exactly below the center of the tool. Worktable is positioned on four die springs, two on each end of the table. This device is located above the slurry bowl and is also the workpiece supporter. Loosening or tightening the four allen bolts can adjust the table adjusted to the desired position. Work samples are pasted by wax on to the workpiece holding plate and magnetized on worktable in a suitable position for machining operations.

The workpiece holding plate (100 × 100 mm square and 10 mm thickness) was made of mild steel specially fabricated with cavity size of 15 × 15 mm square and 8 mm depth at the center for making through hole on the ceramic workpiece [2]. A larger cavity size of 30 × 30 mm square and 8 mm depth at the center was also made for drilling larger holes on workpiece. These work-holding plates were fixed and magnetized on the worktable of the mill module for the specified jobs. The photographic view of the

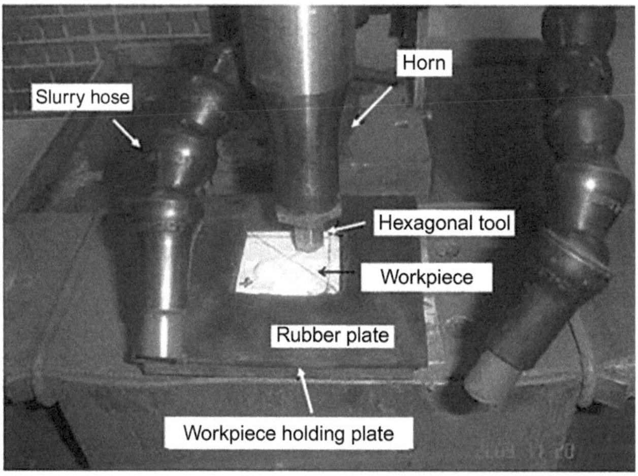

Fig. 3.1.11 Photographic view showing the position of tool and workpiece.

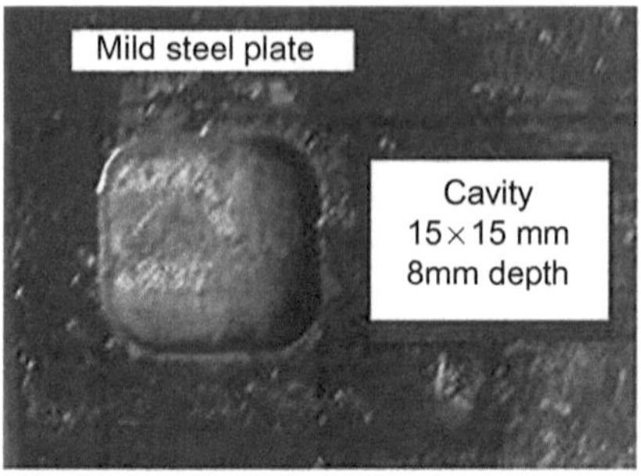

Fig. 3.1.12 Photographic view of workpiece holding plate.

fabricated work holding plate made from mild steel is shown in Fig. 3.1.12. The workpieces were heated up to 70 °C on the hot plate and pasted with pasting wax in a fixed position above the cavity.

3.1.3.3 Abrasive slurry supply unit

An abrasive slurry pump is used as one of the main components of the abrasive slurry circulation system to circulate abrasive slurry solution. The abrasive slurry solution consists of abrasive particles mixed with water, which also functions as cooling system for the tool and workpiece. Varying the control valve attached at the mill-module controls the flow rate of abrasive slurry. The power supply on abrasive slurry pump is controlled from main control unit. The performance of machining operations greatly depends on the slurry concentration, slurry flow rate and cooling system of the ultrasonic machine. Higher abrasive slurry concentration may increase higher material removal rate but lower level of concentration is preferable for better functioning of slurry pump.

The most commonly used abrasive is boron carbide (B_4C), which is one of the hardest materials that can be used as abrasives for USM operations. Abrasive particle is selected for USM based on hardness, size of it and also the type of workpiece materials. The other types of abrasives are aluminum oxide (Al_2O_3), silicon carbide (SiC), boron silicarbide and diamond. Water is used as common liquid media to make abrasive slurry. The other liquids such as oils, benzene and glycerol can also be used.

A re-circulating pump as shown in Fig. 3.1.7 forces abrasive slurry to the work material. Usually, abrasive particles are suspended in a liquid medium, such as water, between the tool face and the work piece. The average grain diameters corresponding to abrasive grit size used for experimentation are listed in Table 3.1.1. Grit numbers above 400 onwards result better surface

Table 3.1.1 Typical grain size of boron carbide

Sl. no.	Grit number	Average grain diameter (μm)
1	220	63
2	325	44
3	400	34
4	500	24
5	600	14

finish but lesser value of material removal rate. Slurry concentration based on selected levels has been taken as 30%, 35%, 40%, 45%, and 50% by volume.

3.1.3.4 Tool feeding and control unit

The tool feed motion is controlled by pneumatic control and solenoid valves. The tool can be fed at different feed rate as per requirement. The feed rate during machining can be measured by dial indicator and stopwatch. The total required feed movement during machining can be set for workpiece of different thickness by using tool height adjusting wheel and micrometer assembly. Figure 3.1.13 shows the principal components of the upper

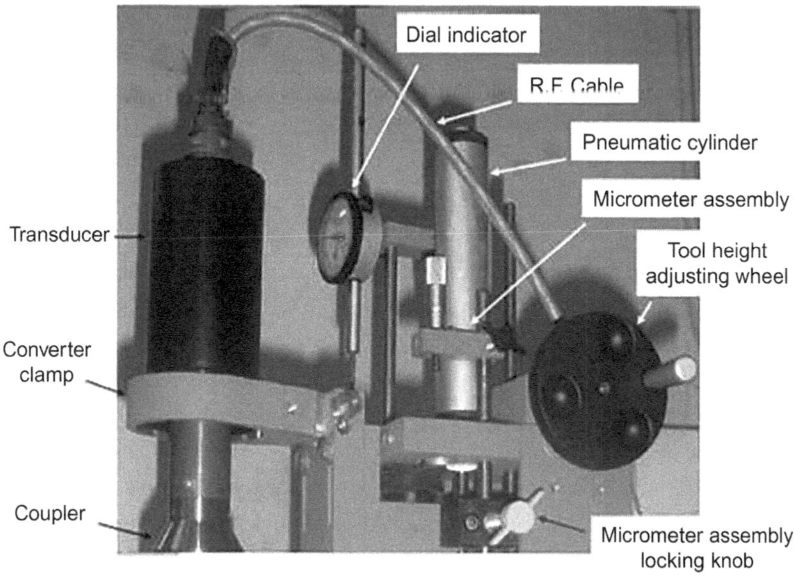

Fig. 3.1.13 Photographic view of upper portion of Mill Module. *(Courtesy Sonic Mill, USA.)*

portion of the mill module comprising dial indicator, micrometer assembly, tool height adjusting wheel and micrometer assembly locking knob.

The photographic view of the electro-pneumatic control unit is shown in Fig. 3.1.7. It is the main control unit, which performs a number of functions for the stationary USM system as listed below:

(i) Provides accurate depth control for the USM process
(ii) Monitors and controls air pressure for raising and lowering *Z*-axis
(iii) Provides a tool lift system to allow slurry under tool
(iv) Controls on/off electrical timing of major functions: slurry, sonic and tool down
(v) Provides manual controls of individual functions in setup and testing
(vi) Provides control for the optional magnetic worktable
(vii) LCD readout timer to monitor cutting time
(viii) Provides control of slurry circulation system
(ix) Optional LCD readout that allows for monitoring of the power supply frequency

Air-compressor is installed to supply required air for this feeding unit at some pressure between 0.41 and 0.82 MPa with a potential volume of 2.5 cubic feet per minute (cfm) or greater. Up air (output) supplies air pressure to raise the tool from the workpiece and feed forced release (output) supplies air pressure to the clutch for tool lift operation. The increase or decrease of air pressure is controlled from the electro-pneumatic control unit. As the air pressure controls the movement of the tool up and down, adequate amount of air must be stored in an air tank for proper functioning of the ultrasonic machine.

3.1.3.5 Power supply unit

The solid-state variable output power supply is equipped with internal or external power control, which converts 50/60 Hz electrical powers into 20 kHz electrical powers. It contains a motherboard assembly, heat sink assembly, fan, system monitor and protection assembly, Sonomax, diode bridge, tuning coil assembly, harness and associated hardware. For optimum efficiency it is equipped with automatic frequency control and automatic load compensation, providing constant output amplitude at desired setting to meet the different energy requirements encountered during operating cycle. The power supply incorporates an overload monitor to protect the system from conditions that could normally result in failure. Normally power supply starts with 20% onwards up to 500 W at 208–240 V I-Phase system. The tuning control is located on the front of the power supply and optimizes the solid-state circuitry for the best possible operation. Tuning control allows for every minor adjustment of the power supply electrical operating frequency to the mechanical resonant frequency of the converter, coupler and horn assembly. Thus, tuning procedure should be performed every time a new horn, new tool tip or a new coupler is used in order to avoid overloading of the power supply. Photographic view of USM power supply unit is shown in Fig. 3.1.7.

3.1.3.6 Specification of ultrasonic machining setup

Typical values of operating process parameters [18] for stationary USM are given in the following Table 3.1.2.

The setup of stationary USM with all its subsystems such as electrical power supply unit, slurry recirculation system, tool feed control device, work holding device and fabricated tools as discussed in this chapter is suitable for machining by setting proper operating conditions.

3.1.4 Details of rotary ultrasonic machining system

Rotary Ultrasonic Machining (RUM) is an upgraded system of stationary USM and RUM has ability to perform various machining operation such as drilling, milling, screw threading, grinding, surface texturing and finishing, etc., at a faster and efficient way. There is simultaneous rotation and ultrasonic vibration of abrasive bonded tool for producing better dimensional accuracy during machining. RUM with workpiece rotation

Table 3.1.2 Typical values for USM operating parameters

Sl. no.	Parameters	Typical values
1.	Power	200 to 1000 W
2.	Frequency	20 to 40 kHz (most frequency 20 kHz)
3.	Abrasive type	Boron carbide, silicon carbide, aluminum oxide, etc.
4.	Abrasive grit no.	100 to 2000
5.	Abrasive slurry concentration	20% to 60% by volume in water
6.	Amplitude of vibration	0.013 to 0.063 mm
7.	Tool tip force	0.45 to 4.5 kg
8.	Tool material	Mild steel, stainless steel, monel, molybdenum
9.	Tool feed rate	0.05 to 2.5 mm/min

arrangement is also another variant for machining. To overcome the drawbacks of stationary USM, i.e., tool wear and slow rate of machining, RUM has been developed to combine conventional diamond grinding and USM process [11].

The Rotary Ultrasonic Machining setup comprises an ultrasonic vibratory-rotary tool system, a data acquisition and control system and a coolant supply system. The schematic representation of details of RUM system is exhibited in Fig. 3.1.14. The photographic view of rotary USM set up is shown in Fig. 3.1.15.

3.1.4.1 Tool vibration unit

The ultrasonic vibratory-rotary tool unit includes an electric motor, transducer, power supply unit and ultrasonic tool spindle. The power supply unit provides ultrasonic frequency electrical power to transducer from low frequency electrical power using high power sine wave generator similar to stationary USM. Like stationary USM, RUM uses either magnetostrictive or piezoelectric transducer to convert high frequency electrical energy into mechanical ultrasonic vibration. Abrasive (diamond) impregnated tool

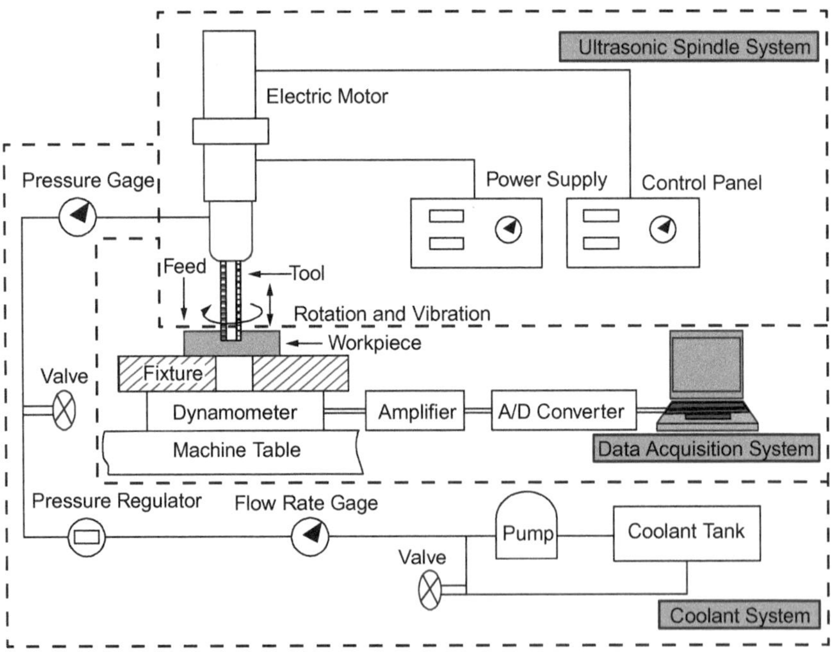

Fig. 3.1.14 Schematic diagram of rotary USM system [1].

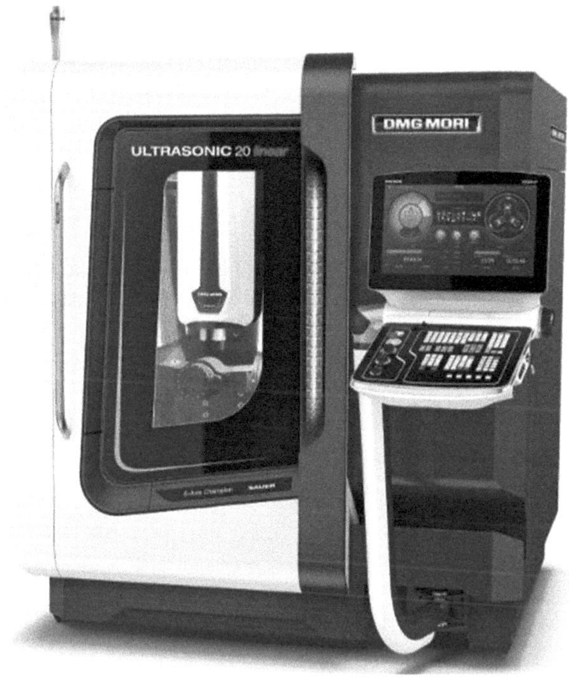

Fig. 3.1.15 Photograph of rotary USM. *(Courtesy DMG MORI USA.)*

receives amplified ultrasonic vibration as per requirement. Electrical motor attached with ultrasonic spindle tool can be rotated at different speed through controller of electrical power supply unit.

3.1.4.2 Data acquisition unit

The data acquisition unit consists of dynamometer along with analog to digital converter which can measure the generated cutting force and torque and computer system to record and plot the measured data for analysis of machining performances. Dynamometer records the cutting force which is one of the important machining criteria of rotary ultrasonic machining process. Less cutting force is obtained at high rotational speed of rotary ultrasonic machining.

3.1.4.3 Coolant supply unit

The coolant supply unit includes coolant tank, pump, flow control valves, pressure gauges to supply the coolant fluid to the tool spindle for efficient internal as well as external cooling during rotary ultrasonic machining.

The coolant supply unit supplies coolant at different pressure as coolant pressure is one of the important process parameter to control MRR, surface roughness and cutting force during rotary ultrasonic machining. For achieving better surface finish cooling pressure should be high to remove the debris from the machining zone.

3.1.5 Influences of process parameters on performance criteria of stationary ultrasonic machining

The ultrasonic machining process parameters such as abrasive grit size, type of abrasives, slurry concentration, power rating, tool feed rate, amplitude and frequency of tool vibration contribute significant role in material removal from workpiece. The machining performance criteria of stationary USM are material removal rate, surface roughness, radial overcut and hole taper, hole circularity and tool wear rate, etc. [2]. An Ishikawa cause and effect diagram as shown in Fig. 3.1.16 represents USM process parameters that may affect USM process criteria.

3.1.5.1 Influences of process parameters on material removal rate

The experiments have been carried out to study the influences of process parameters on material removal rate during ultrasonic drilling on zirconia

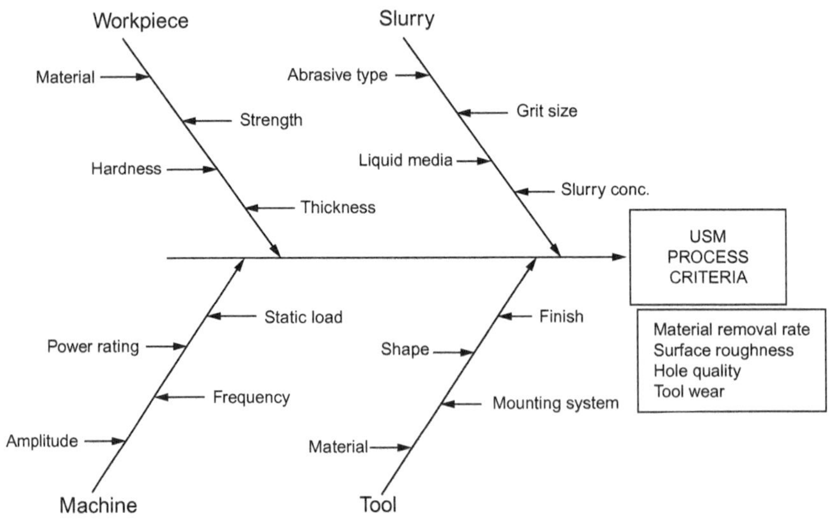

Fig. 3.1.16 Cause and effect diagram for USM process parameters and process criteria.

and alumina ceramics. The abrasive grain diameter is one of the most important parameter in the ultrasonic machining process. Fig. 3.1.17 shows the effect of average diameter of abrasive grain on MRR. From the graph, it is obtained that fine abrasive grain diameter results low MRR but the coarser abrasive grain causes high MRR. Coarser abrasives have sharp corner in every grain particle, therefore MRR is more [19]. The active coarser abrasive grains cut more material from bottom as well as from side of the tool. From this plot, it is obtained that, the zirconia material is removed at higher rate than that of alumina up to certain abrasive grain size.

Figure 3.1.18 presents the effect of abrasive slurry concentration on MRR. The material removal rate depends on the number of effective particles of abrasive grains. If the concentration of abrasive particles in the slurry is more, the more effective particles of abrasive grains come to the working gap [19]. So MRR increases with the increase in abrasive slurry concentration. From this plot, it is obtained that, the higher material removal rate is found for alumina workpiece rather than zirconia up to certain range of abrasive slurry concentration.

Figure 3.1.19 presents the effect of power rating on MRR. From the graph, it is clearly shown that with increase in power rating MRR increases.

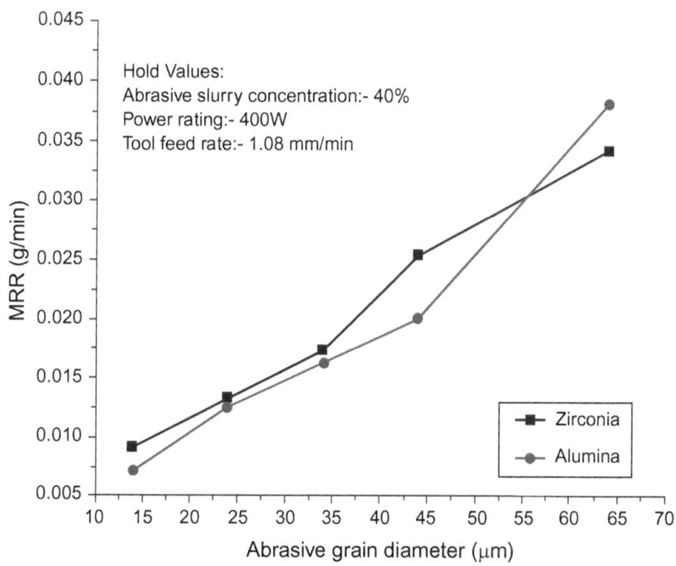

Fig. 3.1.17 Effect of average diameter of abrasive grain on MRR.

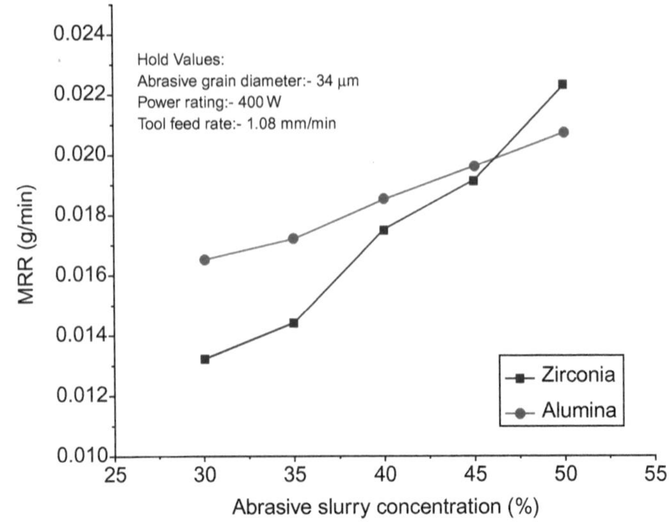

Fig. 3.1.18 Effect of abrasive slurry concentration on MRR.

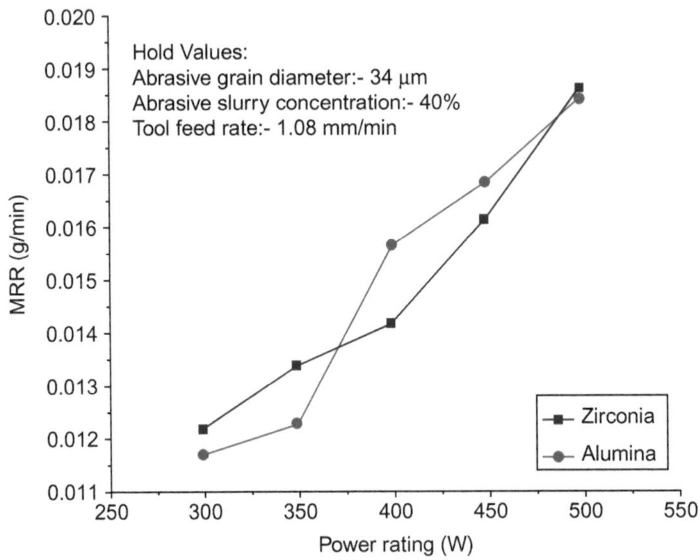

Fig. 3.1.19 Effect of power rating on MRR.

When power rating is higher, then the abrasive particle strikes on the surface of the workpiece with higher value of force. The particles are striking with required momentum that causes rapid fracturing of the work surface due to faster propagation of cracks. For this reason the material removes from the workpiece at faster rate thus promoting more MRR [19]. From this plot, it is

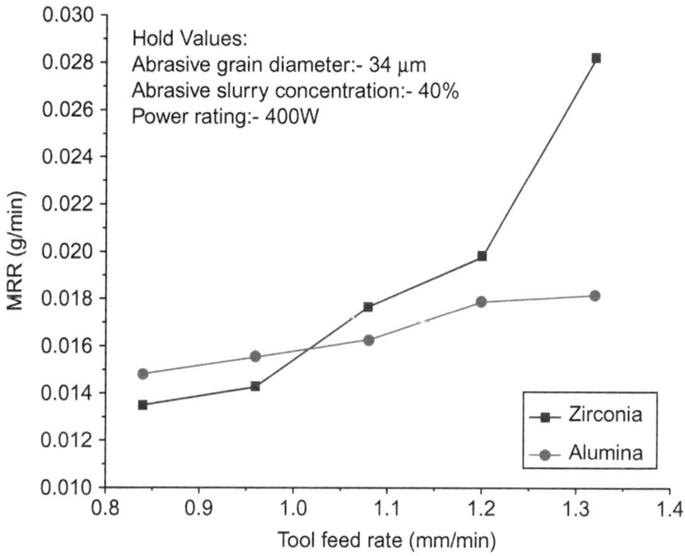

Fig. 3.1.20 Effect of tool feed rate on MRR.

exhibited that, the material removal rate for both materials such as zirconia and alumina are not so much differing at particular value of power rating.

It is observed from Fig. 3.1.20 that the tool feed rate affects the material removal rate during the stepped hole generation by USM on zirconia and alumina workpiece very significantly. While the tool is moved downward direction at high value of tool feed rate, then the crack propagation rate on the workpiece is effectively more. Material is removed from the workpiece at faster rate. For this reason higher value of tool feed rate causes higher MRR [19]. The change of MRR values are more for zirconia by varying tool feed rate but for alumina, these changes are less.

3.1.5.2 Influences of process parameters on overcut criteria

During USM process, the size of generated hole is larger than the size of tool used. For achieving accurate steeped hole ceramics such as zirconia and alumina by USM, the influences of process parameters on overcut criteria have been analyzed.

Figures 3.1.21 and 3.1.22 show the effect of abrasive grain diameter on overcut of larger diameter (OLD) and overcut of smaller diameter (OSD) of stepped hole respectively. From the graph, it is observed that fine abrasive

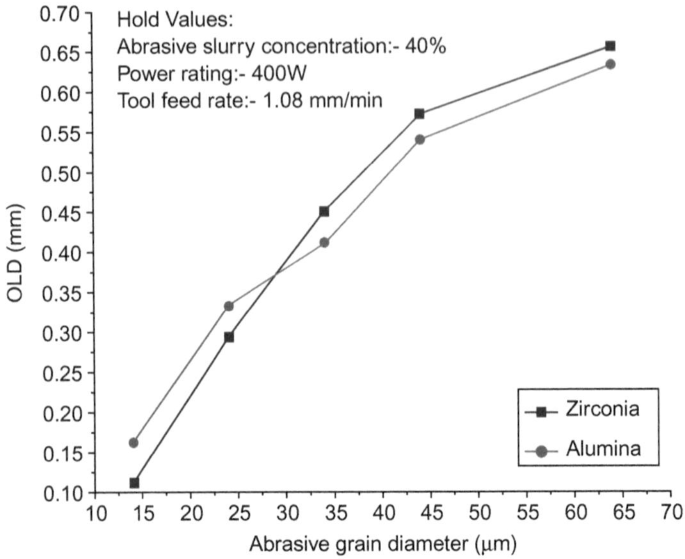

Fig. 3.1.21 Effect of abrasive grain diameter on OLD.

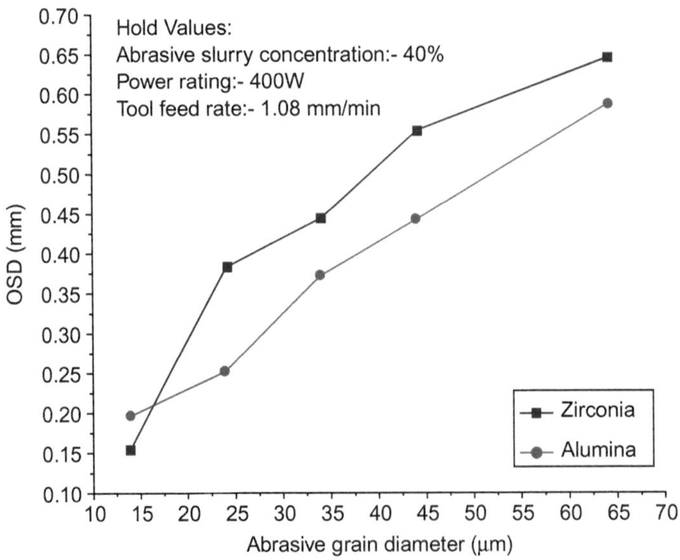

Fig. 3.1.22 Effect of abrasive grain diameter on OSD.

grain produces hole with low overcut, i.e., OLD and OSD but the coarser abrasive grain generates hole with high overcut. Average larger grain diameter is 64 μm and average smaller grain diameter is 14 μm. The contact surface area is less when the fine and smaller abrasive grain is used. Using finer abrasive particles, diametrical overcut of larger hole and smaller hole generated by USM is less [19]. So as to achieve low overcut, i.e., OLD and OSD, the abrasive with low value of grain diameter is preferred. From the graph, it is shown that OLD values are not so much changing for both bio-ceramic materials. From Fig. 3.1.22, it has been found that the OSD of stepped hole produced on alumina is lower than that of zirconia.

Figure 3.1.23 exhibits the abrasive slurry concentration on diametrical overcut of large diameter hole, i.e., OLD of drilled hole. If slurry concentration is low then profile accuracy is good. When abrasive slurry concentration is low, then total mass of abrasive in the working zone is less so hammering force is small. For this reason overcut value is low using low abrasive slurry concentration [19]. Overcut of larger diameter hole for alumina is more than zirconia.

Figure 3.1.24 exhibits the effect of power rating on diametrical overcut, i.e., OLD of large diameter stepped hole. From the graph, it is clearly observed that with the increase in power rating, the overcut, i.e., OLD increases. Actually high value of power rating means the tool vibrates with high power and abrasive particle strikes on the surface of the workpiece with high energy. For this reason the material removes from the workpiece at faster rate [19]. So the overcut is increased. From the graph, it is clearly

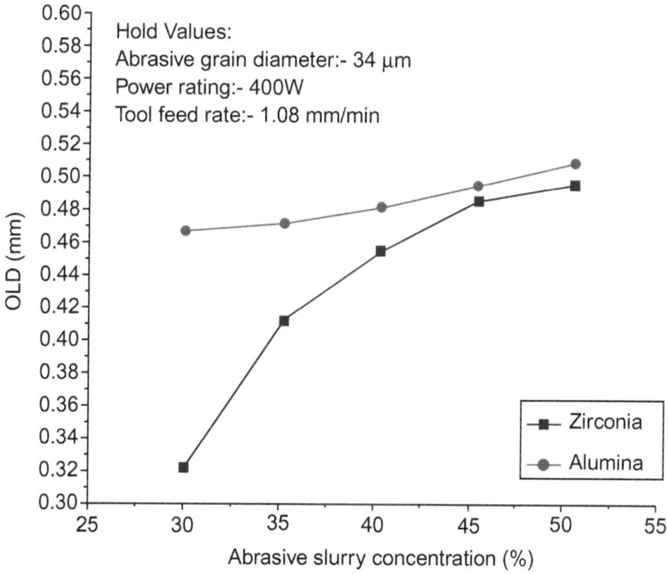

Fig. 3.1.23 Effect of abrasive slurry concentration on OLD.

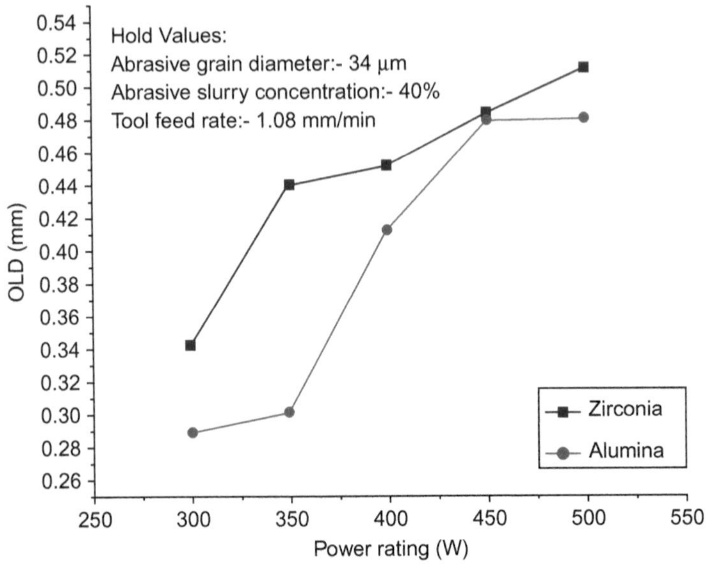

Fig. 3.1.24 Effect of power rating on OLD.

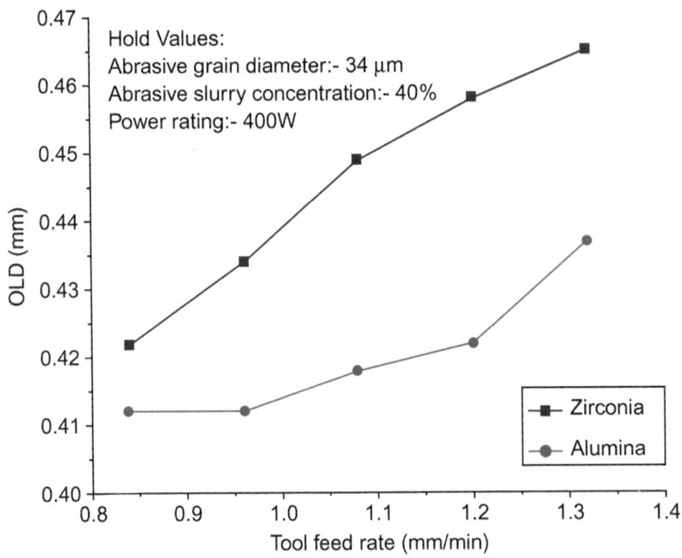

Fig. 3.1.25 Effect of tool feed rate on OLD.

shown that if the power rating is increasing, the overcut is also increasing. So for achieving low overcut lower value of power rating is preferred.

Figure 3.1.25 shows that effects of tool feed rate on overcut, i.e., OLD of USM drilled steeped hole. From the graph, it is obtained that higher value of tool feed rate gives maximum overcut. With higher value of applied tool feed rate, more material is removed by the action of the abrasive particles. Therefore diametrical overcut increases with increase in applied tool feed rate [19]. Low value of tool feed rate is preferred to achieve low overcut, i.e., overcut of larger diameter (OLD) of stepped hole. From the figure it has been observed that overcut is more for zirconia workpiece than that of alumina ceramics during USM stepped hole drilling operation.

3.1.5.3 Influences of process parameters on circularity error

For producing better quality of steeped hole on ceramics, circularity error has to be reduced by controlling the USM process parameters. The circularity error is determined as the difference between the maximum and minimum diameter divided by two. The influences of USM process parameters on circularity error has been analyzed and discussed hereinafter.

Figures 3.1.26 and 3.1.27 show the effect of abrasive grain diameter on circularity error of larger diameter (CELD) and circularity error of smaller diameter (CESD) of stepped hole respectively. From the graph, it is found that fine abrasive grain produces low value of circularity error, i.e., CELD and CESD but the coarser abrasive grain produces high value of circularity error, i.e., CELD and CESD. Average smaller grain diameter is 14 μm. The impacting area of smaller abrasive grain is very low compare to larger

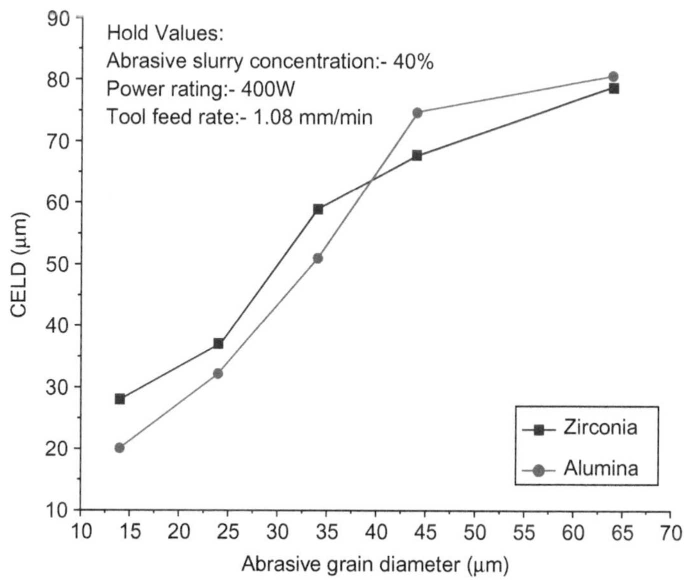

Fig. 3.1.26 Effect of abrasive grain diameter on circularity error, CELD.

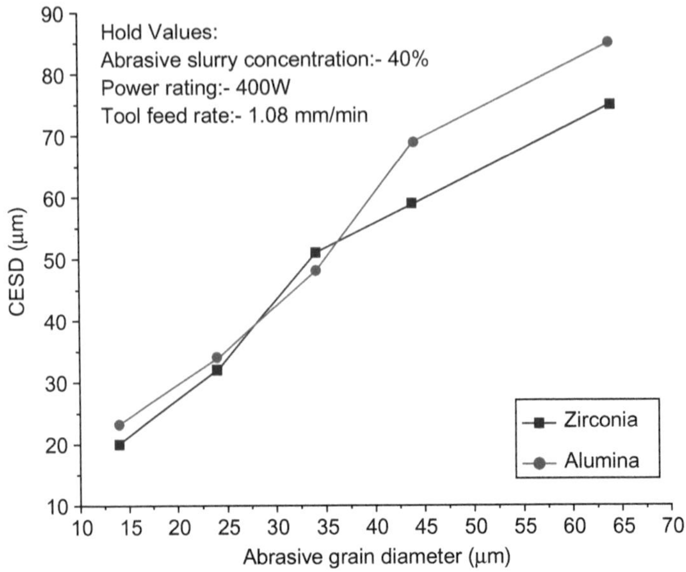

Fig. 3.1.27 Effect of abrasive grain diameter on circularity error, CESD.

abrasive grain. So the circularity error of larger diameter (CELD) hole and smaller hole diameter (CESD) generation by USM is less [19]. Circularity error of larger diameter (CELD) hole and smaller hole diameter (CESD) hole for alumina and zirconia are not so much changing for particular abrasive grain diameter.

Figure 3.1.28 exhibits the influence of abrasive slurry concentration on circularity error, i.e., CELD of USM drilled stepped hole. When abrasive slurry concentration is very low, then total mass of active abrasive grain is low so hammering force is very small. For this reason the value of circularity error, i.e., CELD is low using low abrasive slurry concentration [19]. Circularity error of larger diameter (CELD) hole values for alumina are higher than zirconia with varying abrasive slurry concentration.

Figure 3.1.29 illustrates that effect of power rating on circularity error, i.e., CELD of USM drilled stepped hole. From the graph, it is clearly shown that if the power rating is increasing the circularity error, i.e., CELD is increasing. Actually when high value of power rating is used, the abrasive particle strikes on the surface of the workpiece with high kinetic energy. The material removal is not uniform. So the circularity error is increased [19]. For achieving low circularity error lowers value of power rating is preferred. CELD values for alumina are lower than zirconia with varying power rating.

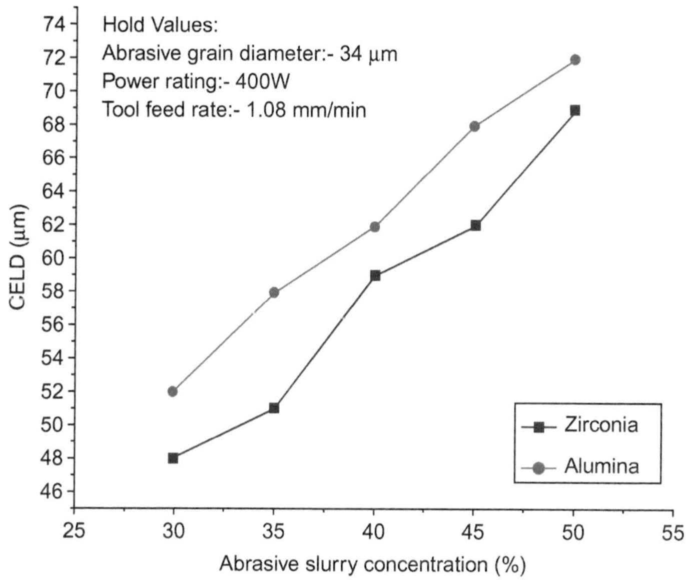

Fig. 3.1.28 Effect of abrasive slurry concentration on circularity error, CELD.

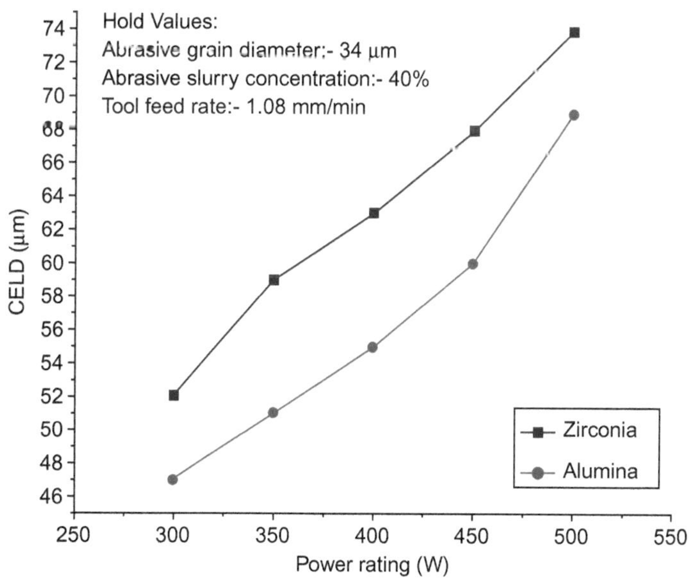

Fig. 3.1.29 Effect of power rating on circularity error, CELD.

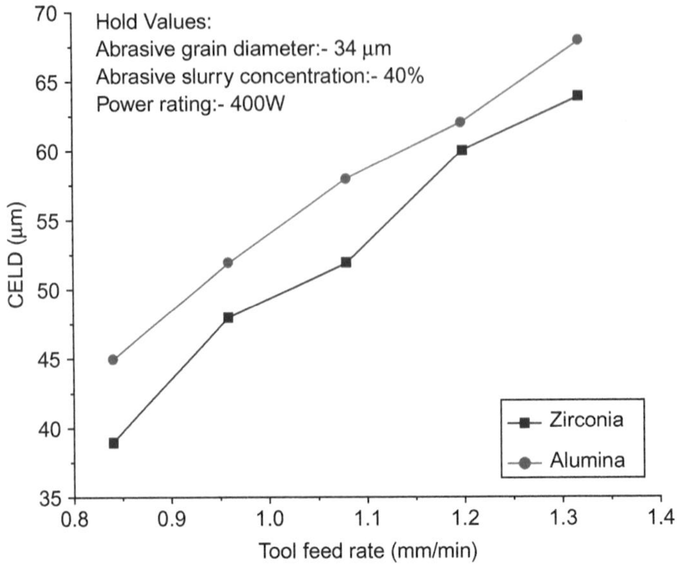

Fig. 3.1.30 Effect of tool feed rate on circularity error, CELD.

Figure 3.1.30 shows that effect tool feed rate on circularity error, i.e., CELD of USM drilled stepped hole. From the graph, it is obtained that higher value of tool feed rate gives higher circularity error. With higher value of applied tool feed rate, non-uniform removal of material are taken place. Therefore circularity error increases with increase in applied tool feed rate [19]. For achieving low value of circularity error of larger diameter and smaller diameter of stepped hole lower value of tool feed rate is preferred. Circularity error of larger diameter (CELD) hole values for alumina are higher than zirconia with varying tool feed rate, but circularity error of smaller hole diameter (CESD) hole values are not so much varying for both alumina and zirconia.

3.1.6 Influences of process parameters on performance criteria of rotary ultrasonic machining

The performance of rotary ultrasonic machining process is influenced by various process parameters related to workpiece, tool, machine, coolant, etc. The important machining process criteria are material removal rate, surface roughness, cutting force, tool wear rate, edge chipping size, etc., for rotary ultrasonic machining for various applications. Figure 3.1.31 shows the cause effect diagram showing process parameters and machining characteristics of rotary USM process.

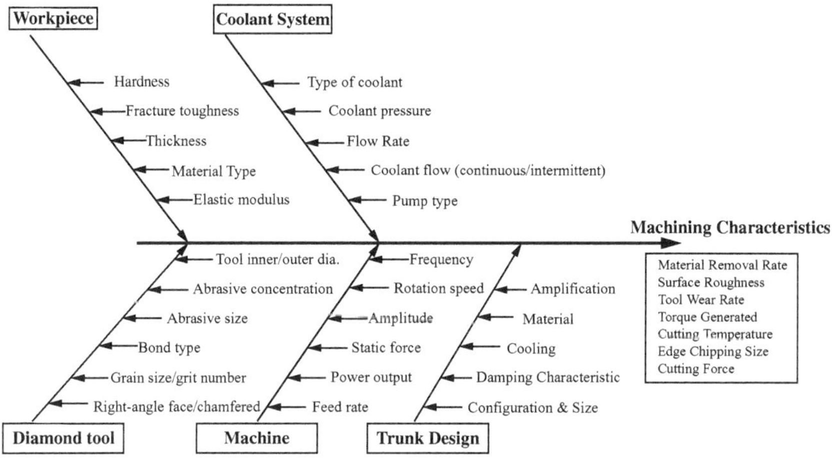

Fig. 3.1.31 Process parameters and machining characteristics of RUM [11].

3.1.6.1 Influences of process parameters on material removal rate

The effects of controllable process parameters of RUM such as static force, amplitude of ultrasonic vibration, rotating speed and grit diameter on MRR during rotatory ultrasonic machining of magnesia stabilized zirconia materials has been analyzed by Pie et al. [12]. The variation of MRR with static force is as shown in Fig. 3.1.32. MRR increases with increase in static force during RUM. From figure it is observed that static force (Fs) is increased from 100 to 600 N at different amplitude of vibration. It was also observed that MRR increases with increase in amplitude of vibration (A) which varies from 0.01 to 0.04 mm. Tool rotational speed (S), average diameter (d) of abrasive, nos. of working abrasive (n) on tool surface are taken as 1000 rpm, 0.05 mm and 100, respectively. The variation of MRR with amplitude of vibration at different static force is shown in Fig. 3.1.33. From the figure it is observed that the rate of increase in MRR with amplitude of vibration is more at higher value of applied static force.

In RUM, tool rotational speed generally varies from 500 rpm to 3500 rpm. The variation of MRR with tool rotational speed is shown in Fig. 3.1.34. From the Fig. 3.1.34 it is observed that while lower static force is applied, effect of tool rotational speed on MRR is very less, but at higher applied static force, MRR increases with the increase in rotational speed.

The variation of MRR with abrasive grit diameter at different static force is shown in Fig. 3.1.35. The average diameter of abrasive generally varies

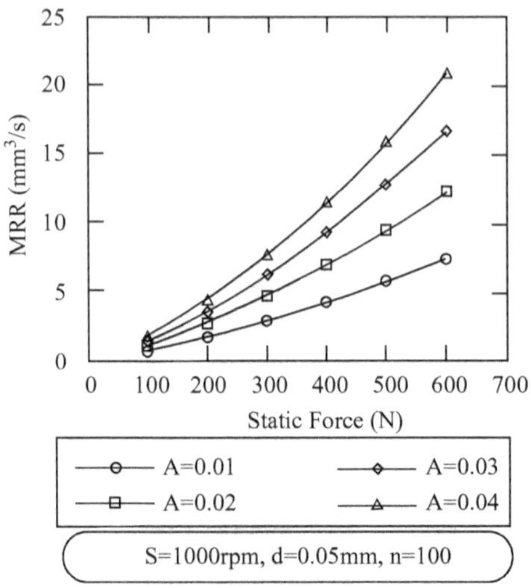

Fig. 3.1.32 Variation of MRR with static force at different amplitude of vibration [12].

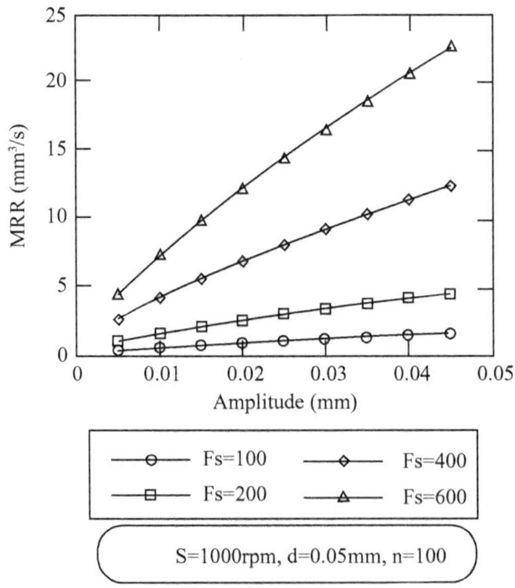

Fig. 3.1.33 Variation of MRR with amplitude of vibration at different static force [12].

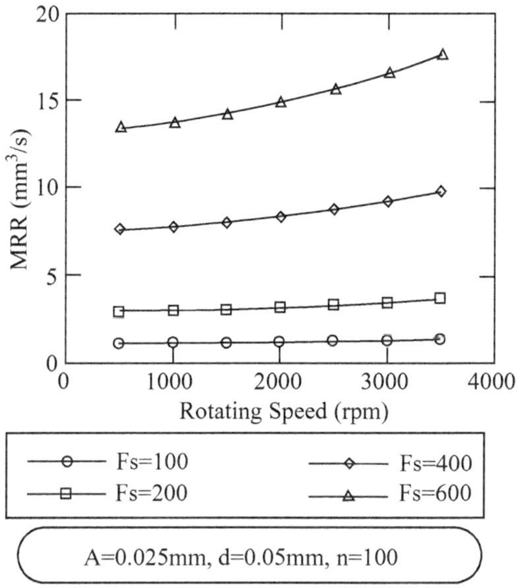

Fig. 3.1.34 Variation of MRR with tool rotational speed at different static force [12].

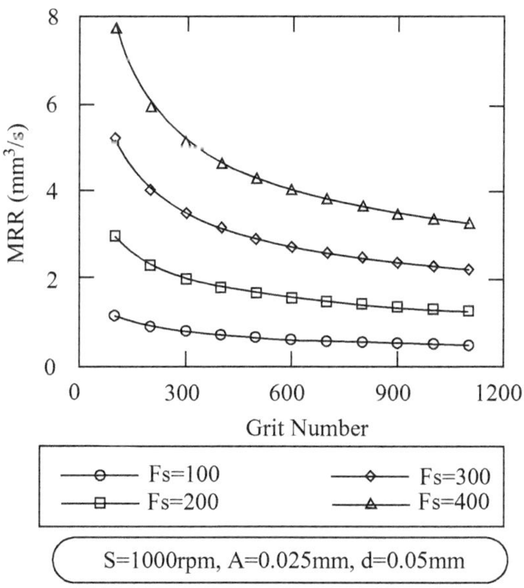

Fig. 3.1.35 Variation of MRR with Abrasive grit diameter at different static force [12].

from 0.05 to 0.4 mm during RUM. From the figure MRR decreases with increase in size of abrasive particle at fixed value of applied static force, because small sharp abrasive particle can penetrate at larger indentation depth into workpiece materials. For RUM, finer abrasive particle with larger applied static force removes more material from workpiece.

3.1.6.2 Influences of process parameters on surface roughness

The influences of ultrasonic power and tool rotational speed on surface roughness during rotatory ultrasonic machining of carbon fiber reinforce plastic (CFRP) composite materials has been analyzed by Cong et al. [20]. The effect of coolant supply of cold air and cutting fluid on surface roughness has also been studied. The variation of surface roughness of machined surface with ultrasonic power and rotational speed at different coolant supply are shown in Fig. 3.1.36. The surface roughness is low while cutting fluid is supplied instead of cold air. The variation of surface roughness with ultrasonic power is less while cutting fluid is applied. But surface roughness increases with ultrasonic power while cold air is supplied as coolant. The surface roughness drastically deceases with increase in tool rotational speed while cold air is used as coolant, but there is less effect of tool rotational speed on surface roughness while cutting fluid is supplied during rotatory ultrasonic machining of carbon fiber reinforced plastic (CFRP) composite materials.

3.1.6.3 Influences of process parameters on cutting force

The effect of tool spindle speed, ultrasonic power and tool feed rate on cutting force during RUM of titanium alloy has been analyzed by Churi et al. [15]. The variation of cutting force with tool spindle speed, ultrasonic power and tool feed rate are shown in Fig. 3.1.37 (a) to (c), respectively.

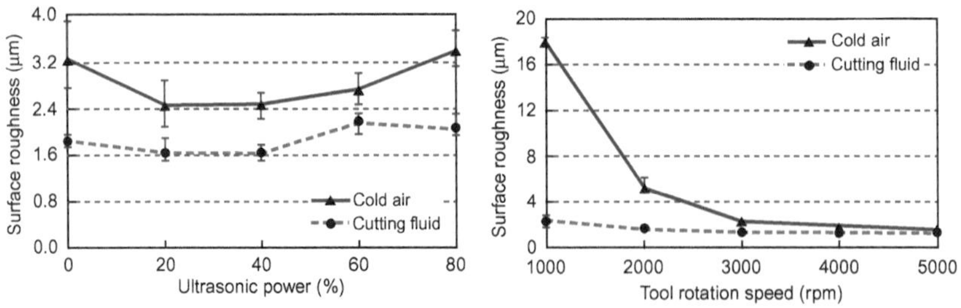

Fig. 3.1.36 Variation of surface roughness with ultrasonic power and tool rotational speed [20].

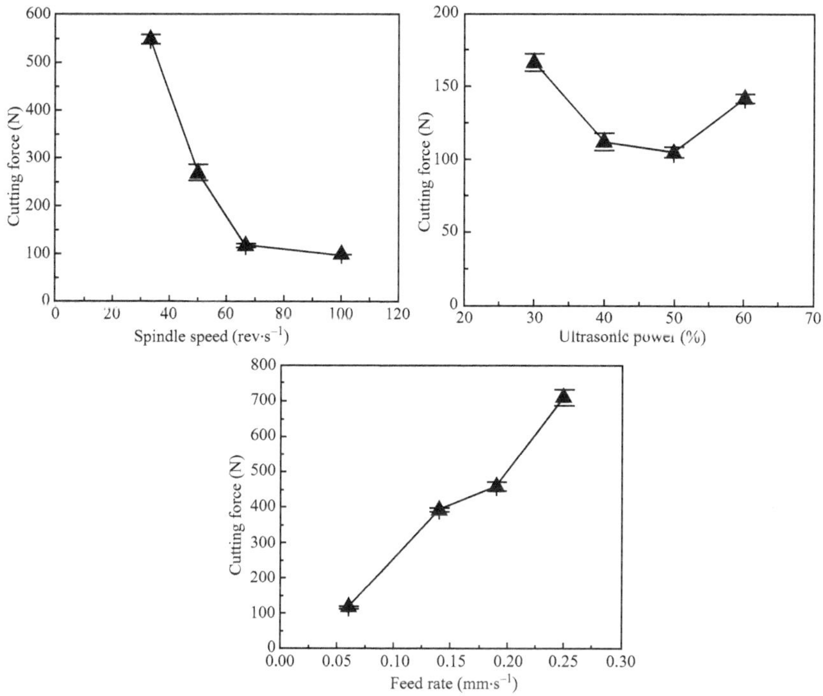

Fig. 3.1.37 Variation of cutting force with tool spindle speed (a), ultrasonic power (b), and tool feed rate (c) [15].

As tool rotational speed increases the cutting force decreases because the coefficient of friction decreases with increase in rotational speed. As ultrasonic power increases the cutting force initially decreases, reaches minimum value and again increases. As tool feed rate increases the cutting force steadily increases. For achieving lower cutting force during RUM of titanium alloy, higher tool rotational speed, medium ultrasonic power and lower tool feed rate are preferred.

3.1.7 Advantages and limitations of ultrasonic machining

3.1.7.1 Stationary ultrasonic machining

Advantages of ultrasonic machining process are discussed here:

(i) The process can machine electrically conductive as well as non-conducting hard and brittle materials.

- **(ii)** There is no direct contact between tool and workpiece therefore, the machined surface is free from excess stress and mechanical damages.
- **(iii)** There is no thermal damage such as thermal cracks and heat affected zone on the machined surface as the water abrasive slurry is used during machining.
- **(iv)** It can produce rough as well as finish surface with good structural integrity.
- **(v)** It can produce burr-free surface.
- **(vi)** It can machine chemically active as well as inert materials.
- **(vii)** It can create any shape depending upon tool shape on very hard and brittle materials, such as glass, silicon, quartz crystal, sapphire, nitride, ferrite and optics.
- **(viii)** Using micro-tool, it can generate micro complex shaped cavity or hole on workpiece.
- **(ix)** Power consumption is less.

Limitations of ultrasonic machining process:

- **(i)** Material removal rate (MRR) is less during finish machining.
- **(ii)** There is change of shape and size of tool after some period of machining due to tool wear and it needs tool changing.
- **(iii)** There is limitation of large depth of small hole.
- **(iv)** There is need of frequent tuning of ultrasonic tool vibration unit by selecting proper weight and shape of horn, tool and tool holding assembly.

3.1.7.2 Rotary ultrasonic machining

Advantages of rotary ultrasonic machining process are discussed here:

- **(i)** The cutting force is reduced significantly compared to conventional diamond grinding process.
- **(ii)** The MRR is increased significantly compared to stationary USM.
- **(iii)** High quality holes are generated by RUM compared to stationary USM.
- **(iv)** Overcut is less and accuracy is high.
- **(v)** Tool life is long.
- **(vi)** The material removal is due to brittle fracture as well as plastic deformation as abrasive bonded tool is used. As a result RUM is applied for brittle as well as ductile materials.
- **(vii)** RUM can perform deep micro-hole drilling operation as there is no problems of availability of abrasive slurry in the gap between tool and workpiece.

(viii) There is no hazard of supply of abrasive slurry as only coolant is used instead of abrasive slurry.

Limitations of ultrasonic machining process:

(i) Rotary USM is expensive;
(ii) Generally high cost diamond dust is used as abrasive for tool;
(iii) The system integration is complex.

3.1.8 Applications of ultrasonic machining

3.1.8.1 Stationary ultrasonic machining

Applications of stationary ultrasonic machining process are given as follows:

(i) This process is mostly suitable for machining hard and brittle materials like as alloys, glass, fiber material, ceramics, carbides, etc.
(ii) This process is applicable for drilling holes in borosilicate glass, quartz for the fabrication of sensors.
(iii) The process can drill very fine holes on shafts and gears which are used in helicopter power transmission.
(iv) It can produce round, square, hexagonal, hemispherical and irregular shaped cavity.
(v) The process is applicable for machining of micro-structured glass wafers for micro electromechanical systems (MEMS) applications.
(vi) The process is applicable for generating stepped hole on bio-ceramics such as zirconia for dental application.
(vii) The process is applicable for generating hemispherical shaped cavity on bio-ceramics such as hydroxyapatite for hip joint application.
(viii) USM process can be applied in aerospace, tool/mold/machine construction, pump and valve industry, time piece industries, etc.

3.1.8.2 Rotary ultrasonic machining

Applications of rotary ultrasonic machining process are given as follows:

(i) Rotary ultrasonic machining (RUM) has a wide range of applications in present day industries such as aircrafts, automobiles, sports goods, medical tools, computers, electronics, optical, etc. [11].
(ii) RUM can machine ceramic materials like zirconia and glass ceramics used for automotive and optical industries.
(iii) It can machine carbon fiber reinforced polymer (CFRP) composites that has wide applications in aircrafts, automotive industries and also in the fabrication of engine blades.

(iv) It can machine titanium and its alloys, which have been used in the fabrication of turbine parts such as rotors, stator and compressor blades, etc.

(v) It can machine stainless steel used for various applications such as in automobile, medical tools manufacturing and house appliances.

(vi) It is capable of machining of engineering metals and alloys, dental ceramics such as macor and dental zirconia as used in dental restorations.

(vii) It can machine sapphire (alumina in single crystal form), generally applicable in the electronics fields, particularly in the fabrication of chips and circuits, etc.

(viii) It can process optical and brittle materials such as K9 and BK7 glass for complex shape generation in various applications.

(ix) RUM process can be applied for various machining operations such as drilling, milling, turning, grinding, surface texturing and finishing operations.

3.1.9 Advancement in ultrasonic machining technology

Ultrasonic Machining (USM) technology has become very much popular because of it advancement in the area of generation of intricate shape on advanced materials such as ceramics and composites for wide applications in the field of bio-medical, aerospace and automobile industries, etc. Some of the outcome of advanced research in USM technology is highlighted here in after.

A stepped hole on zirconia bio-ceramic material has been generated by USM process and it has wide applications in dental implants. Figure 3.1.38 (a) shows developed tool for USM and (b) stepped hole generated on zirconia by USM process. During USM stepped hole generation, boron carbide abrasive of 14 μm, abrasive slurry concentration of 47%, power rating of 500 W, feed rate of 1.03 mm/min were selected as process parameter [21].

Ultrasonic Machining (USM) process can be effectively utilized for fabricating square stepped hole with higher accuracy on bio-ceramics, i.e., zirconia. For achieving the better machining accuracy, i.e., minimum flat to flat overcut, corner to corner overcut and diametrical overcut of square stepped hole, the lower values of abrasive grain diameter, abrasive slurry concentration, power rating and tool feed rate are preferred [22]. The square stepped hole on zirconia ceramics has been produced by

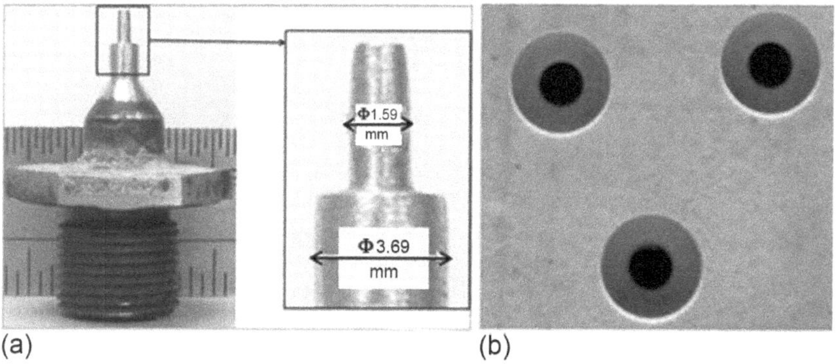

Fig. 3.1.38 (a) Developed tool and (b) step hole on zirconia [21].

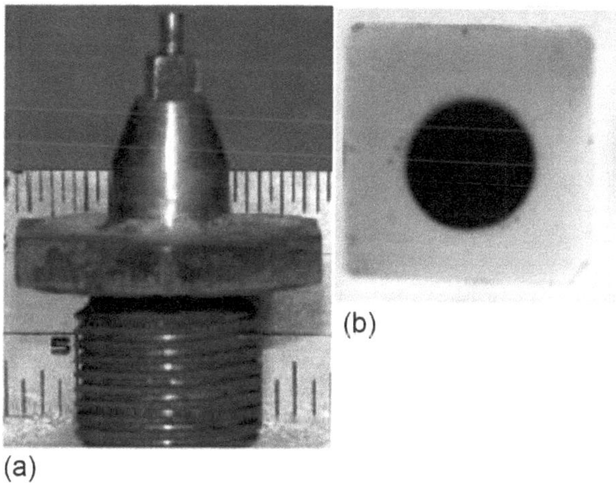

Fig. 3.1.39 (a) Developed square stepped tool and (b) square stepped hole on zirconia.

stationary USM using the developed USM tool as shown in Fig. 3.1.39 (a) and (b), respectively.

An accurate hexagonal hole has been produced on alumina bio-ceramics by USM as shown in Fig. 3.1.40 while boron carbide of 63 μm, abrasive slurry concentration of 50%, power rating of 400 W, feed rate of 1.08 mm/min and slurry flow rate of 10 L/min are selected for USM drilling [23].

To obtain hole with good accuracy, wax coating has been provided on the glass substrate surface before the ultrasonic machining. The coating protects the surface of the glass during USM, and is removed later by clean-up process. Experimental results shows that, the hole accuracy depends on

Fig. 3.1.40 Photographic view of the USM drilled hexagonal hole [23].

coating thickness [24]. The photographic views of holes produced by USM with coating and without coating are shown in Fig. 3.1.41.

A circular through hole on 3.65 mm thick alumina workpiece has been drilled by USM as shown in Fig. 3.1.42 with combination of process parameters such as average grit size of 14 μm, slurry concentration of 40%, power rating of 50% and tool feed rate of 1.20 mm/min at constant flow rate of 8 L/min [26]. The entrance and exit diameter of hole were found as 9.97 and 9.93 mm, respectively, with lesser taper angle of 0.31 degrees [25]. The edge chippings at the exit is observed and it needs precaution and further research to reduce or eliminate the edge chipping which is one of the limitation of the stationary USM process.

Ultrasonic machining has been successfully performed to generate hemispherical shaped cavity on hydroxyapatite bio-ceramics. For achieving the higher material removal rate, the larger abrasive grain diameter, high value of abrasive slurry concentration, high power rating, and high tool feed rate are preferred. For getting better profile accuracy, i.e., minimum diametrical deviation and circularity error, the smaller abrasive grain diameter, low abrasive slurry concentration, low power rating, and low tool feed rate are recommended. Fig. 3.1.43 shows the surface development of hemispherical cavity generated on workpiece by setting USM process parameters such as abrasive grain diameter of 34 μm, abrasive slurry concentration of 40%,

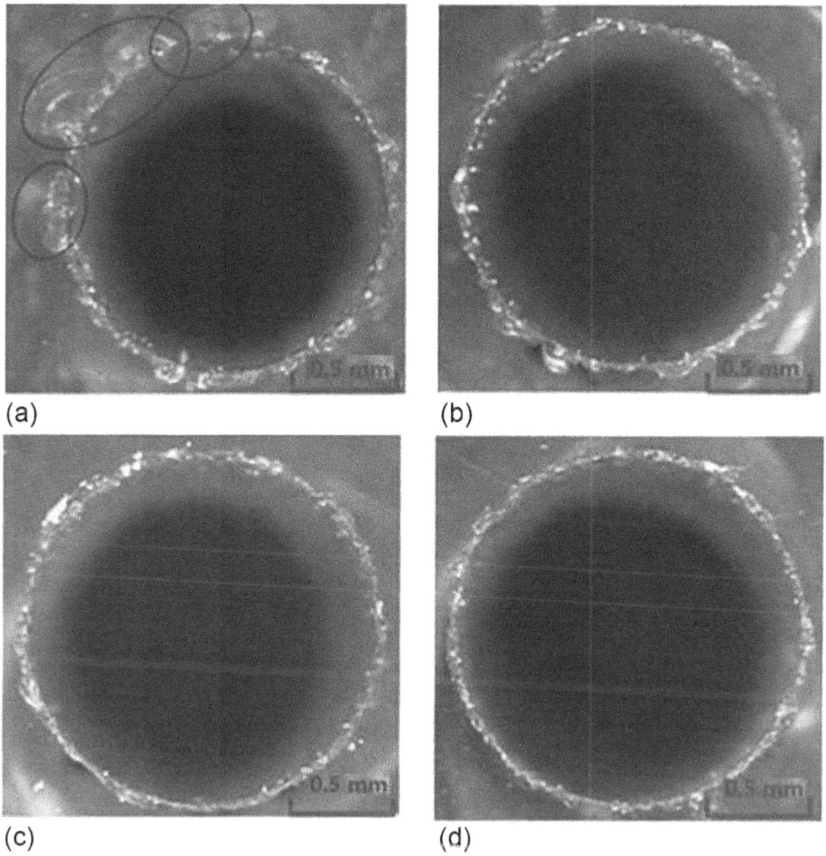

Fig. 3.1.41 Photographic views of holes produced by USM with coating and without coating [24]. (a) without the coating, (b) coating thickness: 0.3 mm, (c) coating thickness: 0.6 mm, and (d) coating thickness: 1.0 mm.

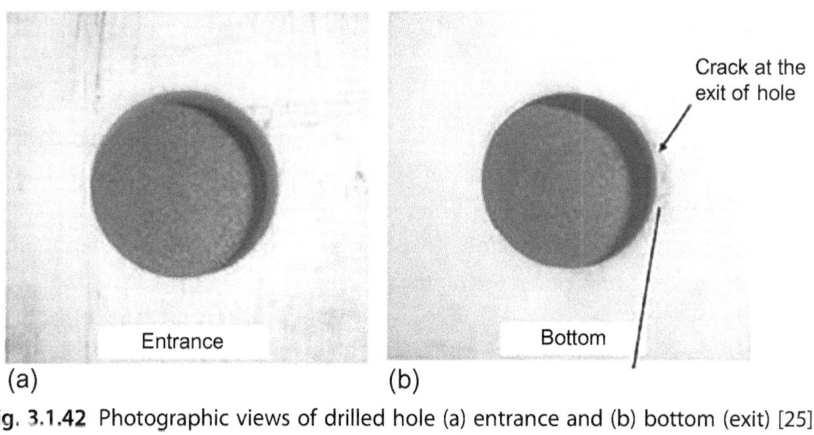

Fig. 3.1.42 Photographic views of drilled hole (a) entrance and (b) bottom (exit) [25].

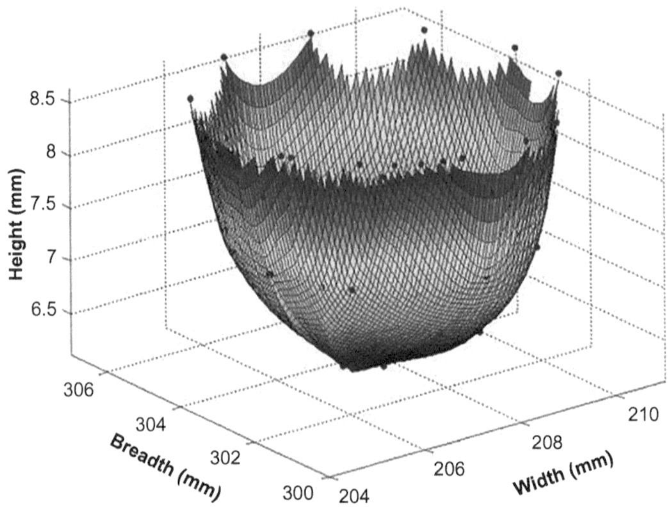

Fig. 3.1.43 Surface development of hemispherical cavity generated on workpiece [27].

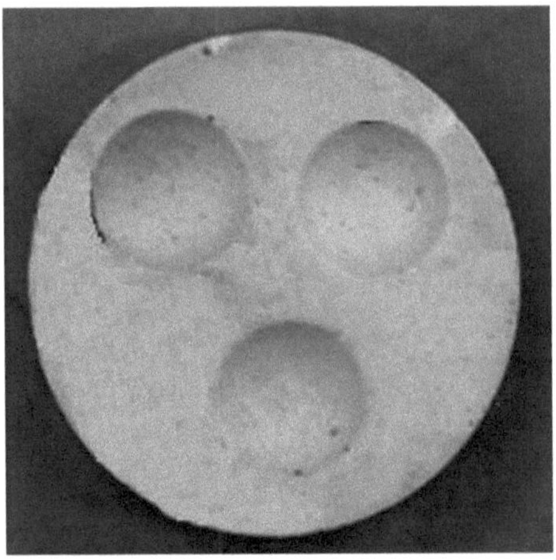

Fig. 3.1.44 Photographic view of hemispherical cavity on hydroxyapatite bio-ceramic [27].

power rating of 400 W and tool feed rate of 1.1 mm/min [27]. Fig. 3.1.44 shows the photographic view of hemispherical cavity generated on hydroxyapatite bio-ceramic workpiece materials.

The accurate tool development for ultrasonic machining to generate the replica of existing hemispherical shape cavity has been done using the concept of reverse engineering (RE) [26]. With the help of this concept of RE, the measurement of data cloud was taken of existing hemispherical cavity of bio-ceramics. Data cloud was imported to RE software and developed 3D surface model of the product with the help of Geomagic Studio12. The CAD model of the cavity was developed after noise reduction and patch filling. Next, the 3D model was validated with the hemispherical cavity. After providing certain tolerances, the CAD model of tool was developed and then the tool was manufactured. The developed tool as shown in Fig. 3.1.45 for generating a hemispherical cavity on the zirconia workpiece is much more accurate. It produces the cavity with less diametrical deviation and less circularity error. Figure 3.1.46 shows the photographic view of developed Tool and Hemispherical Cavity generated during ultrasonic machining. Direct manufacturing of the USM tool using rapid prototyping techniques utilizing a CAD model has the scope of reducing production time for any complex-shaped geometry tool for ultrasonic machining.

Rotary Ultrasonic Machining of Al_2O_3 dental ceramics has been performed to analyze the surface and subsurface damage of machined surface with and without ultrasonic vibration [28]. In comparison to diamond

Fig. 3.1.45 Photographic view of developed tool for USM [26].

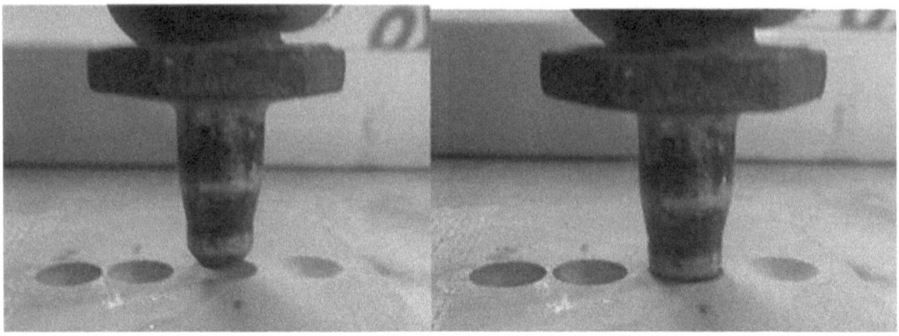

Fig. 3.1.46 Photographic view of developed tool and hemispherical cavity generated during ultrasonic machining [27].

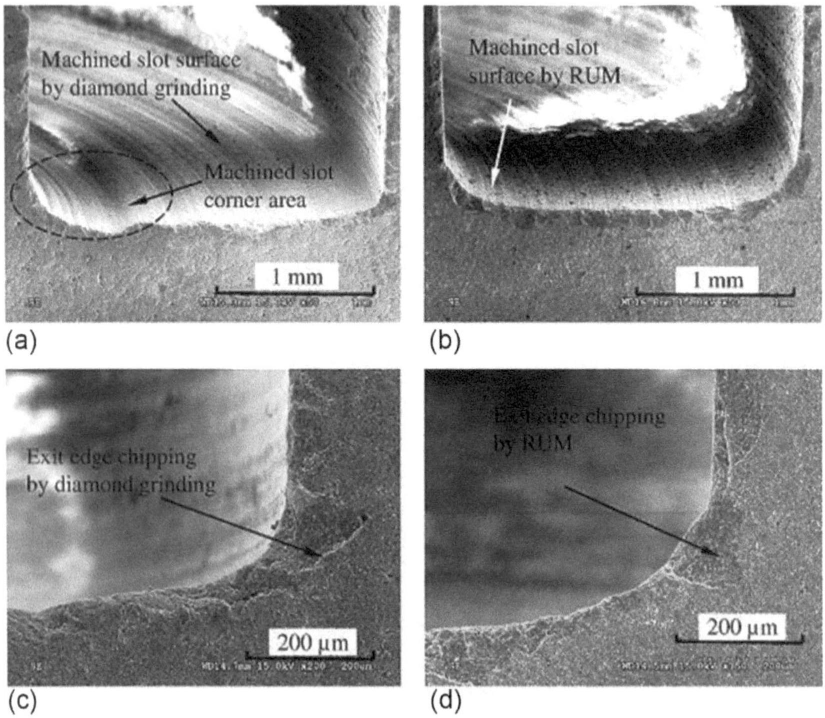

Fig. 3.1.47 SEM micrographs of machined slot produced by diamond grinding and RUM on dental ceramics [28]: (a) surface quality (by diamond grinding), (b) surface quality (by RUM), (c) edge chipping (by diamond grinding), and (d) edge chipping (by RUM).

grinding, RUM generates better quality surface with less subsurface damage (crack of <5 μm depth) and minimum chippings at the exit edge during slot machining. Figure 3.1.47 (a)–(d) shows the SEM micrographs of machined slot produced by diamond grinding and RUM on dental ceramics.

Rotary ultrasonic machining of carbon fiber reinforced plastic (CFRP) and titanium (Ti) stacks have been performed using variable feed rate. Grooving operation has been performed with high feed rate for CFRP plate and low feed rate for Ti plate during rotary ultrasonic machining. Lower cutting force and less tool wear and higher groove depth were reported during rotary ultrasonic machining of CFRP/Ti stacks [29].

Ultrasonic machining technology has great potential in the form of stationary USM as well as rotary USM for processing advanced engineering materials such as glass, ceramics, composites, etc., irrespective of its electrical and thermal properties. Both stationary as well as rotary USM processes are also used for micro-machining applications by selecting the machining conditions in terms of process parameters and tools. USM tool development is a challenging issue as the replica of tool is generated on workpiece. The system details of stationary and rotary USM have been presented in this chapter to understand the working principle of ultrasonic machining process. The material removal mechanism of stationary and rotary USM has been discussed to identify the influencing process parameters affecting the machining criteria of the process. For achieving optimum machining performances such as maximum MRR, minimum overcut, dimensional deviation, surface roughness, edge chipping of workpiece and tool wear, etc., optimization based research studies on stationary and rotary USM has been carried out by various researchers, although experimental based research results are reported and the influences of process parameters on machining process criteria has been analyzed in this chapter.

References

[1] Z.C. Li, Y. Jiaoa, T.W. Deinesa, Z.J. Pei, Rotary ultrasonic machining of ceramics matrix composites: feasibility study and designed experiments, Int. J. Mach. Tools Manuf. 45 (12–13) (2005) 1402–1411.

[2] L.C.H. Vela, Optimal Selection of Process Parameters of Ultrasonic Machining (USM) System (PhD thesis). Jadavpur University, India, 2012.

[3] F. Guiberteau, N.P. Padture, B.R. Lawn, Effect of grain size on Hertzian contact damage in alumina, J. Am. Ceram. Soc. 77 (7) (1994) 1825–1831.

[4] M.C. Shaw, Ultrasonic grinding, Microtechnic 10 (10) (1956) 257–265.

[5] N.K. Jain, V.K. Jain, Modeling of material removal in mechanical type advanced machining processes: a state-of-art review, Int. J. Mach. Tools Manuf. 41 (11) (2001) 1573–1635.

[6] G.E. Miller, Special theory of ultrasonic machining, J. Appl. Phys. (2) (1957) 149–156.

[7] Z.Y. Wang, K.P. Rajurkar, Dynamic analysis of the ultrasonic machining process, J. Manuf. Sci. Eng, Trans. ASME 118 (August) (1996) 376–380.

[8] T.C. Lee, C.W. Chan, Mechanism of the ultrasonic machining of ceramic composites, J. Mater. Process. Technol. 71 (1997) 195–201.

[9] T.C. Lee, D. Jianxin, Ultrasonic erosion of whisker-reinforced ceramic composites, Ceram. Int. 27 (2001) 755–760.

[10] Z.J. Pei, P.M. Ferreira, S.G. Kapoor, M. Haselkorn, Rotary ultrasonic machining for face milling of ceramics, Int. J. Mach. Tool Manuf. 35 (7) (1995) 1033–1046.
[11] R.P. Singh, S. Singhal, Rotary ultrasonic machining: a review, Mater. Manuf. Process. 31 (14) (2016) 1795–1824.
[12] Z.J. Pei, P.M. Ferreira, Modeling of ductile-mode material removal in rotary ultrasonic machining, Int. J. Mach. Tools Manuf. 38 (10) (1998) 1399–1418.
[13] Z.J. Pei, P.M. Ferreira, M. Haselkorn, Plastic flow in rotary ultrasonic machining of ceramics, J. Mater. Process. Technol. 48 (1) (1995) 771–777.
[14] D. Lv, H. Wang, Y. Tang, Y. Huang, H. Zhang, W. Ren, Surface observations and material removal mechanisms in rotary ultrasonic machining of brittle material, Proc. Inst. Mech. Eng. B J. Eng. Manuf. 226 (9) (2012) 1479–1488.
[15] N.J. Churi, Z.J. Pei, C. Treadwell, Rotary ultrasonic machining of titanium alloy: effects of machining variables, Mach. Sci. Technol. 10 (3) (2006) 301–321.
[16] W.L. Cong, Z.J. Pei, T. Deines, Q.G. Wang, C. Treadwell, Rotary ultrasonic machining of stainless steels: empirical study of machining variables, Int. J. Manuf. Res. 5 (3) (2010) 370–386.
[17] G. Ya, H.W. Qin, S.C. Yang, Y.W. Xu, Analysis of the rotary ultrasonic machining mechanism, J. Mater. Process. Technol. 129 (1) (2002) 182–185.
[18] Machining Data Handbook, third ed., vol. 2, Machinability Data Centre (MDC), Metcut Research, Cincinnati, Ohio, 1980.
[19] S. Das, Investigation Into Ultrasonic Machining of Bio-Ceramic Materials (PhD thesis), Jadavpur University, India, 2018.
[20] W.L. Cong, Q. Feng, Z.J. Pei, T.W. Deines, C. Treadwell, Rotary ultrasonic machining of carbon fiber-reinforced plastic composites: using cutting fluid vs. cold air as coolant, J. Compos. Mater. 46 (14) (2011) 1745–1753.
[21] S. Das, B. Doloi, B. Bhattacharyya, Fabrication of stepped hole on zirconia bioceramics by ultrasonic machining, Mach. Sci. Technol. 20 (4) (2016) 681–700.
[22] S. Das, S. Kumar, B. Doloi, B. Bhattacharyya, Parametric studies on profile accuracy of square stepped hole generated on zirconia by USM process, in: Proceedings of 10th International Conference on Precision, Meso, Micro and Nano Engineering (COPEN 10), December 7–9, IIT, Madras, 2017,pp. 599–602, ISBN: 978-93-80689-28-9.
[23] H. Lalchhuanvela, B. Doloi, B. Bhattacharyya, Enabling and understanding ultrasonic machining of engineering ceramics using parametric analysis, Mater. Manuf. Process. 27 (4) (2012) 443–448, https://doi.org/10.1080/10426914.2011.585497.
[24] D.K. Baek, T.J. Ko, S.H. Yang, Enhancement of surface quality in ultrasonic machining of glass using a sacrificing coating, J. Mater. Process. Technol. 213 (2013) 553–559.
[25] H. Lalchhuanvela, B. Doloi, B. Bhattacharyya, Analysis on profile accuracy for ultrasonic machining of alumina ceramics, Int. J. Adv. Manuf. Technol. 67 (2013) 1683–1691.
[26] S. Das, P. Mazumder, B. Doloi, B. Bhattacharyya, Fabricating tool for ultrasonic machining using reverse engineering (RE) method, Int. J. Comput. Integr. Manuf. 31 (3) (2018) 295–305.
[27] S. Das, S. Kumar, B. Doloi, B. Bhattacharyya, Experimental studies of ultrasonic machining on hydroxyapatite bio-ceramics, Int. J. Adv. Manuf. Technol. 86 (4) (2015) 829–839.
[28] Y. Ahmed, W.L. Cong, M.R. Stanco, Z.G. Xu, Z.J. Pei, C. Treadwell, Z.C. Li, Rotary ultrasonic machining of alumina dental ceramics: a preliminary experimental study on surface and subsurface damages, J. Manuf. Sci. Eng. 134 (6) (2012) 064501.
[29] W.L. Cong, Z.J. Pei, T.W. Deines, D.F. Liu, C. Treadwell, Rotary ultrasonic machining of CFRP/Ti stacks using variable feed rate, Compos. Part B 52 (2013) 303–310.

SUBCHAPTER 3.2

Abrasive Jet Machining (AJM)

3.2.1 Introduction

Abrasive Jet Machining (AJM) is one of the mechanical energy based and tool less advanced machining process for removing material from hard and brittle materials by impact erosion caused by the action of high velocity abrasive jet. The fine and sharp edge abrasive particles are driven by high pressure carrier gas and high velocity abrasive jet is formed while the abrasive and gas mixture are passed through the nozzle. A high velocity abrasive jet has great potential for machining applications such as cutting, deburring, polishing, drilling, etching and cleaning of hard and brittle metals, alloys and non-metallic materials such as germanium, silicon, glass, ceramics and mica, etc. The AJM process is not so efficient for soft and ductile materials such as aluminum, rubber, plastics, etc. AJM process is different from sand blasting process where sand blasting is mainly used to clean workpiece surfaces using comparatively larger size sand particles, whereas AJM is used to cut materials with comparatively high finish and accuracy with proper control of process parameters. AJM can machine materials irrespective of electrical and thermal properties. As a stream of abrasive and gas mixture is used in AJM, there is no thermal damage of the machined surface like heat affected zone, recast layer, etc. The cutting action is cool because the carrier gas serves as a coolant. In AJM process no cutting tool is required, so tool related problems are not there, but the nozzle orientation and nozzle wear are to be controlled precisely. AJM is inherently free from chatter and vibration problems. AJM process is also applied for micromachining operation on hard and brittle materials using very fine abrasive particles. The AJM process needs proper control of operating environment to reduce the pollution created by abrasive and gas mixture during and also after AJM operation.

3.2.1.1 Fundamentals of abrasive jet machining

During abrasive jet machining process, material from a surface of workpiece is removed due to impact erosive action of fine-grained abrasive particles impacting at high velocity (200–400 m/s). As the abrasive particle impacts on the surface, it causes a tiny fracture, and the gas stream carries both the abrasive particles and the fractured (wear) particles away. To attain this high velocity, the abrasive particles are driven through a nozzle with compressed carrier gas, usually air. The high velocity stream of abrasives is generated by converting pressure energy of carrier gas or air to its kinetic energy in the nozzle and hence high velocity abrasive jet is produced for machining. Nozzle directs abrasive jet in a controlled manner onto work material maintaining the desired nozzle tip distance from the workpiece. The high velocity abrasive particles remove the material by micro-chipping action as well as brittle fracture of the work material. The fundamental of abrasive jet machining is shown in Fig. 3.2.1.

3.2.1.2 Mechanism of material removal in AJM

In Abrasive Jet Machining Process, a stream of fine abrasive particles and carrier gas with high kinetic energy leaves the nozzle to impact on the work surface with a stand-off-distance of around 0.5–15 mm that causes brittle fractures and the material removal occurs by the erosive action of the abrasive particles striking on the workpiece surface. The fractured (wear) particles are carried away by the gas stream [1]. The AJM process is more suitable

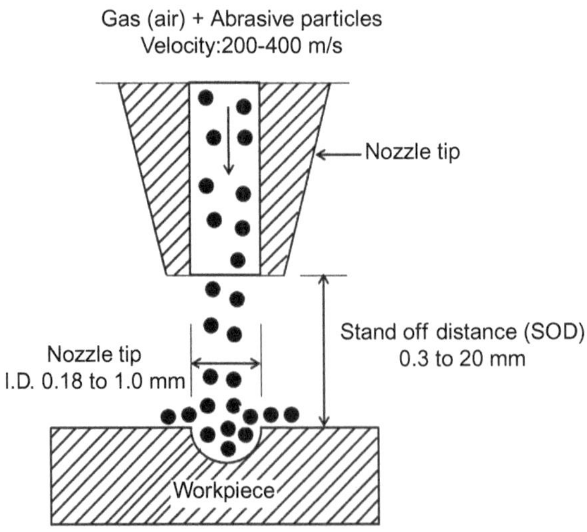

Fig. 3.2.1 Fundamentals of abrasive jet machining.

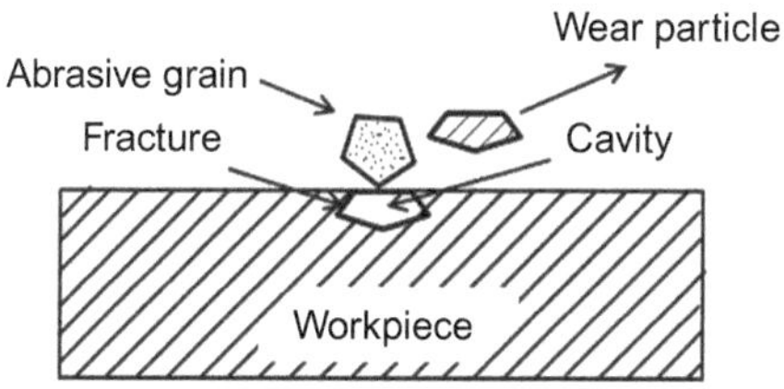

Fig. 3.2.2 Mechanism of material removal in AJM.

for brittle and fragile workpiece materials. The material removal mechanism is represented through the Fig. 3.2.2.

3.2.1.2.1 Modeling of material removal in AJM

As mentioned earlier, material removal in AJM takes place due to brittle fracture of the work material due to impact of high velocity abrasive particles. Figure 3.2.3 shows the interaction of the abrasive particle and the work material in AJM. Modeling has been done with the following assumptions [2]:

(i) Abrasives are spherical in shape and rigid. The particles are characterized by the mean grit diameter.

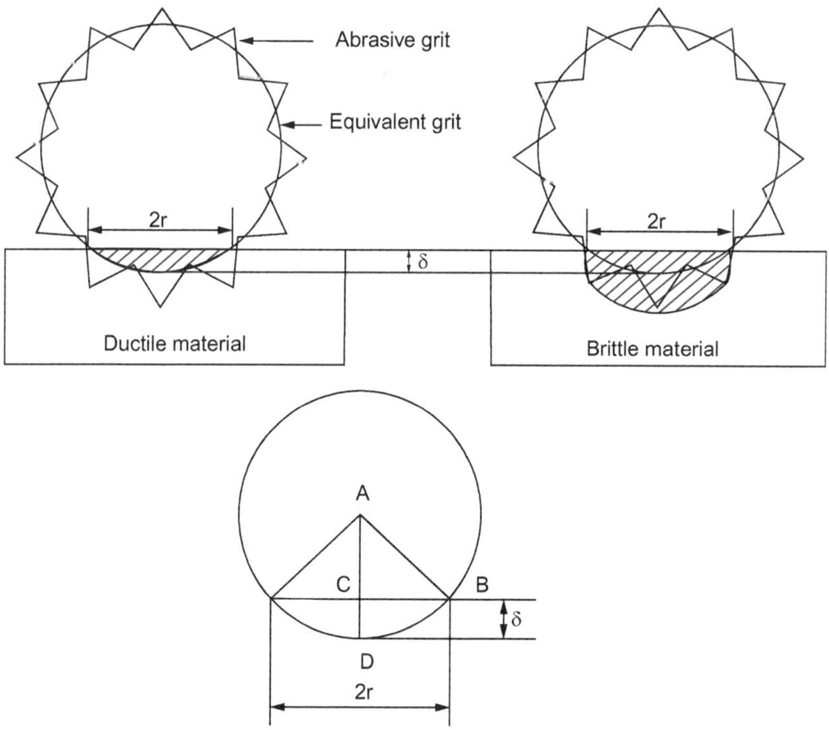

Fig. 3.2.3 Interaction of the abrasive particle and the work material in AJM.

(ii) The kinetic energy of the abrasives are fully utilized in removing material.

(iii) Brittle materials are considered to fail due to brittle fracture and the fracture volume is considered to be hemispherical with diameter equal to chord length of the indentation.

(iv) For ductile material, removal volume is assumed to be equal to the indentation volume due to particulate impact.

Based on the interaction of abrasive particle and work piece and the indentation of abrasive grain on work piece material, the following relations are obtained:

$$AB^2 = AC^2 + BC^2$$

$$BC^2 = r^2 = AB^2 - AC^2$$

$$r^2 = \left(\frac{d}{2}\right)^2 - \left(\frac{d}{2} - \delta\right)^2$$

$$r^2 = -\delta^2 + d\delta \cong d\delta$$

Radius of the crater (r) is expressed as:

$$r = \sqrt{d\delta} \tag{3.2.1}$$

where,

d = average diameter of abrasive grain

δ = depth of the crater

So, volume of material removal in brittle material (V_B) is the volume of the hemispherical impact crater and is given by

$$V_B = \frac{2}{3}\pi r^3 = \frac{2}{3}\pi (d\delta)^{3/2} \tag{3.2.2}$$

For ductile material, volume of material removal (V_D) in single impact is equal to the volume of the indentation and considering small value of δ, V_D is expressed as [2]:

$$V_D = \pi\delta^2 \left(\frac{d}{2} - \frac{\delta}{3}\right) = \pi d\delta^2/2 \tag{3.2.3}$$

Kinetic energy of a single abrasive particle is given by

$$K.E. = \frac{1}{2} m_a v^2 = \frac{1}{2}\left(\frac{\pi}{6} d^3 \rho\right) v^2 \tag{3.2.4}$$

where,

v = velocity of the abrasive particle

m_a = mass of a single abrasive grit

d = average diameter of the grit

ρ = density of the grit

On impact, the work material is subjected to a maximum force F which leads toan indentation of "δ." Thus the work done (W) during such indentation is given by

$$W = \frac{1}{2}\pi r^2 H\delta \tag{3.2.5}$$

where H is the hardness of workpiece materials.

Now, it is assumed that K. E. of the abrasive is fully used for material removal and the work done is considered as equal to the kinetic energy.

From Eqs. (3.2.4), (3.2.5), the depth of crater (δ) is expressed as

$$\delta = dv\left(\frac{\rho}{6H}\right)^{0.5} \tag{3.2.6}$$

Now MRR in AJM of brittle materials (MRR_B) can be expressed as:

$$MRR_B = V_B.n \tag{3.2.7}$$

where n = number of impacts by abrasive grits per second.

The number of impacts by abrasive grits per second,

$$n = \frac{\dot{m}}{m_a} \tag{3.2.8}$$

where,

$\dot{m}$ = mass flow rate of abrasive

m_a = mass of single abrasive grit

From Eqs. (3.2.4), (3.2.8), the value of "n" is expressed as:

$$n = \frac{6\dot{m}}{\pi d^3 \rho} \tag{3.2.9}$$

From Eqs. (3.2.2), (3.2.9), MRR in AJM of brittle materials (MRR_B) is expressed as:

$$MRR_B = 4\frac{\dot{m}}{\rho}\left(\frac{\delta}{d}\right)^{1.5} \tag{3.2.10}$$

From Eqs. (3.2.10), (3.2.6), MRR_B is expressed as:

$$MRR_B = \frac{\dot{m}v^{1.5}}{0.25\rho H^{0.75}} \tag{3.2.11}$$

MRR in AJM of ductile materials (MRR_D) is expressed as:

$$MRR_D = V_D.n \tag{3.2.12}$$

From Eqs. (3.2.3), (3.2.9), MRR_D is expressed as:

$$MRR_D = 3\frac{\dot{m}}{\rho}\left(\frac{\delta}{d}\right)^2 \tag{3.2.13}$$

Putting the value of δ from Eq. (3.2.6) in Eq. (3.2.13) MRR_D is expressed as:

$$MRR_D = \frac{1}{2}\frac{\dot{m}v^2}{H} \tag{3.2.14}$$

The material removal rate in AJM of brittle and ductile materials is estimated from the above mentioned mathematical relations.

The Material Removal Rate (Q) due to the chipping of the work surface for brittle and fragile materials by the impacting abrasive particle scan also is expressed as follows [1]:

$$Q = XZd^3V^{3/2}\left(\frac{\rho}{12H_w}\right)^{3/4} \tag{3.2.15}$$

where,

Z = No. of abrasive particles impacting per unit time
d = mean diameter of abrasive grains
V = velocity of abrasive grains
ρ = density of abrasive grains
H_w = the hardness of the workpiece
X = a constant

Balasubramanium et al. [3] proposed the mathematical model on erosion rate of Abrasive Jet Machining. The erosion rate is considered as proportional to the kinetic energy of the particle. It is also assumed that the particles are spherical in shape and there are no collision between particles within the jet. The maximum mass is considered to be moving at the jet center and minimum mass along the periphery of the jet. The velocity distribution along the jet radius is considered as linear to simplify the analysis.

In this analysis, the radius of jet is taken as R and the radius at any point within the jet is taken as r. The maximum mass of the particle in the jet is m_{max} and the mass of the particle is m_0 corresponding to Q_0, the maximum number density (number of particles per unit volume). The mass of particle at any radial position of the jet is m_r corresponding to Q_r, the number density at any radial position of the jet. V_0 is the maximum velocity of the incoming particle, Vr is the velocity of the incoming particle at any radial position of the jet and V_1 is the velocity at the periphery of the jet (V_R). The variance of the mass distribution is σ_1^2 and σ_2^2 is the variance of number density distribution. Now, number of particles striking the target is $Q_r \cdot V_r$ per unit area per unit time and mass of particles impinging on the target is $m_r \cdot Q_r \cdot V_r$ per unit area per unit time. The erosion rate in AJM is expressed as:

$$Erosion\,rate = p\left(\frac{1}{2} \times mass \times velocity^2\right) \tag{3.2.16}$$

where "p" is a constant and it accounts for all unaccounted factors dissipating the kinetic energy.

The erosion rate (E) in AJM is expressed as [3]:

$$E = 0.5pm_r Q_r (V_r)^3 \tag{3.2.17}$$

Now, m_r, Q_r, and V_r can be expressed as follows [3]:

$$m_r = m_{max} e^{-(r/\sigma_1)^2} \tag{3.2.18}$$

$$Q_r = Q_0 e^{-((m_0 - m_r)/\sigma_2)^2} \tag{3.2.19}$$

Assuming that the velocity distribution is linear along the radius of the jet,

$$V_r = (V_1 - V_0)\left(\frac{r}{R}\right) + V_0 \tag{3.2.20}$$

Therefore the erosion rate at given radius $r \leq R$ is given by the following equation:

$$E = 0.5pm_{max} e^{-(r/\sigma_1)^2} Q_0 e^{-((m_0 - m_r)/\sigma_2)^2} \left((V_1 - V_0)\left(\frac{r}{R}\right) + V_0\right)^3 \tag{3.2.21}$$

The values of erosion rate can be obtained using Eq. (3.2.21) and the erosion profile with different process parameters such as jet radius, jet velocity and abrasive particles mass density can be determined. The analysis on erosion rate and profile is very much effective to generate the cavity of desired shape and size by abrasive jet machining process.

3.2.2 Details of AJM machining system

Abrasive jet Machining (AJM) system consists of gas propulsion system, abrasive feeder and vibratory mixing chamber, nozzle, well ventilated machining chamber and abrasives. Figure 3.2.4 shows the schematic diagram of abrasive jet machining (AJM) system. An abrasive jet machine consists of various components such as compressor; pressure gage; valve; dehumidifier and filter; pressure regulator; flow meter; mechanical vibrator; mixing chamber; working chamber; work holding fixture; vacuum cleaner; nozzle; secondary abrasive chamber; abrasive metering tube; main abrasive chamber and abrasive metering ventury [4].

3.2.2.1 Gas propulsion system

Gas propulsion system includes gas cylinder or air compressor fitted with drier and cleaner and various gauges for detecting pressure, flow rate, etc., as shown in Fig. 3.2.4. It supplies clean and dry air, nitrogen and carbon dioxide, etc., to propel the abrasive particles. Gas may be supplied either from a gas cylinder or air compressor. In case of air compressor, air filter cum drier was used to avoid oil or water contamination of abrasive powder. Filter regulator lubricating (FRL) unit is fitted with outlet of high pressure air compressor to remove any foreign particles and to supply dry and dust

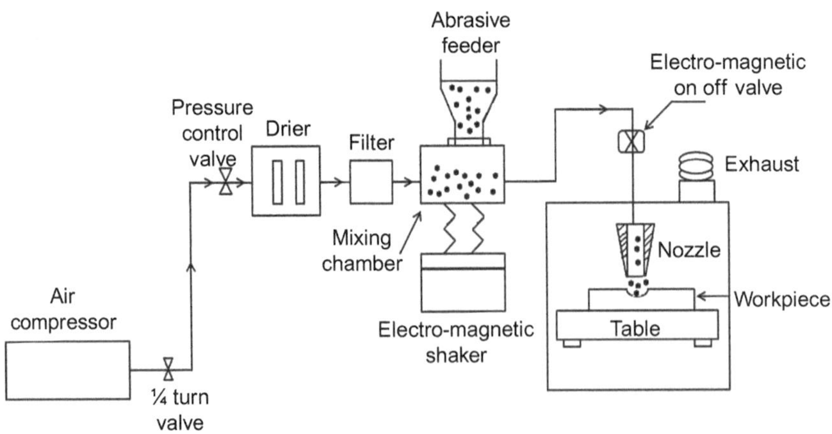

Fig. 3.2.4 Schematic diagram of abrasive jet machining process.

free carrier gas. To avoid coagulation and agglomeration of abrasives during flow through pipelines, the carrier gas should be free from any liquid. The pressure of compressed carrier gas has to be controlled by pressure regulator gauge. The jet velocity in AJM depends on the pressure of carrier gas. The flow rate of carrier gas is also controlled by flow control valve. The mixing ratio of abrasive and gas depends on mass flow rate of carrier gas. Gas should be non-toxic, cheap, and easily available. It should not excessively spread when discharged from the nozzle into the atmosphere. Usually air is taken directly from atmosphere through drier cum filter then compressed to high pressure in the air compressor. The supply pressure of Carrier gas is usually in the range between 15 and 20 bar, based on the compressor capacity and jet velocity requirement. Common carrier gases such as air, pure nitrogen and carbon di-oxide gas are used for better performances of AJM process.

3.2.2.2 Abrasive feeder and mixing chamber

The abrasive feeder supplies the required quantity of abrasive particles to the chamber through a sieve. The sieve is made to vibrate at 50–60 Hz by electromagnetic shaker and mixing ratio is controlled by the amplitude and frequency of vibration of sieve. The particles are propelled by carrier gas to a mixing chamber. There is momentum transfer between gas and abrasives. So abrasives gain momentum in terms of velocity and start flowing with carrier gas through the pipeline. Flow and pressure regulating valves are also employed here to control flow rate and pressure of gas-abrasive mixture. Air abrasive mixture moves further to nozzle. The nozzle imparts high velocity to mixture which is directed at work piece surface.

3.2.2.3 Machining chamber

The machining chamber is well closed to avoid harmful air pollution. A vacuum dust collector cum cleaner is connected with machining chamber to remove tiny suspended abrasive particles and any toxic materials generated during machining from gas and to discharge it to the remote location. Workpiece is mounted by fixture which is placed in the machining chamber. For CNC AJM set up, the machining chamber is fixed with x-y axes CNC machining table for providing motion to the workpiece.

3.2.2.4 Nozzle

High pressure gas and abrasive mixture enters into the nozzle and the hydraulic energy is converted into kinetic energy while the abrasive gas mixture passes through the nozzle. At the outlet of the nozzle high velocity

(200–400 m/s) abrasive jet is obtained. The functions of the nozzle are delivering high velocity jet, controlling jet diameter and regulating inclination angle of abrasive jet. Nozzle is usually made of tungsten carbide or sapphire. The life of sapphire nozzle (usual life = 300 h) is more than that of tungsten carbide nozzle (usual life = 20 to 30 h). High wear resistant material is preferred for nozzle. High wear of nozzle results in more side cutting and high inaccuracy of machining. The cross section of the opening of nozzle is made of either circular or rectangular. For circular nozzle the inner diameter generally varies from 0.2 to 1 mm. For rectangular nozzle the cross section of opening slot generally varies from 0.75 mm × 0.5 mm to 0.15 mm × 1.25 mm. The inner diameter of the nozzle should be more than 20 times of average abrasive diameter. The head of the nozzle can be straight, or at a right angle. To reduce the loss of pressure due to fluid friction, straight head nozzle is preferred. During AJM operation the nozzle tip is kept at a distance from the surface of workpiece. This distance is termed as nozzle tip distance (NTD) or stand off distance (SOD) which is one of the important process parameter for controlling the machining rate and accuracy of AJM process. There should be provision of movement of nozzle head along vertical direction for setting the desired NTD or SOD. The nozzle head is to be moved along horizontal plane as per required path for profile cutting operation. The angle between the workpiece surface and the axis of abrasive jet, termed as inclination angle is another important process parameter which can be set at desired value by orienting the nozzle head for achieving better machining condition. The schematic diagram of a typical nozzle of AJM is shown in Fig. 3.2.5. The photographic view of nozzle fixed in AJM set up is shown in Fig. 3.2.6.

3.2.2.5 Abrasives

The different types and sizes of abrasive particle are selected for AJM operation depending upon the applications, machining rate and surface finish requirements. The commonly used abrasive for AJM process are **Aluminum Oxide (Al_2O_3), Silicon Carbide (SiC), Glass beads, Crushed glass and Sodium Bicarbonate ($NaHCO_3$)**. Abrasive particles must contain sharp edges for easy micro-cutting action. Abrasive grits should have irregular shape but variation of size within the overall mass should be low. Abrasive grits must not contain metal particles as it can cause blockage in the pipeline. It must have good flow characteristics for smooth flow in the pipeline when mixed with carrier gas. It should have high hardness and should be rigid enough so that it does not crumbled into parts when striking the work surface. Abrasives must not be wet. It should be

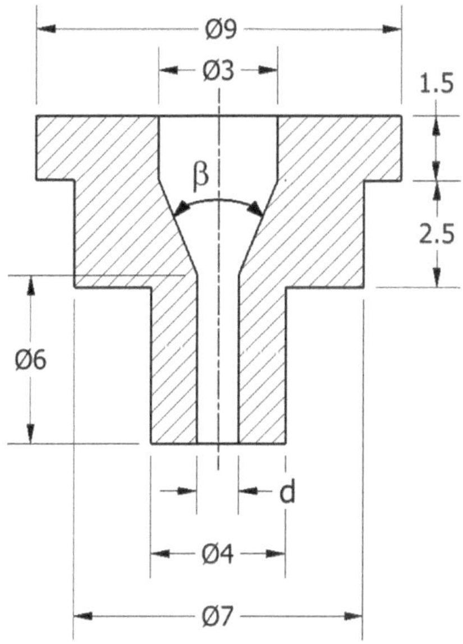

Fig. 3.2.5 A schematic diagram of a typical nozzle of AJM (dimensions in mm) [5].

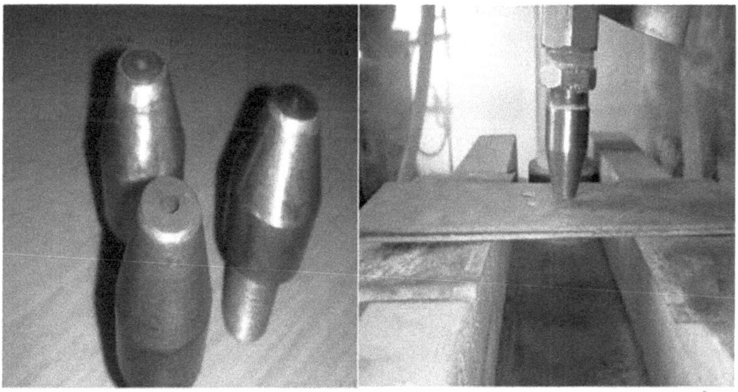

Fig. 3.2.6 Photograph of AJM nozzle fixed to set up [6].

cheap and readily available. Reuse of abrasives is not recommended because cutting capacity decreases after first use and contamination of abrasive particles with metallic chips clogs the small orifices in the nozzle. Silicon carbide is harder abrasive compared to alumina. Silicon carbide (grain size 25–40 μm) is for very hard work materials, but it can substantially reduce nozzle life. Alumina (grain size 12 to 50 μm) is used for cleaning, cutting and deburring purposes for moderate to hard materials. Glass beads (grain size 0.635 to 1.25 mm) are mainly used for

Fig. 3.2.7 Photographic view of vertical AJM set-up [7].

producing matt surfaces for reducing transparency. Crushed glass is used for heavy cleaning, matting and peening operations. Sodium Bicarbonate (grain size 27 μm) or baking soda is commonly used for light duty applications such as cleaning, cutting, etc., on soft materials.

The photographic views of vertical AJM set-up is shown in Fig. 3.2.7.

3.2.3 Influence of process parameters on performance criteria of AJM process

The process parameters of AJM influencing the machining performance of AJM process are given as follows:

(i) Nozzle related Process Parameters: Nozzle Pressure, Nozzle Tip Distance (NTD) or Stand off Distance (SOD), Nozzle Inclination Angle.

For effective machining of AJM, nozzle pressure generally varies from 10 to 30 bar. NTD or SOD varies from 0.75 mm to 20mm. Nozzle inclination angle varies from 60 degree to 90 degree with horizontal.

(ii) Abrasive and carrier gas related Process Parameters: Type and size of abrasive particle, type of carrier gas, mixing ratio, abrasive flow rate, abrasive jet velocity.

The average diameter of abrasive grain varies from 10 to 50 microns. The abrasive flow rate varies from 1.5 to 20 g/min depending on the nature of machining process. The mixing ratio varies from 0.05 to 0.3. The abrasive jet velocity varies from 200 to 400 m/s. The machining performance criteria of AJM process are given as follows:

(i) Material Removal Rate (MRR)
(ii) Nozzle Wear Rate
(iii) Shape and size of machined cavity
(iv) Surface roughness
(v) Taper angle of cut edge or drilled hole
(vi) Kerf Width

3.2.3.1 Effect of abrasive flow rate and grain size and mixing ratio on MRR

The abrasive mass flow rate is one of the important process parameter in AJM process. As the abrasive mass flow rate increases, number of abrasive particles striking the workpiece increases. As a result, MRR increases with increase in abrasive flow rate. But MRR decreases with further increase of abrasive flow rate after reaching optimum value. This is because of the fact that the abrasive flow velocity decreases with increase of abrasive flow rate [8]. Figure 3.2.8 shows the variation of MRR with abrasive flow rate at different grain size of aluminum oxide abrasive while machining of glass plate. As abrasive grain size increases MRR increases because the kinetic energy striking the workpiece increases due to increase of the mass of the abrasive particle.

The material removal rate (MRR) depends on volume flow rate of abrasives as well as carrier gas. Mixing ratio (M) is defined as the ratio of volume flow rate of abrasive particles to the volume flow rate of carrier gas. As the

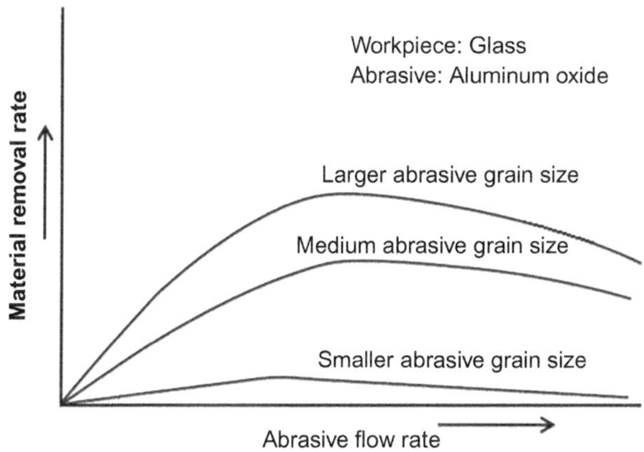

Fig. 3.2.8 Effect of abrasive flow rate and grain size on MRR.

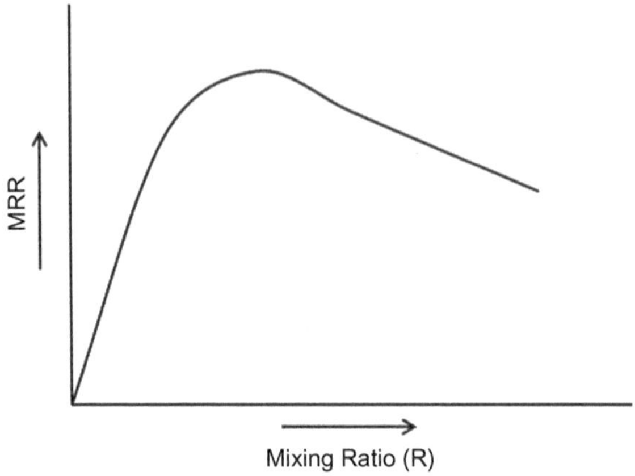

Fig. 3.2.9 Effect of mixing ratio on MRR.

mass flow rate of abrasive grit increases, mixing ratio increases while gas flow rate is constant. As the mixing ratio increases the MRR initially increases and after reaching the maximum value, it decreases because with higher mixing ratio abrasive jet velocity decreases and nozzle may be blocked by larger amount of abrasives particles. Figure 3.2.9 shows the variation of MRR with mixing ratio.

There is another representation of ratio of mass flow rate of abrasive and mass flow rate of abrasive gas mixture which is known as mass ratio. As the mass ratio increases, MRR increases as the mass flow rate of abrasive particle increases. Consideration of mass ratio instead of mixing ratio can be easier for finding an optimum value of mixing ratio that gives maximum MRR.

From Fig. 3.2.10 it is also observed that MRR can be enhanced by proportionally increasing both the abrasive flow rate and gas flow rate at same rate so that mixing ratio remains constant. The mixing ratio is unchanged only by simultaneous increase of both gas and abrasive flow rate.

3.2.3.2 Effect of nozzle pressure on MRR

The abrasive flow rate can be increased by increasing the flow rate of the carrier gas that is only possible by increasing the internal gas pressure. The material removal rate increases with the increase in gas pressure as shown in the Fig. 3.2.11. As the gas pressure increases the velocity of abrasive jet increases. The kinetic energy of the abrasive particles is responsible for the removal of material by erosion process.

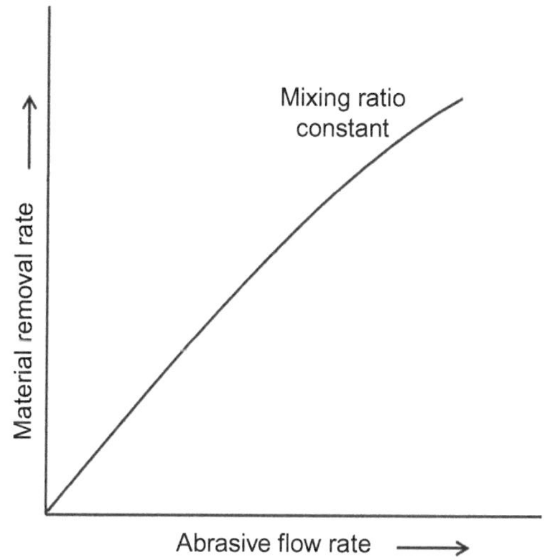

Fig. 3.2.10 Effect of abrasive flow rate on MRR at constant mixing ratio.

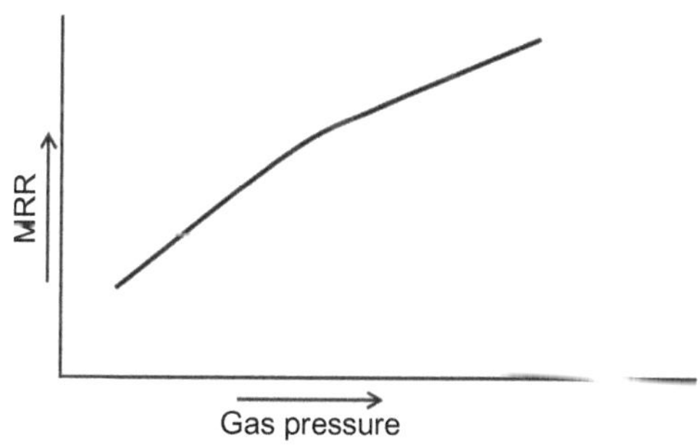

Fig. 3.2.11 Effect of nozzle pressure on MRR.

3.2.3.3 Effect of stand-off distance on width of cut and MRR in AJM

The distance from the work surface to the tip of nozzle in abrasive jet machining is called Stand-Off Distance, abbreviated as SOD. Higher SOD causes spreading of abrasive jet and thus its cross-sectional area increases with the sacrifice of jet velocity. The machining of deeper slots or hole with less width of cut becomes difficult by AJM process. Figure 3.2.12 shows the variation of shape and size of cut with stand off distance of the nozzle. The decrease in SOD improves accuracy, decreases kerf width, and reduces taper in machined groove. However light operations like

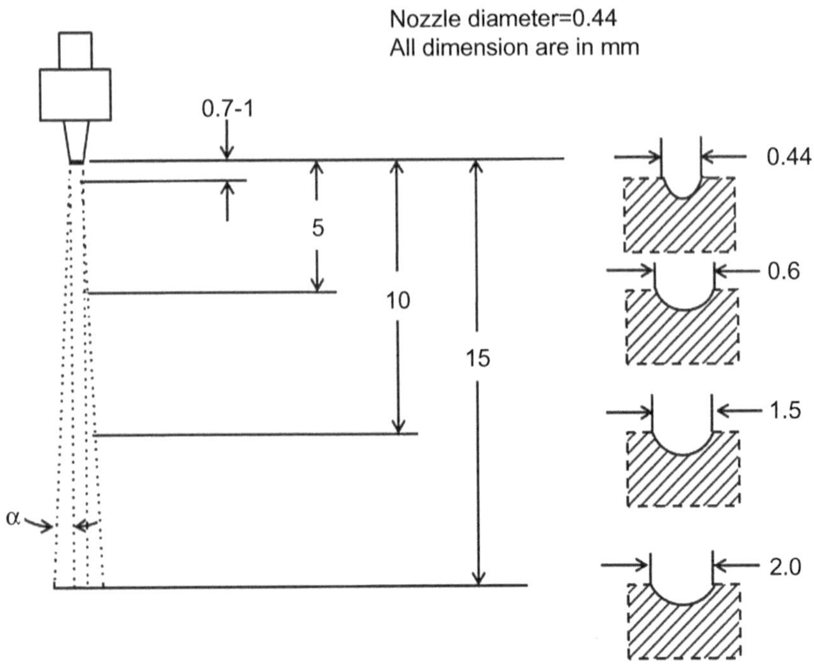

Fig. 3.2.12 Effect of stand-off distance on width of cut.

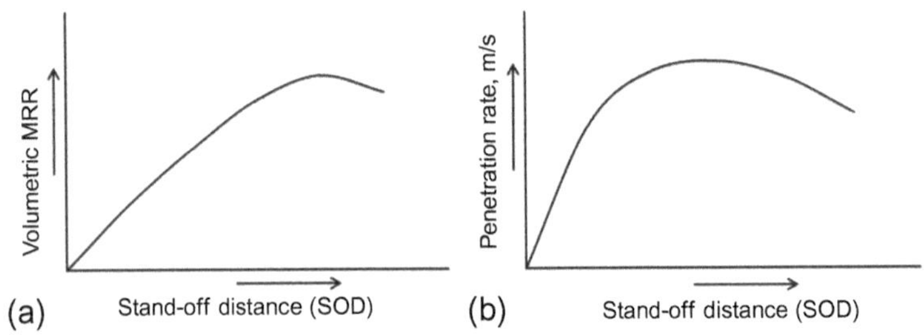

Fig. 3.2.13 Effect of stand-off distance on (a) MRR and (b) penetration rate.

cleaning, frosting, etc., are conducted with large SOD (say 12.5 to 75 mm). Thus an optimum value of stand-off distance is required to be set for obtaining satisfactory performance in abrasive jet machining. Figure 3.2.13 (a) and (b) shows the effect of stand of distance on volumetric material removal rate and penetration rate respectively. The penetration rate of AJM increases with SOD and then decreases, but volumetric material removal rate increases steadily with SOD and remains same with large SOD [9]. The velocity of abrasive jet decreases after certain value of SOD and diameter of abrasive jet increases with SOD. The side and shallow cutting occurs with large SOD. As a result large volume of materials is removed, but penetration rate decreases with large SOD during AJM cutting or drilling operation.

3.2.4 Advantages and limitations of abrasive jet machining processes

The main advantages of AJM are described as below:

(i) Using fine grain abrasive particles, high surface finish of the machined surface can be achieved.

(ii) Due to the supply of carrier gas in the mixture, the heat generation during machining is less. Thermal damage of machined surface is not present. Therefore AJM provides cool cutting action, so it can machine delicate and heat sensitive materials.

(iii) As there is no contact between the tool and workpiece, AJM process is free from chatter and vibration.

(iv) The cost of AJM system is low and it is easy to operate and maintain AJM.

(v) AJM can machine thin sections of hard and brittle materials like germanium, mica, silicon, glass and ceramics, etc.

(vi) The performance of AJM process is independent of electrical or thermal properties of workpiece materials.

(vii) AJM process consumes less power.

(viii) AJM can cut fragile materials with accuracy as the small load is applied on workpiece.

(ix) During AJM process wastage of material is less.

The limitations of AJM are described as below:

(i) AJM is not suitable for soft and ductile materials as abrasive particles are embedded rather than cutting the workpiece materials.

(ii) It is a slower process.

(iii) Due to tapering characteristics of abrasive jet, tapered geometry of machined surface is obtained.

(iv) AJM process can create environmental pollution if abrasives and dust particles are not properly collected after machining.

(v) Side cutting is unavoidable for thicker workpiece materials.

(vi) AJM process can increase noise level at the working area if not properly controlled.

(vii) For machining brittle materials there is chances of generation of micro-cracks at subsurface level.

3.2.5 Applications of abrasive jet machining processes

The main applications of AJM are described as below:

(i) AJM is used for drilling, cutting, de-burring, etching and polishing of hard and brittle materials.

(ii) It is most suitable for machining brittle and heat sensitive materials like glass, quartz, sapphire, mica, ceramics germanium, silicon and gallium, etc.

(iii) AJM process is used for cleaning of metallic smears on ceramics, oxides on metals, resistive coating, etc.

(iv) AJM process is used for abrading and frosting glass more economically as compared to etching or grinding

(v) It is used for deflating small castings, engraving registration numbers on toughened glass used for car windows.

(vi) AJM is useful in manufacture of electronic devices by drilling of glass wafers, deburring of plastics, making of nylon and teflon parts and permanent marking on rubber stencils, cutting titanium foils.

(vii) Register teaming can be done very easily by AJM process.

(viii) Micro module fabrication for electrical contact, semiconductor processing can also be done effectively by AJM micro-manufacturing.

(ix) It is also good method for deburring small hole like in hypodermic needles and for small milled slots in hard metallic components.

3.2.6 Advancement in AJM technology

Abrasive Jet Machining (AJM) has great potential for producing micro-channels on highly curved surfaces of glass rods of 3 mm and 5 mm diameter and PMMA rod of 4.72 mm diameter. The stand off distance and jet incident angle are to be properly selected for producing desired micro-channels on highly curved surfaces. For producing micro-channel on flat surface, axis of jet is generally set at perpendicular to the flat surface [10].

The magnetic abrasives such as iron grits are used to achieve restrained jet for high material removal rate and better surface finish in AJM. The special abrasive is made up of a thermoplastic polymer as base material and thermally coated with SiC particles and magnetic materials. A novel hybrid method that uses magnetic abrasive particles with magnetic action has been adopted to achieve restrained jet, high material removal rate and obtain better surface roughness [11].

Micro-hole drilling on Borosilicate Glass and GFRP Composites (3.5 mm thickness) using AJM process has been performed by Suresh et al. [7]. SiC abrasives of average dia, 200 μm are used during AJM. Carrier gas pressure varies from 4 to 6 bar. Stand off distance varies from 0.5 to 1 mm. Total time to perform the drilling operation varies from 10 to 20 s. High MRR with less taper is obtained for glass compared to GFRP composites. As gas pressure increases MRR increases and taper angle of drilled hole decreases with increase in machining time for both the materials during AJM process.

AJM process has great potential for machining ceramic materials such as zirconium oxide, silicon nitride, aluminum oxide and silicon carbide. Three kinds of common abrasive grits such as aluminum oxide (WA), silicon carbide (GC) and Synthetic diamond (SD) were employed to study the effect of workpiece properties on machinability in AJM of ceramic materials by Wakuda et al. [12]. The appearance of the dimples was observed after 10 s of AJM drilling on different ceramics using various abrasives as shown in Fig. 3.2.14. Using synthetic diamond as abrasive, faster AJM drilling takes place for all the ceramic materials. For machining zirconia ceramics, material removal is independent of abrasive materials. For abrasive jet machining of

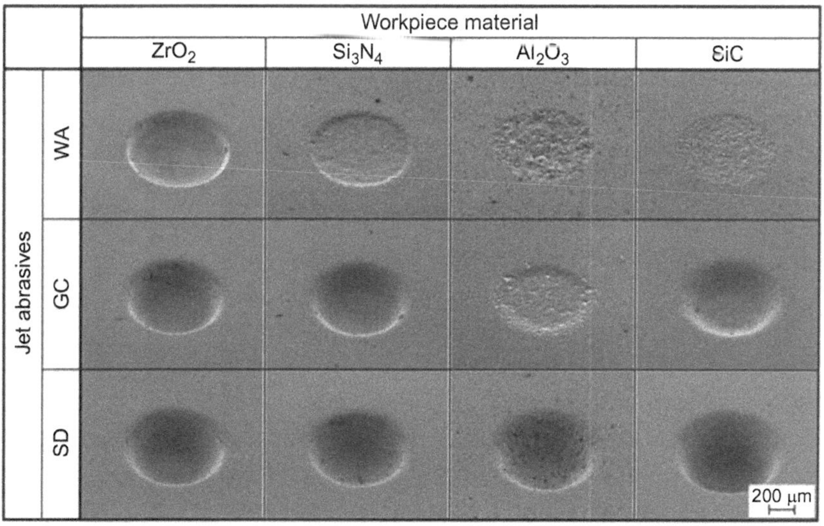

Fig. 3.2.14 Appearance of dimple observed during AJM in different combination of abrasive and workpiece [12].

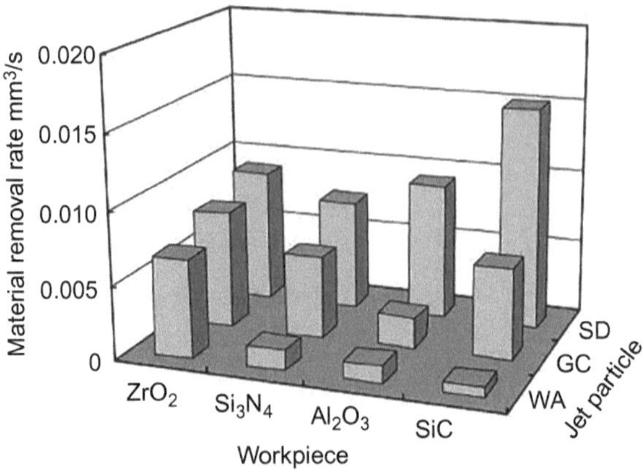

Fig. 3.2.15 Variation of MRR with different combination of abrasives and ceramic workpiece [12].

ceramic like alumina and silicon carbide, synthetic diamond abrasive is the better choice than alumina and silicon carbide abrasives. The MRR were also compared to study the AJM machinability of ceramics through the plot as shown in Fig. 3.2.15.

Abrasive Jet Machining process is applied for turning of glass and PMMA rods to generate helical micro-channels using air driven abrasive jet. These types of micro-channels have applications in micro-fluidic device for mixing of fluids and also separating particles from a flow [13]. Figure 3.2.16 shows the optical microscope images of helical micro-channels produced be abrasive jet turning on (a) 2 mm glass rod, (b) 3 mm glass rod, (c) 5 mm glass rod, and (d) 4.72 mm PMMA rod. The dotting line in the Fig. 3.2.16 represents the wire spring mask used during abrasive jet turning operation. By controlling the exposure time of abrasive jet and using spring masks of various pitches, helical micro-channels of wide range of aspect ratios from 0.09 to 1.2 are generated for micro-fluidic applications. The schematic diagram of abrasive jet turning process is represented in Fig. 3.2.17.

The abrasive jet machining process is applied for producing micro-grooves on glass. The dimension of width of line-type microgroove is 80 μm. The appropriate masking on the workpiece surface is being done to control the dimension of hole or groove during abrasive jet micro-machining operation. The measured surface roughness is about 0.8 μm and 0.6 μm hole type and line type micro-grooves on glass respectively [14].

The application of abrasive jet machining (AJM) technology can be extended in different form such as cryogenic assisted abrasive jet machining,

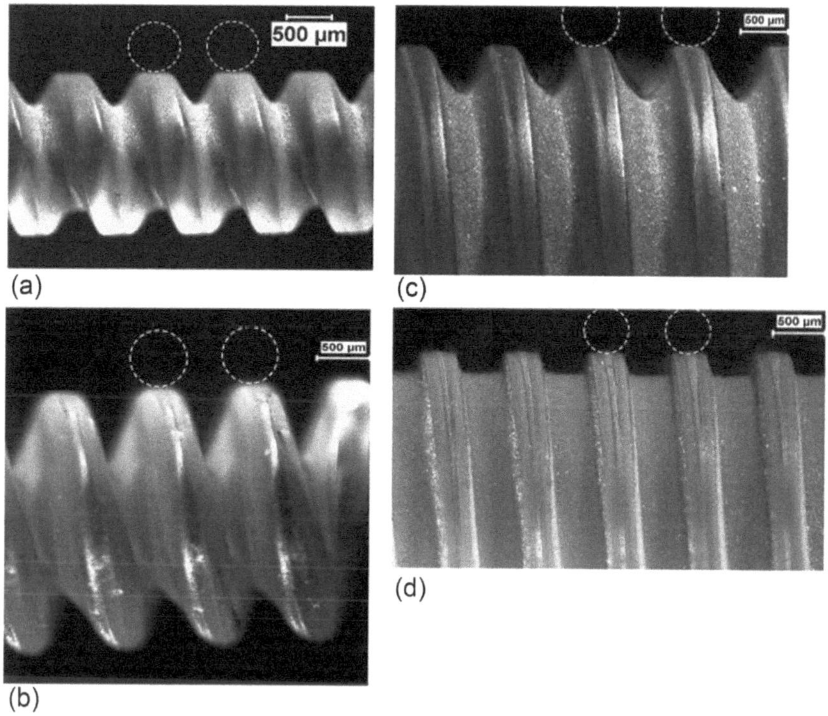

Fig. 3.2.16 Optical microscope images of helical micro-channels produced be abrasive jet turning on (a) 2 mm glass rod, (b) 3 mm glass rod, (c) 5 mm glass rod, and (d) 4.72 mm PMMA rod [13].

thermally assisted abrasive jet machining, ice assisted jet machining, abrasive water jet machining and magneto rheological jet polishing, etc. [15].

The cryogenic abrasive jet machining has been performed on polymer materials like polydimethylsiloxane (PDMS) using aluminum oxide particles as abrasives at cryogenic environment of temperature range between −127 and −178 °C at different angle of attack between 30 and 60 degrees for generating micro-channels for micro-fluidic applications [16]. The PDMS materials under cryogenic environment behave as brittle materials and AJM process can be applied for machining PDMS at this condition and maximum erosion rate can be obtained. Figure 3.2.18 shows the schematic diagram of cryogenic abrasive jet machining (CAJM) System. In this CAJM system the cooled abrasive jet which is the mixture of liquid nitrogen and abrasive jet impinged on the workpiece surface for machining operation.

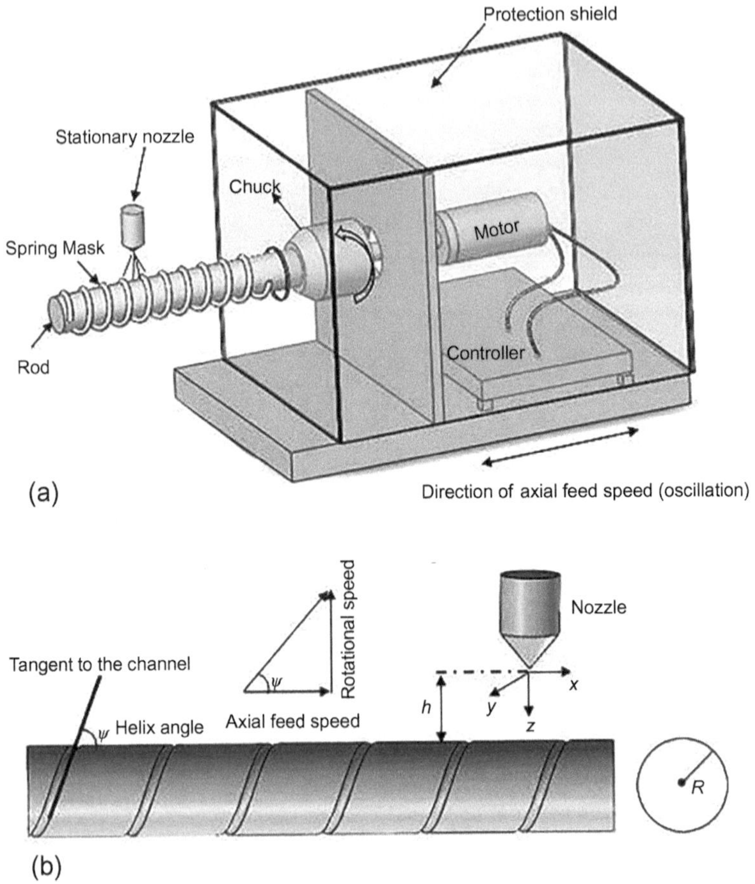

Fig. 3.2.17 Schematic diagram of abrasive jet turning process showing (A) spring mask and (B) helical channel [13].

AJM technology is utilized for micro-machining process in which mask is used to control the effective jet dimension or width of cut in the micron range dimension. The conventional mask is generally fixed to the target surface. The adjustable shadow masking has been implemented for direct writing in abrasive jet micro-machining of glass by Nouhi et al. [17]. The shadow mask is mounted with the nozzle which produces a micro jet of abrasive particles to machine the target workpiece as per the desired pattern. The shadow mask opening were varied from 250 to 300 μm and linear micro-channel of 30 μm width and 30 mm length at a traverse velocity of 1 mm/s.

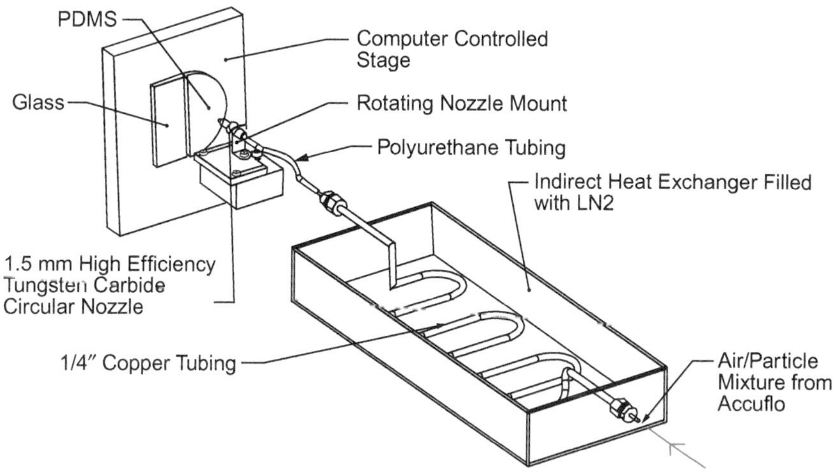

Fig. 3.2.18 Schematic view of cryogenic abrasive jet machining system [16].

Kim et al. [18] developed a magnetic abrasive jet machining system for precision internal polishing of circular tubes such as vacuum tubes, sanitary tubes, etc. of stainless steel. The working fluid mixed with magnetic abrasive particles which are the combination of iron and alumina particles are flowing in the form of jet through the internal surface of tubes under action of magnetic field. As jet pressure increases, MRR as well as surface roughness increases, but with the increase in machining time, surface roughness decreases to the great extent.

Abhishek and Hiramoth [19] utilized a novel approach of feeding the nozzle toward the workpiece during micro-AJM to improve the geometrical accuracy of micro holes machined through micro-AJM set up. The schematic diagram of micro-AJM set up is shown in Fig. 3.2.19.

Using the novel approach of nozzle feeding in AJM, the taper angle and entrance diameter of through hole drilled on 2 mm thick quartz are reduced and hole circularity is also improved. Figure 3.2.20 shows the SEM views of micro-holes machined on quartz by micro-AJM using (a) conventional approach (without feeding of nozzle) and (b) novel approach (with feeding of nozzle). During micro-AJM drilling, alumina abrasives of 320 grit size were used. The air pressure was set as 7 bar. The stand off distance was also set as 2 mm. The inner diameter of nozzle was 0.896 mm [19].

As an alternative to AJM, Barletta et al. [20] proposed the relatively novel hybrid technology of fluidized bed assisted abrasive jet machining (FB-AJM) for internal finishing of tubular components. When the abrasives are jetted

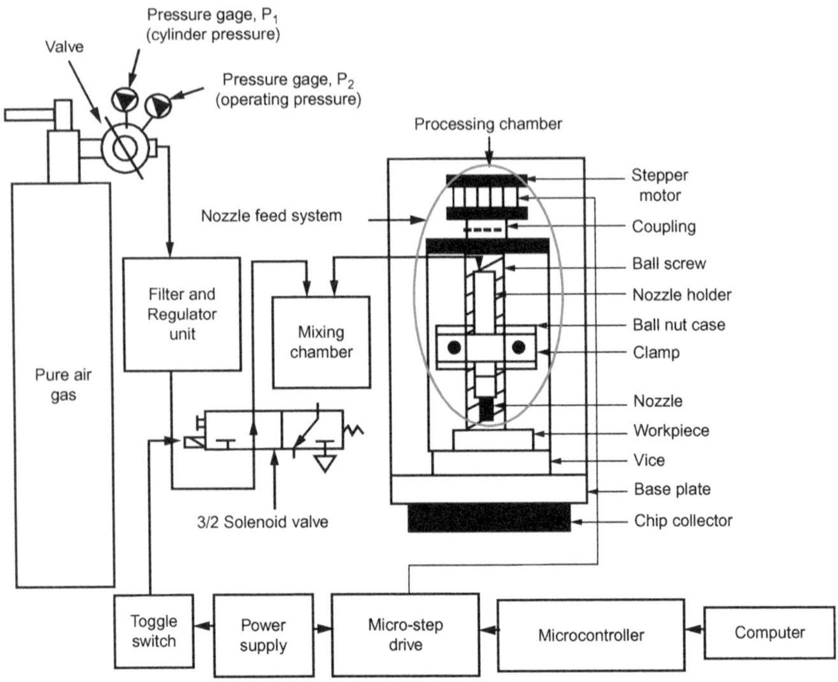

Fig. 3.2.19 Schematic diagram of micro-AJM set up [19].

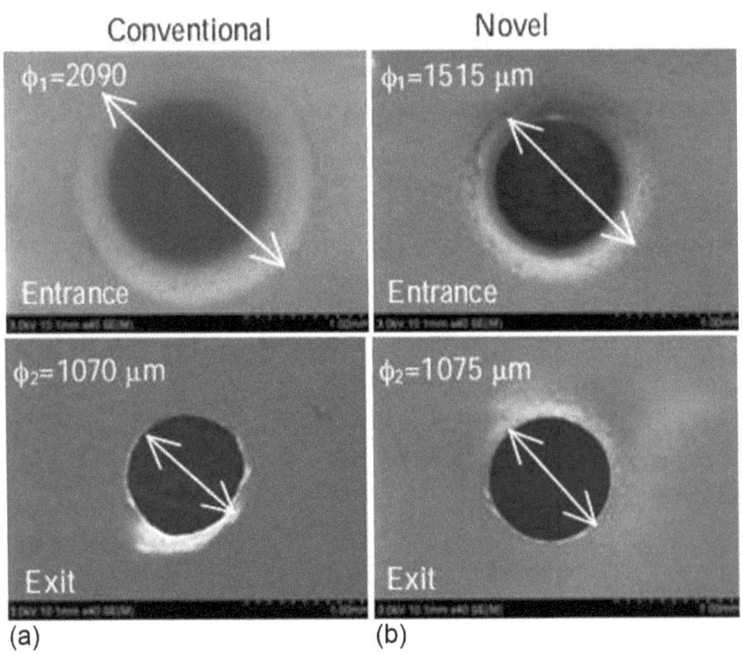

Fig. 3.2.20 SEM views of micro-holes machined on quartz by micro-AJM using (a) conventional and (b) novel approach [19].

onto the workpiece, it machines the metal by striking on it at both high speed and a high impact angle. In particular, the abrasive jet acts preferentially on the peaks and asperities protruding from metal surface by plowing or even cutting them off. As a result, progressive smoothing of workpiece surface associated with remarkable material removal can be expected. In contrast to what happens in standard AJM, the fluidized beds mean a fast transport regime can be established inside the workpiece during machining. This technology uses the peculiar fluidized bed hydrodynamic to improve the abrasive feeding system, the uniformity of abrasive distribution through the workpiece, and the recovery and self-regeneration of abrasive during machining. Consequently, precision and uniform machining as well as accurate surface finishing can be achieved even on inner surfaces of tubular parts. Finally, because of the consistent distribution of abrasive across the section of the workpiece during FB-AJM, good uniformity throughout the entire circular shape of the workpiece was obtained. FB-AJM technology was tested for internal finishing of tubular components with ratio L/D greater than 20 made from a high strength aluminum alloy (AA 6082 T6) using a fluidized bed assisted abrasive jet machining (FB-AJM) system [20]. FB-AJM was also used for finishing tubular workpiece made up of stainless steel.

Fluidized bed abrasive jet machining was also proposed as a method for finishing of complex-shaped cast aluminum components [21]. Abrasive fluidized bed (AFB) machining has wide applications in deburring and finishing of complex-shaped metal components, internal super-finishing of tubular workpiece typically used as dies in injection molding, finishing of axisymmetric metal components, cleaning and polishing of electronic devices and surface preparation of tungsten carbide milling tools. AFB is a flexible finishing technique that is very useful in several industrial fields, yielding end products of a high standard in a short time at low cost using safe and eco-friendly procedures. In AFB machining abrasives are striking on workpiece surface to achieve a smooth and clean finish by progressive removal or displacement of solid material from the surface. This result is achieved by repeated impact on the surface by a discrete number of abrasive particles. The characteristics of impacts between abrasive particles and the workpiece surface are of different types such as (i) rolling impact for a slow impact speed and low impact angle, and (ii) sliding impact for higher impact speeds and angles. However, there are four mechanisms [21] by which abrasives and the workpiece surface can interact such as (i) micro-plowing which is displacement of material involved in impacts with the abrasives without any material removal, (ii) micro-cutting which involves detachment of some material

Fig. 3.2.21 Complex shaped workpiece (a) before finishing (b) after finishing by fluidized bed abrasive jet machining [21].

from the machined surface (iii) micro-fatigue which involves further material removal due to the repeated action of many abrasives impinging on the same area of the workpiece surface and (iv) micro-cracking which involves material removal in the impact area and brittle failure of additional material surrounding the impact area. In general, more of the material directly involved in impacts is removed as result of crack propagation from the impact zone to the surroundings. The complex-shaped cast aluminum components finished by AFB machining is shown in Fig. 3.2.21(b). The finishing effect was observed by comparing the full components in Fig. 3.2.21(a) before finishing and (b) after finishing. It is observed that the workpiece surface became smoother and shinier after fluidized bed abrasive jet machining.

The AJM technology has wide scope of applications for performing various machining operations such as drilling, cutting, cleaning, turning, etc., specially on brittle and hard materials like ceramics, glass, and composites. AJM operations can also be performed on ductile materials in which materials are removed from the workpiece due to ductile mode failure using the mechanical energy of high velocity abrasive jet. AJM technology with assistance of mask is applied for micro-machining and finishing operations. Cryogenic assisted abrasive jet machining is also very much effective for machining polymer materials. Fluidized bed assisted AJM is one of the important finishing method of complex shaped cast components of ductile as well as brittle materials. AJM is applied for producing macro, meso as well as micro dimensional features by controlling various process parameters particularly abrasive size, flow rate, jet velocity. To overcome the pollution problems of AJM process, the AJM system is properly designed to provide protections against any leakage of used abrasive dust and removed particle from workpiece to the working environments.

References

[1] A. Ghosh, A. Kumar Mallik, Manufacturing Science, East-West Press Private Limited, 2014, p. 318, ISBN: 978-81-7671-063-3.
[2] P.K. Mishra, Nonconventional Machining, The Institution of Engineers (India), 2012, pp. 19–20, ISBN: 978-81-7319-138-1.
[3] R. Balasubramanium, J. Krishnan, N. Ramakrishnan, A study on the shape of the surface generated by abrasive jet machining, J. Mater. Process. Technol. 121 (2002) 102–106.
[4] R. Kumar, A.P. Verma, G.K. Lal, Nozzle wear during the flow of a gas particle mixture, Wear 91 (1983) 33–43.
[5] N. Ramachandran, N. Ramakrishnan, A review of abrasive jet machining, J. Mater. Process. Technol. 39 (1993) 21–31.
[6] D.V. Srikanth, M.S. Rao, Metal removal and kerf analysis in abrasive jet drilling of glass sheets, in: 3rd International Conference on Materials Processing and Characterization, Proc. Mater. Sci. vol. 6, 2014, pp. 1303–1311.
[7] R. Suresh, R.K. Sohit, K. Shapur, Abrasive jet machining for mocro-hole drilling on glass and GFRP composites, ICMPC-2017. Mater. Today Proc. 5 (2018) 5757–5761.
[8] G.F. Benedict, Nontraditional Manufacturing Processes, Marcel Dekker, New York, 1987.
[9] A.P. Verma, G.K. Lal, An experimental study of abrasive jet machining, Int. J. Mach. Tool Des. Res. 24 (1984) 19–29.
[10] A. Nouhi, K. Kowsari, J.K. Spelt, M. Papini, Abrasive jet machining of channels on highly curved glass and PMMA surfaces, Wear 356–357 (2016) 30–39.
[11] J.-H. Ke, F.-C. Tsai, J.-C. Hung, B.-H. Yan, Characteristics study of flexible magnetic abrasive in abrasive jet machining, in: 5th CIRP Conference on High Performance Cutting, Procedia CIRP, vol. 1, 2012, pp. 679–680.
[12] M. Wakuda, Y. Yamuchi, S. Kanzaki, Effect of workpiece properties on machinability in abrasive jet machining of ceramic materials, Precis. Eng. J. Int. Soc. Prec. Eng. Nanotechnol. 26 (2002) 193–198.
[13] A. Nouhi, K. Kowsari, J.K. Spelt, M. Papini, Abrasive jet turning of glass and PMMA rods and the micro-machining of helical channels, Precis. Eng. 53 (2018) 151–162.
[14] D.-S. Park, M.W. Cho, H. Lee, W.-S. Cho, Micro-grooving of glass using micro-abrasive jet machining, J. Mater. Process. Technol. 146 (2004) 234–240.
[15] R. Melentiev, F. Fang, Recent advances and challenges of abrasive jet machining, CIRP J. Manuf. Sci. Technol. 22 (2018) 1–20.
[16] A.G. Gradeen, J.K. Spelt, M. Papini, Cryogenic abrasive jet machining of polydimethylsiloxane at different temperature, Wear 274–275 (2012) 335–344.
[17] A. Nouhi, M.R. Sookhak Lari, J.K. Spelt, M. Papini, Implementation of a shadow mask for direct writing in abrasive jet micro-machining, J. Mater. Process. Technol. 223 (2015) 232–239.
[18] J.-D. Kim, Y.-H. Kang, Y.-H. Bae, S.-W. Lee, Development of a magnetic abrasive jet machining system for precision internal polishing of circular tubes, J. Mater. Process. Technol. 71 (1997) 384–393.
[19] K. Abhishek, S.S. Hiramoth, Improvement of geometrical accuracy of micro holes machined through micro abrasive jet machining, in: 7th HPC-CIRP Conference on High Performance Cutting, Procedia CIRP, vol. 46, 2016, pp. 47–50.
[20] M. Barletta, S. Guarino, G. Rubino, V. Tagliaferri, Progress in fluidized bed assisted abrasive jet machining (FB-AJM): internal polishing of aluminium tubes, Int. J. Mach. Tools Manuf. 47 (2007) 483–495.
[21] M. Barletta, Progress in abrasive fluidized bed machining (FB-AJM): internal polishing of aluminium tubes, J. Mater. Process. Technol. 209 (2009) 6087–6102.

SUBCHAPTER 3.3

Water Jet Machining (WJM)

3.3.1 Introduction

High pressure water was used in hydraulic mining in the mid of 1800s. A low pressure water jet nozzle is used for cutting soft materials like paper by paper patents company in 1933. A high pressure water jet was used for cutting plastic by Carl Johnson of Durox International in Luxembourg in 1956. An ultra-high pressure (6900 bar) liquid delivering system to produce hypersonic jet was used to cut stainless steel by B. Schwacha of North American Aviation in 1958. A pulsating water jet was used to cut metals and stone by Philip Rice of Union Carbide in 1962. In the mid of 1960s traditional coal water jet cutting was extended to high pressure water jet cutting of stone by suitable designing of nozzle shape by S.J. Leach and G.L. Walker. In the late 1960s Norman Franz introduced the long chain polymers in the water jet to produce coherent water jet for cutting softer thick materials. Dr. Norman Franz studied the water jet cutting for forestry and wood cutting. Later on lot of research has been conducted to increase the life of nozzle and the machining performance of WJM process for its wide applications.

Water Jet Machining (WJM) is a mechanical energy based advanced machining process in which a very high velocity water jet strikes on the workpiece. The water jet with high momentum erodes the materials from the workpiece surface. The velocity of water jet is increased as high as 1400 m/s. When high velocity jet impinges on the small area of workpiece, then high value of stress is induced which is higher than the mechanical strength of the materials. During WJM process the materials from workpiece are removed based on the principle of mechanical erosion of particles caused by impact of high velocity water jet. The Water Jet Machining (WJM) is non-traditional machining process applied to cut and machine soft and non-metallic materials such as asbestos, wood, paper, foam, leather, etc.

The water jet machining process does not create any pollution as it does not produce any dust and gases during machining. Water jet machining process cleans the surface during cutting. Water jet cutting process is applicable for food industry. Water jet machining is similar to Laser Beam Machining (LBM) and Electron Beam Machining (EBM) in respect of concentrated energy source applied for removing materials [1]. The kinetic energy of water jet is utilized in WJM, whereas thermal energy is utilized in LBM and EBM. WJM process does not produce any heat so it does not create any thermal damage of machined surface. Water jet is used as a cold cutting tool which never becomes dull. Water jet is also used for cutting coal and fracturing rocks in mines. Water jet is used for cutting soft as well as hard tissue like bone in orthopedic surgery. WJM process is also applied for cleaning, polishing and surface engineering applications. Water jet machining process can be applied for various operations such as drilling, grooving, cutting, milling and turning, etc.

3.3.1.1 Mechanisms of material removal

The high velocity water jet of small diameter strikes on the workpiece surface. After striking the surface, the water jet velocity is reduced and the kinetic energy of the jet is converted into pressure energy. After impact of the water jet on the workpiece the generated transient pressure is increased as high as three times the normal stagnation pressure due to water hammering effect [1]. The striking area of the jet is also less. The impact striking pressure of the water jet is very high. The localized stress induced on the workpiece surface due to hammering effect of the water jet exceeds the strength of the materials. As a result material is removed due to mechanical erosion caused by localized compressive failure of the workpiece materials. During compressive failure, the local water pressure is higher than the ultimate compressive strength of the workpiece materials [2]. Due to high pressure impact of water jet in small area, micro cracks are generated specially on brittle materials and micro cracks are propagated until the materials are cut. The high velocity water jet washes away the eroded materials those are formed due to propagation of micro cracks. High velocity water jet produces plastic deformation on the soft materials and finally the materials are eroded by the shearing action of water jet along the edges of the deformed area. The mechanical spalling due to pressure wave of water jet is also another mechanism of material removal from workpiece like ceramics. The cavitation collapse can also increase pressure of water jet which is capable for eroding the materials from workpiece [1]. For high frequency

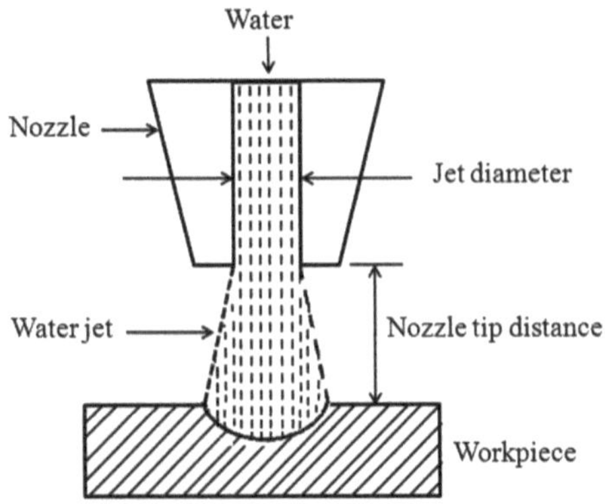

Fig. 3.3.1 Schematic illustration of material removal in WJM process.

pulsating water jet machining the material removal is caused due to fatigue failure in ductile materials and brittle fracture in brittle materials. Figure 3.3.1 shows the schematic diagram of material removal process in WJM.

Material removal rate (MRR) depends on the reactive force F of the jet, where reactive force is the product of mass flow rate (m) of water and water jet velocity (v). Water jet velocity (v) again depends on water pressure (P) and density of water (ρ) and expressed [3] as:

$$v = (2P/\rho)^{0.5} \tag{3.3.1}$$

Reactive force (F) is expressed as:

$$F = m(2P/\rho)^{0.5} \tag{3.3.2}$$

The mass flow rate (m) of water passing through the nozzle is expressed as:

$$m = Cd^2\rho(2P/\rho)^{0.5} \tag{3.3.3}$$

where, d is nozzle orifice diameter and C is a constant.

Therefore material removal rate (MRR) can be expressed as:

$$MRR = K.F = K.m(2P/\rho)^{0.5} = KCd^2(2P) \tag{3.3.4}$$

where, K is a constant of proportionality.

The material removal rate depends on water pressure (P) and nozzle orifice diameter.

3.3.2 Details of water jet machining system

Water jet machining system consists of water reservoir, hydraulic pump, intensifier, accumulator, control valves, flow regulators, high pressure tubing, nozzles, nozzle head motion control unit, catchers, workpiece holding and feeding unit. Figure 3.3.2 shows the schematic diagram of water jet machining system.

(i) WJM pumping unit

The WJM pumping unit supplies high pressure water to the nozzle from water reservoir. The hydraulic oil pump is driven by an electric motor and supplies high pressure oil from oil reservoir to drive a reciprocating plunger pump termed an intensifier. The intensifier delivers ultra-high pressure water to the accumulator by increasing the pressure of water received from low pressure water reservoir through water filtration unit. Accumulator is a reservoir of high pressure water which eliminates the pressure fluctuation of water coming from intensifier. Accumulator provides the continuous supply of high pressure water to the nozzle through control valves and high pressure water tubing.

(ii) Nozzle

Nozzle converts high pressure water into high velocity water jet. The water jet should have minimum lateral expansion and it should remain coherent

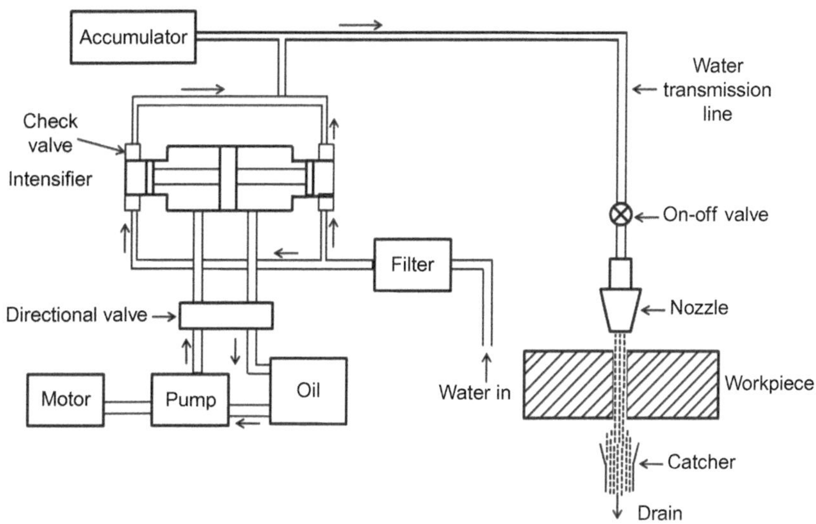

Fig. 3.3.2 Schematic diagram of a water jet machining system.

over the maximum nozzle tip distance. Nozzle should produce better coherent jet to reduce noise level of water jet. Nozzle can produce long coherent jet of length 500–600 times the nozzle diameter with the addition of long chain polymer such as polyethylene oxide to the working fluid due to increase in viscosity of the flowing fluid. Nozzle diameter of 0.05 mm can produce a coherent jet of 25 mm long while a nozzle diameter of 0.35 mm can produce a coherent jet of 175 mm length [4]. The velocity of water jet as high as 900 m/s. The materials of nozzle are of stainless steel, tungsten carbide, diamond and synthetic sapphire, etc. The life of nozzle typically varies from 100 to 500 h. The life of nozzle depends on the nozzle materials, the velocity of fluid flow and the presence of foreign particles in the flowing fluid. Synthetic sapphire is most commonly used nozzle materials for better quality cutting. The inner surface of the nozzle should be very smooth to reduce the level of turbulence of water jet at the exit of the nozzle. The diameter of the nozzle can be varied depending on the application for which the water jet is being used. The nozzle hole diameter typically varies from 0.05 to 0.5 mm. A nozzle has a passage of conical entry of 6 to 20 degrees included angle followed by cylindrical exit having length of 2–4 times end diameter of nozzle opening. For better quality WJM cutting, long exponentially tapered nozzles are also used. The nozzle can be fixed or movable as per requirement. The nozzle can be moved up and down to adjust nozzle tip distance and also to accommodate the thickness of the workpiece. The nozzle tip distance can be typically varies from 2.5 to 50 mm during water jet machining. For CNC WJM machine water nozzle head motion is controlled by CNC X-Y-Z axes drive control system for profile cutting operations.

(iii) Working fluid

For high pressure operation, the mixture of glycerine and water in the ratio 1:1 is used as working fluid. The chances of leakage of fluid are reduced due to the increase of viscosity of mixture of glycerine and water. The presence of glycerine in mixture reduces the compressibility of the working fluid. The materials of the tubing is generally stainless steel to withstand the high stress developed due to high pressure fluid flow. Pure water is commonly used as working fluid as it is cheap and it can be recycled after filtration. Alcohol can be used for cutting meat. For cutting frozen foods the cooking oil can be used [5]. The abrasives such as alumina, zirconia, silicon carbides, glass beads, etc. can be mixed in water jet to improve the machining ability in hybrid abrasive water jet machining process.

(iv) Catcher

In order to reduce the noise level in WJM, the catcher catches the used water jet. After cutting the workpiece, water jet enters into the catcher. A catcher is of long tube or reservoir type. The catcher serves as reservoir for the purpose of collecting the used water jet and also the machining debris entrained in the water jet. The most common type of catcher is the tube type which consists of a 300–600 mm long tube that is attached to a draining hose and the length of the tube is sufficient to allow the water jet to break up completely before it reaches the bottom of the tube [6]. For movable water jet and stationary workpiece, slot type or reservoir type catcher is commonly used, but to reduce noise level it is less efficient. For disposal of waste materials and storage of water a separate catcher basin is used in WJM system.

Figure 3.3.3 shows the photographic view of water jet machining system with single water jet. Figure 3.3.4 shows the photographic view of water jet machining system with multiple water jets for higher productivity.

Fig. 3.3.3 Photographic view of water jet machining system with single water jet. *(Courtesy Flow International Corporation, United States.)*

Fig. 3.3.4 Photographic view of water jet machining system with multiple water jets. *(Courtesy KEMTECH Sweden.)*

3.3.3 Effect of process parameters on machining criteria of WJM

The performance of water jet machining depends on various process parameters such as water pressure, water jet velocity, volume flow rate of water, nozzle diameter, nozzle tip distance and transverse feed rate of nozzle or workpiece and number of passes.

The machining criteria of WJM are material removal rate, nozzle wear rate, kerf width, surface roughness, taper of cut edge and maximum thickness of workpiece that can be cut by water jet cutting process. Table 3.3.1 shows the process parameters and their typical values during WJM process.

Table 3.3.1 WJM process parameters and their typical values

Process parameters	Typical values
Water pressure	1000 to 10,000 bar
Jet velocity	900 to 1400 m/s
Volume flow rate of water	8 L/min
Nozzle diameter	0.075 to 0.5 mm
Nozzle tip distance	2.5 to 50 mm
Traverse feed rate	2 to 8 mm/s
Number of passes	1 or more than 1 depending on thickness of workpiece

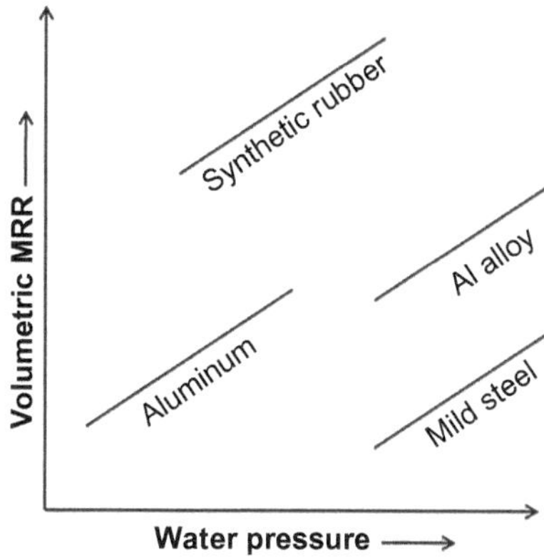

Fig. 3.3.5 Variation of MRR with water pressure.

(a) Effect of water pressure on machining criteria

Material removal rate mainly depends on water pressure. As water pressure increases, material removal rate increases. Figure 3.3.5 shows the variation of MRR with water pressure for different workpiece materials. For machining a particular material by WJM process, there is a threshold pressure below which little or no cutting is observed. The threshold water pressure varies with type of materials to be cut. For cutting mild steel, minimum required water pressure is observed as 6000 bar. For cutting a groove of depth above 1 mm, WJM process needs to set water pressure of 10,000 bar. Depth of groove cut on the workpiece varies linearly with water pressure of the jet. For softer materials like rubber less water pressure nearly 2500 bar to cut groove depth of 50 mm [7]. Figure 3.3.6 shows the variation of depth of groove with water pressure for different workpiece materials.

(b) Effect of transverse feed rate on machining criteria

Material removal rate increases with the increase in feed rate at the lower range from 2 to 4 mm/s during metal cutting by WJM process. For some materials like mild steel and brass, MRR reaches a maximum value at certain feed rate after that it decreases with increase in feed rate at higher range from 4 to 8 mm/s [7]. But for softer metal like copper and aluminum alloy, MRR increases rapidly and then slowly with increase in feed rate. The variation of material removal rate with transverse feed rate is shown in Fig. 3.3.7. For cutting of fiber reinforced plastic by WJM, the width of damaged layer increases with the increase of feed rate after certain limit. The limiting range of feed rate is less for thicker FRP materials. For 5 mm thickness of laminated FRP the feed rate varies from 0.4 to 3 m/min while WJM cutting is

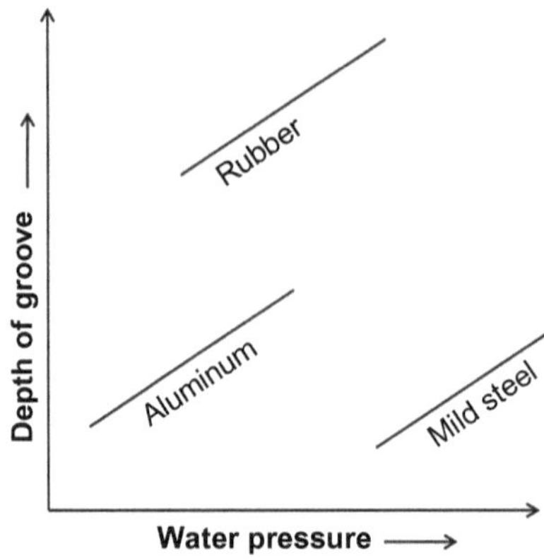

Fig. 3.3.6 Variation of depth of groove with water pressure.

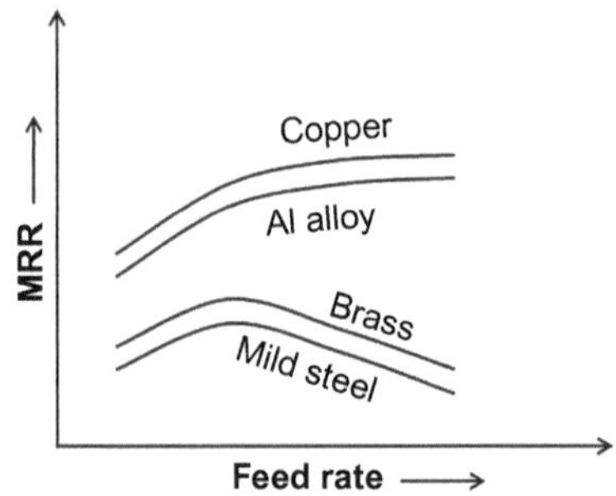

Fig. 3.3.7 Variation of material removal rate with transverse feed rate.

performed at water pressure of 3500 bar and stand off distance of 2 mm and nozzle diameter of 0.18 mm [8].

(c) Effect of nozzle tip distance on machining criteria

As nozzle tip distance (NTD) increases, material removal rate (MRR) increases upto certain level, after that it remains same for certain range of NTD. For maximum MRR, the range of NTD is 3 to 5 mm. With further increase in NTD, MRR decreases. At large NTD, the interaction of atmospheric air with water jet is more, as a result the water jet velocity decreases and diameter of water jet increases. For cleaning and finishing operation by WJM, large NTD is set for better performance. For water jet cutting of materials used in printed circuit boards, NTD is increased upto 19 mm. With the increase in NTD, the accuracy of cut or hole decreases. During WJM cutting the kerf width increases and depth of cut decreases with increase in NTD. With large NTD the hole diameter increases during drilling as

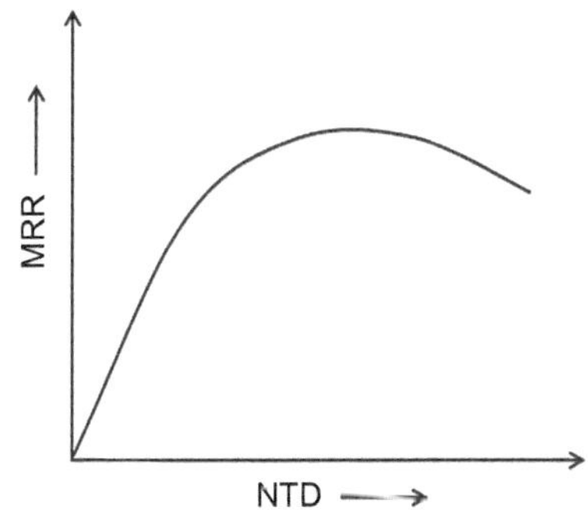

Fig. 3.3.8 Variation of MRR with nozzle tip distance (NTD).

jet diameter increases. At large NTD, the divergent jet with less kinetic energy cut with less accuracy and low machining rate. The variation of MRR with NTD is shown in Fig. 3.3.8.

(d) Effect of nozzle diameter on machining criteria

As the nozzle diameter decreases, the velocity of water jet increases. As a result the kinetic energy of the jet increases. With the increase in kinetic energy of water jet, material removal rate (MRR) of WJM process increases. For generating deeper cut or hole by WJM, smaller nozzle diameter is used. As nozzle diameter increases, jet diameter also increases. With large diameter of water jet, kerf width increases during cutting and hole diameter increases during drilling operation. For increasing the depth of cut or hole, required pressure of WJM process should be high and then only MRR can be increased. For producing micro-hole by WJM, inner diameter of nozzle should be less than 500 μm.

(e) Effect of number of passes on machining criteria

For cutting thicker materials, single pass WJM cutting is not effective. Multi pass WJM cutting operation is to be performed to cut thicker materials. For single pass cutting of materials, proper water pressure of jet is to be selected. For multi pass cutting comparatively lower energy water jet can be utilized by increasing the number of passes of water jet. As number of passes of water jet increases, depth of cut increases and material removal rate increases. The kerf width increases with the increase in number of passes. For controlled machining of thicker material by WJM, the nozzle tip distance and nozzle diameter are to be adjusted along with the number of passes of water jet. For cutting multilayer composite material, proper setting of WJM process parameter for different passes for different layer of materials.

3.3.4 Advantages and limitations of water jet machining

Water Jet Machining Process has wide scope of applications in different fields such as food cutting, electrical and electronics, construction, engineering machinery, mining and biomedical engineering, etc., due to its various advantages, although the process has few limitations.

3.3.4.1 Advantages

The main advantages are described as below:

(i) WJM process eliminates thermal damage on the machined surface as no heat is produced during machining. There is no heat affected zone at the cutting edge which eliminates the inherent problems of thermal cutting process such as Plasma beam cutting, Laser Beam Cutting and Gas Cutting, etc.

(ii) Because of cooling effect, water jet can be used in water jet assisted hybrid machining processes.

(iii) WJM process eliminates tool related problems such as tool wear, tool fabrications, etc.

(iv) Water jet is flexible to utilize at different angles for various operations such as turning, drilling, milling, grooving and cutting, etc.

(v) During WJM better surface finish is obtained and it is also used for deburring purpose.

(vi) WJM produces the cut edge with less taper compared to other cutting process by easy control of the alignment of nozzle with respect to workpiece surface.

(vii) WJM has no hazardous airborne dust contamination and waste disposal problems.

(viii) WJM does not require any post cleaning processes.

(ix) WJM is used for roughing as well as finishing purposes.

(x) WJM can cut thicker materials by increasing the number of passes.

(xi) Water in WJM can be reused after proper filtration.

(xii) WJM can accurately cut softer materials.

(xiii) High pressure and pulsating Water Jet can also cut ductile materials such as copper alloys.

(xiv) Pure water jet machining can be applied for orthopedic surgery specially drilling on soft tissue and hard bones as it has no side effects due to unwanted contaminations.

3.3.4.2 Limitations

The main limitations are described as below:

(i) Equipment cost is moderately high.

(ii) For safety purpose, water treatment is required before disposal.

(iii) WJM requires efficient noise reduction system for safety considerations.

(iv) All the components of WJM are subjected to high pressure, so special attention is needed to avoid any leakage for safety purpose.

(v) Pure water jet is not suitable for hard materials. The use of abrasives mixed in water jet can cut hard materials.

3.3.5 Applications of water jet machining

WJM process has been applied successfully for different machining operations such as cutting, drilling, deburring and wire stripping, etc. It has also been applied in the field of surface engineering such as surface treatment, cleaning, texturing and finishing, etc.

(i) Cutting

WJM process is suitable for cutting of soft, porous, fibrous, granular materials. WJM is applied to cut corrugated asbestos sheet, fiber glass, plywood and marble tiles, etc., without causing airborne dust pollution. Pure Water Jet cutting of food items are also done in food industry.

Water jet Machining process is applied for accurate cutting of contours on fiber-reinforced plastic with negligible thermal damage. Water jet trimming of polycarbonate materials with robotic control of water jet has been performed [9].

WJM process is applied for cutting of unidirectional carbon fiber fabrics. Improved machining process parameters such as nozzle distance, forward speed and water pressure have been determined based on experimental analysis. For best cutting edge quality, low water pressure of 2600 bar, low nozzle distance of 3 mm and high forward speed of 800 mm/min were obtained as optimum parametric combination during water jet cutting of carbon fiber fabrics [10].

Water jet technology is applied for cutting rock in coal mines as it can suppress the dust formed during cutting. Water jet cutting reduces the cutting dust by over 65% on average by increasing the water pressure from 5 to 20 MPa [11].

For cutting thicker materials multiple pass water jet cutting is effective without affecting the cut quality. A slot is produced on materials after single pass and then gradually the cut becomes deeper by increasing the number of passes until the material is completely cut. The maximum thickness of the workpiece to be cut by WJM depends on the materials of the workpiece and also process parametric setting during cutting operation. For cutting very high thickness materials with faster speed, taper edge of cut surface with high kerf width is obtained. Thicker composite materials with multilayer can also be cut by WJM process. CNC WJM system can cut a complex profile with high speed with high accuracy for thinner as well as thicker materials.

Water jet can cut a printed circuit board (PCB) with high speed and acceptable accuracy by using mask and moving small diameter jet as per required profile with the aid of computer numerical control (CNC) technology.

Water jet machining can be applied for cutting laminated glass fiber reinforced composite of thickness 5.0 mm with maximum feed rate of 6.0 m/min setting WJM process parameters such as water pressure of 3500 bar, nozzle diameter of 0.1 mm and nozzle tip distance of 2 mm [5].

WJM has been applied for cutting 12.5 mm thick laminated paper tubes using multiple nozzles at speed of 5.1 m/min. Water jet cutting of difficult to cut materials such as aluminum honeycomb structure of 25 mm thickness has been performed at a cutting speed of 0.85 m/s. Steel and brass can also be cut by high pressure water jet cutting. Water jet cutting is well applied for leather sheet cutting in the shoe industry [1].

(ii) Drilling

WJM process is applied for drilling operation for making a hole on bone tissue for orthopedic surgery. During hole drilling by WJM process, both water jet and workpiece are in fixed position. Generally water jet is kept perpendicular to the workpiece. By tilting and moving the water jet, an inclined and shaped hole can also be created on different materials. WJM process is applied for drilling on rocks and stones in coal mines. For making tunnels of large diameters, multi nozzles high pressure water jet is used along with ratary percussive rock drills. WJM process can also be applied for micro-drilling operation either using mask or micro-water jet.

(iii) Deburring

As better surface finish is observed for WJM cutting, drilling and slotting process, the WJM process is applied for removing burrs of pre-drilled holes and previously cut surfaced of ductile materials using high pressure water jet.

The flexible water jet can be used to remove burrs from narrow opening, blind holes, and complex shaped cavities. The machining or interaction time of water jet controls the size of the burrs to be removed. Both large size and micro-sized burrs can be removed by controlling the process parameters of WJM deburring process. To augment the deburring ability, atomized water jet with induction of compressed air is utilized. Finer abrasives such as silicon carbide, alumina and glass beads mixed with water jet are used to remove burrs efficiently.

(iv) Surface engineering

WJM process has many applications in different area of surface engineering such as cleaning of unwanted coating for surface preparation, cleaning of food utensils, cleaning of rust and residues, polishing, decorative finishing and surface texturing on both metals and non-metals, etc. Water jet technology is applied for cleaning of graffiti which is a damage form of painting, writing or drawing on marble stones [12].

(v) Stripping of cable insulation

WJM process is utilized to remove the insulation material of cable without damaging the metallic core of it. The insulation thickness surrounding the cable is cut by flexible low pressure (690 to 2000 bar) water jet. The water jet rotates surrounding the cable at the required length from the edge of the cable. WJM stripping is faster than conventional thermal or mechanical stripping of cable. Single as well as stranded cable of diameter as large as 46 mm can be stripped by WJM process [6].

(vi) Removing of road stripe

High pressure water jet is applied for removing road stripes of different length, width and types of paints. The problems of traditional grinding method of road stripe removal such as unacceptable air pollution, health hazards, risk of accidents, etc., can be overcome by using high pressure water jet system and automated control system. The water jet system is similar to WJM cutting system which consists of an intensifier and multi nozzles to focus a high pressure water injection on the road stripes. The controllable process parameters are water jet injection distance, injection angle and injection pressure. The injection pressures varies from 1000 bar to 2000 bar. The injection angle varies from 30 to 90 degrees. The nozzle injection distance varies from 40 to 60 mm. The transfer velocity of nozzle is set at 240 mm/min. The advantages of water jet technology such as low dust scattering, environment friendly working conditions and convenient adaptability of automation lead to its application in field of road cleaning and construction [13].

3.3.6 Advancement in water jet machining technology

WJM has wide scope for performing various operations such as drilling, cutting, grooving, surface treatments and cleaning, etc. There are some advancements in water jet technology for improving the machining performances highlighted in the subsequent discussions.

Water jet technology can also be applied in hybrid machining process such as water jet guided laser machining for cutting carbon fiber reinforced plastics. Water jet guided laser machining is very much effective to reduce heat affected zone and improve surface integrity and machining efficiency over simply laser beam machining. The cooling effect of water jet reduces the thermal damage caused by laser [14].

The mechanical effect of water jet in high pressure water jet assisted machining is utilized for fragmentation of the chip in the localized area of cutting. The application of water jet reduces the cutting tool surface temperature and as a result tool life increases due to cooling effect of water jet [15].

High Energy Fluid Jet Machining can be used for micro-machining by reducing the jet diameter which is less than 100 μm. Multi-axis high energy fluid jet machine can generate micro components with complex geometry in a flexible way. By increasing or decreasing the water pressure, the energy of water jet can be applied for cutting and finishing operations. Using harder abrasive media, high energy water jet is useful for deep penetration on hard workpiece materials [16].

The pulsating water jet (PWJ) with high ultrasonic frequency is used to disintegrate copper alloys such as brass and bronze. The PWJ is also used for disintegration of rocks and stones. By increasing the number of passes PWJ can produce a groove of desired depth and size. PWJ can be applied for finishing as an alternative of sand and ball blasting. PWJ technology has wide scope of application in removal of surface deposits and layers for machining as well as cleaning applications specially the cleaning of ships hulls and also in the aircraft industry [17].

Ultra high pressure intensifier is used in Water Jet Machine (WJM) and several components are subjected to very high pressure nearly 600 MPa. The components of ultra high pressure intensifier works under extreme fatigue stress. A multi acoustic emission (AE) sensor based condition monitoring approach is used to check the health and safety of ultra high pressure intensifier used in WJM machine [18].

Water jet Machining has wide scope of application in orthopedic surgical treatments. Water jet drilling in bone tissue has been performed to make holes for fixing screw in bone for bone fracture and dislocation treatment. Single water jet (SWJ) as well as colliding water jets (CWJ) are used to make holes of different depth. The effect of colliding two water jets at different angles can control depth of hole on bone tissue. Using CWJ principle, the hole depth variation caused by inhomogeneity of bone materials can be reduced as compared to SWJ [19].

Water Jet Machining (WJM) is a cold machining process by using high pressure water jet for various operations such as cutting, drilling, cleaning, finishing, polishing, wire striping, removal of road stripes, etc., on varieties of engineering materials. In the area of hybrid machining water jet technology is also applied. There are advancements in the area of development of ultra-high pressure intensifier and generation of high frequency pulsating water jet for efficient WJM system. WJM is also applied for micro-machining operations using high energy fluid jet with micro-hole diameter nozzle. CNC multi axis and multi water jet machining system with robotic control of nozzle heads has been developed to enhance the productivity and flexibility of machining for wide scope of engineering applications. By using abrasives mixed in high pressure water jet, a hybrid machining process like abrasive water jet machining has increased the scope of applications for machining varieties of brittle and ductile materials.

References

[1] P.K. Ray, Water jet machining—a state of art, J. Inst. Eng. 63 (1) (1982) 108–111.

[2] F.J. Lavoi, Water jet machining, Mach. Des. (1973) 89–93.

[3] J.A. McGeough, Advanced Methods of Machining, Chapman and Hall, New York, 1988 ISBN: 0-4123-1970-5.

[4] P.C. Pandey, H.S. Shan, Modern Machining Processes, Tata McGraw Hill, New Delhi, 1980.

[5] H. El-Hofy, Advanced Machining Processes, Non Traditional and Hybrid Machining Processes, McGraw-Hill, New York, 2005, pp. 32–39.

[6] G.F. Benedict, Nontraditional Manufacturing Processes, Marcel Dekker, New York, 1987 ISBN: 0-8247-7352-7.

[7] P.K. Mishra, Noncoventional machining, in: The Institution of Engineers (India), 2012,pp. 45–52, ISBN: 978-81-7319-138-1.

[8] W. Konig, C.H. Wulf, P. Grab, H. Willerscheild, Machining of fibre reinforced plastics, Ann. CIRP 34 (2) (1985) 537–548.

[9] B.K. Engemann, Water jet cutting of fibre reinforced composite materials, Ind. Prod. Eng 3 (1981) 162–164.

[10] M. Voit, G. Reinhart, T. Metzger, Experimental study on water jet cutting of unidirectional carbon fibre fabrics, in: 1st CIRP Conference on Composite Materials Parts Manufacturing, CIRP-CCMPM 2017, Procedia CIRP, vol. 66, 2017, pp. 221–226.

[11] H. Jiang, C. Du, J. Dong, Investigation of rock cutting dust formation and suppression using water jets during mining, Powder Technol. 307 (2017) 99–108.
[12] N. Careddu, O. Akkoyun, An investigation on the efficiency of water jet technology for graffiti cleaning, J. Cult. Herit. 19 (2016) 426–434.
[13] Y.B. Ham, S.W. Kwon, J.H. Noh, J.G. Han, K.T. Kim, Development of road tripe removing equipment using high pressure water jet, Autom. Constr. 15 (2006) 578–588.
[14] D. Sun, F. Han, W. Ying, C. Jin, Surface integrity of water jet guided laser machining of CFRP, in: 4th CIRP Conference on Surface Integrity (CSI 2018), Procedia CIRP, vol. 71, 2018, pp. 71–74.
[15] Y. Ayed, C. Robert, G. Germain, A. Ammar, Development of a numerical model for the understanding of the chip formation in high pressure water jet assisted machining, Finite Elem. Anal. Des. 108 (2016) 1–8.
[16] D.A. Axinte, B. Karpuschewski, M.C. Kong, A.T. Beaucamp, S. Anwar, High energy fluid jet machining (HEFJet-Mach): from scientific and technological advances to niche industrial applications, CIRP Ann. Manuf. Technol. 63 (2014) 751–771.
[17] D. Lehocka, J. Klich, J. Foldyna, S. Hloch, J.B. Krolczyk, J. Carach, G.M. Krolczyk, Copper alloys disintegration using pulsating water jet, Measurement 82 (2016) 375–383.
[18] M. Goletti, M. Grasso, M. Annoni, B.M. Colosimo, Condition monitoring of an ultra high pressure intensifier for water jet cutting machines, in: 8th CIRP Conference on Intelligent Computation in Manufacturing Engineering, Procedia CIRP, vol. 12, 2013, pp. 193–198.
[19] S. Den Dunnen, J. Dankelman, G.M. Kerkhoffs, G. Tuijthof, Colliding jets provide depth control for water jetting in bone tissue, J. Mech. Behav. Biomed. Mater. 72 (2017) 219–228.

SUBCHAPTER 3.4

Ion Beam Machining (IBM)

3.4.1 Introduction

Ion beam machining has become attractive for fabrication of intricate structures especially in micro and nano range with high resolution without altering the properties of parent material. Here, focused beam of ions act as a tool which caused to bombard and eject atoms from the workpiece surface by elastic collisions to crate desired shapes. Thus, this method of machining is commonly known as focused ion beam (FIB) machining. In the history of development of ion beam for industrial applications, there are several milestones. In early 1907, J.J. Thomson established an ion implantation equipment utilizing energetic beam of ions which was an initial breakthrough to utilize ions in this direction. In the early 1960s, an ion implementation technique for semiconductor process was first demonstrated using Van D. Graff-Accelerator which was invented long back by Robert J. Van D. Graff in the year 1931 at MIT, United States. Focused ion beam machining was introduced more than a decade ago and was mainly developed for industrial applications during late 1970s and the early 1980s through the invention of liquid metal ion source [1–3].

In industry, ion beam of single or molecular ions are employed for etching, ion assisted film deposition and implantation, etc. Recent technological advances forced to improve accuracies of manufacturing methods at atomic or molecular level. Ion beam processing can be effectively utilized to fabricate advanced electronic, magnetic and optical devices for achieving this level of accuracy where the characteristic dimensions are less than about a few nanometers. It is important to develop a very low energy ion beam system at high current in order to accomplish these requirements. However, low energy ion beams are very difficult to generate as well as transport. At this juncture, for the development of ion beam processing, Plasma Source Ion Implantation (PSII) in the year of 1987, US and Gas Cluster Ion Beam

(GCIB) in the year 1988, Japan are the two new promising technologies to be noted [4].

Ion Beam Machining is capable to fabricate intricate features in the order of 1 μm or below due to its very short wavelength and extremely large energy density. Thus, IBM becomes a well accepted machining technique for fabricating high quality micro device and micro structures. In the semiconductor industry, IBM is effectively utilized for repairing of mask, modification of devices, analysis of failure and debugging of intricate circuits (IC). FIB has also found its profound utility in areas of advanced manufacturing. FIB can machine any material and can solve micro and nano level machining problems. FIB milling has been successfully utilized for removal of materials from a substrate to fabricate various three dimensional micro structural devices considering a wide range of materials.

3.4.1.1 Basic functions of ion beam processing

Ion beam machining is an atomic bit machining process with very high resolution. Beam of ion consists of highly energetic ions manipulated by electrical means. Material removal is carried out by knocking atoms out of the workpiece surface by electro-elastic collision and recoil actions between the incident ions and the target surface [5]. In this area of machining, there are two processes exist such as ion beam machining (IBM) and focused ion beam machining (FIB) which utilize energetic ions. Material removal mechanisms of both the processes are similar in nature; however there are some minor differences as identified below:

Ion beam machining (IBM)—A stream of charged ions of inert gases, such as argon (Ar^+) is accelerated in vacuum to attend high energies and directed towards workpiece. This broad ion beam removes atoms from the workpiece by transferring its energy and momentum to the atoms on the surface of workpiece.

Focused ion beam machining (FIB)—This system uses beam of different ion species or multiple charged ions and clusters such as gallium ions (Ga^+) from the ion sources, which is focused with the help of deflectors and electrostatic lenses to a small cross-section on the surface and removes atoms from the surface of workpiece. Here, secondary electrons generated due to ions bombardment, are collected to form image of the surface of workpiece.

When energetic ions strike the surface of solid, complicated ion-solid interactions take place which governs the capabilities of ion beam processing. After striking the surface of substrate, ions transfer its energy to the atoms as well as to the electrons. The most vital physical effects of incident ion-solid interactions depending on the ion energy are as follows:

(i) Sputtering of excited atoms from the surface of the substrate which mainly governs material removal in the form of milling. The ion energy ranges from 10 to 100 keV.

(ii) Displacement of atoms in the target surface which may cause implantation of impinging ions and causes induced damage as well as modifies the surface characteristics. The ion energy is higher than 100 keV.

(iii) Chemical reactions causing breaking of chemical bonds and thereby dissociating the gas molecules on the target surface as deposition which employ for thin film deposition. The ion energy ranges from 10 to 50 keV.

(iv) Emission of electrons or photons from the target surface which enables imaging for the purpose of inspection and monitoring. The ion energy ranges from 10 to 100 keV.

(v) Backscattering nuclear reactions occur when ion energy is more than 1 MeV.

Based on these above mentioned interactions, IBM can be used for machining, deposition, implantation and imaging as exhibited in Fig. 3.4.1. High aspect ratio 3D micro/nano features can be generated by FIB milling as shown in Fig. 3.4.1(a). FIB permits controlled maskless deposition of metallic as well as insulting material on the substrate. The deposition process is illustrated in Fig. 3.4.1(b), where precursor gases are sprayed on the surface and incident ion beam decomposes the gases, which allow reaction products remain fixed on the surface as a thin film whereas volatile reaction products de-absorb from the surface and remove through vacuum system. Accelerated ions with high kinetic energy bombard on the surface of the substrate and implant into the substrate as shown in Fig. 3.4.1(c), which can be combined with milling to fabricate fine 3D structures of complex semiconductors. During scanning of ion beam over the substrate, collection of secondary electrons can be utilized as shown in Fig. 3.4.1(d) for characterizing materials as well as failure analysis for different micro and nano scale applications.

FIB technology can be applied for various purposes as mentioned earlier. However, the utilization of ion beam for the purpose of machining of different materials for producing various high quality micro devices and high precision micro components will be highlighted. To achieve desired amount of material removal from a predefined location in a controllable fashion, proper ion species, beam size, shape, current, energy, and incident angle are the key features of IBM.

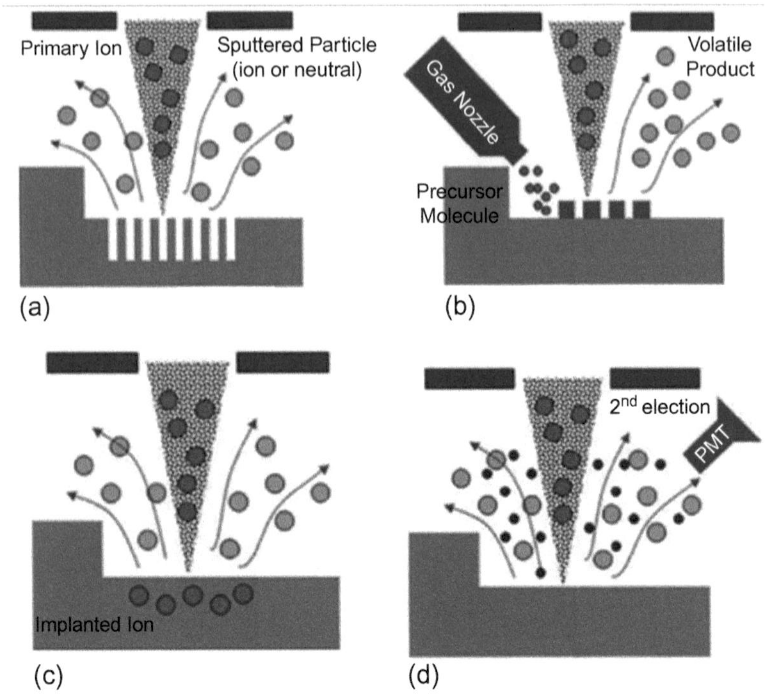

Fig. 3.4.1 Basic four applications of FIB: (a) milling/machining (b) deposition, (c) implantation, and (d) imaging [6].

IBM and electron beam machining (EBM) processes utilize high energy particle beam under vacuum environment. However, IBM has various advantages over EBM. It is better to mention here about the basic differences between IBM and EBM to acquire information about their various characteristics, which will be useful for selection of the process for various modern machining applications. Brief major comparisons between IBM and EBM process characteristics are listed in Table 3.4.1.

3.4.1.2 Material removal mechanism in ion beam machining

During bombardment of energetic ions on the atoms of target surface, various interactions may occur as already stated above. Unwanted side effects may occur due to various interactions that are not entirely separable [7]. To overcome these side effects during a specific machining application, these interactions need to be understood.

For machining applications, incoming ions should interact only with the surface atoms or atoms just beneath the surface layer of the target substrate

Table 3.4.1 Differences between IBM and EBM process characteristics

Characteristics	IBM	EBM
Mechanism	Mechanical sputtering	Melting and vaporization
Energy	Ions are much heavier and strike the target with much greater ion energy, 10–100 keV	Electrons are lighter and strike the target with less energy, upto several hundreds keV
Capability of focusing	5–100 nm	2–25 μm
Varieties of particles	Wide variety of ion masses	Constant electron mass
Interactions	Swelling, deposition, sputtering, implantation, back scattering or nuclear reaction	Excitation of electrons of surface atoms and back scattering
Resolution	Very high (0.005–0.02 μm) due to low lateral scattering for much heavier ions	Moderate (2–5 μm) due to high energy back scattering electrons
Inspection	Possible, in situ	Possible
Surface damage	Lesser or none	Higher (thermal affects)
Surface finish (R_a)	0.01–0.05 μm	0.5–5 μm
Applications	Micro/nano fabrication of 3D features and inspection	Micro hole fabrication, welding and inspection

which lead to a collision cascade on the atoms. The collision can transfer sufficient energy to the surface atoms to overcome its surface binding energy which results in ejection of atoms from the surface of workpiece. This ejection phenomenon of surface atoms is totally dependent on transfer of energy. This ejection phenomenon is known as sputtering which governs the metal removal mechanism during machining. Sputtering is purely physical process which exclusively depends on momentum transfer to remove the atoms from the workpiece surface. Machining mechanism of ion sputter may cause due to various types of ion-atom collisions as shown in Fig. 3.4.2.

Incident ions with low energy may remove single atom from the surface. Impinging ions with higher energy may transfer sufficient momentum to several atoms and results in removal of multiple atoms from the workpiece surface due to a cascade type effect near the surface. The degree of cascading effects occur into the material depend on the energy of the incident ions which drive out several atoms or molecules, ionized or neutral from the target material. As shown in Fig. 3.4.2 incident ions may implant deep into the material as substitutional ions by displacement of atoms from lattice point

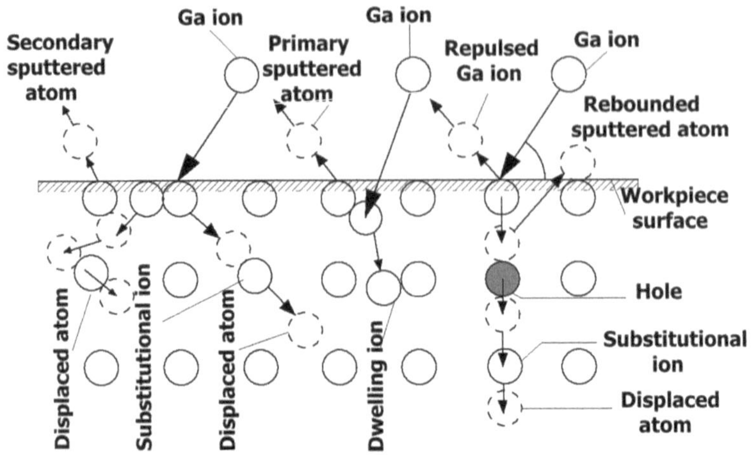

Fig. 3.4.2 Machining mechanism of IBM.

creating holes caused to damage the material. From electron diffraction patterns, it is estimated that the penetration depth of an argon ion (Ar^+) with 1 keV is about 5 nm.

3.4.1.2.1 Mathematical formulation on mechanism of ion beam machining

The mechanism of material removal is known as atomic scale processing because it depends on increasing the internal energy of individual atoms and cluster of ions. In ion beam machining, material removal mechanism greatly depends on individual or cluster of atoms which is utilized to increase the internal energy of the parent substrate. In a thermodynamic point of view, internal energy (U) is the addition of three individual energies, e.g., Gibbs free energy (G), thermal energy (H), linear kinetic energy (E_k), and internal elastic strain energy (E_e).

So,

$$U = G + H + E_k + E_e \tag{3.4.1}$$

However, in practical condition, the required energy known as threshold specific processing energy (E_p) for removal of one atom from the workpiece should be at least greater than sum of two energies, e.g., lattice bonding energy (E_b) and surface barrier energy (E_s).

So,

$$E_p = \frac{(E_b + E_s)N}{m} \tag{3.4.2}$$

where, N is Avogadro's number, 6.023×10^{23} atoms/mol and m is the molar volume of the workpiece, m^3/mol. Threshold specific processing energy is the minimum internal energy required for removing the volume of the material from the workpiece per unit volume. This is the important parameter for ion sputtering in ion beam machining operation. Mechanism of material removal in IBM depends on the presence of defects in the atomic lattice structure of the workpiece material on which focused ion beam strikes. A typical lattice spacing of a solid is as small as 0.3–0.4 nm. However, due to presence of various defects, e.g., vacancies, interstitial defects etc. in the crystal structure suggest that the fracture and failure on the atomic lattice distance may ranging about 1–100 nm. Hence, to enable machining at this scale, typical value of specific processing energy (E_p) for ion sputtering estimated to be 10^{11}–10^{12} J/m^3.

The removal of atoms from the workpiece substrate by the accelerated ions is a physical sputtering process. Through ion sputtering, atoms, positive ions and secondary electrons are removed from the surface of the substrate. From Fig. 3.4.3, relationships can be established between sputtering rate and machining speed with different parameters of ion beam [8].

Sputtering rate (S) is defined as the ratio of number of sputtered atoms to the number of incident ions strike at the workpiece surface.

Thus,

$$\text{Sputtering rate, } S = \frac{\text{Number of sputtered atoms}}{\text{Number of incident ions}} = \frac{Ad\rho NZ}{M} \cdot \frac{I\tau}{\varepsilon} \quad (3.4.3)$$

where, A is the projected area of the focused ion beam is equal to $a.cos\theta$, d is the depth of machining, ρ is the density of the workpiece material, g/cm^3, N is the Avogadro's number, i.e., 6.023×10^{23}, atoms/mol, Z is atom/molecule, M is the molecular weight of specimen, g/mol, I is the Ion beam current,

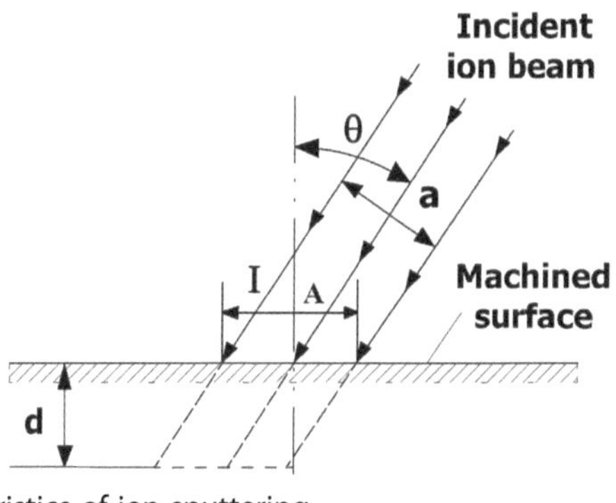

Fig. 3.4.3 Characteristics of ion sputtering.

A, τ is the machining time, s, and ε is the elementary charge, i.e., 1.6×10^{-19}C/ion.

Furthermore, in IBM, machining speed (V) is the ratio of sputter machining depth to current density of the focused ion beam.

Thus,

$$\text{Machining speed, } V = \frac{\text{Sputtered machining depth}}{\text{Beam current density}}$$

Beam current density is the amount of current (I) of the ion beam per cross sectional area of the incident ion beam (a) on the surface of the workpiece. In IBM, incident ion beam strikes the workpiece at an angle known as incident angle (θ).

So,

$$V = \frac{d/\tau}{I/a} \tag{3.4.4}$$

From Eqs. (3.4.3), (3.4.4), machining speed in terms of sputtering rate can be written as:

$$V = S.\frac{M}{NZ\rho\varepsilon}\cos\boldsymbol{\theta} \tag{3.4.5}$$

Hence,

$$\text{Machining speed, } V = S_r.\cos\boldsymbol{\theta} \tag{3.4.6}$$

where, S_r is the sputter machining rate, cm^3/C.

3.4.1.3 Types of ion beam machining processes

Ion beam machining processes can be classified based on ion sources and its applications. A number of ion sources have been developed over the years to generate ions for IBM. There are mainly two types of ion sources available as stated below:

(i) Gas filled ion sources (GFIS)—In this category, various gases are exploited to generate ions by decomposing using different techniques. Gas ions such as H^+, H^{2+}, He^+, Ne^+, and Ar^{++} etc. are employed in earlier days.

(ii) Liquid metal ion sources (LMIS)—Liquid metal ion sources are powerful sources of ions mainly developed in the late 1970. In this category, popular ion species are Ga, As, Be, and Si which are extensively employed for the purpose of microfabrication.

Gas filled ion sources are utilized during the initial stage of development of IBM. However, some of these types of ion sources are still popular for some specific applications in IBM. Some of these ion sources are listed below:

(a) Hot filament Kaufman source
(b) Radio frequency inductively coupled sources
(c) Electron cyclotron resonance (ECR) sources
(d) Duoplasmatron sources

Hot filament Kaufman source is one of the simplest sources of ion utilized in IBM. A cathode filament made of tungsten is heated to the state of thermionic emission by passing high current through it. Electrons are emitted from the tungsten cathode. A solenoid coil provides a magnetic field for the confinement of electrons. Emitted electrons are accelerated towards the anode by high voltage in the order of more than 1 kV. During the movement of high speed electrons, these electrons collide with gas atoms and produce gas ions. These gas ions are extracted into the machining chamber by a grid placed at the front of the source. This type of system is known as ion shower type IBM where argon inert gas is used as shown in Fig. 3.4.4. This type of IBM is used for precision etching of semiconductor. The beam current density of 0.7 mA/cm^2 and energy of 0.6 keV are typical values for etching of

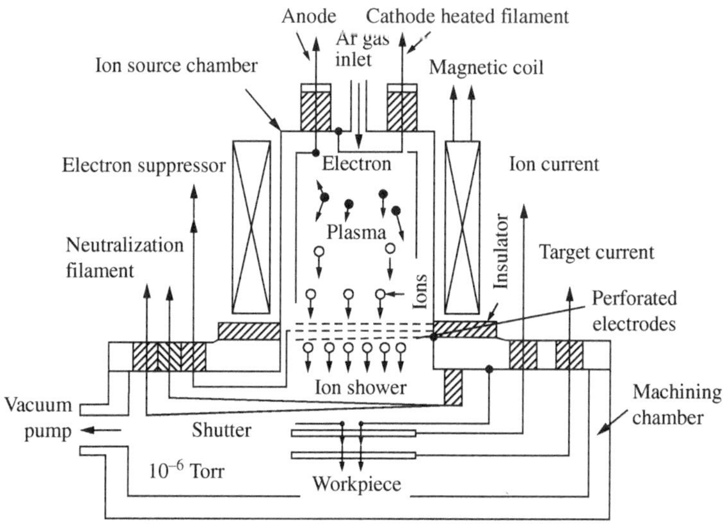

Fig. 3.4.4 Kaufman ion shower type ion beam machining [7]. *(Courtesy AGS-TECH Inc., PO Box 4457, Albuquerque, NM 87196, United States.)*

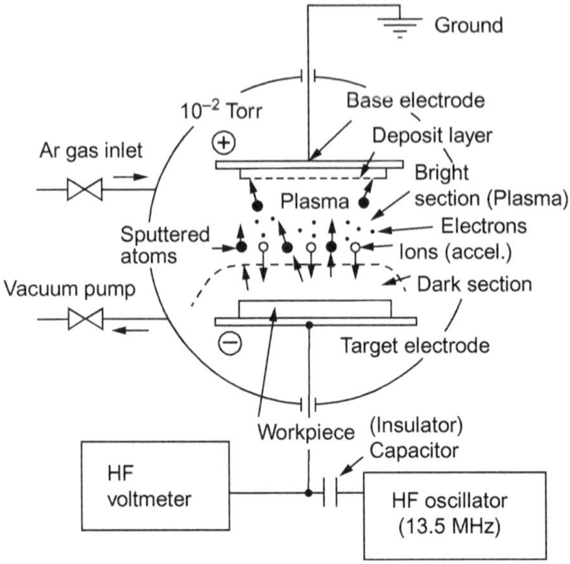

Fig. 3.4.5 High frequency plasma type ion beam machining [7].

semiconductor with photoresist mask. However, the photoresist mask must survive the heat input of ion beam.

Radio frequency inductively coupled sources are used for ionization of gas to form plasma. Here, argon gas is fed in between two opposite parallel electrodes, i.e., base electrode and target electrode and ionized with the help of high radio frequency in the order of more than 10 MHz as shown in Fig. 3.4.5. The wafer to be etched is mounted on the target electrode and anisotropically etched by the high velocity ions. Volatile reaction products generated due to ion beam etching is removed from the chamber by vacuum pump.

Electron cyclotron resonance (ECR) sources are known as hot-plasma ion source which is shown in Fig. 3.4.6. In magnetic field, due to Lorenz force, moving electrons spin around the magnetic field lines. Gyration frequency of electrons is known as cyclotron frequency. The electrons are resonantly accelerated if microwave radiation of the same frequency propagates into the magnetic field. Solenoid or permanent magnet generates axial magnetic field and multiple magnet generates radial magnet field which results in minimum-B-structure. In which magnetic field has a minimum value in the middle of the structure and from there it increases in all direction. Plasma electrons are confined in the minimum-B-structure. This creates a closed

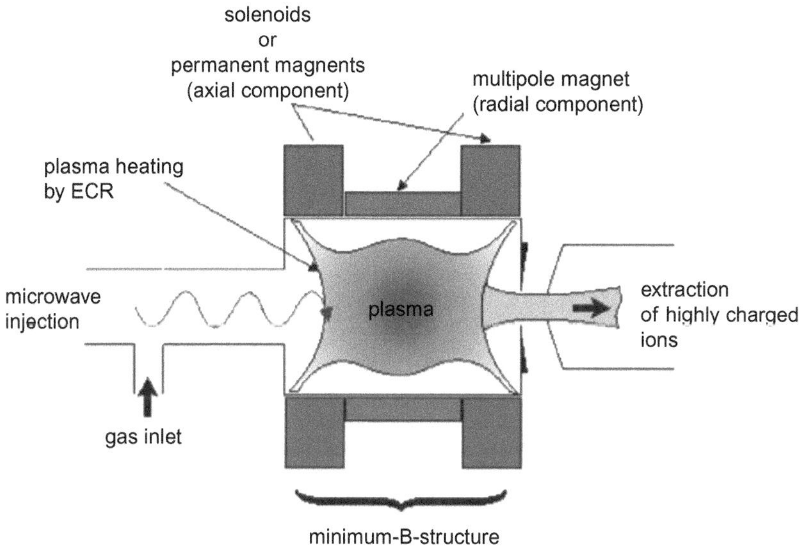

Fig. 3.4.6 Diagram of ECR ion beam source [9].

surface which fulfills the condition of electron cyclotron resonance. Hence, electron passing through this surface can be accelerated resonantly. Electrons pass through the resonance region and gain high energies which intern ionize plasma atoms and ions into high charge state through consecutive single ionization [9].

Duoplasmatron ion source is invented by Manfred von Ardenne [10] which is based on cold cathode source as shown in Fig. 3.4.7. Here, plasma formed is constricted mechanically and magnetically. To generate high

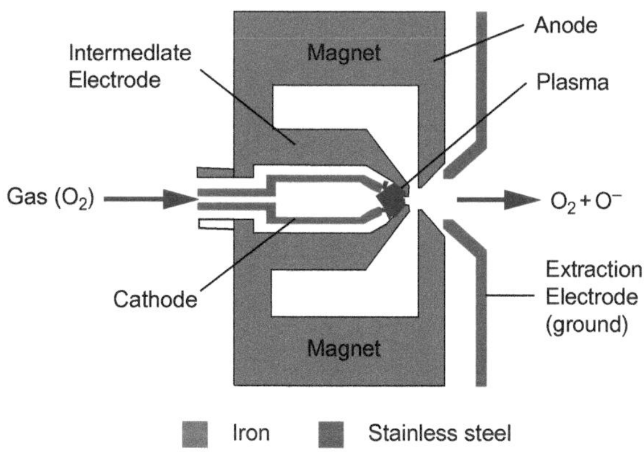

Fig. 3.4.7 Diagram of Duoplasmatron ion beam source [11].

speed ion beam, two highly charged grids, i.e., extraction electrode and acceleration electrode are employed. The ion source includes anode aperture, liquid cooling connection, and source magnet with electrical feed-throughs. Tungsten wire or coated mess filaments are hold by two 50 A feed-throughs. Ion beams of helium, oxygen and argon are commonly produced with currents 2–10 mA, arc voltage of about 100 V and extractor voltage of about 30 kVDC. Duoplasmatron ion source can be used for various gases which form stable ions [12].

3.4.2 Focused ion beam (FIB) machining system setup

FIB machining system is comparatively dynamic process in which atoms from the workpiece surface are removed precisely by an accelerated stream of ions, focused into a tiny area solely by electronic and electrical control [11]. Basic components of FIB system consists of an ion source, an ion optics column and machining chamber with workpiece holding stage through computerized control system. Liquid metal ion sources (LMIS) have been widely used in FIB which provides reliable and steady ion beams for a variety of ion species. FIB systems require vacuum environment to avoid interaction of ion beam with gas molecules and also to avoid contamination and electrical discharge.

3.4.2.1 Liquid metal ion sources (LMIS)

Liquid metal ion sources (LMIS) consists of a capillary tube with a needle through it, an extraction electrode and a shielding as shown in Fig. 3.4.8.

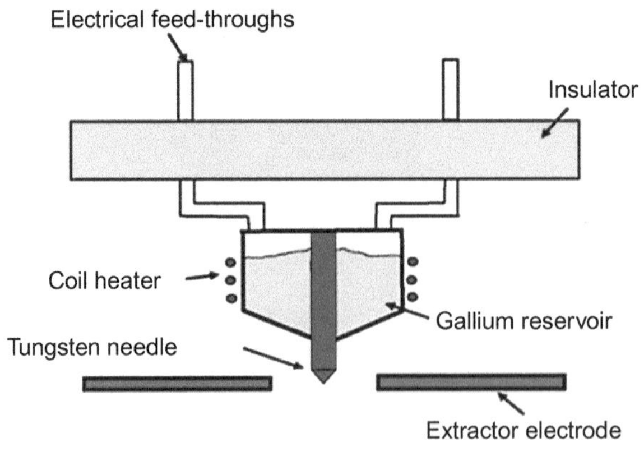

Fig. 3.4.8 Diagram of LMIS [13].

The needle made of tungsten as a tip radius approximately 2–5 μm. Metal, e.g., Gallium is melted with an electrical coil heater. Gallium due to its lower melting temperature, i.e., 29.8 °C remains in molten state for longer period. The capillary tube acts as a reservoir which feeds liquid metal to the tip of tungsten needle. The interactions of strong electrostatic force generated by the extraction electrode and surface tension causes formation of the sharply peaked cone of liquid metal meniscus, which is known as Taylor cone [14]. The application of critical Taylor voltage on the liquid metal cone extracts positively charged ions. Extraction voltage of 7 kV and current of 2 μA are applied under normal condition. The flow of metal ions to the Taylor cone continuously replaces the extracted ions. Extracted metal ions are accelerated utilizing high potential down the ion column and are collimated by the lens electrode. Several alloys can be used in the LMIS to produce various types of ion species for FIB system.

3.4.2.2 Machining system setup

Schematic diagram of FIB system is shown in Fig. 3.4.9 which utilized two lenses, i.e., upper and lower lenses. After the first refinement of the ion beam through the spray aperture, the ion beam is focused further in the first electrostatic lens. Hence, the upper lens which is known as condenser, collimates the diverging ion beams from the source into parallel beams. The focused ion beams is then passed through a mass separator. It allows passing through only required amount of ions with a fixed mass charge ratio. The first octopole adjusts the beam stigmatism. Below this, there is a long and thin drift tube. The function of drift tube is elimination of ions that are not directed vertically. The beam current can be varied ranging from 1.5 nA to 1 pA, utilizing a variable aperture mechanism. For rough milling and high resolution imaging, the heavy and fine beam can be formed with different range of beam current. By blanking deflector and aperture, blanking of the beam is achieved. The second octopole is used for scanning the beam over the workpiece in a pre-planned pattern. Below this, the lower lens is placed which is having short focal length. This lower lens is known as objective lens which helps to improve the focusing to reduce the final spot size of ion beam to nm size at a working distance. The working distance is the gap between lower lens and surface of the workpiece, i.e., substrate. An electrostatic beam deflector is positioned below the objective lens which controls the final landing locations of the ions on the workpiece. Multichannel plate (MCP) is also placed above the target workpiece to view the condition of

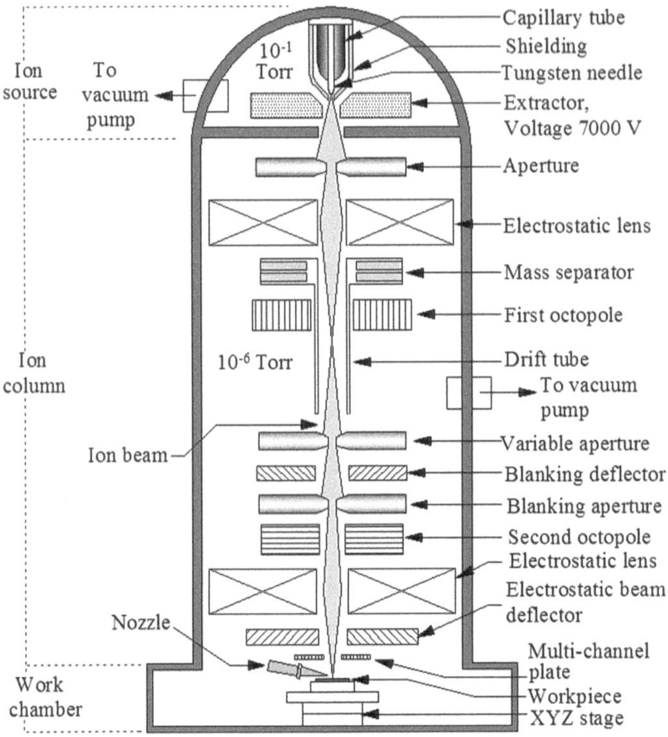

Fig. 3.4.9 Schematic diagram of FIB system setup.

the work surface by receiving and recording the secondary electron emissions during machining.

A nozzle as shown the Fig. 3.4.9 is used to supply gas to the landing spot for FIB induced deposition on the sample surface as discussed earlier. Hence, for some specific applications metal removal by sputtering as well as deposition can also be taken place simultaneously. The FIB setup is placed inside a stainless steel chamber, where low pressure is maintained at 10^{-7} Torr. Vacuum environment will increase mean free-path of the ions as well as strength of the ion beam is not reduced due to interference with gas molecules and other particles in the chamber.

The work substrate is mounted on a CNC stage in the bottom of machining chamber. The stage moves in X, Y and Z directions as well as it can rotate in the X-Y plane and tilt along the X-Z plane with the help of different motors. Thus, complex shape can be machined on the substrate utilizing combined motions of different axes with suitable angular movements which provide enough degree of freedom. With the help of advanced computer controlled technology and graphics, it becomes possible to shape materials for fabrication of complex, high resolution and three-dimensional structures.

All operations, e.g., loading and unloading of workpiece, movement of the stage, control of the gas delivery, control of the vaccum system and manipulating of the ion beam configurations are carried out with the help of an intricate software for which a main frame computer is interfaced with the FIB system.

This type of two lens FIB system generally produces ion energies in the range of 50 to 250 keV. It should be noted that the energy distribution within the spot is highly non-uniform; normally its intensity is close to a Gaussian profile. Thus, produced energy is with the minimum full width at half maximum (FWHM) beam diameter down to 50 nm and can be utilized for both sputtering and implantation [15, 16]. FWHM is the distance between locations on the intensity profile at which the intensity attains half of its maximum value.

Single lens FIB system is also used which provides relatively low energy ions in the range of 10 to 50 keV. Here, ion source is LMIS from where ions are extracted and passed through a beam limiting apparatus to an electrostatic lens. Single lens FIB system does not have objective or lower lens. However, for controlling ultimate beam size, it uses beam limiting aperture which is positioned between extraction electrode and electrostatic lens and it can adjust diameter from 10 to 100 μm. Electrostatic lens is utilized for focusing as well as reducing the ion beam to much smaller than the aperture diameter. It also has beam deflector located above the workpiece. Heavy ion species such as gallium ions are normally used in this low energy type IBM system to improve its sputtering effectiveness.

Single lens IBM system with proper control of beam current and size can be used for fabrication of high precision microstructure with a feature size down to 10 nm. Single lens IBM system is normally utilized for carrying out research in the field of micro-fabrication which provides low quantity production. Two-lens system can handle a variety of ion species and provides high production rate. However, it is much costlier compared to single lens IBM system. Two-lens IBM system is applied in industries for both machining and implantation.

3.4.2.3 Gas cluster ion beam (GCIB)

Concept of Gas cluster ion beam (GCIB) was first established in the year 1998 by Isao Yamada in Japan. In GCIB, electrically charged cluster of ions consisting of few hundred to thousand atoms or molecules of gaseous

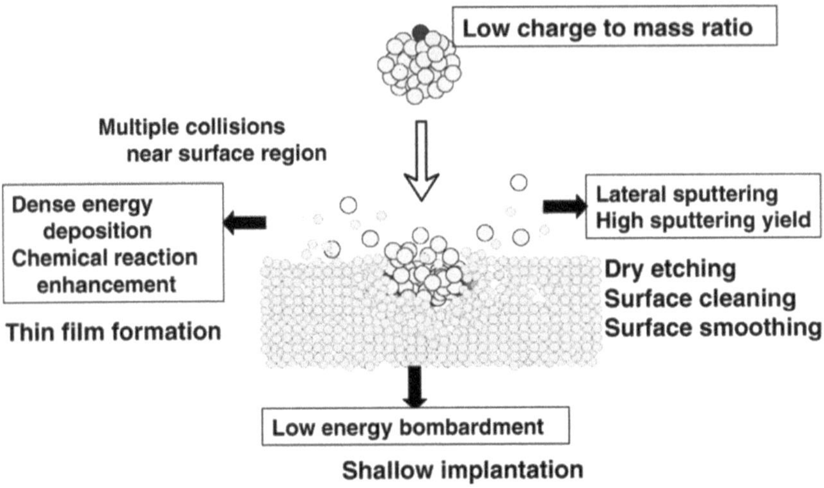

Fig. 3.4.10 Characteristics of GCIB-solid surface interaction and applications [4].

materials are employed as ion beam. It produces very low energy beam at high intensity which presents several unique features of ion-target surface interactions such as effect of low energy bombardment, lateral sputtering and high chemical reaction etc. Impacts of cluster of ions on target surface create concurrent energetic interactions between many atoms comprising the cluster and many atoms of the target which results in non-linear sputtering and implantation effects due to incident of high energy density into a very small volume of the target material. The cluster of ions consists of several thousands of atoms having only single and double ionized atoms in it. Hence, cluster ion beam has very low charge to mass ratio which is one of the significant advantages of GCIB. Cluster ion beam can sputter up to thousands of times more atoms than a monomer ion beam at the same current density. Fig. 3.4.10 shows the characteristics of GCIB-solid surface interaction and their industrial field of applications. Due to lateral sputtering, GSIB produces high sputtering yield which can be utilized for dry-etching, surface cleaning and surface smoothing to atomic levels [4]. Shallow implantation can be effectively made due to low energy bombardment of GSIB. In the field of IC fabrication, thin film at nano-level can be formed utilizing GSIB due to intense energy deposition which enhances rate of chemical reactions because of multiple collisions on the surface region of the workpiece.

3.4.2.4 GSIB processing setup

Schematic of GSIB setup is shown in Fig. 3.4.11 which consists of three main parts such as source of gas cluster formation unit, i.e., nozzle chamber; gas cluster beam formation unit, i.e., ionizer chamber and processing chamber [17]. Supersonic gas expansion from a high pressure source into a vacuum through small nozzle produces gas cluster. Gas cluster formation is achieved due to condensation of gas for adiabatic expansion. Hence, the nozzle plays an important role in GSIB and determines beam flow pattern and cluster size. Gas cluster may consist of 500–10,000 atoms which can be held together by Van Der Waals force. Skimmer, a small aperture collects the primary jet of gas clusters from the source. Electrons emitted and accelerated from a set of filaments, i.e., ionizer, impact on moving beam of neutral clusters and forms positive ion gas clusters. This forward moving cluster ion beam consist usually one electron charge per cluster. Using a series of electrodes, i.e., accelerator, ionized beam of cluster accelerated to a high velocity. Electrostatic lenses are utilized to focus the cluster ions. The beam may consist of some monomers which are filtered out utilizing cluster ion beam transverse magnetic field. The intensity of ion is measured by Faraday Cup. Finally, the focused and accelerated gas clusters ion beam impinges on the workpiece substrate which can be mechanically scanned for desired exposure utilizing a suitable CNC controlled X-Y-Z stage.

Fig. 3.4.12 shows photographic view of FIB machining setup having ability to machining in micro or nano level as well as imaging and analytical performance through SEM with cross beam facility. During FIB-milling,

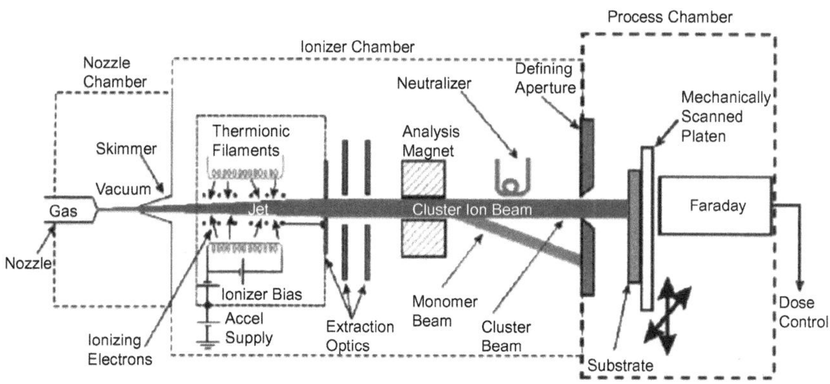

Fig. 3.4.11 Diagram of GCIB setup [17].

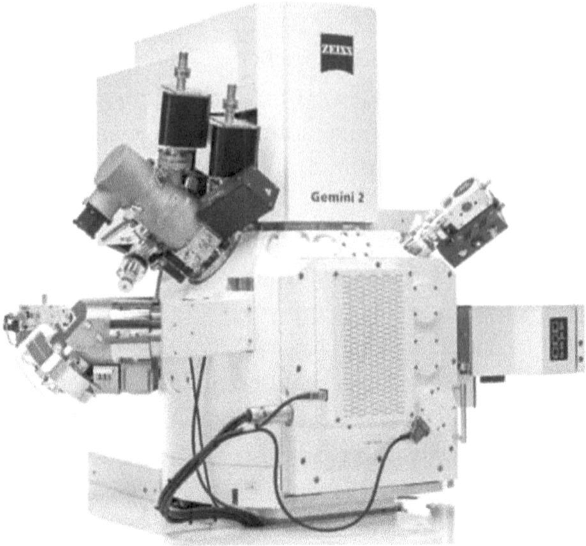

Fig. 3.4.12 Photographic view of FIB with SEMZEISS Crossbeam 340. *(Courtesy ZEISS Germany.)*

imaging or 3D analytics, crossbeam will enhance FIB applications. It has capability of variable pressure as per the user's requirements and fulfills most demanding characterizations.

3.4.3 Focused ion beam milling

FIB milling is one of the techniques for removal of material from the surface of the workpiece to generate different complex profiles. For milling application, the incoming ions should interact only with atoms at the surface or near the surface layer of the workpiece to create collision cascade on the atoms. The collision can transfer sufficient energy to the surface atom to overcome its surface binding energy which results in ejection of atoms from the work surface. Here, sputtering of atoms from the workpiece surface is solely a physical process because the interaction depends purely on momentum transfer. Fig. 3.4.13 shows schematic diagram of FIB milling operation where finely focused ion beam is directed on the workpiece. FIB milling is performed by accurate pixel by pixel movement with precise control of beam diameter, scanning speed, pixel spacing with the help of computer controlled units [18]. The ion beam scans along the programmable path on the workpiece surface to fabricate desired shape.

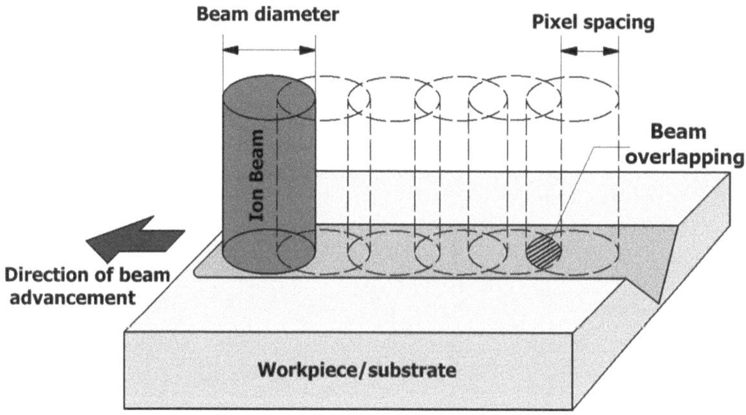

Fig. 3.4.13 Schematic representation of FIB milling operation.

Material removal in FIB milling is purely dependent on sputtering mechanism. Number of atoms ejected per incident ion is the measure of efficiency of material removal and is known as sputtering yield. Sputtering yield is commonly in the range of 1–50 atoms per ion and mainly depends on various variables such as masses of ions, target atom, ion energy, direction of incident, ion flux, etc.

An amount of ejected atoms or molecules is repeatedly deposited into the sputtered region during machining. Hence, it is difficult to control the exact amount of material removal during sputtering due to redeposition of material. Thus, it is important factor during IBM milling to control the amount of redeposition during sputtering to achieve accurate amount of material removal for precise shape generation. FIB milling is a combination of physical sputtering and material redeposition. The material removal can be achieved in the range of sub micron utilizing high energy density and small size of the ion beam as well as suitable scanning strategy.

3.4.4 Influence of process parameters on performance criteria of IBM process

Ion beam processing can be achieved by controlling three factors, e.g., beam parameters, operation parameters and scan strategy. To achieve higher accuracy and efficiency during FIB fabrication, proper setting of ion beam parameters are to be carefully considered. Operation parameters can be appropriately controlled to achieve suitable machining condition during FIB milling to enhance accuracy and efficiency. The scanning

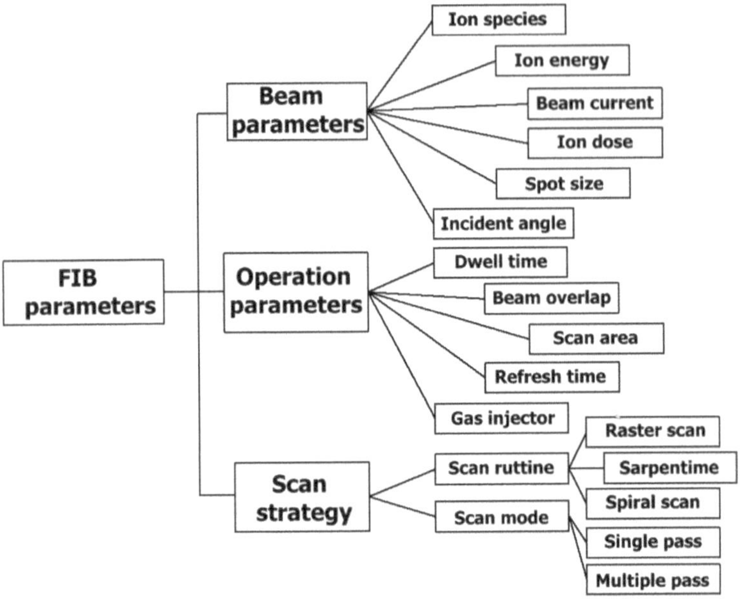

Fig. 3.4.14 FIB controlling factors and parameters.

strategy of the ion beam movement for fabricating the desired shape should also be appropriately selected. Fig. 3.4.14 represents meticulous identification of FIB controlling factors and parameters.

Major performance criteria of IBM are material removal rate, i.e., sputtering yield; profile and shape accuracy, i.e., length, depth, width, etc. and surface quality, e.g., surface nature, finish, integrity, etc. Important FIB controlling factors and parameters should be controlled to achieve desired performance criteria for fulfilling diverse design requirements. Detailed analysis of various controlling factors and parameters and their influences on performance criteria is discussed here under.

3.4.4.1 Beam parameters

(i) Ion species—In the beginning, gas ions such as H^+, H^{2+}, He^+, Ne^+, and Ar^{++} etc. were employed in FIB. However, presently metal ion species commonly used in FIB for micro-fabrication are Gallium (Ga), Arsenic (As), Beryllium (Be), and Silicon (Si). In the case of implantation of semiconductor materials, Argon (Ar), Boron (B), and Phosphorous (P) are utilized in FIB processing. Depending upon the type of applications, selection of the ion species plays an important role in FIB. Range of materials being used as ion species in FIB systems is growing to expand its field of applications. Generally, liquid metal ion source is utilized for generation of ion species. However, it is difficult to produce ions from some of the pure metal species due to their reactivity and volatility at high melting temperature. Hence, in those cases, liquid metal alloys of these materials are utilized as source in FIB.

The preference of ion species is mainly dictated by the application and availability. Choice of ion species is depended on the target material to be processed utilizing FIB. It also influences by the type of processing, e.g., sputtering, implantation, and deposition, etc. to be performed by FIB. The energy of ions varies as atomic weight of ion species differs depending on types of materials. Hence, heavier ions are having higher energies and thus produce higher sputtering yields.

(ii) Ion energy—Sputtering yield directly depends on ion energy. It increases with the increment of the ion energy. However, rate of increase of sputtering yield reduces as the ion energy increases. Fig. 3.4.15 depicts the relationship between ion energy and sputtering yield. As shown in figure, heavier ions produce higher sputtering yield due to higher kinetic energy. Normally, ion energy varies from 10 to 250 keV. Furthermore, it can be stated that that sputtering yield decreases for ion energies higher than 100 keV because at this

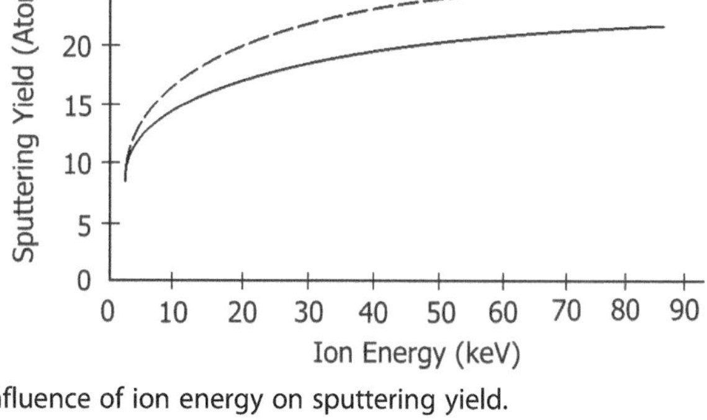

Fig. 3.4.15 Influence of ion energy on sputtering yield.

relatively higher energy-interaction, implantation of the ions into the target workpiece occur instead of sputtering.

(iii) Incident angle—Sputtering yield is greatly influenced by incident angle of ion beam striking the target substrate. Sputtering yield is governed mainly by surface collision cascades. As shown in Fig. 3.4.16, sputtering yield increases with the increment of incident angle. Maximum sputtering yield can be achieved at near about 80 degrees of incident angle after that it decreases very rapidly as the incident angle come close to 90 degrees [19]. As the angle of collision between the ions and target atoms increases leads to increase the chance of escaping more atoms from the substrate which results in increment of sputtering yield. As the incident angles approach to 90 degrees, the sputter yield decreases due to the termination of collision cascade at the target surface due to reflection of the ion. Surface smoothing can also be achieved by varying the angle of incident [20]. As shown in Fig. 3.4.17, surface roughness remains same at the lower incident angle and increases up to the incident angle 80 degrees due to the higher sputtering yield. Beyond this incident angle, surface roughness drastically reduces due to lower sputtering yield as well as lateral sputtering.

(iv) Ion dose—During FIB, imparted and absorbed quantities of ions on to the target substrate over a particular area is known as ion dose. It is also

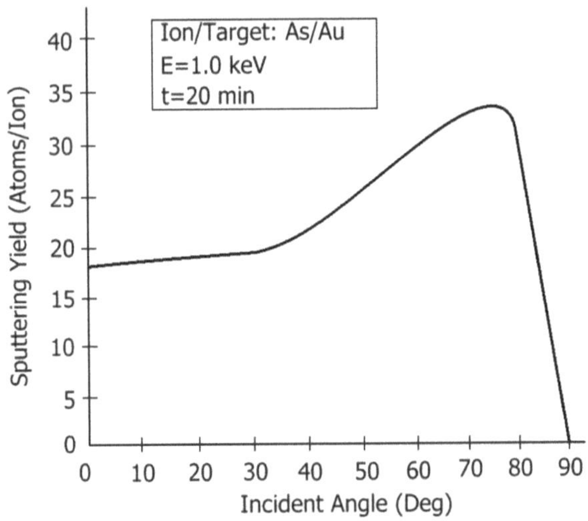

Fig. 3.4.16 Influence of incident angle on sputtering yield.

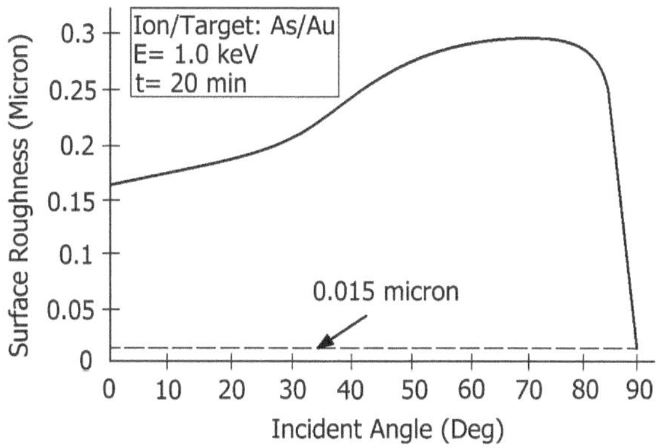

Fig. 3.4.17 Influence of incident angle on surface roughness.

termed as ion-fluence or ion-flux. Ion dose increases delocalization which affect deposition of the material at the edge of the machined profile. Amorphization of crystalline substrate may occur in the area of bombardment if the level of ion dose is not high enough during FIB milling due to induced swelling.

(v) Beam current—It determines by the aperture size and condenser lens placed in ion beam column. During FIB operation, beam current can be varied from pico amps to nano amps depending on the requirements. It is also related to the spot size and current density [20]. Beam current increases sputtering yield as well as material redeposition. Hence, selection of optimum beam current for constant sputtering yield for particular combination of ion-substrate is most important. However, surface roughness decreases with the increment of ion beam current at a constant ion dose due to generation of repelled surface structure because of redeposion effect. Small beam current allows machining of materials at a higher resolution which is in the few nanometer range.

(vi) Spot size—In FIB system, the beam is focused by the ion column to a finer spot diameter ranging from 10 to 50 nm known as spot size which enables to achieve best resolution during machining. Energy distribution within the spot is Gaussian. Due to this type of distribution, energy at the center of the spot is 2–3 times higher than the edge leads to several primary defects during fabrication. The intensity at the beam tail region is much smaller than the core which is not strong enough to sputter atoms from the substrate. Lower ion dose at the beam tail leads to amorphizing or swelling in the substrate by displacing the target atoms from their lattice side so that it redeposites on the neighboring surface [18].

3.4.4.2 Operation parameters

(i) Dwell time—The amount of time that the beam remains on a target pixel is called dwell time. Hence, it is the exposure time of the beam on each pixel which is expressed in μs or ms. Increase in dwell time causes angle dependent milling where, material redeposition occurs and results in slant wall or tapering effect. Decreasing dwell time causes reduction in sputtering yield however, due to reduction in redeposition, more vertical side walls and smooth surface can be achieved. Resultant sputtering yield reduces due to the redeposition effect with higher dwell time. At a particular machining condition, machining time reduces as dwell time increases which lead to better surface finish as shown in Fig. 3.4.18.

(ii) Beam overlap—During FIB milling, as shown in Fig. 3.4.13 common area between two consecutive spot is known as beam overlap. With the increment of scanning speed, the percentage of overlap reduces. Hence, smaller overlap leads to lesser machining time. However, for maintaining the same ion dose, smaller overlap is responsible for the increment of processing time due to longer travel distance. Beam overlap is considered to be one of the important parameters for achieving uniform ion flux to fabricate consistent structure. However, at higher overlapping, cracks produce on the milled surface due to the rapid change of temperature.

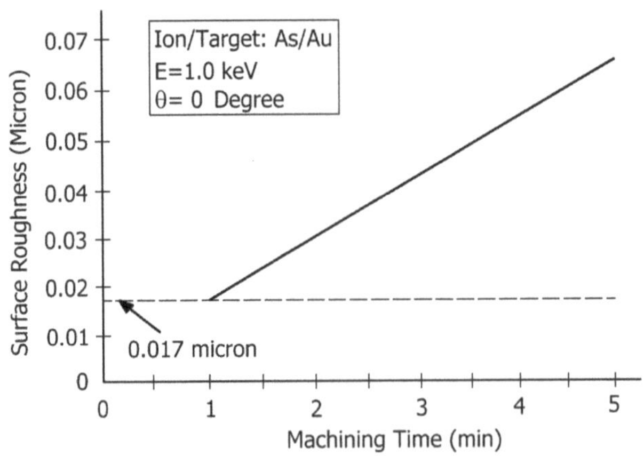

Fig. 3.4.18 Influence of machining time on surface roughness.

(iii) Gas injection—The precursor gas is delivered through the gas nozzle in the gas injection system in order to spray gas molecule onto the local target area. Impinging gas distribution depends on the size, angle, distance and the height of gas injection system in FIB deposition.

(iv) Refresh time—It is the time interval of each scan loop to replenish the gas molecule during FIB deposition. Refresh time allows to supply sufficient gas or molecules on the target surface to deposit the material.

3.4.4.3 Scan strategy

Milling is performed by a precise pixel by pixel movement in computer control FIB machine which is shown in Fig. 3.4.13. It consists of movement pattern of ion beam on the target as well as the number of passes to achieve the desired depth of the profile.

(i) Scan type—Different types of scan strategy is used in FIB system, e.g., raster scan, serpentine scan and spiral scan as shown in Fig. 3.4.19 and discussed below [21]. Raster scan is widely used in FIB milling compare to other two types.

- **(a)** Raster scan—In this scanning type, ion beam moves in the same direction throughout the scan as shown in Fig. 3.4.19A. These scans are carried out in a sequential manner one scan line after another.
- **(b)** Serpentine scan—In this scanning type ion beam moves in a reverse direction after each pass as shown in Fig. 3.4.19B. Here also, scanning is done one line after another as in the case of raster scan.
- **(c)** Spiral scan—The movement of the ion beam is circular instead of straight line as in the above two types as shown in Fig. 3.4.19C. It is mainly used during the fabrication of holes by FIB milling. This scan types improves side wall angle and symmetry of holes by removing localized deposited materials.

(ii) Scan mode—The material removal rate in FIB milling cannot be directly computed by sputtering yield due to redeposition factor. Using ion beam, micro channels are initially milled with one pass to measure the milling rate for quantifying the effect of redeposition, accordingly number of passes is selected. Hence, to achieve precise channel featuring flat channel bottom, vertical sidewalls and high aspect ratio profile etc. in the micron and nano range for fabrication of micro devices need single as well as multi pass scan mode during FIB milling. It should be noted that direction of scanning path should be started at a region where the smallest milling depth is needed and ended at the region where the largest milling depth is desired.

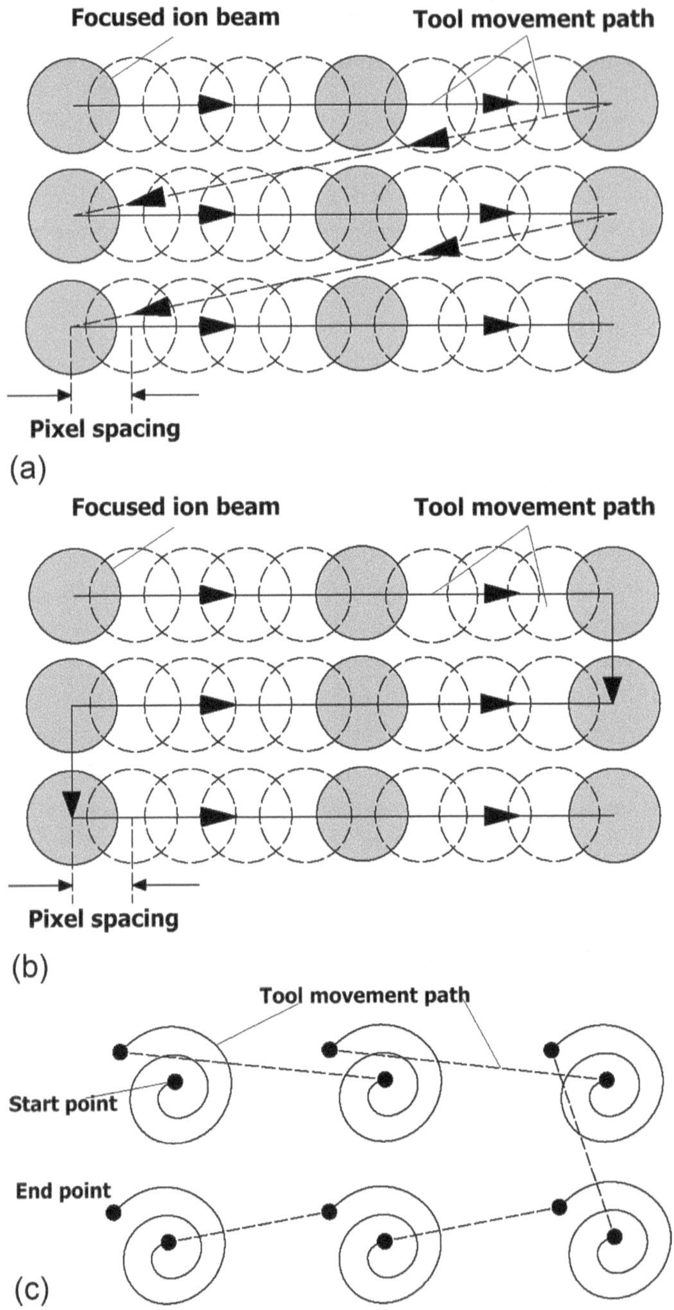

Fig. 3.4.19 (a) Scanning strategy of Raster scan, (b) scanning strategy of Serpentine scan, and (c) scanning strategy of Spiral scans.

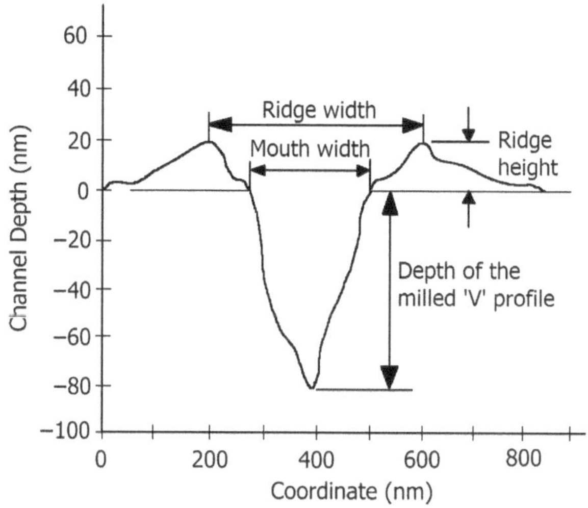

Fig. 3.4.20 Features of the "V" shaped milled profile.

(a) Single pass—The ion beam moves once over the substrate to achieve desired profile. Due to redeposition effect, the milled channel profile will be "V" shape with maximum depth at the centre and maximum width at the mouth as shown in Fig. 3.4.20. Four features of the "V" shaped milled profile can be identified as follows:

I. Ridge width, i.e., the distance between the ridge picks form at the mouth of the channel

II. Mouth width which is the channel width at the original substrate surface

III. Depth of the milled "V" profile which is the depth of the profile from the original substrate surface

IV. Ridge height, i.e., height of ridge from the original substrate surface which is form due to the redeposition

FIB milled channel is the relatively shallow due to redeposition not only outside of the channel, i.e., the formation of the ridge but also inside the channel, i.e., formation of tapering effect which results in "V" shaped profile [22]. However, this effect can be controlled by proper selection of FIB milling parameters, e.g., dwell time, beam current, spot size, etc.

(b) Multiple pass—Repetitive pass of ion beam is inevitably needed to overcome the effect of redeposition, which creates tapered or slanted side wall. To fabricate the channel with flat bottom which is important for various micro devices, multiple pass of ion beam on the target area is needed for removing redeposited material. Here, in each passes small amount of materials is removed as well as smaller amount of redeposition takes place which is also removed in the subsequent passes. Hence, repetitive pass mode can lead to more uniform milling where, nearly flat bottom and almost vertical sidewalls can be achieved.

3.4.4.4 FIB simulation software

For predicting sputtering yield of different ions with wide range of energy, software packages are available. Monte-Carlo simulation is utilized to calculate the transport of ions in matter as well as stopping range of ions into the substrate by the quantum-mechanical treatment of ion atom collisions [23]. Widely used software package in this area are TRIM (Transport of Ions in Matter) and SRIM (Stopping and Range of Ions in Matter). Predictions trough these software are very efficient and very close to experimental values. These software packages are commonly used to analyze the influence of various ion beam parameters on the major outcomes of focused ion beam machining.

3.4.5 Advantages and limitations of ion beam machining process

Focused ion beam machining has become increasingly attractive and receiving great attention from the manufacturing engineers due to its several advantages for fabrication of precise structures to fulfill the demands in industry and research. Some of the major advantages of FIB are identified hereunder:

(i) FIB can be successfully utilized to fulfill various purposes, e.g., fabrication, implantation and deposition thus offering unique and potential techniques for fabrication of three dimensional micro structures.

(ii) It has high flexibility and extremely high features resolution with low lateral scattering.

(iii) FIB is capable of maskless fabrication of complex micro/nano features on substrate with higher aspect ratio.

(iv) Any hard metal as well as non metals can be machine by FIB.

(v) Cluster ion beams can be effectively utilized for surface smoothing utilizing its advantage of lateral sputtering.

However, inspite of its various advantages, it has few limitations also which are listed as follows:

(i) Cost of the FIB machine is very high compare to the other non-traditional machines.
(ii) Substrate or workpiece size is limited for FIB machining.
(iii) It needs to be carried out in the clean and vacuum environment.
(iv) Time taken for FIB machining is extremely high compare to the other advanced machining processes.

3.4.6 Applications of ion beam machining process

Recently focused ion beam machining technique owns the attention of the manufacturing engineers because of its capability to machine various high quality devices and high precise components at the micron range. With either FIB assisted sputtering or deposition technique, several micro 3D structures as well as micro components, e.g., honey comb and nano-hand structure, micro-actuators, sensors, SPM probe, field emission device, micro fluidic channel, lens molds, stamps and free space nano-wiring can be fabricated [24]. Some broad area of FIB applications are discussed below.

(a) Cutting tools of micro/nano scale

Micro or even nano scale mechanical cutting tools for micro-machining can be successfully machined by FIB technique. Using FIB sputtering process, researchers have already fabricated tool for micro-milling with a diameter less than 25 nm of cobalt M42, high speed steel, and C2 micro-grain tungsten carbide. For the fabrication process, at first 20 keV Ga ion beam is focused on the circular facet followed by different end mills with various metal alloys are fabricated which are having two, four, five cutting edges. The radii of the curvature of the cutting edge are less than or equal to 0.1 mm [25]. An advanced tungsten carbide turning micro-tool has also been manufactured by the researchers with HSS and single crystal diamond tool as shown in Fig. 3.4.21. This micro-tool is having different shapes such as rectangular, triangular, and other complex shape with the dimensions of 10–150 nm range and cutting edge radii with a curvature of 40 nm.

(b) Different types of manipulators

Using an exclusive FIB-CVD technique, various complex micro/nano-structures, different types of manipulators such as two-pillar, 3D lamented pleats structure, nano-actuator with a coil structure and 3D nano-rotors have been fabricated as shown in Fig. 3.4.22. These rotors were fabricated by shaping in the free-space using 100 nm sheet. Additionally, with this technique a four fingers nano-manipulator has been manufactured which is able to catch the target and an electrostatic 3D nano-manipulator acts as a micro-gripper which can hold a 1 mm micro-sphere [28, 29].

(c) Biological tools

FIB has plenty of applications in the field of biotechnology. A tissue stabilizer cuts with the help of FIB milling for microsurgical operation. In bio-experiments and medical treatment, the cell wall cutting tool (CWCT) was fabricated by the researchers on a glass capillary. It has a needle that penetrates the cell, a retention table to hold the organelles, and a nano-net structure to filter them. The wire size of the net-structure is about 200 nm. Fig. 3.4.23 shows special type of nano-tool often called "nano-net" and using this polystyrene microsphere can be captured.

(d) Microstructures with different patterns

(i) Annular profiles

Fabrication of micro-channels with curved profiles can be possible by controlling the dwell time at each pixel during FIB milling. However, due to the redeposition and the instability or drift of the beam current, there is chance of deviation between the required and the obtained milled profiles. This fabricated micro channels are often used in the areas where geometrically precise arrays are required, e.g., optical interconnects, multiple spot scanning microscopy, laser-to-fiber coupling etc. Axisymmetric sinusoidal wave channels can be fabricated by FIB in a Si substrate, as depicted in Fig. 3.4.24.

(ii) Non planar surface

FIB technique can also be utilized to mill patterns on non-planar surfaces as shown in Fig. 3.4.25. However, it is practically impossible to pattern fine features on non-planar surfaces. This becomes possible if the FIB system is capable of a depth of focus an order of magnitude

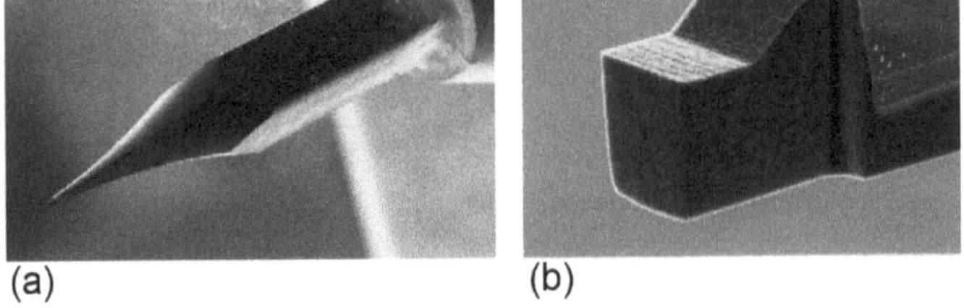

Fig. 3.4.21 Mechanical cutting tools: (a) low magnification view of a single crystal diamond tool shank and junction with mandrel, (b) high magnification view of (a) [26].

(a) (b) (c)

Fig. 3.4.22 Various types of nano-rotors [27].

higher with a 50 nm resolution [32]. Therefore, under proper working conditions, FIB can fabricate patterns over practically 180 degrees of a cylinder with diameter of a few 100 nm without the need for "rotation." For this, it urgently needs computer-controlled stage which is not only move in the *x*- and *y*-direction but also rotates in the θ direction.

(iii) High aspect ratio microchannels

Due to the increasing trend of manufacturing micro-devices, fabrication of microstructures with high-aspect-ratio is becoming increasingly essential and a leading factor. Recently, researchers have used a 50 keV Ga + FIB to mill nanochannels of various aspect ratios to investigate the impact of shrinking feature size on milling efficiency [33]. At relatively low dose intensities ranging from 1013 to 6 × 1015 ions cm^{-2}, swelling in the Si substrates was observed because of amorphization. The maximum swelling size of 3 nm was found in the neighborhood of 1015 ions cm^{-2}. Higher dose intensity leads to material removal with a yield of 2.5 Si atoms per Ga ion. At higher aspect ratios, redeposition is again the dominant effect. The higher the dose, the deeper the channel can be milled.

3.4.7 Advancement in IBM

Current demands for advanced micro engineering parts have influenced the promotion of FIB in the area of micro and nano fabrication. However, till date FIB is not fully utilized due to its low productivity, complexity as well as high cost factor. As the improvement in FIB equipments continues, the ability to achieve higher depth with tolerance in the order

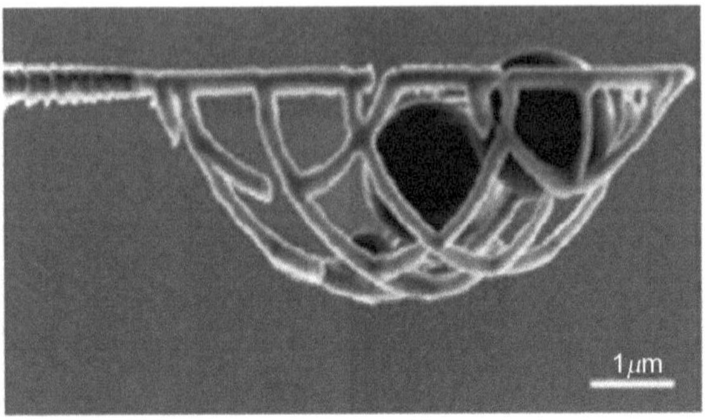

Fig. 3.4.23 SEM image of nano-net [30].

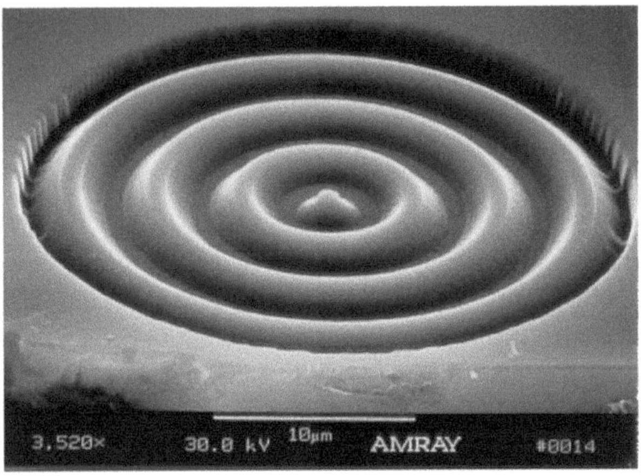

Fig. 3.4.24 SEM image of sinusoidal annulus microchannels viewed at 60 degrees [31].

of nanometer range is still the most critical factors during milling of high precision components.

Development of machining strategies based on mathematical formulations in FIB milling is needed for the successful fabrication of predetermined curved shapes 3D structure [34, 35]. Researchers have introduced control of pixel dwell time thus achieving different sputter yield for the fabrication of curved microstructures [31]. The FIB milling of curved shapes, 3D structure has also been achieved by introducing 2D slice-by-slice method considering thin and constant thickness [36]. Additionally, continuous slicing method has been modified by combining the spiral scan as a part of vector scan [37]. The redeposition effects of milling BK7 glass curved substrates have been studied by considering different scanning sequences [38].

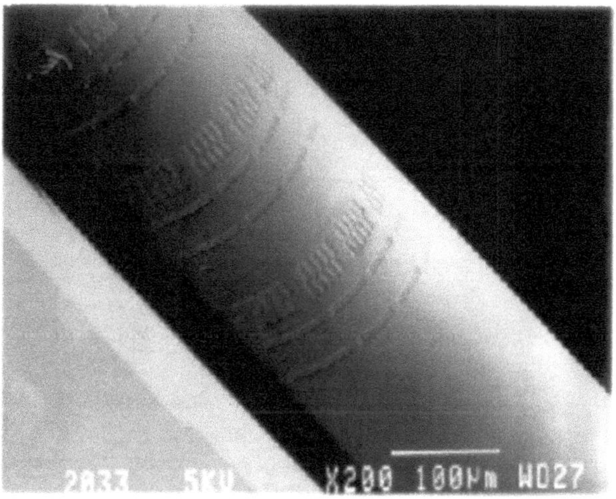

Fig. 3.4.25 Flexible cylindrical fiber-based neural recoding and stimulating electrode implants [32].

New concept is developed for fabrication of Si island arrays which are possible by focused ion beam milling though stigmating and defocusing the beam intentionally to form honeycomb shape [39]. It is important to drift the beam with and without distortion of the shape of beam spot to generate the island formation on Si. Fig. 3.4.26 shows the fabricated islands by this process.

Fabrication of 3D structures can also be achieved by precision turning utilizing focused ion beam, as illustrated in Fig. 3.4.27. Rotating the job by a precision motor-driven system with a very small run-out, the

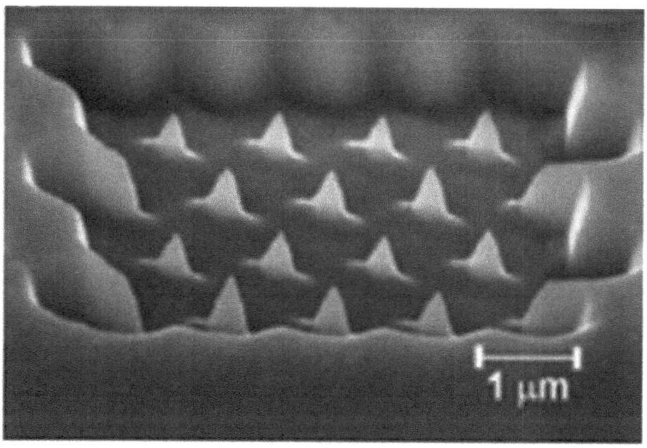

Fig. 3.4.26 SEM image of Si island arrays by highly defocused beam [39].

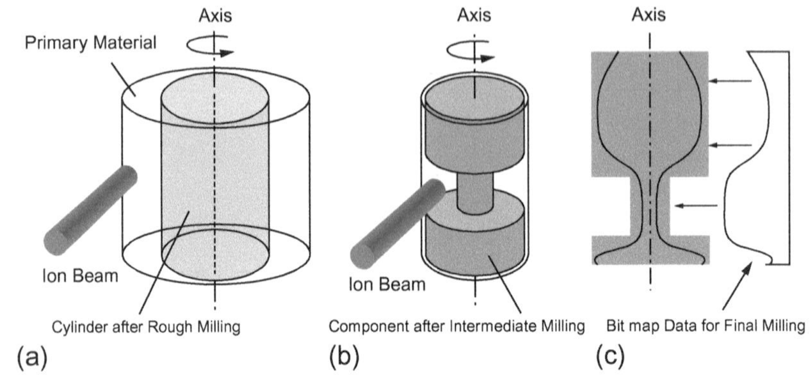

Fig. 3.4.27 Process of forming an outline of the wine glass with the nano lathe: (a) rough milling, (b) intermediate milling, and (c) final milling [40].

ion beam can be operated as tool in a nano-lathe to shape the axisymmetric structures.

Scientists are trying to developed FIB system with feedback control that detects as well as monitors in-situ fabricated shapes with high accuracy which is in the order of nanometer range that may further improves the quality of the products, e.g., flatness, texture and surface finish etc.

Attempts have been made to achieve high throughput of FIB system which is a major challenge for all micro and nano fabrication technologies. Milling rate of the existing FIB system has to be increased to achieve higher throughput and thus leading to higher applicability in the advanced manufacturing system.

Development of variable beam diameter FIB system is also essential to obtain multi resolution milling for achieving different levels of accuracy requirements. During FIB milling, larger diameter beam can be utilized for rough milling to improve the machining rate and smaller diameter beam can be applied to fulfill higher resolution requirements. Heavy duty FIB system with superior automation should be developed to improve its utility in production of micro and nano components thus transform it as the mainstream tool in future.

3.5 Model Questions

Ultrasonic machining (USM)

Q1. Explain the principle of stationary ultrasonic machining with sketch?
Q2. Explain the principle of rotary ultrasonic machining with sketch?

Q3. Discuss on material removal mechanism of rotary USM.
Q4. Discuss on various components of USM system with sketch.
Q5. Discuss on various elements of RUM system with sketch.
Q6. Mention process parameters and machining criteria of USM and RUM with the aid of fishbone diagram.
Q7. What are advantages and limitation of USM and RUM?
Q8. State the various shapes can be generated on different materials by USM.
Q9. What are the applications of USM and RUM?

Abrasive jet machining (AJM)

Q1. What is the mechanism of material removal in AJM?
Q2. Discuss various elements of abrasive jet machining system with sketch?
Q3. Discuss on the effects of process parameters on MRR of AJM processes.
Q4. What are the advantages and limitations of abrasive jet machining process?
Q5. What are the various applications of abrasive jet machining process?
Q6. What are the advancements in abrasive jet machining technology?

Water jet machining (WJM)

Q1. Explain the mechanisms of material removal of water jet machining with sketch.
Q2. Explain various components of water jet machining system with sketch.
Q3. Discuss on the effect of process parameters on machining criteria of WJM.
Q4. State the advantages and limitations of water jet machining process.
Q5. What the various applications of water jet machining technology?

Ion beam machining (IBM)

Q1. Define ion beam machining (IBM) and differentiate it from EBM.
Q2. Depending on the ion energy, classify different application areas of FIB.
Q3. What is sputtering rate and how it can be calculated?
Q4. What is the different ion sources utilized in FIB?
Q5. Explain with reasons why vacuum in FIB system is essential?
Q6. What is liquid metal ion source (LMIS)? What are the advantages for using Gallium in LMIS of FIB system?

Q7. Describe in brief different sub units of FIB system setup.

Q8. Discuss about the differences in between electrostatic lens and electromagnetic lens?

Q9. What is gas cluster ion beam (GCIB)? How it can be utilized for different applications?

Q10. Discuss in brief about implantation, back scattering, swelling and redeposition occurring in FIB machining.

Q11. Describe about FIB-milling and its applications in precision fabrication.

Q12. Discuss about the controlling factors and influencing parameters in FIB operation.

Q13. Identify the different scanning strategies during FIB milling.

Q14. In FIB what should be the strategy for minimizing surface damage in nano level machining?

Q15. How microstructuring can be done by FIB?

Q16. Can FIB be used for additive micromanufacturing? Justify your answer with proper reasons.

References

[1] R.E.J. Watkins, P. Rockett, S. Thoms, R. Clampitt, R. Syms, Focused ion beam milling, Vacuum 36 (1986) 961.

[2] J. Orloff, High-resolution focused ion beams, Rev. Sci. Instrum. 64 (1993) 1105.

[3] J. Orloff, M. Utlaut, L. Swanson, High Resolution Focused Ion Beams: FIB and Its Applications, Kluwer Academic. Plenum Publisher, 2003.

[4] I. Yamada, N. Toyoda, Nano-scale surface modification using gas cluster ion beams—a development history and review of the Japanese nano-technology program, Surf. Coat. Technol. 201 (2007) 8579–8587.

[5] N. Taniguchi, N. Taniguchi, M. Ikeda, I. Miyamoto, T. Miyazaki (Eds.), Energy Beam Processing of Materials, Oxford University Press, 1989,p. 38 ISBN: 0198590059, (Chapter 1).

[6] C.-S. Kim, S.-H. Ahn, D.-Y. Jang, Review: developments in micro/nanoscale fabrication by focused ion beams, Vacuum 86 (2012) 1014–1035.

[7] N. Taniguchi, Current status in, and future trends of, ultraprecision machining and ultrafine materials processing, Ann. CIRP 32 (2) (1983).

[8] I. Miyamoto, Ultra fine finishing of diamond tools by ion beams, Precis. Eng. 9 (2) (1987) 71–78.

[9] R. Trassl, Development of ECR Ion Sources, Institut für Atom- und Molekülphysik, niversita, Giessen, Germany, 2002. Accessed at: http://www.strz.uniiessen.de/ezr/english/prinzip.html.

[10] M. Von Ardenne, Tabellen der Electronenphysik, in: Ionen-physik und Ubermikroskopie, VEB-Deutscher Verlag der Wissenschaften, Berlin, 1956.

[11] D.M. Allen, P. Shore, R.W. Evans, C. Fanara, W.O. Brien, S. Marson, W. O'Neill, Ion beam, focused ion beam, and plasma discharge machining, CIRP Ann. Manuf. Technol. 58 (2009) 647–662.

[12] Pelletron, http://www.pelletron.com/negion.htm.

[13] L.A. Giannnuzzi, F.A. Stevie, Introduction to Focused Ion Beams; Instrumentation, Theory, Techniques and Practice, Springer Science + Business Media, New York, USA, 2005, pp. 3–5, ISBN: 0-387-23116-1.
[14] I. Brodie, J.J. Muray, Particle beams: sources, optics, and interactions, in: The Physics of Micro/Nano-Fabrication, Plenum, New York, 1992 (Chapter 2).
[15] L.R. Harriott, Technology of finely focused ion beams, Nucl. Inst. Methods Phys. B55 (1991) 802–810.
[16] J. Bi, P.W.H. de Jager, J.E. Barth, P. Kruit, Influence of Coulomb interactions on current density distribution in a two-lens focused ion beam system, Microelectron. Eng. 42 (1998) 249–252.
[17] A. Kirkpatrick, Gas cluster ion beam applications and equipment, Nucl. Inst. Methods Phys. Res. B 206 (2003) 830–837.
[18] A.A. Tseng, Recent developments in micro-milling using focused ion beam technology, J. Micromech. Microeng. 14 (2004) R15–R34.
[19] M.J. Vasile, R. Nassar, J. Xie, H. Guo, Microfabrication techniques using focused ion beams and emergent applications, Micron 30 (1999) 235–244.
[20] D.Z. Xie, B.K.A. Ngoi, Y.Q. Fu, B.H. Lim, Etching characteristics of TiNi thin film by focused ion beam, Appl. Surf. Sci. 225 (2004) 54.
[21] W.C.L. Hopman, F. Ay, W. Hu, V.J. Gadgil, L. Kuipers, M. Pollnau, et al., Focused ion beam scan routine, dwell time and dose optimizations for sub-micrometer period planar photonic crystal components and stamps in silicon, Nanotechnology 18 (2007) 195305.
[22] A.A. Tseng, I.A. Insua, J.S. Park, C.D. Chen, Milling yield estimation in focused ion beam milling of two-layer substrates, J. Micromech. Microeng. 15 (2005) 20.
[23] J.P. Biersack, L.G. Haggmark, A Monte Carlo computer program for the transport of energetic ions in amorphous targets, Nucl. Inst. Methods Phys. Res. B 174 (1980) 257–269.
[24] S. Reyntjens, R. Puers, A review of focused ion beam applications in microsystem technology, J. Micromech. Microeng. 11 (2001) 287–300.
[25] D.P. Adams, M.J. Vasile, G. Benavides, A.N. Campbell, Micromilling of metal alloys with focused ion beam-fabricated tools, Precis. Eng. 25 (2001) 107.
[26] Y.N. Picard, D.P. Adams, M.J. Vasile, M.B. Ritchey, Focused ion beam beam-shaped microtools for ultra-precision machining of cylindrical components, Precis. Eng. 27 (2003) 59.
[27] J. Igaki, R. Kometani, K. Nakamatsu, K. Kanda, Y. Haruyama, Y. Ochiai, et al., Three-dimensional rotor fabrication by focused-ion-beam chemical-vapor-deposition, Microelectron. Eng. 83 (2006) 1221.
[28] R. Kometani, T. Hoshino, K. Kondo, K. Kanda, Y. Haruyama, J. Kaito, et al., Performance of nanomanipulator fabricated on glass capillary by focused-ion-beam chemical vapor deposition, J. Vac. Sci. Technol. B 23 (2005) 298.
[29] S. Matsui, Focused-ion-beam deposition for 3-D nanostructure fabrication, Nucl. Inst. Methods Phys. Res. B 257 (2007) 758.
[30] R. Kometani, T. Hoshino, K. Kanda, Y. Haruyama, T. Kaito, J. Fujita, et al., Three-dimensional high-performance nano-tools fabricated using focused ion beam chemical vapor deposition, Nucl. Inst. Methods Phys. Res. B 232 (2005) 362.
[31] M. Vasile, J. Xie, R. Nassar, Depth control of focused ion-beam milling from a numerical model of the sputter process, J. Vac. Sci. Technol. B 17 (1999) 3085–3090.
[32] Mitan M M and Pivin D P 2003, http://www.eas.asu.edu/~fib/projects.htm
[33] A. Lugstein, B. Basnar, J. Smoliner, E. Bertagnolli, FIB, processing of silicon in the nanoscale regime, Appl. Phys. A A6 (2003) 545–548.
[34] M.J. Vasile, Z. Niu, R. Nassar, W. Zhang, S. Liu, Focused ion beam milling: depth control for three-dimensional microfabrication, J. Vac. Sci. Technol. B 15 (1997) 2350.

[35] D.P. Adams, M.J. Vasile, T.M. Mayer, Focused ion beam sculpting curved shape cavities in crystalline and amorphous targets, J. Vac. Sci. Technol. B 24 (2006) 1766.
[36] Y. Fu, N.K.A. Bryan, Fabrication of three-dimensional microstructures by two-dimensional slice by slice approaching via focused ion beam milling, J. Vac. Sci. Technol. B 22 (2004) 1672.
[37] C.S. Kim, J. Park, W.S. Chu, D.Y. Jang, S.D. Kim, S.H. Ahn, Fabrication of silicon micro-mould for polymer replication using focused ion beam, Microelectron. Eng. 86 (2009) 556.
[38] Y. Fu, N.K.A. Bryan, Experimental study of microcylindrical lenses fabricated using focused-ion-beam technology, J. Vac. Sci. Technol. B 19 (2001) 1259–1263.
[39] S.E. Wu, C.P. Liu, Direct writing of Si island arrays by focused ion beam milling, Nanotechnology 16 (2005) 2507.
[40] T. Fujii, M. Iwasaski, M. Munekane, T. Takeuchi, M. Hasuda, T. Asahata, et al., A nanofactory by focused ion beam, J. Micromech. Microeng. 11 (2005) S286.

Further reading

[41] N. Taniguchi, Research on, and development of energy beam processing of materials in Japan, Bull. Jpn Soc. Prec. Eng. 18 (2) (1984).

CHAPTER FOUR

Machining processes utilizing thermal energy

Contents

Modern Machining Technology
https://doi.org/10.1016/B978-0-12-812894-7.00004-9
© 2020 Elsevier Inc.
All rights reserved.

4 Introduction

Thermal energy can be directly utilized for removing material to overcome the limitations of different conventional machining processes where cutting tool or abrasives are directly employed. For generating desired shape, control removal of material from the work piece is achieved by melting, vaporization or ablation. Over the year's different forms of thermal energy, e.g., electrical spark, plasma arc, laser beam and electron beam etc. are used by the engineers for machining advanced materials to fulfill the needs of modern manufacturing industries. Some of the modern machining processes, e.g., Electro-Discharge Machining (EDM), Plasma Arc Machining (PAM), Laser Beam Machining (LBM) and Electron Beam Machining (EBM) uses different forms of thermal energy and are become popular now-a-days due to their inherent advantages over conventional machining processes. Most of the above mentioned machining processes are still developing to overcome there's various limitations. Scientists are involved to improve the method of applications of thermal energy for removal of material to generate desired shaped with high accuracy and better surface quality. As properties of materials are improving as well as design considerations are becoming more complicated in different modern engineering applications, complexity of machining is also escalating. To solve these challenges of machining, engineers are trying hard to understand the basic mechanisms of material removal and to control precisely the above mentioned machining processes. Still there are some areas where improvements are needed which necessitates in depth study of advance machining processes which utilizes thermal energy for removal of material. Hence, the details of various thermal based advanced machining processes are discussed here under.

SUBCHAPTER 4.1

Electro Discharge Machining (EDM)

4.1.1 Introduction

Presently Electro Discharge Machining (EDM) is most widely used advanced machining process where material removal is achieved utilizing thermal energy. It is one of the popular nonconventional machining processes that uses spark discharges to machine electrically conducting materials regardless of its mechanical properties. There is no contact between tool electrode and workpiece. Hence, poor machinable materials and complex shaped jobs can be economically machined by EDM which provides enormous support to the manufacturing industries.

The method of controlled material removal from the workpiece utilizing a series of small electrical sparks is commonly known as electrical discharge machining. In the year 1770, English chemist Joseph Priestly discovered the erosive effects of electrical discharge which was the point of origin of the EDM process. However, only in 1943 during World War II physicist B. R and N. I. Lazarenko at the Moscow University investigated into sparking phenomena occurring between electric power contacts to minimize the erosion due to electrical discharges and consequently exploited the destructive properties of electrical sparks for productive uses [1]. This historical break through initiated the development of electrical discharge machining utilizing controlled discharge condition to achieve precision machining by melting and vaporization from the surface of difficult-to-machine materials [2]. Till 1950, EDM machine utilized the power supply as developed by Lazarenkos and afterwards it served as the foundation for the successive developments. Since then, EDM technology has improved rapidly by the various researchers and become indispensable in manufacturing applications such as die and mold making, prototyping and micro machining etc.

Fig. 4.1.1 shows the basic scheme of EDM. Pulse discharge takes place in the gap filled with an insulating medium, i.e., dielectric liquid between tool electrode and workpiece. The initiation of the discharge is started by a high voltage due to the breakdown of the dielectric in the small gap. Generation of electrical discharge produces intense heat in the sparking zone which melts and evaporates materials from the workpiece as well as tool. Spark discharge of shorter duration and high repetition rate have the ability which needed for effective machining. The spark energy can be concentrated in a tiny area by submerging the discharge in the liquid dielectric. Negative tool electrode shape is copied on the surface of the workpiece with an offset equal to the gap size. Hence, for achieving precise shape generation the gap between tool and the workpiece should be maintained in the range of 10–100 μm. Discharge should occur at a high frequency in the range of kHz to MHz because the metal removal per discharge is very small. Each pulse creates discharge at a single spot location where gap between the tool and workpiece surface is shortest. Due to the discharge, materials are evaporated or ejected in the molten phase from the spot which creates a small crater both on the tool electrode and the workpiece surfaces. Molten materials are cooled and resolidified in the dielectric media forming spherical debris which are then removed from the gap by the flowing dielectric. The discharge spot shifted after each pulse to the next shortest gap thus sparks travels throughout the surfaces. These repetitive discharges all over the workpiece surface results in uniform metal removal and at last the mirror image of the tool electrode shape is created on the workpiece surface. Servo control tool feeding arrangements is utilized to maintain the small spark gap between tool and workpiece.

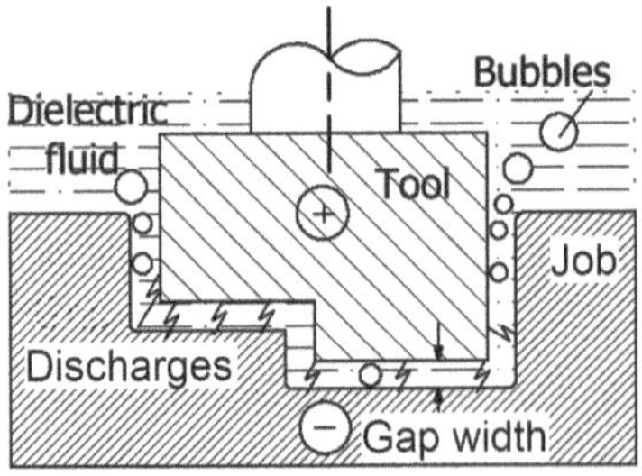

Fig. 4.1.1 Basic scheme of EDM.

4.1.2 Mechanism of material removal

EDM is a thermal process that uses spark discharges to erode electrical conducting materials. The gap between tool and the workpiece is filled up by dielectric liquid. Pulse DC power is applied across tool and the workpiece. Suitable voltage is build up across the electrodes. An electro static field of sufficient strength is established between the smallest gap region which causes cold emission of electron from cathode. Liberated electrons are accelerated by the electric field and moves towards anode and in-between collide with dielectric molecules thus dislodge more electrons resulting in an electron avalanche in the narrow gap. In due course a plasma channel is established in the form of electrical spark. EDM gap phenomena are very complicated and not yet completely understood. However, attempts have been made to describe sequence of events which occurred during single discharge of an EDM cycle.

Figure 4.1.2 depicts a model of gap phenomena when a pulse DC is applied in between tool and the workpiece. As discussed earlier a plasma channel is created at the narrowest gap due to presence of surface irregularities on workpiece and tool. Because of the electric field, positively charged dielectric ions migrate to the cathode and simultaneously negatively charged particles, i.e., electrons moves towards anode which forms highly conductive bridge across the gap.

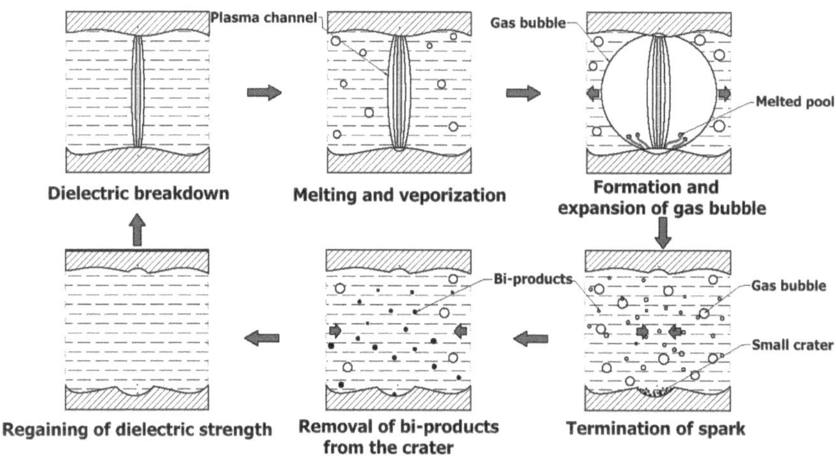

Fig. 4.1.2 Gap phenomena in EDM.

As voltage increase between two electrodes, dielectric breakdowns continue which establish a strong plasma channel. In the plasma channel, dielectric fluid and charged particles vaporizes and ionizes and resulting in the formation of the spark discharge between the two electrode. The diameter of the plasma channel increases with the passage of time [3]. At this juncture due to formation of highly conductive channel, the voltage decreases and amount of current delivery increases to a higher level which cause the increment of temperature and pressure in the spark channel. Due to extreme temperature (approximately 12,000–20,000 °C) of the spark, materials from the surface of workpiece and tool electrode melt and/or vaporize at the contact points of spark. Because of vaporization of the liquid dielectric, a bubble forms around the spark channel. This bubble expands rapidly due to the further evaporation of material of electrodes and dielectric liquid as well as dissociated molecules and ionized atoms. The expansion of gas bubble is constrained by the inertia and viscosity of the liquid which causes extremely high pressure inside the bubble. The boundary between the bubbles and the liquid expands with a very high velocity [4]. Inner detail of discharge spot is shown schematically in Fig. 4.1.3 [5]. This extremely high pressure and velocity in the bubble plays an important role for material removal in EDM.

When the electrical pulse terminates which results in end of the discharge duration as well as heating action instantaneously. As a result, spark channel and gas bubbles collapse inside the gap. Due to the contraction of channel results in violent inrush of dielectric fluid as well as high pressure and velocity of collapsing gas bubble, explosive ejection of molten material occurs from the workpiece and tool surface which causes

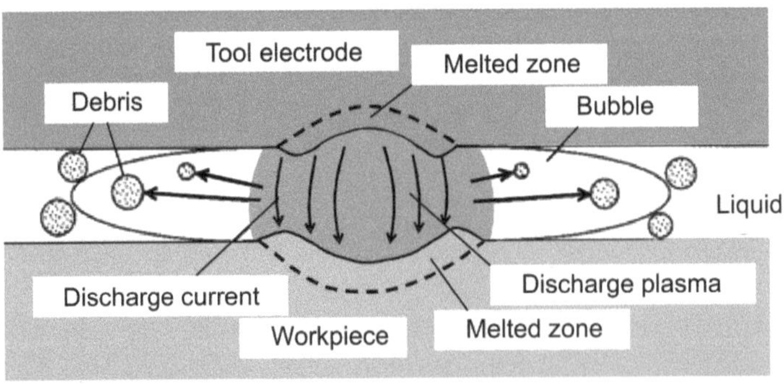

Fig. 4.1.3 Inner detail of discharge spot [5].

creation of small crater in both the surfaces. Evaporated and molten materials are cooled and solidified by the dielectric liquid to form the spherical debris particles as well as gas bubbles are also generated by dissociation of the dielectric liquid. Dielectric fluid plays another important role in flushing debris particle and gas bubbles from the narrow gap. At the end of the cycle, dielectric strength of the liquid is recovered and ready for the next spark. However, hundreds of tiny debris particles and gas bubbles are generated per single pulse discharge which contaminated dielectric in the gap and make the process more complex.

4.1.2.1 Energy distribution

Spark energy is responsible for material removal in EDM. Distribution of energy between anode, cathode and dielectrics indicates the removal amount which greatly depends on discharge duration as well as electrode materials [6]. As shown in Fig. 4.1.4 energy distribution to workpiece and tool electrode is approximately 34% and 48% respectively. However, remaining energy, i.e., 18% goes into the spark gap which is responsible for the formation of plasma. Hence, maximum energy is dissipated into the anode. Out of 34% of the discharge energy which goes into the workpiece, 3% energy loss due to convection and radiation remaining 1% energy carried away by the debris. 30% energy is conducted into the workpiece which is also responsible for material removal.

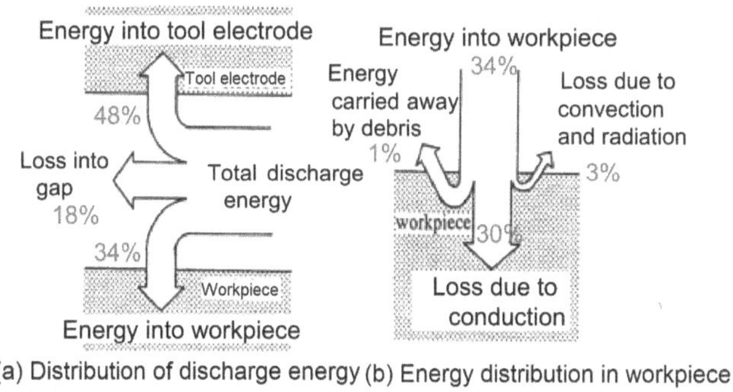

Fig. 4.1.4 Energy distribution in EDM process. (a) Distribution of discharge energy. (b) Energy distribution in workpiece [5].

4.1.3 EDM equipment

In EDM, workpiece can be machined either by complex shaped or simple shaped tool electrode. Complex profile is generated on the workpiece by replication of shaped tool and known as die-sinking EDM. 3D movement of simple shaped electrode can also be generated complex profile and known as EDM milling. EDM systems are available in different configurations, e.g., sinking EDM, EDM drilling, EDM milling center, ED Grinding, wire EDM, hybrid EDM, micro EDM and innovative EDM applications etc. which are operated occasionally manual or by sophisticated computer controlled. Fig. 4.1.5 represents diverse type of EDM systems which mainly utilize electro discharge phenomena to remove metal for different manufacturing applications. Fig. 4.1.6 shows schematic diagram of different major sub components of EDM system. It consists of power supply unit, dielectric supply system, tool electrode and gap control by servo systems etc. which are discussed in detail in the subsequent sub chapter here under. Die sinking EDM is most popular and widely used in manufacturing industry. Detail classifications of die sinking EDM machine based on performance measures, machining capacity and auxiliary facilities is shown in Fig. 4.1.7.

4.1.3.1 EDM power supply

For any EDM system power supply is one of the most vital units which mainly control the machining characteristics. EDM power supply transforms main supply of alternating current (AC) into pulsed direct current (DC) which requires for generating spark discharges at the machining zone.

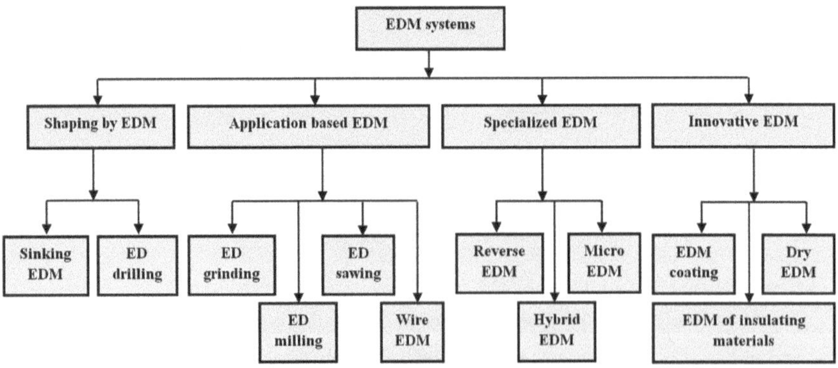

Fig. 4.1.5 Different types of EDM systems.

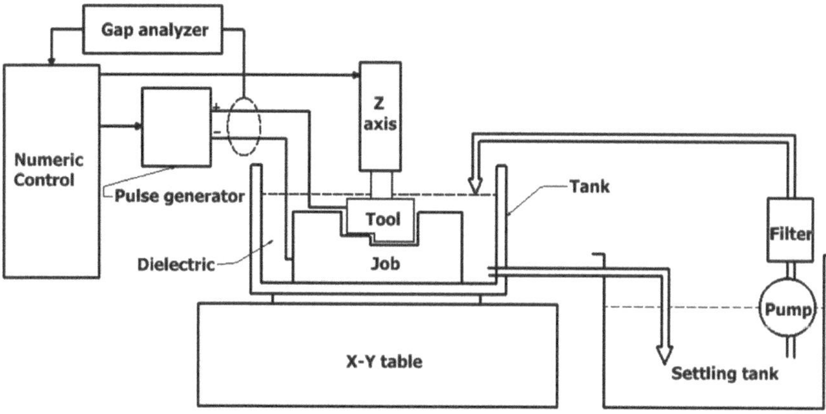

Fig. 4.1.6 Schematic of different major sub units of EDM system.

Initially, input AC power is converted into continuous DC power by full wave solid state rectifiers. Different electric power circuits are existing to generate pulsesetting DC between the gap of tool and workpiece. Basic principles of generating pulsed DC for EDM can be classified into three broad categories as follows:

(i) Resistance-Capacitance (RC) relaxation circuit,
(ii) Rotary impulse circuit.
(iii) Transistor type controlled pulse circuit.

According to the machining application requirements suitable EDM power circuits are applied. Between these three categories of power circuits, at the initial stage of development RC circuit was invented then Rotary impulse circuit was developed to generate pulsesetting of DC for creating repetitive discharge for EDM operations. However, rotary impulse circuit is obsolete now a day due to different limitations. Presently, controlled pulsed circuit is most popular for EDM operations due to its several advantages. RC circuit is used in EDM for some special applications, e.g., micro-EDM; however, this circuit have some limitations over controlled pulsed circuit.

(i) Resistance-Capacitance (RC) relaxation circuit

At first, RC type relaxation circuit was used when pulse discharges are generated for removal of metal by EDM. Resistance and capacitance are connected in series and parallel to the electrode gap respectively as shown in Fig. 4.1.8. In this circuit main function of the capacitance is to accumulate charges along with time and at breakdown voltage, the discharging

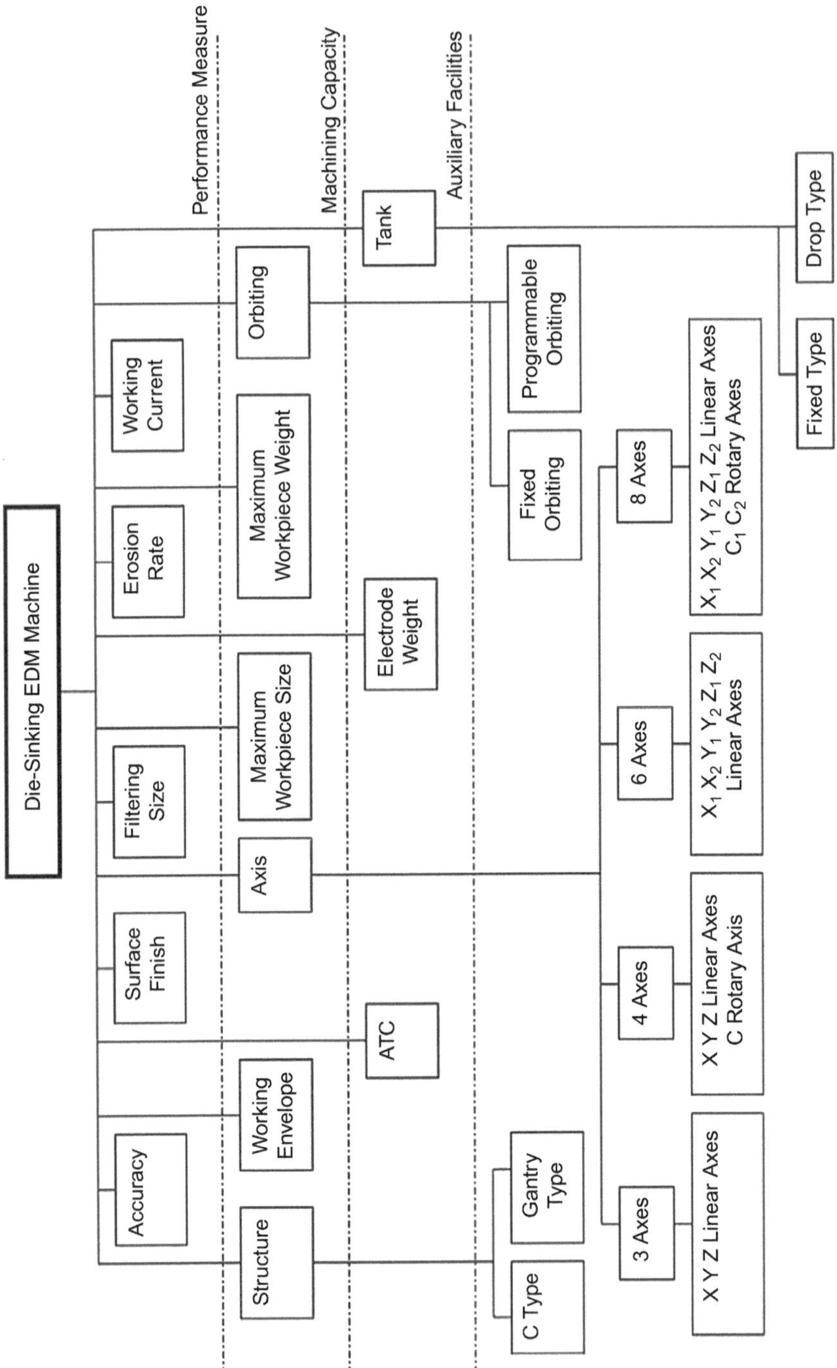

Fig. 4.1.7 Classifications of die sinking EDM machine [7].

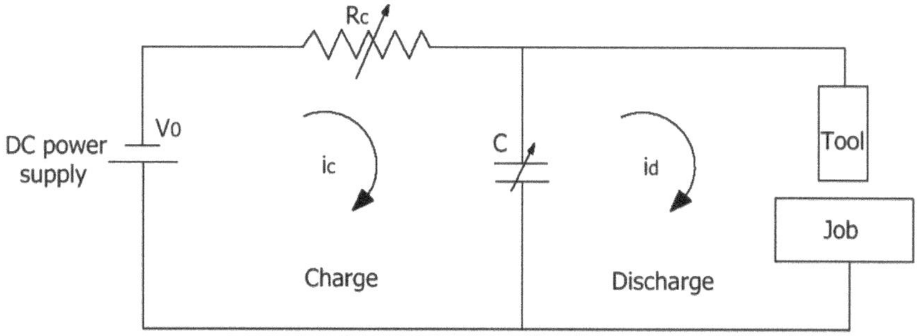

Fig. 4.1.8 RC type relaxation circuit.

phenomena occurs and creates required sparks across the gap between tool and the job which are immerged in dielectric fluids. The pulse pattern of voltage and current in the RC circuit is shown in Fig. 4.1.9. In the figure, it is clearly shown that voltage is increasing with respect to time, i.e., charging time. Then voltage decreases sharply with time when it reaches to the breakdown level, i.e., discharge time. At this period of discharging time, current flow increases across the gap sharply and creates required discharge. In this cycle of pulse period, charging and discharging occurs simultaneously

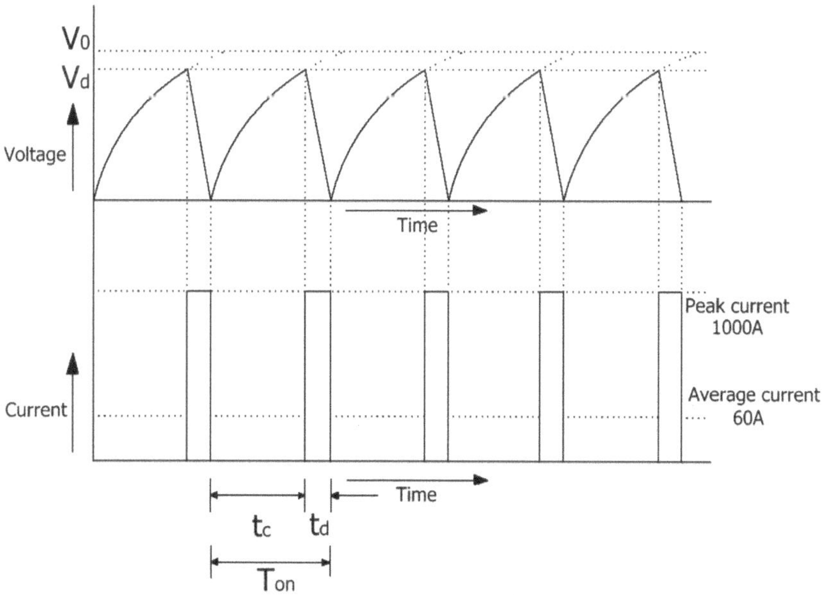

Fig. 4.1.9 Gap voltage and current wave form for RC circuit.

and act as ON and OFF time of each pulse. Major characteristics of this circuit which influences the machining performances are as follows:

(i) Due to longer charging time, i.e., OFF time which is ideal time of the cycle, the material removal rate is low,

(ii) Shorter discharging time, i.e., ON time leads to improvement of the surface quality.

(iii) High peak current is responsible for higher temperature which creates thermal damage and higher tool wear.

(ii) Rotary impulse circuit

To increase MRR, rotary impulse circuit is utilized for spark generation which is shown in Fig. 4.1.10. In the circuit capacitor is charged by the diode for the first half cycle and next half cycle voltage is generated by the generator. The sum of the voltage thus generated by the generator and the capacitor is applied for spark generation between the gap of tool and workpiece. The motor sped is mainly responsible for frequency of the sine wave generation which influences the operating frequency. However, surface finish is not so good in spite of higher MRR.

(iii) Transistor type controlled pulse circuit

Two circuits as discussed above have no control over current flow and its duration as well as frequency. Hence, power parameters, e.g., pulse duration, on time and off time, frequency and amount of current flow which mainly influence the sparking characteristics during EDM operations must be controlled to achieve better machining performances, i.e., MRR, surface finish and accuracy etc. The transistor type pulse circuit is developed which automatically control the power parameters by introducing advanced switching device in the circuit. Fig. 4.1.11 shows the details of the controlled pulse circuit utilized for modern EDM machining system.

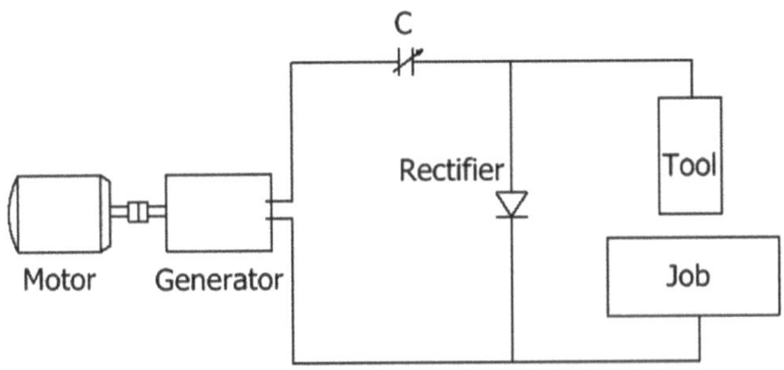

Fig. 4.1.10 Rotary impulse circuit.

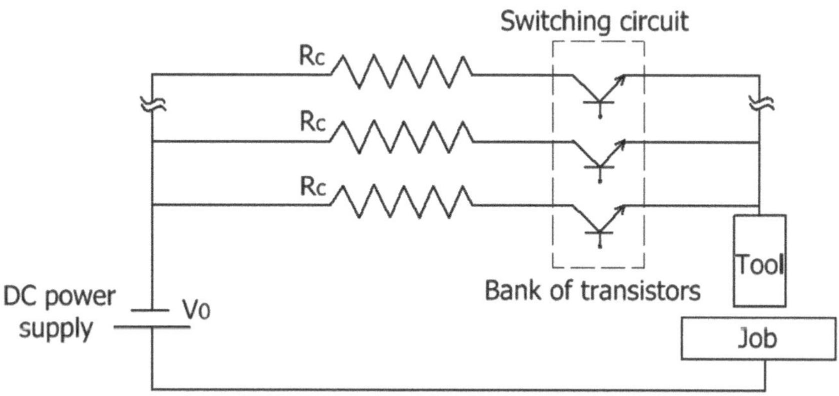

Fig. 4.1.11 Transistor type pulse generator circuit.

The circuit consists of a series of resistances and transistors which are connected in parallel between machining gap and DC power source as shown in figure. A high frequency pulse generator is utilized to generate square wave signals with precise timing. These signals are used to trigger a group of power transistors which acts as a high speed switching device to regulate the flow of current to the machining gap. The numbers of transistors utilized in the circuit which are switched on at the same time as a result discharge current increases proportionally. Advance high power transistor, e.g., Metal Oxide Semiconductor Field Effective Transistor (MOSFET) is utilized for the purpose of high speed switching of high DC power. This advanced circuit has provision for an automatic prevention from short circuit which may occur during machining. Thus, with this advanced controlled pulse circuit, DC power characteristics, e.g., current, pulse pattern, frequency and pulse duration can be effectively controlled to achieve high energy pulses which are responsible for material removal by generating controlled discharges in the gap.

Fig. 4.1.12 shows a typical voltage and current waveforms when controlled pulse circuit is applied at the gap during EDM operation. The open voltage U_0 is applied between the tool electrode and workpiece when bank of transistors are switched on. The discharge at the gap occurs after the ignition delay time (μs-ms), i.e., T_d which is needed to create plasma channel between the tool electrode and workpiece by accelerating electrons to generate electron avalanche which is required for dielectric breakdown. This ignition delay or charging time may differ depending on the condition of the gap, dielectric, amount of contaminated electrically conductive debris particle, etc. Dielectric breakdown strength decreases with the increment of debris particles in the gap

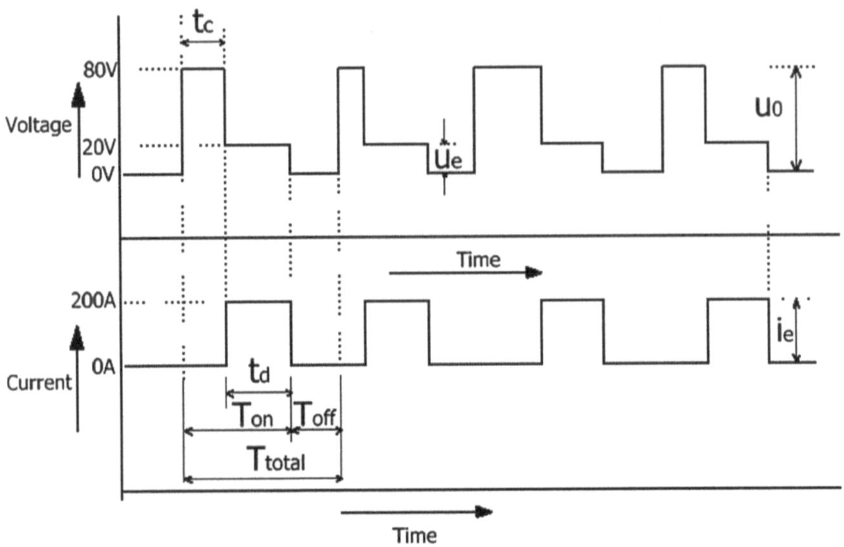

Fig. 4.1.12 Voltage-current waveform for transistor type pulse generator.

as well as decrease of gap width. A discharge current i_e is passed through the gap during sparking which is occurred due to breakdown of dielectric. Switching circuit is controlled by pulse generator which allows the bank of transistors to on for the discharge duration, i.e., T_{on} after the dielectric breakdown resulting in uniform crater size. Thus, the discharge duration is independent of ignition delay time which varies for each discharge due to the variation of gap conditions. Discharge current, i_e and discharge duration, t_e can be set by controlling the pulse generator for different types of EDM operations, e.g., roughing and finishing operation. After each discharge, there is a pulse interval, i.e., OFF time of the cycle. OFF time of the pulse cycle should be adequately large to allow regaining of the dielectric strength after each discharge. Major characteristics of controlled pulse circuit which influences the machining performances are as follows:

(i) Discharge duration, i.e., ON time and pulse interval, i.e., OFF time can be controlled hence; the machining can be controlled depending on the design requirements, e.g., metal removal rate, accuracy, surface quality etc.

(ii) Peak current can be reduced resulting in improvement of the machining accuracy as well as reduced tool wear and thermal damage.

(iii) By monitoring the gap voltage (U_e), discharge interval can also be controlled adaptively.

4.1.3.2 Control system for EDM process

Monitoring and controlling of suitable spark gap with a servo control system and discharge pulse parameters with power generator have greatly influence the EDM process performance. Hence, performance of the EDM process can be improved by the appropriate integration of the state-of-the-art monitoring and control technologies. Attempts have been made by the researchers around the world to develop on-line monitoring and control strategies to identify the state of the spark gap and regulate the machine parametric settings in real-time to continue the EDM process at optimal performance level [8–11]. To develop the control strategy for automation of the EDM process, in order to make the decision when control interactions should be taken, it is very crucial to get accurate process information during machining with the help of different process sensing devices. An EDM unit mainly consists of main machine, servo control system, pulse power generator and dielectric system as shown in Fig. 4.1.6. All these subunits are interconnected for which robust feedback control strategies should be developed to achieve an efficient overall control system for EDM process.

In this regard, most important control parameters of EDM process are servo reference voltage and ignition delay time. In EDM, servo reference voltage as well as ignition delay time are proportional to gap width. Fig. 4.1.13 also shows different feedback loops which should be utilized for process control in EDM. First servo feedback loop is utilized for maintaining a proper spark gap

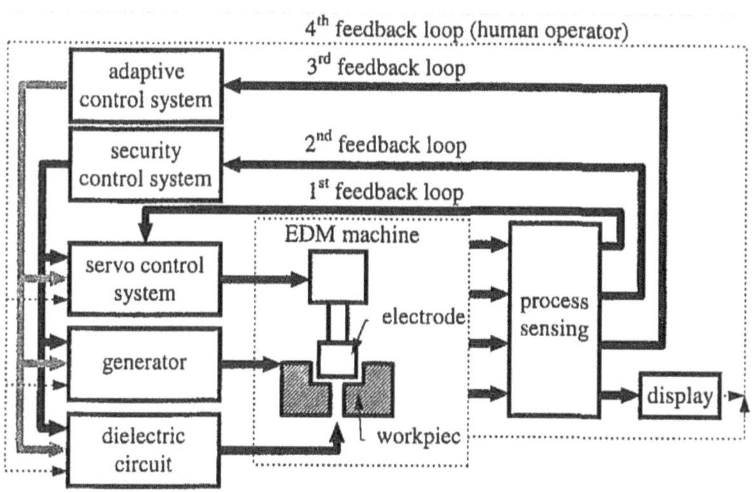

Fig. 4.1.13 Feedback loops for adaptive control in EDM [8].

by servo control system in EDM machine. In security control system, second servo feedback loop is utilized when major disturbance occurs which may damage the vital parts of the machine unit. Third feedback loop is utilized for adaptive control system to improve the degree of automation. Process optimization can also be achieved by adaptive control system. For man machine interaction fourth feedback loop is employed. Amongst above mentioned control systems, servo control system is most crucial for the machining performance in EDM which needs to be discussed in detail.

4.1.3.2.1 Gap control by servo feed mechanism

Unlike conventional machining methods, tool electrode feed in EDM is not constant. The gap between tool and workpiece is continuously monitored and control with the help of a servo control system. To match precisely the rate of material removal, tool electrode feed is controlled such that the gap remains same during machining. The servo system is commanded by the signals received from the sensors of the gap voltage in the power supply. A servo reference voltage is preset by the operators considering the gap to be retained during machining. As shown in Fig. 4.1.14 during machining, a proper gap width has to be maintained for controlled machining with average gap voltage. Larger gap widths are responsible for longer ignition delays result in higher average gap voltage which is larger than the preset servo reference voltage accordingly the tool feed increases to regain the proper gap. However, in case of comparatively lower gap, the ignition delay will be smaller as a result average voltage will be lower compared to reference preset

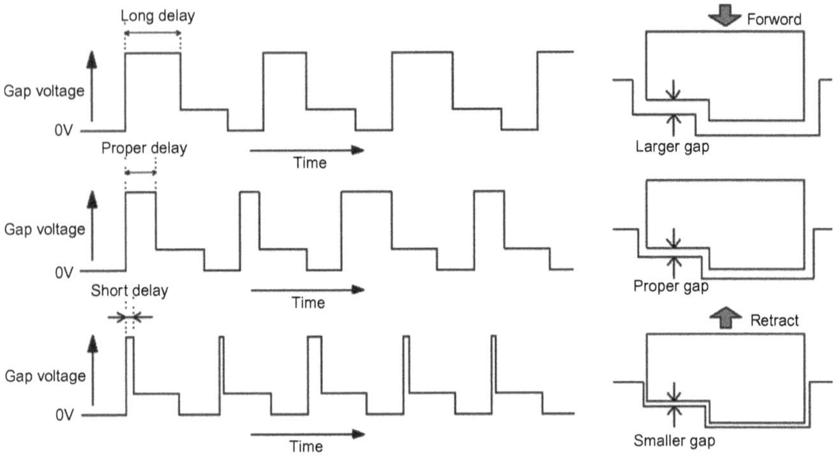

Fig. 4.1.14 Principle of servo feed control.

servo voltage. Hence, to maintain the preset value of the servo voltage, the tool feed decreases or is retracted as shown in Fig. 4.1.14. In case of short circuits which may cause due to presence of debris particles, the servo system will retract the tool automatically sensing the gap voltage. This operation continuous until the dielectric flushes the gap by removing the debris and the tool in feeding starts and machining operation continuous.

To monitor the gap width, the average ignition delay time may also be used in place of average gap voltage. Advance gap monitoring and control systems have also been developed to improve stability and the quality of machining. Five types of discharge states can be occurred during machining which must be monitored to achieve stability [8,12]. Fig. 4.1.15 shows voltage and current waveforms of the five discharge states, e.g., (i) Open circuit, (ii) Normal discharge, (iii) Transient arcing, (iv) Harmful Arcing and (v) Short circuit.

EDM gap monitoring system identifies the above mentioned major state of the gaps. The sate of gaps can be identified by detecting the level of pulse voltage during ON time. The normal discharge and harmful arcs which can also be differentiated by observing the ignition delay time. Gap control system provides duty factors of various gap states to monitor machining condition. Normal discharge ratio, i.e., duty factor during normal discharge has good correlation with actual material removal rate during machining which

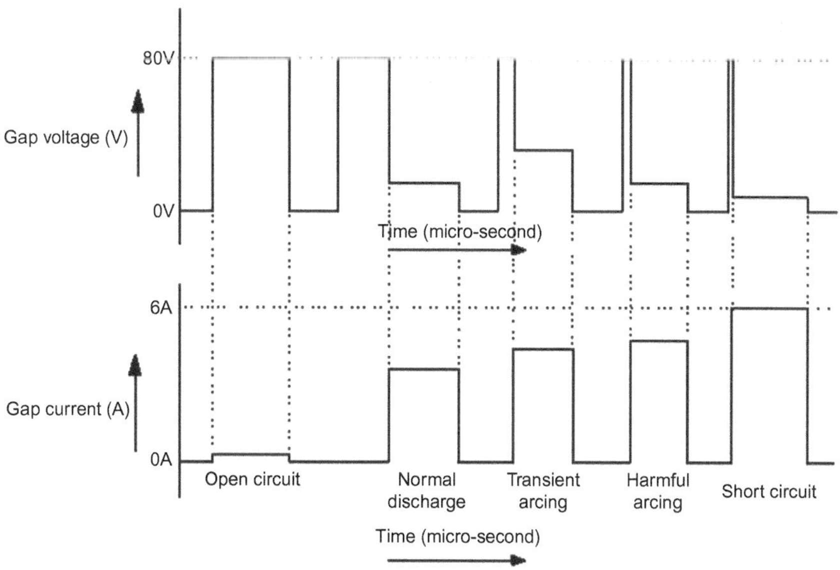

Fig. 4.1.15 Voltage and current waveforms of five basic EDM gap states.

can be utilized effectively for adaptive control system. Ignition delay is also an important indicator to maintain the stability of machining under above mentioned situations. Various online gap control strategies can be developed to monitor and control the above mentioned discharge states.

During machining of deep and narrow cavity, tool jumping operation is adopted. During tool jumping operation, the tool retracts from the machining zone to improve the flushing condition. The ratio of sum of abnormal discharge time to the total pulse on time is utilized for adaptive jumping control system [13,14]. Jump down time should be reduced in comparison with tool lifting time for avoiding harmful arcing during machining of larger depth and narrow cavities.

4.1.3.3 Dielectric system

One of the major units of EDM is dielectric system. It consist of dielectric fluids, its circulating system, e.g., pump, filters, reservoir and flow control unit and different types of delivery devices to guide dielectric flow efficiently to the narrow gap in the machining zone.

Dielectric fluids are needed in the gap between tool electrode and workpiece to initiate controlled electrical discharges for achieving removal of material during EDM operation. These dielectric fluids possess high dielectric strength to remain electrically nonconductive. It acts as an insulating medium but at the particular voltage level, i.e., breakdown voltage, it behaves like a conductive medium and helps to create concentrated sparks to a tiny area in tool electrode and workpiece. Just after the completion of the spark, these fluids regain its insulating property rapidly which facilitates further sparking in a repeated manner. Thus, high spark repetition rate can be maintained in a small machining area. Additionally, these fluids should have high specific heat which helps to extract heat from the machining zone and act as effective coolants. Moreover, it should also have low viscosity thus it can enables to remove efficiently the debris particles from the narrow machining gap. Selection of appropriate dielectric depends on the main following functions which should have to perform during machining:

(i) High dielectric strength for appropriate ignition delay,
(ii) Faster retention of the insulating property after breakdown,
(iii) High specific heat to serve as a coolant,
(iv) Low viscosity to remove debris particles,
(v) High flash point for safety, and.
(vi) Good oxidation stability to prolong permanence.

4.1.3.3.1 Types of dielectrics

Hence, considering above mentioned properties commonly used dielectric fluids in EDM are hydrocarbon oil, e.g., transformer oil, paraffin oil, kerosene; lubricating oil, gas, air and also deionized water. Dielectrics can be classified into the following three groups as follows:

(i) Hydrocarbon oil based dielectrics: This type of dielectrics of EDM is normally used in sinking EDM. Oxidized dielectric oil, i.e., mixture of fatty acid and hydrocarbon oil perform most stable machining due to dissolve of suspended debris by the formation of organometallic compounds. However, the oil based dielectrics are not environmentally friendly. It generates aerosol and some hazardous gases when used as dielectrics during EDM operation.

(ii) Water based dielectrics: This type of dielectrics widely used in wire EDM and micro-hole EDM drilling. These dielectrics are more environmentally friendly compared with the oil based dielectrics. Investigation proved that the poor performance of water based dielectric is found in the case of sinking EDM due to higher tool wear.

(iii) Gaseous based dielectrics: Presently, the use of gaseous dielectrics draws more attentions from the researchers. Gaseous dielectrics are more environmental friendly and more economical compared with the liquid dielectrics. However, there are still some technical problems need to be solved before it can be applied commercially in the industry.

Hydrocarbon oils are more popular in case of die sinking EDM; however it breaks down at high temperature and forms carbon which deposit on the electrodes as well as reacts with molten metal to form different alloying elements. In case of deionized water, MRR as well as tool wear are high and coolant effect is also more. Deionized water is mainly used in Wire EDM as well as micro EDM for precision machining, however, it causes corrosion during machining of ferrous workpiece. Different physical properties and compositions of dielectrics are listed in Table 4.1.1 [15].

4.1.3.3.2 Flushing of dielectric

At the initial stage of machining, there are no debris particles in the die electric. As machining continues, the concentration of debris particles in the dielectric increases, consequently dielectric strength reduces which reduces the ignition delay time. Hence, these debris particles should be removed from the dielectric, for this efficient filtration of dielectric is needed in the dielectric system. The amount of debris will be large in the narrow machining gap if dielectric flushing pressure is not high enough for which a proper pumping unit is required. Electrical conductive path generates

Table 4.1.1 Physical properties of the five dielectrics [15]

Dielectrics	Thermal conductivity λ (W/m K)	Specific heat C (J/kg °C)	Viscosity μ (Pa-s)	Density ρ (kg/m³)	Composition (vol%)
Oxygen	0.027	0.913	–	1.43	99.9%
Air	0.016	1.005	–	1.29	99.9% N_2, 21% O_2, 1% other gases
Deionized water	0.62	4200	1×10^{-6}	1000	100% H_2O
Kerosene	0.14	2100	2.3×10^{-6}	860	100% hydrocarbon
W/O emulsion	0.22	2747	8.7×10^{-4}	890	30.8% H_2O, 68.7% hydrocarbon, O_2, 0.5% surfactant

between electrodes due to large amount of debris concentration in the machining gap leads to undesirable sparking as well as arcing results in damage of tool and work piece. Hence, proper flushing of dielectric in the machining zone must be maintained. Different types of dielectric flushing techniques are utilized in EDM for effective removal of unwanted accumulation of debris particles from the machining zone. During machining of high aspect ratio and complicated shapes, flushing plays a crucial role in EDM for achieving desired accuracy. Different techniques are used for flushing dielectrics fluid through the machining zone for removing debris particles. Generally used techniques are listed as follows:

(i) Suction flushing,
(ii) Pressure flushing,
(iii) Jet flushing,
(iv) Immersion flushing, and
(v) Assisted flushing.

(i) *Suction flushing*: Dielectric flow forcefully through the machining zone by suction effect as shown in Fig. 4.1.16(a). It may enter in the machining zone by the side gap between tool and the workpiece and flow out through the tool electrode or workpiece as shown in Fig. 4.1.16(b). The holes in the workpiece for flushing of dielectric should not harm the machined product. This technique is useful to avoid any tapering effect due to secondary sparking between debris and side walls of the tool electrode.

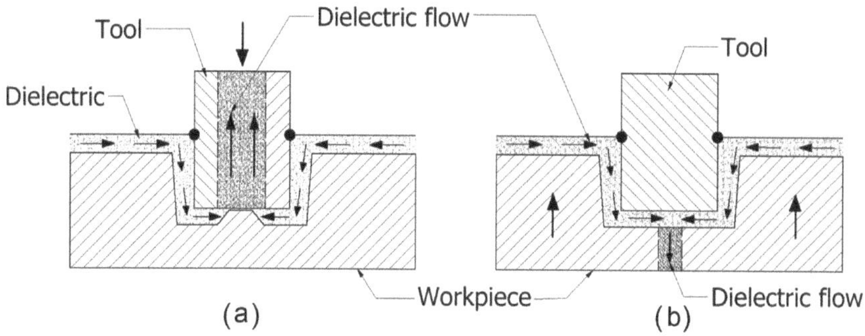

Fig. 4.1.16 Suction flushing through (a) tool electrode and (b) workpiece.

(ii) *Pressure flushing*: Dielectric being fed by pressure through the machining zone as shown in Fig. 4.1.17(a). It may enter either through the tool electrode or workpiece as shown in Fig. 4.1.17(b). Holes provided in the workpiece for flushing of dielectric should not damage the machined product. However this technique may generate tapering effect on the workpiece arising due to secondary sparking in the side gap because of accumulation of more debris particles in the side gap which acts as an exit path for the dielectric.

(iii) *Jet flushing*: This type of flushing uses nozzle through which dielectric can be directed to the adjacent of the machining zone. Supply of dielectric fluid through one direction results in increased concentration of debris particles in the another direction causes deterioration of the machining accuracy. Hence, multiple nozzles from different directions can be used to guide the flow of the dielectric for uniform removal of debris from the machining gap as shown in Fig. 4.1.18. It can also be used combined with suction or pressure flushing technique to achieve better flushing effect.

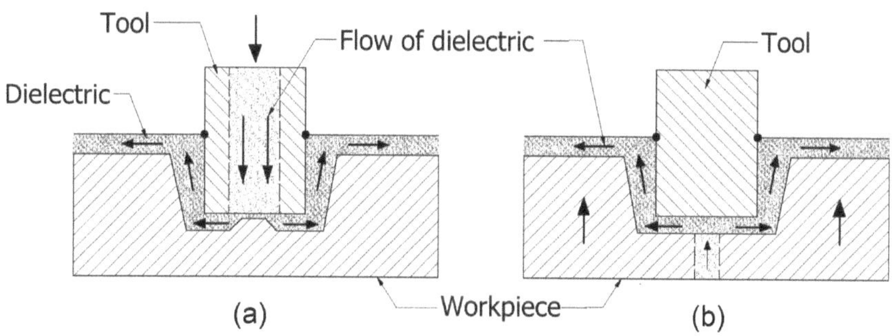

Fig. 4.1.17 Pressure flushing through (a) tool electrode and (b) workpiece.

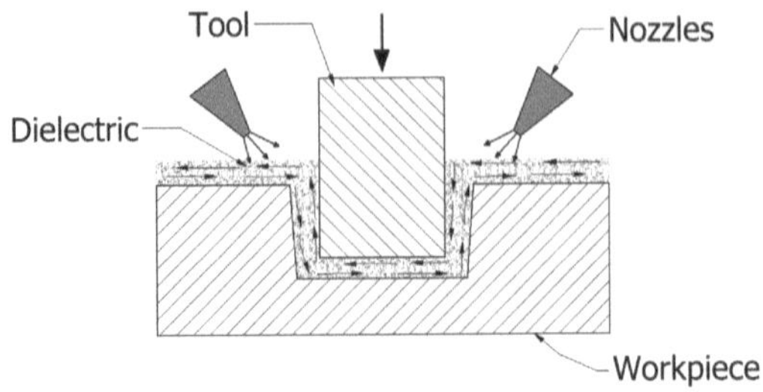

Fig. 4.1.18 Jet flushing.

(iv) *Immersion flushing*: Simple immersion of the discharge gap in the dielectric is sufficient for shallow cuts or perforations of thin sections. This technique can be effectively utilized when the cutting requirement of the depth is very low. This technique is more simple and economical.

(v) *Assisted flushing*: This is one type of emersion flushing where machining zone submerged in dielectric and effective flushing can be achieved with the assistant of other means as shown in Fig. 4.1.19. The tool electrode can be lifted at regular intervals to replace the contaminated dielectric fluid in the gap with fresh dielectric. This technique of lifting of tool periodically is known as jumping operation which can be controlled by adaptive jumping system. Jump down time in one jumping cycle is reduced to avoid abnormal sparking during machining of deep and narrow cavity. The tool electrode can also be rotated to generate forced circulation of the dielectric in the machining gap. This technique is effective during machining of high aspect ratio components and can also be utilized combining with other flushing techniques. Orbital motion can also be imparted to tool electrode to facilitated proper circulation of dielectric fluid in the machining zone.

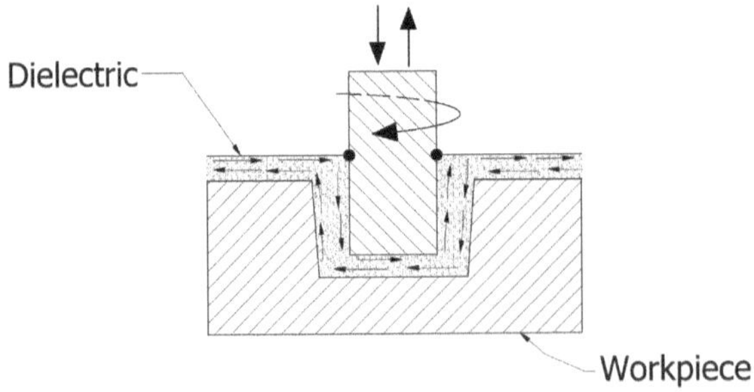

Fig. 4.1.19 Assisted flushing.

4.1.3.4 Tool electrode

Tool electrodes play a key role in EDM operations. The shape of tool electrode finally determines the profile of the cavity or hole generated during EDM. Hence, different characteristics of the tool electrode, e.g., material, tool wear as well as design considerations are essential factors to achieve better results.

4.1.3.4.1 Tool electrode materials

Theoretically, any conducting materials can be used as tool in EDM. Materials with the highest melting point and lowest electrical resistively are, in theory, the most appropriate for tool. Several materials are used for tool electrodes in EDM applications. However, selection of tool electrode materials are depended on various factors, e.g., higher melting point, higher boiling point, higher conductivity, lower cost, higher availability, good machinability, low wear and ability to generate good surface finish on the job etc.

Copper, brass and graphite are most commonly used materials for tool as they have fulfilled the above mentioned criteria. Copper and brass exhibits stable machining with high metal removal but produce high tool wear rate. Graphite is the most useful tool material in EDM because it vaporizes at very high temperature as a result it exhibits extremely low tool wear rate. However, fabrication of tool electrodes with copper and brass is much easier compare to graphite due to it's brittleness. Mainly, molding or powder metallurgy is used for fabrication of graphite tool electrodes. Hence, grain size of graphite can be controlled which mainly influences surface finish, tool wear and metal removal rate. Very small graphite grains improve surface finish, metal removal rate and decrease tool wear.

Some alloy materials, e.g., aluminum alloys, copper tungsten alloys and silver tungsten alloys are also used as tool electrode material due to their excellent properties as needed during EDM operation. Copper tungsten and silver tungsten alloys exhibit extremely low tool wear with high metal removal. These alloys are quite expensive, that is why their use is limited to very small and high precision parts. However, machinability of these alloys is poor compared to other tool materials as mentioned above.

4.1.3.4.2 Tool wear

As already mentioned the tool electrode gets eroded due to extensive heat generation during sparking. The shape of the tool electrode deteriorates due to tool wear as EDM operation continues which results in poor machining accuracy. Hence, for achieving better machining accuracy, tool wear must be minimized. Tool wear depends on various factors like voltage, peak current, duty ratio, electrode material, types of dielectric and polarity etc. Tool wear of different electrode materials can be quantified by a number known as wear ratio. It is the ratio of material removed from the workpiece to the material removed from the tool. The wear rate of the electrode tool material W_t and the wear ratio R_w, are as follows [16]:

$$W_t = \left(11 \times 10^3\right) i T_m^{-2.38} \tag{4.1.1}$$

$$R_w = 2.25\,\mathrm{T}_r^{-2.3} \tag{4.1.2}$$

Where, W_t, wear rate of the tool, mm^3/min, i, EDM current, A, T_m, melting point of the tool electrode, °C and T_r, ratio of the workpiece and tool electrode melting points.

Fig. 4.1.20 shows various features of EDM tool wear. Tool gets eroded at a faster rate from the sharp corners and edges due to higher current concentration as shown in figure. Hence, during EDM operation shape of the front portion changes as well as length of the tool becomes shorten. Due to this, after a certain time of machining tool must be replaced by a fresh tool. In some CNC controlled EDM system, automatic tool changing facility is available. To compensate the length of the tool during machining, tool wear compensation factor is incorporated in the programme of CNC EDM system. Till date limited information are available regarding tool wear of various tool materials under different parametric conditions as well as in the context of various tool shape features. Hence, there is an urgent need to develop expert system to assist CAD programming for CNC EDM based on knowledge base system supported by in-depth database.

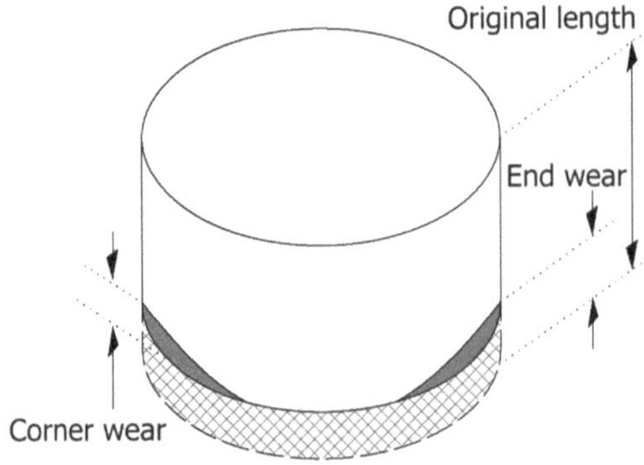

Fig. 4.1.20 Various characteristics of EDM tool wear.

4.1.3.4.3 Fabrication and design aspects

Various fabrication methods as well as design aspects of tool electrode play an important role during EDM operation. Extensively adopted tool fabrication methods are conventional machining, e.g., such as turning, milling, shaping, grinding, engraving etc. used for copper, brass, Cu-W, Ag-W, graphite etc. Casting method can also be used for Al alloys and Zn-Sn alloy. Metal forming as well as spraying is also utilized for the fabrication of tool electrode. Advanced methods like powder metallurgy (PM) technology as well as rapid prototyping (RP) are becoming popular now a day for fabrication of tool electrodes.

Allowances must also be included during design of tool electrode considering various tool wear features. To achieve desired shape of the machined components, this tool wear factor must be considered for allowance during tool design.

The composite electrodes fabricated using powder metallurgy from special powders is used to modify EDMed surfaces for achieving improved wear and corrosion resistance. Special powders, e.g., chromium in the Cu-Cr composite electrode migrated to the work piece, resulting in good corrosion resistance of the machined surface after EDM [17].

Special tool electrode having micro holes or micro channels for jetting dielectric liquid over the working surface has been developed to achieve high speed and high precision EDM of deep slots. Supply of dielectric liquid through the micro holes and channels using the newly developed tool electrode enabled stable machining of deep slots without jump flushing [18].

New type of bundled electrode has been developed by discretizing the solid die-sinking electrode into many smaller cell electrodes. A bundled die-sinking electrode is fabricated by combining large number of thin tubular cell electrodes and adjusting the length of each cell electrode [19]. This unique fabrication method generates coaxial hole-matrix structure that provides intensive inner flushing. It is evident that the special structural feature of the bundled electrode will improve the machining conditions in the gap through multi-hole inner flushing during EDM operation. Bundled electrodes can achieve much higher peak current, which results in higher metal removal and lower tool wear as compared with solid die-sinking electrode [20].

4.1.4 Modeling of EDM processes

EDM is basically a thermal process where removal occurs at the discharge location. During discharge, various factors play important roles which must be simulated to understand the mechanisms of metal removal.

To start with development of models for EDM, it is considered that material removal is a collective result of single discharges.

4.1.4.1 Material removal at single discharge in EDM

The quantity of material removal due to a single discharge can be determined by considering the diameter of the crater and the depth to which the melting temperature is reached. To do this some assumptions have to be made which are as follows [21,22]:

(i) The spark is a uniform circular heat source on the electrode surface and the diameter (*2d*) of this circular source remains constant.
(ii) The electrode surface is a semi-infinite region.
(iii) The electrode surface is insulated except the portion of the heat source.
(iv) Throughout the discharge duration the rate of heat input is remain constant.
(v) Electrode material properties do not change with the temperature,
(vi) The vaporization of the electrode material is neglected.

If the diameter of the crater (D_c), i.e., the spark diameter is assumed to be $2d$ (cm) then,

$$D_c = kE^{k_1} t_d^{\,k_2} \tag{4.1.3}$$

Where, E is the total pulse energy, (Joules), t_d is the discharge duration, (*sec*) and k, k_1, k_2 are constants depending on properties of the electrodes and dielectric medium.

Then, crater volume will be

$$V_c = \frac{\pi}{6} d_c \left(\frac{3}{4} D_c^{\,2} + d_c^{\,2} \right) \tag{4.1.4}$$

Where, V_c is the crater volume, (cm^3) and d_c is the crater depth, (cm) which is related to the melting temperature.

From this equation it can be estimated that the material removal is very low for small discharge time and increases with t_d. It can also be stated that the material removal per discharge greatly depends on the melting point of the material.

Since, the size of the crater depends on the spark energy; the depth and the diameter of the crater are considered as follows for a rough estimation of the material removal rate.

$$d_c = k_3 E^{\frac{1}{3}} \tag{4.1.5}$$

$$D_c = k_4 E^{\frac{1}{3}} \tag{4.1.6}$$

Where, E is the spark energy, (Joules) and k_3, k_4 are the constants depending on properties of the electrodes and dielectric medium.

Combining Eq. (4.1.4), Eq. (4.1.5) and Eq. (4.1.6), the crater volume can be expressed as

$$V_c = \frac{\pi}{6} k_3 \left(\frac{3}{4} k_3{}^2 + k_4{}^2 \right) E \tag{4.1.7}$$

Hence, approximate relationship of material removal rate in terms of melting point of the workpiece material is given by

$$Q \approx 4 \times 10^4 T_m{}^{-1.23} \tag{4.1.8}$$

Where, Q is material removal rate, (mm^3/amp-min) and T_m is the melting temperature of the material, (°C).

4.1.4.2 Estimation of material removal and surface finish for RC circuit

The RC circuit was used in the first developed electric discharge machine. The details of the circuit is already discussed and shown in Fig. 4.1.8

$$i_c = C \frac{dV_d}{dt}$$

$$i_c = \frac{V_0 - V_d}{R_c}$$

Where, i_c is the charging current, V_0 is the open voltage and V_d is the discharge voltage.

The capacitor 'C' is charged through a variable resistor 'R' by the DC source of voltage V_0. The voltage across the gap varies with time according to the following equation:

$$V_d = V_0 \left(1 - e^{t_c/RC} \right) \tag{4.1.9}$$

Where, t_c is the charging time starting at the moment V_0 is applied.

Sparks occurs when the voltage across the gap reaches a certain value also called discharge or breakdown voltage V_d then the frequency of the sparking (f) will be as follows:

$$f = \frac{1}{t_c} = \frac{1}{RC\log_e\left(\frac{V_0}{V_0 - V_d} \right)} \tag{4.1.10}$$

Energy released per spark,

$$E = 1/2\, C V_d^{\,2}$$

The average value of power delivered is:

$$P_{av} = \frac{E}{t_c + t_d}$$

As discharge time, t_d is negligible compare to charging time, t_c so, $t_d \approx 0$ and then average power becomes:

$$P_{av} = \frac{1}{2}\frac{1}{R\xi} V^{02}\left[1 - e^{-\zeta}\right]^2$$

Considering, $\zeta = \frac{t_c}{RC}$

For maximum power delivery,

$$\left[\frac{\partial P_{av}}{\partial \zeta}\right]_{\zeta=\zeta_{opt}} = 0$$

Using this condition in equation

$$\left(2\zeta_{opt} + 1\right)e^{-\zeta_{opt}} = 1$$

And solving this equation ζ is found as 1.26 so,

$$\left(\frac{V_d}{V_0}\right)_{opt} = 0.72$$

For maximum power delivery, the discharge voltage should be 72% of the supply voltage.

If material removed per spark is proportional to the energy released per spark then MRR can be expressed as:

$$Q \approx k_5\left(\frac{1}{2} C V_d^{\,2}\right) f \qquad (4.1.11)$$

where k_5 is the proportionality constant indicates the power fraction which is effectively used in material removal. Using Eq. (4.1.10) it can be expressed as:

$$Q \approx k_5 \frac{V_d^2}{2R\log_e\left(\frac{V_0}{V_0 - V_d}\right)} \qquad (4.1.12)$$

From the Eq. (4.1.12) it can be postulated that MRR increases with the decrement of 'R' but up to a certain level, below a certain value of 'R', i.e., R_{cr}, arcing is occurred as shown in Fig. 4.1.21.

The R_{cr} depends on inductance L and capacitance C of the circuit. However, for a purely inductive discharging circuit, that critical value is as follows:

$$R_{min} = \sqrt{\frac{L}{C}}$$

However, the practical limit of R_{min} considered as 30 L/C.

In case of machining of steel under normal condition, the removal rate can be roughly expressed as:

$$Q \approx 27.4W^{1.54}$$

4.1.4.3 Surface finish

Material removal in EDM is achieved due to the formation of craters by sparks so, larger crater sizes deteriorates surface quality of the machined surface. Crater size or crater depth, d_c (mm) depends on energy released per spark E. Hence, it can be expressed as:

$$d_c \approx n_1 E^{0.33} \quad (4.1.13)$$

d_c can be expressed in terms of C and Vd as follows:

$$d_c \approx n_1 \frac{1}{2} C^{0.33} V_d^{0.66}$$

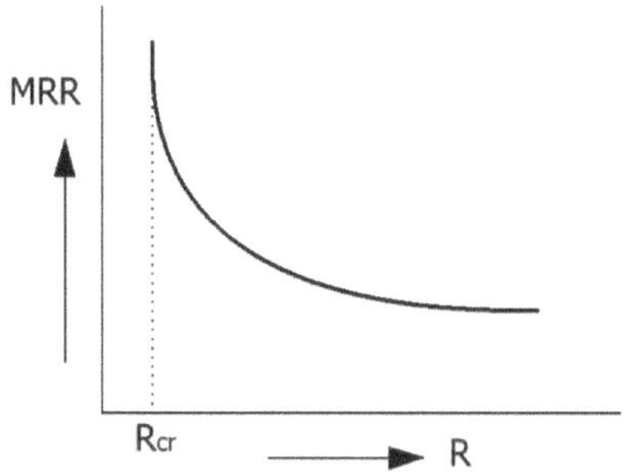

Fig. 4.1.21 Variation of MRR with resistance.

As the surface is greatly dependent on spark energy when machining steel under normal condition the relation between MRR and surface roughness can be approximately expressed as:

$$R_{rms} \approx 1.11 Q^{0.384} \tag{4.1.14}$$

where R_{rms} is the root means square value of the surface roughness in microns and Q is material removal rate in mm^3/min.

4.1.4.4 Material removal for pulse generator circuit

EDM is a thermal process. Hence, thermal modeling plays an important role in EDM. Major factors which help to estimate MRR in single spark EDM model comprise of the amount of heat input, radius of plasma spark, and the thermo-physical behaviors of material as well as dielectric fluid. The material removal in successive discharges is a collective result of single pulse discharges.

Gaussian distribution of heat flux input [23] can be used for rough estimation of the heat from the plasma. Due to EDM spark, heat, q enters into the workpiece expressed as follows:

$$q(r) = q_0 e^{-4.5\left(\frac{r}{R_{pc}}\right)^2}$$

where r is spark radius at the work surface (mm).

Using this equation, the maximum heat flux q_o can be postulated as follows [24]:

$$q_0 = \frac{4.57 F_c V_d i_d}{\pi R^2{}_{pc}}$$

where F_c is fraction of total EDM spark power entering into the workpiece and V_d is discharge voltage and i_d is discharge current.

Spark radius is an important factor in the thermal modeling of EDM process. Because, it is assumed that the crater radius is approximately equal to the spark radius.

The spark radius is also roughly considered as equivalent heat input radius. Spark radius can be derived as a function of discharge current, I (A) and discharge duration, t_{on} (μs) [25] as follows:

$$R_{pc} = (2.04e-3) I^{0.43} t_{on}{}^{0.44}$$

Equivalent heat input into the workpiece can be calculated with the help of Eqs. (4.1.4), (4.1.6). Hence, the heat flux equation is as follows:

$$q(t) = \frac{3.4878 \times 10^5 FcVI^{0.14}}{t_{on}{}^{0.88}} e^{-4.5\left(\frac{t}{t_{on}}\right)^{0.88}} \quad (4.1.15)$$

The transient heat transfer can be solved by applying the above heat flux equation at the spark location using the discharge duration as the time step for the further analysis.

A typical crater cavity is generated by single spark due to transient heat transfer. The crater volume V_c (mm^3) has already been expressed in Eq. (4.1.7) and can be utilized for further analysis.

However, the actual material removal during EDM process will influence by various factors such as ignition delays, sparks frequency, flushing efficiency, polarity of electrodes, behavior of debris particles, etc. [26,27]. It is difficult to include all these factors into the mathematical models. As a result, ideal MRR can be computed for a particular process parametric condition assuming all sparks are equally effective. Hence, the material removal rate can be computed by the equation as follows.

$$MRR = \frac{60 \times V_{ct}}{t_{on} + t_{off}} \quad (4.1.16)$$

Total cycle time includes on time, t_{on} and off-time, t_{off}. t_{on} is also summation of ignition delay time t_c and discharge time t_d.

4.1.4.5 Generalized approaches for EDM modeling

In EDM metal removal occurs at the discharge position where small size crater is produce. At the discharge locations thermal, hydrodynamic and electromagnetic behaviors of the workpiece, tool materials and dielectric fluids make the process of material removal more complex and stochastic in nature. The spark gap consists of dielectric fluids, gas bubbles as well as debris particles. Modeling of EDM process becomes more difficult due to variation of composition of gap, gap width, pressure and temperature in the narrow spark gap [28]. Fig. 4.1.22 exhibits comprehensive approach of modeling for various factors in EDM.

Development of model in EDM initiates with the determination of discharge location. It considers probability of discharge, discharge delay time, theory of determination of discharge location taking into account the influence of solid debris particles and bubbles. The next stage of modeling

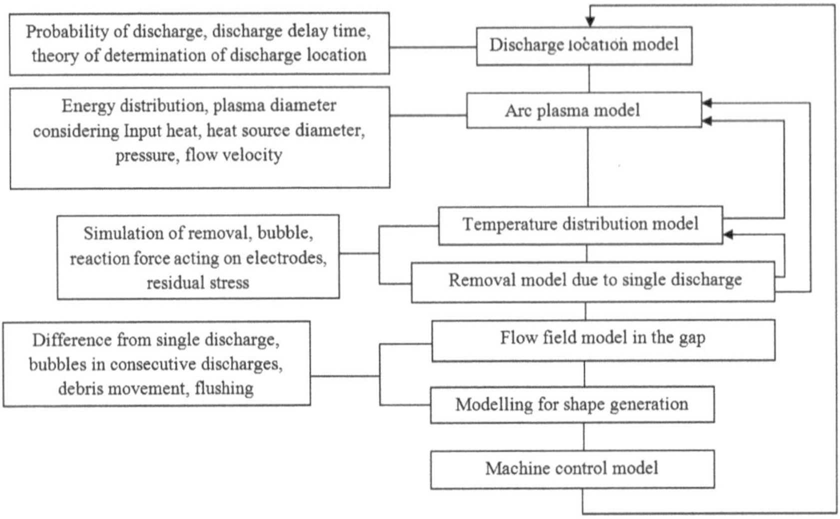

Fig. 4.1.22 Comprehensive approach of modeling for various factors in EDM.

consists of simulation of EDM arc plasma. It is essential for determining temperature distribution to evaluate the amount of material removal during sparking. It consists of modeling for energy distribution, plasma diameter considering input heat, heat source diameter, pressure, flow velocity etc. which provides necessary boundary conditions. By knowing the heat input and diameter of arc plasma, simulation of temperature distribution and metal removal for single discharge can be performed at the subsequent stages.

Volume of metal removal due to single spark depends on the size of crater formation and ejection of molten pool from the crater. At the end of the discharge, it is found that some amount of molten material ejected from the crater and it depends on various factors, e.g., flow field of dielectric liquid in the gap and different forces acting on electrodes. Flow field are produced by the blast of the bubbles at the discharge spot, flow of dielectric due to flushing as well as jumping motion of tool. Hence, development of model related to flow field in the spark gap for multiple discharges considering bubbles in consecutive discharges, debris movement and nature of flushing are essential.

Finally, modeling for workpiece shape generation is developed which depends on tool shape. The volume of material removed per unit area of tool is known as 'load of removal' which is not uniform and depends on various factors as mentioned above as well as the curvature of tool electrode surface. Most significant model in EDM is machine control which depends on nature of discharge. It is mainly related to the gap control by servo mechanisms. Average gap voltage and average discharge delay time play significant role in actual servo control mechanisms in machine control.

4.1.5 Influence of EDM parameters on machining characteristics

Various parameters of EDM play important role during machining operations. Major influencing parameters are broadly classified into two categories, electrical parameters and nonelectrical parameters. Nonelectrical parameters are tool materials, flushing pressure and type of dielectrics. Electrical parameters significantly influence the machining performance. Major electrical parameters are peak current, open circuit voltage, pulse frequency, pulse on time, pulse off time, duty cycle and polarity etc. Machining output criteria, e.g., material removal rate (MRR), tool wear ratio and surface roughness are the key machining characteristics of EDM. Material removal rate represents the average amount of material removed from the workpiece per unit time. Higher MRR, low tool wear ratio (TWR) and good surface finish are contradictory to each other and cannot be achieved at the same time utilizing a particular combination of parametric setting. Hence, optimum machining performance can be achieved taken care of the objective requirements separately in different segments of operation with diverse considerations. Different segments of operation are considered mainly as rough and finish machining. Maximum material removal is of prime importance in case of rough machining. Whereas, accuracy and surface finish are the most important criteria for finish machining. Accuracy also depends mainly on tool wear. Hence, surface finish and TWR must be lower in case finish machining. Role of various EDM parameters on major machining performance criteria during machining of tungsten carbide are discussed hereunder [29] due to its wider range of applications in advanced manufacturing.

Discharge energy is directly proportional to the peak current. Hence, peak current significantly influences the machining criteria, e.g., MRR, TWR and surface quality. As shown in Fig. 4.1.23 MRR increases sharply with the peak current up to a certain value then remains almost same when machining with higher peak current. As peak current increases, it increases the spark energy which intern increases the heating effect results in higher MRR due to melting. MRR remains constant at high peak current region due to uncontrolled discharges. Fig. 4.1.23 also shows that MRR increases following same pattern with the increment of pulse on time. At particular

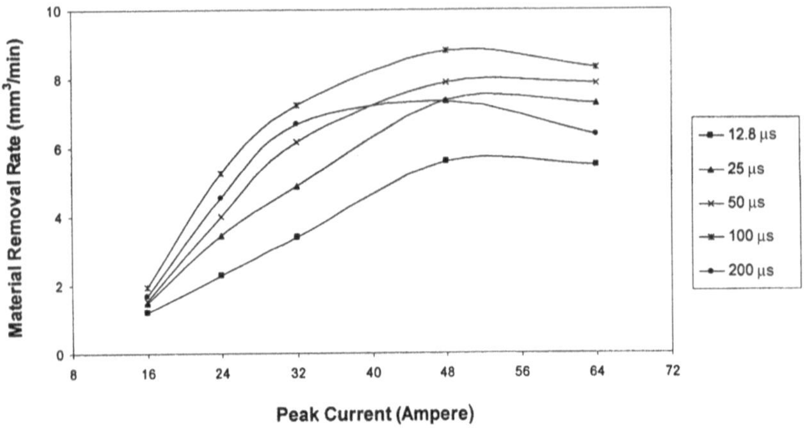

Fig. 4.1.23 Influence of peak current on MRR [29].

peak current, higher energy is available during machining for longer on time results in higher MRR.

The influence of peak current on TWR is shown in Fig. 4.1.24. The TWR decreases gradually with the increment of peak current up to a certain level then increases sharply due to availability of higher sparking energy which results in higher TWR due to more heating effect. Fig. 4.1.24 exhibits that TWR increases following the same pattern with the increment of pulse on time. At particular peak current, higher spark energy is available during machining for longer on time results in higher TWR.

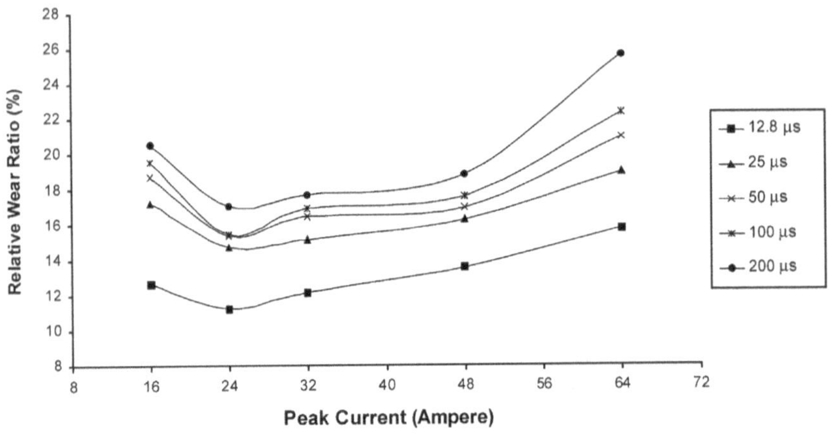

Fig. 4.1.24 Influence of peak current on relative wear ratio [29].

The influence of peak current on surface finish is shown in Fig. 4.1.25 Surface finish decreases with the increase in peak current. High peak current generates sparks with higher energy in turns creates larger and deeper crater size, for which surface finish deteriorates. Surface finish decreases following same pattern with the increment of pulse on time. Surface finish is a function of two major power parameters, i.e., peak current and pulse on time. Higher spark energy is available during machining for longer on time at particular peak current which results in poor surface finish.

Hence, peak current is one of the most significant parameters together with pulse on time in EDM. Peak current and pulse on time mainly control MRR along with surface roughness, overcut and TWR. Fig. 4.1.26 indicates relation between peak current and pulse on time with machining characteristics. Larger peak current with longer pulse on time results in higher MRR, lower TWR and suitable for rough machining. Carbon layer which is deposited on the anode tool electrode due to thermal dissociation of hydrocarbon oil results in lower TWR. Carbon layer is thicker when pulse on time is longer which reduces TWR. However, surface roughness is not good due to larger crater size as generated by each longer pulse discharge with higher peak current.

As shown in Fig. 4.1.26, combination of longer pulse on time and lower peak current results in lower TWR and good surface finish. However, this combination generates lower MRR due to lower discharge energy which creates lower heat flux. Combination of higher peak current and shorter pulse on time results in higher MRR and better surface roughness due to

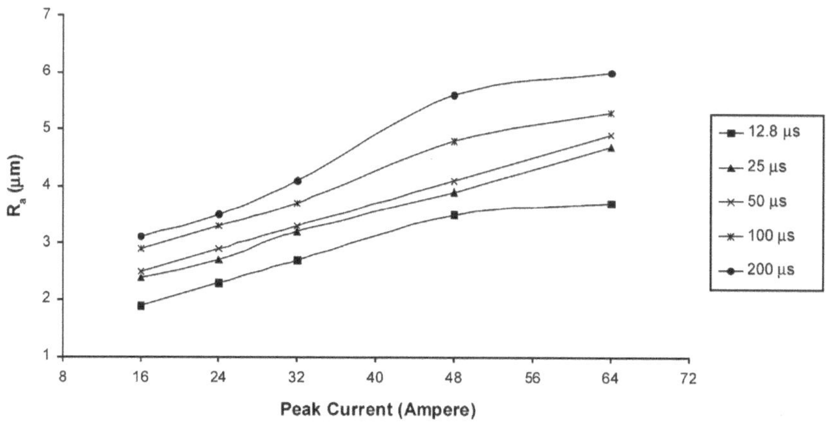

Fig. 4.1.25 Influence of peak current on surface roughness [29].

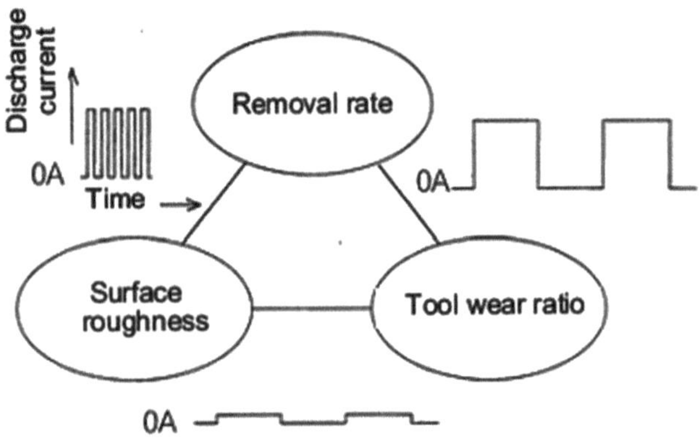

Fig. 4.1.26 Relation between discharge current waveform and machining characteristics [30].

smaller crater size. This combination is favorable for finish machining. Here, higher repetition rate of discharge per unit time also increases the MRR. However, TWR is high due to thinner carbon layer as deposited on the tool electrode because of shorter pulse duration.

Pulse off time is the time period when no energy is available during one pulse duration. During this period, dielectric fluid re-establishes its strength at the end of each discharge duration, i.e., on time and establishing the machining process for the next phase of cycle. Hence, at a particular duty cycle, longer off time reduces MRR, TWR and improves surface finish and overcut due to better removal of debris particles from the narrow gap. Shorter pulse off time with higher peak current results in high concentration of debris leading to uncontrolled sparking, which is unfavorable for EDM.

Pulse frequency is also another major parameter in EDM. Keeping duty ratio as constant, with the increment of frequency, material removal as well as surface roughness reduces. Due to shorter pulse duration at higher frequency, effective spark energy reduces which results in lower MRR and TWR as well as lower surface roughness because of smaller crater size.

Open circuit voltage is also followed the same pattern as peak current as discussed above. Increase of voltage raises peak current, which produces higher MRR, TWR and surface roughness during EDM operation.

Improper flushing can also reduces metal removal due to accumulation of higher concentration of debris in the gap leading to unstable sparking. Low flushing pressure cannot remove the gas bubbles and solid debris particles after each discharge. However, higher flushing pressure also disturbs the ionization effect of sparking operation. Flushing pressure has minimum effect on surface roughness. Proper flushing technique with appropriate pressure is essential for stable machining.

4.1.5.1 Influence of polarity

In straight polarity or normal polarity of EDM, workpiece is positive and tool is negative. When workpiece is negative and tool is positive, this type of polarity is commonly known as reversed polarity. As discussed earlier, energy dissipation into the anode is greater than cathode during sparking. In case of die sinking EDM, reverse polarity, i.e., positive tool electrode is commonly employed for reduction of tool wear during longer machining operation. Hydrocarbon oil breaks down at high temperature and produces carbon particles. Due to higher energy dissipation into the anode, thicker carbon layer deposited on anode tool electrode which protects the tool surface from wear.

In case of wire EDM and micro EDM, straight polarity, i.e., negative tool electrode is employed. Deionized water as dielectric is commonly used in wire EDM and micro EDM. However, during finish machining and micromachining utilizing hydrocarbon oil, tool wear is smaller due to application of lower energy pulses with extremely shorter pulse on time, which also reduces carbon deposition.

Keeping in mind, higher energy dissipation and carbon layer deposition on the anode as well as thermal properties and microstructure of electrode materials, selection of appropriate parametric combination is important to achieve desired performance criteria in EDM.

4.1.6 Machining accuracy and surface integrity

In EDM operations, various factors influence the machining accuracy. Inaccuracy mainly takes place due to machining instability because of different factors, e.g., improper tool feeding, occurrence of secondary sparking as well as gradual change of shape of tool due to tool wear.

During EDM operation, tool shape changes as machining progress because of removal of the material from the tool. This change of tool shape influences the machining accuracy. This factor has already been discussed at the earlier section. Tool wear compensation during tool feeding as well as replacement of tool improves the machining accuracy.

4.1.6.1 Taper

During machining hole or cavity, tapering effect is observed at the side walls of the machined profiles as shown in Fig. 4.1.27. Taper is the deviation between enlarged entry shape compare to bottom shape of the cavity generated during EDM operation.

In case of hole, it can be measured by the following equation:

$$\text{Taper angel}\,(\theta) = \tan^{-1}(D_i - D_o/2d) \tag{4.1.17}$$

where *Di,* entry diameter, mm, D_o, exit diameter, mm and d, depth of the hole, mm.

During machining of hole or cavity with a certain depth, as tool enters into the cavity, more number of sparks occurs from the side wall of the tool at the entrance area compare to the front portion of the tool at the bottom area of the machined hole or cavity. Machining occurs for longer period at the entry portion compare to bottom portion which causes more removal at the entrance area results in tapering effect. This taper angle should be minimized for improving accuracy of the machined job. Taper allowances can be incorporated in the tool during tool shape design as well as improved flushing can also reduce taper.

4.1.6.2 Overcut

Overcut is the deviation between shape feature of the cavity or hole and shape feature of the tool. In case of hole, it can be measured by the following equation:

$$\text{Radial Overcut} = D_i - D_t/2 \tag{4.1.18}$$

where D_t, Tool diameter, mm.

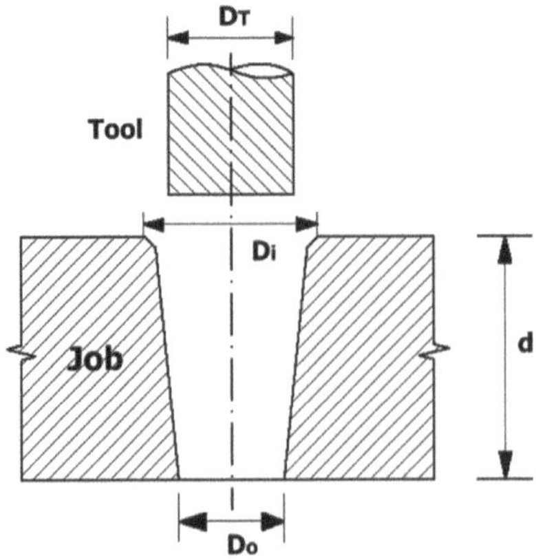

Fig. 4.1.27 Taper of the machined hole.

The scale of overcut depends on spark energy which can be controlled by monitoring different parameter of EDM. Hence, overcut for a particular machining operation influences by the spark gap. Presence of debris particles in the side gap during machining is also responsible for the overcut due to generation of secondary sparks and this can be minimized by proper flushing.

As shown in Fig. 4.1.28, to maintain proper spark gap between tool electrode and work piece for a particular parameter settings, overcut is produced in the workpiece. Spark gap varies depending on different parametric conditions of EDM operations hence, overcut also varies. Overcut can be minimized by considering the spark gap factor during tool design for a particular parametric combination.

4.1.6.3 Surface integrity

EDM is a thermal process where electrical sparks produce small craters on the workpiece surfaces for achieving desired shape. Hence, the machined surface is covered with tiny craters of different sizes. The depth and diameter of these craters depend on spark energy which relies on various parameters of EDM, e.g., mainly peak current, pulse on time, frequency, etc. As discussed earlier, crater sizes are influenced by not only discharge action but also dielectric fluid as well as electrode materials. Final surface finish of the machined job depends on the size and shape of the craters. Surface finish improves when crater size is smaller due to application of higher pulse

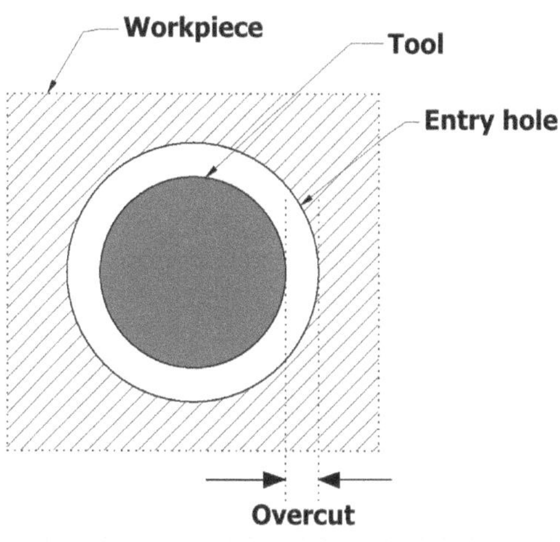

Fig. 4.1.28 EDM electrode and resulting overcut in workpiece.

frequency, smaller pulse on time and peak current which is already explained through Fig. 4.1.26.

Material is removed from the workpiece by melting and vaporization due to high thermal action of sparking which generates temperature ranging from 8000 °C to 12,000 °C. This high temperature is responsible for metallurgical change in the surface of the workpiece. Three distinctive layers can be identified, i.e., recast layer, heat affected zone (HAZ) and conversion zone on the machined surface as shown in Fig. 4.1.29.

At the end of each discharge, molten material from the crater ejected to the dielectric and after solidification forms tiny debris particles. However, some of the molten material rapidly solidifies in the crater due to faster cooling and gets adhered to the machined surface. A thin re-solidified layer is produced throughout the machined surface which is very hard and brittle due to its faster cooling and known as white layer or recast layer. Its thickness varies from 2.5 μm to 25 μm depending on the parametric combinations utilized during EDM operation. Recast layer is normally porous in nature and consists of micro cracks. The thickness of the recast layer should be minimized by controlling EDM parametric combinations. After machining, removal of recast layer from the machined product is essential before industrial applications and it may be done mechanically and electrochemically.

Beneath the recast layer, a heat affected area is produced due to cyclic heating and cooling action which causes the thermal stress and metallurgical damage. The thickness of this zone is less than 250 μm with the hardness of around HRc 65 and consists of grain boundary cracking and grain boundary weaknesses. These factors cause reduction in fatigue strength. Chemical composition may also changes due to formation of carbide because of migration of carbon from the hydrocarbon oil dielectric at high temperature.

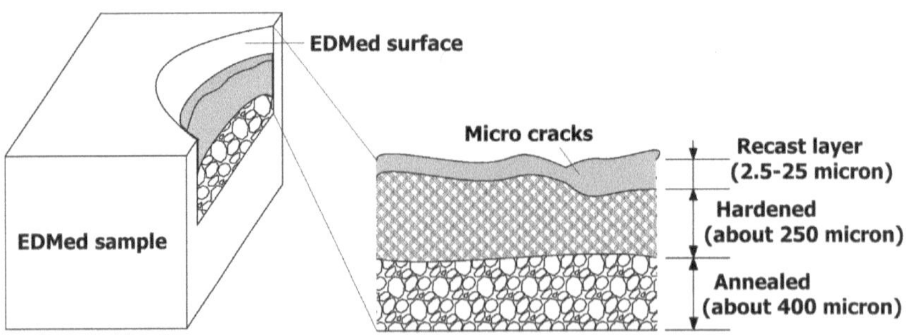

Fig. 4.1.29 Recast layer, heat affected zone (HAZ) and conversion zone on the machined surface.

Below the heat affected zone, conversion zone exists where grain structure is changed from its original structure. Depth of this zone may go up to 400 μm. By selecting the appropriate parameter setting, thickness of this layer can be reduced to some extent. Fatigue strength can be regained by removing the recast layer and heat affected zone.

4.1.7 Wire Electro Discharge Machining (WEDM)

In the late 1960s, wire EDM was developed for replacing the tools of different shapes which were needed in EDM for the fabrication of products with diverse configurations. In 1974, D.H. Dulebohn developed an optical system to control automatically the geometry of machined component utilizing WEDM process. WEDM process popularity was growing rapidly, as its capabilities were better understood by the industries. It can machine material that is electrically conductive regardless of the hardness, from relatively common materials such as tool steel, aluminum, copper, and graphite, to most advanced space-age alloys including hastaloy, waspaloy, Inconel, titanium, carbide, polycrystalline diamond compacts and conductive ceramics. In the end of 1970s, immense invention in the WEDM process was the introduction of computer numerical control (CNC) system. The accuracy, cutting speed and surface finish are continuously improving since the inception of wire EDM process. Since newer and more exotic materials are being developed, as well as design requirements of more complex shapes are increasing, conventional machining reach their limitations; hence the uses of WEDM in manufacturing are becoming more popular and continues to grow at an accelerated rate.

4.1.7.1 Basic principle

The fundamental principle of wire EDM is basically the same as that of the conventional EDM process. In wire EDM, the materials are machined with a series of discharges that are generated between a precisely positioned moving wire and the workpiece. High frequency pulses are utilized to create discharges from the moving wire to the work piece within a very small spark gap which is filled by insulated dielectric fluid, i.e., de-ionized water. The heat generated by sparks, erodes away a minute amount of material by melting and vaporization from the workpiece. During the process, some of the wire material is also eroded. These debris particles are flushed away from the machining zone with the aid of stream of de-ionized water which flows with high pressure through the top and bottom flushing nozzles. Basic scheme of WEDM is shown in Fig. 4.1.30.

4.1.7.1.1 Features of WEDM

In WEDM, the electrode is a continuously moving electrically conducive wire of diameter normally ranging from 150 to 300 μm. The mechanism of material removal in WEDM involves the complex erosion effect by rapid and repetitive (discrete) spark discharges between the wire tool and job immersed in a dielectric medium. A D.C. power supply is used to generate a very high frequency pulses. The wire electrode is threaded through a starting hole or pilot hole which is drilled at the desired location of the workpiece before machining. The wire is unwounded from the feeding spool by a drive unit which moves at constant speed through the tension mechanism and finally through the machining zone. The used wire which is deformed due to thermal effect because of sparking is either rewound on a take up spool or sliced into small pieces and collected in a bin. Hence, fresh wire electrode is continuously fed to the machining zone. The dielectric fluid is injected through nozzles into the working area of WEDM. Flowing dielectric flushes away the eroded particles from the machining zone. De-ionized water is used as the dielectric medium because of the fact that it not only provides better flushing and cooling but also it is environment friendly and noninflammable. It also provides large spark gap due to low conductivity. The wire electrode under tension is also vibrates due to

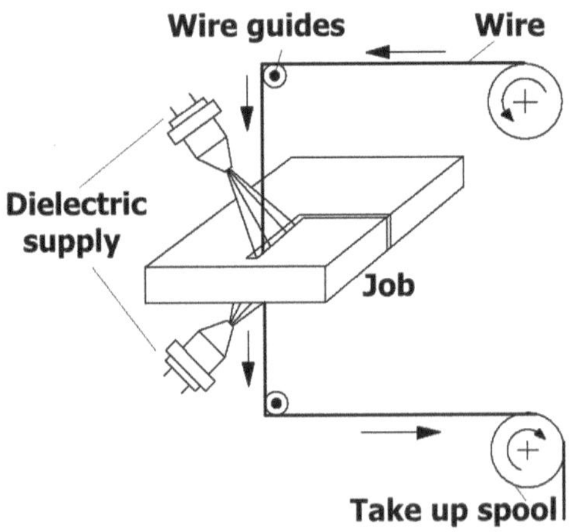

Fig. 4.1.30 Basic scheme of WEDM.

various forces which act on it during sparking. De-ionizing units are provided as standard accessories in WEDM system.

The typical features of a WEDM set up include the wire, working table, electrical and electronic control system, power supply, wire feed drive unit and dielectric supply system etc. The WEDM system are shown in Fig. 4.3.31 which consist of mechanical, power supply and computerized numerical control (CNC) system. Functions like wire feed, table movement, dielectric circulation and power is controlled by the CNC system. The quill advance is affected by a servo control system which maintains an optimum spark gap between the wire and the workpiece. The electro hydraulic servo control is usually preferred to electromechanical servo system because of its superior response time. The difference between a selected reference voltage and the actual voltage across the gap serves as the input signal for the servo control system. This signal is utilized to activate the wire movement accordingly. In case of a short circuit across the gap, servo control system is capable to control the tool position until the correct gap is established. A power supply unit capable of supplying 50–300 V, DC with the pulse frequency ranging from 50 to 300 kHz, and a current of 1–30 A is used in WEDM. A part program or CAD drawing of the job profile can directly be interfaced to the computer.

4.1.7.2 Major components of WEDM systems

Wire EDM systems are available in various configurations as well as styles of flushing or submerged type machines. Large scale WEDM system can accommodate workpieces weighing over 4500 kg and can cut over 50 cm thick job.

Automatic Wire Threader (AWT) is usually standard accessories in most of the models. In addition to the X-Y table travels for job, wire EDM system can generate taper profile in the job by utilizing U/V travels of the wire. Most machines can cut tapers of 10–40 degrees depending on thickness of the workpiece. The wire EDM system with all configuration detail is shown in Fig. 4.3.31. The system consists of CNC controller, power supply with anti-electrolysis circuitry, automatic wire threading, hand held pendant, programmable X-Y-Z as well as U-V-axis, dielectric supply system along with water chiller and filtration system, etc.

Out of above mentioned subsystem of WEDM, major components can be identified as follows:

(a) Mechanical Structure,

(b) Power Supply unit,

(c) Dielectric supply unit,

(d) Computerized Numerical Control (CNC) system.

As exhibited through Fig. 4.1.31, the wire EDM machine tool comprises of a main work table, i.e., x-y table, an auxiliary table for wire tilting, i.e., u-v table and a wire drive mechanism. The workpiece is mounted and clamped on the main worktable. This table as well as the u-v table move along x and y axes, in steps of 0.5 μm usually by means of servo motors. U and V axes are parallel to x and y axes respectively.

A traveling wire which is continuously fed from wire feed spool is caused to travel through the workpiece. The wire is supported along its traveling path under tension between a pair of guides, which are located at the upper and lower portion of the workpiece. The lower wire guide is permanently stationary whereas the upper wire guide is supported by the u-v table. The upper wire guide can traverse along a predetermined path as the machining process continues. The CAD program of the profile to be generated can be supplied to the CNC controller. Straight cut profile is generated utilizing only the movement of x-y table moves along the predetermined path keeping the u-v table as stationary.

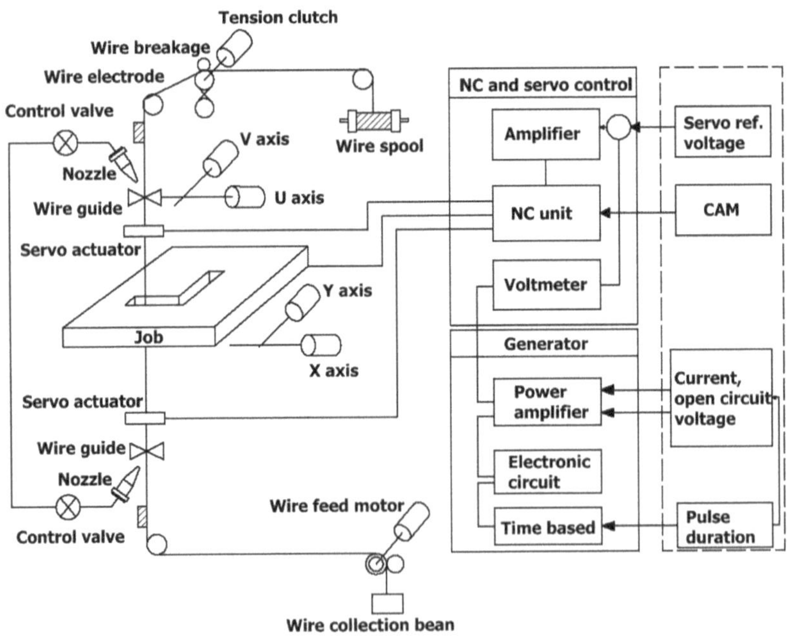

Fig. 4.1.31 The mechanical, power supply and CNC system of WEDM.

Fig. 4.1.32 shows the details of the dielectric system which consists of tanks for clean as well as dirty deionized water, various pumps, e.g., filter pump, flushing pump, etc. cartridges for resin, conductivity meter, filter units, control valves and nozzles. Conductivity and temperature of the deionized water are controlled to fixed values during machining. Parameters like pH of the deionized water are also to be controlled to restrict chemical or electrochemical reactions such as corrosion. Filtration is important in the dielectric supply system because debris particles create a bridge when an electric field is applied between wire and the workpiece. Hence, filtration unit with nominal particle pore size of 5 to 3 μm is common today. The used deionized water with the debris generated during machining is collected in dirty water tank and pumped through filters to the clean water tank. After that temperature stabilization and conductivity controls of the cleaned dielectric are performed by cartridges filled with resin and supply to the machining zone through upper and lower nozzles. The flushing pressure of dielectrics in the machining zone can be controlled by control valves.

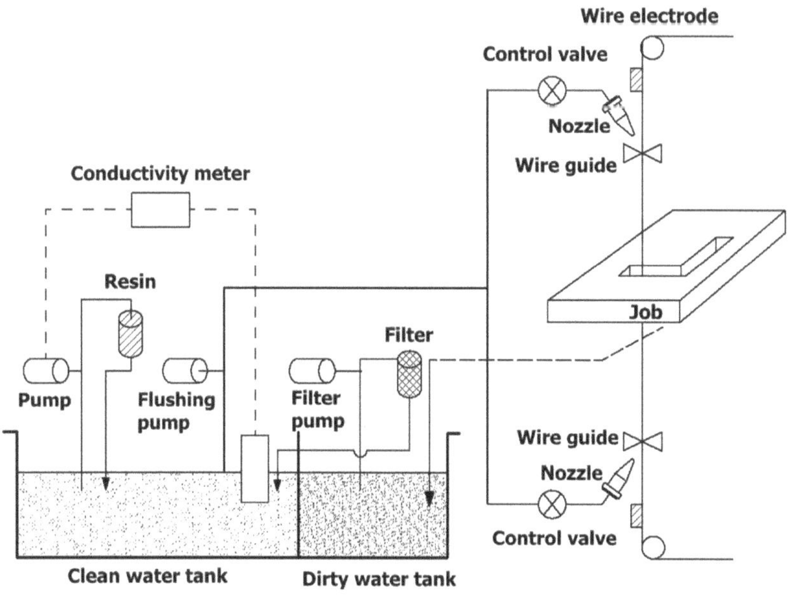

Fig. 4.1.32 The details of dielectric supply system.

For taper cut machining, the wire electrode has to be tilted by shifting the upper wire guide along u-v axes with respect to the lower wire guide. A desired taper angle is achieved by simultaneous control of the movement of x-y axes and u-v table along their respective predetermined paths as stored in the CNC controller. To achieve high order of accuracy, WEDM system usually uses an extremely low friction slide drive systems. The slides for movement of x and y axes can be mounted on hardened precision guide ways which are suitably protected from dust by using bellows. The slide drive is achieved by using either close loop DC motor drive system or DC stepper motor drive system.

The power supply unit comprises of three phase voltage stabilizer, electric pulse generator, motor drives and servo control unit for x, y, u and v axes and CNC controller. For the electro-erosion of the workpiece material, a series of electrical pulses generated by the high frequency pulse generator unit is applied between the workpiece and the traveling wire electrode.

The machining zone is continuously flushed with de-ionized water passing through the nozzles on both sides of the workpiece to ensure coaxial flushing. The spark discharge across the workpiece and wire electrode creates debris particles which increases the conductivity of the de-ionized water which is used as a dielectric medium. Ion exchange resin bowels can be used in order to prevent the increase of conductivity as well as the conductivity can be maintained constant with the help of a controller unit. To maintain the temperature of the de-ionized water dielectric, chiller unit is incorporated within the WEDM system.

CAD program consists of the geometry of the profile of the required job and the motion of the wire electrode along the desired path of the profile should be fed to the machine controller. The wire compensation for wire diameter and overcuts and taper angle can be specified for the total path or each path element separately. The entire wire travel path can be verified on the display screen and corrected, if necessary. After successful completion of the tool path profile, it can be stored in the computer and can be executed by the CNC controller during machining operation.

4.1.7.3 Influence of WEDM process parameters

Emphasize on realization of higher machining productivity with desired accuracy and surface finish is becoming a major issue in Wire electrical discharge machining. Though, even a highly skilled operator with a state-of the art WEDM is rarely able to achieve the optimal performance because of involvement of a large number of variable parameters as well as other controllable factors in WEDM. The optimum utilization of the WEDM process needs appropriate selection of machining parameters. WEDM is complex in nature and can be controlled by large number of parameters and other factors as shown in Fig. 4.1.33.

Major variable parameters of WEDM can be identified as pulse peak current, voltage, pulse on time, pulse off time, duty factor, flushing pressure of dielectric, types of wire electrodes, wire feed rate and wire tension etc. Other major factors of WEDM which are indirectly influencing the machining responses are properties and thickness of job materials, types of dielectric, construction and controlled features of the machine tools. Research is going on to investigate the effect of various process parameters on desirable outputs. Major outputs which play a crucial role in modern manufacturing applications are cutting speed (MRR), accuracy, overcut and surface integrity etc. However, the different parameters controlling the machining performance can be described here-in-under:

(i) Pulse-on time: During this period the voltage is applied across the wire electrode and job. The discharge energy contains in a single pulse, increases with the increase of T_{on}, resulting in a higher cutting rate.

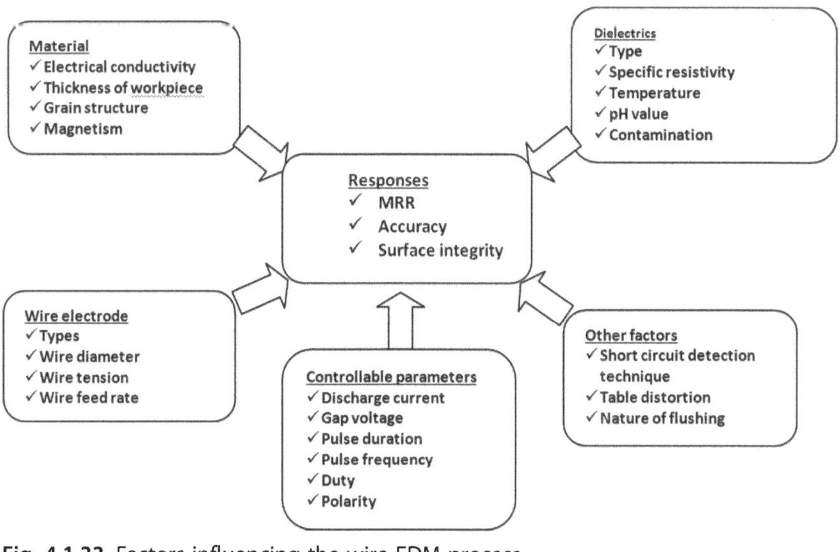

Fig. 4.1.33 Factors influencing the wire EDM process.

(ii) Pulse-off time: During this period the voltage across the electrodes is absent. With a lower value of T_{off} for a particular pulse period increases pulse on time, results in more number of discharges which increases sparking efficiency and cutting rate. However, very low value of T_{off} causes wire breakage and an unstable discharge condition. When sparking becomes unstable, it is better to increase T_{off} which allows lower pulse duty factor thereby reduces the average gap current.

(iii) Peak current: The increase in the peak current, I_p increases the pulse discharge energy. At higher value of I_p, the gap condition may become unstable with improper combination of T_{on}, T_{off}, voltage and servo feed settings. Reducing the I_p value helps to attain an unstable discharge condition into a stable condition.

(iv) Flushing pressure of dielectric: Generally, a high flushing pressure of dielectric is necessary for cutting with higher values of pulse power and, especially, while cutting the jobs of higher thickness. However, low input pressure should be used for thin jobs and precise cutting.

(v) Wire feed rate: This is a feed rate at which the fresh wire is fed continuously into the sparking zone during machining. For working with higher pulse power, higher values of wire feed rate are required.

(vi) Wire tension: This is a gram equivalent load with which the wire being continuously fed to the system such that wire can be kept under tension and it remains straight between the wire guides. Due to spark induced reactive forces and flow of water dielectric in the machining zone, the wire vibration and deflection occur which deteriorate the machining accuracies. The wire tension makes the wire straight between the wire guides during machining.

(vii) Servo feed setting: The parameter decides the servo speed of the worktable, which can vary in proportion with the gap voltage or can be held constant during machining.

(viii) Threshold setting voltage: This is threshold setting for corrective action in abnormal discharge condition. At gap short or short circuit condition, the controller will reduce the current drastically and will retrace the sparking path in backward direction so as to overcome the short circuit condition.

Pulse on time, pulse off time and peak current are most influencing parameters for cutting speed during rough cutting. Pulse on time is the most significant parameter during finish cutting with constant cutting speed. The surface roughness is influenced mostly by peak current during rough cutting. During finish cutting, most influencing parameters for surface roughness are pulse on time, peak voltage, gap voltage, dielectric flow rate, wire tool offset and constant cutting speed. Highest productivity with best surface finish and least geometrical inaccuracy can never be achieved by a single set of parametric combination. A proper trade-off is essential during selection of parameter setting to satisfy all the above-mentioned three machining responses simultaneously.

4.1.7.4 Mathematical modeling for material removal rate (MRR) in WEDM

The metal removal rate in WEDM may be evaluated through development of a suitable mathematical model. In WEDM, the metal is removed from the workpiece through repetitive spark discharges leading to generation of craters in the work surface. The volume of a single crater depends on the pulse energy for each discharge when all other conditions are considered to be unchanged. The depth and diameter of a crater are expressed by the following relations:

$$h_c = K_1 W_p^{\frac{1}{3}} \tag{4.1.19}$$

$$D_c = K_2 W_p^{\frac{1}{3}} \tag{4.1.20}$$

where hc, height of crater, mm, Dc, diameter of the crater, mm, Wp, pulse energy or energy contained in a pulse, Joule, K1, K2, constants for a particular combination of electrode, dielectric and workpiece.

Thus, considering the crater shape as approximately spherical, the volume of a crater may be calculated as:

$$\begin{aligned} V_c &= \frac{\pi}{6} h_c \left(\frac{3}{4} D_c^2 + h_c^2 \right) \text{mm}^3 \\ &= \frac{\pi}{6} K_1 W_p^{\frac{1}{3}} \left(\frac{3}{4} K_2^2 W_p^{\frac{2}{3}} + K_1^2 W_p^{\frac{2}{3}} \right) \\ &= \frac{\pi}{6} K_1 W_p \left(\frac{3}{4} K_2^2 + K_1^2 \right) \\ &= K W_p \end{aligned} \tag{4.1.21}$$

where $K = \frac{\pi}{6} K_1 \left(\frac{3}{4} K_2^2 + K_1^2\right)$, a constant which may be calculated for the known values of K_1 and K_2.

Thus, if the discharge frequency is γ (Hz, cycles/s)
The metal removal rate,

$$MRR_v = 60bv_c\gamma \ \text{mm}^3/\text{min} \tag{4.1.22}$$

Also, the MRR in WEDM can be expressed in another form:
Metal Removal Rate,

$$MRR_v = hv_cb \ \text{mm}^3/\text{min} \tag{4.1.23}$$

where v_c, cutting speed or work-table feed, mm/min, h, workpiece height, mm, b, width of cut, mm, i.e., $d+2g$, d, wire-tool diameter, mm, g, spark gap, mm.

Although, the spark gap is highly sensitive to most of the machining parameters, the domain of the variation of spark-gap values for a particular machining set up is very small. The cutting speed drastically varies with the thickness of the workpiece even at the same parameter setting. Therefore, the MRR is more conventionally expressed as follows:

$$MRR = v_ch\,\text{mm}^2/\text{min} \tag{4.1.24}$$

However, from the above expressions, it is clear that for a particular job-tool combination, v_c is a fair indicative of MRR in WEDM process.

From equations Eq. (4.1.21), Eq. (4.1.22), Eq. (4.1.23) following relationship can be made:

$v_chb = 60v_c\gamma = 60 \text{ K } W_p\gamma$

$$\text{or,} \ v_c = \frac{60KW_p}{hb(T_{ON} + T_{OFF})}\text{mm/min} \tag{4.1.25}$$

The pulse energy may be obtained by two ways. For a R-C relaxation circuit,

$$W_p = \frac{1}{2}CVd^2\,\text{Joule} \tag{4.1.26}$$

where C, Capacitance, F, Vd, discharge voltage, V, and, for other electronic controlled pulsed circuits generating high frequency, Wp may be obtained from the pulse parameters as below:

$$W_P = \int_{t_1}^{t_2} I\,V\,dt \tag{4.1.27}$$

where, t_1 = discharge start time, *sec*, t_2 = discharge finish time, sec, I = discharge current, A, V = gap voltage, V.

Thus, for a perfectly square shaped voltage and corresponding square current pulse, I and V are constant and W_p is determined as,

$$W_p = I_p V_p T_{on} \tag{4.1.28}$$

Thus, cutting speed, v_c may be expressed in general as follows:

$$v_c = \frac{60K}{bh} \cdot \frac{T_{on}}{T_{on} + T_{off}} \cdot I_p \cdot V_p$$
$$= \frac{60\mathrm{K}}{\mathrm{bh}} \cdot \mathrm{I_p} \cdot \mathrm{V_p} \cdot D_f \ \mathrm{mm/min} \tag{4.1.29}$$

where, Duty Factor $D_f = \frac{\text{pulse on time}}{\text{cycle time}}$

Therefore, for a known value of K in a particular machining condition and set up, the cutting speed as well as the volumetric metal removal rate may easily found out.

4.1.7.5 Taper cutting system in WEDM

Wire EDM can also be utilized for generating tapered jobs. Tapered jobs are often needed in the die and mold making industries. During cutting of straight job, upper guide remains stationary in u–v plane with respect to lower guide. The lower guide remains permanently stationary during machining. The worktable carrying the job can be moved along the programmed path instead of movement of both upper and lower guides.

During taper cutting, the upper wire guide moves along u and v axes with respect to the lower guide at a predetermined speed with the worktable movement. Hence, the wire does not remain vertical due to the generation of motion of the top slide in u and v axes. The angle produced with the vertical axis can be changed at various points and controlled and monitored by the CNC controller through different path programming. Hence, desired taper angle can be achieved by simultaneous control of the movement of the x–y table and the u–v table along their respective predetermined path as designed. The path information of such movements can be stored in the CNC controller. The machine controller makes it possible to move and control all the four axes simultaneously to generate different profiles with tapered surfaces. However, during tapper cutting in WEDM, machining should be carried out very carefully and at a considerably low cutting speed as compared to straight cutting. Tapper cutting system may come across following difficulties during machining which must be properly addressed:

(i) Proper flushing is extremely difficult as flushing direction is vertical and the wire remains tilted during tapper cutting. This causes to be severe heating of wire creating thermal load result in the wire rupture. This limits the cutting speed in taper cutting.

(ii) Wire tilting causes instability in wire due to fluctuation of wire tension;

(iii) Possibility of the wire breakage increases due to deformation of the wire which becomes inevitable at the exit of the upper guide and at the entry of the lower guide.

To overcome above difficulties, the upper and lower flushing valves are kept fully open to make sure of proper flushing as far as possible. Also the distance between the upper flushing nozzle and the top surface of the workpiece should be kept minimum in the range of 0.5–1 mm.

4.1.7.6 Forces acting on the wire

During machining, mechanical behavior of the wire electrode is extremely complicated. This is due to the occurrence of sparks between wire and workpiece, which is highly stochastic in nature and also the magnitudes and directions of various forces acting on the wire are not constant. The Wire EDM process is also become complicated due to a combination of other factors which includes fluctuation in voltage and current, flow distribution and decomposition of dielectric, random ionic migration, interaction of two successive discharges and presence of gas bubbles as well as debris particles in the narrow machining zone.

However, the main forces acting along or upon the wire are forces from gas bubbles formed by the plasma of the erosion mechanism, hydraulic forces due to flushing, electrostatic force and electrodynamic force. Though, it is very difficult to estimate quantitatively any of these forces, researchers have been able to determine experimentally the forces acting perpendicular to the wire during the WEDM operation for particular machining conditions. However, it is considered that there is slight effect of electrodynamic force on the wire. Instead, another force acts perpendicular to the wire, which is impact. However, forces acting on the wire electrode during machining can be categorized by four types, i.e., discharge reaction force, electrostatic force, hydrodynamic force and electromagnetic force as shown in Fig. 4.1.34.

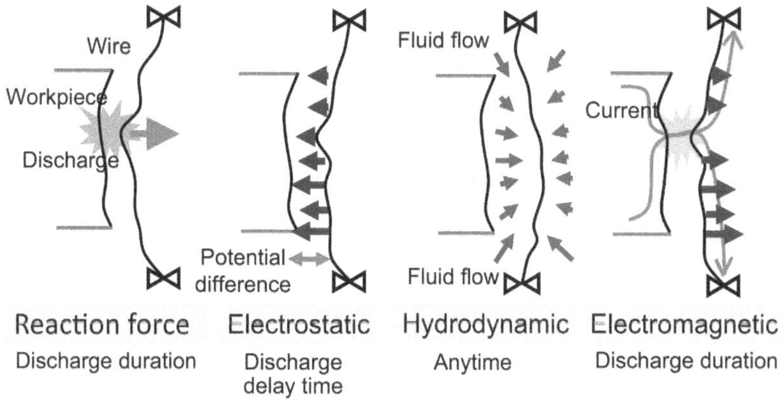

Fig. 4.1.34 Forces acting on wire electrode [31].

Discharge reaction force or impact force mainly occurs due to rapid expansion of gas bubbles at the discharge spot as shown in Fig. 4.1.34. During every individual spark discharge, the wire experiences reaction force, which acts in the reverse direction of the discharge occurrence. Electrostatic force occurs between the work electrode and the workpiece during the ignition delay time when an open voltage is applied. Compare to reaction force, the electrostatic force is small per unit length of the wire. However, the influence of this force is significant considering the total length of the wire which facing the workpiece during machining as shown in Fig. Hydrodynamic force is the drag force generated by the flow of dielectric through the tiny spark gap. This force is active on any point of the wire throughout the machining time as shown in Fig. Electromagnetic force generates at any location of the wire which is facing the workpiece. An electric current flow produces on the workpiece surface due to presence of high frequency current pulses during WEDM. The flow of current on the workpiece is in parallel with the wire electrode due to skin effect and generates electromagnetic repulsive forces on the wire. All these above-mentioned forces along with the axial tensile force applied to the wire is responsible for the wire vibration and wire bending in the reverse direction of cutting when discrete sparks are generated between the wire electrode and workpiece.

4.1.7.7 Finishing and accuracy

Compared to EDM, WEDM is more complex machining process which depends on large number of machining parameters and factors. Finishes and accuracies greatly depend on the influence of these process parameters with some other major control factors. Machining accuracy of WEDM is in the order of ±0.2 μm which depends on position accuracy of the machine elements, e.g., X-Y and U-V table, wire feeding arrangement and dielectric flushing system etc. Surface finish obtained is moderately good which depends on craters size of the sparks of the eroded surface. However, surface finish (Ra) can be improved to the range of 0.8–0.1 μm by selecting suitable combinations of parameter settings. Single cuts generate a good finish; however, more finely finishes with better accuracies are essential for fulfilling stringent design requirements. To achieve these strict design requirements, trim cuts or skim cuts are used in wire EDM. Trim cutting generates excellent finishes due to application of less energy during machining which produces smaller sparks energy and thus smaller craters on the cut surfaces. Control sparks with energy produce enormously good surface finishes with superior accuracies. However, not only surface finish but also accuracy depends on some other output factors as discussed bellow which also needs to be taken care of.

(i) Kerf Width

The wire electrode cuts the workpiece with the help of controlled and repetitive electrical discharges. The workpiece moves along the predefined path and maintain constant distance, i.e., spark gap from the workpiece such that it never contact with the wire electrode. Spark gap depends on different process parametric combinations during WEDM. Always, there will be a chance of occurring short circuit due to the contact between workpiece and wire electrode because of several disturbances which may arise in the tiny spark gap during machining. Due to short circuit, wire may rupture and no further cutting is possible. Frequent wire rapture intern deterioration of the machining accuracy. To continue stable machining, a spark gap has to be maintained for which the kerf width is always slightly larger than the wire diameter. As shown in Fig. 4.1.35 the cut width or kerf width generates due to machining is the summation of spark gap, wire diameter as well as crater depth. A normally used wire diameter of 250 μm, usually produces a 350 μm kerf width. Hence, the diameter of the wire enforces a restriction on the minimum achievable kerf width which may be expressed as follows:

$$\text{Minimum kerf width} = d + 2g + 2d_t \tag{4.1.30}$$

where, d, diameter of the wire, g, spark gap and d_t, average crater width.

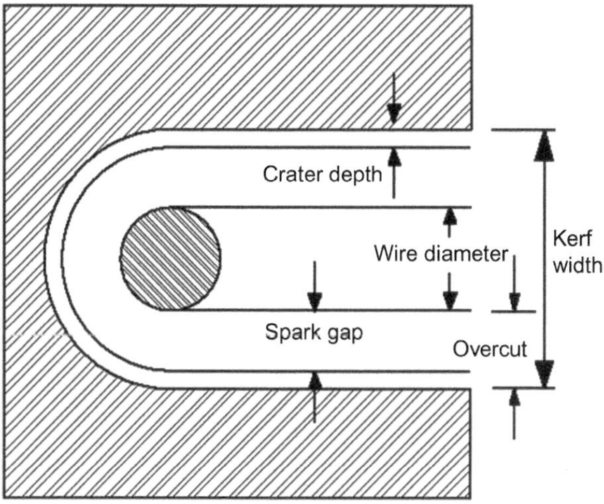

Fig. 4.1.35 Generation of Kerf width during WEDM.

Generated overcut is the summation of spark gap and average crater depth. To minimize the kerf width, spark gap and average crater depth must be minimized. This objective can be attained by minimizing the spark energy with the stable machining condition. However, smaller wire diameter makes the kerf width thinner.

(ii) Corner radius

It is difficult to produce sharp corners on the cut profile by WEDM. During complex profile cutting, when the wire turns a corner along the desired profile, it generates a sharp edge on the inside corner and always produces a small radius on the outside corner. Fig. 4.1.36 shows this effect of corners and radii during corner cutting. Wire diameter, spark gap and cutting speed influence the degree of radius generated during corner cutting. For some practical application such as dies for stamping, small outside radius is intentionally provided which increase the life of dies because sharp corners generally wears first. Trim cuts in essential in WEDM for generating extremely sharp outside corners. Rough cutting operations must be followed by multiple trim cutting runs as per the design requirements for achieving sharp outside corners with greater accuracy which requires more machining time in turn increases the overall cost. However, small diameter wire can also produce sharp corners with small outside radius by single run with much lower cutting speed. To have faster machining, it is advantageous to use larger diameter wires to reduce the chances of wire rapture. Hence, always there will be chances of generating radius during sharp corner cutting in WEDM. It is

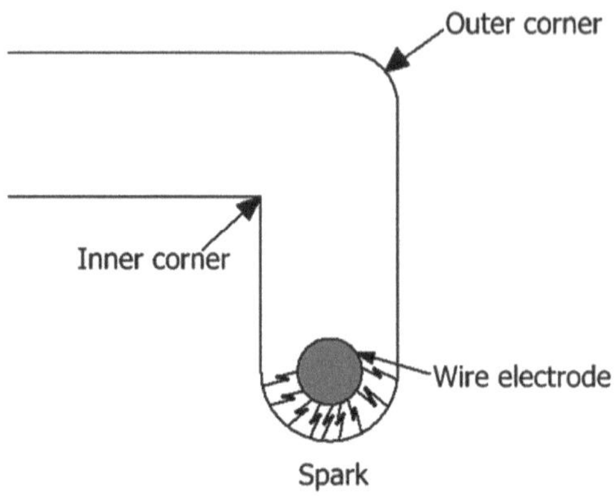

Fig. 4.1.36 Effect of corners and radii during corner cutting.

observed that the minimum outside corner radius for wire diameter of 250–300 μm is ranging from 12 to 16 μm.

(iii) Barreling effect and wire trail-off

The gap between wire and workpiece during WEDM operation is very narrow which is generally in the range of 20–30 micron. The gap depends on various process parametric combinations and mainly on intensity of the spark energy. In this narrow gap, controlled localized material removal takes place due to concentrated high heat energy which results in melting and vaporization. Due to the sparking forces and generation of gases, the proper flushing of dielectrics along the gap is not possible. The effect of flushing is good around the entrance and the exit portion, i.e., upper and lower portion of the workpiece where, material removal is more effective compare to the middle portion. Hence, metal removal at the middle potion of the workpiece is less which causes slight bulging at the middle zone and known as barreling effect. This effect is more prominent in case of thicker workpieces. Barreling effect on the cut surface of the workpiece can be minimized by selecting appropriate parameters setting or utilizing trim cutting.

Wire trail off is generated at the end point of cutting a profile of the workpiece. During any profile cutting, a small portion of the material is left behind for a short distance at the end point followed by the wire returns to its predefined path to improve the accuracy. This undercut portions is trail off at the end to separate the cut portion from the stock which creates an impression on the cut profile. This slight undercut is negligible for most of the jobs. However, undercut is more prominent for profile with sharp corner which can be improved by trim cutting. Barreling effect and wire trail-off can be avoided by slowdown the cutting speed during corner cutting.

4.1.7.7.1 Trim cutting features in WEDM

The advanced wire electrical discharge machining systems are provided with the arrangement for trim cutting. The trim cut also called skim cut is an operation where the wire electrode traces back the same path after the first cut is over. Obviously, this is used for machining of job profile in multi-pass cuts. Multi pass cutting of jobs is usually done in two or three cuts depending on the specific requirement. While trim cutting, very low energy pulses are applied to obtain a very less amount of material removal. The wire tension should be kept high as well as high pressure flushing is avoided in case of trim cutting to obtain better cut quality in respect of dimensional accuracy and surface integrity. However, the main objectives of using trim cuts are as follows:

(i) To achieve better dimensional accuracy;
(ii) To achieve improved surface finish;
(iii) To reduce inaccuracies produced by deformations after the first cut due to residual stresses in the job material;
(iv) To reduce the bowing effect on the cut surface produced in the first cut due to wire vibration or improper flushing;
(v) To improve the life of the job by minimizing heat affected layer formed on the machined surface in the first cut or previous cut.

For multi pass cutting of a job, it is necessary to hold the job in place for each cutting pass. Usually, it is done by leaving a portion of path, about 2 to 4 mm, at the end of profile during cutting. This uncut length of the profile holds the job in place and subsequent trim cut can be completed. After carrying out the trim cutting, the uncut path length is cut in a single cut with the proper wire offset. The amount of material to be cut in second and third pass depends on the job height, size and shape of the job and cutting speed etc. as selected for the first cut.

4.1.7.8 Coated wire electrode

Suitability of wire materials for WEDM depends on the electrical as well as mechanical properties of the materials. After the introduction of WEDM in 1969, during the first 10 years standard materials like copper and brass were used as wire electrode materials. Wire should be kept straight under applied tension during machining which is essential for achieving high machining accuracy and it is easy to keep a copper straight due to its higher ductility compare to brass. However, brass wire has more erosion resistance compared to copper wire. Thermal conductivity of the copper is slightly higher than that of brass. But, brass wire can withstand higher tensile stress than that of copper wire. The diameters of the wire-tool used commercially are in the range of 100–350 μm. To achieve sharp corners and high precision during material machining utilizing WEDM much smaller diameter is required. In such cases use of tungsten and molybdenum wire, diameter in the range of 25–40 μm are recommended. Resistance to corrosion and tensile strength are much higher in case of tungsten and molybdenum wire as compared to copper and brass wires for the same size. Although, tungsten and molybdenum wires are costlier compare to copper and brass wires. However, use of tungsten and molybdenum wires of lower diameter reduces the chances of wire rupture during high precision machining.

Since 1980, due to further development in technology of WEDM, there was exponential increment of cutting speed which necessitates development of newer type of wire electrode, e.g., coated wire electrode of different compositions. To enhance the machining accuracies, the development of new wires with higher conductivity and good sparking ability was needed. These new wires have a core of copper alloy with thick coating of zinc or zinc alloy and known as stratified wire is shown in Fig. 4.1.37.

During machining, upper layer of zinc vaporizes extracting latent heat from the core of the wire thus helps to lower down the overall temperature of the wire which reduces the chances of wire rapture during carrying more current for generation of powerful sparks to achieve higher cutting speed. However, to meet the superior machining accuracy, the wire with lower diameter have to be kept straight by tension mechanism which needs wire core with high tensile strength regardless of electrical conductivity. To meet the current trends of WEDM, composite wires of several compositions have been introduced with high tensile strength core and several types of coatings.

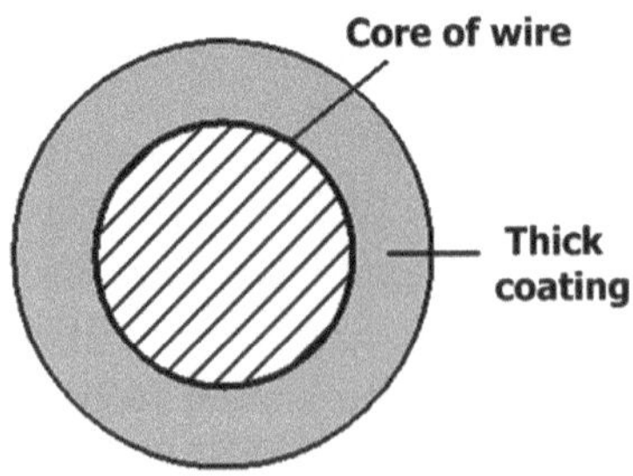

Fig. 4.1.37 Stratified wire for WEDM.

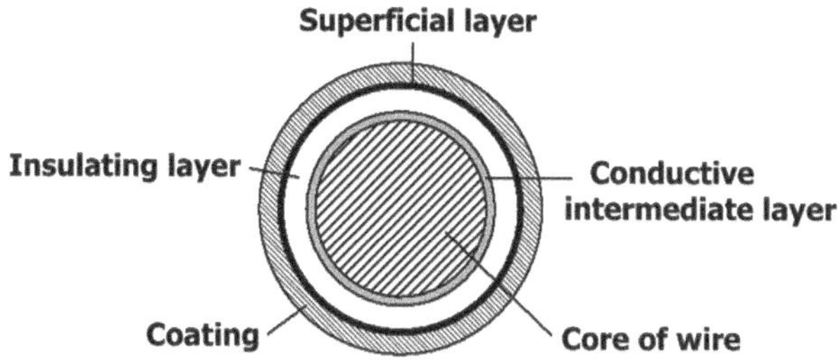

Fig. 4.1.38 Composite wire for WEDM.

Wire electrode with a core of pearlitic steel along with zinc coating can fulfill the requirements of high tensile strength for the improvement of accuracy during corner cutting. To improve electrical conductivity, an intermediate conductive layer between the core and the outer coating is recommended. Mostly copper is used for the intermediate layer with thickness in the range of 10–4 μm for wire diameter of 250–100 μm on core of pearlitic steel, which can attain higher peak current. The enhancement of cutting speed by improving the sparking ability can be possible by zinc coating in the range of 1–2 μm thickness at the outer portion of the wire. The wire performance can be further enhanced by applying a very thin superficial layer in the order of few nanometers thickness on zinc coating on the top of wire. As well as, thermal insulation of the core is also needed to prevent thermal damage. Hence, a good composite wire preferably should have five elements as shown in Fig. 4.1.38, which contains a high tensile core, an insulating layer, a conductive intermediate layer, a coating and a superficial layer [32].

4.1.7.9 Wire path strategy

The limit on machining accuracy in WEDM depends to the dimensional consistency of the wire and the positional accuracy of elements of the work-table. However, there are other factors which prevent this theoretical precision from being achieved. Most of the uncertainties arise because the working region is an unsupported section of the wire, remote from the guides. It is necessary to hold the wire in a designated position against the object because the wire generates complex oscillations due to electrical discharge between wire and the work piece. The unsupported length of wire changes with thickness of the job which may also alter the wire vibration frequency. The CNC positioning system continuously maintains the gap between the wire and the work piece. Disturbances from the external and internal sources generate vibrations in the wire, which ultimately influence the repetitive sparking process. This deviation of the wire electrode from its mean position therefore has substantial influence on the occurrence of the succeeding discharge. It also changes the breakdown voltage of the discharge and the discharge energy since the gap is changing constantly during the wire vibration.

The amplitude of the vibrations is in the order of 10 μm or more, and cannot be neglected [33]. WEDM machines are specially designed for rigidity to minimize the wire vibration. However, it is not possible to completely reduce the influences of wire vibration, which always have a tendency to dislocate the wire from its mean position.

4.1.7.9.1 Wire tool vibration

The most important performance factors effecting WEDM are discharge current, pulse duration, pulse frequency, wire speed, wire tension, type of die electric fluid and dielectric flow rate. However, wire EDM owing a large number of variables and the stochastic nature of the process, even a highly trained operator will still find it difficult to attain an optimal processing and avoid wire breakage. The wire running vertically between two guides and can vibrate in forward or backward directions as shown in Fig. 4.1.39. The vibrations of the wire can be divided into two components: vibrations parallel to the cutting direction and vibrations perpendicular to the cutting direction. However, wire vibrations perpendicular to the cutting direction is more disturbing effects on machining, due to constriction of the two sidewalls of the work piece inside the machined kerf. Vibration is relatively small and symmetrical. Vibration and deflection of wire electrode during Wire EDM machining process was studied by few researchers [34–36]. The force components that influence the wire during machining were analyzed. Researchers focused on the measurement of static deflection of the wire, i.e., wire lag. Mathematical modeling of vibration behavior of the wire needs to be studied in details. Vibration behavior plays an important role in deciding precision and accuracy of wire EDM products. Wire vibration during wire EDM process was directly observed by using a high-speed video camera [37,38]. The effects of machining conditions such as wire tension, wire speed, flow rate of flushing and servo voltage on the wire vibration amplitude and machined kerf width were investigated by analyzing the recorded images. Response surface methodology can be used to optimize machining conditions based on experimental results.

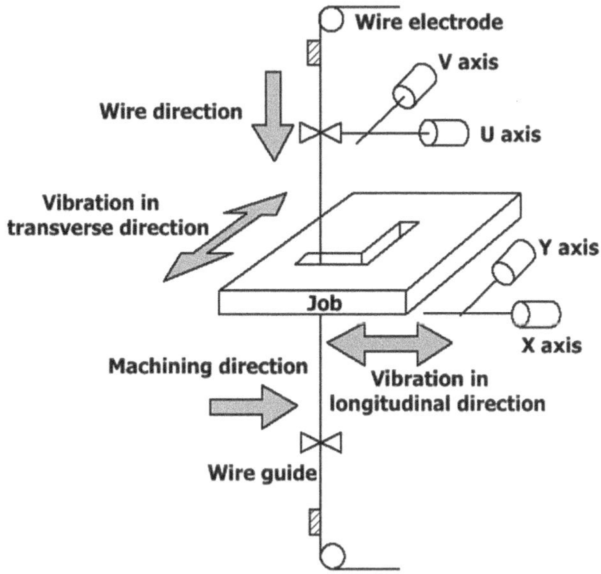

Fig. 4.1.39 Vibrations of the wire during WEDM.

Methods for minimizing wire vibration

(i) Improved wire transportation system

Wire tension and speed are significant factors on nature of wire vibration. Wire tension and velocity changes within some range due several factors e.g., inertia of wire reel, fluctuation of load and nonuniformity of unreeling wire, etc. Improved wire transportation system mainly intends to maintain the wire velocity and tension at a steady value. Some improvements can be incorporated in wire transportation system. A closed-loop wire tension control system can be developed by introducing a tension sensor. Through comparison of responsive tension and control signal, tension motor can adjust the wire tension. Dynamic absorbers can be designed to absorb fluctuation of wire tension and speed from the source [39].

(ii) External vibrational excitation on work piece

The period, amplitude and phase position of wire vibration characteristic can be calculated. An external vibrational excitation can be applied on work piece. Lateral vibration of wire can be considerably controlled if the frequency of external vibrational excitation of the work piece is same as the lateral vibrations and the differences of two phase positions are maintained as half period.

(iii) WEDM parameters optimization
The machining parameters, e.g., power settings, wire tension and speed, etc. have an immense influence on the wire vibration. Hence, selecting optimal parameter setting for minimum wire vibration during machining may provide a useful method to restrain wire vibration. Lower discharge energy per pulse considering smaller pulse-on time, larger pulse-off time and lower peak discharge voltage combination produces smaller discharge impulsive force. By increasing wire tension, stiffness of wire electrode can be increased. Also medium wire velocity may generate wire vibration frequency which is different from wire natural frequency [40].

4.1.7.10 Wire-lag phenomenon during machining

Wire bending is a main cause of cutting imprecision and it is hindering the part accuracy for various applications. This wire bending or wire deflection is called wire lag. When a corner or curved profile is being machined, the wire lag creates a geometrical inaccuracy on the product. This error can be of the order of a few hundred microns; which for some applications becomes unacceptable. For proper dimensional control, the knowledge of wire lag value is extremely essential to achieve exact dimension of job and for precision contour generation.

The mechanical behavior of the wire during machining is extremely complicated in nature. This is because the magnitudes and directions of various forces acting upon the wire are not always constant as the occurrence of sparks is highly stochastic in nature. The stochastic nature of the Wire EDM process is also attributed to a combination of factors including fluctuation in voltage and current, decomposition and distribution of dielectric, random ionic migration, interaction of two successive discharges and presence of debris particles in the machining zone. Also, several of the above-mentioned factors vary noticeably changes with time and space in the discharge zone [41].

Several repulsive and attractive forces act on the wire while cutting operation continues in WEDM and those have been described in details. When these forces are not perfectly balanced, the deformation or deflection of wire may occur. Mechanical force due to pressure generated from these gas bubbles formed by plasma of the erosion mechanism and impact on the wire due to the generation of sparks act always in the same direction, i.e., in the reverse direction of cutting. These forces contribute the majority of the total force acting perpendicular to the wire. Experimental results reveal that while rough cutting, the wire lag behind with respect to designed cutting path. This fact confirms the idea that there must be a static deflection of the wire and the wire vibrates randomly around this shifted mean position of the wire.

4.1.7.10.1 Mathematical modeling of wire lag in WEDM

The static deflection of a stretched wire supported by the two wire guides at opposite ends may be mathematically modeled with the assumptions as stated below:

(i) The axial tensile force (Γ) remains constant between the wire guides;
(ii) The wire is static;
(iii) The wire is perfectly flexible;
(iv) The wire mass is uniformly spread along its length;
(v) The viscous force resisting the deflection of wire in the dielectric medium is negligibly and.
(vi) The disturbing forces acting per unit length of the wire perpendicular to the axial force is constant over the span of the wire between guides.

Thus, the wire bending or the static deflection (y) of the stretched wire of length 'L'in a plane (along z-axis) may be expressed as [42]:

$$F\frac{\partial^2 y}{\partial x^2} = q(z) \tag{4.1.31}$$

As has been discussed in the previous section that the deviation of the wire just before sparks takes place or the mean shift of the wire during the machining, is solely dependent on the body force. Again, the majority of the body force is constituted by the formation of pressure wave front from the gas bubbles and impact. So, the body force itself is highly dependent on the energy contained in pulses. Thus, for small energy pulses, generally used in finish cutting operation, it is possible to obtain very small amplitude of vibrations. Selecting suitable pulse parameters, significant wire vibrations may be eliminated. But, the existence of static deflection of wire still persists and may be described as in the Eq. (4.1.31).

However, the Eq. (4.1.31) may be further simplified considering the above assumptions:

$$F\frac{\partial^2 y}{\partial x^2} = q_0 = \text{Constant, within the workpiece} \tag{4.1.32}$$

And, $F\frac{\partial^2 y}{\partial x^2} = 0$, in the rest of the span of the wire except over the workpiece height.

Where, F = wire tension, N; y = y(z) = Wire deflection, m.

q_0 = a constant specific external load or body force per unit length of the wire, N/m.

Fig. 4.1.40 shows the schematic representation of the static deflection of wire in WEDM process.

The solution of the equation Eq. (4.1.32) with the boundary conditions $z=0$, $y=0$ and $z=1$, $y=0$ will appear as follows:

$$y = -\frac{q_0}{2F} Z(L - Z) \tag{4.1.33}$$

Eq. (4.1.33) describes a parabola which indicates that the shape of the wire within the workpiece will be parabolic during the machining.

If, the deflection of the wire is y_1 at $Z = H$ and $Z = H + h$ (Fig. 4.1.40), then

$$y_1 = -\frac{q_0}{2F} H(L - H) = -\frac{q_0}{2F} H(H + h) \quad [L = 2H + h]$$

Also, $y_{max} = y_{z = \frac{L}{2}} = -\frac{q_0}{2F} \cdot \frac{L}{2} \left(\frac{L}{2}\right)$ [from Eq. (4.1.33)]

$$= \frac{q_0}{2F} \cdot \frac{1}{4} \cdot (2H + h)^2 = -\frac{q_0}{8F}\left(4H^2 + h^2 + 4hH\right)$$

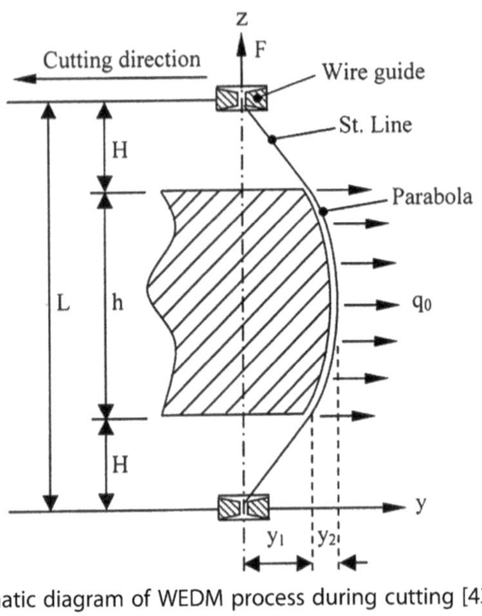

Fig. 4.1.40 Schematic diagram of WEDM process during cutting [43].

$$= -\frac{q_0}{2F}H(H+h) - \frac{q_0 h^2}{8F}$$
$$= y_1 + y_2 \tag{4.1.34}$$

where $y_1 = -\frac{q_0 h^2}{8F}$

The negative sign in Eq. (4.1.33), indicates that the wire deflection occurs in the reverse direction of cutting. The maximum wire deflection comprises of two components, namely y_1 and y_2. y_1 is the displacement of the wire outside the workpiece and y_2 is the further displacement within the workpiece.

Again, solving the Eq. (4.1.32) with the boundary conditions, $Z=0$, $y=0$ and $Z=H$, $y=y_1$, the solution appears as:

$$y = \frac{y_1}{H}.Z = -\frac{q_0 H}{2F}(H+h).\frac{1}{H}.Z$$
$$= -\frac{q_0}{2F}(H+h).Z \tag{4.1.35}$$

Eq. (4.1.35) describes a straight line; i.e., the wire outside the workpiece will be straight up to the guides during machining [43]. The geometrical error resulting out of the wire lagging phenomenon affects a corner cutting as depicted in Fig. 4.1.41.

From Fig. 4.1.41, it is shown that the geometrical accuracy of the machined contour is affected mostly by the magnitude of y_1. The

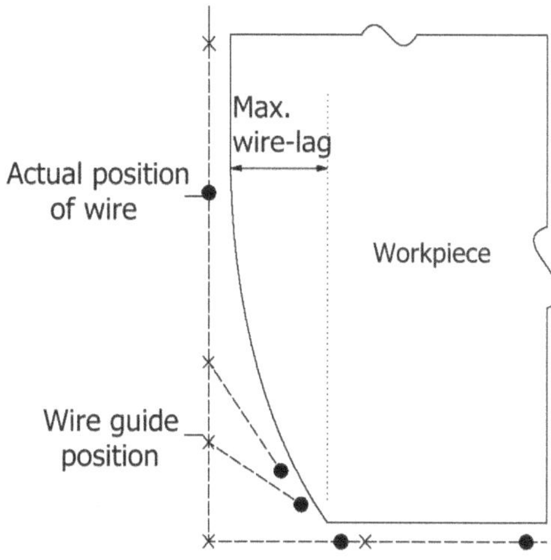

Fig. 4.1.41 Wire lag corner cutting in WEDM.

significant factors for geometrical inaccuracy due to wire lag are pulse on time, pulse off time and pulse peak current during rough cutting; and pulse peak voltage, wire tension, servo spark gap set voltage, wire tool offset and constant cutting speed during trim cutting. The experimental investigation further confirms that the difference between the offset values for the rough cut and the trim cut should be maintained as high as possible till the surface finish criterion permits. This will help in minimizing the geometrical inaccuracy in the most effective way. Thus, the minimization of the geometrical inaccuracy due to wire lag will find tremendous potentiality in industrial applications for manufacturing jobs with ultraprecision contours.

4.1.7.11 Advancement in WEDM

Research and development has been taken place during the last few years in the field of wire EDM, to achieve improvement in the area of metal removal rate, tool life, cutting speed, surface finish and accuracy etc. Some of the recent developments are identified here under:

(i) Newer wires with different multi layered coating having high strength, rigidity as well as high conductivity and better flushibility have been developed and used for machining, resulting in better surface finish and improved cutting speeds [33].

(ii) Development of advance power supply for Wire EDM is to provide the functions of anti-electrolysis, high frequency and very low energy pulse control. This can be used for restricting titanium's bluing and rusting effects and capable for reducing micro cracking in tungsten carbide caused due to electrolysis and oxidation and also for achieving fine surface finish of the order of 0.22 μm Ra [44].

(iii) Recent development of advanced control system for improving machining accuracy in WEDM. Dynamic performance of the closed loop wire tension system is introduced utilizing Proportional Integrate (P.I.) controller as well as dynamic dampers can be employed to reduce the vibration of the wire electrode [45].

(iv) Development of new wire guides to reduce wire bending defects as well as advancements in taper cutting are the focused areas of latest research in WEDM.

4.1.7.12 Applications of wire-cut EDM

Wire EDM is used for cutting aluminum, brass, copper, carbides, graphite, steels and titanium. The wire material varies with the application requirements. Example: for quicker cutting action, zinc-coated brass wires are used while for more accurate applications, molybdenum wires are used. The process is used in the following areas:

(i) Aerospace, Medical, Electronics and Semiconductor applications,
(ii) Tool and Die making industries,
(iii) For cutting the hard Extrusion Dies,
(iv) In making Fixtures, Gauges & Cams,
(v) Cutting of Gears, Strippers, Punches and Dies,
(vi) Manufacturing hard Electrodes,
(vii) Manufacturing micro-tooling for Micro-EDM, Micro-USM and such other micromachining application,
(viii) WEDM process is able to machine exotic and high strength and temperature resistive (HSTR) materials and eliminate the geometrical changes occurring in the machining of heat-treated steels, and
(ix) WEDM is also used for varying degree of taper ranging from 15 degree for a 100 mm thick to 30 degree for a 400 mm thick workpiece, etc.

4.1.8 Electro discharge grinding

Electro discharge grinding (EDG) is a thermal process based on spark erosion employed for precision machining of electrically conducting material. In EDG a rotating disk electrode is utilized in place of stationery electrode as in the case of sinking EDM. This process is known as electrical discharge grinding (EDG), because of the similarities of process with conventional grinding. However, material is removed from the job due to the electrical discharge, In this process, the spark is generated between rotating wheel and workpiece.

Flushing of debris from the narrow inter electrode gap is a major difficulty in EDM. Improper flushing in EDM results in short circuits followed by arcing between two electrodes due to accumulation of debris particles which terminates the machining operation. Compare to EDM, rotating electrode in EDG provides effective flushing of the working gap which improves the productivity and accuracy remarkably with a better finish. There is no mechanical contact between rotating electrode and the work piece; hence, it can be effectively utilized on very fragile parts for producing thin section without any damage.

4.1.8.1 Process mechanism

Basic mechanism of metal removal in EDG is melting and vaporization which is very much similar in principle with EDM. However, there are major differences in EDG setup configuration with that of EDM instead of mechanism of material removal. Fig. 4.1.42 shows basic differences between EDG and EDM.

In EDG, an electrically conductive rotating wheel is used as a tool electrode instead of stationary tool electrode as used in EDM. The wheel and workpiece are submerged into the dielectric liquid where the wheel is connected to the negative terminal and workpiece is connected to the positive terminal of a pulsed DC power supply. During EDG operation, the wheel electrode is rotating and simultaneously fed towards workpiece. There is no contact with workpiece and tool electrode, i.e., rotating wheel during material removal by repetitive electric sparks. Peripheral speed of the rotating wheel electrode transmitted to the dielectric into the very thin gap between wheel electrode and workpiece results in improvement of flushing, thus efficiency of process is enhanced. The molten material is effectively ejected from the craters created by the electrical sparks on the workpiece. As well as no debris particles accumulation take place into gap which intern

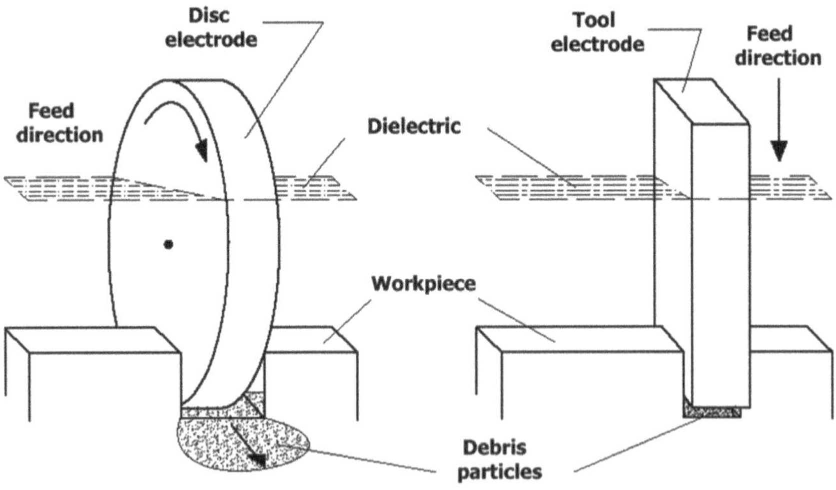

Fig. 4.1.42 Differences between EDG and EDM.

improves the flushing efficiency [46]. Due to the better flushing, higher material removal and enhanced surface finish can be obtained in EDG as compare to the conventional EDM process [47]. A moderate rate of rotating speed of the wheel electrode is maintained during EDG because the high speed of wheel is not favorable and after a certain limit of speed, the electrical discharges becomes instable and produces adverse effect on performance [48]. In EDG, wheel electrode wear is not localized as in the case of EDM. Here, tool wear is evenly spread throughout the circumference of the wheel electrode; hence, shape deterioration of the wheel electrode is not severe which may leads to better production of corners on the workpiece [49].

4.1.8.2 Setup detail

EDG setup is having similarities with EDM setup and conventional abrasive grinding setup. The detail of EDG setup is illustrated in Fig. 4.1.43. In this process, an eclectically conductive metallic wheel is used which is known as rotating tool electrode. The rotating wheel used in this process, having no abrasive particles and rotates at its horizontal axis. Various sub-units of EDG setup are power supply, dielectric delivery system, filtration system and working chamber, which are more or less similar with conventional EDM system. However, overall appearance of the EDG setup structure is

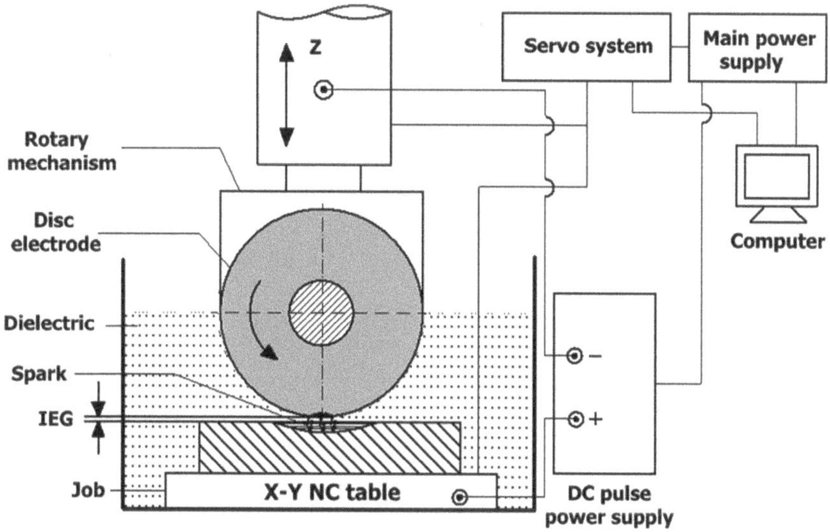

Fig. 4.1.43 EDG setup details.

similar to a conventional surface grinder considering the wheel mounting arrangements and machine tool axis motions.

The rotating wheel and job both are separated by dielectric fluid. During machining, both are submerged in dielectric fluid. Kerosene oil, Paraffin oil, Transformer oil or de-ionized water can be utilized as dielectric fluids for EDG. A pulse generator is used for maintaining the DC pulse power supply with capabilities of variable current and voltage along with variable pulse parameters.

Wheels used for EDG are commonly made of graphite. In some cases, other materials such as copper, brass and mild steel are also be used for wheels. Commonly used range of wheel diameter is between 100 and 300 mm and range of wheel width is between 0.25 and 150 mm. In EDG, shape generated on the workpiece is just the opposite shape of grinding wheel. Copper rings can also be used as rotating disc electrodes by mounted on mandrels made of medium carbon steel. For machining complex profiles on the workpiece, EDG wheels with complex surface are required. Fabrication of graphite wheel is easy and can be made by high speed steel tools. Form tool or tracer lathe can be utilized for fabrication of EDG wheels with complex surfaces.

The servo control mechanism is utilized to maintain the constant gap between workpiece and rotating wheel. EDG servo drive system senses and automatically monitors the gap between wheel and workpiece and simultaneously fed the wheel into the workpiece. But, in some setups, the rotating wheel is fixed at a position and workpiece is fed into the wheel. Control of feed rate of wheel or workpiece by the servo control mechanism is based on metal removal rate from the workpiece. During operation, if contact occurs between wheel and workpiece, the protective circuit dynamically senses it and retracts the wheel or workpiece to avoid such situations and subsequently make the machine ready for further operation.

4.1.8.3 Types of EDG

There are many developments found in EDG process, depending on the type of wheel, rotation, relative position, electrode feed directions, etc. Normally, wheels rotate on the workpiece and feed is provided to the wheel as shown in Fig. 4.1.44(a). However, modified EDG process has been developed by placing workpiece above the wheel and feed is provided to the job as shown in Fig. 4.1.44(b).

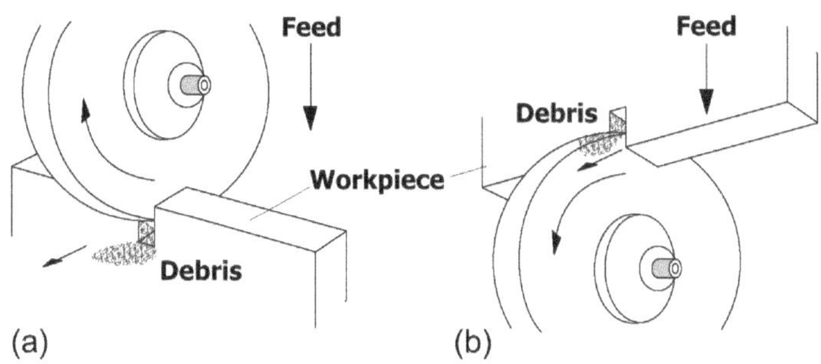

Fig. 4.1.44 EDG process (a) Conventional, (b) Modified.

In this developed modified EDG set up, gravitational force also contributes to the improvement of debris removal from the gap compared to conventional EDG setup. The modified EDG setup is capable of precision micromachining due to higher MRR with better finish [50]. Electro discharge milling is another new development of EDG process. Here, tool electrode rotates and workpiece fed horizontally below the rotating wheel as shown in Fig. 4.1.45. In ED milling, MRR and surface quality is much better and machining time can be reduced up to 60% compare to sinking EDM. Insulating ceramics such as Al_2O_3, SiC, Si_2N_3 etc. can be machined by ED milling process. Thin copper sheet as assisting electrode is placed in between insulating ceramics and rotating wheel. During development of EDG process, toothed steel wheel as rotating electrode has also been utilized during ED milling which further improves MRR and surface quality [51].

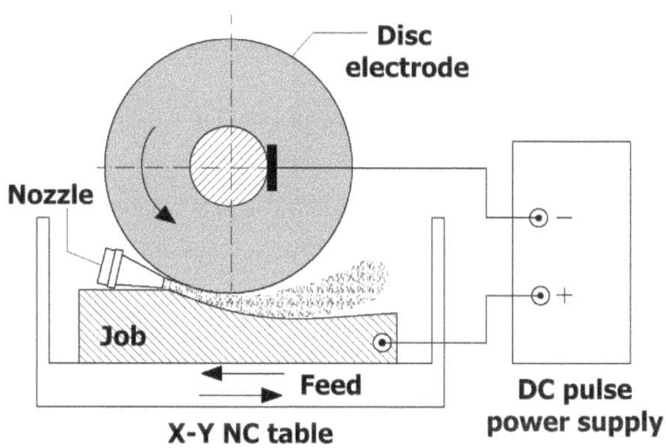

Fig. 4.1.45 Milling by EDG.

4.1.8.4 Influence of process parameters

Process parameters involve in EDG are voltage, current, frequency, rotating speed of wheel electrode, types of dielectric fluid and flushing direction, etc. The power parameters also consist of polarity, pulse duration and duty ratio, etc. The range of voltage, current and frequency normally utilized in EDG are 30–100 V, 1–150 A and 5–200 kHz respectively. Wheel electrode speed in EDG is in the range of 50–650 RPM, considerably lesser compare to conventional grinding. Higher speed of wheel electrode is not at all suitable for EDG. Dielectric cannot enter into the discharge gap due to the centrifugal effect of higher rotating speed of wheel electrode results in discharge instabilities which adversely affect the performance capability of EDG process.

For higher metal removal during EDG, main influencing parameters are positive polarity, high peak current and longer pulse duration. For achieving good surface finish during EDG, negative polarity, lower peak current and shorter pulse duration needs to be considered [52]. Higher current leads to faster metal removal, however surface roughness and the depth of heat affected zone increase. Higher current also creates problem when grinding Tungsten Carbide (WC) material which contains cobalt. Surface finish can be also improved during EDG by utilizing higher frequency.

Wheel wear rate is proportional to current. Positive electrode polarity causes higher wheel wear rate [52]. Longer pulse duration leads to lower wheel wear rate. Wheel electrode may gain weight due to carbon deposition on the wheel surface which also protects the wheel from wear. Vaporized debris from the workpiece during EDG may also splash onto wheel and form alloy clot on the wheel surface which also increases the weight of the wheel.

4.1.9 Advantages and limitations of various EDM processes

Electro Discharge Machining processes are most popular and acceptable machining processes due to its various advantages as listed below. However, these processes have also some limitations as mentioned below which should be overcome by controlling the processes more effectively.

4.1.9.1 Advantages

(i) Material can be machined without generating any cutting forces,
(ii) High accuracy with no burrs,
(iii) Complex features can be machined in a single run,

(iv) No shaped electrode for WEDM leads to smaller production lead time,
(v) Unmanned machining for WEDM,
(vi) Effective flushing in case of EDG, and
(vii) Fragile parts can be grind by EDG, etc.

4.1.9.2 Limitations

(i) High machining time due to low metal removal rate,
(ii) Electrode wear,
(iii) Thermal effects on the machined surface, e.g., recast layer, micro cracks and heat affected zone (HAZ) etc.,
(iv) Tool fabrication requires for complex shapes generation leads to longer lead time,
(v) Possibilities of health hazards,
(vi) High capital cost for WEDM,
(vii) Large and thicker job cannot be machined by WEDM, and
(viii) Electrolysis occurs for some material, etc.

4.1.10 Environmental impacts of EDM

Electro Discharge machining (EDM) is one of the most popular and important machining processes and become indispensable in manufacturing application such as die and mold making, prototyping, micromachining, etc. EDM is based on thermal effect. EDM has various advantages over other machining processes; however it has several hazards possibilities. Melting and vaporization of material from the workpiece are caused due to creation of highly plasma between tool and workpiece electrodes separated by non-conductive dielectric. The high temperature in inter electrode gap results in generation of solid, liquid and gaseous by-products, which makes the process rather hazardous and having appalling environmental impact.

Various hazard potentials of EDM process are identified in Fig. 4.1.46. Generation of gaseous by-products due to high temperature mainly consists of hazardous smoke, vapor and aerosols. Hydrocarbon dielectrics have harmful effect on the human skin. These types of dielectrics also having significant fire hazard and explosions may occur under hostile circumstances. Decomposition of products and heavy metals occur during material removal at high temperature, which accumulates in the dielectric and erosion slurry. Sharp edge metallic particles generated during machining may also damage

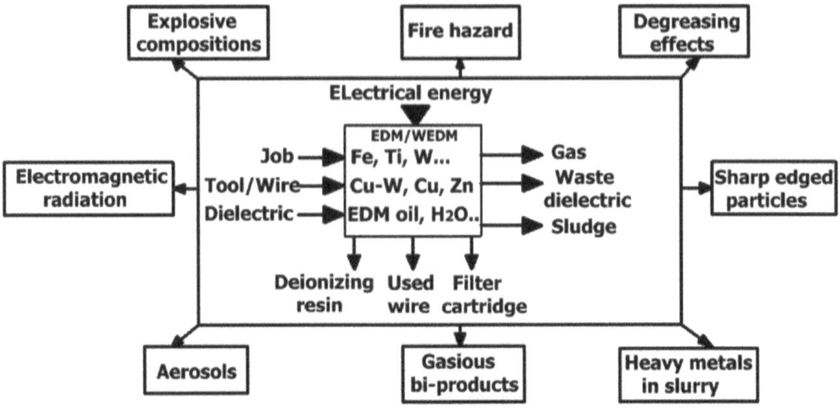

Fig. 4.1.46 Various hazard potentials of EDM process.

human skin. Due to high temperature plasma, generation of electromagnetic radiation is inevitable, which may create harmful impacts.

The amount of gaseous products, aerosols and heavy metals depends on the type of EDM process; dielectrics and work materials used as well as depend on rough and finish operation. The composition of dielectric influences the generation of fume and vapor. Its viscosity also plays an important role in the machining zone under extremely high temperature. EDM die sinking uses mineral oil and organic fluid, which may produce hazardous fumes such as polycyclic aromatic hydrocarbons, benzene, vapor of mineral oil, mineral aerosols and other by-products. However, water based dielectrics used for wire-EDM and micro-EDM produce hazardous fumes such as carbon monoxide, nitrogen oxide, ozone, aerosols and harmful by-products. Fumes may also consist of eroded metal and oxide particles of very small diameters from the tool and workpiece like tungsten carbide, titanium carbide, nickel, molybdenum, and chromium, etc. De-ionized water as dielectric is most environmental acceptable compared with other dielectrics. Erosion slurry is another important area of interest regarding environmental impact of EDM. Erosion slurry contains eroded work and tool materials as well as solid decomposition products of dielectric [53].

4.1.10.1 Protective measures

At present, there are extensive research activities going on to monitor, control and minimize the level of pollution generated due to EDM process. The type and composition of dielectric at very high temperature influence the generation of fumes in EDM. Dielectrics with lower viscosity generate lower fume and vapor. A flashing point of dielectric is also important and should be above 100 °C. The height of dielectric level in machining chamber plays an important role in this context. The upper level should be higher than 80 mm from the sparking zone, which allows condensation and absorption of maximum amount of vapor and fume in the dielectric. Fume extractor must be installed for reducing level of pollution with the help of multi-stage filtration systems, which may consist of mesh filter, electrostatic filter and chemical filter. Erosion slurry mixed with dielectric must be filtered before recirculation during EDM operation. Each EDM setup must have a dielectric cleaning system for removal of erosion slurry. Protective measures regarding location of equipments, setup design, operation, maintenance, selection of dielectrics are some of the important issues to be considered in order to reduce the health risk as well as undesirable environmental impact of EDM [54]. Different hazardous substances as mentioned above generate during EDM operation. Hence, careful treatment and appropriate disposal of wastes are challenging tasks for manufacturing engineers.

4.1.11 Applications of EDM

EDM is used for machining any materials irrespective of its mechanical properties. It can machine efficiently harden materials without generating shape distortion. EDM process in terms of the type of applications can be significantly extended with continuous advancement in the machining efficiency and the integration of advanced pulse generator with adaptive control. Some of the application areas are listed below:

(i) EDM process is able to machine exotic and high strength and temperature resistive (HSTR) materials,
(ii) It is extensively used in tool and die making industries,
(iii) It is useful for fabrication of various machine elements, e.g., fixtures, gauges, cams, gears, strippers and punches etc.,
(iv) It is capable of fabrication of various tools as well as micro tools for Micro-EDM, micro-ECM, Micro-USM and such other micro-machining applications,
(v) EDM can drill holes of various shapes on contour surfaces,
(vi) WEDM can be used for varying degree of taper ranging from 15 degree for a 100 mm thick to 30 degree for a 400 mm thick workpiece,

(vii) Most attractive features of WEDM are the ability to operate without any human interaction for several hours or even days,
(viii) WEDM is capable to produce extrusion dies and dies for powder metallurgy and enhance the die life as well as reduce the machining time compare to conventional machining,
(ix) EDG is useful for grinding of very fragile parts and fabrication of micro structures,
(x) EDG is popular for form grinding,
(xi) EDG fabricates tooling for semiconductors industries, and
(xii) Different types of EDM can be effectively utilized for automotive, aerospace, medical, electronics and semiconductor applications etc.

4.1.12 Advancement in EDM

Continuous research and developments have been carried out during last few decades in the field of EDM. Compared to other advanced machining processes, physics involved in EDM is so complicated that its metal removal mechanism is yet to be fully understood. Research areas involving EDM mainly focused on the increase in metal removal rate, design of tool and its life, surface finish and productivity of the process etc. Enormous efforts are being made to bring further developments in the EDM technology for its utmost utilization in the future. Some promising advancements in the field of EDM are briefly discussed below.

4.1.12.1 Advance power supply and control system

Advanced power supply and control system is urgently needed to make EDM technology more acceptable and competitive for industrial applications. Researchers are attempting to develop advanced pulse generator integrated with control system. Adaptive control systems based on fuzzy logic provides reliable pulse control mechanisms during EDM process [55]. Genetic algorithm is also utilize to synthesize the membership functions of fuzzy sets automatically [56]. Attempts have also been made to utilize radio frequency (RF) signal generated during sparking for monitoring and controlling of the EDM process [57]. Attempts have also been made to integrate EDM within the computer integrated manufacturing (CIM) environment [DeVries]. However, standardization of EDM process parameters and related controlling factors are still not exist which obstruct the integration of EDM within CIM environment. It has also been considered to integrate EDM within the next generation agile manufacturing environment. In this direction, data created during monitoring of different EDM systems has been integrated with communication function through internet which can be recorded, displayed for off line analysis [58,59].

4.1.12.2 Tool design and fabrication

Tool design and development in EDM has also been progressed utilizing different advanced computer aided systems. Generation of complex 3D shaped cavities consisting of sharp corners is difficult in EDM considering complexity of measurement of curvatures along the edges of the surface. Tool as well as holder design can be possible utilizing CAD system from the complex shaped 3D geometry of the workpiece [60]. Attempts have been made to utilize rapid prototyping (RP) techniques for the fabrication of tool electrode. Extensive research in the area of tool fabrication involving RP is required to make it successful considering the influence of porosity factor on different EDM performance criteria. Advancement in CAD/CAM systems has also provided a systematic integration towards the design and fabrication of tools by choosing essential machining parameters prior to the machining operation [61].

4.1.12.3 Dry EDM

To minimize hazardous effects of hydrocarbon dielectric in EDM, researchers are attempting to make EDM more environment friendly. In this direction, development of dry EDM is a noble step. In this EDM method, air or air mixed with water mist is supplied to the discharge zone as dielectric for machining purpose, demonstrates almost same machining capabilities as conventional EDM. Dry EDM process utilizes thin walled pipe electrode and argon or helium gas as dielectric medium. High velocity gas effectively flushes out molten workpiece material from the working gap. The discharge plasma formed during pulse-on time is blown away during pulse-off time by the high velocity gas resulting in decrease in the temperatures of the discharge zones [62]. Various advantages of dry EDM include enhanced material removal rate during machining of steel utilizing oxygen as gas, negligible tool wear, thinner heat affected zone, narrower machining gap, etc. compared to conventional EDM.

4.1.12.4 Alloying and coating with EDM

Another innovative development in the EDM technology is surface treatment, which is an additive process compared to removal process in conventional EDM. At extreme temperature, hydrocarbon dissociates and forms carbon. This carbon reacts with the molten metal and resolidifies as metal carbides forming layer on the surface of the workpiece, when the polarity of the workpiece is positive. This inturn results in increased resistance to abrasion and corrosion of the machined product [63]. The workpiece EDMed utilizing Si, SiC or Al powder mixed dielectric improves surface roughness as well as increased corrosion resistance compared to the surface machined without powder [64]. Tungsten (W) powder mixed dielectric fluid during EDM forms an alloyed layer on the positive workpiece and enhances surface roughness, hardness, corrosion resistance. Moreover, tribological properties of the machined workpiece are also improved [65]. Metal carbide coating on steel rolls, molds, dies, cutting tools, turbines and aeroengine components can be successfully accomplished through EDM. Negative polarity of the tool and long discharge duration is utilized in the said process. In this process, more material removed from the tool reacts with carbon dissociated from the hydrocarbon dielectric, gets deposited over the surface of anode workpiece forming carbide coating [66,67].

4.1.12.5 Machining of nonconducting materials

Machining of electrically nonconducting materials can also be possible by EDM. In this regard, a thin metal sheet is placed on the nonconductive material connected with positive polarity. During machining, sparks generate between the metal sheet and tool electrode. Due to extreme heat generation, hydrocarbon oil is thermally decomposed and generates carbon which deposited on the work piece. The surface of the nonconductive workpiece is covered with the thin carbon layer even after the metal sheet eroded out. Sparks continue to occur between electrically conductive carbon layer over the workpiece. Hence, machining of nonconductive materials can be possible so far carbon layer exists. By such process machining of complex shapes on various nonconductive material, e.g., Si_3N_4, ZrO_2, SiC, Al_2O_3, glass, and diamond can be possible [68].

Due to the stochastic nature of the sparking phenomenon involving a large number of process parameters, EDM process is obviously more complicated. To understand the EDM phenomena clearly, it needs in depth further investigations, without this the potentials of EDM technology may not be fully realized. It also necessitates optimizations of the process parameters which contributes for the overall machining efficiency for achieving improvement in the machining performance, e.g., MRR, TWR and SR, etc. Some new advancements in EDM, e.g., hybrid EDM, powder mixed EDM and micro-EDM may lead to further developments of the EDM technology in the near future. In this regard, details of hybrid EDM as well as micro-EDM have been discussed in the consequent chapters. Thus, the EDM process needs to be relentlessly rejuvenated to remain competitive for achieving important role in the field of manufacturing of difficult-to-machine materials by fulfilling stringent design requirements.

References

[1] B.R. Lazarenko, To invert the effect of wear on electric power contacts, Dissertation of the All-Union Institute for Electro Technique in Moscow/CCCP, 1943 (in Russian).

[2] S. Webzell, That first step into EDM, in: Machinery, 159, (4040), Findlay Publications Ltd, Kent, UK, 2001, p. 41 November.

[3] R. Snoeys, F. Van Dijck, Plasma channel diameter growth affects stock removal in EDM, Ann. CIRP 21 (1972) 39–40.

[4] P.K. Eckman, E.M. Williams, Plasma dynamics in arc formed by low-voltage parkover of a liquid dielectric, Appl. Sci. Res. Section B 8 (1960) 299–320.

[5] B.M. Schumacher, About the role of debris in the gap during electrical discharge machining, Ann. CIRP 39 (1) (1990) 197–199.

[6] S. Hayakawa, M. Yuzawa, M. Kunieda, N. Nishiwaki, Time Variation and mechanism of Determining Power Distribution in Electrodes during EDM Process, IJEM 6 (2001) 19–26.

[7] K.H. Ho, S.T. Newman, State of the art electrical discharge machining (EDM), Int. J. Mach. Tool Manuf. 43 (2003) 1287–1300.

[8] R. Snoeys, D. Dauw, J.P. Kruth, Improved adaptive control system for EDM processes, Ann. ClRP 29 (1) (1980) 97–101.

[9] K.P. Rajurkar, W.M. Wang, A new model reference adaptive control of EDM, Ann. CIRP 38 (1) (1989) 183–186.

[10] K.P. Rajurkar, W.M. Wang, Real-time stochastic model and control of EDM, Ann. ClRP 39 (1) (1990) 187–190.

[11] W.M. Wang, K.P. Rajurkar, Modeling and adaptive control of EDM systems, J. Manuf. Syst. I (5) (1995) 334–345.

[12] W.M. Wang, Investigations of Monitoring Discharge Parameters and Adaptive Control for EDM, Ph. D Dissertation, Harbin Institute of Technology, Harbin, China, 1985.

[13] W.S. Zhao, T. Masuzawa, Adaptive control of EDM-jump with self-tuning approach, Bull. Jpn. Soc. Precis. Eng. 24 (1) (1990) 45–50.

[14] W.M. Wang, K.P. Rajurkar, K. Akamatsu, Digital gap monitor and adaptive integral control for auto-jumping in EDM, ASME J. Eng. Ind. 117 (1993) 253–258.

[15] Y. Zhang, Y. Liu, Y. Shen, R. Ji, Z. Li, C. Zheng, Investigation on the influence of the dielectrics on the material removal characteristics of EDM, J. Mater. Process. Technol. 214 (2014) 1052–1061.

[16] S. Kalpakjian, Manufacturing Processes for Engineering Materials, third ed., Addison Wesley, Reading, MA, 1997.

[17] H.C. Tsai, B.H. Yan, F.Y. Huang, EDM performance of Cr/Cu-based composite electrodes, Int J Mach Tool Manu 43 (2003) 245–252.

[18] T. Shibayama, M. Kunieda, Diffusion bonded EDM electrode with micro holes for jetting dielectric liquid, Ann. CIRP 55 (1) (2006).

[19] W.S. Zhao, L. Gu, L. Lei, Bunched- electrode for electrical discharge machining, in: Proceedings of the ISEM XV, Pittsburgh, USA, 2007, pp. 41–44.

[20] L. Gu, L. Li, W. Zhao, K.P. Rajurkar, Electrical discharge machining of Ti6Al4V with a bundled electrode, Int J Mach Tool Manu 53 (2012) 100–106.

[21] A. Singh, A. Ghosh, Thermo-electric model of material removal during electric discharge machining, Int. J. Mach. Tools Manuf. 39 (4) (1999) 66982.

[22] R. Snoeys, F.S. Van Dijck, Investigation of electro discharge machiningoperations by means of thermo-mathematical model, CIRP Ann. 20 (1) (1971) 357.
[23] M.R. Patel, M.A. Barrufet, P.T. Eubank, D.D. DiBitonto, Theoretical models of the electrical discharge machining process. II. The anode erosion model, J. Appl. Phys. 66 (9) (1989) 410411.
[24] H.K. Kansal, S. Singh, P. Kumar, Numerical simulation of powder mixed electric discharge machining (PMEDM) using finite element method, Math. Comput. Model. 47 (1112) (2008) 121737.
[25] T. Ikai, K. Hashigushi, Heat input for crater formation in EDM, in: Proceedings of International Symposium for Electro-machining ISEM XI, EPFL, 1995, p. 16370.
[26] S.H. Yeo, W. Kurnia, P.C. Tan, Critical assessment and numerical comparison of electro-thermal models in EDM, J. Mater. Process. Technol. 203 (2007) 24151.
[27] S. Das, M. Klotz, F. Klocke, EDM simulation: Finite element-based calculation of deformation, microstructure and residual stresses, J. Mater. Process. Technol. 142 (2003) 43451.
[28] S.H. Lee, X.P. Li, Study of the effect of machining parameters on the machining characteristics in electrical discharge machining of tungsten carbide, J. Mater. Process. Technol. 115 (2001) 344–358.
[29] M. Kunieda, B. Lauwers, K.P. Rajurkar, B.M. Schumacher, Advancing EDM through fundamental insight into the process, CIRP Ann. 54 (2) (2005) 64–87.
[30] S. Hinduja, M. Kunieda, Modelling of ECM and EDM processes, CIRP Ann. Manuf. Technol. 62 (2013) 775–797.
[31] J.P. Kruth, B. Lauwers, B. Schacht, J. Humbeek, Composite wires with high tensile core for WEDM, CIRP Ann. Manuf. Technol. 53 (1) (2004) 171–174.
[32] B. Patric, L. Michel, S. Gerald, Composite Wire for Electrical Discharge Machining, U.S Patent No. 20080061038, (2008).
[33] A. Herrero, S. Azcarate, A. Rees, A. Gehringer, A. Schoth, J.A. Sanchez, Influence of force components on thin wire EDM, in: 4th International Conference on Multi-Material Micro Manufacture, 4 M 2008, Cardiff, UK, 2008.
[34] S. Tomura, M. Kunieda, Analysis of electromagnetic force in wire-EDM, Precis. Eng. 33 (3) (2009) 255–262.
[35] H. Obara, M. Abe, T. Ohsumi, Control of wire breakage during wire EDM, 1st report, IJEM 4 (1999) 53–58.
[36] S. Habib, A. Okada, Study on the movement of wire electrode during fine wire electrical discharge machining process, J. Mater. Process. Technol. 227 (2016) 147–152.
[37] A. Okada, Y. Uno, M. Nakazawa, T. Yamauchi, Evaluations of spark distribution and wire vibration in wire EDM by high-speed observation, CIRP Ann. 59 (1) (2010) 231–234.
[38] Z. Chen, Y. Huang, H. Huang, Z. Zhang, G. Zhang, Three-dimensional characteristics analysis of the wire-tool vibration considering spatial temperature field and electromagnetic field in WEDM, Int J Mach Tool Manu 92 (2015) 85–96.
[39] Y.S. Liao, Y.Y. Chu, M.T. Yan, Study of wire breaking process and monitoring of WEDM, Int J Mach Tool Manu 37 (1997) 555–567.
[40] Y.S. Liao, J.T. Huang, H.C. Su, A study on the machining para-meters optimisation of the wire electrical discharge machining, J. Mater. Process. Technol. 71 (1997) 487–493.
[41] I. Betrami, A. Bertholds, D. Dauw, A simplified post process for wire cut EDM, J. Mater. Process. Technol. 58 (1996) 385–389.
[42] A.B. Puri, B. Bhattacharyya, An analysis and optimization of the geometrical inaccuracy due to wire lag phenomenon in WEDM, Int J Mach Tool Manu 43 (2003) (2003) 151–159.
[43] M.-T. Yan, Y.-P. Lai, Surface quality improvement of wire EDM using a fine finish power supply, Int. J. Mach. Tools Manuf. 47 (11) (2007) 1686–1694.
[44] N. Kinoshita, M. Fukui, T. Fujii, Study on wire-EDM: accuracy in taper cut, CIRP Ann. Manuf. Technol. 36 (1) (1987) 119–122.

[45] K.P. Rajurkar, S.M. Pandit, Formation and ejection of EDM debris, Trans. ASME J. Eng. Ind. 108 (1986) 22–26.
[46] K.M. Shu, G.C. Tu, Study of electrical discharge grinding using metal matrix composite electrodes, Int. J. Mach. Tools Manuf. 43 (2003) 845–854.
[47] H.E. Hofy, Advanced Machining Processes: Nontraditional and Hybrid Machining Processes, McGraw-Hill Companies, New Delhi, 2005.
[48] P. Koshy, V.K. Jain, G.K. Lal, Experimental investigations into electrical discharge machining with a rotating disk electrode, Precis. Eng. 15 (1993) 6–15.
[49] H.M. Chow, B.H. Yan, F.Y. Huang, Micro slit machining using electro-discharge machining with a modified rotary disk electrode (RDE), J. Mater. Process. Technol. 91 (1999) 161–166.
[50] Y. Liu, R. Ji, Q. Li, L. Yu, X. Li, Electric discharge milling of silicon carbide ceramic with high electrical resistivity, Int. J. Mach. Tools Manuf. 48 (2008) 1504–1508.
[51] H.R. Shih, K.M. Shu, A study of electrical discharge grinding using a rotary disk electrode, Int. J. Adv. Manuf. Technol. 38 (2008) 59–67.
[52] C. Frei, et al., A new approach for contamination measurements for EDM dielectrics, Ann. ClRP 3611 (1987) lll–113.
[53] N.N, Electro-discharge machining, Industrial and Process Fire Safety, FS 6035.
[54] Y.S. Tarng, C.M. Tseng, L.K. Chung, A fuzzy pulse discrimi-nating system for electrical discharge machining, Int. J. Mach. Tools Manuf. 37 (4) (1997) 511–522.
[55] Y.S. Tarng, J.L. Jang, Genetic synthesis of a fuzzy pulse discri-minator in electrical discharge machining, J. Intell. Manuf. 7 (1996) 311–318.
[56] S.K. Bhattacharyya, M.F. El-Menshawy, Monitoring the EDM process by radio signals, Int. J. Prod. Res. 16 (5) (1978) 353–363.
[57] A. Kapoor, K.P. Rajurkar, R.E. Williams, R. Bishu, Internet-based real time remote monitoring system for machining processes, Trans. NAMRI/SME XXVII (1999).
[58] K. Kobayashi, T. Nagao, G. Li, K. Kakiuchi, T. Noda, K. Nakamura, T. Inoue, The study on the methods for applying Internet-use technology to EDM fields, in: Proceedings of ISEM XIII, 2001, pp. 523–533.
[59] X.M. Ding, J.Y.H. Fuh, K.S. Lee, Y.F. Zhang, A.Y.C. Nee, A computer-aided EDM electrode design system for mold manu-facturing, Int. J. Prod. Res. 38 (13) (2000) 3079–3092.
[60] J.P. Kruth, J. Peters, Steps towards an integrated CAD/CAM system for mould design and manufacture: anisotropic shrink-age, component library and link to NC machining and EDM, Ann. CIRP 35 (1) (1986) 83–88.
[61] NASA, Inert-Gas Electrical-Discharge Machining, (1985)NTIS Tech Note.
[62] J.P. Kruth, H.K. Toenshoff, F. Klocke, Surface and sub-surface quality in material removal process for tool making, Proc. ISEM 12 (1998) 33–64.
[63] H. Narumiya, N. Mohri, N. Saito, H. Ootake, Y. Tsunekawa, T. Takawashi, K. Kobayashi, EDM by powder suspended working fluid, Proc. ISEM 9 (1989) 5–8.
[64] K. Masui, K. Demizu, Y. Sato, T. Sone, Surface modification of tool steels by alloying method using EDM process, Proc. ISEM 11 (1995) 419–426.
[65] J.P. Kruth, L. Froyen, L. Stevens, P. Dejonghe, In-process alloying of the white layer of a workpiece machined by die-sinking EDM, IJEM 3 (1998) 33–38.
[66] Y. Fukuzawa, H. Katougi, N. Mohri, K. Furutani, T. Tani, Machining properties of insulating ceramics with an electric discharge machine, Proc. ISEM 12 (1998) 445–453.

Further reading

[67] S. Hayakawa, M. Yuzawa, M. Kunieda, N. Nishiwaki, Time variation and mechanism of determining power distribution in electrodes during EDM process, IJEM 6 (2001) 19–26.
[68] D.K. Aspinwall, R.C. Dewes, H.G. Lee, J. Simao, Electrical discharge surface alloying of Ti and Fe workpiece materials using refractory powder compact electrodes and cu wire, Ann. CIRP 52 (1) (2003) 151–154.

SUBCHAPTER 4.2

Plasma Arc Machining (PAM)

4.2.1 Introduction

Plasma is an ionized gas which is of ions, electrons, and neutral particles. Plasma is also called as fourth state of matter. When the electric arc is passed through the certain gas such as air, hydrogen, nitrogen and argon etc., the temperature of a gas is raised to the higher value and the gases become ionized. The gas in this stage is termed as plasma. The plasma gas strikes on the workpiece materials with high heat and kinetic energy. The temperature of the workpiece surface increases above the melting point and the material is melted and vaporized. Machining by plasma was adopted in the early 1950s as an alternative method for oxy-fuel flame cutting. In oxy-fuel cutting the materials are oxidized and the heat of exothermic reaction of oxidation is utilized. But for plasma arc machining the energy of plasma arc is utilized to melt the workpiece and the material is removed by the jet of plasma gas during machining. Figure 4.2.1 shows

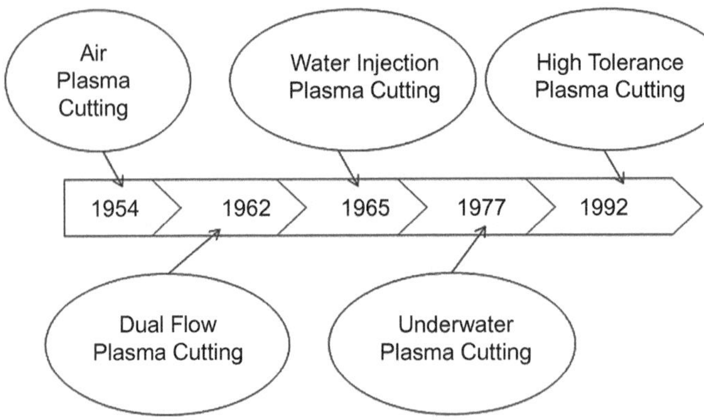

Fig. 4.2.1 Historical development of plasma arc cutting method [1].

various development of Plasma arc cutting with the passage of time [1]. Air plasma cutting technology was developed in the year 1954. After that Dual gas flow plasma arc cutting method was developed in 1962 in which plasma forming gas and shielding gas flow arrangements are used. Water injection plasma cutting method was developed in 1965. After that under water plasma cutting technique was in use in 1977. High Tolerance Plasma cutting method also developed in 1992. Plasma Arc Cutting (PAC) process can cut at a rate faster that abrasive water jet cutting and PAC can cut thicker materials compared to laser beam cutting [2]. Plasma arc machining can be applied for both electrically conducting as well as non-conducting materials such as stainless steel, aluminum, mild steel, ceramics and composites etc.

4.2.1.1 Mechanism of material removal and machining rate in PAM

In plasma arc machining a continuous arc is generated between a cathode and an anode. A gas is introduced around the cathode and flows through the anode. The temperature, in the narrow orifice around the cathode, reaches 28,000 °C, which is enough to produce a high-temperature plasma arc. Various physical phenomena such as heat conduction, convection, radiation, mechanical deformation and phase transition etc. occurred in PAM process. The energy of plasma arc is absorbed by the workpiece materials and a very high temperature above the melting point of workpiece material is raised. Under these conditions, the metal being machined is very rapidly melted and vaporized. The high velocity stream of ionized gases flushes away the machining debris from the surface of the workpiece. For organic materials machined by PAM, volatile products are formed and removed by the action of high velocity plasma jet and also vaporized due to heating by plasma arc jet.

For plasma arc machining of workpiece materials of density, p (kg/ m^3), kerf width, w (m) and thickness, t (m) with cutting speed v (m/s), the required rate of energy transfer from plasma jet to workpiece, E (J/s) is expressed [3] as:

$$E = w.t.p.v.L \tag{4.2.1}$$

where L is the energy for converting unit mass of workpiece materials to effluent (J/kg).

The energy, E can also be expressed as the difference between input energy of plasma arc and loss of energy of outgoing plasma and removed materials.

$$E = P.K.\eta - H.V \tag{4.2.2}$$

where P is input electrical power of plasma torch (W); K is the constant to account the spreading energy losses of plasma jet (<1); η is the efficient of plasma torch (<1); H is the heat energy content of plasma and effluent per unit volume (J/m^3) and V is the volume flow rate of plasma gas (m^3/s).

The machining speed in terms of cutting speed can be expressed as:

$$v = (P.K.\eta - H.V)/w.t.p.L \tag{4.2.3}$$

The machining speed decreases with increase in width of cut as well as thickness of cut of workpiece material. The machining speed increases with the input electrical power of plasma arc. The material removal rate is directly proportional to the cutting speed for constant cross sectional area of the workpiece materials. The material removal rate (MRR, kg/s) in PAM can be expressed as:

$$MRR = (P.K.\eta - H.V)/L \tag{4.2.4}$$

The material removal rate varies with input power of plasma arc and also plasma gas flow rate during plasma arc machining process.

4.2.1.2 Types of plasma arc machining system

There are two types of Plasma Arc Machining (PAM) Systems such as Transferred Arc PAM system and Non-Transferred Arc PAM system. In Transferred Arc PAM system, Plasma arc is transferred between electrode and workpiece. The positive terminal of D.C. Power source is connected with workpiece materials and the negative terminal of power source is connected with electrode. The workpiece should be electrically conductive for transferred Arc PAM system. High power can be achieved with high electrothermal efficiency up to 90% for transferred Arc PAM system. The transferred arc PAM system is shown in Fig. 4.2.2.

In non-transferred arc PAM system, the arc is not transferred from electrode to the workpiece. The arc is maintained between the electrode and nozzle. The negative terminal of D.C. power source is connected with electrode and positive terminal is conned with nozzle. The nozzle materials should be

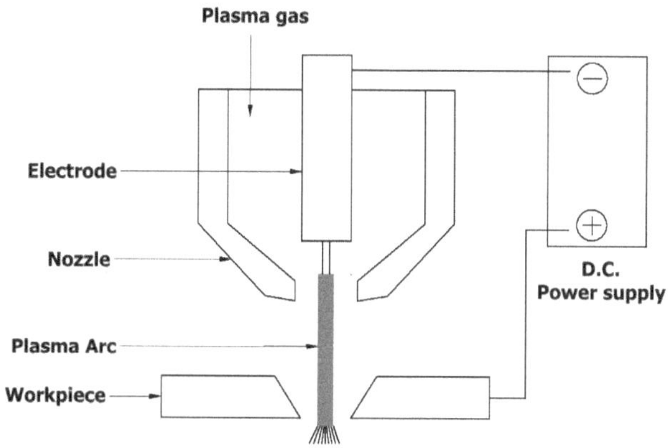

Fig. 4.2.2 Transferred arc PAM system.

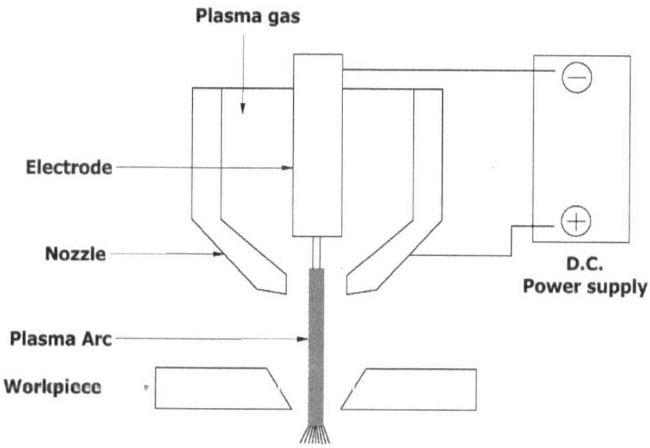

Fig. 4.2.3 Non-transferred arc PAM system.

electrically conductive and this non-transferred arc PAM can be applied for both electrically conducting as well as non-conducting materials. The arc ionizes gas and plasma arc is formed between electrode and nozzle. Plasma arc strikes on workpiece with high velocity. The electro-thermal efficiency for non-transferred arc PAM is less up to 75% compared to transfer arc PAM system. The non-transferred arc PAM system is shown in Fig. 4.2.3.

4.2.2 Details of plasma arc machining system setup

Plasma Arc machining system setup consists of different subsystems such as power supply unit, plasma gas and shielding gas supply unit, plasma

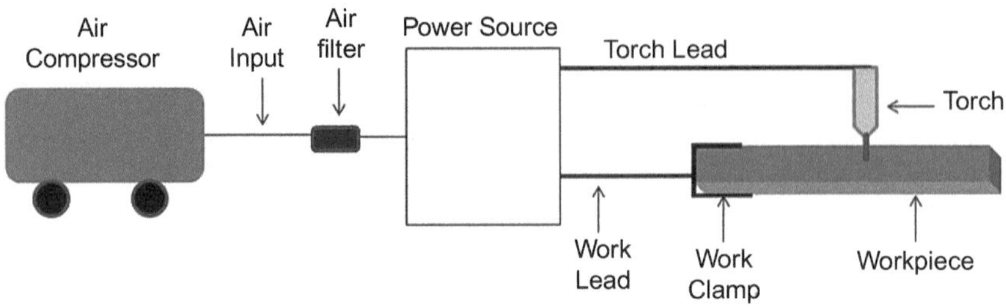

Fig. 4.2.4 Schematic diagram of plasma arc machining system.

Fig. 4.2.5 The photographic view of 80 A plasma arc cutting system. *(Courtesy: LOTOS Model LTP 8000.)*

torch, cooling unit and control unit. The schematic diagram of plasma arc machining system is shown in Fig. 4.2.4. The photographic view of Plasma arc cutting machine is shown in Fig. 4.2.5.

4.2.2.1 Power supply unit

Smooth and constant DC Power supply is used to connect the negative terminal with electrode and the positive terminal with either workpiece or nozzle. The supply voltage varies from 200 to 400 V to generate arc across the gas flow and ionize the gas to produce plasma arc. The required arc current varies based on type and thickness of material of workpiece. High frequency generator circuit is used as an arc staring circuit which generates an AC voltage of 5000 to 10,000 V at 2 MHz. The high voltage is to be supplied to initiate a high intensity arc to ionize the plasma forming gas in the plasma torch.

4.2.2.2 Plasma gas supply unit

Individual gases like air, nitrogen, argon, helium, hydrogen and steam are used as plasma forming gas. The mixture of gases like argon and hydrogen, nitrogen and hydrogen, nitrogen and oxygen can also be used as plasma forming gas. Plasma forming gas must have high ionization energy, high thermal conductivity to supply high heat energy to workpiece and high atomic weight to provide sufficient energy to remove material from the workpiece. The gas is pressurized by compressor and high pressure gas is supplied into the nozzle to increase its velocity. The gas is passed through filter cum drier to separate moisture and foreign particles etc. before entering into the nozzle of the plasma torch. The gas supply pipes should withstand the high pressure and carry large volume with high speed.

4.2.2.3 Shielding gas supply unit

Air and carbon dioxide gases are commonly used as shielding gas for PAM. Oxygen, nitrogen and hydrogen can also be used as shielding gas in PAM. High pressure shielding gas is supplied through the concentric nozzle around the plasma gas forming nozzle. The shielding gas is used to improve the cut quality by providing effective cooling and reducing the oxidation effects and formation of smokes during machining. The shielding gas also increases the arc constriction to blow away the dross during cutting. It also reduces the risk of double arcing during high speed cutting using non-transferred plasma arc system. Various combinations of shielding gas and plasma forming gas can be selected for efficient plasma cutting operation. Air is used as shielding or secondary gas for air plasma as well as nitrogen plasma for achieving better quality cut for some ferrous materials. Carbon dioxide gas is used as shielding gas while nitrogen plasma or Argon/hydrogen (mixture of 65% argon and 35% hydrogen) plasma for good cooling effect and less smoke formation during plasma arc cutting.

4.2.2.4 Plasma torch

Plasma torch is the device in which high velocity plasma arc is generated and focused on the workpiece. Plasma torch consists of torch body cover, electrode, gas distributer inner nozzle for plasma gas flow, outer nozzle for shielding gas flow or cooling medium, swirl ring and an additional shielding cap. Plasma torches have self-adjusting and self-aligning consumables parts such as electrodes and nozzles as wear of these parts takes place during PAM operation. The distance between tip of the nozzle and the workpiece is called as stand-off distance or nozzle tip distance. The position of the Plasma torch with respect to workpiece is to be controlled for better cutting action. For profile cutting operation the movement of plasma torch is provided by CNC control. The mechanized torch is mounted on robot or CNC controlled tractor.

The electrode in the plasma torch is made up of copper with an insert of hafnium. Hafnium alloyed electrode has wear resistant properties to increase the life of the electrode. Tungsten can be also used as electrode, but it has shorter life. The electrode is connected with negative terminal of DC power source and it supplies the electricity to ionize the plasma forming gas.

The nozzle is placed at the tip of the plasma torch and the nozzle tip constricts and focuses the plasma arc jet on the workpiece. The tips are made up of copper materials with orifice at the centre. For non-transferred arc, the nozzle should be of electrically conductive materials. The size of the nozzles depends on the current rating of the plasma torch which is selected based on thickness of the work piece. For transferred arc PAM system the nozzle materials can be made up of electrically non-ducting materials such as ceramics.

Plasma torch also sometimes consists of swirl ring in the path of the plasma forming gas flow. The swirl ring is used to spin the plasma gas in a vortex for controlling the constriction of plasma arc jet. The swirl ring is made up of high temperature plastic with angled holes at the periphery to spin the plasma gas.

Depending upon gas (plasma forming gas and secondary/ shielding gas) supply requirements plasma torch shapes are of different types such as Air plasma torch, Oxygen injected nitrogen plasma torch, Dual gas torch, Water injected torch etc.

In Air plasma torch as shown in Fig. 4.2.6, compressed air is used as the plasma forming gas instead of nitrogen or argon. When air is passed across high temperature of the electric arc it breaks down into its constituent gases. Since the oxygen, in the resulting plasma, is very reactive especially with ferrous metals, machining rate is increased by 25%. The main problem of this method is that the machined surface becomes oxidized, which is frequently obtained in case of stainless steel and aluminum. As tungsten is reactive with oxygen, hafnium copper (Hf-Cu) or hafnium-zirconium (Hf-Zr) alloys is as electrode instead of tungsten electrode. Stainless and chrome nickel alloy steels, aluminum, and copper can be machined by air plasma. Because of less constriction of air plasma arc jet, high taper of cutting edge is obtained.

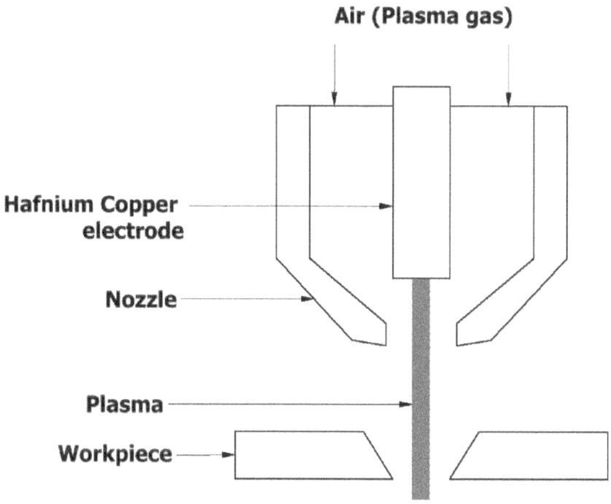

Fig. 4.2.6 Schematic view of air plasma torch.

In oxygen injected nitrogen plasma torch, nitrogen is used as plasma forming gas and oxygen is injected downstream of the plasma jet. For mild steel cutting with higher material removal rate, this PAM system is used. As a result the life of electrode is increased avoiding the oxidation of the electrode because the plasma forming gas is nitrogen instead of air.

In dual gas plasma torch, nitrogen is used as plasma forming gas which is flowing through the inner nozzle and the secondary shielding gas such as oxygen for steel, carbon dioxide for stainless steel, argon hydrogen mixture for aluminum etc. depending upon the workpiece materials are supplied through co-axial outer nozzle for cooling and protecting the cut surface from thermal damage as well as oxidation.

In water injected plasma torch as shown in Fig. 4.2.7, nitrogen is used as plasma gas and water is injected downstream of plasma arc for constriction of the plasma jet. The water vapor is so formed surrounding the plasma jet to provide insulating layer over the nozzle surface. Water constriction of plasma jet takes place at the lower portion of the nozzle which can be of electrically non-conducting materials such as ceramics so that double arcing effect can be reduced. Less heat affected zone, reduction in smokes, less formation of oxide layers and increase in nozzle life are the advantages of water injected plasma torch. Water muffler which covers the plasma torch is used to reduce smoke and noise above the workpiece and water table place below the workpiece is used to reduce the effects of sparks and noise. The best way of reduction of smoke and noise is under water Plasma Arc Cutting.

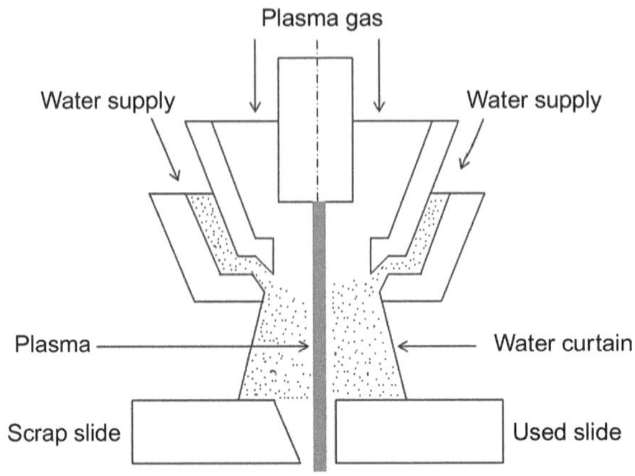

Fig. 4.2.7 Schematic view of water-shielded plasma torch.

4.2.3 Influence of process parameters on performance criteria of PAM process

Various process parameters of PAM are type of plasma forming gas and shielding gas, gas flow rate, gas pressure, arc current, arc voltage, cutting speed and stand-off distance or cutting height. The machining performances of PAM are represented in terms of Material Removal Rate (MRR), Surface Roughness (Ra Value), Taper or conicity, Heat Affected Zone (HAZ) and Dross formation etc.

The performance of Plasma arc machining is influenced by thermo-physical and chemical characteristics of plasma forming gas [4]. Individual as well as mixture of gases such as nitrogen, hydrogen, oxygen and argon are used as plasma forming gas. Hydrogen plasma has maximum heat capacity and high heat conductivity, therefore for same arc power, cutting speed is higher. The hydrogen plasma can be used for cutting thicker plate, but the chances of wear and thermal damage of nozzle is very high. The gas mixture of hydrogen and argon can be used as plasma forming gas, the layer of argon gas can protect the nozzle to some extent. Nitrogen and air are commercially used as plasma cutting gas where the cutting speed is lower than that for hydrogen as plasma cutting gas, but these are used for better quality cut. Air plasma is most economical and air plasma jet has higher concentration of energy. Air plasma is used specially for non-ferrous alloy. Oxygen can only be used as plasma forming gas where heat is applied for cutting from oxidation reaction as well as plasma arc. Steam, a combination of hydrogen and oxygen can also be used as plasma gas, the application of it is still limited. Argon is not efficient plasma forming gas, but it can be used for thinner plates of non-ferrous metals and alloys.

The range of process parameters and experimental results during CNC Plasma cutting of 15 mm thick mild steel using oxygen as primary plasma gas and air as shielding gas are given in the following Table 4.2.1 and 4.2.2, respectively [1]. From the analysis it was observed that the cutting height is the most influencing process parameters for controlling surface roughness and conicity of the cut edge. The cutting current is the most influencing process parameter for heat affected zone of the cut edge.

During Plasma arc cutting of 12 mm thick plate of Hardox 400 materials, process parameters such as cutting speed, arc voltage and plasma gas flow rate are varied to study the unevenness of the cut surface. The oxygen was used as plasma gas and water cooled nozzle of outlet diameter 2.5 mm was used for plasma cutting operation. The experimental results are given in Table 4.2.3 [5].

From the experimental investigation and analysis it has been observed that very good quality plasma cut with minimum unevenness can be achieved by reducing cutting speed to some extent, but arc voltage and gas flow rate are to be controlled. The minimum unevenness of 420 μm

Table 4.2.1 Range of process parameters during CNC plasma cutting [1]

Parameters	Lower limit	Upper limit
Cutting speed-CS (mm/min)	1365	3365
Cutting current-CC (A)	160	280
Cutting height-CH (mm)	3.7	9.9
Plasma gas pressure-PGP (psi)	60	99

Table 4.2.2 Experimental results of CNC plasma cutting of mild steel [1]

	CS (mm/min)	CC (A)	CH (mm)	PGP (psi)	Ra (μm)	Conicity (%)	HAZ (μm)
1	1820	180	4.1	60	0.820	11.84	400
2	1820	210	6.1	75	0.912	7.85	500
3	1820	240	8.1	90	1.178	7.43	550
4	2275	180	6.1	90	0.888	9.30	425
5	2275	210	8.1	60	0.976	9.46	425
6	2275	240	4.1	75	0.772	10.56	400
7	2730	180	8.1	75	1.022	8.80	350
8	2730	210	4.1	90	0.724	11.29	400
9	2730	240	6.1	60	0.908	8.08	325

Table 4.2.3 Experimental results of CNC plasma cutting of Hardox 400 [5]

Trial no.	Cutting speed (mm/min)	Voltage (V)	Plasma flow Rate (l/h)	Unevenness (μm)
1	2100	125	70	420
2	2100	125	70	422
3	2100	125	80	654
4	2100	125	80	660
5	2100	135	70	570
6	2100	135	70	565
7	2100	135	80	610
8	2100	135	80	600
9	2300	125	70	421
10	2300	125	70	423
11	2300	125	80	537
12	2300	125	80	535
13	2300	135	70	495
14	2300	135	70	490
15	2300	135	80	575
16	2300	135	80	580

is achieved at the cut surface of 12 mm Hardox plate at optimal setting of process parameters such as plasma gas flow rate of 70 l/h, arc voltage of 125 V and cutting speed of 2100 mm/min [5].

During plasma cutting of 16 mm thick mild steel plate the effect of nature of plasma gas has been studied. Plasma forming gas such as steam, air, nitrogen and oxygen were used to study the effects of plasma gas nature and flow rate on plasma arc characteristics such as arc voltage, arc temperature, energy input and also material removal rate (MRR) during plasma arc cutting operation. The variations of arc voltage with gas flow rate for different plasma gases are shown in Fig. 4.2.8.

For higher plasma gas flow rate, the arc voltage is also higher. As gas flow rate increases, more energy is needed to ionize the gas, therefore the arc voltage should be higher. The variation of arc voltage with gas flow rate is almost similar for different plasma forming gases such as air, nitrogen, oxygen and steam as shown in Fig. 4.2.8. The voltage level for steam is higher because the enthalpy and thermal conductivity of steam are much higher compared to other gases.

The nature of variation of temperature profile of the plasma jet with radial distance from centre of the jet is similar for different plasma forming gases such as nitrogen, air, oxygen and steam as shown in Fig. 4.2.9. The

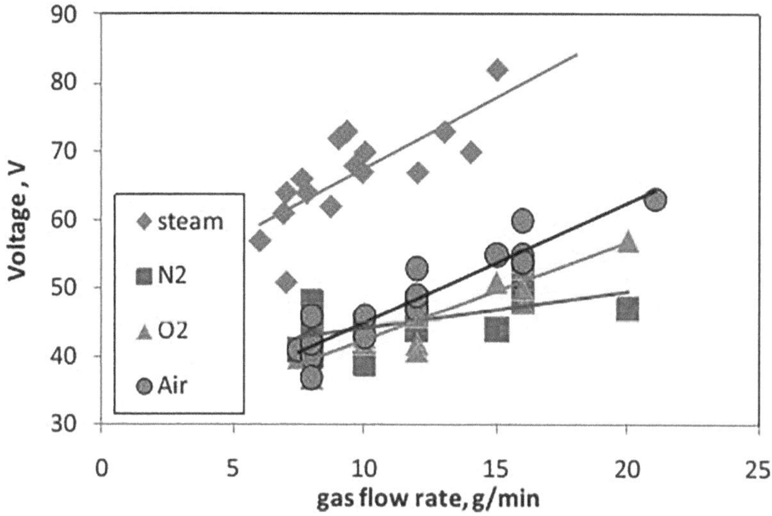

Fig. 4.2.8 Variation of arc voltage with gas flow rate [6].

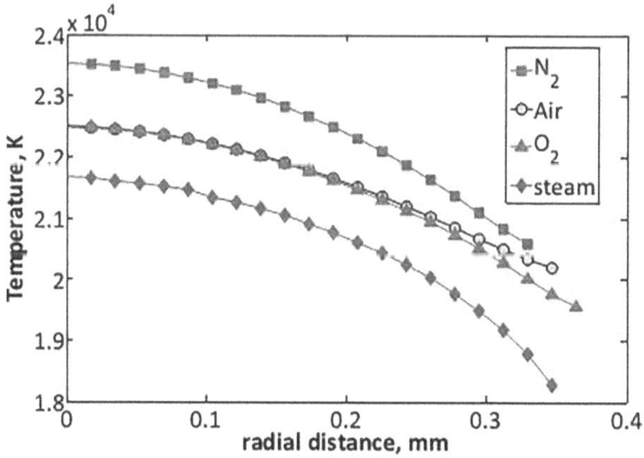

Fig. 4.2.9 Variation of temperature with radial distance of nozzle [6].

temperature is maximum at the entre of the jet and decreases with radial distance towards the edge. For nitrogen plasma temperature is highest and for steam the temperature is lowest. Temperature distribution for oxygen and air plasma is exactly same.

The effects of gas flow rate for different gases on material removal rate were studied and shown in Fig. 4.2.10. For nitrogen, air and steam plasma, as gas flow rate increases the MRR decreases because heat transfer rate from hot plasma to material decreases due to either higher arc constriction or the

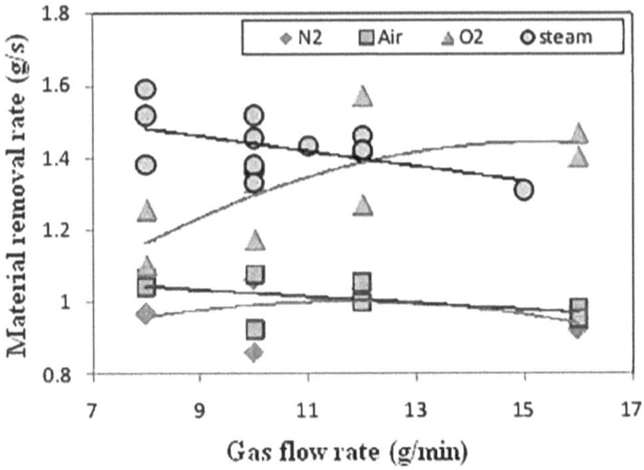

Fig. 4.2.10 The effect of gas flow rate on MRR [6].

formation of cold gas layer surrounding the plasma jet. For oxygen plasma cutting, MRR increases with the gas flow rate because there is additional heat release due to exothermic reaction of oxygen with iron which increases the rate of melting of work piece materials.

The effect of plasma gas on kerf geometry was shown in Fig. 4.2.11. The cutting speed of 30 cm/min, stand- off distance of 2 mm and gas flow rate of 8 g/min were set during plasma cutting of 16 mm thick mild steel plate using different plasma gases such as nitrogen, air, oxygen and steam. From the kerf geometry as shown in figure it was observed that kerf is wider at top and narrower at bottom for nitrogen as well as air plasma gas. The taper cut is observed. The kerf is narrower at the top, then it widens at the middle and then again narrower at the bottom for oxygen and steam. From the kerf shape variation it is concluded that heat distribution along the cut is not uniform.

The machining speed during PAM depends on the thickness of work-piece materials as well as type of materials to be cut. Figure 4.2.12 shows

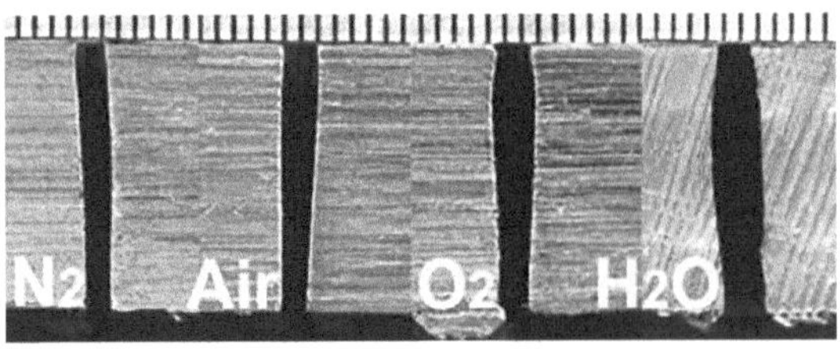

Fig. 4.2.11 The effect of plasma gas on Kerf geometry [6].

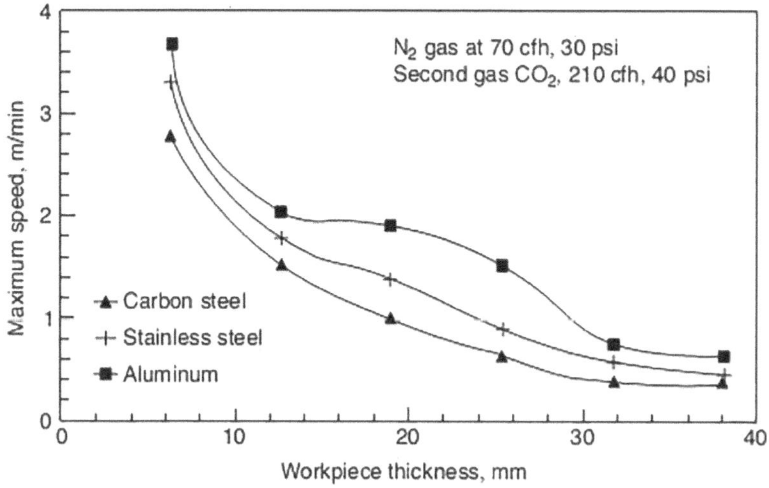

Fig. 4.2.12 Variation of machining speed with thickness of different material in gas-shielded plasma [7].

the variation of machining speed with thickness of workpiece for carbon steel, stainless steel and aluminum using nitrogen as plasma gas and carbon dioxide as shielding gas during PAM. As workpiece thickness increases the machining speed decreases keeping other process parameters remains same. The nature of variation is same for different materials. For aluminum cutting speed of PAM is higher compared to the cutting speed of stainless steel and carbon steel. The machining speed of PAM for carbon steel is very low while workpiece thickness is about 40 mm. Figure 4.2.13 shows the power consumption factor during plasma beam rough turning of different alloys. A low power consumption factor indicates either low energy required for machining or high removal rates. As the power is increased, the efficient removal of melted metal is found to need a corresponding rise in the gas flow rate. During plasma machining of 12-mm-thick steel plate using 220 kW the machining speed is 2500 mm/min, which is 5 times greater than that for oxy-gas cutting.

During plasma arc cutting, dross formation at the bottom is to be minimized or eliminated by controlling the process parameters such as cutting speed, arc power, pressure of nozzle etc. At low speed, input energy to the workpiece is high and it causes melting of more materials. Dross is formed at the bottom if adequate force of plasma jet is not available. To obtain a dross free cut surface, plasma force and energy input to workpiece are to balanced properly. The plasma power increases with plasma gas flow rate and arc

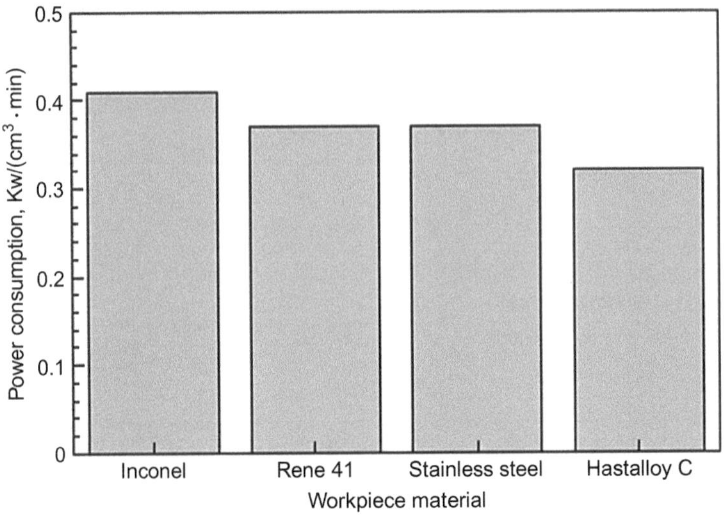

Fig. 4.2.13 Power consumption in plasma beam rough turning [7].

current. In order to achieve a square cut of narrow kerf without dross, a narrow plasma jet with high plasma power density and arc force is required. The nozzle bore should be narrow. Design and selection of nozzle materials should be such that it can withstand the thermal and electrical stresses in terms of excessive nozzle erosion and plasma instabilities due to double arcing [8].

For Plasma arc cutting of stainless steel, SS 304, arc current from 30 to 40 A, stand off distance from 2.5 to 3.5 mm, nozzle pressure from 4 to 5 bar and cutting speed from 0.2 to 0.46 m/min are varied. With high arc current, material removal rate initially increases with increase in stand off distance and then MRR decreases at higher stand off distance. Higher current and lower pressure of nozzle gives higher MRR. With increase in pressure, MRR decreases due to the cooling effect on workpiece materials. The selected cutting speed is not so high that exposure of heat energy at the specific point on the workpiece materials for less time is insufficient for plasma cutting. However with setting of high speed, medium stand off distance, high current and medium pressure, good quality cut surface with less dross appearance is observed during plasma arc cutting of stainless steel [9].

Plasma arc cutting of EN 31 steel has been performed and optimal setting of process parameters such as gas pressure, arc current and torch height has been determined for maximum MRR and minimum surface roughness characteristics using Gray Relational Analysis. The middle level of gas

Table 4.2.4 Typical values of operating parameters of PAM [11]

Power supply	Type: Direct current Wattage: up to 200 kW Current: 50 to 1000A
Primary gas	Type: Argon–nitrogen (80%–20%), hydrogen Flow: 0.42 to 5.66 m^3/h
Shielding gas	Type: Nitrogen, oxygen, water, compressed air, CO_2 Flow: up to 11.3 m^3/h
Water shield flow (in place of shielding gas)	up to 11.3 m^3/h
Plasma	Temperature: 11100^0 to 27800^0C Cutting speed:50 to 6100 mm/min Standoff distance: 6.4 to 9.5 mm Nozzle orifice size: 1.59 to 6.35 mm (diameter)
Kerf width	1.52 to 2.29 mm (thin plate) 4.76 mm (for 25.4 mm plate) 19 mm (for 152 mm plate)
Kerf angle	Normal: 2° to 7° Special: 1° to 2°
Accuracy	0.81 mm on 6.4 to 35 mm plate 3.2 mm on 152 to 203 mm plate
Corner radii	3.9 mm to 38 mm (increases with increasing cutting speed)

pressure (6 bar), higher level of arc current (200 A) and higher level of torch height (6 mm) are obtained for optimal performance of Plasma cutting of EN31 steel. From analysis it is obtained that MRR is proportional to gas pressure and surface roughness is proportional to arc current [10].

Based on the research findings on the parametric influences of power supply, plasma forming gas, shielding medium, cutting speed, stand off distance etc. on machining criteria such as kerf width, kerf angle and accuracy etc., the typical values of operating parameters of PAM are given in Table 4.2.4.

4.2.4 Advantages and limitations of plasma arc machining process

The Advantages of Plasma Arc Machining Process are given as follows:

(i) PAM process can be applied for cutting electrically conducting as well as non-conducting materials;

(ii) PAM can cut materials faster than Oxy fuel cutting;
(iii) PAM can be used for thicker materials;
(iv) Using high power of plasma torch, high material rate can be obtained.
(v) PAM process can be utilized for high speed profile cutting of any kind of engineering materials such as stainless steel, super alloy, aluminum and ceramics etc.;
(vi) No cutting tool is required for this machining process;
(vii) PAM process does not require any chemicals. So there is no such harmful effect due to corrosion.
(viii) Automated Plasma cutting can be performed using CNC technology;
(ix) Underwater cutting is also possible by PAM process. There are certain advantages of under water PAM such as less noise, less fumes, high solubility of removed materials with water and less effect of oxidation by selecting proper plasma forming gas with respect to workpiece materials.

The Disadvantages of Plasma Arc Machining are given as follows:
(i) The large power supplies is required to cut thicker materials at high speed;
(ii) High heat can damage the cut edge of workpiece;
(iii) Toxic fumes are produced;
(iv) Dimensional accuracy of cutting is less and difficult to generate sharp corners;
(v) Generally kerf width is more, but narrow kerf width can be obtained by High Tolerance Plasma Arc Cutting (HTPAC) system;
(vi) The cutting edge is taper.

4.2.5 Applications of plasma arc machining process

The applications of Plasma Arc Machining Process are given as follows:
(i) Computer Numerical controlled (CNC) PAM is used for profile cutting of various shapes on thicker sheet metals such as steels and aluminum etc. those are difficult to cut by oxyacetylene gas cutting technique. CNC Plasma Arc Cutting system provides automatic cutting of large number of parts from large sheet.
(ii) Simultaneous cutting of different shapes from one plate can be done using multiple torch PAM system.
(iii) PAM is applied for turning of difficult-to-machine materials. In this regard, cutting speeds of 2 m/min and a feed rate of 5 mm per

revolution produced a surface finish of 0.5 mm *Ra*. The depth of cut can be controlled through the machining power or surface speed.

(iv) PAM can cut 1.5-mm-deep, 12.5-mm-wide grooves in stainless steel at 80 mm/min, using 50 kW as the cutting power. Lower machining rates are obtainable when these grooves are cut in nonconductive materials. The groove dimension however depends on the traverse speed, arc power, and the angle and height of the plasma arc [12].

(v) PAM system can produce a bevel edge of the end of pipe at a certain angle before welding. It cut tubes of wall thickness of up to 50 mm.

(vi) PAM system is used to cut 6 mm Aluminum oxide ceramic plate with cutting speed 1.2 m/min with kerf width of less than 4.6 mm using hydro-magnetically confined plasma arc.

(vii) Plasma Arc machining technology can be applied for under water cutting of plates and tubes with machining accuracy of ±0.2 mm. Under sea water PAM technology has potential applications for cutting mild steel plates and pipes etc.

(viii) Fine plasma cutting process can be applied to cut advanced aerospace materials like aluminum alloy in both a monolithic and a metal matrix composite form using highly concentrated High Tolerance Plasma Arc Cutting (HTPAC) method. Less kerf width (less than 1 mm) and small HAZ can be observed during plasma cutting of these materials with high cutting speed (55–60 mm/s) [13].

4.2.6 Advancement in PAM technology

The plasma arc machining technology can be applied for cutting laminar composite materials which is composed of two aluminum plates with a polyethylene core. The optimal process parameters setting are: gas pressure 0.41 MPa, Feed 50 m/min, stand off distance 5 mm, cutting current 30A and cutting power 7KW. High power plasma with high cutting speed is recommended for cutting this sandwich panel of this composite materials to avoid the thermal damage of polyethylene core at high speed cutting. With the use of CNC technology for precise motion control of plasma torch or workpiece, high quality and complex shape cutting can be done [14].

Hydro-magnetically confined plasma arc was used to cut engineering ceramic (aluminum oxide) plate of 6 mm to reduce the kerf width and also to improve the cut quality. Figure 4.2.14 shows the principle of hydro-magnetically confined PAM [15]. Both water constriction and magnetic constriction are combined to improve the plasma arc stability to cut ceramic

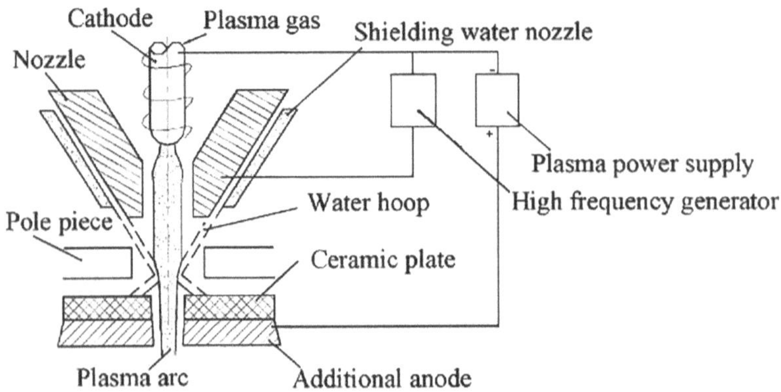

Fig. 4.2.14 Schematic diagram of the principle of hydro-magnetically confined PAM [15].

materials. A high quality cut can be produced by using lower arc current and life of nozzle is increased due to reduction in heat load by applying hydro-magnetic constriction.

High Tolerance Plasma Arc Cutting (HTPAC) has been performed on titanium sheets of 5 mm thick [16]. Better cut quality in terms of reduced unevenness, kerf width and HAZ was achieved at dross free and higher feed rate condition due to reduced interaction of material with high energy plasma beam during HTPAC process. During HTPAC process nitrogen or oxygen was used as plasma forming gas and nitrogen was used as shielding gas. HTPAC system operates highly constricted plasma and sometimes magnetic field is applied surrounding the arc to stabilize the arc. HTPAC system offers the advantages such as high energy density plasma beam due to less diameter of beam, high cutting speed, precision cutting with narrow kerf width, less unevenness and less heat affected zone. HTPAC process can also be used for cutting of carbon steel, stainless steel and aluminum etc. with better cut edge quality.

High tolerance plasma arc cutting (HTPAC) operation has also been performed on 25 mm thick mild steel plate using 200 A DC arc current [17]. The cutting speed and arc voltage affect the kerf formation mechanism. During plasma cutting the range of cutting speed (1890–2310 mm/min), arc voltage (139–149 V), plasma gas flow rate (2495–2873 l/h), shield gas flow rate (2590–3140 l/h) is selected for HTPAC process. From analysis it is found that arc voltage is the major influencing parameter for controlling the plasma cut quality.

A new hybrid plasma torch has been developed for material processing using two plasma generating technologies, non-transferred arc plasma and magneto plasma dynamic thruster (MPDT) to generate high energy plasma jet [18]. Argon gas is used as plasma forming gas. The hybrid plasma torch consists of primary and secondary cathodes and also primary and secondary cylindrical anodes. Initially the gas is ionized by electrical discharge between primary cathode and anode. The plasma gas is further ionized in the passage by electrical discharge between secondary cathode and anode. Two power sources (primary and secondary) are used in the plasma torch. The circumferential magnetic fields are generated in the primary as well as secondary cathode to propel the plasma jet with high velocity from the torch. The secondary cathode is tungsten electrode and other electrodes are made up of copper materials. The schematic diagram of hybrid plasma torch is shown in Fig. 4.2.15. Figure 4.2.16 shows the schematic diagram of experimental setup of Plasma Arc Machining using Hybrid plasma torch.

An intelligent control strategy of power supply based on fuzzy neural network has been utilized for plasma arc cutting process [19]. During plasma cutting using conventional constant current closed loop power supply control, cutting surface finish becomes poor and deeper ripples are observed. Using fuzzy neural network based intelligent power control strategy, air plasma with carbon dioxide shielding gas can cut 20 mm thick low carbon steel with better finish and uniform and even ripples by CNC automatic controlling the cutting speed and torch height.

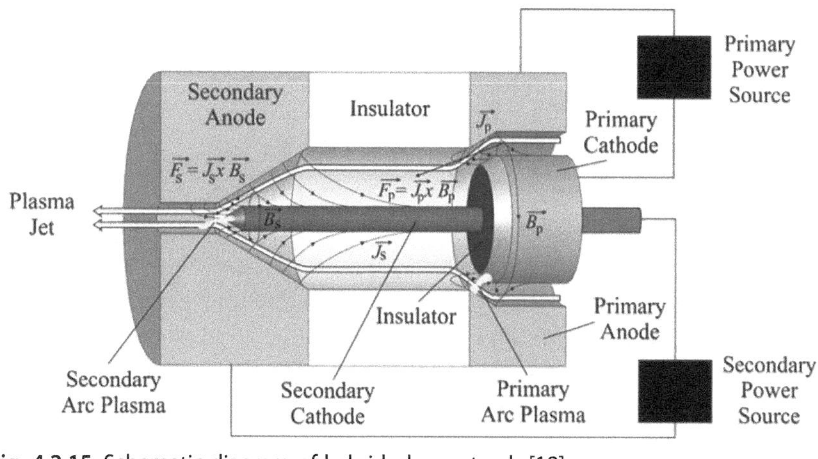

Fig. 4.2.15 Schematic diagram of hybrid plasma torch [18].

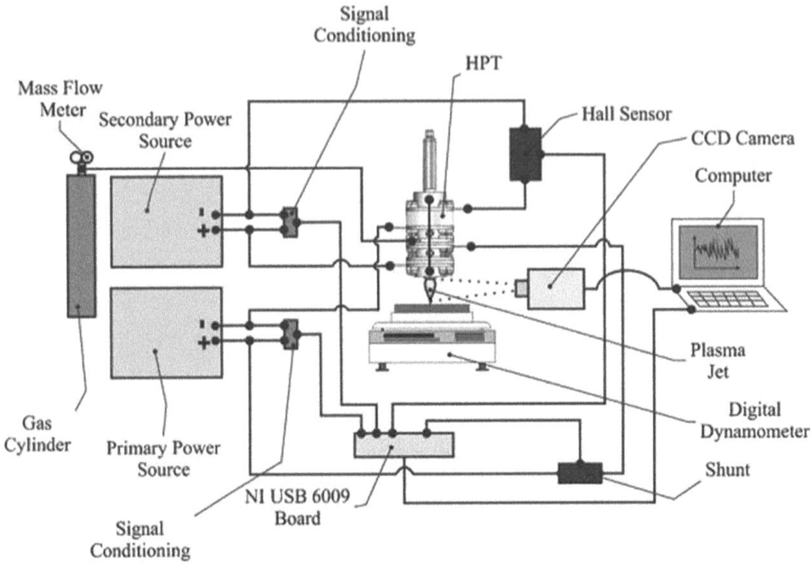

Fig. 4.2.16 Schematic diagram of experimental setup of plasma arc machining [18].

A D C double anode plasma torch has been devised which generates a highly stable argon plasma jet at atmospheric pressure [20]. Using Fast Fourier Transform (FFT) and time correlation function method, the effects of plasma gas flow rate and arc current on arc fluctuation and dynamic behavior of double arc argon plasma jet were investigated. Low frequency (150 Hz) electrical signal is attributed to the inherent characteristics of the torch power source and it is independent of any change in the operating parameters.

Plasma arc machining is becoming a popular and common modern machining process for machining electrically conducting as well as non-conducting thicker materials. There are lots of advances in plasma arc machining technology for efficient machining in terms of high speed, better geometry of kerf and less thermal effect on varieties of materials. CNC Plasma machining has wide applications for cutting different shapes and contours with better accuracy. Plasma beam with high current density and hypersonic velocity combined with CNC automation and Robotic system can perform 3 D cutting operation for its well acceptance in present fabrication industry.

References

[1] K. Salonitis, S. Vatousianos, Experimental investigation of the plasma Arc Cutting Process, Proc. CIRP 3 (2012) 287–292.
[2] D. Karjcarz, Comparison metal water jet cutting with laser and plasma cutting, Proc. Eng. 69 (2014) 838–843.
[3] J.A. McGeough, Advanced Methods of Machining, Chapman and Hall, 1988. ISBN 0412 319 705.
[4] P.K. Mishra, Nonconventional machining, in: The Institution of Engineers (India) Text Book Series, Narosa Publishing House, 2011. ISBN 978-81-7319-138-1.
[5] S. Chamarthi, N.S. Reddy, M.K. Elipey, D.V.R. Reddy, Investigation analysis of plasma arc cutting parameters on the unevenness surface of Hardox-400 material, Proc. Eng. 64 (2013) 854–861.
[6] T. Kavka, A. Maslani, M. Hrabovsky, T. Stehrer, H. Pauser, Experimental study of the effect of gas nature on plasma arc cutting of mild steel. J. Phys. D: Appl. Phys. 46 (2013), 224011 https://doi.org/10.1088/0022-3727/46/22/224011.
[7] H. El-Hofy, Advanced Machining Processes: Nontraditional and Hybrid Machining Processes, McGraw Hill, 2005. ISBN 0-07-145 334-2.
[8] S. Ramakrishnan, M. Gershenzon, F. Polivka, T.N. Learney, M.W. Rogozinski, Plasma generation for the plasma cutting process, IEEE Trans. Plasma Sci. 25 (5) (1997).
[9] A.H. Patel, A.B. Pandey, Effect of process parameters on MRR and dross formation in plasma arc cutting of SS304, in: 4th International & 25th All India Manufacturing Technology, Design and Research Conference (AIMTDR 2012), December 14–16, 2012, Jadavpur University, Kolkata, India, 2012.
[10] M.K. Das, K. Kumar, T.K. Barman, P. Sahoo, Optimization of process parameters in plasma ARC Cutting of EN 31 steel based on MRR and multiple roughness characteristics using grey relational analysis, Proc. Mater. Sci. 5 (2014) 1550–1559.
[11] Machining Data Handbook, 3rd ed., Vol. 2, Machinability Data Centre (MDC), Cincinnati, Ohio, 1980. ISBN 0-936 974-02-8.
[12] G.F. Benedict, Non-traditional Manufacturing Processes, Marcel Dekker Inc, New York, 1987.
[13] A.P. Hoult, I.R. Pashby, K. Chan, Fine plasma cutting of advanced aerospace materials, J. Mater. Process. Technol. 48 (1995) 825–831.
[14] A. Iosub, G. Nagit, F. Negoescu, Plasma cutting of composite materials, Int. J. Mater Form 1 (Suppl 1) (2008) 1347–1350.
[15] W.J. Xu, J.C. Fang, Y.S. Lu, Study on ceramic cutting by plasma arc, J. Mater. Process. Technol. 129 (2002) 152–156.
[16] E. Gariboldi, B. Previtali, High tolerance plasma arc cutting of commercially pure titanium, J. Mater. Process. Technol. 160 (2005) 77–89.
[17] R. Bini, B.M. Colosimo, A.E. Kutlu, M. Monno, Experimental study of the features of the kerf generated by a 200 A high tolerance plasma arc cutting system, J. Mater. Process. Technol. 196 (2008) 345–355.
[18] R.T. Lermen, I.G. Machado, Development of a new hybrid plasma torch for materials processing, J. Mater. Process. Technol. 212 (2012) 2371–2379.
[19] J. Deli, Y. Bo, An intelligent control strategy for plasma arc cutting technology, J. Manuf. Process. 13 (2011) 1–7.
[20] X. Tu, J.H. Yan, B.G. Cheron, K.F. Cen, Fluctuations of DC atmospheric double arc argon plasma jet, Vacuum 82 (2008) 468–475.

SUBCHAPTER 4.3

Laser Beam Machining (LBM)

4.3.1 Introduction

LASER is an acronym of Light Amplification by Stimulated Emission of Radiation. Laser light is coherent and monochromatic in nature. Laser beam can be focused to a very small spot size in the order of few microns only. Because of very small spot size, the energy intensity is therefore very high. Because of high energy intensity at the focus spot, laser can be applied as energy source in many areas of manufacturing. Laser light is different from ordinary torch light. Ordinary torch light cannot be focused down to a very tiny spot. These common light sources, such as the incandescent light bulb, emit incoherent photons in almost all directions, usually over a wide spectrum of wavelengths. The laser, a beam of photons can perform better than electrons because photons are electrically neutral and the size is much smaller than electrons. Photons have no volume and these can be concentrated to a tiny space. Photons do not repulse each other like electrons.

Einstein laid foundation of laser theory based on the concept of stimulated emission. Planck has given the theory of Quanta in 1900 and in 1920 it was well accepted that laser shows particle like characteristics while interacting with matters and exchange energy in the form of photons. The first laser built in 1962 was a Ruby Crystal Laser by T. H. Maimann. Since then, a large variety of lasers were produced. The CO_2 laser was invented by C.K.N Patel in 1964 at Bell Laboratories. Nd:YAG laser was invented by J.E. Guesic, H.M Marcos and L.G Van Uitert in the same year at Bell Laboratories with power in the order of 100 W. CO_2 lasers (developed by Dr. C.K.N. Patel) in the order of 1000 W were produced. In the late 1980s the Fiber Laser technology was available using erbium-doped single-mode fibers. In the early 80s lasers were no longer treated as laboratory research equipment. Since that time, laser manufacture has become a multi-billion dollar industry. Industrial applications of laser in the manufacturing include cutting, drilling, marking and engraving, welding, surface hardening, surface alloying and cladding, dressing and cleaning etc. for processing of diverse range of materials-plastics, metals, ceramics, diamonds and composites etc.

Laser Beam Machining process is a thermal process in which a laser beam of high energy density generates a highly localized heating on a workpiece, causing the material to melt and vaporize or ablate, so that the material can be removed in the micron level from the workpiece. It is a non-contact type tool less energy beam based thermal machining process with several unique advantages such as potential for micro-machining, elimination of finishing operation, adaptability of automation and greater material utilization and faster machining. Laser has long history for machining application such as cutting, slicing, drilling, grooving, turning, marking, cleaning, dressing etc. and non-machining applications such as welding, cladding and surface modification etc. for different manufacturing applications of varieties of materials. The lasers used for machining of materials are CO_2 Laser, Nd: YAG laser, Fiber Laser, Diode Laser, Metal Vapor Deposition Laser and Excimer lasers, etc. Among these, Nd:YAG laser provides superior micro-machining characteristics with high accuracy and surface finish. CO_2 Laser has a wavelength of 10640 nm. Fiber Laser and Nd:YAG laser have a wavelength of 1064 nm the infrared (IR) range. Excimer laser is a gas laser with a wavelength of ultraviolet (UV) range (ArF:193 nm; KrF:248 nm; XeCl:308 nm; XeF:351 nm). Laser technologies are utilized in various fields like manufacturing, instrumentation, imaging, medicine, communications, etc. Although each of these fields is as important as others, here the discussion is on various types of laser beam machining technologies and their applications.

4.3.1.1 Types of lasers in machining applications

Lasers in machining applications are classified based on lasing medium such as Solid State lasers and Gas Lasers. Solid state lasers are sub-classified as Nd: YAG Laser, Fiber Laser, Diode laser and Ti: Sapphire Laser. Depending upon the type of excitation source of energy, solid state lasers are classified as lamp pumped and diode pumped lasers. Gas lasers for machining applications are sub-classified as CO_2 Laser, Excimer Laser and Copper Vapor Laser. Depending upon wave characteristics of laser beam, laser is classified as Continuous Wave (CW) and Pulsed Wave (PW). Depending upon the duration of pulse, lasers are classified as long pulse laser (pulse duration in the range of milliseconds), short pulse laser (pulse duration in the range micro-seconds and nanoseconds), ultra short pulse laser (pulse duration in the range of picoseconds and femtoseconds). For laser machining short and ultra short pulse lasers are useful because of high beam quality with high peak power but less thermal effect during laser material interactions. Depending upon the wavelength of laser beam, lasers are classified as Infrared (IR) laser and Ultra-violet (UV) laser. Depending upon the average power of laser beam, lasers are classified as low power laser and high power laser. Table 4.3.1 shows various types of lasers in machining applications.

4.3.2 Fundamental of laser generation

Laser is a coherent, monochromatic, highly directional energy beam of photon. For generation of laser beam, lasing medium is to be excited by a source of energy. Lasing medium cannot stay for long time at excited state. It comes to the normal state after emitting the energy in the form of beam of photon known as laser beam.

Mechanism of light emission: An atom consists of a positively charged nucleus surrounded by orbiting electrons. When the electrons are excited by the supplied energy, it changes its orbit. After short periods of time, it returns to the ground state and it emits the energy in the form of light/photon. This follows the Planck's law ($E_2 - E_1 = h\gamma$, γ = frequency of photon and h = Planck's constant). Light radiation is the part of an electromagnetic radiation whose wavelength occupies the range from 0.1 μm to 100 μm. Emission of photon can be either spontaneous or stimulated process:

Spontaneous emission: Normally an atom will adsorb energy, raising electrons to higher energy levels and will remain in that energy state for some period of time (nano-sec/milli-sec.). Within this period the atom will spontaneously emit a photon of light in a random direction and return to the ground state. This process is called spontaneous emission process.

Stimulated emission: For stimulated emission, the atoms are to be stimulated to release its photon of light by interacting with another passing photon. The passing photon must be of same energy as the difference of energy of upper and lower energy levels. The photons are generally emitted from lamp or semiconductor diode. The photon wavelength (λ) is given by its energy ($E = hc/\lambda$, c = velocity of light). The passing and emitted photon will be of same wavelength. Both photons will be in same phase. This equal property of wavelength, direction and phase constitutes this light as LASER light, which is coherent light.

Table 4.3.1 Type of lasers used in machining application [1]

Type of laser	Laser material	Wavelength	Pulse length	Frequency
Solid state laser	Nd:YAG (second harmonic)	532 nm	100–10 ns	50 Hz
	Nd:YAG (third harmonic)	355 nm		
	Nd:YAG (forth harmonic)	266 nm		
	Nd:YVO	1064 nm	2.8–7.9 ps	84 MHz–77 GHz
	Nd:GdVO	1053 nm	37 ps	100 MHz
	Nd:BEL	1070 nm	2.9–7.5 ps	250 MHz–20 GHz
	Nd:LSB	1062 nm	1.6–2.8ps	177–240
	Nd:glass	1054 nm	7 ps	
	Nd:VAN	750–870 nm		
	Nd:YLF	1047–1053 nm	1.5–37 ps	76 MHz–2.85GHz
	Yb:YAG	1030 nm	340–730 fs	35–81 MHz
	Yb:glass	1025–1082 nm	58–61 fs	112 MHz
	Yb:GdCOB	1045 nm	90 fs	100 MHz
	Yb:KGW	1037 nm	176 fs	86 MHz
	Ti:sapphire	750–880 nm	6–150 fs	15 MHz–2 GHz
	Cr:LiSAF	800–880 nm	12–220 fs	82–200 MHz
	Cr: LiSAF	800–820 nm	20–170 fs	90–95 MHz
	Cr: LiSAF	830–895 nm	14–100 fs	71–119 MHz
	Cr: LiSAF	860 nm	90 fs	140 MHz
	Cr:Forsterite	1.21–1.29 μm	14–78 fs	81–100 MHz
	Cr:YAG	1.52 μm	44–120 fs	81 MHz–1.2 GHz
	Fiber lasers	1064 nm	100 ns	20–50 Hz
	Diode lasers	0.8 μm		
	Microchip lasers	1064 nm	<100 ps	100 kHz
Gas laser	ArF	193	5–25 ns	1–1000 Hz
	KrF	248	2–60 ns	1–500 Hz
	XeCl	308	1–250 ns	1–500 Hz
	XeF	353	0.3–35	1–1000 Hz
	CO_2 laser	10.600	200 μs	5 Hz
	Copper vapor lasers	611-578	30	4–20 Hz

Population inversion: Atoms are in the ground state at normal room temperature and few atoms are in the higher excited state. The population distribution from thermal equilibrium can be changed by supplying the energy externally to the atoms such that selectively there will be large number of atoms in higher excited state than in the lower energy level. This situation is called the population inversion. The external energy source which generates this situation is called energy Pump. As the population inversion is non-equilibrium state, the situation will not exist for long time and the atoms from the higher excited state soon transferred to lower energy levels following the thermal equilibrium population distribution. Therefore to get continuously population inversion and subsequently stimulated emission, excitation by external energy source is the basic requirement of laser generation process.

An atom consists of a positively charged nucleus surrounding by orbiting electron. These electrons are revolving in definite orbits called the energy levels. When the energy is supplied to an electron in an orbit, it changes its orbit (energy level) and it subsequently emits the energy by returning to the ground state.

Lasers convert electrical energy into a high energy density beam of light through stimulation and amplification. The energy required to raise an electron from one energy state to another is provided by an excitation process or pumping. This is achieved by the lasing medium's absorption of energy from mechanical, chemical, electrical or light sources. The lasing medium typically contains ions, atoms or molecules which have discrete electron energy level. The wave length produced is a characteristic of the lasing medium. At the beginning of the lasing process, photon emissions are random in nature. As each photon stimulates other excited electrons to emit photons, however the new photons will have similar wave length, direction and phase characteristics as the initial photon. Eventually the stream of photons with identical wavelength, direction and phase will be produced.

The amplification of light in a laser is accomplished by an optical resonator, which is composed of a cavity with the lasing medium set between two high precision, aligned mirrors. One is fully reflective and the other is partially transmissive to allow for the beam output. The mirrors channel light back into the lasing medium; as the photons pass back and forth through the lasing medium, they stimulate more and more emissions. The mirrors to stimulate more emissions, so that the cavity will only amplify those photons with the proper orientation, do not redirect photons, which are not aligned with the resonator. Figure 4.3.1 shows the schematic diagram of stimulated emission and photon generation principle.

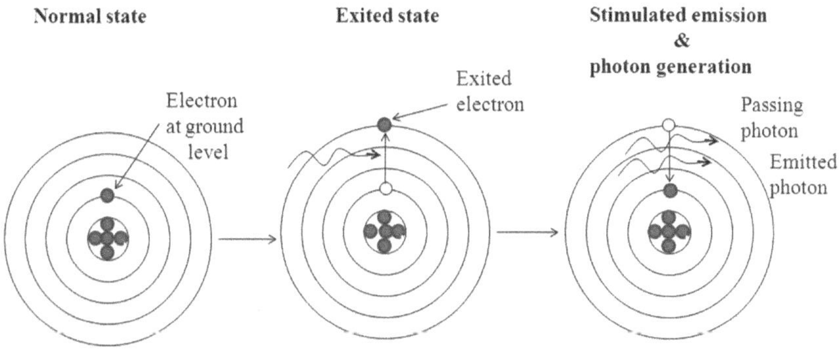

Fig. 4.3.1 Schematic diagram of stimulated emission and photon generation principle.

4.3.2.1 Fundamentals of Nd: YAG laser generation

Nd:YAG laser consists of crystalline YAG with a chemical formula $Y_3Al_5O_{12}$ as a host material. The Nd^{3+} ions substitute yttrium ion sites in the lattice with a maximum doping level of around 2%. This is a typical four-level energy laser system illustrated in Fig. 4.3.2.

Such systems offer significant advantages such as ease of achieving population inversions. Hence, simple designs of flash lamps with modest amount of pumping energy are sufficient to achieve the efficient population inversions. Due to splitting of initial and final energy levels, several lasing wavelengths are possible, 1.064 μm being the strongest one. The output of the

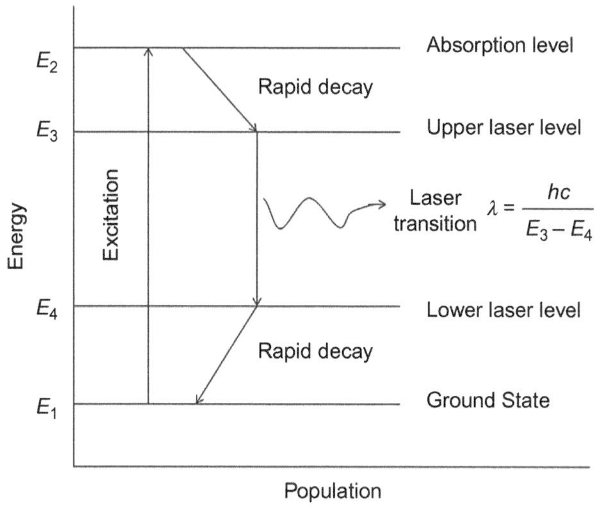

Fig. 4.3.2 Energy transitions in a four-level laser system [2].

Nd:YAG laser can be continuous or pulsed. The light source for pumping depends on the absorption characteristics of the crystal. For continuous operation, the laser is excited by continuous krypton-filled or xenon-filled arc lamps or semiconductor diode lasers. Krypton lamps are efficient pumping sources for continuous Nd:YAG laser because the emission lines from Krypton lamps agrees better with absorption lines in Nd:YAG. For pulsed operation, flash lamps are generally used. If the pulses of relatively large-pulse energy are desired, the laser is excited by a flash lamp, which gives pulses at relatively low pulse repetition rates. Nd:YAG laser is also available in frequency-doubled mode in which the output of the laser is in the green portion of the visible spectrum at 532 nm. In addition to frequency doubled operation, the laser is also available in frequency-tripled (355 nm) and frequency-quadrupled (266 nm) modes. The schematic diagram of Nd:YAG laser generation principle is shown in Fig. 4.3.3. The detail specification of Nd: YAG laser beam is given in Table 4.3.2.

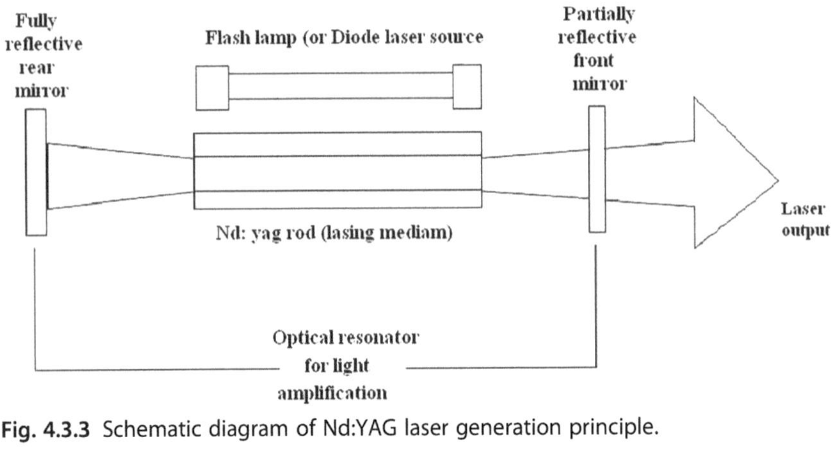

Fig. 4.3.3 Schematic diagram of Nd:YAG laser generation principle.

Table 4.3.2 Details of Nd: YAG laser beam

Laser type and wavelength	**Nd: YAG laser, 1064 nm**
Mode of operation	Q-Switched (pulsed)
Type of Q-switch used	Acousto optic Q-switch
Mode of laser beam	Fundamental mode (TEM_{00})
Mirror reflectivity	Rear mirror 100% Front mirror 80%
Beam diameter	1 mm
Pulse width	120 ns
Laser spot diameter at focal point	100 μm

4.3.2.2 Fundamentals of fiber laser generation

The fundamental principle of fiber laser generation can be identified as excitation of a rare earth element erbium or ytterbium as active laser medium within the fiber optic cable from its ground state to an excited state (population inversion) and stimulated emission of photons in that medium. The source of excitation energy is generally diode laser. The number of laser diodes is depended on the application of the fiber laser and the output power required. Laser light is amplified by "stimulated emission" along the axis of the resonator cavity. An optical fiber is a cylindrical dielectric waveguide made of material such as a silica glass. It has a central core with refractive index n1 in which the light is guided, embedded in an outer cladding of slightly lower refractive index n2. The small particles of rare earth elements like Ytterbium are mixed into the core of the fiber when it is made (Doping) Diode laser as a pump source. Ytterbium atoms act as laser medium. Pump radiation is launched into the fiber core and gradually absorbed over the fiber length and converted into the laser radiation. In general, both the pump and the laser radiation are guided in an active doped waveguide structure. Emitted photons are trapped inside the fiber core. Light rays are guided through the core without refraction into cladding. It is important that light which incidents on the boundary between fiber core and cladding, should be at angles greater than critical angle and must undergo a total internal reflection The cavity mirrors consist of two primary optics, front and rear mirror; reflect the photons to and fro through the laser medium to increase amplification. Through the pumping process, only a fraction of energy passes through the front mirror [3]. Fiber Bragg Gratings or dielectric mirrors are used to create the laser cavity. The gain of the laser medium is determined by the product of a pump light intensity and an interaction length with the laser radiation in the gain medium. The choice of rare-earth doping is primarily decided by the required emission wavelength. If high average power levels are needed, the ytterbium is the first choice among all the rare-earth ions. In conventional fibers doped single-mode core and a cladding, the laser power is restricted to about 1 W. To achieve higher output powers, fiber lasers use double-clad fibers as the amplifying medium [4]. Most of fiber lasers employ Q switch mode of operation though introduction of "master oscillator, power amplifier" (MOPA) design which enables independent control of pulse width, pulse repetition frequency, and average power. In Q switched fiber laser systems, pulse energy is independent of the fiber length. A schematic diagram of multidiode ytterbium doped fiber laser is shown in the Fig. 4.3.4. The specification details of fiber laser are given in Table 4.3.3.

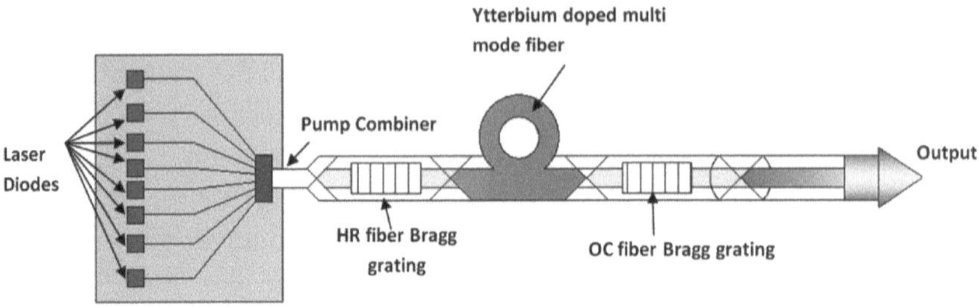

Fig. 4.3.4 Schematic diagram of multidiode ytterbium doped fiber laser [5].

Table 4.3.3 Specification of the fiber laser

No.	Specification	Description
1.	Laser type	Pulsed
2.	Wavelength	1064 nm
3.	Mode of operation	Pulsed
4.	Mode of laser beam	TEM_{00}
5.	Beam diameter	8.8–9 mm
6.	Laser beam spot diameter	18–20 μm (after focus)
7.	Average power	50 W
8.	Pulse width	120 ns
9.	Pulse repetition rate	50–120 kHz
10.	Output fiber length	3 m

4.3.2.3 Fundamentals of CO_2 laser generation

CO_2 laser is an electrically excited gas discharge laser and it makes use of three gases: carbon dioxide (2–5%) as the active medium, nitrogen (10–55%) as a charging medium and helium (40–88%) as coolant. Nitrogen gases are charged by means of high voltage direct current (DC) or RF charging. When the nitrogen molecules are charged, they have to transfer energy to the carbon dioxide molecules by means of collisions. This process is only effective if the gases are cold enough. The carbon dioxide molecules are excited to higher energy level and then come to ground state by emitting laser light. CO_2 lasers operate in the infrared region, 10.6 μm. The amplification of laser light takes place in the optical resonator which consists of two cavity mirrors such as fully reflective concave mirror and partially transmitting output mirror. The output power of carbon dioxide laser beam varies from 10 W to 45 kW. CO_2 laser can be obtained in the continuous wave (CW) form as well as pulsed wave (PW) form. For PW CO_2 laser, the peak power is very high, although the average power is low. The pressure in the discharge chamber is usually depending on the type, size, and operating mode of the laser. Pulsed lasers usually operate at higher pressures than continuous-wave (CW) lasers, with some pulsed laser pressures as high as 1 atm. CO_2 laser can produce outputs with conversion efficiency of 10–15%. Because of the high power levels available (combined with reasonable cost for the laser), Transverse excited atmospheric pressure operated (TEA) CO_2 laser is generated for high power laser applications. CO_2 lasers are frequently used in industrial applications for cutting and welding and - surface engineering etc. Figure 4.3.5 shows the schematic diagram of CO_2 laser generation principle.

4.3.2.4 Fundamentals of excimer laser generation

The name Excimer is an acronym for Excited Dimer. Two gas molecules are excited by energy released by electrical discharge. Excited Dimer or excimer molecules are formed by the electronic charging of a mixture of halide gas (fluorine or chlorine) with a rare gas such as argon, krypton or xenon. No excited dimer can stay for long time at this excited state. After some time, the excited molecules are separated out with releasing some energy in the form of beam of photon which has called as excimer laser. Excimer laser is also generated due to stimulated emission process. The excimer laser is a gas laser. The excimer laser is an ultraviolet (UV) laser. The operating medium is a

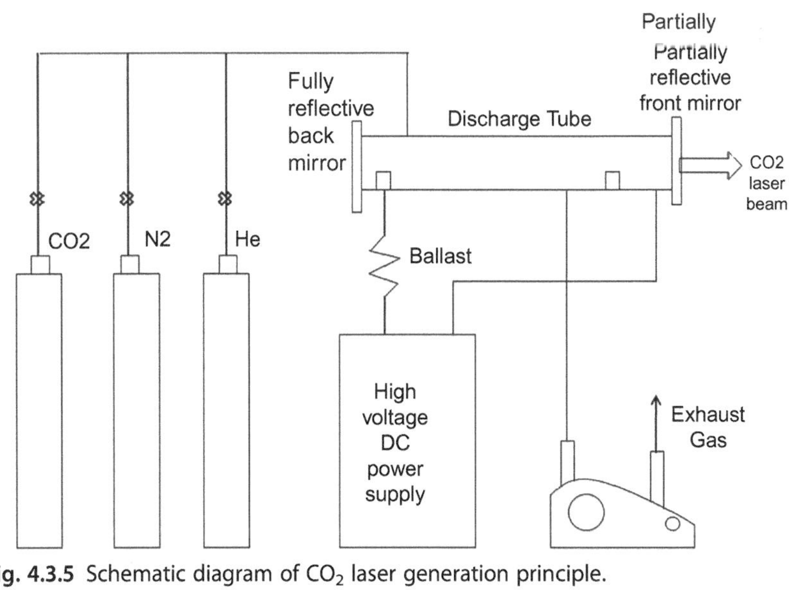

Fig. 4.3.5 Schematic diagram of CO_2 laser generation principle.

gaseous mixture of a noble gas and a halide gas mixed with either He or Ne as a buffer gas, with a total gas pressure in the chamber of about 3 bar. The wavelength produced by the laser depends upon the gas mixture used, but all wavelengths are in the ultra-violet part of the spectrum. In a typical gas mixture, 2% noble gas, 0.2% halogen gas and balance of buffer gas are used.

During generation of excimer laser, ArF, the mixture of argon and fluorine is excited due to electrical discharge, the following processes occur:

1. Electron attachment: $e + F_2 \rightarrow F^- + F^-$
2. Formation of excited molecule: $Ar^+ + F^- \rightarrow (ArF)$
3. Dissociation of excited molecule: $(ArF) \rightarrow Ar + F +$ Photon (193 nm)

For excimer laser generation, the gases are excited by a high voltage (around 23 kV) pulsed electrical discharge switched by a thyratron trigger circuit. The Laser gas must be circulated and cooled using an internal water heat exchanger. An excimer laser generation system consists of a sealed tube filled with laser gas through which an excitation pulse is passed. The electric pulse is perpendicular to the laser beam axis. Laser cavity can be repeatedly refilled, as the laser gas degrades during use. It is necessary that the laser cavity, optics, and electrodes must be designed to resist corrosion by halogen in the laser gas. Passive components typically are coated with teflon, and electrodes are made of halogen-resistant materials such as solid nickel or brass. Figure 4.3.6 shows the scheme of excimer laser generation principle.

The maximum repetition rate for existing industrial excimer laser generally varies between 100 and 250 Hz, although some units may work outside these limits. Each Laser pulse has an energy of between 300 to 1000 mJ depending upon the choice of laser and gas mixture. This translates to average powers of between 30 to150 W. However, since the pulse lengths are

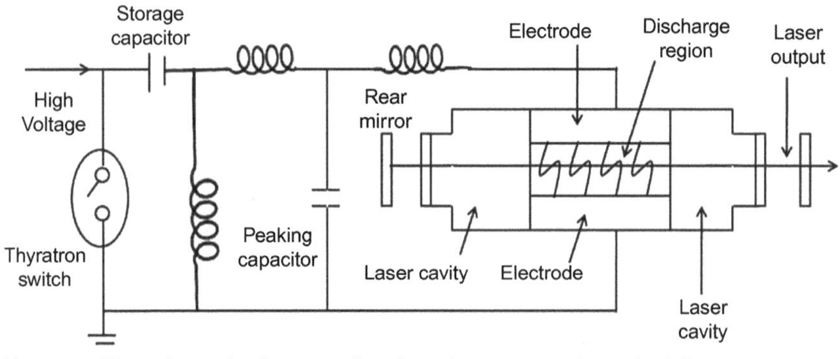

Fig. 4.3.6 The schematic diagram of excimer laser generation principle.

short, around 25 ns, peak powers can be as high as 25 MW. Moreover, the energy is delivered in very short periods of time (typically 25 ns), making it possible to remove material in small increments. In comparison to other lasers, excimer laser beams are short pulsed, shorter wavelength, have small coherence and high divergence. Figure 4.3.7 shows the photographic view of 500 W Excimer Laser Set-up for machining. The typical values of excimer laser parameters are given in Table 4.3.4.

4.3.2.5 Fundamentals of diode laser generation

A diode laser is a laser where the active medium is a semiconductor similar to that found in a light-emitting diode. The most common and practical type of diode laser is formed from a p-n junction and powered by injected electrical

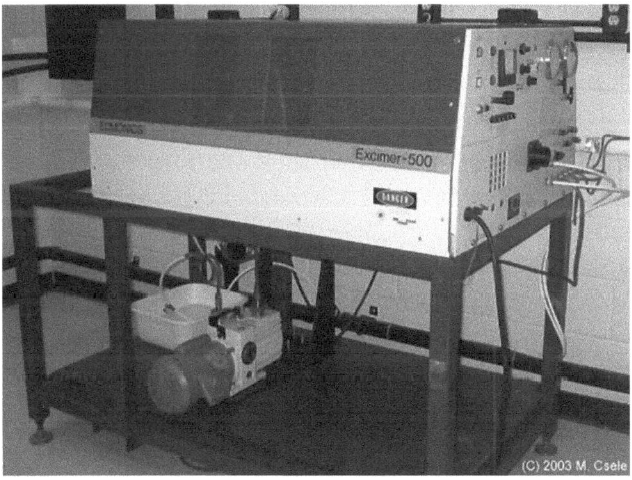

Fig. 4.3.7 Photographic view of excimer laser system. *(Courtesy: Lumonics Inc.)*

Table 4.3.4 Typical excimer laser parameters [6]

Parameters	Value
Wavelength, nm	193 (ArF); 248 (KrF); 308 (XeCl); 351 (XeF)
Pulse energy, mJ	400–500
Repetition rate, Hz	100–200
Average power, W	40–100
Pulse duration, ns	15–30
Peak power, MW	10–30
Peak fluence, mJ/cm^2	100–200
Beam divergence angle, mrad	1–3
Beam size, mm	8–15 × 25–30

current. These devices are sometimes referred to as injection laser diodes to distinguish them from optically pumped laser diodes, which are more easily produced in the laboratory.

When a voltage is applied in a p-n junction diode in the forward direction, electrons are injected from the n-region into the depletion region of the junction and holes are injected from the p-region. The electrons and holes co-exist in the same region for some time. When injection current density is sufficiently high, population inversion and lasing are induced. The electrons and holes are recombined by stimulated emission. The photons are emitted with recombination energy with same phase difference and same wavelength. The beam of photons generated from diode is laser beam. AlGaAs and InGa As diodes produce Continuous Wave (CW) laser with output power in the range of several watts in the wavelength rage 0.8–1.0 μm. InGaAsP diode produces laser in the wavelength range of 1.3–1.8 μm. AlGaInP diode can produce laser at wavelengths as short as 640 nm. The electrical to optical power conversion efficiency is greater than 50%. The multiple arrays of laser diodes generate high power laser for machining operation. Semiconductor diode lasers are widely used as optical pump source for generating solid state laser. Figure 4.3.8 shows the diode laser generation principle.

High power direct diode laser can be used for machining of polymer and composites and highly reflective materials like copper and aluminum. Direct diode laser are applied for non-machining applications such as soldering, brazing, welding, hardening, melting and surface modifications of softer

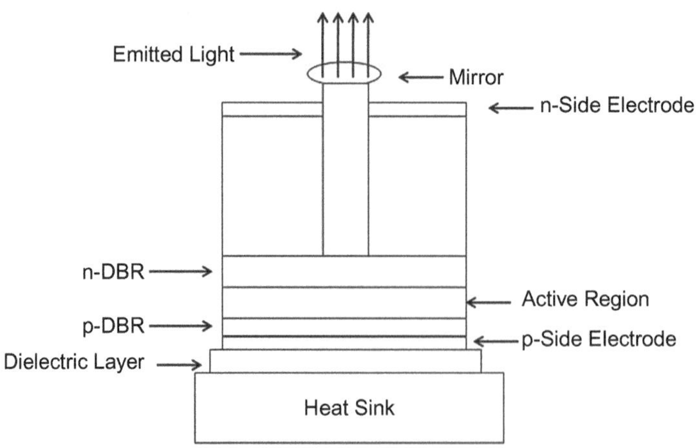

Fig. 4.3.8 Schematic diagram of diode laser generation principle.

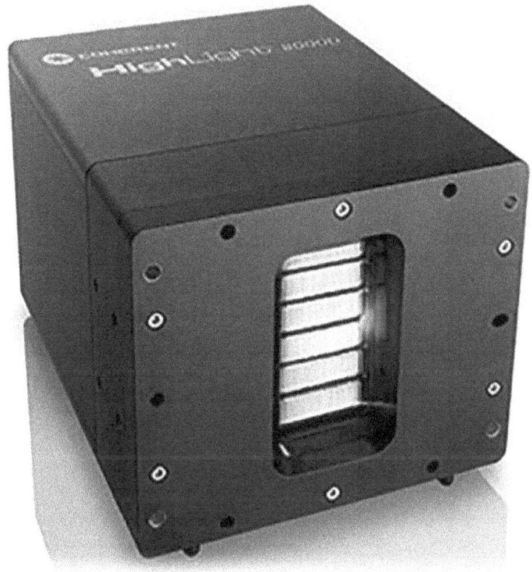

Fig. 4.3.9 Photographic view of high power direct diode laser. *(Courtesy: Coherent Inc.)*

materials like plastics as well as metals. High power diode laser have many advantages like high energy efficiency (>50%), portable, more reliable, longer service life, less maintenance, less heat affected zone and better surface finish. The limitation of diode laser includes high beam divergence, inability to produce very high peak power with short pulse laser beam. Figure 4.3.9 shows the photographic view of high power direct diode laser.

4.3.3 Material removal mechanism in laser beam machining

The energized laser beam is focused to a point after passes through a lens. As a result, a high energy concentration is obtained at that point. Laser radiation is first absorbed by the material surface where the optical energy is converted into heat. Amount of absorption of laser radiation by a work piece mainly depends on wavelength of laser, temperature, reflectivity of the work piece, angle of incidence of the laser beam on the work piece and intensity of the focused laser beam. This intense heat melts and vaporizes the material. A portion of the material vaporizes so rapidly that a substantial mechanical impulse is generated to remove a large amount of molten material from the cutting zone. Laser ablation occurs for short pulse laser. Thermal spalling takes place for ceramic materials. An additional assist gas jet of very high velocity is used which exerts momentum on the material leading to the ejection of molten material.

There are various effects of laser material interactions such as heating, surface melting, vaporization, plasma formation and ablation as shown in Fig. 4.3.10 during material removal process of laser beam machining. While long pulse laser beam is used in machining, there are thermal effects such as formation of melting zone, heat affected zone, recast layer, thermal damages of the adjacent layers etc. during material removal process as shown in Fig. 4.3.11. But while short pulse or ultra-short pulse laser beam is used during machining the materials are removed without melting and therefore there are no such thermal damages during material removal process.

When laser beam is incident on the surface of the materials, there are changes of states such as heating of room temperature to melting point, melting, heating of melt to vaporization temperature and then vaporization of molten materials. The required energy (E) for these changes is given as:

$$E = m\left[Cp(Tm - Tr) + Lm + Cp\,(Tm - Tv) + Lv\right]/At \tag{4.3.1}$$

where E = required energy of laser beam to remove mass (m) of materials; Cp is the specific heat of the materials; Tr is the room temperature of the

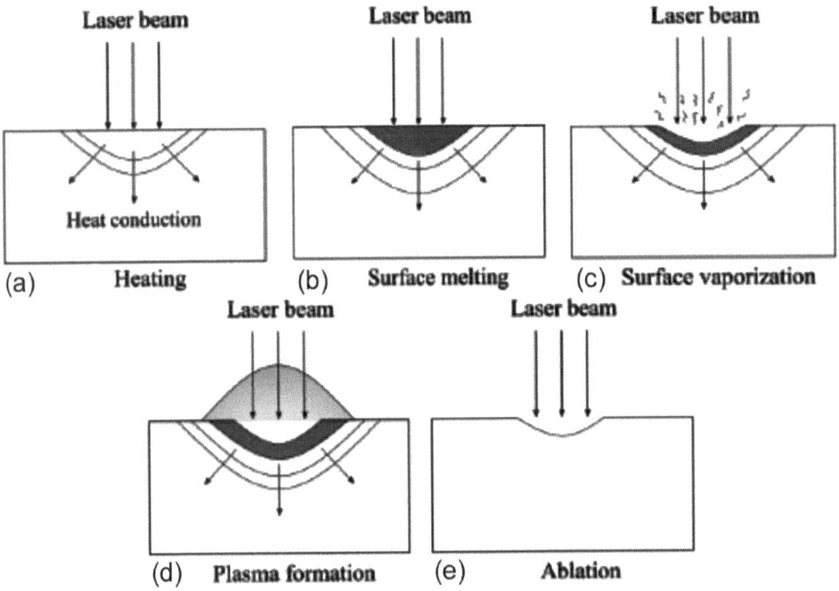

Fig. 4.3.10 Various effects of laser material interactions such as (a) heating, (b) surface melting, (c) vaporization, (d) plasma formation and (e) ablation [7].

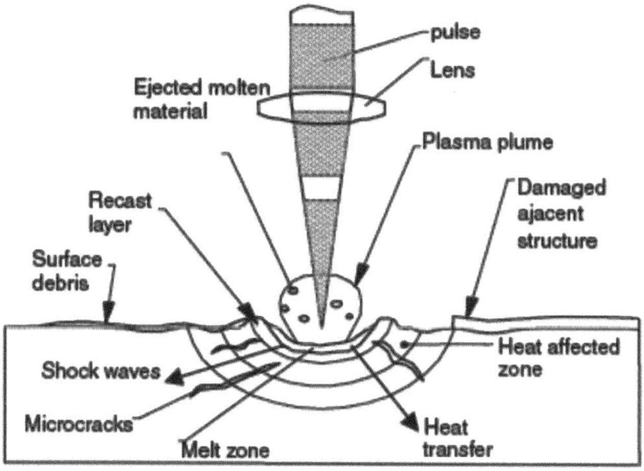

Fig. 4.3.11 Schematic representations of thermal effects of long pulse laser material interactions [6]. *(Reproduced from D.T. Pham, S.S. Dimov, P.V. Petkov, Laser milling of ceramic components, Int. J. Mach. Tools Manuf. 47 (2007) 618–626.)*

materials; Tm and Tv are melting point and vaporization point temperature respectively. Lm and Lv are latent heat of melting and vaporization respectively. The machining time is represented as t. The percentage of incident energy absorbed is represented as A.

From the energy balance equation, mass of material removed per unit time (MRR) can be expressed as:

$$MRR = m/t = EA/[Cp(Tm - Tr) + Lm + Cp\,(Tm - Tv) + Lv] \quad (4.3.2)$$

The machining speed (V) during laser cutting of a slot of width, w and depth d of material with density p is represented as:

$$\begin{aligned} V &= m/(w.d.p.t) \\ &= EA/[Cp(Tm - Tr) + Lm + Cp\,(Tm - Tv) + Lv]w.d.p \end{aligned} \quad (4.3.3)$$

4.3.4 Laser beam machining systems details

Laser Beam Machining system consists of various sub-systems such as laser generation and beam delivery unit, cooling unit, assist air supply unit and CNC controller for X-Y-Z axis movement. Figures 4.3.12 and 4.3.13, respectively, show the schematic and photographic view of the typical Nd: YAG laser beam machining system indicating various sub-systems.

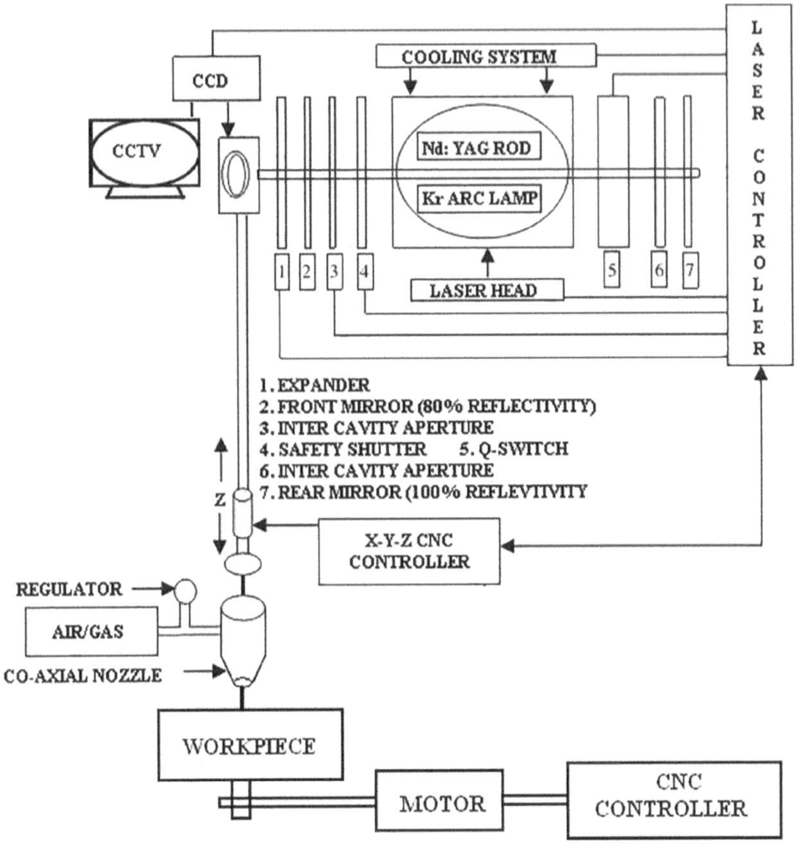

Fig. 4.3.12 Schematic view of the typical Nd: YAG laser beam machining system indicating various sub-systems.

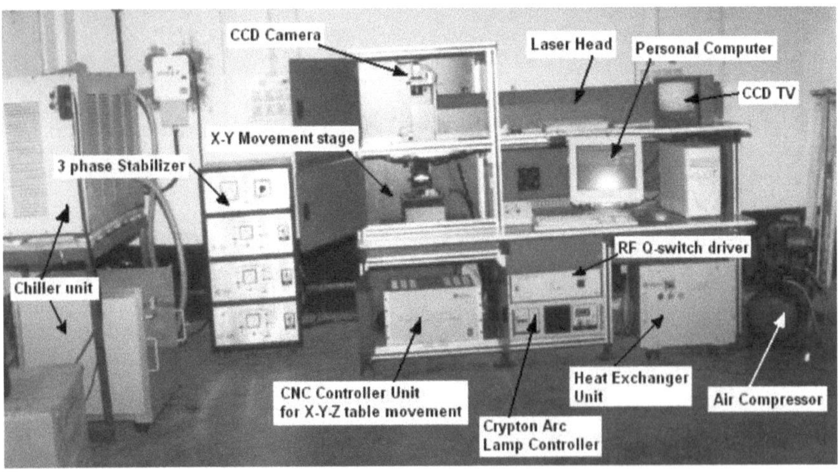

Fig. 4.3.13 Photographic view of the pulsed Nd: YAG laser marking system indicating various sub-systems. *(Courtesy: M/S Sahajanand Laser Tech. Ltd. India.)*

4.3.4.1 Laser generation and beam delivery unit

Laser Generation and Beam Delivery Unit consists of various sub-systems such as

(a) Laser head
(b) Optical excitation unit
(c) Cavity mirrors
(d) Q switch and RF drive unit
(e) Laser beam delivery and focusing unit

(a) Laser head

Laser head for lamp pumped Nd: YAG Laser consists of mainly Nd: YAG rod, krypton arc lamp and gold plated elliptical cavity. The krypton arc lamp is placed at one focus of the ellipsoid, and the Nd: YAG rod of the active medium at another focus. The inner surface of the cavity is coated with a reflective coating usually gold, such that all the radiation emitted from the lamps ended at the active medium. The active medium in the Nd: YAG laser is neodymium ions (Nd^{3+}) which is doped in a solid host material of yttrium aluminum garnet (YAG). The host is a synthetic crystal with a garnet-like structure and the chemical formula $Y_3Al_5O_{12}$. Neodymium ions take the place of yttrium ions in the garnet lattice—they are roughly the same size. The main advantage of YAG compared with other host materials is its good thermal stability. Laser light is generated through transitions between energy levels of the neodymium ion. Most solid-state lasers are derived from optically pumped transitions among electronic levels of ions in a crystal or other solid state host.

Laser Head for diode pumped fiber laser system is comprised of a fiber itself doped with Yb^{3+}, fiber Bragg grating and number of diodes. The number of laser diodes is dependent on the total output power of the system. A total number of 8 diodes are utilized for optical pumping and generating average fiber laser power of 50 W. Fiber Bragg gratings are sliced into two ends of the fiber optics so that the laser can be generated. The total length of the fiber can be up to 3 mm depending upon the output power generated by the fiber laser. After the generation of the laser within the optical fiber, the laser is transferred into a collimator. In the collimator, two prisms are fitted at the two ends of it. This collimator acts as a beam expander, i.e., the laser can propagate to the required distance. The typical value of diameter of the laser beam is 9 mm at the end of the collimator.

(b) Optical excitation unit

Optical excitation unit for Nd:YAG Laser machining system mainly consists of krypton arc lamp and its power supply unit that controls the laser output. It mainly ignites and controls the intensity of light emitted by the Krypton arc lamp. The intensity of light produced by the lamp is used for pumping the Nd atoms in Nd: YAG rod. To start the lamp, which is having very high impedance, first it is to be triggered with high voltage pulse. Once the discharge in lamp is produced, then by changing the current flowing through the lamp, the intensity of the light emitted by the lamp can be controlled. Now in many applications laser is not used continuously and therefore the power supply is provided with a special feature of standby mode which keeps alive the discharge in lamp. This arrangement is very useful in increasing the operational life of the lamp and also that of power supply, because this will save the whole process of generating trigger pulse for igniting the lamp. The power supply unit of the lamp is coupled with internal chiller unit by interlocking arrangement. This is because, if suddenly, cooling system stop working, then the lamp will get shutdown and thus, this arrangement saves both very expensive lamp and Nd: YAG rod. The lamp current is varied by rotating a knob in the power supply system of lamp and is the most important controllable process parameter for laser machining process. Figure 4.3.14 shows the schematic representation of Optical Pumping of Solid State Laser.

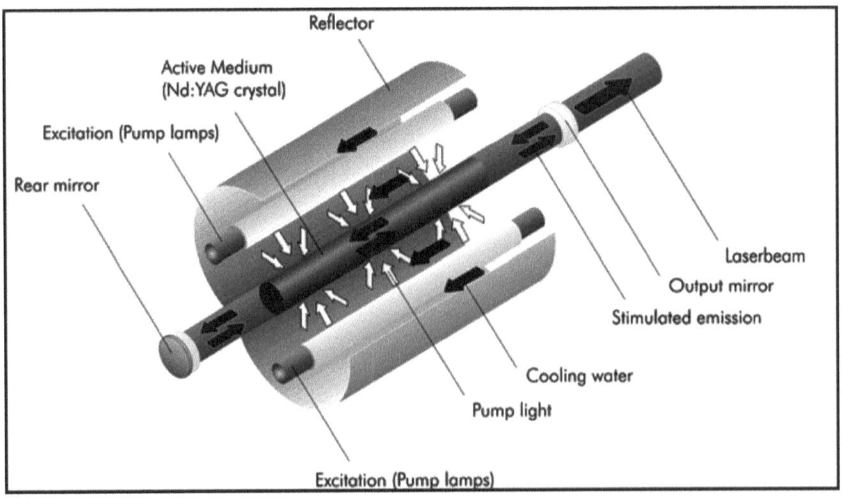

Fig. 4.3.14 Schematic diagram of optical pumping of solid state laser [8].

(c) Cavity mirror

A fully reflecting mirror is located at one end of the rod with 100% reflectivity which is called back mirror or rear mirror, and a partially reflecting output mirror of reflectivity of about 80% at the other is called as front mirror. To get amplified light, feedback is essential which is provided by two mirrors. This arrangement of light application by cavity mirrors is common for most of the laser generation unit.

(d) Q-switch and RF drive unit

The output of a gain switched, pulsed solid-state laser is generally a train of irregular pulses—irregular in peak power, pulse width and repetition frequency. It is possible to remove these irregularities and at the same time greatly increase the peak power by a technique called Q switching. Q-switched lasers normally emit giant pulse in an operational cycle. The pulse length is typically less than a microsecond down to several nanoseconds and peak power increases from 10^6 W to 10^9 W. Q switching is a mode of laser operation in which energy is stored in the lasing medium during pumping in the form of excited atoms and suddenly released in a single, short burst. This is accomplished by changing the optical quality of the laser cavity. The quality factor Q is defined as the ratio of the energy stored in the cavity to the energy loss per cycle. Figure 4.3.15 represents the position of the acousto-optic Q switch. A Q-switch is essentially a shutter like arrangement placed between the active medium and the highly reflective (HR) mirror. With this shutter is closed, the HR mirror is blocked preventing oscillation. When the amplifier gain reaches a predetermined value, the shutter is opened to increase the cavity quality. Several techniques can be used for Q-switching of lasers. Within acousto-optic Q-switches a transparent element is placed in the cavity. This

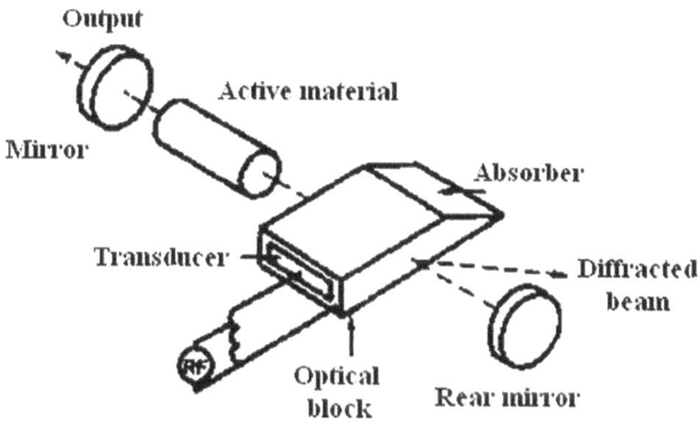

Fig. 4.3.15 Schematic representation of the acousto-optic Q switch [9].

transparent device, when excited with intense, standing, acoustic waves by piezoelectric crystals, exhibits a diffraction effect on the intra cavity laser beam and diffracts part of the laser beam out of the cavity alignment. This results in a low feedback. When the acoustic wave is removed, the diffraction effect disappears, the cavity is again aligned, and the system emits a giant pulse.

To get the pulsed laser output with high peak power, the laser is operated in Q-switched mode. This is done by using an acousto-optic Q-switch. This Q-switch requires a special power supply to operate it. An acousto-optic Q-switch makes use of RF source at 27.2 MHz that is modulated with a frequency ranging from 0.1 kHz to 19.9 kHz. For this radio frequency, RF generator is used. As it is being switched with a high frequency, it is also cooled by chilled water for repetitive operations and therefore the unit is provided with over temperature interlock.

(e) Laser beam delivery and focusing unit

The laser beam delivery system consists of beam bender, intra cavity shutter, intra cavity aperture and a focusing lens. The laser beam from the beam expander of the laser source is to be sent to the work-surface. Beam bender accomplishes this by bending the beam at 90 degree, which is then focused by the focusing lens.

An important component in the laser cavity is the electrically activated safety shutter. It also blocks the path of refraction between the optical feedback mirrors. Unlike the Q-switch, the safety shutter is a mechanical device designed to block the lasing action for long period of time. The shutters are designed in such a way that gravity will drop the shutter into the closed or non-lasing position in the event of an electrical failure. The safety shutter is located inside the laser head assembly and is actuated by toggle switch. The shutter terminates lasing by blocking the laser beam path and preventing emission of laser radiation out of the head assembly enclosed.

Aperture restricts the amplification to occur along the off axis of the resonator as this apertures are used inside the cavity so the name intra cavity aperture. For light to get amplified, feedback is needed which is provided by two mirrors.

Focusing units consists of CCD camera and CCTV. The CCD camera operating at 12 V is provided on top of the laser head for capturing the image during laser beam machining and the beam has to be properly focused on the work surface for effective machining. The position of focused beam can be adjusted with the help of CCTV which is connected to the CCD camera in order to obtain accurate focusing of laser beam. There will not be any material removal if the focal point of the beam is well above the job surface.

Moreover, if the beam is focused well below the surface then it may lead to thermal shock which may lead to the cracking of the material. Once laser beam properly focused the material starts to melt and evaporate, hence position of focal point of laser beam has to be adjusted perfectly. The experimental set up had a CNC controlled arrangement for effective adjustment of the focal point by controlling the movement of lens position along Z-axis.

4.3.4.2 Cooling unit

The cooling system is used to cool the laser head and Q-switch. The cooling unit consists of the subunits such as three-phase chiller unit, heat exchanger and pump unit. Three-phase chiller unit is used for providing the chilled water to the heat exchanger in order to maintain the desired temperature during machining operation. Heat exchanger and pump unit are used for circulating the de-ionized water through resin filter via water to water heat exchanger, laser head & Q-switch. When the laser is produced inside the laser head cavity, a large quantity of heat is generated inside the pumping cavity due to very low electro-optic efficiency. Therefore if the heat is not removed from the cavity then it will cause both lamp and rod to get damaged. So, a proper arrangement of heat removal should be made. For this purpose chilled deionized water supply and cooling unit are provided. Chiller system supplies the chilled water to the heat exchanger. Pump system is responsible for circulating de-ionized water via water-to-water heat exchanger through resin filter to the laser head and Q-switch. Inside the laser cavity both lasing medium and the lamp are immersed in flowing cold de-ionized water. The de-ionized water is used because it has high optical transparency and low electrical conductivity. The heat from de-ionized water is subsequently removed by water-to-water heat exchanges connected to an outside water source. The de-ionized water temperature is controlled by means of a solenoid, which turns the outside water flow "ON" and "OFF" as required. Regulation of de-ionized water temperature to its flow degree further contributes to the stability of laser output. For high power laser system, water cooling is an essential unit, but for low power laser system air cooling arrangement is generally provided.

4.3.4.3 Assist gas supply unit

Air is supplied by the compressor and the compressed air is chosen as the assist gas. Primary function of assist gas is to assist in removing the molten material from the ablated surface and it is supplied to the laser marking zone through a co-axial air flashing nozzle. With the change of supply pressure of assist air the quality of machining surface can be improved. The supply line of the compressed air passes through moisture separator and connected to pressure regulatory valve then it supplies dry pressurized air to the laser machining zone through the co-axial nozzle attached to the beam delivery system. Hence, for experimental purpose the range of air pressure of 0.3–2.3 kgf/cm^2 is used for experimentation. Moreover, gas pressure has little significance at high laser power when instantaneous vaporization occurs.

4.3.4.4 CNC controller for X-Y-Z axes movement

The work piece movement along the X-Y axis and the focusing lens movement along Z axis are CNC controlled and can be operated via a computer for machining by proper focusing of laser beam by means of focusing lens. The feed rate is controlled by X-Y table movement so that the machining is done appropriately. The CCD camera together with CCTV monitor is used for viewing the location of work piece and also for checking the proper focusing condition.

Different units of diode pumped fiber laser micro-machining system of a 50 W ytterbium doped fiber laser system with a wavelength of 1064 nm working in TEM00 mode of operation, made by M/S Sahajanand Laser Technology Limited (Model: AKSHAR Fiber Pro-309) are discussed here. A diode pumped fiber laser micro-machining system is comprised of (a) power supply unit, (b) laser head, (c) collimator, (d) beam bender, (e) beam delivery unit and focusing lens, (f) assist gas supply unit and (g) computer numerical control (CNC) controller for X-Y-Z movement. The schematic diagram of the fiber laser machining system is shown in Fig. 4.3.16. In the aforesaid fiber laser system, the working range of pulse frequency is 50–120 kHz, whereas pulse width is in the regime of 120 ns. The laser finally propagates through a F-θ lens of 71 mm of diameter (focus lens) which is protected by a nozzle for preventing against dust and other contaminations. The spot diameter of the fiber laser beam is 21 μm. The laser beam falling on the surface from the focusing lens has to be précised and appropriately aligned. If the center of the lens does not coincide with the beam center, then the beam that will pass through the lens will not be linear, resulting in low energy of the laser beam and less micro-machining efficiency. CCTV and CCD camera are the two units used for focusing. At the top of the laser head, the CCD camera is placed in order to capture the image during the operation of the laser micro-machining. Further, the beam should also be focused on the surface so as to receive optimum micro-machining operations. The function of the CCTV is to adjust the position of the focused beam on the

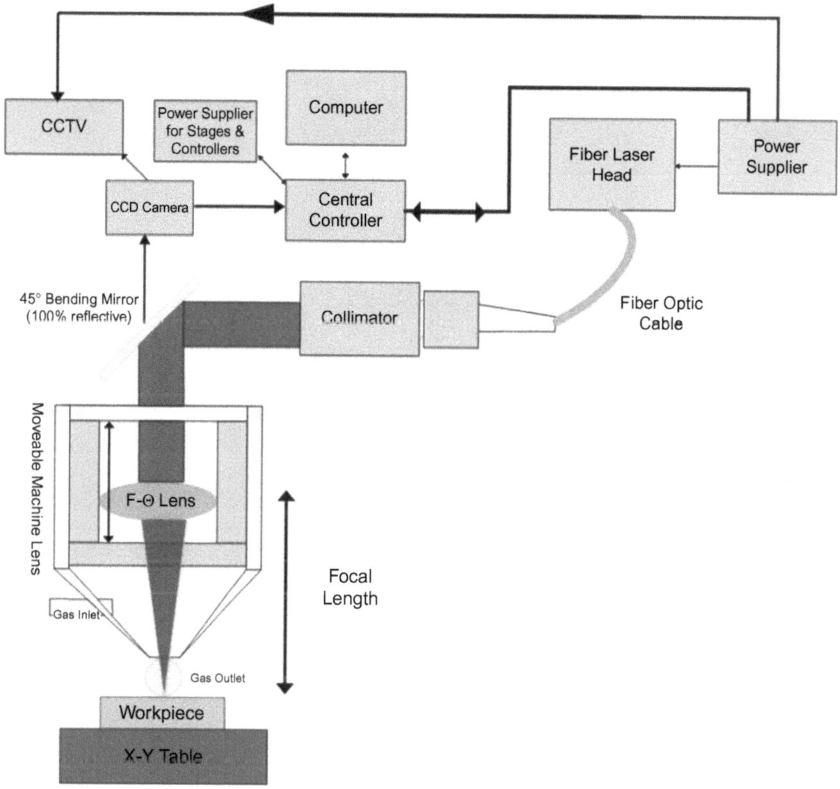

Fig. 4.3.16 Typical fiber laser beam machining system [10].

workpiece. This device is connected to the CCD camera so as to obtain précised laser beam focusing condition. After the laser beam is appropriately focused, the material melts at a faster rate and thereby evaporates. This is the reason why the focal point position of the laser beam should be properly adjusted. A CNC controlled arrangement is provided in the experimental setup, in order to effectively adjust the focal point of controlling the lens movement position along the Z axis. The details of the 50 W diode pumped fiber laser beam machining setup is shown in the Fig. 4.3.17.

4.3.5 Process parametric studies on performance criteria of LBM

The important process parameters those are to be controlled during laser beam machining are Pulse Frequency, Pulse Duration, Peak Power,

Fig. 4.3.17 Photographic view of 50 W multimode ytterbium doped fiber laser beam machining setup. *(Courtesy M/S Sahajanand Laser Tech. Ltd, India.)*

Scanning Speed, Focused Spot Size, Air Pressure etc. The process parameters of laser beam machining are discussed here-in-after:

(a) Pulse Frequency, Pulse Duration and Peak Power

The pulse repetition rate determines the amount of time between laser's bursts allowed to strike a surface. The ability to Q switch is to provide short pulses with very high peak powers that enable the laser to machine wide range of materials. The laser parameters such as pulse duration, peak power and average power can be varied as per requirement according to the material and type of machining process. The average power, peak power, pulse frequency and pulse duration are related as follows:

$$\text{Peak Power} = \text{Average power}/(\text{Pulse frequency} \times \text{Pulse duration}) \quad (4.3.4)$$

(b) Scanning speed

In laser beam machining process the scanning speed of laser beam with respect to the work piece is another important variable and the scanning speed must be set to achieve the desired machining performance during various machining operations. For deeper machining (typically >0.05 mm) each point on the material will require exposure to several pulses to achieve depth. The scanning speed must be reduced until the desired depth is achieved. The result of pulse frequency and beam speed is the laser spot overlap. As a general rule, pulses should overlap at least 50% to give the appearance of a continuous laser beam machining and around 70% to 90% to ensure a good machining.

(c) Focal spot size

Focal spot size determines the maximum energy density that can be achieved when the laser beam power is set, so the focal spot size is very important for material processing. The diameter of the focused laser beam on the work surface determines both the width and depth of machining.

The focal spot diameter (d) depends on the focal length (f) of the focusing lens and laser beam divergence angle (α). The beam divergence angle (α) depends on the wavelength of laser and the beam diameter (D) at the entry to the lens and the relation is given as follows:

$$d = f.\alpha = f.2\lambda/D \quad (4.3.5)$$

Figure 4.3.18 shows the laser beam profile with respect to position of lens and workpiece.

Therefore the focal spot diameter depends on type of laser. For example, wavelength of CO_2 laser ($\lambda = 10.64\,\mu m$) is ten times greater than that of Nd: YAG laser, therefore the laser spot size is 10 times greater than that of Nd:

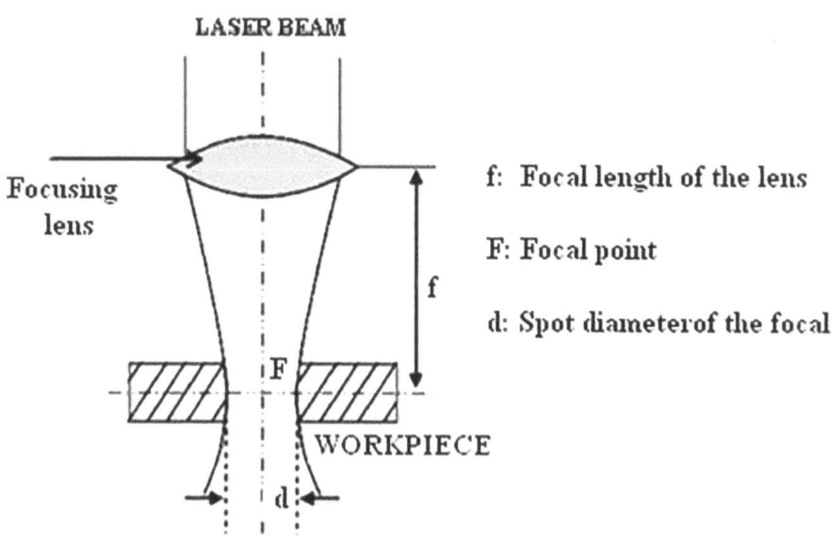

Fig. 4.3.18 Schematic diagram of laser beam profile [11].

YAG laser. As wavelength of fiber laser is same as Fiber laser, so focal spot size is almost same for both fiber laser and Nd:YAG laser considering other parameters as same. Focal spot size can be well below than 500 μm for laser beam and the energy intensity of laser beam beams very high in the range of 10 GW/cm^2. The laser beam can be applied for micro-machining applications.

(d) Air pressure

Air pressure plays an important role during laser beam machining. It mainly assists to remove the material and to minimize the heat affected zone during laser beam machining. Moreover, the air protects the focusing optics against vapor or spatter emitted due to high heat generation during machining. The air pressure during laser machining affects the machining surface quality characteristics. However, the pressure should not be so high such that it will create vibration of machine parts.

The performance criteria of Laser Beam Machining for various operations are given as follows:

(i) For laser cutting operation: cut or kerf width, depth, taper formation and surface roughness, Heat Affected Zone (HAZ) and dross formation;

(ii) For laser drilling operation: hole diameter (entry and exit side), depth, surface characteristics, HAZ, aspect ratio and taper;

(iii) For laser marking operation: mark width, mark depth, mark continuity and mark intensity or contrast.

(iv) For laser turning operation: depth, length, surface finish and HAZ.

(v) For laser grooving operation: groove dimensions, groove geometry, HAZ and surface quality.

The cause effect fish bone diagram which represents the input process parameters and machining responses is given in Fig. 4.3.19.

4.3.5.1 Laser drilling

Laser drilling produces micro-hole as well as large size hole on workpiece. Laser percussion drilling is the micro-machining process for generating micro-hole of diameter less than 500 μm. During laser drilling the beam is normally stationary with respect to the work piece. The beam heats the material to vaporization temperature, after which it penetrates to form a cavity. The assist gas with high pressure helps to remove molten material from the cavity wall to its outer rim where it is expelled. Laser hole drilling can be done by four practical techniques i.e. single pulse drilling, percussion drilling

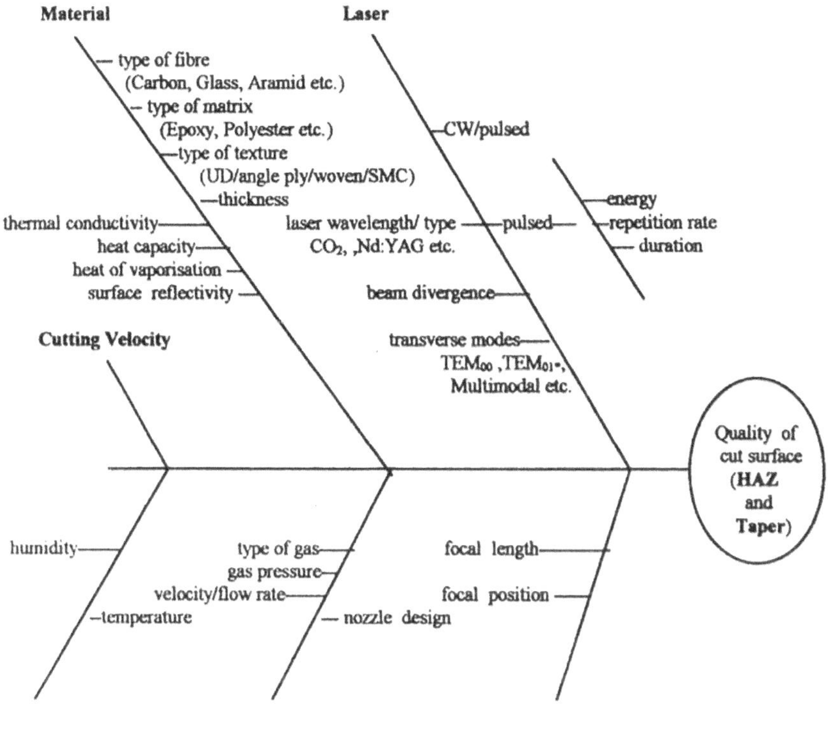

Fig. 4.3.19 A cause effect diagram showing the various parameters affecting the quality of the laser beam machining [12].

and trepanning and helical drilling. In single pulse drilling a hole is drilled with a single pulse with energy of tens of joules and duration in the order of 1 ms. A large quantity of molten and vaporized material is produced, which causes difficulties in maintaining smaller dimensional tolerance. In percussion drilling technique, several lower energy pulses are used to drill each hole, thus limiting the amount of molten and gaseous material production, enabling tighter tolerances. Here the diameter of the hole is smaller than the laser focused beam. Trepanning is a means of cutting holes using one of the above mechanisms by guiding the beam around pre-defined locus. It is used when the diameter of the hole to be cut is considerably larger than the laser spot diameter. Material is first pierced at the center of the hole, and the cutting front guided out to the periphery. In trepanning, the movement of beam determines the accuracy of hole diameter. This is a sharp contrast to the other two types of drilling techniques, where the beam characteristic mostly influences hole quality. In helical drilling, laser beam follows the helical path for movement and also it rotates about its own axis with respect to workpiece. The rotational movement of laser beam is controlled by Dove prism. Laser beam helical drilling can produce high quality hole with high aspect ratio and minimum heat affected zone. The schematic diagram of various types of laser hole drilling operation is shown in Fig. 4.3.20.

Nd: YAG Laser drilling can be used to make a micro hole on ceramics such as alumina and zirconia etc. By controlling the process parameters such as lamp current (19–25 A), pulse frequency (0.2–4.2 kHz), assist air pressure (0.5–2.5 kgf/cm^2), duty cycle (2–22%), and micro hole of less than 100 μm was drilled on alumina by CNC Pulsed Nd: YAG Laser drilling machine. The cross sectional view of laser drilled hole on alumina is shown in Fig. 4.3.21.

The micro-drilling of zirconium oxide ceramics has been carried out by CNC Pulsed Nd:YAG Laser beam. The minimum half taper angle of 0.04 radian and minimum HAZ thickness of 0.1296 mm of micro-hole on zirconia have been achieved at Nd: YAG laser beam machining parameter setting of lamp current, 17 amp; pulse frequency, 1.624 kHz, assist air pressure, 1.074 kgf/cm^2 and pulse width, 2% of duty cycle [14].

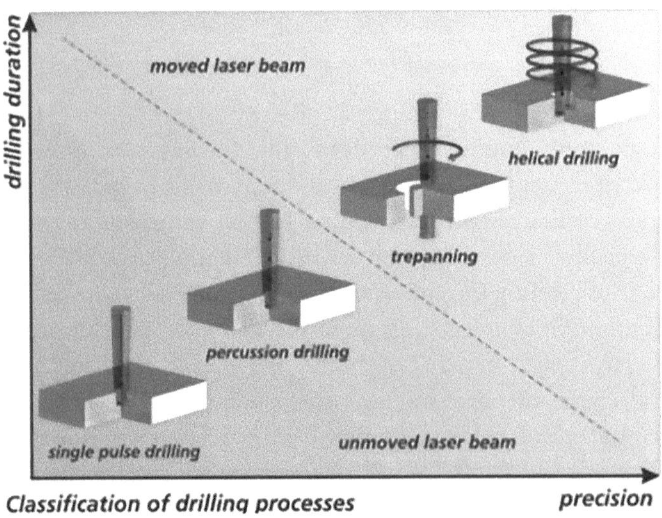

Fig. 4.3.20 Schematic diagram of various types of laser hole drilling operation [13].

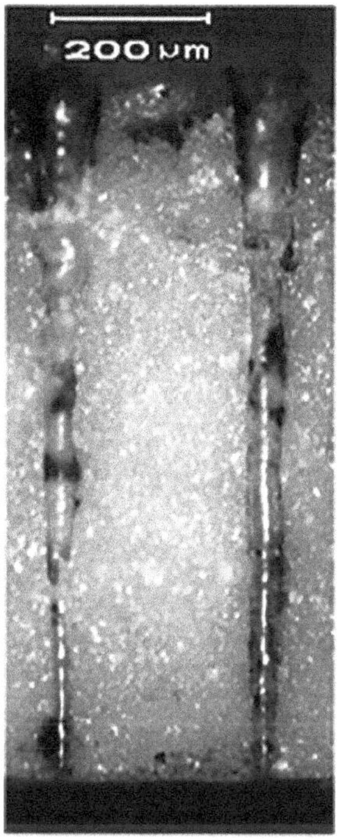

Fig. 4.3.21 Cross sectional view of laser drilled hole on alumina.

Laser drilling operation has also been performed on Ti 6Al 4V sheet of 8 mm thickness and nickel alloy of 4 mm thickness and taper hole with half taper angle 0.5–2 degree was obtained. The hole taper decreases with the increase in pulse frequency irrespective of material thickness combinations as shown in Fig. 4.3.22. The hole taper also varies with depth of hole as shown in Fig. 4.3.23. As depth increases, hole taper decreases and becomes almost constant. The variation of hole taper with pulse energy as shown in Fig. 4.3.24 is very less while pulse energy varies from 7.5 to 15 J for fixed values of pulse width 0.7 ms during laser drilling of 8 mm thick titanium alloy [15].

The HAZ and recast layer formation were also observed during Nd: YAG micro-hole drilling on titanium alloy and also on nickel alloy sheet. The recast layer thickness decreases with increase in pulse frequency as shown in Fig. 4.3.25.The variation of HAZ thickness with pulse frequency is very less for laser drilled hole on titanium alloy as shown in Fig. 4.3.25. The recast layer thickness decreases with increase in pulse energy,

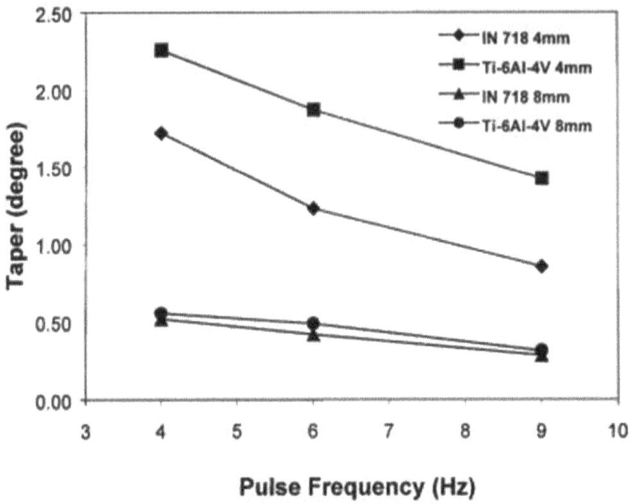

Fig. 4.3.22 Variation of taper angle with pulse frequency [15].

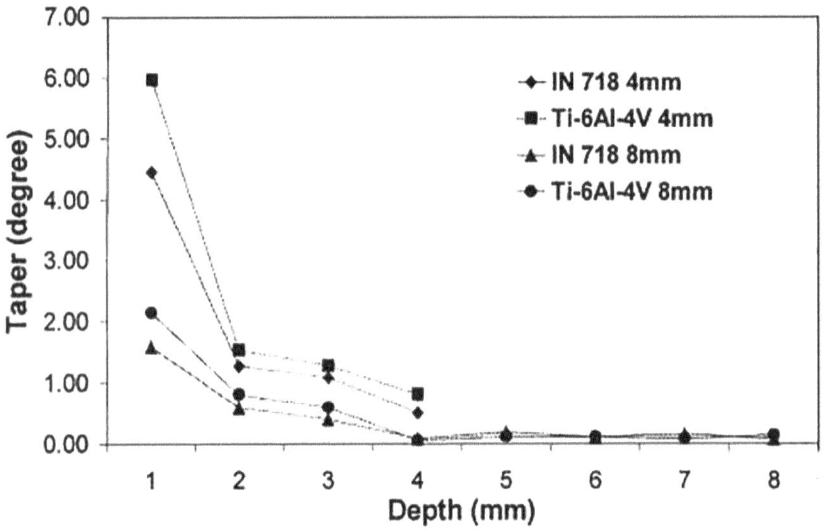

Fig. 4.3.23 Influence of depth on taper angle [15].

but HAZ thickness slightly increases with pulse energy as shown in Fig. 4.3.26.

A 300 W CO_2 laser percussion drilling operation has been performed on alumina ceramics of 2 mm thick plate. For drilling 1 mm hole diameter, entrance hole circularity of 0.963, exit hole circularity of 0.965, minimum

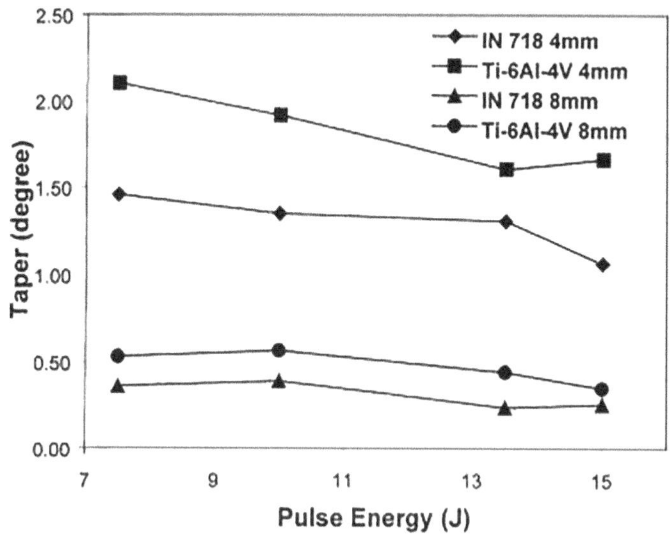

Fig. 4.3.24 Variation of pulse energy on taper angle [15].

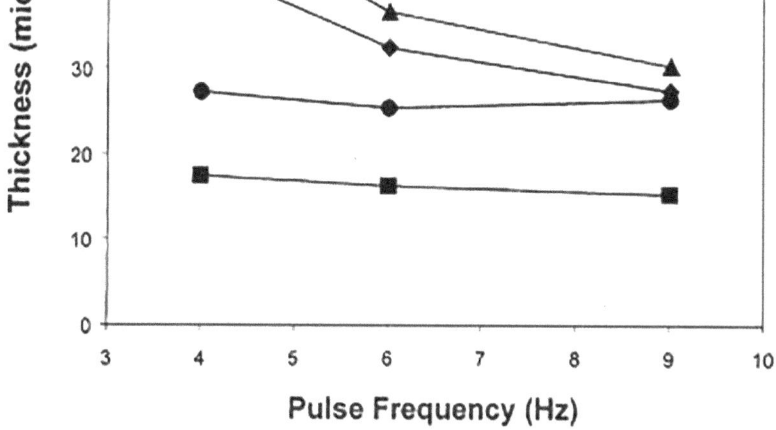

Fig. 4.3.25 Variation of recast layer and HAZ thickness with pulse frequency [15].

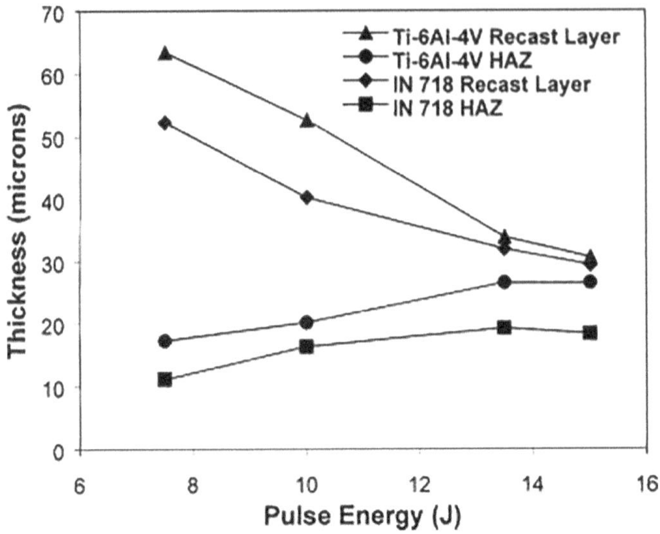

Fig. 4.3.26 Variation of recast layer and HAZ thickness with pulse energy [15].

HAZ of 0.55 mm, minimum taper of 0.357 degree was achieved by setting the CO_2 laser process parameters such as high pulse frequency of 7.5 kHz, high laser power of 240 W, moderate scanning speed of 3.85 mm/s. Figure 4.3.27 shows CO_2 laser drilled hole and observed HAZ at the entry side on alumina [16].

The advanced laser trepanning system can produce a large hole with positive taper, taper less and reversed taper by controlling the beam alignment, rotation and displacement of optics as shown in Fig. 4.3.28. Laser trepanning

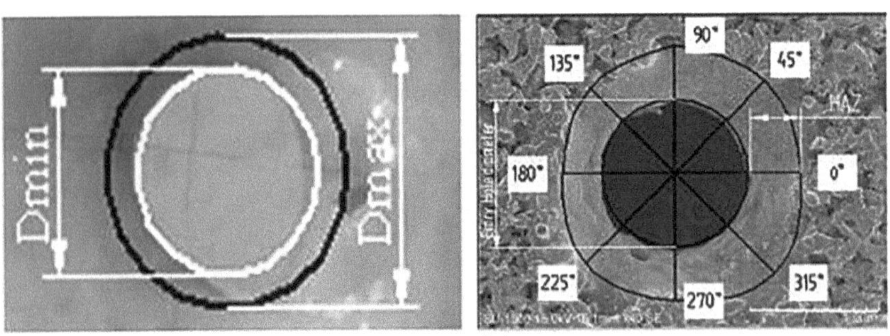

(a) (b)

Fig. 4.3.27 (a) CO_2 laser drilled hole and (b) observed HAZ at the entry side on alumina [16].

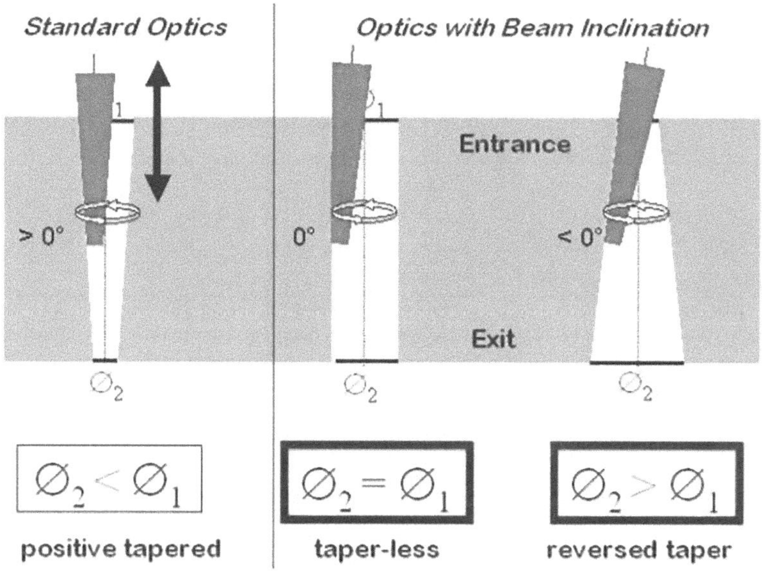

Fig. 4.3.28 Various taper geometries of hole [17].

on stainless steel of 500 μm has been performed using femtosecond laser system [17].

Laser beam helical drilling operation has been performed on stainless steel of 1mm using diode pumped frequency doubled Nd: YVO_4 laser with wavelength of 532 nm, average power 18 W, pulse frequency 20 kHz, pulse duration 20 ns. Figure 4.3.29 shows the illustration of rotating motion during laser helical drilling operation [13].

Excimer Laser KrF was utilized to drill an array of 300 micro-holes of diameter 55 μm with 27 degree taper on 50 μm thick polyamide for injet nozzles. The frequency of pulse laser was 300 Hz and pulse energy density was 0.6–0.8 J/cm^2. The total drilling time is about 1 s. Figure 4.3.30 shows array of holes on ink jet nozzles by KrF excimer laser.

4.3.5.2 Laser cutting

Laser cutting is a material removal process by which shaping and separating of parts into pieces is being done by utilizing the thermal energy of laser beam. Laser cutting process is commonly used for cutting difficult to cut materials in sheet metal industries. For laser cutting either workpiece or laser beam moves over the surface as per pre-defined path as per CNC program

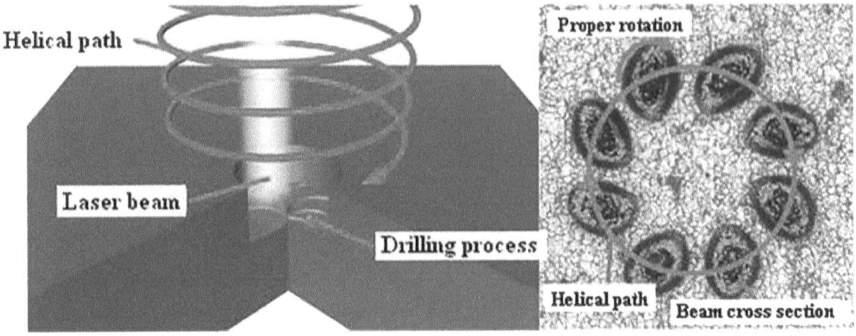

Fig. 4.3.29 Illustration of rotating motion during laser helical drilling operation [13].

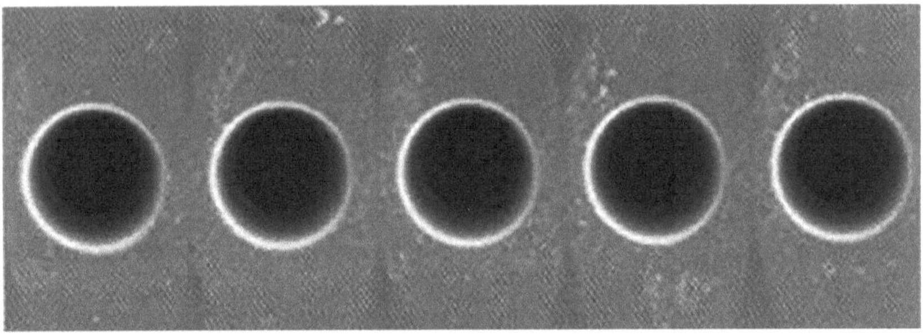

Fig. 4.3.30 Array of holes on ink jet nozzles by KrF excimer laser [18].

generated from CAD drawing. The principle components are the laser itself with some shutter control, beam delivery unit and focusing optics. The focused beam then passes through a coaxial nozzle with flow of gas jet. The gas jet is needed for cutting operation and to protect the optics from spatter and assist the ejection of materials from cutting zone. Laser cutting has an excellent performance with respect to kerf width, surface finish, generation of contour surface and variety of work piece materials over other cutting process such as plasma cutting, flame cutting and water jet cutting. Laser cutting has also good performance with respect to cutting speed and width of HAZ over other cutting process.

Laser cutting has been classified according to material transformation process into four types such as laser fusion cutting, laser oxidation cutting, laser vaporization cutting and mixed cutting process. In laser fusion cutting, materials of workpiece is transformed into molten state after absorbing the heat energy of laser beam and ejected out by the high pressure jet of inert gas of nitrogen and argon. Stainless steel, titanium alloy, aluminum alloy, polymer and ceramics are cut by laser fusion process with oxide free cut edges. In laser oxidation cutting, the cutting takes place by melt ejection of oxidized layer formed due to exothermic reaction of material with oxygen supplied along with laser beam as heat source. Mild steel, alloy steel and metal etc. are cut by this laser oxidation cutting process. In laser vaporization process, the cutting takes place due to vaporization of the materials directly from solid by minimizing or eliminating the melting process. For laser vaporization cutting, high power and short pulse laser is used so that the peak power intensity is very high. A gas jet is supplied at the cutting zone to assist the ejection of the material vapor. Acrylic, polymer, wood, paper, leather and ceramics etc. are cut by laser vaporization process for producing complex cut geometries on thin job. In mixed laser cutting process, the laser cutting occurs due to melting and vaporization process. Materials are transformed into molten state and then vaporized. During mixed process if air is used as assist gas then partial burning or oxidation of materials also takes place during mixed laser cutting. The supply of inert gas can prevent the oxidation during laser cutting process. Under liquid or water cutting is also another variant of laser cutting process to reduce the thermal damage of the workpiece.

Nd: YAG laser beam can be used for cutting metals like tool steel, aluminum, titanium alloys and ceramic materials such as silicon nitride, aluminum oxide and composite materials like carbon fiber reinforced plastics etc. As pulse frequency increases, surface roughness decreases because the peak power decreases and the resultant surface is generated due to overlapping of small craters. As lamp current increases, the surface roughness of cut edge increases because the applied power of laser beam increases. With higher cutting speed better surface finish of cut edge is achieved. During lamp pumped pulsed Nd:YAG laser cutting of silicon nitrides of 2 mm thickness, pulse frequency varies from 6 kHz to 8.5 kHz, lamp current varies from 22 A to 27 A and cutting speed varies from 17 to 22 mm/s.

The Nd:YAG laser machining operation on silicon nitride has been successfully carried out. From the experimental observation it is found that to perform cutting operation on silicon nitride successfully with better surface finish (Ra Value: 3 μm), the parametric settings are Lamp Current 25 amps; Q-switch frequency 7.0 kHz and cutting speed 20.5 mm/sec. Figure 4.3.31 shows the photographic view of cut edge of silicon nitride ceramics.

The profile cutting on high speed steel of thickness 0.7 mm has been performed using CNC Pulsed Nd:YAG Laser beam machining and the minimum Ra value of 4 μm at the cut edge is obtained while pulse frequency of

Fig. 4.3.31 Photographic view of cut edge of silicon nitride ceramics [19].

5 kHz, lamp current of 22 A and cutting speed of 10 mm/s have been set and minimum angular deviation of 0.03 degree is obtained while pulse frequency of 4.5 kHz, lamp current of 21 A and cutting speed of 10 mm/s have been set during laser profile cutting an operation. Figure 4.3.32 shows photographic view of hack saw tooth profile cut on HSS materials.

Fig. 4.3.32 Photographic view of hack saw tooth profile cut on HSS [20].

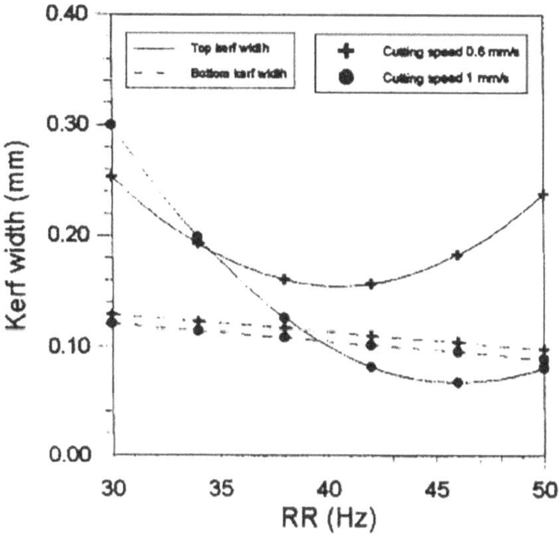

Fig. 4.3.33 Effect of RR and cutting speed on the kerf width [12].

Pulsed Nd:YAG laser cutting of Carbon fiber reinforced plastic composites (CFRP) of 2 mm thickness has been performed by setting the cutting speed ranging from 0.5 to 0.8 mm/s, pulse energy ranging from 1.5 to 2 J, pulsed duration from 0.4 to 0.7 ms, pulse repetition rate from 35 to 45 Hz and gas pressure ranging from 5 to 7 kg/cm^2 [21]. The variation of kerf width (top and bottom) with pulse repetition rate (R.R.) at different cutting speed is shown in Fig. 4.3.33 while pulse energy of 1.8 J; pulse duration of 0.6 ms and gas pressure of 6 kg/cm^2 have been set as fixed process parameters. As the pulse repetition rate (RR) increases the peak power decreases, the kerf width decreases, but kerf width increases with decrease in cutting speed. At lower cutting speed the laser materials interaction time is more for wider cut.

CO_2 laser is used to cut 3 mm thick sheet of carbon fiber reinforced plastic (CFRP) composites. A minimum HAZ thickness of 540 μm was achieved by using pulsed wave CO_2 laser. CO_2 laser cutting process can replace mechanical cutting of CFRP composites in terms of productivity and accuracy [21]. The values of process parameters during CO_2 laser cutting using continuous wave mode and pulsed wave mode are given in Table 4.3.5.

During CO_2 laser cutting, kerf width varies with duty cycles as shown in Fig. 4.3.34. The entry side kerf width is more than exit side kerf width. The both the variation is less but similar in nature.

Table 4.3.5 Range of values of process parameter of CO_2 laser cutting [21]

Process parameters	Value for continuous wave	Value for pulsed wave
Average laser power, W	200–2000	300–3000
Pulse frequency, Hz	–	10–4000
Duty cycle, %	–	20–100
Cutting speed, mm/min	200–4000	200–4000
Focal length, mm	127	127
Focal position, mm	On the surface	On the surface
Assist gas	Argon	Argon
Gas pressure, bar	5	2–6
Nozzle diameter, mm	2	2
Stand off distance, mm	1.5	1.5

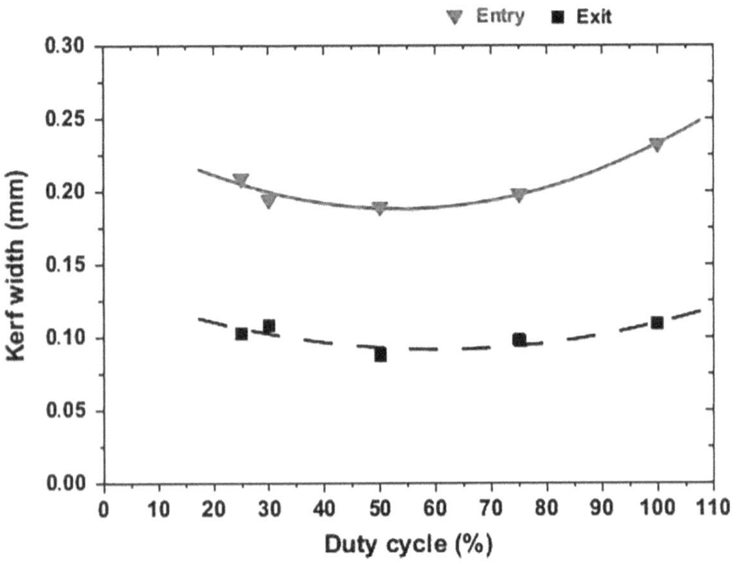

Fig. 4.3.34 Variation of kerf width with duty cycle [21].

The taper angle of cut edge also depends on pulse energy, frequency and duty cycle while pulsed wave CO_2 laser is applied for cutting operation. The variation of taper angle with pulse energy, pulse frequency and duty ratio are shown in Fig. 4.3.35(a)–(c) respectively. Taper angle increases with pulse energy and duty ratio, but it decreases with pulse frequency during laser cutting.

The assist gas pressure plays an important role for controlling kerf width, taper angle and kerf HAZ during cutting operation. As pressure of assist gas

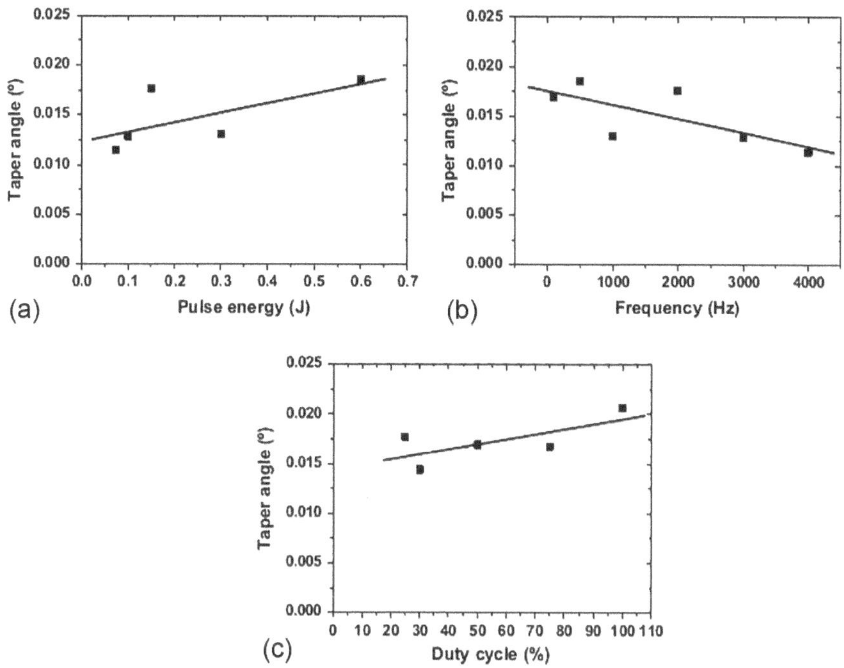

Fig. 4.3.35 Variation of taper angle with (a) pulse energy, (b) pulse frequency, and (c) duty cycle [21].

increases, kerf width at entry side of cut increase at a faster rate than that at exit side because more materials are removed from the entry side. As a result the taper angle of cut is also increased with the increase in assist gas pressure. The observed variation of HAZ width is less with gas pressure during CO_2 laser cutting operation. Figure 4.3.36 shows the influences of pressure of assist gas on (a) kerf width, (b) taper angle and (c) HAZ width.

The laser power is an important influential process parameter for controlling kerf width, taper angle and kerf HAZ during cutting operation. As laser increases, kerf width first decreases and the increases at entry side as well as exit side of cut, but it is more at entry side of cut than that at exit side because more materials are removed from the entry side. The taper angle of cut decreases with the increase in laser power. The observed HAZ width is more while laser power is more. The HAZ width increases with increase in laser power because more heat is generated during CO_2 laser cutting operation with high laser power. Figure 4.3.37 shows the influences of laser power on (a) kerf width, (b) taper angle and (c) HAZ width.

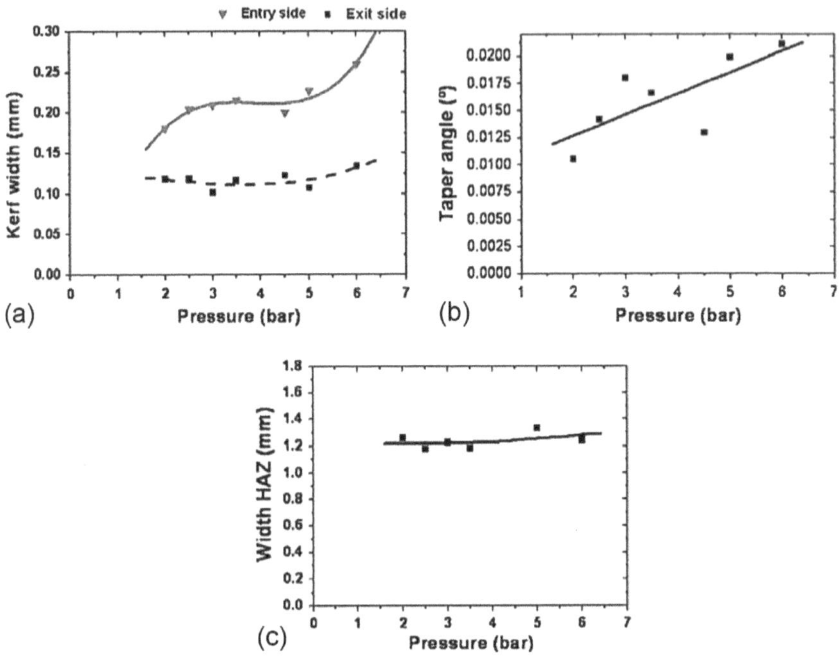

Fig. 4.3.36 Effects of pressure of assist gas on (a) kerf width, (b) taper angle, and (c) HAZ width [21].

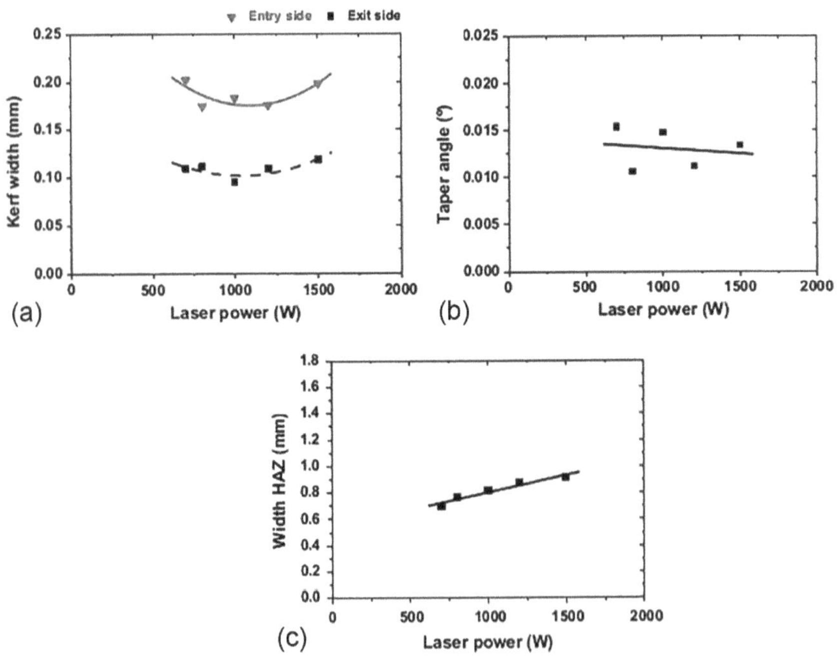

Fig. 4.3.37 Effects of laser power on (a) kerf width, (b) taper angle, and (c) HAZ width [21].

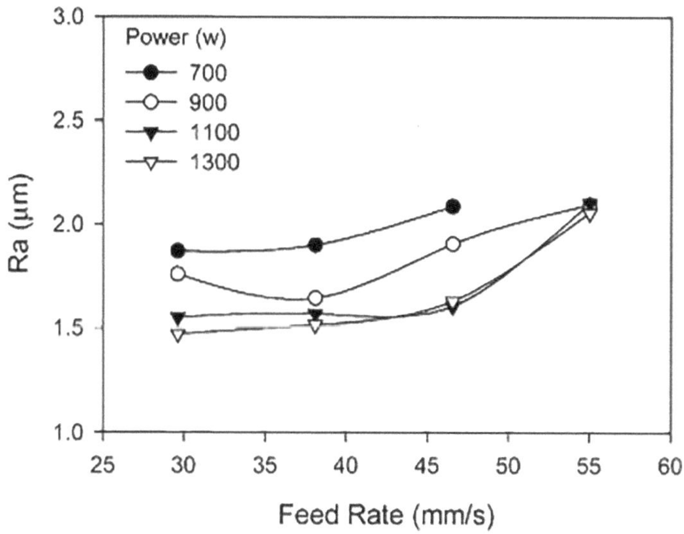

Fig. 4.3.38 Variation of surface roughness with feed rate [22].

CO_2 laser cutting of 4130 steel of thickness 1.27 mm has been done by Rajaram et al. [22] and it has been observed that the feed rate has significant influence on surface roughness of cut edge as shown in Fig. 4.3.38. The laser power has less effect on surface roughness. The feed rate has also large effect on Kerf width and HAZ. For achieving lower kerf width and HAZ, the lower feed is preferred.

4.3.5.3 Laser grooving

Laser grooving is one type of laser beam machining in which grooves are cut as per required geometry. Two laser beams can be used simultaneously as shown in Fig. 4.3.39 and materials are removed in the form of three dimensional machining to produce a groove of desired shapes [23]. Single laser beam can also generate a groove by removing materials layer by layer while laser beam is scanned over the workpiece by number of passes. Laser grooving operation can be performed on flat surface as well as cylindrical surface of the workpiece. By controlling the laser machining process parameters such as number of passes, scanning speed, pulse frequency, pulse duration, average power etc. the desired groove dimensions such as width, depth and taper etc. can be achieved. The groove depth defines the boundary of the volume of material removed. The scanning velocity defines rate at which this volume is formed on the work piece. The surface quality is related to heat affected zone and surface roughness at the groove surface during beam/material interaction.

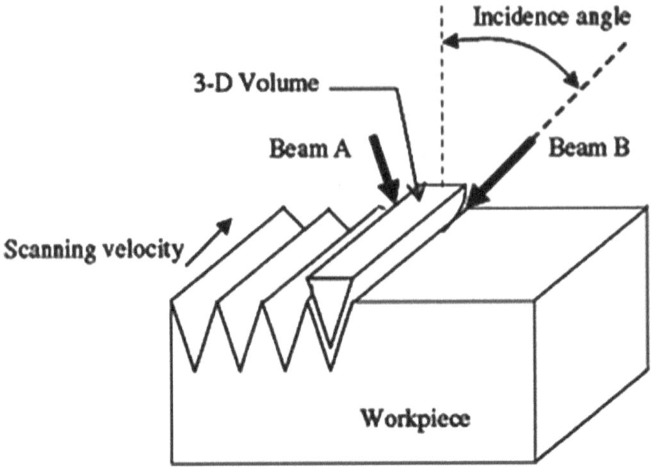

Fig. 4.3.39 Three dimensional laser grooving by two laser beam [23].

Fig. 4.3.40 SEM micrograph of micro-grooves cut on aluminum titanate [24].

The Nd:YAG laser micro-grooving operation on flat plate of aluminum titanate ceramics has been performed at optimal parametric setting of lamp current 20.4 A; pulse frequency 1 kHz, pulse width 10% of duty cycle, cutting speed 10 mm/s and assist pressure of 1.3 kgf/cm^2 for achieving trapezoidal micro-groove of 300 μm upper width and 100 μm lower width and 500 μm depth. The minimum percentage of deviation of taper angle and depth of laser micro-groove were obtained as 0.71% and 1%, respectively. Figure 4.3.40 shows the SEM view of trapezoidal micro-groove cut on aluminum titanate ceramics. [24].

Pulsed Nd:YAG laser micro-grooving on alumina ceramics has been performed and Fig. 4.3.41 shows SEM view of micro-groove cut on

Fig. 4.3.41 SEM view of micro-groove cut on alumina by Nd:YAG laser.

alumina by Nd:YAG Laser. For achieving minimum deviation in upper width of 0.002 mm the optimal combination for minimum deviation of upper width of the micro-groove is: Air pressure of 0.5 kg/cm^2, Lamp current of 17.5 Amp., Pulse frequency of 4 kHz, Pulse width of 8% of duty cycle and Cutting speed of 14 mm/sec. The minimum deviation in lower width of micro-groove is observed as 0.0343 mm and the optimal combination is set as Air pressure of 2 kg/cm^2, Lamp current of 17.5 Amp., Pulse frequency of 2 kHz, Pulse width of 6% of duty cycle and Cutting speed of 26 mm/sec. The optimal parametric combination for minimum deviation of depth 0.0066 mm of the micro-groove is as Air pressure of 0.5 kg/cm^2, Lamp current of 20 Amp., Pulse frequency of 3 kHz, Pulse width of 8% of duty cycle and Cutting speed of 18 mm/sec [25].

Fiber laser micro-machining of Ti-6Al-4V of 1.1 mm thickness has been performed [5]. Figure 4.3.42 shows micro-groove cut onTi-6Al-4V by CNC diode pumped pulsed fiber laser beam machining while process parameters are set at 8numbers of passes. scanning speed of 40 mm/sec, Average power of 30 W and pulse frequency of 50 kHz. Fiber laser beam is also utilized to cut a groove on PMMA setting process parameters at scanning speed of

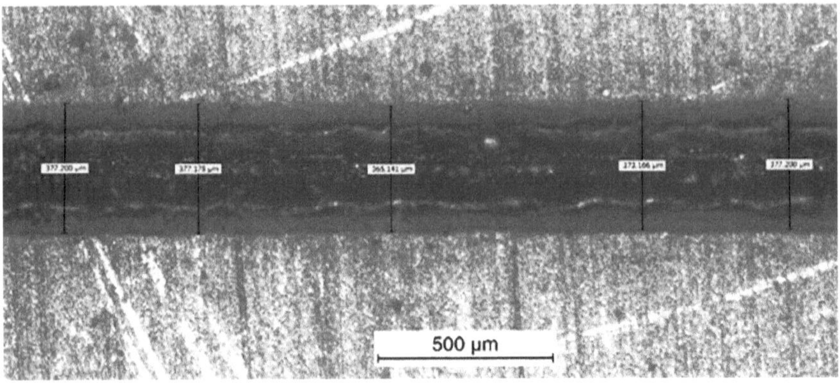

Fig. 4.3.42 The micro-groove cut on Ti-6Al-4V by fiber laser beam [5].

50 mm/s, average power of 15 W, pulsed frequency of 50 kHz and single pass of laser beam.Figure 4.3.43 shows the micro-groove cut on PMMA by fiber laser beam [26].

Shashi Prakash et al. [27] fabricated micro-channel on PMMA using CO_2 laser beam machining method. The width of microchannel varies from 171 to 192 μm while average laser power is 1 W. Figure 4.3.44 shows width variation with laser power in different passes. As laser power increases the width of microchannel increases with number of passes.

The depth of micro-channel cut on PMMA by CO_2 laser increases with number of pass as well as laser power. The depth variation with number of pass and laser power is shown in Fig. 4.3.45.

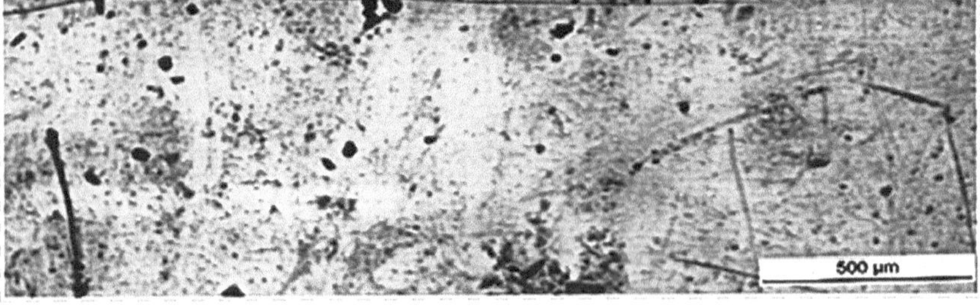

Fig. 4.3.43 The micro-groove cut on PMMA by fiber laser beam [26].

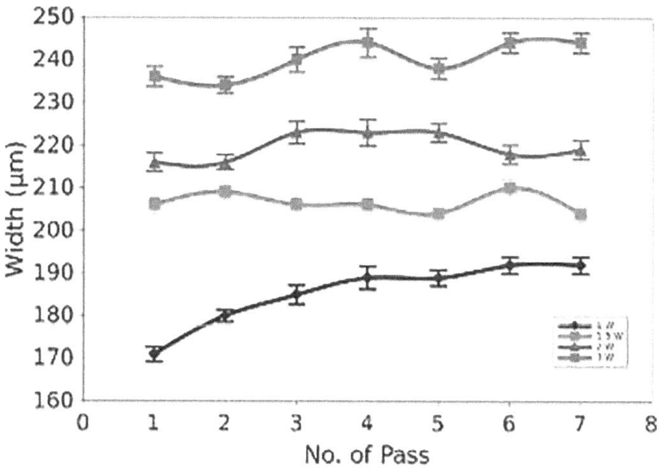

Fig. 4.3.44 Cut width variation with power in different passes [27].

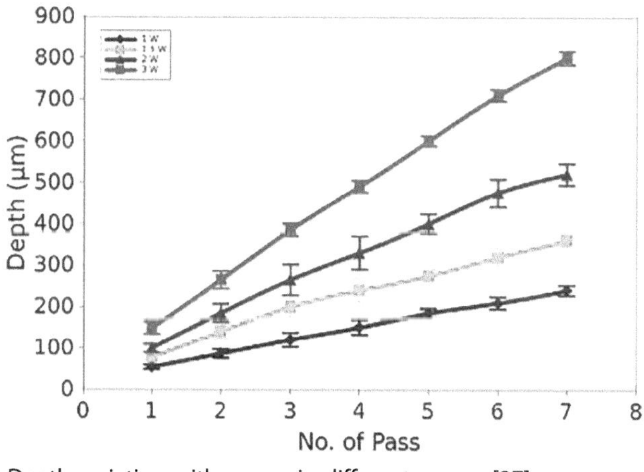

Fig. 4.3.45 Depth variation with power in different passes [27].

Figure 4.3.46 shows photographic view of micro-channel cut on PMMA by CO_2 laser at different number of passes. The variation of width is less but more depth of micro-channel is obtained with increasing the number of passes of laser beam.

The surface roughness decreases with the increase in number of pass. With more number of pass, smooth surface is achieved and there is less variation of surface roughness with laser power while number of pass is more than 5 during micro-channeling on PMMA by CO_2 laser beam. Figure 4.3.47 shows variation of surface roughness with number of pass at different laser power settings.

Fig. 4.3.46 Microchannel depth in different passes at 1.5 W and 50 mm/s [27].

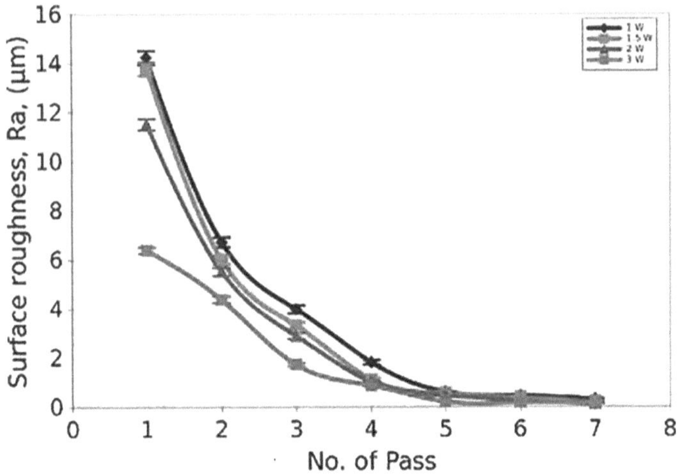

Fig. 4.3.47 Surface roughness (Ra) variation with power in different number of passes [27].

The heat affected zone (HAZ) on surface of micro-channel varies and increases with no. of pass of laser beam at higher laser power. With lesser no. of passes and lower laser power the HAZ becomes narrow (50 µm) and it becomes almost constant with number of passes up to 5. Figure 4.3.48 shows the variation of HAZ with no. of passes and laser power during fabrication of micro-channel on PMMA by CO_2 laser beam machining.

4.3.5.4 Laser turning

Laser turning operation is performed on cylindrical job to remove materials at micron level. The workpiece rotates about its own axis like conventional turning operation. Instead of cutting tool laser beam is used for turning operation. Thc job is also movcd along the axial direction to set the length of turning. The depth of turning is controlled by the number of passes of laser beam as well as the position of focal point of laser beam. For removal of bulk

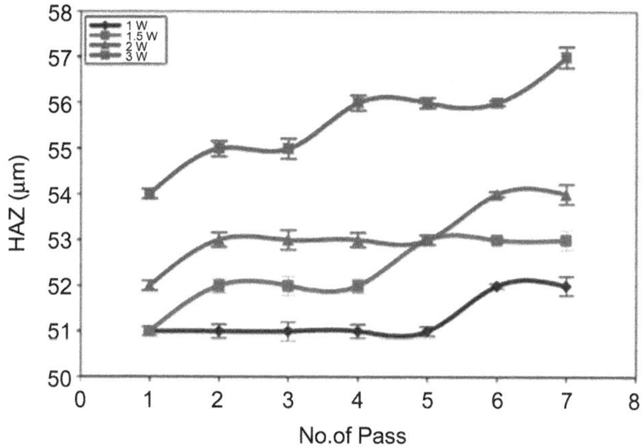

Fig. 4.3.48 HAZ variation with power in different passes [27].

form (like a ring shape or tapered groove) of material from workpiece, laser micro-turning process is applied by using two intersecting laser beams. However, for carrying out laser micro-turning, a single laser beam can be used to remove very thin layer from work material surface by irradiating and focusing the laser beam onto the machining surface of rotating workpiece through the desired length of turn along the axis of work sample. Laser micro-turning is the process for generating a continuous micro-groove (spiral type) over the cylindrical surface of rotating workpiece. The main important criterion of the process is the degree of overlap between two adjacent micro-grooves generated due to movement of work surface under the high intense laser beam. In practice, higher amount of overlap produces better quality turned surface. Figure 4.3.49 shows the schematic diagram of workpiece feeding and rotating arrangement for Nd:YAG laser micro-turning operation.

Nd: YAG laser turning process generates micro-groove on cylindrical parts of alumina ceramics at optimal parametric setting of lamp current 19A; pulse frequency 3.2kHz, pulse width 6% of duty cycle, rotational speed 22 rpm and assist pressure of 0.13N/mm^2 for achieving square micro-groove of 200 μm width and 200 μm depth. The minimum deviation of upper width, lower width and depth of laser micro-groove was obtained as 10.1, 9.8 and 6.9 μm, respectively. Figure 4.3.50 shows the SEM views of cross section of micro-groove generated on alumina by laser turning operation [29].

For laser micro-turning operation spot overlapping and circumferential overlapping are the two phenomena for generating better laser turned

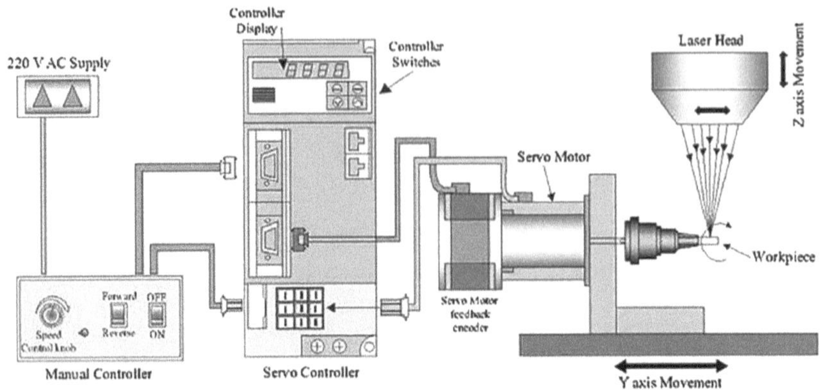

Fig. 4.3.49 Schematic representation of workpiece feeding and rotating system for Nd: YAG laser micro-turning operation [28].

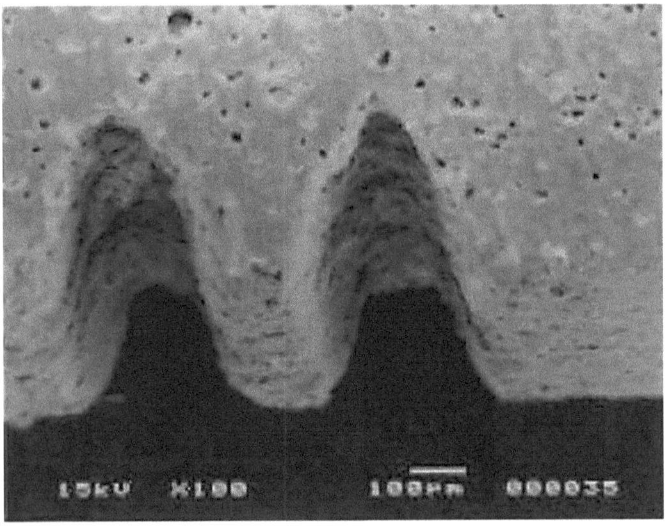

Fig. 4.3.50 SEM views of cross section of micro-groove generated on alumina [29].

surface. The spot overlapping phenomena is schematically represented in Fig. 4.3.51. Fig. 4.3.52 shows the SEM views of laser micro-turned surface generated on alumina ceramics at different magnifications.

Fiber laser micro-threading operation can also be performed by generation spiral micro groove on cylindrical workpiece. The basic concept of laser micro-threading process is to produce multiple side-by-side spiral micro-grooves on the cylindrical workpiece by irradiating the focused laser beam on the continuous rotating workpiece. Workpiece is being fed along

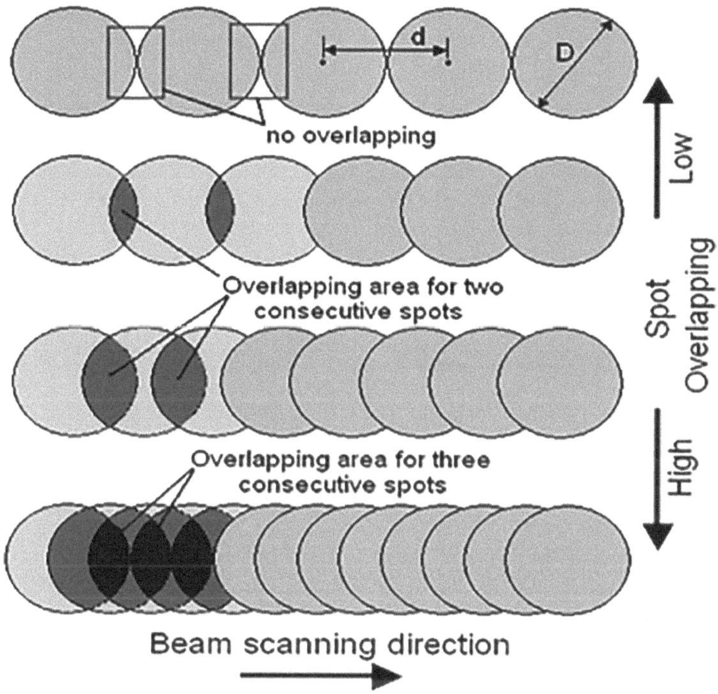

Fig. 4.3.51 Schematic representation of the spot overlapping phenomena [7].

axial direction for maintaining a laser scan lines while rotating the workpiece. As a result, spiral micro groove of required length on the workpiece surface is generated. Then, it is repeated from reversed direction. This process is continued until the adjacent two peak touch each other. The process parameters involve in this process are laser pulse frequency, pulse duration, axial feed rate and work piece rotation. The speed of rotation and axial movement of workpiece is so controlled that multiple spiral micro-groove is produced on workpiece surface and there will be no overlap between consecutive groove widths. Thus, a spiral micro-groove is generated keeping a little unmachined gap between two consecutive micro-groove widths. To achieve required amount of micro-groove depth, multi-pass laser scanning can be carried out on the same laser scan track of previous one (multi-pass process). For achieving better quality laser spiral micro-groove, spot overlap plays an important role during the process. The schematic representation of laser spiral micro-groove is shown in Fig. 4.3.53. A single laser beam can be used simply to carry out laser spiral micro-grooving operation by properly controlling the speed of rotation and also the axial feed rate of the workpiece and also to ensure that there will be no overlap between two consecutive

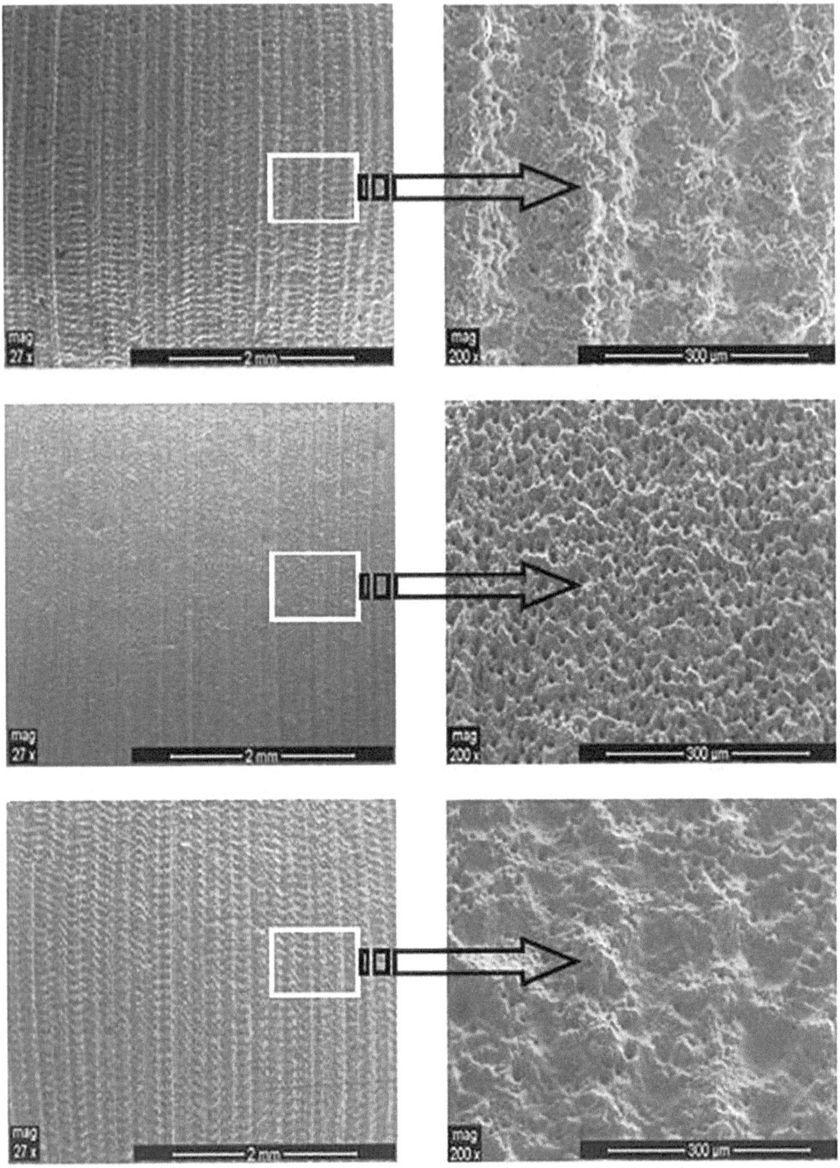

Fig. 4.3.52 SEM views of laser micro-turned surface generated on alumina ceramics at different magnifications (27 × and 200 ×) [28].

micro-grooves produced during two consecutive rotation of cylindrical work sample. In this process, high intense laser beam interacts with material and due to generation of large amount of heat at the focused spot, the material melts and evaporates. The amount of material removes from the focused

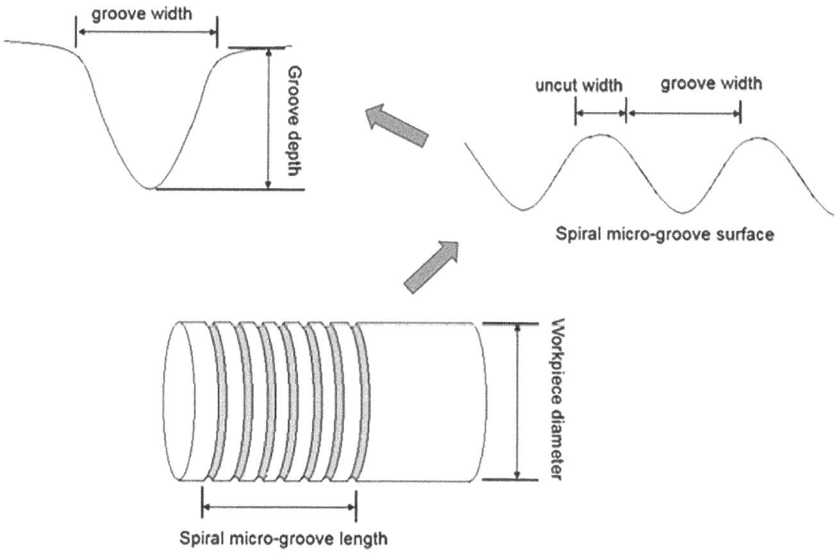

Fig. 4.3.53 Schematic diagram of laser spiral micro-groove and its dimensions [30].

spot mainly depends upon the average power of laser beam, pulse frequency and time duration of laser-matter interaction.

4.3.5.5 Laser marking

Laser marking is one of the well developed technologies of materials processing. Laser marking is the best and most applied permanent marking method. Laser marking is flexible, programmable and environmentally clean method which employs a thermal process that utilizes high intensity beam of focused laser light to create a contrasting mark on the material surface and is primarily used to mark or print bar codes, matrix codes, universal product codes (UPC) and serial numbers, etc. The main application of laser marking is for the purpose of product identification and traceability. Compared to traditional marking techniques, such as inkjet, hot stamping or mechanical scribing, laser marking offers many advantages such as flexibility, speed, indelibility, reliability, free of consumables and high cleanliness. Nd:YAG laser marking has great potential in diverse range of applications, i.e., electronics components, automotive products, jewelry, medical implants and surgical instruments, etc. The line width produced by Nd:YAG laser is narrower than that of CO_2 lasers, which is a major factor for higher resolution. In some cases, second and third harmonic Nd:YAG lasers at 532 and 355 nm, as well as ultraviolet excimer lasers are the optimum choice to mark highly transparent materials. The application of laser technology is pacing the demands from industry for smaller geometries and increasingly tighter specifications.

Figure 4.3.54 shows a typical laser marking system which includes galvanometric scanner for scanning the laser beam over the workpiece mounted on CNC x-y table for marking various features on different materials.

Technical demands from the electronic, medical, aerospace and other industries are pushing the boundaries into micron level for smaller components which need fine marking with more precise tolerances. Nd: YAG laser has some unique properties for using it in marking processes. The advantage of laser systems is the small beam width, allowing manufacturers to mark particularly small parts, an issue of increasing concern as the quest for miniaturization is being increasing very rapidly.

Nd: YAG lasers are typically found in industrial applications as they are known for their ability to mark metal, either by ablation or by creating a color change through annealing. The major concern in the marking process is the number of laser parameters involved. Hence there is a requirement to explore it in detail on laser marking of ceramics, which has different applications in various fields. In order to attain clear and excellent mark, it is most important that proper laser marking methods and the suitable process parameters are to be selected for the specific material, which leads to increase in the

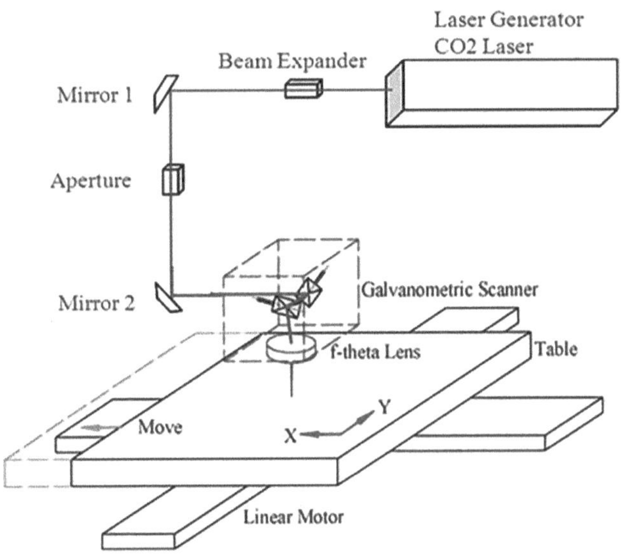

Fig. 4.3.54 A typical laser marking system [8].

marking quality. Laser marking characteristics may be altered due to a small change in the marking parameters; therefore, monitoring and control of the laser marking parameters are necessary

Mechanisms for marking can broadly be divided into two categories such as (i) Marking by surface modification and (ii) Marking by material removal from the surface. In the first category of marking by surface modification, the laser radiation affects the material composition to create a high contrast mark without material removal. It may locally melt the material, causing it to oxidize or chemically alter to form a visible mark. Foaming, carbonization, annealing and coloring of materials are the methods of laser marking under first category. Laser etching, surface melting, ablation and engraving are the methods of laser marking under second category.

The key parameters of the laser marking process that define a given laser's marking ability are the pulse frequency, lamp current, scanning speed, spot size, the peak and average power of the pulse.

Evaluation of laser marking quality characteristics is done by estimating mark width, mark depth and mark intensity. The quality of a mark is assessed and evaluated using complementary techniques such as optical microscopy, surface roughness measurement, and contrast evaluation.

The marking intensity or the mark contrast (c) can be determined by the following expression:

$$c = \frac{g_{background} - g_{mark}}{g_{white} - g_{black}} \tag{4.3.6}$$

where $g_{background}$ is the grey level value of unmarked area and g_{mark} is grey level value of marked area. $g_{black}$is the grey level of black color and g_{white} is the grey level of white color. After taking the optical photographs using optical microscope, the image is analyzed in a PC, which computes the gray level of mark of different portion.

Laser marking process can be effectively applied for permanent fine marking on varieties of materials from softer to harder, electrically conducting to non conducting, brittle to ductile and metal to non-metals etc. Various kinds of lasers such as Nd:YAG, Fiber, Excimer and CO_2 laser can be selected for marking depending upon their thermal properties rather than their mechanical and electrical properties. Various methods or mechanism of laser marking can also be applied as per requirement and constraints of the materials point of view. Controlling and optimizing the laser process parameters and over lapping characteristics of laser beam, desired mark width, mark depth and mark intensity can be achieved.

The parametric effects on laser marking characteristics such as mark width, mark depth and mark intensity are discussed through the graphical representation. Figure 4.3.55 shows the variation of mark width with pulse frequency with different lamp current. As pulse frequency increases the peak power decreases at fixed lamp current as a result mark width decreases. As lamp current increases, the laser power increases, hence mark width increases as shown in Fig. 4.3.56. Again at high scan speed of 9 mm/s and lamp current 24 A, mark width is nearly 95 μm for fine marking on ceramics, the position of focal point should be placed 0.3 mm below the top surface of the workpiece as shown in Fig. 4.3.57. For deeper marking, the laser power should be more and scan speed should be low. The variation of mark depth with lamp current at different scan speed is shown in Fig. 4.3.58. The mark

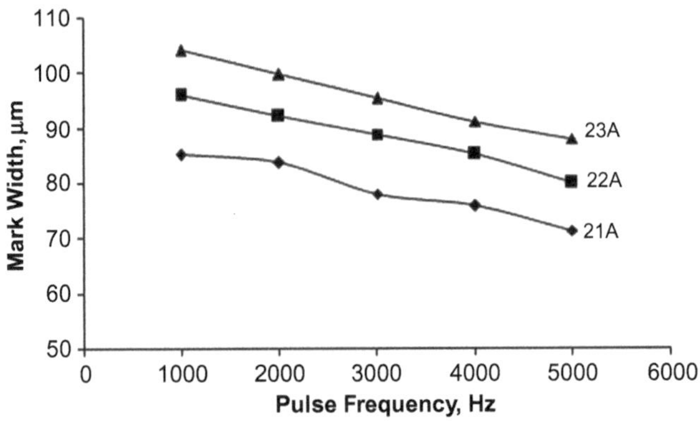

Fig. 4.3.55 Variation of mark width with pulse frequency with different lamp current [31].

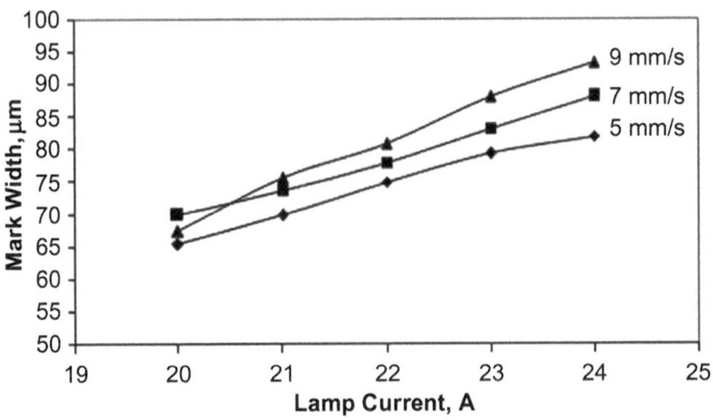

Fig. 4.3.56 Variation of mark width with lamp current with different scanning speed [31].

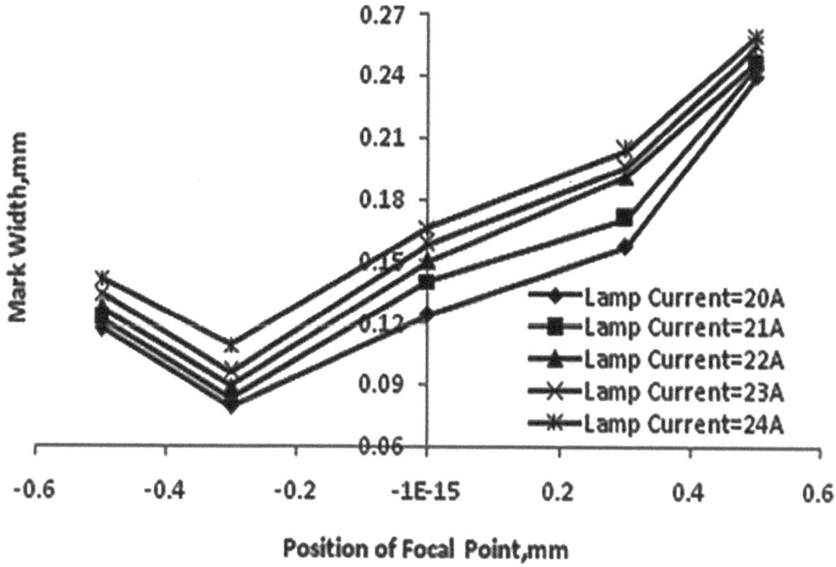

Fig. 4.3.57 Variation of mark width with position of focal point with different lamp current [31].

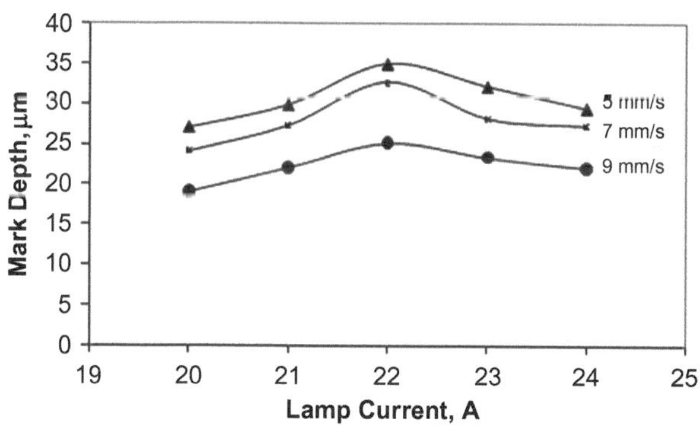

Fig. 4.3.58 Variation of mark depth with lamp current with different scanning speed [31].

intensity increases with pulse frequency as shown in Fig. 4.3.59. At high frequency, low lamp current and high scanning speed, the mark intensity becomes high as shown in Fig. 4.3.60.

Nd: YAG laser marking on ceramics has been done as shown in Fig. 4.3.61(a) and (b) by Peter et al. [8]. The optical parametric setting are pulse frequency of 6.8 kHz, lamp current of 26.2 A, pulse width of 3%, scanning speed of 1 mm/s and air pressure of 2.3 kgf/cm^2.

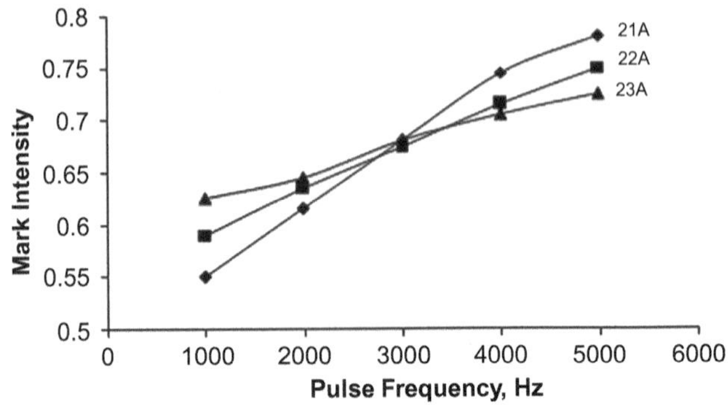

Fig. 4.3.59 Variation of mark intensity with pulse frequency at different lamp current [31].

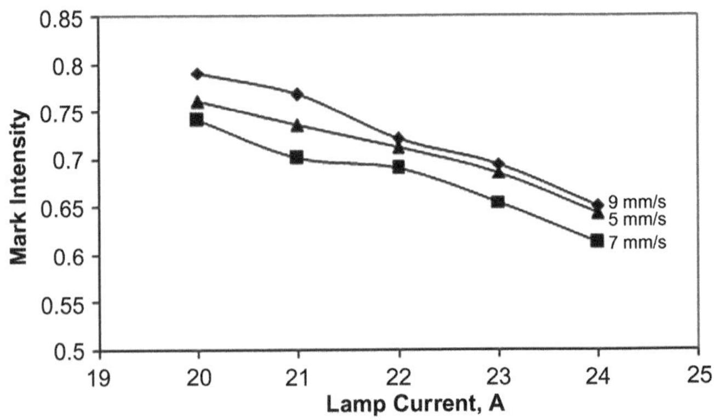

Fig. 4.3.60 Variation of mark intensity with lamp current at different scan speed [31].

At the optimal setting of process parameters, different types of mark as shown in Fig. 4.3.62(a) logo and (b) number are marked on alumina by Nd:YAG laser.

Ricciardi et al. [32] described an excimer laser marking process for marking on polymer material (ABS and PC) of standard computer keyboards. A laser marking cell has been developed to meet all the marking requirement of mark contrast and color quality with a throughput of 1 keyboard every 15 seconds. Li et al. [33] used 60 W fiber delivered diode laser for marking and engraving on building materials such as marble, granite, clay tiles, ceramic tiles, roof tiles, cement bricks, glass and wood. A good quality mark with high contrast was produced on these building materials by melting and material removal mechanisms.

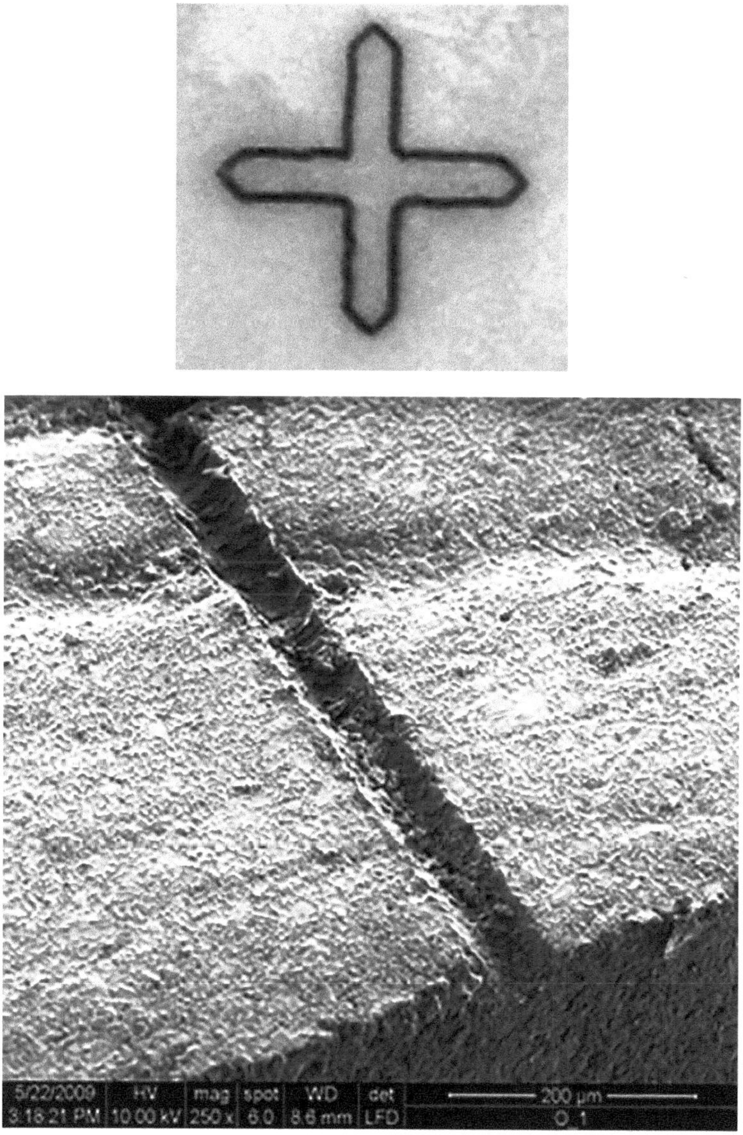

Fig. 4.3.61 (a) Photographic view of mark on alumina. (b) SEM view of marked line on alumina [8].

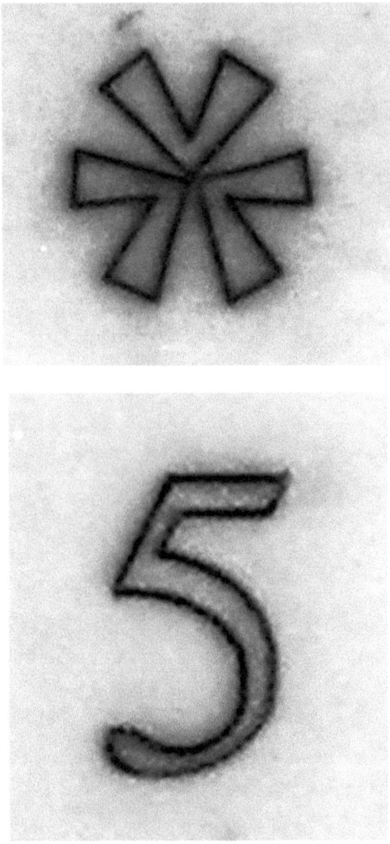

Fig. 4.3.62 (a) Laser mark of logo on alumina. (b) Laser mark of number on alumina [8].

4.3.6 Advantages and limitations of laser beam machining processes

Laser beam machining (LBM) is an excellent material processing technique to machine a wide range of advanced materials, especially difficult-to-machine advanced ceramics and composites. LBM can be used to process these materials with intricate shapes and sizes with greater precision even difficult to reach surfaces. Laser beam machining can replace the traditional machining processes for various industrial applications due to several following advantages.

(a) LBM can produce better finished machining surface and hence post finishing operation can be eliminated;

(b) Laser beam can be transported to remote place for machining hazardous materials like nuclear fuel rod etc.

(c) LBM can remove materials from the selective regions of the workpiece by focusing the beam;

(d) No solvents/chemicals are used during machining, therefore it is a clean machining process;

(e) LBM has adaptability to automation for high productivity and high speed machining;

(f) LBM is non-contact type processing and there is no physical tool. Thus, there is no machining force or wear of the tool takes place. There is no mechanically induced material damage and vibration of machine components,

(g) Complex profiles with high degree of accuracy can be easily cut with laser,

(h) Laser beam machining process does not depend on mechanical properties (hardness, brittleness, etc.), rather depends upon thermal and to some extent optical properties of the material to be machined. Therefore, the material of any degree of brittleness or hardness can be machined. Materials of low thermal diffusivity and conductivity are particularly well suited for laser machining,

(i) Laser beam machining is independent of electrically conductive nature of the workpiece. It can machine wide range of materials from plastics to diamonds.

(j) Processing of materials with laser is very fast, thus, it produces very small heat affected zone (HAZ),

(k) Large aspect ratio micro-features can be produced along with acceptable accuracy and dimensions,

(l) In LBM, less material removal is removed from selected portions of material. Thus, it reduces material wastage,

(m) Laser beam machining is highly flexible machining system when it combines with multi-axis work positioning system or robot. Thus, LBM can be adapted with automation, and

(n) Laser beam machining provides a wide range of material processing techniques ranging from drilling, cutting, grooving, scribing, marking, cleaning and turning etc.

As laser beam micro machining is a thermal type material removal process, there are also some limitations as follows.

(i) It requires high initial investment cost,

(ii) Maintenance cost of LBM system is high and to carry out various operations, it requires skilled operators,
(iii) The efficiency is low when compared with the laser beam power with respect of input electrical energy to the LBM machine. However, development and research issues of various industrial lasers may increase the efficiency of converting electrical energy to beam energy, and
(iv) Highly reflective and transparent materials at the machining condition cannot be processed by laser beam, etc.
(v) It has limited applications for thicker materials but highly applicable to thin cross sectional materials.
(vi) Metal removal rate is very low.

4.3.7 Applications of laser beam machining processes

Q-switched Nd:YAG laser with TEM00 mode is mostly used in micro-machining application due to their high energy density and small focused spot up to 2 μm diameter [6]. In general, pulsed laser beam machining produces micro parts with high dimensional accuracy and surface quality as compared with conventional micro machining process. Laser drilling can produce micro holes as small as 0.05 mm in diameter in work piece at a rate up to 1 ms/hole. Presently, it is used in industry for producing holes in turbine blades, combustion chamber and aerosol nozzles also in other areas like environment sensing, laser marking, microsurgery, measurement techniques, range finding and frequency conversion etc. Some more applications of various lasers are as follows:

(i) Drilling micro as well as macro holes in hard metals such as tungsten and ceramics.
(ii) Trimming of sheet metal, plastic parts and carbon resistors and wafer labeling in microelectronics.
(iii) Machining of complicated geometrical profiles geometry on thin and hard metals.
(iv) Engraving pattern on thin films;
(v) Fabrication of integrated chips (IC) and computer peripherals.
(vi) Communication and space applications.
(vii) Aeronautics and guided missiles in defense.
(viii) Medical device fabrication, micro holes in nozzles and artificial ceramic hip joints.
(ix) Applications in non-machining processes like heat treatment, surface hardening, alloying, cladding, welding, etc.

(x) Laser beam applied for various machining processes like cutting, grooving, drilling, trepanning, turning, milling, cleaning, dressing and marking etc.;
(xi) Laser Assisted machining.
(xii) Laser Assisted Oxygen Flame Cutting (LASOX)
(xiii) Laser Assisted Hybrid Machining

The various types of machining applications on different materials by different types of laser beam are listed in Table 4.3.6.

Table 4.3.6 List of laser beam machining applications [34]

Laser	Applications	Material
Micro-electronics packaging Excimer Lamp-pumped solid-state Diode-pumped solid-state CO_2 sealed or TEA	Via drilling and interconnect drilling Via drilling and interconnect drilling High volume via drilling, tuning quartz oscillators Excising and scribing of circuit devices, large panel via drilling	Plastics, ceramics, silicon Plastics, metal, ceramics, silicon Plastics, metal, inorganic Ceramics, plastics
Semiconductor manufacturing Excimer Solid-state CO_2 or TEA	UV-lithography IC repair, thin films, water cleaning IC repair, thin films, bulk machining resistor and capacitor trimming Excising, trimming	Resist, plastics, metals, oxides silicon Plastics, silicon, metals, oxides silicon, thick film Silicon
Data-storage devices Excimer Diode-pumped solid-state CO_2 or TEA	Wire stripping air bearings, heads micro via drilling Disk texturing servo etching micro via drilling Wire stripping	Plastics, glass silicon ceramics plastics Metal, ceramics metals, plastic Plastics
Medical devices Excimer Solid-state CO_2 or TEA	Drilling catheters balloons, angioplasty devices. Micro-orifice drilling Stents, diagnostic tools Orifice drilling	Plastics, metals ceramics, inorganics Metals Plastics
Communication and computer peripherals Excimer Solid-state CO_2 or TEA	Cellular phone, fiber gratings, flat panel annealing, ink jet heads Via interconnect coating removal tape devices Optical circuits	Plastics, silicon, glass, metals, inorganics Plastics, metals, oxides, ceramics Glass, silicon

4.3.8 Advancements of laser beam machining technology

A new development of application of laser beam is laser polishing, where melting of material is done by CW Laser and finishing is done by short pulse laser.

Direct diode laser is recently used for micro-machining applications such as scribbing on softer materials.

Laser beam machining of transparent materials is also a challenging area. The very high power laser beam of intensity about 400 GW/cm^2 at the focal spot engraves inside the transparent materials by breakdown materials and generating the micro deformations which can scatters the light.

There has been a lot of advancements in the area of laser cutting, drilling, micro-machining and marking already being made, but 3 D laser machining like turning, milling and grooving are still waiting for industrial use. The use of laser beam machining for complex profile cutting of thicker materials are limited. Future research work on laser beam machining can also be directed in this direction.

Under Liquid Laser Beam Machining (UL-LBM) is an advanced machining process in which thermal effects of LBM can be minimized [35]. Under water LBM reduces HAZ, recast layer and prevents formation of micro-cracks and dross during machining of difficult to cut materials. The Material removal is assisted by the cavitation effects of bubble collapse in under water laser drilling process [36]. The water jet assisted laser ablation is also another process variant of hybrid laser beam machining process.

Laser cleaning, laser dressing and laser assisted dressing are the advanced applications of LBM where the controlling of process parameters such as laser power, focal point location, laser spot size etc. are to be precisely controlled to achieve the optimum performance of LBM. There are lot of advancements in the area of laser assisted machining for micro tool fabrication and grinding applications.

There are advancements of lasers for its applications in biomedical engineering, machining of metal stent and bio polymer stent and micro-drilling on catheters etc. Femtosecond laser using Chirped Pulse Amplification (CPA) techniques can be applied for micro cutting of silicon for fabrication of micro-chips, sensors and actuators utilized in semiconductor industry. Femtosecond laser has ultra-short pulse duration which is less than 100 fs, as a result very high peak power more than 15 GW can be obtained which causes ablation mechanism of material removal.

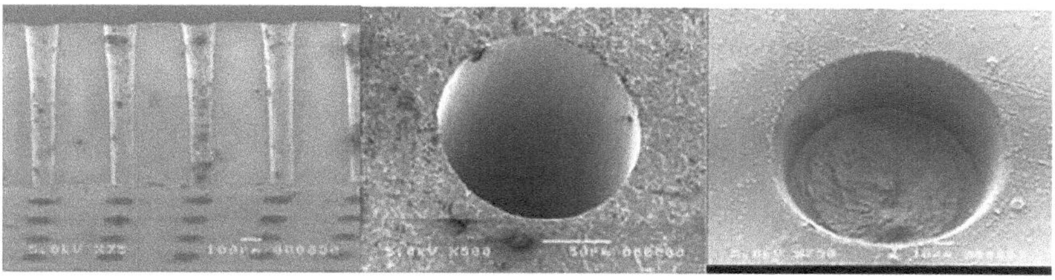

Fig. 4.3.63 Micro-hole drilled on medical filters and close-up view of hole and blind hole by CVL [18].

Copper Vapor Laser (CVL) has proved its potentials for machining applications such as drilling, cutting and milling of varieties of materials such as metals, silicon, diamonds, ceramics and polymers. As CVL is strongly absorbed in metals than IR laser, deep hole on stainless steel with small HAZ, large aspect ratio (>40) and surface roughness in the order of 1–2 μm are reported by Allen et al. [37]. CVL is a faster process compared to micro-EDM for generating micro-structures with better surface profile on electrically conducting materials. CVL can create micro-channel, engrave and mark on hard and electrically non-conducting materials like glass. Fig. 4.3.63 shows micro-hole drilled on medical filters and close-up view of hole and blind hole by CVL [18].

References

[1] S. Mishra, V. Yadav, Laser beam micromachining (LBMM): a review, Opt. Lasers Eng. 73 (2015) 89–122.

[2] J.C. Ion, Laser Processing of Engineering Materials: Principles, Procedure and Industrial Application, Butterworth-Heinemann, Oxford, 2005.

[3] F.D. Teodoro, Pulsed Fiber Laser, High Power Laser Hand Book, McGraw-Hill, New York, USA, 2011.

[4] R.D. Schaeffer, Fundamentals of Laser Micromachining, CRC Press, Taylor & Francis Group, 2012, pp. 5–36.

[5] A. Sen, B. Doloi, B. Bhattacharyya, Fibre laser micromachining of Ti 6Al 4V, in: Laser Based Manufacturing, Springer, 2015.

[6] J.F. Ready, D.F. Farson, Hand Book of Laser Materials Processing, LIA Magnolia Publishing Inc., USA, 2001

[7] G. Kibria, Experimental Investigation into Nd:YAG Laser Micro-turning of Engineering Ceramics, PhD Dissertation thesis, Jadavpur University, India, 2014.

[8] J. Peter, Experimental Investigations into Nd:YAG Laser Marking on Engineering Materials, PhD Dissertation thesis, Jadavpur University, India, 2015.

[9] O. Svelto, D.C. Hanna, Principles of Lasers, fourth ed., Springer, 1998. ISBN 0-306-45748-2.

[10] A. Sen, B. Doloi, B. Bhattacharyya, Fibre laser micro-machining of engineering materials, in: Non-traditional Micromachining Processes–Fundamentals and Applications, Springer, 2016.

[11] A.K. Dubey, V. Yadava, Laser beam machining: a review, Int. J. Mach. Tool Manuf. 48 (6) (2008) 609–628.

[12] J. Mathew, G.L.G. oswami, N. Ramkrishnan, N.K. Naik, Parametric studies on pulsed Nd:YAG laser cutting of carbon fibre reinforced plastic composites, J. Mater. Process. Technol. 89-90 (1999) 198–203.
[13] C. Fornaroli, J. Hltkamp, A. Gillner, Laser beam helical drilling of high quality micro holes, in: Laser in Manufacturing Conference 2013 Physics Procedia, vol. 41, 2013, pp. 661–669.
[14] A.S. Kuar, B. Doloi, B. Bhattacharyya, Modelling and analysis of pulsed Nd:YAG laser machining characteristics during micro-drilling of zirconia (ZrO2), Int. J. Mach. Tools Manuf. 46 (12-13) (2006) 1301–1310.
[15] S. Bandyopadhyay, J.K.S. Sundar, G. Sundarajan, S.V. Joshi, Geometrical features and metallurgical characteristics of Nd:YAG laser drilled holes in thick IN718 and Ti6Al 4V sheets, J. Mater. Process. Technol. 127 (2002) 83–95.
[16] A. Bharatish, H.N. Murthy, B. Anand, C.D. Madhusoodana, G.S. Praveena, M. Krishna, Characterization of hole circularity and heat affected zone in pulsed CO_2 laser drilling of alumina ceramics, Opt. Laser Technol. 53 (2013) 22–32.
[17] D. Jahns, T. Kaszemeikat, N. Mueller, D. Ashkenasi, R. Dietrich, H.J. Eichler, Laser trepanning of stainless steel, Phys. Proc. 41 (2013) 630–635.
[18] J. Meijer, Laser beam machining (LBM), state of the art and new opportunities, J. Mater. Process. Technol. 149 (1-3) (2004) 2–17.
[19] A.S. Kuar, B. Doloi, B. Bhattacharyya, Experimental investigations on Nd: YAG laser cutting of silicon nitride, Int. J. Manuf. Technol. Manag. 7 (2-4) (2005) 181–191.
[20] A.S. Kuar, Parametric Analysis on Nd:YAG Laser Machining System for Micro-Machining Applications, PhD Dissertation thesis, Jadavpur University, India, 2005.
[21] A. Riveiro, F. Quinero, F. Lusquinos, J. del Val, R. Comesana, M. Boutinguiza, J. Pou, Experimental study on the CO_2 laser cutting of carbon fibre reinforced plastic composite, Compos. Part A 43 (2012) 1400–1409.
[22] N. Rajaram, J.S. Ahmad, S.H. Cheraghi, CO_2 laser cut quality of4130 steel, Int. J. Mach. Tools Manuf. 43 (2003) 351–358.
[23] G. Chryssolouris, Laser Machining: Theory and Practice, Springer-Verlag, New York, 1991.
[24] D. Dhupal, B. Doloi, B. Bhattacharyya, Parametric analysis and optimization of Nd: YAG laser micro-grooving of aluminum titanate (Al_2TiO_5) ceramics, Int. J Adv. Manuf. Technol. 36 (2008) 883–893.
[25] D. Dhupal, Experimental Investigations into Nd:YAG Laser Micro-machining on Engineering ceramics, PhD Dissertation thesis, Jadavpur University, India, 2009.
[26] A. Sen, B. Doloi, B. Bhattacharyya, Fibre laser microchanneling of polymethyl methacrylate (PMMA), Lasers in Engineering 35 (1-4) (2016) 123–138.
[27] S. Prakash, S. Kumar, Experimental investigations and analytical modeling of multi-pass CO_2 laser processing on PMMA, Precis. Eng. 49 (2017) 220–234.
[28] G. Kibria, B. Doloi, B. Bhattacharyya, Predictive model and process parameters optimization of Nd:YAG laser micro-turning of ceramics, Int. J. Adv. Manuf. Technol. 65 (2013) 213–229.
[29] D. Dhupal, B. Doloi, B. Bhattacharyya, Modeling and optimization on Nd:YAG laser turned micro-grooving of cylindrical ceramic material, Opt. Lasers Eng. 47 (2009) 917–925.
[30] O.F. Biswas, Experimental Investigation on Spiral Micro-grooving on Aluminium using Fibre Laser, PG Dissertation thesis, Aliah University, India, 2016.
[31] J. Peter, B. Doloi, B. Bhattacharyya, Parametric analysis of Nd: YAG laser marking on alumina ceramics, Inst. Eng. (India): Ser. C 94 (4) (2013) 287–292.
[32] G. Ricciardi, M. Cantello, G.S. Aira, Marking of computer keyboards by means of excimer lasers, Ann. CIRP 45 (1) (1996) 191–196.

[33] L. Li, J. Lawrence, J.T. Spencer, High power diode laser marking and engraving of building materials, SPIE Proc. 3097 (1997) 600–611.
[34] G. Ogura, J. Angell, D. Wall, Applications test potential of laser micro-machining, Laser Focus World (1998) 117–123.
[35] R.R. Behera, M. Ravi Sankar, State of the art on under liquid laser beam machining, in: 4th International Conference on Materials Processing and Characterization, Materials Today: Proceedings, vol. 2, 2015, pp. 1731–1740.
[36] A.K. Nath, D. Hansdah, S. Roy, A.R. Chowdhury, A study on laser drilling of thin sheet in air and underwater, J. Appl. Phys. 107 (2010) 123103 (1-9).
[37] D. Allen, H. Almound, P. Logan, A technical comparison of micro-electrodischarge machining, micro-drilling and copper vapour laser machining for the fabrication of ink jet nozzles, Proc. SPIE 4019 (2000) 531–540.

SUBCHAPTER 4.4

Electron Beam Machining (EBM)

4.4.1 Introduction

At present, one of the most advanced material processing techniques is Electron Beam Machining (EBM) which involves the use of focused beam of high velocity electron. EBM is a thermal material removal process where material from the workpiece is removed by means of melting and vaporization to achieve desired shape and size. The first application of electron beams for machining and welding was reported through pioneer work of Dr. K·H Steigerwald and his team [1]. In the year 1952, it culminated with the first development of prototype of EBM machining setup in Germany. The growth of EBM technology during the last couple of decade has been very fast due to its processing speed and easy control mechanism. The electron beam can be utilized not only for machining but also for welding, heat treatment, deposition, purification and inspection of materials. EBM machine has been adopted commercially in industry for manufacturing since 1960, primarily for welding applications in the field of nuclear and aerospace. Subsequently, EBM is also used for high speed drilling of small hole, cutting, and milling of any kind of materials particularly on semiconductor in the area of micro electronics. High power density, precise controllability of spot diameter as well as adjustability of parameters are the key features of EBM. It is also a promising technology for micro-machining applications due to its very small focused spot as well as flexibility and high level of accuracy. Among various thermal manufacturing processes, the application range of electron beam is the largest as shown in Fig. 4.4.1, which exhibits various combination of power density and spot diameter of different heat sources.

Energy density of electron beam can be controlled from low to very high power density according to the needs of various applications, e.g., hardening, surface modification, welding and different types of machining, e.g., perforation, drilling, milling of metals as well as plastics. Fig. 4.4.2 represents different application range of electron beams. For various applications requirements of power density of the electron beam on the spot and time of interaction are shown in the figure [1].

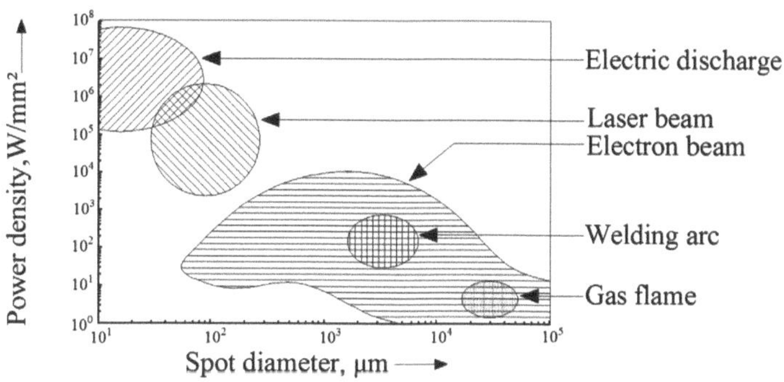

Fig. 4.4.1 Application range of various thermal manufacturing processes.

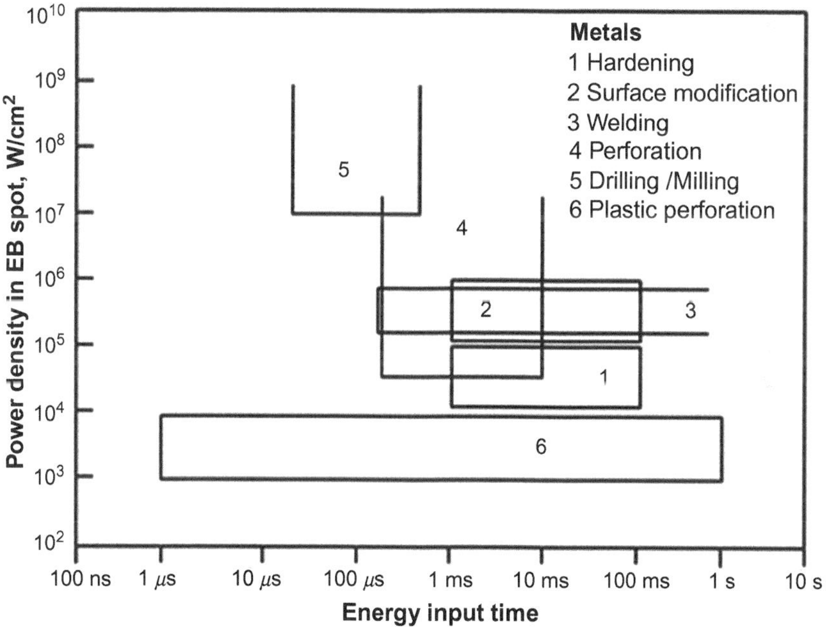

Fig. 4.4.2 Different application range of electron beam.

4.4.1.1 Fundamental of electron beam generation

Free electron can be generated by heating a Tungsten or Tantalum filament at a high temperature. In order to emit focused beam of high velocity of electrons, electron beam gun is utilized. Hence, electron beam gun is the key component of any electron beam machining facility. Electron beam gun is basically consists of two parts. First part produces electrons and second part mainly generates focused beam of electrons which is utilized for different purpose.

Fig. 4.4.3 represents detail configuration of electron gun. Electron gun is mainly consisting of a cathode which generates free electron, a negatively biased grid cup with respect to cathode and an anode at ground potential. The source of electron is highly negatively biased cathode cartridge which consists of tungsten hair pin (filament). Tungsten hair pin is heated up with the help of cathode heater supply which leads to thermo ionic emission of electrons. Cathode cartridge can be changed in case of failure due to the prolong heating. The temperature of tungsten hair pin goes around 2500 to 3000 °C due to emission current with the magnitude of 25–100 mA by maintaining the range of current densities in between 5 and 15 A/cm^2 depending on cathode material and its cross sectional area. Electron emission is further enhanced by keeping low vacuum in the chamber of electron beam gun. Moreover, thermo ionic electrons are repealed away from the cathode due to the negative biased. Generated electrons do not diverge and approach in the form of beam towards highly negative annular bias grid which is placed just below the cathode. The beam passes through the central hole of the annular bias grid. This annular bias grid is also used for controlling the beam current. This electron beam is attracted with the help of annular

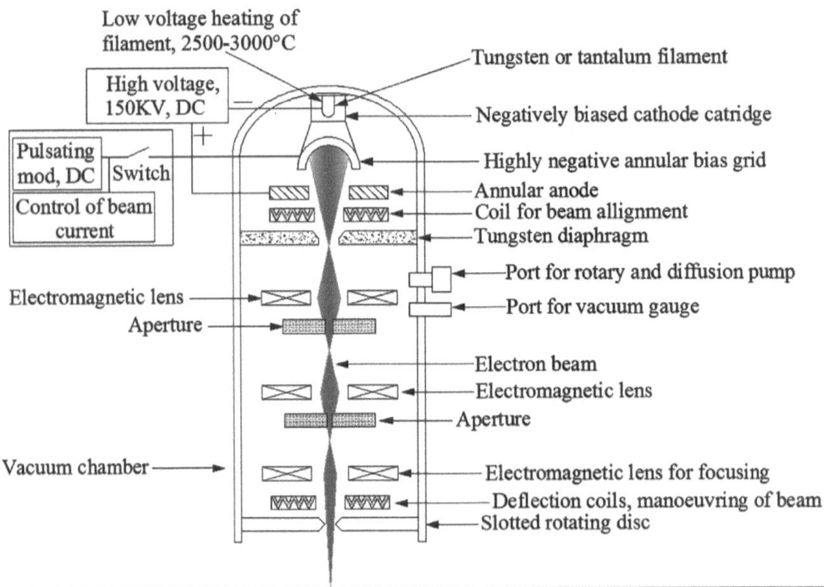

Fig. 4.4.3 Detail configuration of electron gun.

anode maintaining the accelerating voltage in the order of 100 to 150 kV. This amount of high voltage also accelerates electrons to a high velocity. Tungsten diaphragm having central hole separates this part of the electron beam generation unit from the rest of the beam focusing unit of electron beam gun. The gun needs to be maintained vacuum in the range of 10^{-4} and 10^{-6} Torr. This is essential for retaining the energy of electrons which leads to increase the life of cathode cartridge.

Different types of electron gun configurations are available for generation of electron beam. Electron beam accelerates due to the high voltage of the annular anode. Three types of electron gun systems can be used for different purposes. Fig. 4.4.4 shows the line diagram of three different electron guns. Normal triode system of electron gun has cylindrical annular grid and annular anode. It has two type of configurations, i.e., closed anode and open anode. However, normal triode system is widely utilized closed anode type configuration. Pierce type electron gun system utilized concentrically spherical electrodes which is indirectly heated [2]. Steigerwald type electron gun has a hairpin shaped filament surrounded by conically shaped annular bias grid [3]. Pierce and Steigerwald type systems form crossover spots after the anode which requires a long machine column. Due to the longer column lengths, these types of guns are critically dependent on perfect alignment. However, in case of normal triode system the column length is comparatively smaller. Since alignment is most important, normal triode systems are preferred for the electron gun.

After annular anode the electron beam passes through magnetic lenses and apertures. Magnetic lenses reduces the divergences of the beam and controls

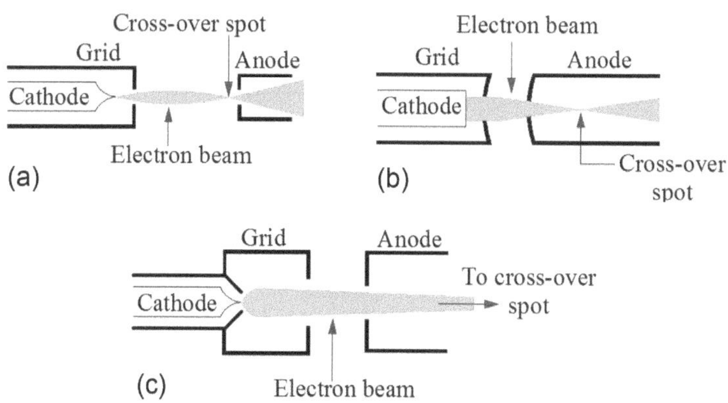

Fig. 4.4.4 Line diagram of three different electron guns. (a) Normal triode system, (b) Pierce gun configuration, (c) Steigerwald gun configuration.

the shape of the beam. The low energy stray electrons from the fringe of the beam are restricted by the apertures. Apertures make the beam more narrower by allowing the convergent electron only to pass through it. With the simultaneous help of the magnetic lenses and apertures, the quality of the electron beam is enhanced. The focused beam achieved its spot size with the help of final set of electromagnetic lens. With the help of deflection coil, the focused beam with desired spot size can be precisely maneuvered within small range. At the exit of the gun, a series of slotted disks are mounted which protects the electron beam gun from damage by metal spatter and vapor during various operations. Typical triode Electron beam gun is generally used for various machining applications for generation of different high accuracy features.

4.4.1.2 Material removal mechanism in electron beam machining

Electron beam with high velocity as generated from the electron gun strike the workpiece. Electron is the smallest stable elementary particle of the atom having a negative charge of 1.602×10^{-19}C and mass of 9.109×10^{-31}kg. As the mass of electron is negligible; the kinetic energy of electron mainly depends on its velocity. Hence, the velocity of the electrons of the focused beam should be kept very high utilizing the electron gun. The electron beam attain high velocity which is imparted by acceleration voltage of annular anode and subsequently focused by set of electromagnetic lenses of the electron gun and impinge on to the workpiece to a very small well defined area in the order of 2 to 5 μm in diameter. Power of the electron beam is adjusted so that a high power density is available on the workpiece surface which is in order of 10^8 W/cm^2. Focused beam of that high power density electron strikes the workpiece with high kinetic energy. The kinetic energy is then rapidly translated into heat which results in rapid increase in temperature of workpiece in a localized spot. The instant rise in temperature is more than sufficient for melting and vaporization of any material regardless of its melting point and thermal conductivity. Hence, rapid melting and vaporization causes metal removal from the workpiece. Due to the high power density of EBM all engineering materials can be machined.

Initially high velocity electron penetrates into the surface of the work material without disturbing the molecules up to certain depth often called transparent layer. After that electrons start colliding with the molecules and subsequently lose its kinetic energy which converted to heat energy through thermal agitation that is taking place below the transparent layer.

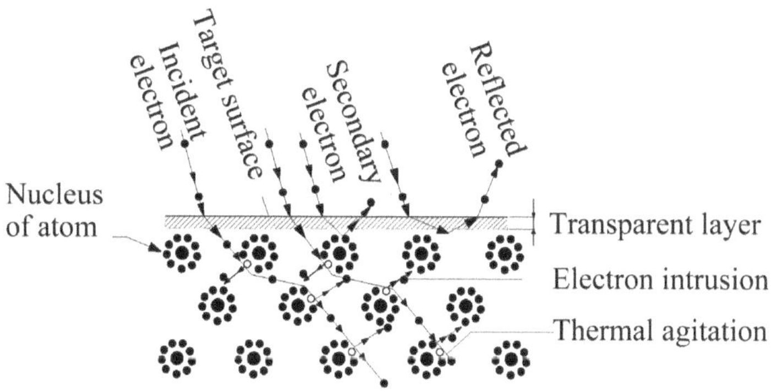

Fig. 4.4.5 Colliding phenomena of incident electrons.

Fraction of energy of the incident beam also decreases due to back scattering effect of incident electrons due to the collisions which produce secondary electrons as shown in Fig. 4.4.5. Some electrons go out of the target material and some electron also retain in the lattice of the material known as reflection and intrusion electron respectively. The degree of penetration of the beam into the material depends on the energy of the incident electron and atomic number of work piece material [4]. Because of generation of high temperature due to the collisions of incident electron with the molecules of the workpiece material, melting and vaporization occurs at the colliding spot. The beam as well as evaporating molecules exerts pressure on the liquid interface as melting occurs at that spot. The rise of high force due to this interaction is known as recoiling pressure. Complicated dynamic problem that occurs at the liquid interface is not properly understood till date. As well as the way in which the focused beam penetrates the workpiece is not accurately understood due to the involvement of its complex mechanisms. However, the incident beam and recoil pressure have a remarkable effect on the rate of vaporization on the workpiece material. Hence, machining mechanisms mainly occurs due to high recoiling pressure and surface tension at the liquid interface. Due to rapid ejection and vaporization of the liquid molten material, material removal takes place from the striking spot of the workpiece.

Machining is achieved using controlled high power density electron beam by instant local heating, melting and vaporization. EBM is extensively utilized for the generation of multiple high aspect ratio holes of different cross sections. In case of drilling, workpiece is supported by an auxiliary backing plate which may be of organic or synthetic material. During the

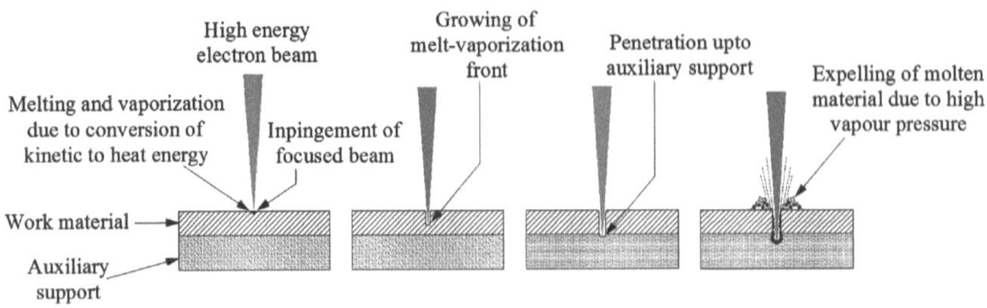

Fig. 4.4.6 Mechanism of perforation by EB.

process, very small amount of material actually vaporized which form a capillary channel inside the molten material. The beam and the capillary channel instantly penetrate into the workpiece as well as up to the certain depth of backing plate. The backing material which is come under the direct contact of the beam is almost vaporized leads to violent discharge of backing material vapor. This high pressure vapor of backing material forced out the molten material wall of the generated capillary. The vapor pressure is so high that the molten material leaves the channel as a shower of sparks forming a hole in the workpiece and form a small cavity in the backing material [5]. These subsequent phenomena of mechanism of perforation by EBM are exhibited in the Fig. 4.4.6.

4.4.2 Electron beam machining system setup

The electron beam machining setup produces high energy focused beam of electrons for machining any type of materials. To fulfill the requirement, the construction of EBM set up should have some basic requirements which include electron gun, focusing system, vacuum system, X-Y work-table, viewing unit and main machine control system as shown in Fig. 4.4.7. Electron gun produces electron beam in different ranges of high power with various spot sizes combinations. Construction details of electron gun and its operation have already been discussed. Focusing is needed for reducing the beam diameter to achieve necessary power densities for the purpose of machining. To maintain low pressure for allowing the generated electrons to travel a longer path, vacuum system is essential. On the X-Y work table, job is mounted for machining. Any profile or feature can be machined by the precise movement of the job with the help of CNC X-Y work table. Main machine control system integrates all the subsystems of EBM setup to achieve desired machining performance.

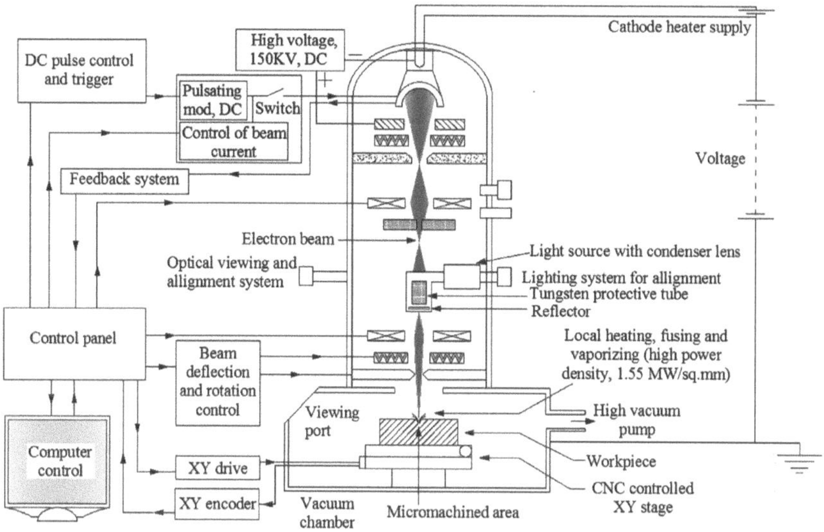

Fig. 4.4.7 Construction details of EBM set up.

For the construction of EBM machine, electron gun of triode system type is widely used. In electron gun the emission of electron from the tungsten hair pin cathode are normally depends on the filament temperature. The most advantage factor is that by controlling the temperature, amount of electron emission can be easily controlled. In the gun, electrons are accelerated with the help of annular anode. As per the requirements of machining applications, acceleration voltages are to be considered. EBM machine is operated only in the pulse mode for material removal applications. Electron beam travels through a high negative annular bias grid before passing through annular anode. This annular bias grid acts as a switching device for generating pulsed beam by controlling the flow of electrons. The beam current can also be controlled adequately for the machining purpose with the help of annular bias grid.

Focusing of the highly accelerated beam can be obtained utilizing electrostatic or electromagnetic lenses. Electromagnetic system of focusing is widely used for the construction of EBM machine because it requires low voltage compare to electrostatic system. High voltage of electrostatic system of focusing leads to non-uniform field in the region of machining area. Magnetic lens consists of coil and wire surrounded by iron shroud. An intense electromagnetic field is generated in the narrow gap and pole piece of the shroud. Normally two electromagnetic lenses are utilized in EBM system. The focused electron beam can be preferentially deflected without defocusing for a very small amount utilizing beam deflection coil. For ensuring a round beam spot at the work surface, stigmator is used which correct minor shape irregularities of the beam, i.e., aberration. At the exit of the electron gun a series of rotating slotted disks are mounted for protecting the electron gun from the metal spatter and vapor. The electron beam only passes through these slots. By synchronizing the rate of rotation of disk with beam pulse, spatter and vapor are not allowed to enter into the gun.

The designed profile on the workpiece can be machined by moving the point of incidence of the electron beam on the workpiece. This can be achieved by moving the workpiece which is mounted on the XY stage as well as deflecting the beam itself electromagnetically. The beam can be deflected very precisely within 1 mm in most cases without considerable defocusing. Hence, for machining of larger areas along with beam deflection, movement of worktable is needed. Movement of job as well as worktable is achieved by utilizing various programs through CNC controller.

During machining, it is necessary to observe the machining zone. For this, optical viewing unit is utilized in the machining chamber. However, during machining interval the same electron beam with lower power and smaller spot size can also be used as a beam of scanning electron microscope to inspect the actual condition of the machined profile [6].

All the major parameters of electron gun, e.g., beam current, acceleration voltage, power, pulse frequency etc. can be controlled with the help of control panel which is inter connected with computer controlled system as shown in Fig. 4.4.7. Precise movements of the workpiece with X-Y drive work table can also be possible with the help of control panel interfaced with central computer unit. Accurate focusing and alignment of the beam with the help of magnetic lens and electromagnetic coil can also be achieved with the help of control panel. With this, beam deflection and rotation of slotted disk can be controlled and monitored. Hence, adequate controlling system for EBM must be supported by a high end CNC system. It provides necessary flexibility and makes control of machining parameters much easier. Beam parameters along with workpiece movements are simultaneously controlled so that any minor deviation of mechanical movements is automatically rectified by the beam adjustment [1]. Fig. 4.4.7 shows detail block diagram of a CNC system indicating all the control features of an ideal EBM system. Any type of complicated shapes can be produced in the workpiece through proper programming of the CNC controlled EBM system. Fig. 4.4.8 shows photographic view of EBM system setup.

Fig. 4.4.8 EBM unit. *(Courtesy: pro-beam AG & Co. KGaA, Germany.)*

4.4.3 Theoretical consideration for EBM

Theoretical study of EBM mainly centered around calculation of maximum current density on the focused object and how the energy transferred to the work material to achieve desired material removal. Influences of current density and other factors on machining performances of EBM are discussed as follows.

4.4.3.1 Current density

The temperature of the cathode tip mainly influences the emission of electrons as well as Maxwellian spread in energy. By the electron gun, electrons from the tip of the cathode are focused into a crossover spot which is smaller than the cathode tip. Langmuir [7] has shown that cathode temperature enforces limitations on the maximum obtainable current density at the focused spot. According to the Brown and Nickols [8], the maximum current density in the focused image is as follows:

$$j = j_c\left(\frac{qV}{kT}\right)\alpha^2 \tag{4.4.1}$$

where j_c is the current density at the cathode and q is the charge of electrons. α is the semi angle, q_V is the energy, k is the Boltzmann's constant and T is the cathode temperature, K.

Demagnified image of the crossover spot is form on the workpiece utilizing a series of magnetic objective lenses. The distribution of the beam current within this image is approximately Gaussian in the absent of abrasion. This image is termed as Gaussian spot and can be described by an RMS diameter d_O. Hence, the maximum current in the final image is given by:

$$I = j_c\left(\frac{qV}{kT}\right)\alpha^2 \times \frac{\pi}{4} d_0{}^2 \tag{4.4.2}$$

However, $d_O = d_g \times M$, where, d_g is the diameter of the crossover spot and M is the magnification factor of the magnetic lens which is less than one. Current density at the final spot can be made sufficiently large when d_0 is very small.

Maximum cathode current density J_c is limited to about 2 A/cm^2 by the maximum temperature at which cathode can be operated, i.e., 2800 K which leads to shorter life of the cathode. Hence, there is a limitation on the rise of the cathode temperature. The maximum semi angle α is also limited due to spherical aberration of the objective lens. Hence, current density can only be increased by increasing accelerating voltage V. Due to difficulty faced with the working at very high voltage and electron penetration effect, the maximum value of V is limited.

4.4.3.2 Effects of aberrations on maximum current density

Spherical aberration in the objective lens which focuses the beam as well as aberrations associated with beam deflection limits the maximum current density. Other aberrations in the EBM system are not significant due to sufficient stability of accelerating voltage and lens current. In the image, crossover spot is focused not to a point but which form like a disc of approximately Gaussian form whose diameter is as follows [9]:

$$d_s = 1/2\, C_s\alpha^3 \tag{4.4.3}$$

where C_s is a constant of the objective lens for a constant focal length which depends on the construction of the lens [10]. Due to spherical aberration, effective diameter, d of the image can be determined as follows [11]:

$$d^2 = {d_0}^2 + {d_s}^2 \tag{4.4.4}$$

Hence, considering the value of α and d as constant and utilizing Eqs. (4.1.2), (4.1.3), and (4.1.4), maximum current density, $J_{\max}$ occurs when, $d = 2d_s$ and $d_0 = 3^{1/2} d_s$ and can be expressed as follows [12–14]:

$$J_{\max} = 0.75 j_c (qV/kT)(d/c_s)^{2/3} \tag{4.4.5}$$

4.4.3.3 Material removal

Generally, rates of material removal in EBM are determined in terms of number of pulse required for melting and vaporization of certain amount of material. The depth of machining can be achieved by controlling the machining time which can be adjusted by monitoring the number of pulses.

In EBM, two types of pulse numbers are used. Firstly, a linear number n_{pL}, is adopted for hole sinking by EBM. Secondly, the volume number n_{pV}, is used for evaluating slotting by EBM, that is parting off or cutting into the materials.

So, for drilling, the number of pulses required for removing a hole of depth d can be described by

$$n_{pL} = \frac{d_h}{d_p} \tag{4.4.6}$$

where d_h is the required depth of the hole, mm and d_p is the depth of the hole per pulse, mm.

The machining time t_m can be given by

$$t_m = \frac{n_{pL}}{f} \tag{4.4.7}$$

where n_p is the number of pulses and f is the frequency, Hz.

Therefore the drilling rate λ, mm/min, can be calculated by

$$\lambda = \frac{d_h f}{n_{pL}} \tag{4.4.8}$$

According to Kaczmarek [15], the number of pulses n_{pL} can be described as a function of accelerating voltage V, Volt and emission current I, A.

So,

$$n_{pL} = \frac{1}{KIV} \tag{4.4.9}$$

where K is a constant.

Hence, drilling rate λ, mm/min will become

$$\lambda = Kd_h fIV \tag{4.4.10}$$

And volumetric removal rate VRR_d, mm^3/min for drilling will become

$$VRR_d = \frac{\pi}{4} Kd_b d_h fIV \tag{4.4.11}$$

where d_b is the beam diameter in contact with the workpiece, mm.

Similarly, in case of slotting, the number of pulses required for removing material from the work surface can be described by

$$n_{pV} = \frac{m}{m_p} \tag{4.4.12}$$

where m is the mass of material to be removed and m_p is the material removed by a single pulse.

So, the slotting time will be

$$t_m = \frac{n_{pV}.L}{f.d_b} \tag{4.4.13}$$

where L is the length of the slot, mm.

And volumetric removal rate VRR_s, mm^3/min for slotting will become

$$VRR_s = Kd_b{}^2 mfIV \tag{4.4.14}$$

Putting the value of I in Eq. (4.4.2), the volumetric removal rate VRR_d, mm^3/min for drilling will become:

$$VRR_d = \left(\frac{\pi}{4}\right)^2 Kj_c \left(\frac{qV}{kT}\right) \alpha^2 d_0{}^2 d_b d_h fV \tag{4.4.15}$$

Penetration rate of electrons influence the depth of machining and intern volumetric material removal. High energy electrons penetrate the surface layer to the depth of a few microns. Electron impinging range (R) can be correlated with acceleration voltage (V) and density of the work materials (ρ) as follows:

$$R = 2.2 \times 10^{-12} V^2/\rho \qquad (4.4.16)$$

Electron impinging range is 7 μm and 10 μm for steel and aluminum at 50 kV accelerating voltage. Hence, acceleration voltage is most influencing factor in EBM. Energy is transferred to the atoms of the material to be machined over a relatively large zone which is not controllable hence; ultra precision machining is not achievable by EBM [4].

The relationship between parameters of beam, the nature of work material and the depth and width of the cut produced has not been accurately established till date. Depth and width of cut produced during EBM is not only dependent on vaporization of the material but also other effects of non thermal nature.

4.4.4 Influence of process parameters on performance criteria of EBM processes

The control parameters for Electron Beam Machining are beam current, accelerating voltage, power, pulse width, pulse frequency, cutting speed as well as lens current, beam deflection signal and position of the focus with respect to work piece etc. which influence the performance criteria of EBM processes [16]. However, major influencing parameters for machining are pulse width for generation of depth and beam current for achieving the desired width. Lens current determines the size of focus spot which also controls the diameter of the hole. Machining characteristics also depends on the power distribution within the beam as well as focusing condition.

4.4.4.1 Process parameters

Beam current influences the energy of the pulse which can be utilized during machining. Beam current increases energy per pulse delivered to the workpiece during machining. Beam current can be in the range of 100 μA to 1 A. Thus, EBM can generate pulse energies in the order of 150 joules/pulse which are much more comparing to available energies of laser machining system. Hence, EBM is capable of machining high aspect ratio features at a very faster rate utilizing its enormously high pulse energy.

Pulse width influences depth of machining as well as diameter of the hole. The width and depth of machined features can be increased by utilizing longer pulse width during EBM operation. However, longer pulse width will increase the interaction time of machining which may leads to higher degree of thermal damage during machining. Hence, to minimize thermal damage shorter pulse duration is preferable. Pulse duration as short as 30 μs and as long as 20 ms can be produced by EBM systems which indicates the wider range of machining capability.

Number of pulse is directly proportional to pulse frequency for a particular machining duration. Larger number of pulse is required at a particular voltage to achieve higher depth of machining. Increasing pulse duration and accelerating voltage at a particular pulse frequency also rises the pulse energy which can reduce the number of pulses needed to obtain required machining depth.

The size of the focused spot of the electron beam can be varied by controlling the focusing power of the electromagnetic lenses through scheming of the lens current. The lens current also determines the working distance, i.e., the distance between focused spot and electron beam gun. The diameter of the machined hole depends mainly on the beam spot size. Power density depends on the spot size which must be sufficient to promote machining and should be in the range of 10^7 to10^9 W/cm^2 in EB spot.

Another important consideration during machining is the positioning of the EB spot on the workpiece by controlling the lens current. There are mainly three types of focusing conditions, e.g., under focus, just focus and over focused [17]. Under and over focused condition is known as defocusing conditions during machining which affects the machining capabilities. By selecting different focusing conditions, various types of metals and nonmetals can be machined successfully as well as the shape accuracy is also improved. Beam deflection coil is also used to control the shape of the machined profile in the smaller range. Beam deflection signal can be programmed such a way that the beam can move along the predefined path to machine desired profile. Beam deflection coil can be used to scan the workpiece during operation and accordingly machining can also be monitored.

4.4.4.2 Process capabilities related to various performance criteria

By controlling various parameters of EBM, different achievable output criteria can be controlled during machining. Some of the important output criteria are material removal rate, penetration rate, surface finish and accuracy etc. EBM is a thermal process hence, it produces heat affected zone on the machined surface. Damage layer thickness and micro cracks produce

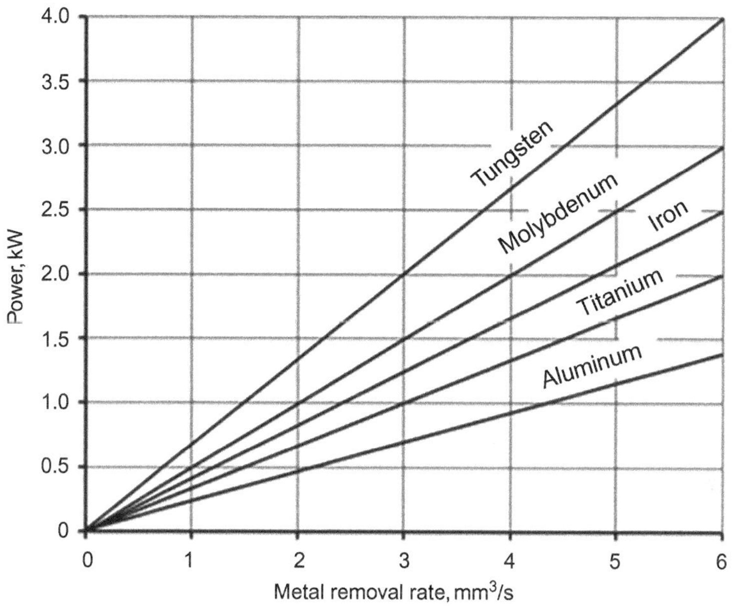

Fig. 4.4.9 Power requirements for different metals during EBM.

during machining can be controlled by selecting proper combination of process parameters.

EBM can successfully machine a wide range of metals, e.g., stainless steel, nickel and cobalt based alloy, tungsten, molybdenum, titanium, tungsten, tantalum, copper and aluminum etc. [18]. Fig. 4.4.9 shows power requirements to achieve different range of metal removal rate for different metals during EBM. Some material, e.g., aluminum and titanium require less power to achieve certain amount of metal removal compare to steel, molybdenum and tungsten. This is due to differences of melting temperature of the materials. Boiling point and thermal conductivity of different metals have major influences in determining how easily they can be machined.

Nonmetals, e.g., ceramics, quartz, silicon, plastics and leather can also be successfully machined by EBM [18]. Fig. 4.4.10 depicts a typical representation of the influence of beam power during machining of nonmetals, e.g., quartz crystal, silicon wafers, alumina, sapphire crystal and alumina wafers. Hence, it reveals the fact that low energy beam will be most effective during precise machining of non metals compare to the metals.

One of the most important capabilities of EBM is hole drilling. It has the ability to drill deep and high aspect ratio holes utilizing its high power density. It can also produce different cross sectional holes, e.g., conical, barrel as

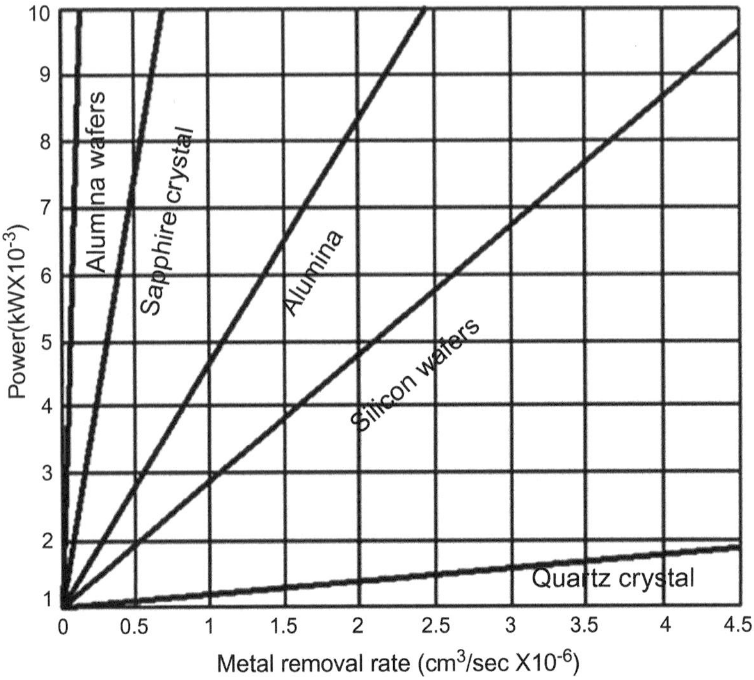

Fig. 4.4.10 Power requirements for different non-metals during EBM.

well as various shaped holes like square, elliptical etc. Aspect ratio in the range of 15:1 can be machined with consistent accuracy and high speed at the rates of several thousand holes per second. EBM can drill several numbers of holes in the range of 10^3 to 10^5 holes per second. Hence, it can be considered one of the most economical machining processes for perforation.

From the Fig. 4.4.11, it can be seen that drilling rate decreases with the increment of thickness of the workpiece. Drilling rate also increases with the decrease of diameter of the hole which is most acceptable criteria for perforation [19]. Multi pulse drilling is required for higher thickness workpiece. However, for thinner workpiece up to the range of 1 mm, single pulse drilling is sufficient. Fig. 4.4.12 shows capability of EB perforation typically for steel and nickel alloy. Wider range of hole diameters can be drilled in thinner workpiece, i.e., in the order of 20 μm. However, when workpiece thickness is >5 mm, hole diameter range is drastically reduced [19].

Quality of machined surface in EBM depends on various process parameters, e.g., pulse energy, spot diameter, focal position as well as type of materials. In EBM, machined surface defects, e.g., recast layer, micro cracks and heat affected zone due to thermal effect is less because of its extremely high

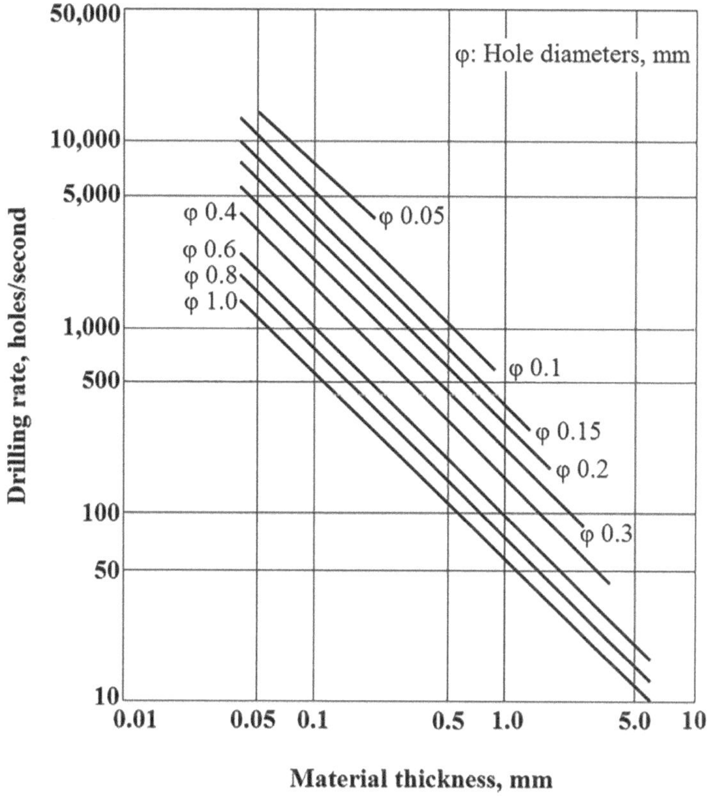

Fig. 4.4.11 Influence of thickness of the workpiece on drilling rate.

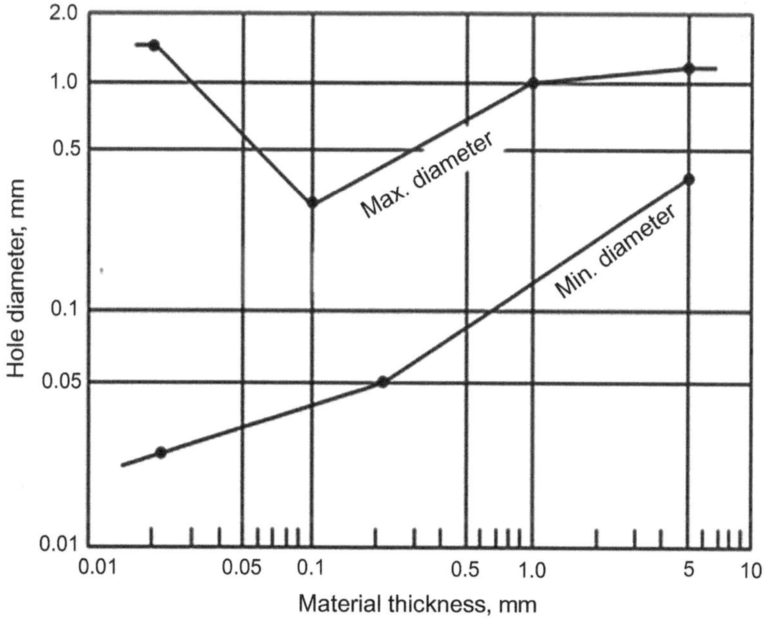

Fig. 4.4.12 Capability of EB perforation for steel and nickel alloy.

beam power density and shorter interaction time during machining. Surface roughness increases with pulse energy for materials like titanium, nickel, tungsten, gold and carbon. The range of achievable surface roughness in EBM can vary from 1 μm to 5 μm depending on the combination of different parameters utilize during machining. Dimensions of heat effected zone and its nature depends on beam energy, diameter and distribution of the beam in the focus spot and target material. During EB drilling, the damage layer thickness increases with mainly pulse duration and hole diameter. A typical thickness of heat affected layer can be up to 0.25 mm in EBM.

The best machining could be achieved on material of high atomic number, low specific heat and low thermal conductivity. However, selection of optimal combination of parameter setting of EBM for different materials is essential to achieve most effective machining performances. Table 4.4.1 represents specific application data of various EBM parametric combinations which can be useful for different types of machining operations applicable to a range of materials [20].

4.4.5 Advantages and limitations of electron beam machining processes

EBM provides very high drilling rates when small holes with large aspect ratio are to be drilled. Moreover it can machine almost any material irrespective of their mechanical properties. However, EBM has its own share of limitations.

4.4.5.1 Advantages

Various advantages of the EBM are as follows:

(i) Very high drilling rate ranging from 1000 to 10,000 is possible depending on the hole diameter and the thickness of the workpiece.
(ii) Less work holding and fixturing cost due to zero mechanical cutting force during EBM operations.
(iii) Fragile and brittle materials of thin workpiece can also be machined by EBM process.
(iv) Due to shorter pulses in EBM, the heat affected zone is rather less.
(v) EBM is capable to generate holes of any shape with high accuracy by combining beam deflection using electromagnetic coils and the CNC table.
(vi) The process can achieve high accuracy and repeatability of 0.1 mm for position of holes and 5% for the hole diameter.

Table 4.4.1 Specific applications data for electron beam machining

Materials	Shape produced	Cutting speed/time	Voltage (kV)	Beam current (μA)	Pulse width (μs)	Pulse frequency (Hz)
Alumina 0.76 mm thick	0.30 mm dia. hole	30s	125	60	80	50
Alumina wafers	0.76 mm dia. hole	10s	125	60	80	50
0.25 mm thick	0.10 mm wide slot	304 mm/min	90	150	80	150
Ferrite wafers 0.25 mm thick	0.025 mm dia. holes	<1 s	140	25	5	50
Mylar tape 0.038 mm thick	Cut	117 m/min	110	600	–	continuous
Sapphire crystal 0.66 mm thick	0.038 mm dia. hole, taper approx. 2°	<30s	110	20	9	50
Silicon wafers 0.254 mm thick (gold deposited)	Scribed to 0.05 mm depth	127 mm/min	130	70	4	3000
Steel, hardened 3.175 mm thick	Rectangle −0.45 by 1.82 mm	10 min	140	150	80	50
Stainless steel 6.35 mm thick	0.508 to 1.016 mm dia. holes	3 min	145	4000	2100	12.5
Stainless steel 1.57 mm thick	Rectangle −0.2032 by 6.35 mm ±0.00635 mm	5 min	140	120	80	50
Hastelloy 11.43 mm thick	2.54 mm dia. hole	70s	130	5000	5300	100
Molybdenum 0.254 mm thick	<0.05 mm dia.	<1 s	140	20	20	50
Tantalum alloy 1.016 mm thick	0.127 mm dia. hole	<1 s	140	100	80	50
Tungsten 0.254 mm thick	0.0254 mm dia. hole	<1 s	140	50	20	50

(vii) Compare to other non conventional machining processes, EBM produces moderate level surface finish with high level of accuracy.
(viii) Mechanical or thermal distortion is not produce during machining, and.
(ix) Controlling of machining parameters is very easy due to interfacing of computer to the EBM system etc.

4.4.5.2 Limitations

The primary limitations are the high capital cost of the equipment and necessary regular maintenance applicable for any equipment using vacuum system. Moreover in EBM there is significant amount of non-productive pump down period for attaining desired vacuum. However this can be reduced to some extent using vacuum load locks. Though heat affected zone is rather less in EBM but recast layer formation cannot be avoided. Other disadvantages of the EBM are listed as follows:
(i) EBM requires high level of operator skills,
(ii) >10 mm material thickness cannot be machined by EBM,
(iii) Long production time due to the time needed to generate a vacuum and
(iv) EBM needs auxiliary backing material during perforation etc.

4.4.6 Concise summary of EBM characteristics

Accelerating voltage: 50–150 kV
Beam current: 25 mA–1 A
Current density: 5–15 A/cm^2
Pulse time: 4–64,000 μs
Pulse frequency: 0.1–16 kHz
Vacuum: 10^{-4} and 10^{-6} Torr
Power: 0.5–60 kW
Power density: 10^7–10^9 W/cm^2
Beam intensity: 1.55×10^5–1.55×10^9 W/cm^2
Spot size: 2–25 μm
Material application: All materials
Maximum cutting ability for material thickness: 10 mm
Maximum material removal rate: 40mm^3/s
Penetration rate: 0.25 mm/s
Maximum tolerances: ± 0.03 mm of hole size
Heat affected zone: <0.25 mm deep
Surface roughness (R_a): 0.5–5 μm

4.4.7 Applications of electron beam machining processes

Electron beam machining is used for various applications, e.g., milling, cutting, welding and heating materials for vacuum deposition, vacuum purifications of materials, and for inspecting, examining and analyzing materials. For examining and analyzing materials, electron microscopy and X-ray micro-analysis can also be considered. EBM is more accepted in industries like aerospace, insulation, food processing, chemical, clothing etc. The most popular use of EBM process is drilling thousands of holes which diameter is <1 mm in very thin plates used for turbine engine combustor domes. The applications of EBM also cover the areas like making of fine gas orifice in space nuclear reactors, holes in wire drawing dies, cooling holes in turbine blades, metering holes in injector nozzles of diesel engines etc. Main areas of applications include drilling, perforation and pattern generation for fabrication of intricate circuit etc. Some specific applications of EBM are described as follows.

4.4.7.1 Drilling

An electron beam machine fitted with a system for numerically controlling the beam power, focus and pulse duration, and mechanical motion can be effectively used for various types of drilling. Several thousand of cylindrical, conical, and barrel-shaped holes of various diameters can be drilled with consistent accuracy [21]. Holes with an inclination angle of about 15 degrees can also possible with the application of EBM. Drilled holes with diameter of 1.5 mm and 10 mm of depth can be accurately drilled by EBM and aspect ratio which is depth-to diameter ratio is normally in the range of 1:1 to 1:15 as shown in Fig. 4.4.13. For deeper holes, in the range of 2.5 to 7.5 mm, Steigerwald and Mayer recommended the need for a stable power supply that can produce required groups of pulses with well-controlled beam of closely defined diameter. Angle of aperture and role of beam deflection coil have strong influence on the shape of the hole produced [3]. Holes of about 19 mm were produced. By EB-drilling, holes at an angle ranging from 60 to 90 degrees in a turbine blade of super alloy can also be possible.

4.4.7.2 Perforation of thin shit

By EB perforation 10^4 to 10^5 holes per second can be produced which is economically acceptable. Hence, single pulses with few microseconds are needed for perforation. In some typical applications, sheet or foil is stretched on a rotating drum, which is simultaneously shifted in the direction of its axis which produced rows of perforations following a helical line. Manipulators capable of linear and rotating movements in four axes are used for EBM perforation of jet engine components. For fabrication of filtration unit, foil made of a synthetic material has been perforated with 620 holes per square millimeter at a rate of one hole for every 10 s. EBM perforation can be applied to the production of masks of color television tubes, sieve and also for sound insulation and in glass fiber production. EBM is also used for perforation in clothing and shoe industries and capable of drilling these materials at a rate of 5000 holes/s [22].

4.4.7.3 Slotting

Different cross sectional slots can be successfully fabricated utilizing EBM. Rectangular slots of 0.2 by 6.35 mm are produced on 1.57 mm-thick stainless steel plate within 5 min utilizing EBM with the parameter settings, e.g., 140 kV, 120 μA, a pulse width of 80 μs, and a frequency of 50 Hz. Workpiece thickness mainly influences slotting speed. For example, using similar machining conditions, 0.05 mm-thick stainless steel was cut at a speed of 100 m/min, while 0.18 mm-thick stainless steel was cut at 50 m/min.

Fig. 4.4.13 Electron beam drilling. *(Courtesy: AGS-TECH Inc., AGS-TECH Inc., PO Box 4457, Albuquerque, NM 87196, USA.)*

4.4.7.4 Integrated circuit fabrication

Integrated circuit fabrication is usually made by anisotropic etching technique which consists of mask preparation utilizing photo lithography. The application of photolithography is restricted due to constrains of generation of minimum dimensions of the pattern utilizing comparatively long wavelength of visible light [23]. It also requires several hours to accomplish the task of masking for creation of final patterns. As well as, there are certain geometrical constraints of using anisotropic etching technique. Some advanced materials do not exhibit anisotropic etching property.

Electron beam with its very short wavelength, large frequency and beam deflection capacity offers high throughput for volume production and flexibility in much finer pattern generation for integrated circuit fabrication. First attempt was made by Birnie and Champney to use of electron beam technology in scribing thin film circuits for the electronic industry [24]. Further, detailed account for the manufacture of integrated circuits has been presented by Yew et al. [25].

For the production of the lithographic definition, the beam should be positioned accurately by means of deflection coils at the spot where a pattern needs to be written by exposing a film of electron resist coated on either a chrome mask blank or wafer. To break the bonds between the polymer molecules or to cross-linking in the polymers, an electron beam of energy about 10 to 20kV is required. The solubility changes when the resist film is immersed in the developer, typically a solvent for the resist. A differential material removal occurs because of the difference in solubility between the original and exposed resist polymers. To avoid unwanted etching of the integrated circuit mask, a fine pattern of polymer is thus obtained which is used as an active mask pattern generation and this can be carried out by a vector scan. Thus, the electron beam is deflected accordingly only to the locations at which the electron resist is to be exposed. The electron beam action is started as soon as the deflection system completes the positioning of the beam at the required place. With the raster scan method, described by McGeough, the chip pattern is divided into subfields [26]. The electron beam in a raster which is employed with the television scans each subfield. The electron beam is then turned on and off along each raster line as needed. By the combined effects of electron beam exposure and subsequent resist development, the required pattern is fully formed. Fig. 4.4.14 shows a typical application where EBM is used for a miniature electronic hybrid circuit engravement with a 40 micron-wide trace. Direct wafer processing by EBM

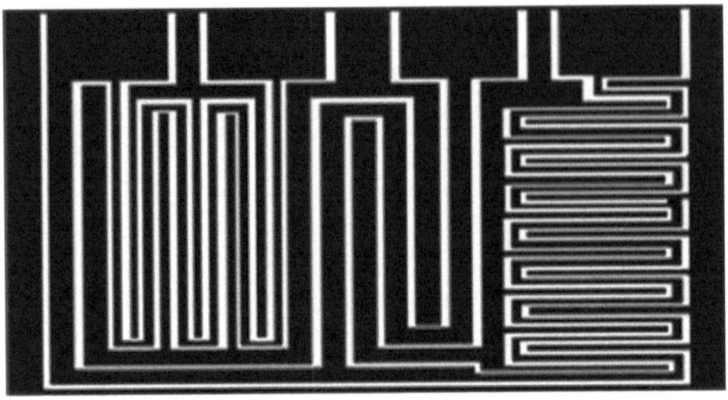

Fig. 4.4.14 Hybrid circuit engraved with 40 μm traces at speed >5 m/s [27].

is utilized for feature sizes below one or two microns. With electron beam techniques at a speed as high as 20 MHz, line width as small as several hundredth of an angstrom can also be written. Producing 22 wafers per hour can be possible by direct wafer processing electron beam systems. A typical intricate circuit fabrication utilizing electron beam is shown which refers to the engraving of a hybrid circuit with 40 micron width traces at a machining speed of 5 m/s [27].

4.4.8 Advancements in EBM

Electron beam has been exploited as a heat source for several years. However, its uses in various technological applications, especially in the area of machining, are still developing not only in the research stage but also in various engineering applications. The advancement of electron beam machining and its further applications in the area of welding, surface modification, rapid prototyping, cladding, engraving as well as texturing etc. have been reported through various research works around the world [28].

For EBM operation, attempt has been made to use acoustic emission signals which may act as effective tools for monitoring and control [29]. During EBM operation, two different types of acoustic emission are observed. Signal level is low while molten metal is removed outside the workpiece during EB drilling. However, burst signals of high level are generated intermittently during electron beam bombardment. Occurrence of the burst signals indicates about the fact of molten metal quantity which is remaining in the machined portion. The occurrence frequency increases with the quantity of existing molten metal. Hence, this signal indicates that the EB drilling proceeds to the welding process. This phenomenon has been utilized for effective monitoring of the EBM process and for also determining the appropriate condition for EB drilling. Hence, the occurrence elastic emission frequency is strongly related to the molten metal quantity. This measuring method can be utilized successfully for monitoring and controlling the electron beam machining.

Influence of beam de-focusing during drilling of alumina green sheet with an electron beam has been investigated by the researcher. High speed machining tests with a deflection coil have been carried out. The diameter increases because of the beam defocus effect. The beam defocus effects also depend on the time interval of the pulse. The dynamic focal control during EB drilling is very effective for hole diameter variation [17].

Cross-section of the rapidly solidified layer of a graded steel structure formed by low energy high-current pulsed electron beam treatment has been successfully analyzed [30]. The results of electron backscattering diffraction measurements revealed that the microstructures and phase components vary significantly through the melted layer and heat-affected zone. The characteristics of such a graded structure also changes with the number of pulses. Phase transformations are determined both by temperature gradients and composition gradients in the surface layers induced by EB treatment.

Electron beam is presently utilized in Rapid Prototyping (RP) for the source of controlled heating. Due to the high temperature of electron beam melting, EB prototype services produce parts with lower densities without any internal stresses. Hence, EB prototyping can be utilized successfully to produce both prototypes and low cost, high quality parts for various engineering applications. Fig. 4.4.15 shows schematic of rapid prototyping

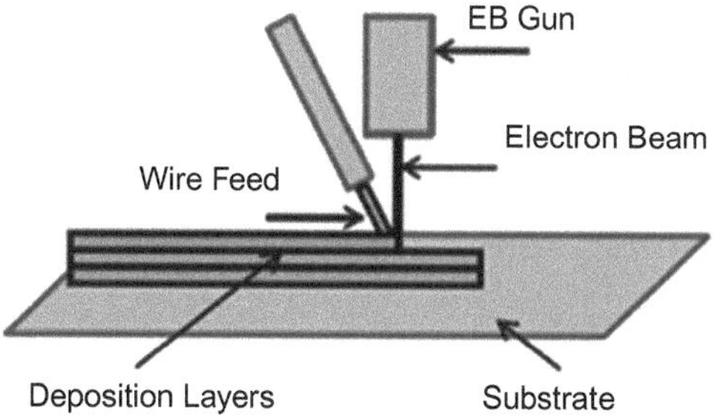

Fig. 4.4.15 Schematic of rapid prototyping process with wire and electron beam as a heat source [31].

process with wire and electron beam as a heat source [31]. EBM is one of the 3D printing techniques which can be used successfully for aerospace and biomedical applications. In EB rapid prototyping, different titanium, nickel, aluminum alloys as well as cobalt chrome wire are used primarily for those applications where electron beam is used as a heat source [32]. Recently, it has been reported that General Electric Company utilizes EB 3D printing for the turbines for new passenger aircraft.

Electron beam is presently utilized as a tool for repairing technology. It can be adopted for repairing of welding for the parts that have worn out or are broken which are critical for the system and may bring to a halt the entire production. Applications of electron beam for repairing technology can be proved as economical, which provides satisfactory solution [33].

A direct metal rapid prototyping (RP) fabrication technique utilizing electron beam melting process has been used to fabricate successfully porous Ti6Al4V parts with fully interconnected, controlled internal pore architecture. As shown in Fig. 4.4.16, Ti6Al4V parts with highly uniform internal honeycomb-like structure, controlled pore morphology and complete pore interconnectivity had been effectively fabricated by the researchers utilizing EBM process in direct metal layered fabrication

Fig. 4.4.16 Highly uniform internal honeycomb-like structure on Ti6Al4V parts [34].

technique. The compressive strength and Young's modulus of parts proved to be well-matched with those of natural bone. The potential of EBM process for fabricating Ti6Al4V parts with controlled internal pore architecture fulfilling the requirements of orthopedic implants is well established [34].

EBM needs further developments for effective utilization of this technology in the manufacturing fields. Attempt is to be made for the development of EB gun, where the requirements are to extend cathode life, improve beam consistency, and simplify the equipment structure as well as its operation. Other area of EBM which are to be developed such that it can operate in reduced pressure which avoids the need for large vacuum chambers with associated sealing and pumping provisions. It has been observed that research in electron beam machining and its applications in the area of fabrication, welding and surface engineering is progressing well and still growing.

4.5 Model questions

Electro discharge machining (EDM)

Q1. In EDM, which factors influence the generated spark?

Q2. Explain different phenomena occur in the spark gap during machining.

Q3. Discuss about the different variants of EDM system.

Q4. Explain the nature of waveform in pulse generator circuit for EDM.

Q5. Explain the role of ignition delay during spraking in pulse generator circuit

Q6. What should be the ratio of breakdown voltage to supply voltage if the power delivery through the circuit is maximum in RC circuit.

Q7. Identify nature of energy distribution in the gap and which one is most responsible for metal removal?

Q8. In EDM, what are the reasons for using graphite electrode?

Q9. What do you understand by (ii) wear ratio, (iii) heat affected zone (HAZ), (iv) critical resistance in RC circuit of EDM.

Q10. In EDM, explain mechanism of servo feed control.

Q11. What are the different dielectric flushing techniques applied in EDM?

Q12. How do you compensate the tool wear during EDM operation?

Q13. The role of influencing parameters on various machining performance criteria.

Q14. Why reverse polarity is generally employed during die-sinking EDM?

Q15. How the EDM parameters affect the surface integrity of machined components?
Q16. Explain in brief the working principle of wire-EDM.
Q17. Describe different sub components of WEDM system.
Q18. Identify the different factors influence the WEDM performance.
Q19. How taper cutting is performed in WEDM?
Q20. Explain different forces act on the wire during WEDM operation?
Q21. What are the different of accuracy factors associated with WEDM?
Q22. Explain the role of trim cutting on the performances of WEDM.
Q23. In WEDM, how the stratified wire works?
Q24. Describe about advanced wire utilized in WEDM.
Q25. How wire-tool vibration affects the profile accuracy in WEDM?
Q26. Deduce an expression for wire lag phenomena and how it can be minimized?
Q27. Differentiate between EDG and EDM.
Q28. Explain about the different types of EDG.
Q29. Identify major environmental impacts of EDM and how it can be minimized?
Q30. Describe about the dry EDM?
Q31. How nonconductive materials can be machined utilizing EDM?

Plasma arc machining (PAM)

Q1. What is the mechanism of material removal in PAM?
Q2. Distinguish between transferred arc and nontransferred arc PAM system.
Q3. Discuss on various elements of plasma arc machining system setup with sketch.
Q4. State the various process parameters and performance criteria of PAM process.
Q5. Mention the advantages and limitations of plasma arc machining process.
Q6. What are the various applications of plasma arc machining process?
Q7. Discuss on various advancements in PAM technology.

Laser beam machining (LBM)

Q1. How is laser beam different for ordinary torch light?
Q2. What are the various types of laser in machining applications?
Q3. Discuss on principle of laser generation for following types of laser:

(i) Nd:YAG laser;
(ii) CO2 laser;
(iii) Excimer laser;
(iv) Diode laser.

Q4. Discuss the material removal mechanism of laser beam machining.
Q5. Discuss on the various elements of laser beam machining system with sketch.
Q6. State the various process parameters and performance criteria of LBM process.
Q7. What are the various types of laser micro-machining processes?
Q8. Mention the advantages and limitations of laser beam machining process.
Q9. What are the various field of applications of laser beam machining process?

Electron beam machining (EBM)

Q1. Why the application range is maximum for EBM over other thermal manufacturing process?
Q2. What type of manufacturing applications can be performed by EBM in the context of power density and interaction time?
Q3. Explain different types of electron guns and which one is mostly used?
Q4. Give explanation about the mechanism of material removal in EBM process?
Q5. Describe mechanism of EB perforation.
Q6. Explain in details different units of EBM system setup.
Q7. In EBM, why vacuum environment is needed?
Q8. Discuss about the major process parameters of EBM.
Q9. In EBM, how the maximum current density can be generated?
Q10. Role of pulse width and beam current on machining performance in EBM.
Q11. Differentiate between EBM and laser beam machining in the context of machining rate.
Q12. Discuss about the differences in between electrostatic lens and electromagnetic lens.
Q13. In EBM elucidate about the voltage and current requirement in electrostatic lens.
Q14. Justify the power requirements for different metal during EBM operation.

Q15. Justify the power requirements for different nonmetal during EBM operation.

Q16. Explain the role of beam deflection during EBM operation.

Q17. Why HAZ is much less in EBM?

Q18. How EBM can be utilized as heat source in RP?

References

[1] K.H. Steigerwald, Material processing with electron beams, in: Proc. Fourth Congress on Electron Microscopy, Berlin 1958, 1960, p. 276 (in German).

[2] J.R. Pierce, Rectilinear electron flow in beams, J. App. Phys. 11 (1940) 548.

[3] K. Steigerwald, E. Meyer, New developments in electron beam machining methods, in: Electrical Methods of Machining and Forming IEE, Conf. Publ. No. 38, 1967, pp. 252–258.

[4] N. Taniguchi, Current status in, and future trends of, ultra-precision machining and ultrafine materials processing, Ann. CIRP 32 (2) (1983) 573–582.

[5] G.F. Benedict, Nontraditional Manufacturing Processes, Marcel Dekker Inc, New York, 1987.

[6] S. Kelly, H.N.G. King, The use and design of a high-voltage electron beam machine, Microelectron. Reliab. 4 (1965) 85.

[7] D.B. Langmuir, Theoretical limitations of cathode-ray tubes, Proc. IRE 25 (1937) 977.

[8] G. Brown, K.G. Nichols, A review of the use of Electron beam machines for thermal milling, J. Mater. Sci. 1 (1966) 96–111.

[9] M.E. Haine, The Electron Microscope, Spon, 1961.

[10] G. Liebmann, E.M. Grad, Imaging properties of a series of magnetic electron lenses, Proc. Phys. Soc. Lond. 64 (1951) 956.

[11] O.C. Wells, Calculation of the heat-affected zone during pulsed electron-beam machining, IEEE Trans. Electron Dev. (4) (1965) 224.

[12] T.E. Everhar, Berkeley University, California, A SD Technical Documentary Report, (1962), p. 63.

[13] H. Schwarz, Remarks on power density, total power, and pressure of electron beams as a welding, cutting, and surfacing tool, Rev. Sci. Inst. (USA) 33 (1962) 688.

[14] P.A. Einstein, Design of an experimental electron beam machining apparatus, in: Proc. Solid Circuits and Microminiaturization, West Ham, 1963, p. 239.

[15] J. Kaczmarek, Principles of Machining by Cutting, Abrasion and Erosion, Peter Pergrenius Ltd, Stevenage, UK, 1976.

[16] J.A. McGeough, Advanced Methods of Machining, Chapman and Hall, London, New York, 1988.

[17] K. Horio, T. Terabayashi, Beam defocus effect in electron beam machining of green ceramic sheet, Ann. CIRP 36 (1) (1987) 95–98.

[18] R.K. Springbone (Ed.), Nontraditional Machining Processes, American Society of Tools and Manufacturing Engineers, Dearborn, MI, 1967, p. 141.

[19] J. Drew, Electron Beam Drilling, Influence of Metallurgy on Hole Making Operations, American Society for Metals, Metals Park, OH, 1978, p. 119.

[20] Machining Data Handbook, 3rd ed., vol. 2, Metcut Research Associates Inc, Ohio, USA, 1980.

[21] G. Bellows, J.B. Kohls, Drilling without drill, in: American Mechanist, 1983, pp. 183–185. Special Report 743.

[22] K.H. Steigerwald, Electron Beam Machining – The Process and Its Applications, Farrel Company Division, USM Corporation, Ansonia, Conn, 1978.
[23] W. Kulcke, Metals and Materials in Micro-Electronic Technology, Plenum Press, New York/London, 1982, pp. 127–145.
[24] J. Birnie, M. Champney, The contribution of electron beams to machining and forming in electrical methods of machining and forming, in: IEE Conf., Publ. No. 36, 1967, pp. 210–221.
[25] Yew. N.C, Electron Beam – Now a Practical a Lesser Production Tool, Solid State Technology, August, (1977) pp. 86–90.
[26] J.A. McGeough, Micromachining of Engineering Materials, Marcel Dekker, New York, 2002.
[27] R. Snoeys, F. Staelens, W. Dekeyser, Current trends in non-conventional material removal processes, Ann. CIRP 35 (2) (1986) 467–480.
[28] M.S. Węglowski, S. Błacha, A. Phillips, Electron beam welding techniques and trends – review, Vacuum 130 (2016) 72–92.
[29] T. Miyazaki, et al., Monitoring electron beam machining by generated acoustic emission, Ann. CIRP 34 (1) (1985).
[30] J.X. Zou, T. Grosdidier, K.M. Zhang, C. Dong, Cross-sectional analysis of the graded microstructure in an AISI D2-steel treated with low energy high-current pulsed electron beam, Appl. Surf. Sci. 255 (2009) 4758–4764.
[31] W.E. Frazier, Metal additive manufacturing: a review, J. Mater. Eng. Perform. 23 (2014) 1917–1928.
[32] P. Wanjara, M. Brochu, et al., Electron beam free forming of stainless steel using solid wire feed, Mater. Design 28 (2015) 2278–2286.
[33] D.v. Dobeneck, Examples of 30 Years Jop Shop Experience, Pro-beam AG& Co. KGaA, 2007, pp. 1–60.
[34] X. Li, C. Wang, W. Zhang, Y. Li, Fabrication and characterization of porous Ti6Al4V parts for biomedical applications using electron beam melting process, Mater. Lett. 63 (2009) 403–405.

CHAPTER FIVE

Machining processes utilizing chemical and electrochemical energy

Contents

Modern Machining Technology
https://doi.org/10.1016/B978-0-12-812894-7.00005-0
© 2020 Elsevier Inc.
All rights reserved.

SUBCHAPTER 5.1

Chemical Machining (CM)

5.1.1 Introduction

Chemical Machining (CM) is one the oldest and still useful machining method in which material is dissolved and removed from the workpiece by controlled chemical reaction using reactive chemical solution, i.e., strong alkaline or acidic reagent. Chemical machining method was initially applied to etch copper jewelry by citric acid in the Ancient Egypt in 2300 BC [1]. Until the 19th century this process was generally used for decorative etching. The development of photography opens up a new dimension to chemical machining and in 1826, J.N. Niepce was the first to exploit a photoresist made from bitumen of Judea asphalt for etching pewter (an alloy of 80–90% of tin and 10–20% of lead). William Fox Talbot (1852) patented a method for machining copper with ferric chloride using a photo-resist generated from bichromated gelatin (GB Patent No: 565). In 1888, John Baynes described a process for etching material on two sides using a photoresist which was patented in the United States (US Patent No: 378423). In 1953, North American Aviation Inc. (California, United States) applied the process to etch aluminum components for rockets. In 1956, the company named the process "chemical milling" and patented it (US Patent No: 2739047) [2]. This machining method is known in different names such as chemical etching, chemical milling, chemical blanking, chemical engraving, photochemical machining, wet etching, and etching, etc.

It has the capability to generate precise and accurate features on workpieces by controlled chemical reactions. In ancient times, artisans used the chemical machining method to etch metals. But in recent times, CM is widely used for milling of pockets and for generating intricate geometric features where no thermal distortion and cutting forces are concerned. The process is broadly applied to fabricate intricate and precision components from the advanced engineering materials in aerospace, electronics and automotive manufacturing [3, 4]. Multiple précised parts such as miniaturized microelectronics components and deep internal cavities may only be manufactured by chemical machining process [5, 6]. In chemical machining, special insulation called maskant, which protects derived areas from which metal is not to be removed. CM is utilized to fabricate contours and pockets by removing the metals from components having a high strength-to-weight ratio. Additionally, this method is extensively employed to produce the précised components for different industrial applications such as semiconductor industries and microelectromechanical systems (MEMS) [7, 8].

5.1.2 Working principle

In chemical machining, the material is removed from specified areas of metal workpiece by chemical etching solution. Here, the metal part is immersed in strong etchant solutions like alkaline and acid solutions, $FeCl_3$, HF, etc. which depends on types of materials to be machined such as copper, aluminum, etc. Chemical reagents react with the metal in the solution and produce the required features. Before machining, the workpiece is cleaned properly and coated with chemically inert maskant apart from the specified areas onto where the etching is to be occurred. Coating materials allow the chemical solution for dissolving and penetrating the required specified areas of workpiece. The etchant solution removes the material by controlled chemical reactions from whole exposed areas of metal where the material removal rate differs from 0.0024 to 0.01 mm/min. But, the main limitation is observed in the characteristics of isotropic etching where the etchant will remove the material not only from downwards of the metal but also sideways below the resist layer. The ratio of the depth to the undercut is termed as the "etch factor" and can be determined in case of one and two sided CM operation as shown in Fig. 5.1.1.

$$\text{Etch factor} = \frac{\text{Depth of cut}}{\text{Undercut}} \tag{5.1.1}$$

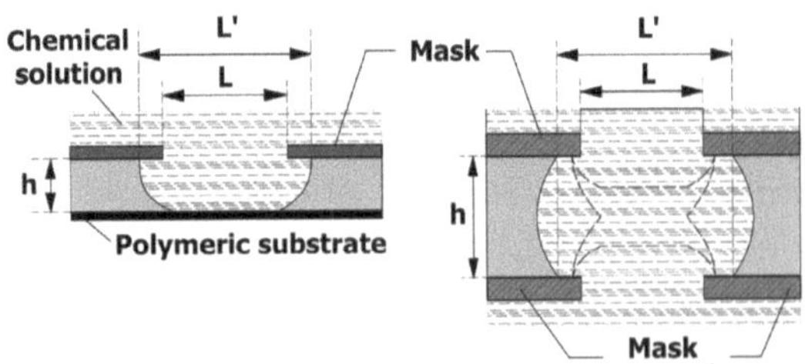

Fig. 5.1.1 Schematic representation of etch factor.

The amount of generated undercut in the etched surface is a function of many factors including type and strength of the etchant and workpiece material and depth of cut. For compensation of undercut, it may be recovered during selective masking of workpiece and controlling process variables to obtain the proper final size of component. The depth of cut can be accurately controlled by immersion time in the etchant solution. During etching time, gas bubbles are generated. If the gas bubbles are trapped on workpiece, these bubbles may insulate the work surface during etching resulting non-uniform material removal from workpiece. Therefore, to avoid these difficulties, workpieces are tilted for escaping the bubbles from the workpiece.

Chemical machining is a prolonged process, but it can produce contours, complex shapes, pockets, etc. Chemical machining process includes chemical blanking where chemical reaction is performed on thin sheets, chemical milling, photochemical machining, etc. [9]. The process parameters such as type and method of applying maskant, method for circulation of etchant, type and temperature of etchant, etc. can affect the performance of chemical machining process. The material removal mechanism has been shown in Fig. 5.1.2.

CM process has different steps, i.e., job preparation, coating with masking material, scribing of the mask, etching and cleaning masking material for generating the précised parts and these different steps of CM discussed here under:

(i) Job preparation: The workpiece is cleaned properly at the starting of chemical machining process. The grease, oil, rust, dust or any substance are removed from the surface of workpiece material. Proper cleaning operation creates the better adhesion bonding between the job and masking material. Two types of cleaning methods are available, one is chemical and other is mechanical methods. The chemical method is most extensively used as a cleaning procedure due to it produces less damage comparing to mechanical method.

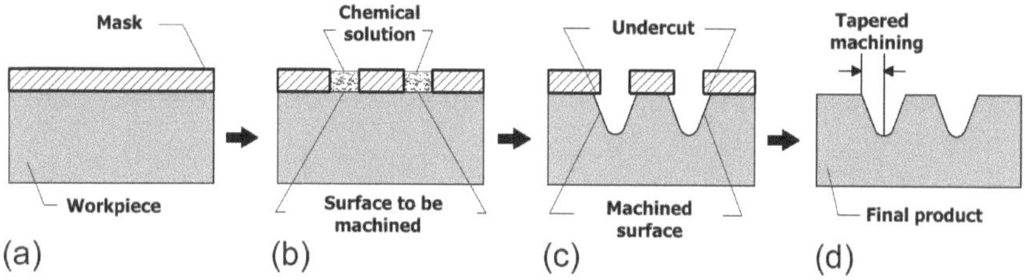

Fig. 5.1.2 Outline of mechanism of material removal by CM.

Ultrasonic cleaning machine is also utilized for cleaning the workpiece using special cleaning solution and heating is advantageous during the cleaning operation.

(ii) Coating with masking material: The insulation operation is carried out in the next step on cleaned workpiece. The selected coating material must be readily strippable insulation, which has enough adhesion strength to withstand the chemical abrasion during reactions and it is chemically impregnable.

(iii) Scribing of the mask: This step is directed by templates to expose the unmasked areas for chemical etching. The selection of mask material depends upon some specific factors such as the number of parts to be produced, the desired geometry and the size of the workpiece material. Silk-screen masks are chosen for shallow cuts requiring close dimensional tolerances.

(iv) Etching: This step is the most vital stage to generate the required parts from the workpiece. The immerse type etching machine is used to carried out this operation as shown in Fig. 5.1.3. The workpiece is immersed into selected etchant and the unexposed areas are machined to produce the required shape. The etching operation is conducted in the specific temperature which depends on the etched material. Then the machined workpiece is cleaned for removing the etchant from the machined zone.

(v) Cleaning masking material: The last step is to remove the masking material from machined workpiece. Before packaging the finished part, the inspections of the surface quality and dimensions are accomplished [10].

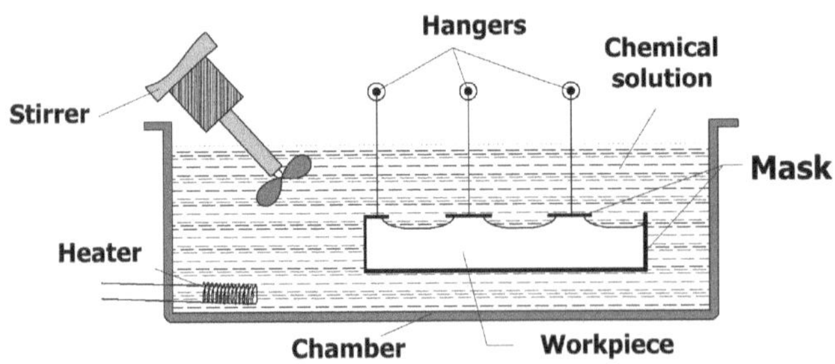

Fig. 5.1.3 Chemical machining setup.

5.1.3 Process parameters of CM

The major process parameters for chemical machining consist of type of etchant, temperature of the etchant, composition of etchant, method of circulating the etchant, type of maskant and method of applying the maskant that affect onto the work surface during machining time.

5.1.3.1 Parameters related to etchants

Various chemical etchants such as ferric chloride, nitric acid, chromic acid, etc. are utilized for dissimilar materials for chemical reactions. The etchant solution selection depends on some particular aspects such as types of workpiece and maskant, material removal capability, desired surface finish, cost, and required depth of machined workpiece, etc. Ferric chloride is applied for copper, nickel, aluminum and their alloys. Ferrous nitrate etchant is utilized for silver. For chemical machining of titanium, hydrogen fluoride is employed. Nitric acid is applied for tool steel and chromic acid is used for phosphor bronze. The combined mixture of etchant and material forms surface oxides, which damage the surface finish. Faster material removal rate reduces the cost of etchant solution. Due to higher material removal rate, the maskant of workpiece may deteriorate resulting lower surface finish and higher heat generation. Sometimes, etchant solution removes the corrosion from the workpiece surface. Some chemical reagents produce good surface finish, however that may diminish the etch depth. The cost, maintenance and disposal of chemical etchants are also considered during selection of etchant solution in CM.

5.1.3.2 Parameters related to masks

Generally maskants are categorized into three types, i.e., cut and peel, photoresist, and screen print. In cut and peel maskants, the entire workpiece is dipped in the coating solution or spray the coating and then cut and peeled off from the specified areas required to be uncovered for chemical reaction. The dimensional accuracy of this method is lower but very deep etching can be achieved. The photoresist maskant is applied in photochemical machining process where machining is performed by photosensitive resist material and can fabricate the intricate shapes. In screen printing technique, a screen maskant is employed to press on the work material surface. The screen has the aperture on to which the chemical reaction has to be carried out.

Chemical reaction occurs in the uninsulated area when the maskant is rolled over. This technique is utilized for mass production but the machining accuracy of the product is low.

The selection of proper resisting material depends upon the job material. Mask selection for different materials depends on various factors such as part configuration, chemical resistance, quantity of parts, cost of mask, ease of removal of mask and required resolution, etc. [9]. The thicker maskant is suitable for longer exposure time before demolishing. Therefore, thicker masks are appropriate for deeper etching. Some special masks are pertinent to flat parts. The masking process should be less manpower intensive for higher productivity. Generally the actual cost of resist material is varying as per requirements. Higher priced masks are usually simple to remove. The accuracy of different machined parts depends on maskant and its application method. Thicker mask is not suitable for higher accuracy. Maskant should not have any chemical effect on the metal surface and it should be stable at high temperature of etchant bath during CM.

Various methods like cut and peel, screen printing, and photoresist masks are available for depositing the maskants on the workpiece surface. Generally for cut and peel method, the maskant materials are butyl, vinyl based material, neoprene, etc. These masks normally prepared by flow, spray and dip coating. These masks are used in thickness ranging from 0.025 to 0.13 mm. Thicker masks can endure prolonged etching time and etch much higher depth. Hand scribing using a template as a guide is the most common technique for the removal of maskant. The undesired mask is manually peeled away after scribing for etching. The accuracy level is not better than ±0.13 to ±0.75 mm due to manual method which depends on type of feature and size being produced. Cut and peel masks are only the simple type of mask for rescribing produced step etching. So it is better for the batch production and produces with a large depth normally more than 1.5 mm. Typical applications include in large manufacturing industries such as chemical, aircraft and missile industries.

In screen printing technique, a fine mesh silk or stainless steel screen is used for masking. The back pattern corresponds to the image that is to be etched. The screen is printed against the workpiece surface and the mask is rolled on. When screen is removed, the mask contains on the part in the required pattern. After dry of maskant using baking method, the maskant is used for etching. It is a rapid and economical masking process for higher production volume where better accuracy is not required. The tolerance level of image varies from ±0 to ±0.18 mm depending upon the application. Typically mask thickness is less than 0.05 mm in this method. So its lifetime is less and limiting to the etched depth of 1.5 mm. The technique is suitable for the part size less than 1.2 × 1.2 m. It is also desirable for flat surfaces with moderate contours, etch depth less than 1.5 mm per sides and lower degree of accuracy.

Photoresist masking method is much versatile known as photochemical machining (PCM). Ii is utilized to produce complex shapes using photographic image on a workpiece. In this method, the artwork is used to make the required image on the workpiece. The artwork is either computer generated 2 to 20 times life-size or hand drawing. An oversized original drawing is utilized due to better accuracy upon reduction. An alternative method for producing a photomaster from a drawing, a laser pattern generator is used. Compensation values in terms of undercut are accounted during the design of artwork. After designing, the artwork is photographed and produced with higher accuracy. Laser pattern generator is an alternative process for producing a photomaster from a CAD drawing. This generator receives the data from the CAD system and produces the images and features on the photomaster film. Accuracies of ±0.0028 mm and repeatability of ±0.0006 mm are achievable. This generator produces a 0.45 m × 0.6 m photomaster within 6 min. After creating the photomaster, the workpiece surface is cleaned properly to remove all dirt and oxides. After cleaning, the workpiece is insulted with thin layer of mask by spin-coating process. Then a strong ultraviolet light source is utilized to expose the photoresist while photomaster and workpiece are kept together properly. Then, the photoresist is developed to remove the insulation from entire surface of workpiece where etching is required. After developing the photoresist, the parts are etched and inspected. PCM process is applicable for high production volume and requires tolerance level better than ±10% of the part thickness [11]. To avoid small parts from prematurely dropping out of the sheet metal which are being blanked, small attachment tabs are connected the design to ensure the components for easy handling. This process is known as tabbing. After completion of all PCM operations, the components are cut from their tabs. Generally the maskants should be resistant to the etchant. These should be simply removable after etching. The maskants must not have any chemical effect on the workpiece and must be stable at high temperature of etchant bath during CM.

5.1.4 Chemical blanking

Chemical blanking process is a prominent nontraditional machining process for producing machined features on thin plates and foil material using a suitable etching solution, which were generally produced by mechanical blanking presses. Chemical blanking penetrates the job material entirely such as features like slots, holes, etc. The method blanks total parts of the desired profile from a sheet of material by chemical etching. In this process, the blanking operation is performed by utilizing different masking techniques, e.g., photoresist, screen printing masking and cut and peel methods. Various difficulties such as backlash, part distortion and vibrations have been occurred during production of smaller parts using mechanical blanking presses. But, these types of problems are not occurred during chemical blanking using a favorable solution. This process has mainly four steps, i.e., pre-process, masking, through-material etching, and post-process. In pre-process, material surface is cleaned by pickling, degreasing, and grinding. In masking process, unmachined portion of metal part is covered with chemical resistant insulation. For extra precise applications, photo-resist material is used. In through-material etching, material is to be penetrated through etching solution from both sides at the same time. Through-material etching is two types, i.e., spray type and paddle type. In spray type method, spray nozzles spray the etching solution on the material using a rotating wheel. In paddle type, the material is immersed in the corrosive liquid and the liquid is continuously stirred using a paddle wheel. Air injection is extensively utilized in stirring method. In post process, resist material is removed by washing off etchant. If the coating material is non-water soluble, organic solvents are applied.

5.1.5 Chemical milling

Chemical milling is the subtractive machining process using baths of temperature-regulated etching chemicals to remove material for producing the required shape [11]. It is defined as the method of chemically corroding of material to produce blind features like pockets, channels, etc. It also removes material from the entire surface of components for the purpose of weight reduction. This method is extensively used on metals, although other different materials are gradually becoming more essential [9]. During the Renaissance as alternative for etching on metal, it was developed from printing etching and armor-decorating processes. The process basically dips the cutting areas in a corrosive chemical known as an etchant, which etches with the material in the area to be cut and causes the solid material to be dissolved. The inert substances, i.e., maskants in the workpiece are used to shield the constricted areas of the material as resists [10]. Some important organic chemical compounds, i.e., citric and lactic acids had also been utilized to react the metals and produce the products as early as 400 BCE. At that time, vinegar was applied to oxidize the lead metal and generates the pigment ceruse, which is known as white lead. In recent times, various important alkaline and acid etchants are utilized in modern chemical milling methods to acquire the précised parts. Until the 15th century, Armor etching process was not invented using strong mineral acids. At that time, different etchants mixed from charcoal, vinegar and salt were employed to etch the plate armor that had been decorated with a maskant of linseed-oil paint. The etchant would react into the unexposed areas and the painted areas are to be protected from chemical reactions. This controlled etching in this manner allowed armor plate to be decorated as if with accurate engraving without the existence of formed burrs. It also protected the inevitability of the armor being softer than an engraving tool. In the 17th century, this method was utilized to improve the quality of measuring instruments gradually. This etching process could produce the thinner lines for the production of more precise and accurate instruments than were possible before. Not long after, it became applied to engrave the trajectory information plates for artillery and cannon operators and the etched plates could be reasonably robust during the rigors of combat. Regularly such information (normally ranging marks) was etched onto equipment such as stiletto daggers.

In 1782, John Senebier discovered that the certain resins lost their solubility to turpentine when light is applied, i.e., they hardened. This matter helps to develop the photochemical milling process where a liquid maskant is used to the whole material surface and the outline of the area is to be masked produced by exposing it to UV light. Photochemical milling was widely applied in the improvement of photography methods, allowing light to produce the impressions on metal surface. The earliest use of chemical etching was started in 1927 to mill commercial parts. In that time, Swedish company Aktiebolaget Separator patented an important method for producing edge filters by chemical milling. Later, around 1940s, it is widely used to etch thin samples of very hard metal.

Chemical milling process is generally carried out in a series of five steps: cleaning, masking, scribing, etching, and demasking [10]. Fig. 5.1.4

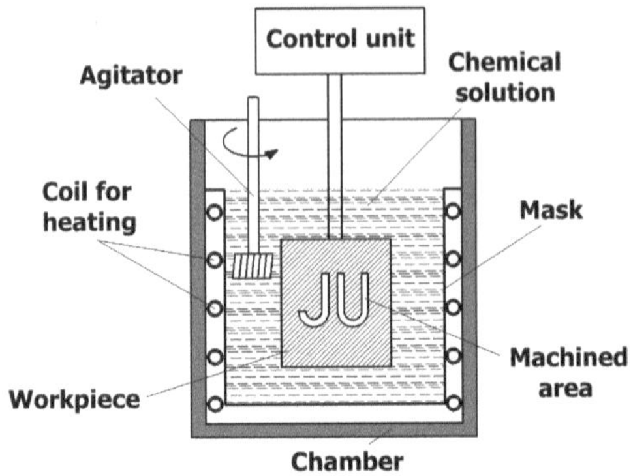

Fig. 5.1.4 Schematic diagram of chemical milling.

represents the schematic diagram of chemical milling process. At first, the workpiece surface is cleaned properly for removing the contaminants which could negatively impact the quality of the finished part. An improper cleaned surface could result non-uniform etching and poor adhesion of the maskant causing inaccurate final dimensions of workpiece surface. The workpiece surface must be remained free from grease, primer coatings, oils, markings, scale (oxidation), and any other foreign contaminants. For most of metals, cleaning is carried out by applying a solvent substance on the surface for etching and washed away foreign particles. The job material may also be cleaned in alkaline or specific de-oxidizing solutions.

Masking is the procedure of applying the maskant material to the workpiece surface to ensure that only desired areas are etched [10]. Maskant material must also be chemically inert enough with respect to the etchant to protect the workpiece surface. Dip-masking process is also utilized for masking on workpiece surface using liquid maskants. In this process, the workpiece surface is dipped into open tank of liquid maskant and then it is dried. Liquid maskant may be used on workpiece surface by the flow of liquid maskant. Electrostatic deposition may also be applied for surface coating onto the surface of the material using certain conductive maskants. Here, electrical charges are applied to conductive maskant particles for generating the coating. The charge causes the conductive particles of maskant to stick to the workpiece surface. Most modern chemical milling processes exploit the maskants with an adhesion around 350 g/cm. If the adhesion strength is too strong, the scribing procedure may be too complicated to perform. If the adhesion strength is too low, the etching area may be inaccurately defined. Most industrial chemical milling methods utilize the maskants based upon isobutylene-isoprene copolymers or neoprene elastomers. Scribing is the elimination of maskants on the etched. Sometimes, an etching needle, scribing knife or similar tool is often used for decorative applications. In modern industrial applications, an operator scribing may be used with the help of a template or computer numerical control process. Complex templates using color codes may be used for different parts scribing with the aid of multiple stages of etching. Etching is the actual immersion technique of the workpiece material into the chemical bath and the action of chemical is to be milled on the part. The immersing time in the chemical bath decides the depth of the resulting etch which can be determined by the formula as mentioned below:

$$E = \frac{s}{t} \tag{5.1.2}$$

where, E is the etching rate of workpiece material, s is the depth of the resulting etch, and t is the total immersion time. Higher agitation of the etchant may generate the cavitations or stagnation resulting ridges, weavings, and grooves in the etched surfaces. Etching rate of chemical milling differs based on various factors including the material to be etched, temperature and type of composition and concentration of the etchant. Magnesium is commonly etched at rates around 0.46 cm/h and aluminum about 0.178 cm/h [12]. Sodium hydroxide and Keller's reagent are the common etchants for aluminum. For steels, hydrochloric and nitric acids etchants are used. Cupric chloride, ferric chloride, ammonium persulfate, nitric acid, hydrochloric acid and hydrogen peroxide are the normal etchants for copper. Hydrofluoric acid is the general etchant for silica.

Demasking is the combined method of removing the part of maskant and etchant. Etchant is normally flushed away with a wash of clear cold water however other substances may also be used in specialized processes. Sometimes, a de-oxidizing bath may be used in the normal case that the chemical process left an oxide film on the workpiece surface. Other methods used to eradicate the maskants, the most familiar being simple hand removal using scraping tools. This is simultaniously both laborious and time-consuming. But, this step may be automated for large-scale processes.

Chemical milling process has applications in semiconductor fabrication industries and the printed circuit board . This process is also utilized in the aerospace industry to eliminate the shallow layers of material from missile skin panels, large aircraft components and extruded parts for airframes.

It is also broadly applied to produce Microelectromechanical systems and integrated circuits.

5.1.6 Photochemical machining

Photochemical machining (PCM) is also popularly known as photo etching or photochemical milling process which is applied to produce sheet metal components using a photoresist and etchants to corrosively machine away selected areas. This process had been emerged as an offshoot of the printed circuit board industry in the 1960s. This process can fabricate highly intricate parts with very fine detail accurately and economically [13]. This technique is economical alternative to the modern manufacturing processes such as punching, laser or water jet cutting, stamping, wire electrical discharge machining (WEDM) for thin gauge precision parts. It retains dimensional tolerances and does not generate sharp edges or burrs. PCM can be applied almost on any commercially accessible alloy or metal. It is also suitable to material thickness ranging from 0.012 to 2.042 mm. PCM can be effectively used for metals which include brass, nickel, copper, silver, manganese, steel, zinc, aluminum and titanium [14]. Photochemical machining is a type of photo engraving and a similar method in microfabrication is called photolithography.

PCM procedure begins by printing the profile of the part onto dimensionally stable and optically clear photographic film. The "phototool" contains two sheets of this film showing negative images of the parts. Fig. 5.1.5 shows the outline of photochemical machining procedure [11]. The metal sheets are cut to the required size and shape and cleaned properly to remove dirt, grease or any foreign particles. Then the metal piece is spin-coated with the required mask and laminated on both sides with a UV-sensitive photoresist. The coated metal is placed between the two sheets of the phototool and a vacuum is drawn to ensure intimate contact between the metal plate and phototool. Then the plate is exposed in UV light that permits the areas of resist that are in the clear sections of the film to be hardened. After exposure, the metal plate is developed under the etching solution and the required pattern is obtained. Sometimes, the etching line is a multi-chambered machine in which the driven-wheel conveyors are used to move the plates and arrays of spray nozzles above and below the plates. The etchant is naturally an aqueous solution of alkaline, acid, and ferric chloride. These are heated and directed under pressure to both sides of the plate. The etchant starts the reactions with the unprotected metal, basically corroding it away quite rapidly. After neutralizing and rinsing, the remaining resist is eliminated from the plates and the plates are cleaned and dried.

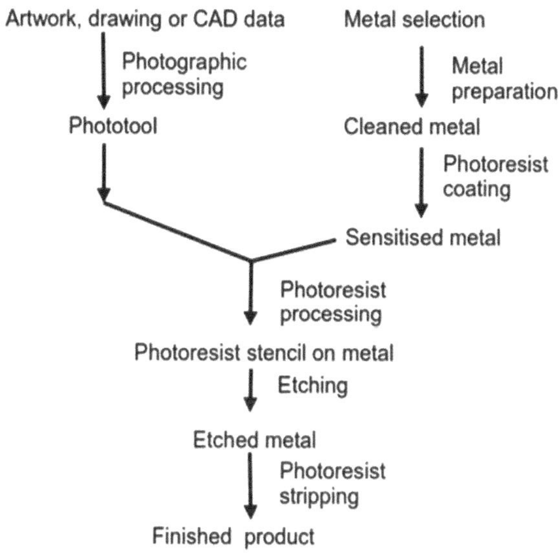

Fig. 5.1.5 Outline of photochemical machining [11].

This process can be utilized in various industrial applications like apertures and masks, fine screens and meshes, heat sinks, sensors, springs, RF and microwave circuits and components, battery grids, fuel cell components, pressure membranes, metal gaskets and seals, flexible heating elements, semiconductor leadframes, shields and retainers, encoders and light choppers, motor and transformer laminations, electrical contacts, jewelry and washers [11].

5.1.7 Advantages and limitations of CM

The chemical machining provides several advantages as follows:

(i) No effect of workpiece materials properties such as hardness.
(ii) No stress introduction and burr formation.
(iii) Low capital cost of equipment and tooling costs.
(iv) Requirement of less skilled worker.
(v) Easy and quick design changes and weight reduction.
(vi) Good surface quality of machined surface.
(vii) Simultaneous removal operations on several workpieces, etc.

However, chemical machining has also some limitations:

(i) Difficult to get sharp corners.
(ii) Limited depth of cut, hence difficult to chemically machine thick material.

(iii) Less dimensional accuracy.
(iv) Workpiece must have flat surface and consists of homogeneous structure.
(v) Etchants are very hazardous for workers.
(vi) Etchant disposals are very expensive, etc.

5.1.8 Applications of CM

CM has wide range of applications ranging from small to very large workpieces for generating desired shape profiles. Some of the typical applications are identified below:

(i) CM is greatly applicable for generating the metallic parts like enclosure screens, recording heads, instrument panels, semiconductor device, printed circuit boards (PCB), integrated circuit (IC) and other circuit devices to be utilized in electronics industry.
(ii) Chemical machining is applied to etch different metals such as titanium, copper, nickel, aluminum and its alloys as well as various types of non metal components like glass, plastic, ceramics, etc. [15].
(iii) This process is employed for producing shallow cavities and holes over large surface area [16], contours, pockets, depression [17], machining of edges of sheet metal, and engraving on metal workpiece.
(iv) Machining of huge work surfaces like airplane wing to tiny workpiece like IC chips [18], and fabrication of stainless steel edge filter by photochemical machining [19], batch size production of various other components in chemical and missile industries.
(v) Other popular applications of chemical machining engage minimizing thickness of ribs, webs and walls of components produced by conventional process, narrow cuts in large thin sheets applicable in aerospace industries, etc.
(vi) Photochemical machining has provided a quick response service to deliver the components to electrical, electronic and mechanical engineering industries. The demand is growing day by day for utilization of CM or PCM to fabricate on thin, complex and precision parts due to its economical consideration. Some prime examples are TV shadow aperture masks, integrated circuit leadframes, disk drive suspension head assemblies and etched gaskets in mobile telephones as shown in Figs. 5.1.6–5.1.9, respectively [11].

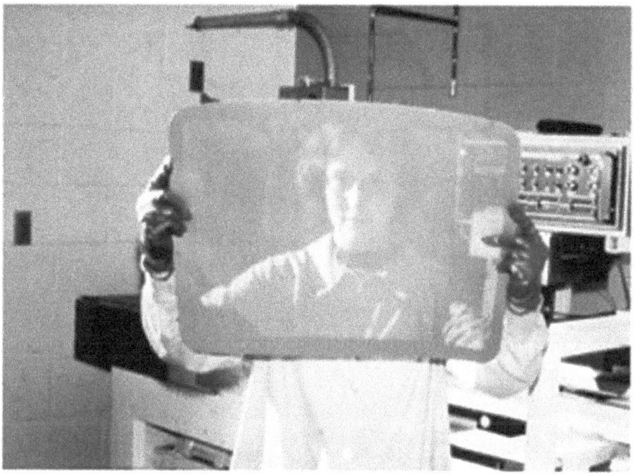

Fig. 5.1.6 Low resolution TV shadow aperture mask [11].

Fig. 5.1.7 60 mm wide integrated circuit leadframes [11].

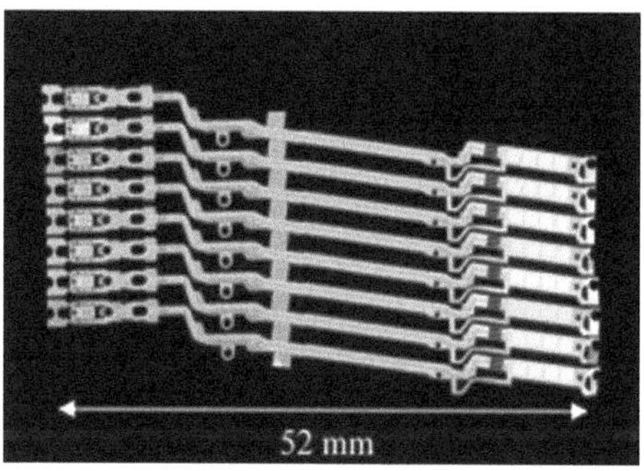

Fig. 5.1.8 Etched suspension head assemblies [11].

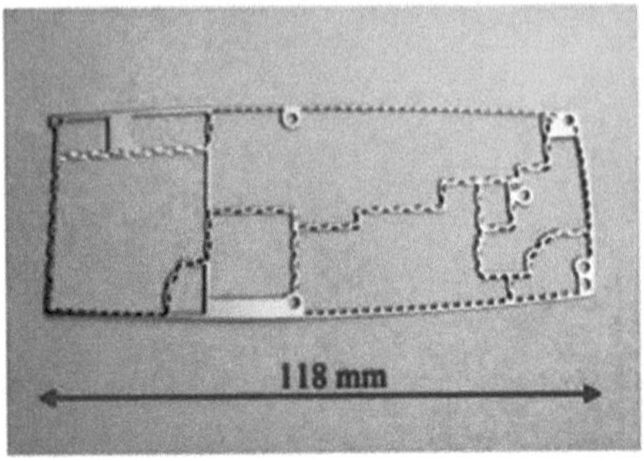

Fig. 5.1.9 Etched mobile telephone gasket [11].

5.1.9 Advancement in chemical machining

Chemical machining is one of the important nontraditional machining processes in which material is removed using strong etchant solution. This is basically the accelerated and controlled corrosion method. This process has several advantages over traditional machining processes [20, 21]. Copper is the most commercial engineering materials which is widely applied in different industries such as electronics, chemical, and automotive industries due to its excellent thermal and electrical conductivities, high corrosion resistance, ease for which CM can be effectively utilized of fabrication, good strength and fatigue resistance [22].

The chemical machining is also more efficient and generates superior etching properties on Stainless Steel and aluminium. The efficient chemical reagent can produce better surface finish and higher etching rate. So, the selection of significant etchant is most likely the important parameter. $FeCl_3$ is the important chemical etchant for most of the engineering materials such as steels, metals and alloys [23–25]. It is inexpensive and simple to control during the etching process and ferric chloride or nitrate solution is also suitable for various etchant regeneration systems from the industrial point of view. The best possible etchant has some better advantages like high dissolved-material capacity, high etching rate, compatibility with used maskants, and easy control of process, good surface finish, economic regeneration and personal safety maintenance [26, 27]. The good relationship between etching rate and the sufficient concentration of etching solution has been established by a practical etchant composed of ferric nitrate during the chemical milling process [28]. Ferric chloride is also used at different operating temperatures during chemical milling of aluminum alloys. In the acidic solution, chemical milling of aluminum alloys have been done at low temperature (20–50 °C) and have higher surface finish (about 7–12 μm). In the alkaline system, aluminum has dissolved in sodium hydroxide with the addition of Na_2S, $NaNO_3$, Na_2CO_3, etc. [29]. Some chemical solutions such as sodium nitrate, hydroxide and ethylene glycol are suitable in chemical milling of aluminum alloys with high compositions of zinc or copper [30]. The alkaline solution results a better surface quality by adjusting other parameters at a higher temperature of 55–65 °C compared to the acidic system in the chemical milling process [31]. The mixed solution of NaOH, Na_2S, triethanolamine (TEA) and Al^{3+} is used for better chemical milling process of 2219 aluminum alloy at high reaction temperature of 80 °C [32].

Environmental issues are the most important matter in chemical machining which affects the machining process for the use of various chemical reagents. Some hazardous liquids are poisonous etchants, cleaning solutions, strippers, etc. The disposal and handing of these hazardous and poisonous etchants are very expensive matter. Industrial trends are increasing for environmentally suitable etchants for chemical machining process. Etched metal recovery from waste reagents and regeneration of waste etchant are increasing in the modern chemical industry. Some etchants like alkaline etchants, $CuCl_2$ and $FeCl_3$ could be recovered and regenerated [33].

Due to further advancement in chemical machining process, it can also be used to fabricate components used for microengineering products, e.g., microsystems technology (MST), micro-electromechanical systems (MEMS), medical diagnostic equipment and biomedical engineering applications such as body implants. There has been a remarkable rise in the number of non-silicon MEMS applications. It can also be used to fabricate three-dimensional products required for many technical applications as well as in decorative applications. CM will also be suitable process to the future demands of technology especially for rapid prototyping and miniaturization.

References

[1] W.T. Harris, Chemical Milling, Oxford University Press, Oxford, UK, 1974.
[2] M.C. Sanz, Process of Chemically Milling Structural Shapes and Resultant Article, USA Patent No: 2739047, 1956.
[3] G.F. Benedict, Nontraditional Manufacturing Processes, Mercel Decker, New York, USA, 1987.
[4] J.A. McGeough, Advanced Methods of Machining, Chapman and Hall, London, UK, 1988.
[5] K.P. Rajurkar, Nontaditional manufacturing processes, in: Handbook of Design, Manufacturing and Automation, John Wiley & Sons, New York, USA, 1992.
[6] K.P. Rajurkar, R.F. Ross, B. Wei, J. Kozak, R.E. Williams, The role of nontraditional manufacturing process in future manufacturing industries, in: The Proceedings of the International Conference "Manufacturing International" Dallas, USA, 1992, pp. 23–27.

[7] J.W. Dini, Fundamentals of chemical milling, Am. Mach. 768 (1984) 113–128.
[8] H.A.-G. El-Hofy, Advanced Machining Processes, McGraw-Hill Companies, Blacklick, OH, USA, 2005.
[9] G.F. Benedict, Nontraditional Manufacturing Processes, Vol. 19 (Manufacturing Engineering and Materials Processing), CRC Press, 1987, pp. 190–196.
[10] O. Çakir, A. Yardimeden, T. Özben, Chemical machining, Arch. Mater. Sci. Eng. 28 (8) (2007) 499–502.
[11] D.M. Allen, Photochemical Machining: from 'manufacturing's best kept secret' to a $6 billion per annum, rapid manufacturing process, CIRP Ann. 53 (2) (2004) 559–572.
[12] David Fishlock, New ways of cutting metal, New Sci. 8 (212) (1960) 1535.
[13] D.M. Allen, The Principles and Practice of Photochemical Machining and Photo-etching, Adam Hilger, IOP, Bristol, 1986. ISBN 0-85274-443-9.
[14] P. Greiner, Results of PCMI industry trends survey 2000, PCMI J. 84 (2002) 27–35.
[15] J.T. Blak, et al., DeGarmo's Materials and Processes in Manufacturing, 10th ed., John Wiley & Sons, Hoboken, USA, 2007.
[16] T.J. Drozda, Tool and Manufacturing Engineers, Handbook (Chapter 14: Nontraditional Machining), SME Publishing, Michigan, USA, 1989.
[17] T.A. Fadaei, A new etchant for the chemical machining of St304, J. Mater. Process. Technol. 149 (2004) 404–408.
[18] A.H. Al-Ethari, K.F. Alsultani, N. Dakhil, Variables affecting the chemical machining of stainless steel 420, Int. J. Eng. Innov. Technol. 3 (2013) 210–216.
[19] D.M. Allen, T.N. Talib, Manufacture of stainless steel edge filters: an application of electrolytic photopolishing, Precis. Eng. 5 (2) (1983) 57–59.
[20] E. Paul DeGarmo, J.T. Black, R.A. Kohsern, Materials and Processes in Manufacturing, eighth ed., Prentice-Hall, Englewood Cliffs, NJ, 1997.
[21] M. Langworthy, "Chemical Milling Nontraditional Machining Process Machining" Handbook, ASM, Ohio, USA, 1994.
[22] D.M. Allen, Progress towards clean technology for photochemical machining, CIRP Ann. Manuf. Technol. 42 (1) (1993) 197–200.
[23] D.M. Allen, The state of the art of photochemical machining at the start of the twenty-first century, Proc. Inst. Mech. Eng. B J. Eng. Manuf. 217 (2003) 643–650.
[24] O. Çakir, A. Yardimeden, T. Özben, Chemical Machining, J. Mater. Sci. Eng. 28 (2007) 499–502.
[25] O. Çakır, H. Temel, M. Kiyak, Chemical etching of Cu-ETP copper, J. Mater. Process. Technol. 162–163 (2005) 275–279.
[26] O. Çakır, Chemical etching of aluminium, J. Mater. Process. Technol. 199 (2008) 337–340.
[27] A. Fadaei Tehrani, E. Imanian, A new etchant for the chemical machining of St 304, J. Mater. Process. Technol. 149 (2004) 404–408.
[28] B. Chambers, Etching of aluminum alloys by ferric ion, Met. Finish. 98 (2000) 26–29.
[29] O. Çakir, Chemical etching of aluminum, J. Mater. Process. Technol. 199 (2008) 337–340.
[30] D.W. Gross, Chemical Milling Processes and Etchants Therefore, (1986)US Patent No.: 4, 588, 474 pp.
[31] W. Chandler, Pros and cons of alkaline vs. acid etching of aluminum method chosen boils down to desired appearance and physical attributes of finished product, Met. Finish. 106 (15) (2008) 18.
[32] Q. Li, J. Wang, W. Hu, Optimization of chemical milling solution for 2219 aluminum alloy, Trans. Mater. Heat. Treat. 36 (2015) 243–249.
[33] D.M. Allen, O. Çakır, H.A.J. White (Almond), The photochemical machining of brass with cupric chloride etchant and a technique for the partial recovery of dissolved zinc, in: Proceedings of the Symposium High Rate Metal Dissolution Processes, The Electrochemical Society 95-19, 1995, pp. 305–315.

SUBCHAPTER 5.2

Electrochemical Machining (ECM)

5.2.1 Introduction

Electrochemical machining (ECM) has tremendous potential on account of versatility of its applications and it is expected that it will be one of the most promising, successful and commercially utilized machining processes in the modern manufacturing industries. ECM is an anodic dissolution process where work piece and tool are made anode and cathode respectively and separated by an electrolyte solution. When the electric current is passed through the solution of electrolyte, the anode work piece dissolves locally. The principle of electrochemical anodic dissolution was discovered long back in the 19th century by Michael Faraday (1791–1867). In 1833, Faraday established the two fundamental laws of electrolysis, which are the foundation stones of the electrodeposition and electro dissolution techniques. In 1929, the Russian researcher W. Gusseff first developed a process to machine metal anodically through electrolytic process and patented the technique. Much later in 1941 C.F. Burgess published a paper in the electrochemical society, United States regarding the application of anodic metal removal as a machining technique. Later, in 1959, Anocut Engineering Company of Chicago developed and commercially established the anodic metal machining technique. Subsequently, based upon research by the Battelle Memorial Institute, Steel Improvement and Forge Company utilized this technique for commercial application. Significant advances during the 1950s and the 1960s developed ECM into a major technology in the aircraft and aerospace industries for shaping, finishing, deburring, and milling operations [1]. However, initially ECM was applied mainly for machining of large components made of advanced and difficult-to-cut metals particularly for the gas turbine industry. Faster development of a wide range of electrochemical machines by several manufacturers in the United States, Europe, and Japan made the process more popular and acceptable due to its numerous advantages such as quality, flexibility, and cost. Only from the 1980s, ECM was employed in several machining applications in automotive, offshore petroleum, and medical engineering industries, as well as by aerospace firms, which are its principal user still now [2]. In recent years, ECM has also received much attention in the fabrication of smaller parts for various micro engineering applications [3].

5.2.1.1 Principles of electrolysis

Electrolysis is a chemical process that occurs when an electric current passes between two conductors submerged in a solution. As shown in Fig. 5.2.1, two copper conductors dipped in $CuSO_4$ solution and connected to a DC power source. As current passes through the $CuSO_4$ solution in between two conductors and thus established an electrolytic cell. This phenomenon proves that $CuSO_4$ solution has the property to conduct electrical current. The two conductors are called electrodes. The conductors which are connected to the positive terminal and negative terminal are known as anode and cathode respectively. The solution in between the cathode and anode is known as electrolyte. Due to the flow of current between two conductors, chemical reactions occur at the electrodes which are called anodic and cathodic reactions simultaneously electrochemical reactions also take place in the electrolyte itself. Current is carried in the electrolyte by the atoms which have either gained or lost electrons to form ions. Positively charged ions are known as cations and move toward cathode. Similarly, negatively charged ions are known as anions and move toward anode.

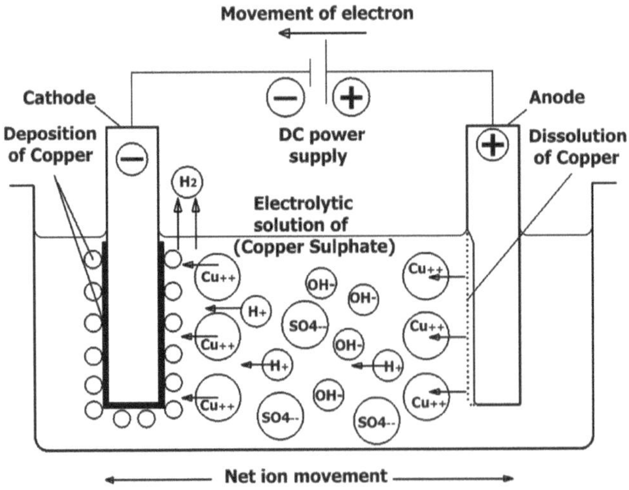

Fig. 5.2.1 Principles of electrolysis.

In the electrolytic cell, electrons are always move round to the external circuit to maintain the cathodic and anodic reaction thus anode copper breaks to positively charged ion by the following anodic reaction:

$$Cu \rightarrow Cu^{++} + 2e \quad (5.2.1)$$

Cation is neutralized by the electrons on the cathode by the following reactions:

$$Cu^{++} + 2e \rightarrow Cu \quad (5.2.2)$$

$$2H^{+} + 2e \rightarrow H_2 \uparrow \quad (5.2.3)$$

To maintain the neutrality of the bulk electrolyte, equal amount of reaction should occur at both the electrodes. So, electrolysis in copper sulfate solution with copper electrodes looks like just transfer of copper metal from cathode to anode. Therefore, positively charged copper ions continuously moves from anode and deposited on cathode. Finally at the end of electrolysis, the amount of weight which anodic conductor looses is the same amount which cathodic conductor gains. Electrolysis is mainly governed by Faraday's two laws which are as follows [3]:

(i) Faraday's first law of electrolysis: The mass of a substance altered at an electrode during electrolysis is directly proportional to the quantity of electricity transferred at that electrode. Quantity of electricity refers to electrical charge, typically measured in coulombs, and not to electrical current.

(ii) Faraday's second law of electrolysis: For a given quantity of electricity (electric charge), the mass of an elemental material altered at an electrode is directly proportional to the element's equivalent weight. The equivalent weight of a substance is its molar mass divided by an integer that depends on the reaction undergone by the material.

These above mentioned laws are utilized for determination of the amount of metal removed or deposited during electrolysis process.

5.2.1.2 Working principle

The basic principle of electrolysis is utilized for electroplating. However, the objective of electrochemical machining is to remove metal from the workpiece by utilizing the same principle. Hence, the workpiece is connected to the positive terminal and tool is connected to the negative terminal in the electrolytic cell. Thus, metal ions remove from the anodic workpiece by means of anodic dissolution. Basic mechanism of metal

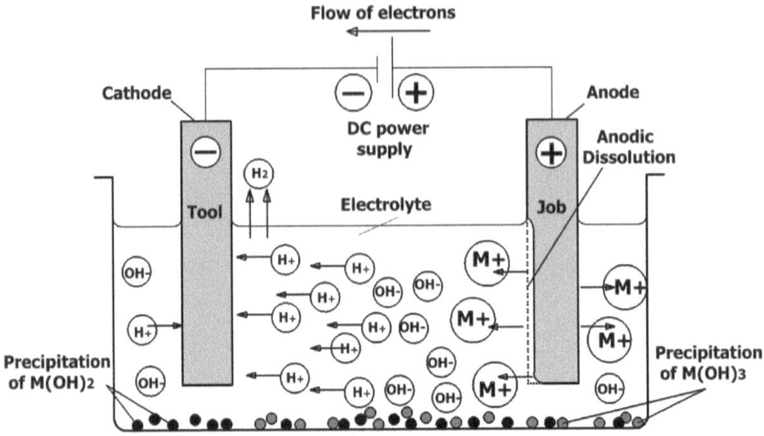

Fig. 5.2.2 Basic mechanism of metal removal during ECM.

removal during electrochemical machining is shown in Fig. 5.2.2 where, dissolution of metal is taking place in an electrolytic cell with electrolyte solution. As mentioned above, iron ion dissolves from the anodic workpiece and before precipitate to cathodic tool it reacts with hydroxyl ions of electrolyte and form metal hydroxide. Metal hydroxide settles down at the bottom surface of the machining cell due to gravity. Hence, due to the flow of current in the machining cell, more and more metal ion dissolves into the electrolyte and results in removal of metal from the anodic workpiece.

When a potential difference is applied across the electrodes, different electrochemical as well as chemical reactions occur in the electrolytic cell. At the cathode, the reaction having the smallest oxidation potential will take place, and at the anode, the reaction having the largest oxidation potential will occur first. The factors that influence the oxidation potential and determine the kind of reactions those will occur which depends on nature of metal being machined, type of electrolyte, current density and temperature, etc. [4].

The cathodic and anodic reactions with anodic workpiece and electrolyte solution are as follows:

(i) Anodic reactions

At the anode two possible reactions can occur

(a) Evolution of oxygen or halogen gas.

(b) Dissolution of metal ions.

The reactions leading to oxygen evolution are as follows:

$$2H_2O \rightarrow O_2\uparrow + 2H^+ + 2e^- \text{ (In acidic electrolyte)} \tag{5.2.4}$$
$$(OH)^- \rightarrow 2H_2O + O_2\uparrow + 4e^- \text{ (In alkaline electrolyte)} \tag{5.2.5}$$

In the hydrolysis of water, oxygen is liberated and hydrogen ions are formed, which increases electrolyte acidity locally.

The reaction which governs the dissolution of metal is as follows:

$$M \rightarrow M^+ + e^- \tag{5.2.6}$$
$$M^+ + (OH)^- \rightarrow M(OH) \tag{5.2.7}$$

(ii) Cathodic reactions

At the cathode two possible reactions can occur

(a) Evolution of gas bubbles (hydrogen).

(b) Neutralization of positively charged metal ions.

The reactions causing the evolution of hydrogen gas at the cathode are as follows:

$$2H^+ + 2e^- \rightarrow H_2\uparrow \text{ (In acidic electrolyte)} \tag{5.2.8}$$
$$2H_2O + 2e^- \rightarrow 2(OH)^- + H_2\uparrow \text{ (In alkaline electrolyte)} \tag{5.2.9}$$

Sometimes, metal ions may also reach the cathode and deposited by neutralization of positively charged metal ions is caused by the following reaction:

$$M^+ + e^- \rightarrow M \text{ (Metal) (In acidic electrolyte)} \tag{5.2.10}$$

As example, anodic dissolution of iron utilizing NaCl electrolyte solution involves following reactions:

$$\text{At anode: } Fe \rightarrow Fe^{++} + 2e \tag{5.2.11}$$
$$\text{At catode: } 2H_2O \rightarrow H_2\uparrow + 2(OH)^- \tag{5.2.12}$$

Electrochemical reaction:

$$Fe^{++} + 2(OH)^- \rightarrow Fe(OH)_2\downarrow \tag{5.2.13}$$
$$4Fe(OH)_2 + 2H_2O + O_2 \rightarrow 4Fe(OH)_3\downarrow \tag{5.2.14}$$

As per the above electrochemical reactions, material removes from the anode workpiece and hydrogen gas bubbles generate as well as ferrous hydroxide ($Fe(OH)_2$) and ferric hydroxide ($Fe(OH)_3$) form in the electrolyte and precipitates as sludge.

Hence, following observations pertinent to ECM can be drawn:

(i) Due to electrochemical reactions, metal dissolves from the anode where rate of dissolution depends on its atomic weight (M), it's valency (Z) and the current (I) which is passed and time (t) for which the current passes. Hence, dissolution or machining rate is not influenced by the hardness or other characteristics, e.g., physical, mechanical properties of the metal to be machined.

(ii) Only hydrogen gas is evolved at the cathode tool, the shape of the tool remained unchanged during machining. Therefore, any soft electrically conductive metal can be chosen as tool during machining.

Fig. 5.2.3 shows machining of a workpiece utilizing shaped tool during ECM where, the gap in between tool and the workpiece is filled with the electrolyte. When current passes, dissolution of the anode occurs. The dissolution is more at the smallest gap between tool and the workpiece due to flow of higher current as shown in the Fig. 5.2.3 As tool moves toward the workpiece, gradually work surface profile tends to become the shape of tool. After sometime, at steady state condition the gap becomes uniform and profile of tool surface is replicated on the workpiece as shown in Fig. 5.2.3.

Considering Faraday's two lays of electrolysis, rate of metal removal can be expressed as follows:

$m \alpha M/Z$ (from the First Law), $m \alpha I.t$ (from the Second Law) and combining these:

Metal removal rate, $\dot{m} = MI/ZF$, g/s.

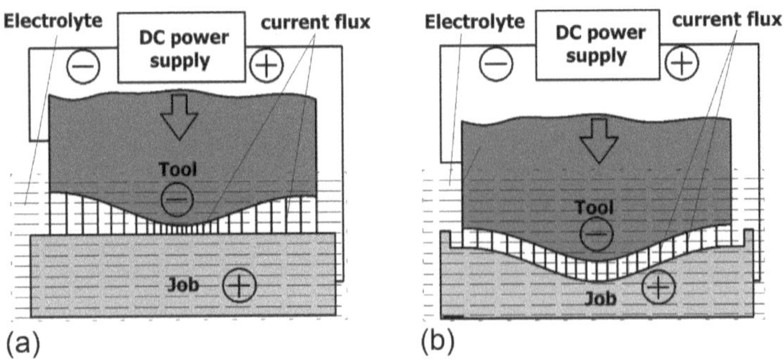

Fig. 5.2.3 ECM utilizing shaped tool (a) Initial condition (b) end of ECM operation.

where, M is the atomic weight of the element in g, Z is the valance of the element.

Volumetric metal removal rate, $\dot{w} = MI/ZF\rho_{alloy}$ cm^3/s.

In case of alloy, M/Z and ρ can be calculated as:

If the workpiece material is an alloy then density of the alloy ρ_{alloy}, can be calculated from the following equation:

$$\rho_{alloy} = \frac{1}{\sum_{i=1}^{n} \frac{\alpha_i}{\rho_i}} \quad (5.2.15)$$

And electrochemical equivalence for an alloy Ex_{alloy} can be calculated from the equation as follows:

$$Ex_{alloy} = \frac{1}{\sum_{i=1}^{n} \frac{\alpha_i Z_i}{M_i}} \quad (5.2.16)$$

where, α is the fraction of weight percent of individual element in an alloy expressed in decimal value, M is the atomic weight of the element in g, Z is the valance of the element and i is the number of elements in that alloy.

Hence, the metal removal mainly depends on the amount of current flowing in between tool and the workpiece which depends on electrolyte conductivity as well as the inter electrode gap width. Considering Ohm's law,

V = IR again, $R = r_e.h/A$.

Where, r_e is the resistivity, h is the gap thickness and A is the cross sectional area of the tool.

So,

$$V = (I/A).(r_e.h) \quad (5.2.17)$$

where, J = I/A which is the average current density.

Hence,

$$J = V/r_e h = k_e V/h \quad (5.2.18)$$

Where, k_e is the conductivity of electrolyte.

In practical ECM operation, various factors influence the metal removal process, e.g., flow of current, mass transfer, electrolyte type and density, temperature, types of tool as well as workpiece material, etc.

Flow of current is governed by [5]:

(i) Mass transfer from the anode surface to the bulk solution.
(ii) Electron transfer at the electrode surfaces.
(iii) Chemical reactions preceding or following the electron transfer.
(iv) Other surface reactions, etc.

Out of these above mentioned factors during ECM, mass transfer plays an important role during machining. Mass transfer is the movement of material from one location to another, in electrolytic solution, and is achieved through the different modes as follows:

(i) Migration is the movement of ion (M^+) under the influence of an electrical potential.
(ii) Diffusion is the movement of a species under the influence of concentration gradient.
(iii) Convection is related to hydrodynamic transport. Generally fluid flow occurs because of natural convection and forced convection.

Due to the above mentioned reasons, actual rate of metal removal during machining differs from the metal removal which is computed theoretically. Hence, ratio of the amount of actual metal removal to the theoretical metal removal computed from Faraday's law is known as Current Efficiency.

$$\text{Current Efficiency} = \frac{\dot{m}}{\left(\frac{M}{ZF}\right)I} \tag{5.2.19}$$

During machining, current is also utilized for other anodic reactions which occur simultaneously, i.e., oxidation of water, release of oxygen, hydrogen and other gas, etc. As well as selection of wrong valency of the metal may also lead to an incorrect estimation of theoretical metal removal, this results in inaccurate evaluation of current efficiency. Hence, the current efficiency is always <100%. Formation of non-conducting oxide layer on anode surface will reduce the metal removal rate, which also diminishes actual metal removal and leads to reduction in current efficiency. Sometimes high-velocity electrolyte removes grains of metal from the anode surface due to the traction forces which may also increase the actual metal removal rate compare to theoretical and leads to increase of current efficiency. The current efficiency greatly depends on the material of workpiece, electrolyte, and potential difference as well as current density.

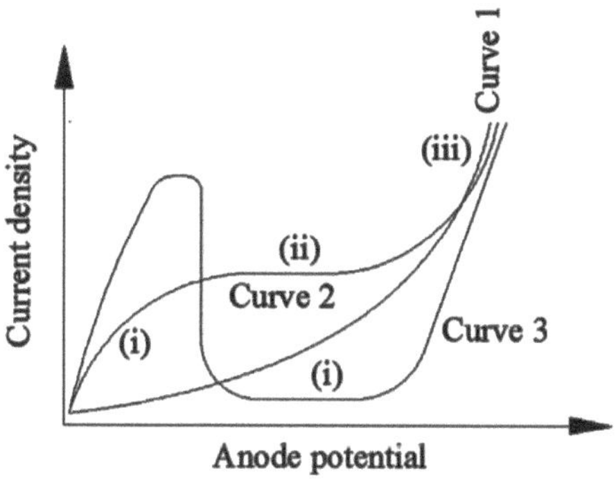

Fig. 5.2.4 Polarization curves.

5.2.1.2.1 Polarization curves

In ECM, anode potential and current density play major roles and governs the manner in which the metal is removed from the anode, which in turn affects the surface finish. Therefore, to understand the role of anode potential and current density in ECM, analysis of polarization curves is greatly needed. Various polarization curves and their behavior on the nature of metal dissolution are shown in Fig. 5.2.4.

Curve 1 represents the etching phenomena in which metal dissolves more rapidly from areas that have a wider atomic space and attacks the grain boundaries and leads to formation of an uneven surface.

Curve 2 represents the phenomena of polishing. It is subdivided into three regions. In region (i) etching still occurs due to non generation of anodic film which required for polishing. In region (ii) at high current density, polishing occurs due to the generation of essential anodic film. The process of anodic dissolution in region (ii) is diffusion controlled over the current density plateau region. In segment (iii), at higher potential due to evolution of gas at the anode and subsequent rapture of the anodic film, polishing action combined with selective dissolution, i.e., pitting of the surface occur.

Curve 3 represents occurrence of passivation which results in the formation of passive oxide film in segment (i) and diminishes the current density, which leads to lower dissolution. As potential further increases, transpassive phenomenon occurs due to breakdown of the passive film and dissolution becomes uniform resulting in efficient machining during ECM.

The polarization curves present the relationship between anode potential and current density and provide essential information about the electrochemical characteristics of metal dissolution in ECM. The pattern of the curve varies according to the metal and electrolyte combinations and other machining parameters.

5.2.1.2.2 Overvoltage

In ECM metal dissolution is achieved by the application of an external potential difference, which gives rise to a corresponding current flow. The greater the current flow, the greater is the anodic dissolution due to irreversible electrode reaction for polarization which is measured by the overpotential. The overpotential is the deviation of the electrode potential from its equilibrium value and can be determined as follows:

$$\Delta V = V - V_{eq} \tag{5.2.20}$$

Where, V is the electrode potential and Veq is the electrode potential at equilibrium condition.

Total voltage profile considering various overpotentials that influence the total potential drop in the ECM cell are shown in Fig. 5.2.5 and are as follows:

(i) Activation overpotential: In the no current flowing condition, the electrochemical changes occurring at an electrode are in equilibrium. The electrode potential between the interface of the electrode and the electrolyte acts as an obstacle to increase the rate of reaction. Additional energy must be supplied to activate the ions discharged at the required rate to promote flow of current.

(ii) Concentration overpotential: The movement of ions in the electrochemical cell is controlled by migration, convection, and diffusion. Migration means movement of ions under the influence of potential difference. In convection, ions move due to physical movement of

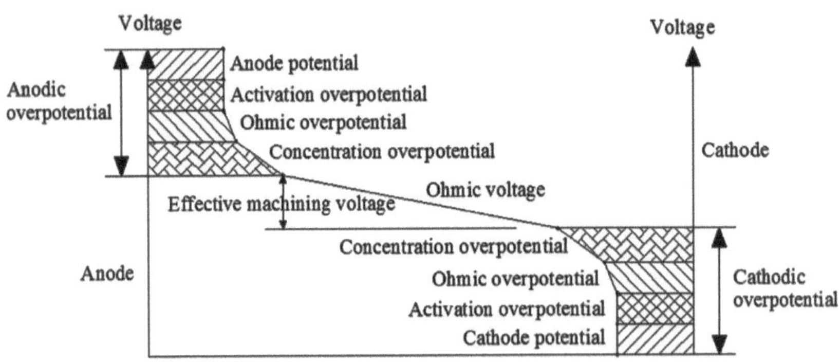

Fig. 5.2.5 Total potential drop in ECM [6].

the electrolyte solution. In diffusion, ions move due to difference in gradient of ion concentration in the solution. Ions migrate toward the electrode surface and form a layer of concentrated ions which acts as concentration barrier. The extra energy required for the movement of the ions through this concentration layer is known as concentration overpotential.

(iii) Resistance overpotential: Films of oxide layer forming on the electrode surface offer resistance to the passage of current. Resistance overpotential is generally regarded as the potential drop across this thin layer on the electrode surface.

(iv) Decomposition potential: For initial values of potential difference, the current is low. When the value of the potential difference increases, the current rises sharply. The current increases appreciably when the potential difference further increases. At this stage, the efficient anodic dissolution started and corresponding potential is known as decomposition potential.

Hence, considering total overvoltage, i.e., ΔV between tool and the workpiece, the current flow for machining can be estimated by $I=(V-\Delta V)/R$. Where, R is the Ohmic resistance of the electrolyte in the gap between tool and the workpiece. However, R is not constant which depends on the various properties of the electrolyte, e.g., type, concentration, temperature, presence of sludge produced during machining as well as amount of gas bubbles evolve in the electrolyte.

Hence, to maintain steady conductivity of the electrolyte which leads to constant resistance of the gap for achieving uniform machining necessitates flowing electrolyte through the narrow gap in between tool and the workpiece. The flow of electrolyte is necessary due to the following facts:

(i) to avoid the metal ions deposition of the tool.

(ii) for removing the sludges, gas bubbles generated during ECM, etc.

(iii) for avoiding the ion concentration to overcome the layer of concentrated ions which acts as concentration barrier at anode surface.

(iv) to control the effect of overheating of the electrolyte due to high current flow during machining.

5.2.2 Kinematics and dynamics of ECM

ECM can be undertaken without any feed to the tool or with a feed to the tool so that a steady machining gap is maintained. Figure 5.2.6

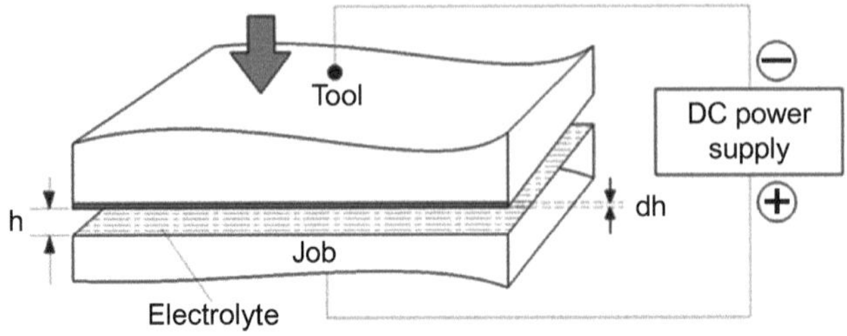

Fig. 5.2.6 Schematic of the machining with plain and parallel electrodes.

schematically shows the machining (ECM) with plain and parallel electrodes and an instantaneous gap between the tool and workpiece of "h."

Now, over a small time period "dt" current of I is passed through the electrolyte and that leads to a electrochemical dissolution of the material of amount "dh" over an area of A.

$$I = \frac{V}{R} = \frac{V}{\frac{r_e h}{A}} = \frac{VA}{r_e h} \tag{5.2.21}$$

Volumetric material removal rate can be expressed as:

$$\dot{\mathrm{w}} = \frac{dv}{dt} = A\frac{dh}{dt} = \frac{\mathrm{MI}}{\mathrm{FZ}\rho_a} \tag{5.2.22}$$

Then,

$$\frac{dh}{dt} = \frac{\mathrm{MI}}{\mathrm{FZ}A\rho_a} \tag{5.2.23}$$

Substituting the value of I,

$$\frac{dh}{dt} = \frac{1}{F}.\frac{M}{Z}.\frac{\mathrm{V}}{r_e h \rho_a} = \frac{G_e}{h} \tag{5.2.24}$$

Where,

$$G_e = \frac{1}{F}.\frac{M}{Z}.\frac{\mathrm{V}}{r_e \rho_a} = constant \tag{5.2.25}$$

During actual ECM, tool is continuously moves toward the workpiece at a constant feed f. Then, the effective rate of change of IEG can be written as:

$$\frac{dh}{dt}=\frac{G_e}{h}-f \tag{5.2.26}$$

This is the basic equation which represents the dynamic of ECM process. Now, three particular cases which occur during ECM process can be considered as follows:

(i) Under no feed condition

This is the case when there is no movement in between tool and the workpiece, i.e., feed, f=0.

Hence, Eq. (5.2.26) becomes

$$\frac{dh}{dt}=\frac{G_e}{h} \tag{5.2.27}$$

Or,

$$h.dh=G_e.dt \tag{5.2.28}$$

Integrating both side with time of machining, following equation is obtained:

$$h^2=2G_et+K \tag{5.2.29}$$

Initially, when $t=0$, then $h=h_0$ and the above equation gives the value of K as h_0^2.

Then,

$$h^2-{h_0}^2=2G_et \tag{5.2.30}$$

Or,

$$h=\sqrt{h_0^2+2G_et)} \tag{5.2.31}$$

Eq. (5.2.31) shows under no feed condition the gap in between tool and the workpiece increases gradually with time and follows a parabolic curve as shown in Fig. 5.2.7.

In practice, this particular situation is intentionally created and utilized during EC deburring operation. By measuring the burr's height, machining time can be calculated for which ECM will be carried out for removal of the burrs.

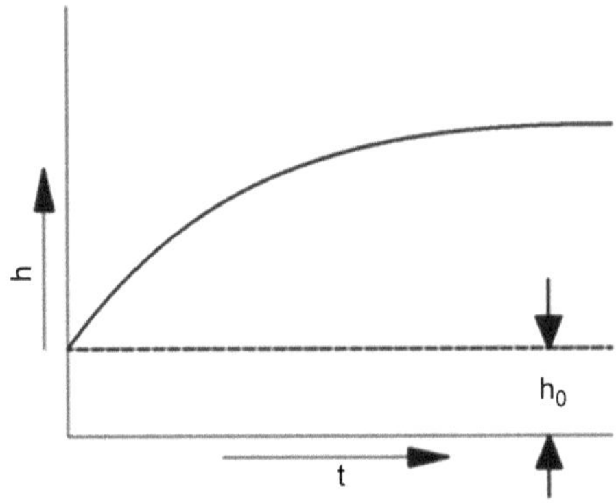

Fig. 5.2.7 Change in gap with time at no feed condition.

(ii) At a constant tool feed condition

This is the case when tool moves toward the workpiece at a constant feed rate f units/s.

Then,

$$\frac{dh}{dt} = \frac{G_e}{h} - f \tag{5.2.32}$$

If the rate of dissolution at anode is same as the tool feed rate then the gap between tool and workpiece is always uniform. Thus with respect to the tool, the workpiece seems to stationary and this situation is known as steady state condition. In this condition, the amount of tool moves toward the workpiece is always compensated by dissolution of the anode workpiece.

Thus, when $\frac{dh}{dt} = 0$, h will become h*, i.e., h = h*.

At steady state condition h* is known as equilibrium gap and $h^* = \frac{G_e}{f}$,

Hence, Equilibrium gap,

$$h^* = \frac{1}{F} \cdot \frac{M}{Z} \cdot \frac{V}{r_e \rho_a f} \tag{5.2.33}$$

In practice, at the beginning of ECM operation it is very difficult to set the value of h* as the initial gap. Thus it is required to be analyzed if the initial gap value would have any effect on progress of the process.

Now,

$$\frac{dh}{dt} = \frac{G_e}{h} - f$$

For making the analysis more generalized, the parameters, e.g., machining time and IEG need to make dimensionless.

Thus,

$$h' = \frac{h}{h^*} = \frac{hf}{G_e} \tag{5.2.34}$$

And,

$$t' = \frac{t}{(h^*/f)} = \frac{tf^2}{G_e} \tag{5.2.35}$$

Then,

$$\frac{dh'}{dt'} = \frac{1}{f}\frac{dh}{dt} \tag{5.2.36}$$

So,

$$\frac{dh'}{dt'} = \frac{1 - h'}{h'} \tag{5.2.37}$$

Then,

$$dt' - \frac{h'}{1 - h'} dh' \tag{5.2.38}$$

Now integrating both side when, $t' = 0$ to $t' = t'$ when, h' changes from h_0' to h_1', the equation will be as follows:

$$t' = h_0' - h_1' + \ln \frac{h_0' - 1}{h_1' - 1} \tag{5.2.39}$$

Now, for different values of h_0', i.e., initial gap and h_1', i.e., steady state gap seems to approach equilibrium gap as shown in Fig. 5.2.8.

After non dimentionalized the parameters, the final equation can be written as:

$$t = \frac{1}{f}\left[h_0 - h_t + h_e \ln \frac{h_0 - h^*}{h_t - h^*}\right] \tag{5.2.40}$$

where, h_t is the instantaneous IEG at time t. This indicates when the given feed rate is same as the metal removal rate, ECM process is self regulating. At a particular parametric combination, various initial gaps approach to equilibrium gap after sometime depending on the values of initial gap. At

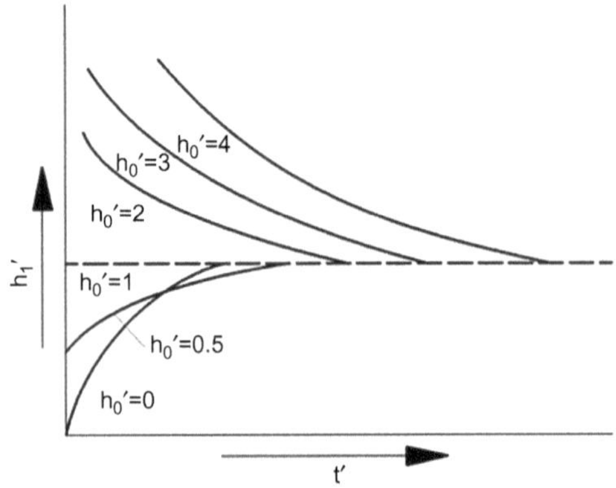

Fig. 5.2.8 Realization of equilibrium gap for various initial gaps at constant feed rate.

this equilibrium gap, stable machining takes place due to machining rate becomes equal to the tool feed rate.

(iii) Under inclined tool feed condition

This is the case when tool moves toward the workpiece at a constant feed rate f units/s and the feed velocity vector is inclined to the surface at an angle θ as shown in Fig. 5.2.9. Where θ is measured between a normal to the workpiece surface and the tool feed direction.

In this case, feed rate f is replaced by f cosθ in the above analysis. Hence, at inclined tool surface, the equilibrium gap between tool and workpiece is as follows:

$$h^* = \frac{G_e}{f\cos\theta}, \text{where, } G_e = \frac{1}{F}\cdot\frac{M}{Z}\cdot\frac{V}{r_e\rho_a}$$

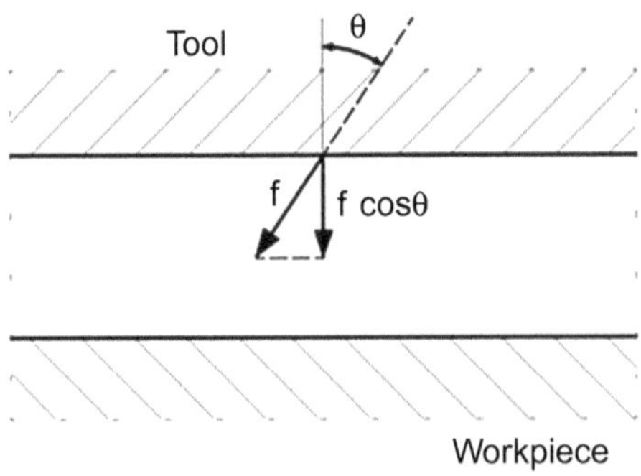

Fig. 5.2.9 Feeding of tool at inclined plane.

When $\theta = 90^0$, the feed rate is perpendicular to the workpiece and only then the situation, i.e., at constant feed rate condition can be applied and the equilibrium gap h* will be $\frac{G_e}{f cos\theta}$. However, this condition can be utilized during tool design for generation of complex job profiles.

5.2.3 Effects of heat and bubble generation

During the machining operation, metallic compounds and gaseous products accumulates in the small machining gap. Due to the flow of high machining current through the narrow machining gap, large amount of heat generated which increases the temperature of electrolyte. Hence, as mentioned before to avoid this crisis, the electrolyte is supplied in the gap with high pressure to dissipate effect of heating as well as removal of the gas bubbles and sludge. To estimate the typical flow velocity of electrolyte through the gap, effect of Joule heating can be considered.

The Joule's effect is a physical relationship between the heat generated by the current flowing in the gap through the electrolyte medium between tool and workpiece can be expressed as:

$$H = I^2.R.t \tag{5.2.41}$$

where, H is the heat generated by current I flowing through the gap with electrical resistance R, for time t. Joule heating is important in ECM; due to the current flow, the electrolyte gets heated, which changes its conductivity. Total amount of heat dissipates in the electrolyte can be expressed as:

$$H = mC_e\,\delta T \tag{5.2.42}$$

where, m is the mass flow rate of electrolyte, g/min, C_e is the specific heat of the electrolyte, J/K.

And mass flow rate of electrolyte, $m = A.U.\rho_e$.

Considering plane and parallel electrodes as shown in Fig. 5.2.10 where, electrolyte passes through the gap, h between tool and workpiece from one to another end for a length, L. During the flow velocity analysis, it is assumed that all the heat generated due to Joule's heating by the flow of current remain in the electrolyte and dissipates in the electrolyte due to the electrolyte mass flow rate through the gap.

Hence, from Eqs. (5.2.41), (5.2.42), $H = I^2 R\delta T = mCe\,\delta T$ as $R = \frac{r.h}{A}$

Then,

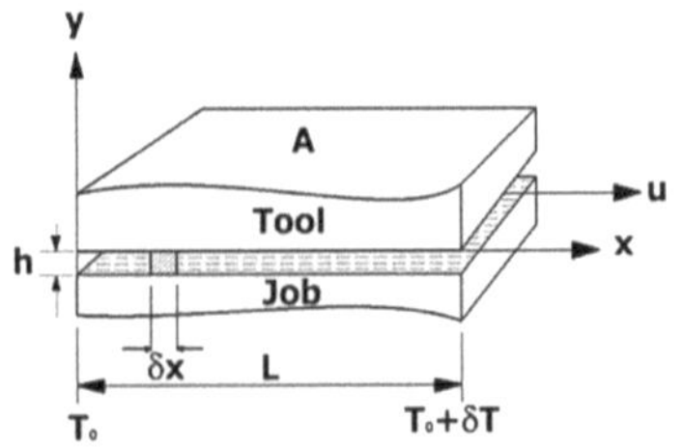

Fig. 5.2.10 Flow velocity of electrolyte through plane and parallel electrodes.

From Fig. 5.2.10

$$AU\rho_e\,C_e\,\delta T = \frac{I^2\delta x}{k_e A} \tag{5.2.43}$$

$$\delta T = \left(\frac{I}{A}\right)^2 \frac{\delta.x}{k_e\rho_e c_e U} \tag{5.2.44}$$

$$\delta T = \frac{J^2}{k_e\rho_e c_e U}\delta x \tag{5.2.45}$$

Again,

$$K_e = K_0[1 + \alpha\Delta T]$$

Thus,

$$\delta T = \frac{J^2}{k_0[1 + \alpha\Delta T]\rho_e c_e U}\delta x \tag{5.2.46}$$

$$\int_T^{T+\Delta T}[1 + \alpha\Delta T]\delta T = \int_0^L \frac{J^2}{k_0\rho_e c_e U}\delta x$$

$$U = \frac{J^2 L}{k_0\left[\Delta T + \frac{\alpha}{2}\Delta T^2\right]\rho_e c_e} \tag{5.2.47}$$

Due to electrochemical reaction at cathode tool evaluation of hydrogen takes place that changes the conductivity of the electrolyte which intern varies the local anodic dissolution rate at exit region of the electrolyte flow path. Hence, the amount, size as well as location of the gas bubbles in the machining gap should be studied.

Considering the effect of temperature as well as hydrogen gas bubbles on the electrolyte conductivity, this can be expressed utilizing Bruggeman's equation as follows:

$$k_m = k_e(1-\upsilon)^n \tag{5.2.48}$$

where, k_m is the conductivity at the electrolyte-gas medium, k_e is the conductivity of the electrolyte and υ is the void fraction and generally n is considered as 1.5. For a sufficiently small υ, say < 0.2 the above equation can be written as:

$$k_m = k_e(1-1.5\upsilon) = K_0[1+\alpha\Delta T](1-1.5\upsilon) \tag{5.2.49}$$

or,

$$K_0 = \frac{k_m}{[1+\alpha\Delta T](1-1.5\upsilon)} \tag{5.2.50}$$

Then, flow velocity through the gap can be estimated as:

$$U = \frac{J^2 L[1+\alpha\Delta T](1-1.5\upsilon)}{k_m\left[\Delta T + \frac{\alpha}{2}\Delta T^2\right]\rho_e c_e} \tag{5.2.51}$$

where, h is the gap and R_e is the Reynolds number. Flow of electrolyte is usually turbulent due to the narrow gap between tool and the workpiece. Hence, the Reynolds number should be >2300.

Reynold's number,

$$Re = \frac{\rho_e \ddot{u} d_0}{\mu} \tag{5.2.52}$$

where, $\ddot{u}$ is the mean velocity, d_0 is the hydraulic mean diameter of the electrolyte flow path and μ is the absolute viscosity of the electrolyte. Hence, pressure drop in the electrolyte flow path of length, L for turbulent flow can be expressed as:

$$\Delta p = \frac{1}{2}\frac{fL\rho_e \ddot{u}^2}{d_0} \text{ where, friction factor, } f = \frac{0.3164}{R_e^{1/4}}$$

Hence, total pressure required for maintaining the flow velocity, U through small IEG, h across the length, L can be expressed as follows:

$$p = \frac{0.3146\rho_e U^2 L}{4h Re^{0.25}} + \frac{\rho_e U^2}{2} \tag{5.2.53}$$

where, hydraulic mean diameter, $d_o = 2h$.

In the above mentioned equation first term denotes the amount of pressure required to overcome the viscous friction which acts on the tool face. The second term denotes the pressure required to overcome the inertia.

5.2.4 ECM equipment details

To apply controlled anodic dissolution for the ECM in practice for industrial applications, desired metal removal with high accuracy is needed. To achieve this, the ECM setup requires several subunits. Presently, electrochemical machines have been developed by various machine tool manufacturers with excellent standard of performance and reliability. ECM units are available of many types and wide range of sizes as standard models. A typical ECM system setup consists of four major subunits which are as follows:

(i) Mechanical unit
(ii) Electrolyte flow system unit
(iii) Power supply unit
(iv) Machine control unit

Fig. 5.2.11 shows the above mentioned basic sub units of a complete ECM system setup.

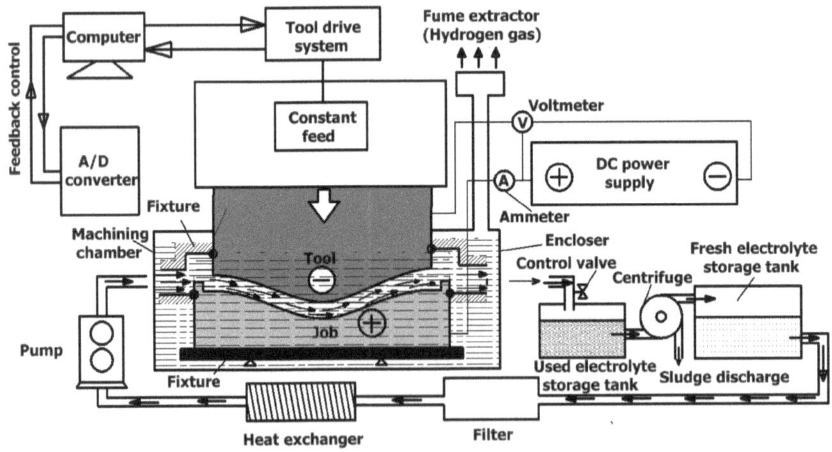

Fig. 5.2.11 Scheme of ECM setup with different sub units.

(i) Mechanical unit

This unit mainly consists of machine structures which are of various type, e.g., vertical frame, horizontal frame or combined configuration, etc. depending on the various applications. However, vertical frame type structure as shown in Fig. 5.2.11 is widely used. The structure of the machine should be robust to counter forces generated by high electrolyte flow. To avoid corrosion caused by electrolyte, generally non metallic materials and proper coating should be utilized for fabrication of different components of the mechanical units. It should have proper arrangements for connections of electrical power terminals to tool and job. All parts of the mechanical unit should also be isolated properly from electrical current by stable insulation.

Ideal machine work chamber should consist of job holding device, job and tool feeding units, proper sealing arrangements as well as online observation unit. It should be properly enclosed by plexiglass shields during ECM operation for protecting the operators from any kind of health hazards. Explosion proof blowers are fitted with the work chamber to remove hydrogen gas and fumes produced during ECM operation. Arrangement for forced circulation of air through the work enclosure in addition to an exhaust duct is needed to keep the hydrogen concentration below 0.5% by volume.

Gap between tool and the workpiece is very small as well as that should be maintained constantly throughout the ECM operation. To avail this, a tool feeding unit is used along with the feed axis. To move the job precisely, X-Y stage is utilized. The slide ways need corrosion protection utilizing grease coating and specially designed cathodic protection. Different drive units, e.g., motors, coupling, ball-screws and guide ways for the movement of tool and the job must be protected by sealing.

(ii) Electrolyte flow system unit

Electrolyte should flow with high velocity through the narrow gap between tool and the workpiece for the purpose of effective machining. For maintaining stable and continuous flow of electrolyte, separate electrolyte flow system is needed in ECM setup. It mainly consists of pumps, filters, pipe lines, control valves, storage tanks, clarification system as well as heat exchanger for temperature control. All subunits are made of non corrosive materials. Special types of pumps are needed to maintain uniform high pressure. As per the requirements, single as well as multi stage pumps are used which runs at a high speed with proper sealing arrangements for shafts. Generally, stainless steel or monel mesh filter screens are utilized for proper online filtering which are periodically clean.

From the main machining chamber, used electrolyte collects in a storage tank and it further moves to clean electrolyte storage tank through centrifuge. Storage tank of used electrolyte should have much larger floor area for settling down the metal hydroxide, i.e., sludge due to gravity. Centrifuge is used to further separate the sludges and precipitates from the electrolyte. Pump is utilized to supply clean electrolyte with high pressure to the machining zone through filter and heat exchanger. During machining, it is needed to control the temperature of electrolyte as well as its Ph value for maintaining stable conductivity to achieve better machining.

(iii) Power supply unit

Generally, direct current power unit is used for ECM setup to supply smooth direct machining current at a constant voltage. It also provide protection against damage during malfunctioning of the machining process, e.g., generation of sparking, arcing and short circuiting. A typical power unit consists of step down transformer, rectifiers, control and protection unit circuits and cooling system. To the ECM power supply, 3 phase AC input supply is stepped down to low voltage and high current utilizing the step down transformer. Then, it is converted to smooth DC output utilizing silicon-controlled rectifiers (SCR).

Power supply unit has also electronic protective device for detecting electrical transient during machining. The electrical power is cutoff during occurrence of electrical transient and protects the tool and the work surface from damage. The response time of this protection circuit should be very small in the order of 10–20 micro-second to prevent severe spark damage to tool and workpiece.

(iv) Machine control unit

Machine control plays an important role in ECM which is monitored and controlled through computer interfacing. Various sub units of ECM system setup, e.g., power supply, tool feeding system, job moments, and different units of electrolyte flow system can be controlled manually as well as automatically depending upon the machining requirements as well as different conditions. It has already been informed that main power supply should have protection unit to safeguard the ECM setup from the undesirable conditions which may occur during machining. With the help of machine control unit, all the other subunits are automatically shut-off during such situations. Closed loop control system is essential for feedback control depending upon various signals which are sensed from various subunits utilizing different transducers. Machining can be performed utilizing DC power source through two modes, i.e., constant voltage (CV) mode and constant current (CC) mode. CV mode is mostly utilized during ECM during its simplicity in operation.

However, for online monitoring and controlling of the gap between tool and the workpiece, major task in machine control is to choose the appropriate sensors or sensing parameters. The gap can be sensed directly with the help of eddy current. The micro fluctuations of the machining current due to micro variation of the gap can also be sensed utilizing the current sensors. This variation in gap current signals can be used for gap identification. Depending upon the variation of the gap, the current signals can be utilized to move the tool forward or backward directions to maintain the desired gap.

5.2.5 Electrolyte flow paths and insulation

Availability of the electrolyte in the machining gap is most important in ECM. Hence, correct electrolyte flow across the machining gap is the basic requirement of ECM process. Depending upon the requirements, different mode of electrolyte flow path can be considered as follows:

(i) Straight flow path
(ii) Reverse flow path
(iii) Crossed flow path
(iv) Mixed flow path

(i) Straight flow path

Fig. 5.2.12 shows the arrangement of straight flow path where, electrolyte flows with high pressure through the tool and reaches to the machining zone. Discharge of electrolyte takes place from the machining zone to the work chamber at atmospheric pressure. Hence, due to this pressure difference different flow disturbances may occur which leads to formation of boss or ridges occur on the work surface in the machining zone which hamper the accurate shape generation.

(ii) Reverse flow path

Fig. 5.2.13 shows the arrangement of reverse flow path, where electrolyte coming out through the tool from machining zone with high pressure.

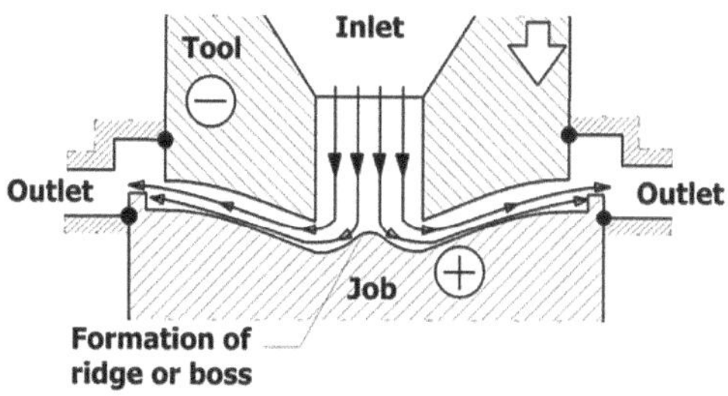

Fig. 5.2.12 Straight flow path.

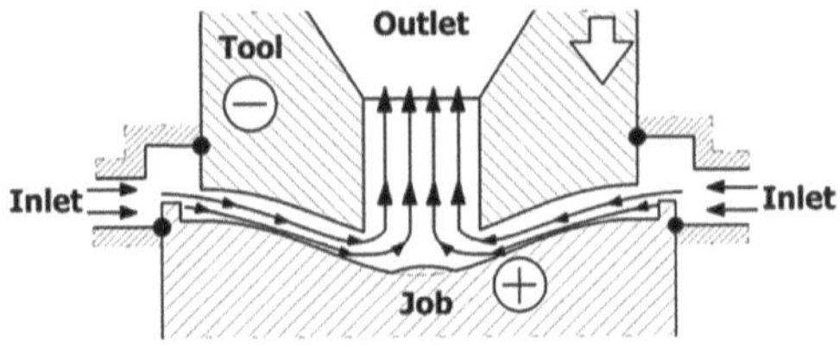

Fig. 5.2.13 Reverse flow path.

Here, back pressure at the exit of the electrolyte flow path can be controlled as per the requirements. It increases the chances of availability of electrolyte at every corners of the machining zone due to prevailing back pressure. Reverse flow path reduces non-uniform material removal as well sludge induced sparking. It also decreases the size of the produced gas bubbles which increases the conductivity of the electrolyte thus helping the efficient machining.

(iii) Crossed flow path

Fig. 5.2.14 shows the arrangement of crossed flow path, where electrolyte incoming to machining zone from entry side with high pressure, moves along the machining area and leaves at the exit side. This type of flow path is generally utilized for machining of large surface components. Drawback of this mode of electrolyte flow path is the requirement of high inlet pressure. It also increases presence of sludges and gas bubbles at the outlet zone which leads to uneven machining at exit machining area.

(iv) Mixed flow path

It is a combination of straight and crossed flow path as shown in Fig. 5.2.15. During large surface machining utilizing cross flow, electrolyte at the exit area becomes impure due to the accumulation of sludges and gas bubbles. This can be overcome by introducing further flow of fresh electrolyte at the exit area with the help of straight flow through the tool as shown in

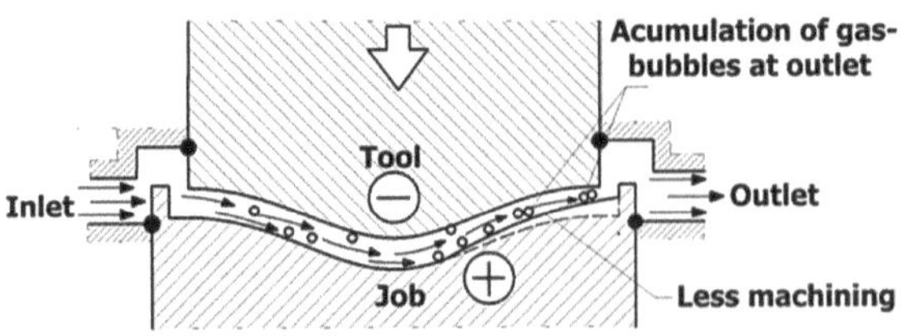

Fig. 5.2.14 Crossed flow path.

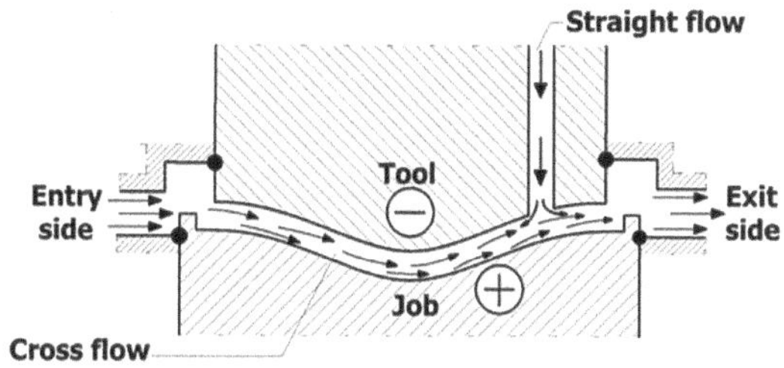

Fig. 5.2.15 Mixed flow path.

Fig. 5.2.15. This type of mixed flow path helps the conductivity of the electrolyte almost same throughout the machining surface.

Design of electrolyte flow path is important for reducing the chances of cavitation, stagnation as well as the formation of vortex which results in uneven machining and poor surface finish. During tool design electrolyte flow path plays an important role. Correct electrolyte flow is most important for effective machining. Tools with narrow slots for supplying electrolyte to the machining zone are effective however, these may leave small ridges on the machined surface.

Fig. 5.2.16 shows different slots as designed for supplying electrolyte to ensure smooth flow condition for overcoming generation of passive areas. Slot width should be twice the estimated machining gap size for supplying

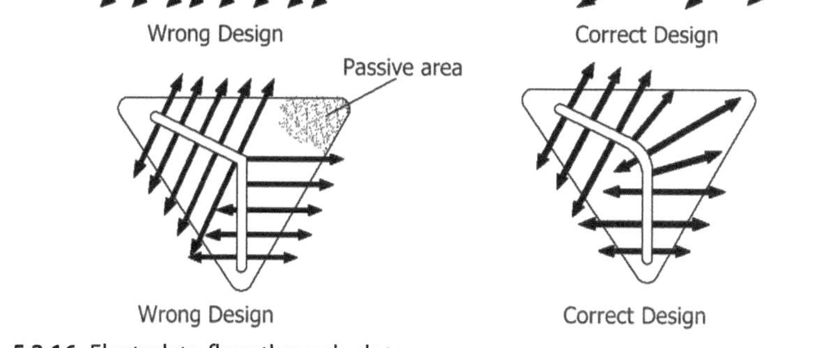

Fig. 5.2.16 Electrolyte flow through slots.

sufficient electrolyte. Flow takes place in a direction perpendicular to the slot and deteriorates at the end sharp or rounded corner which can be improved by introducing modification in slots design as shown in Fig. 5.2.17.

Quality of the machining can be improved by controlling undesirable stray current flow through the circumference of the tool. Fig. 5.2.18 shows the effect of stray machining and improvement of machining through insulation. The areas of the tool as well as the work surface where, machining is not desirable should be insulated for reducing flow of stray current. Selection of insulation material as well as the technique

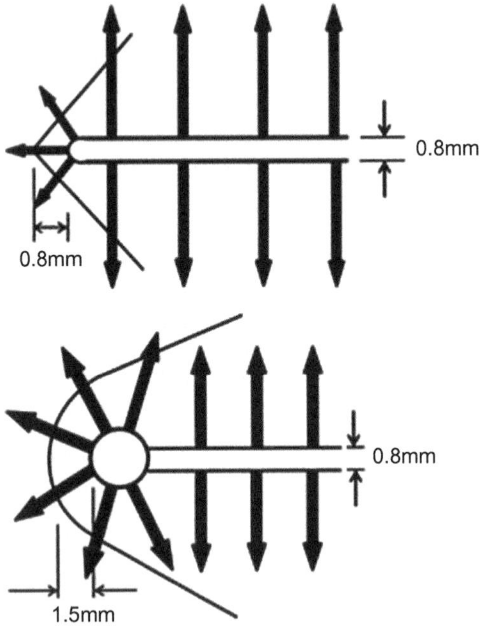

Fig. 5.2.17 Modification in slots design.

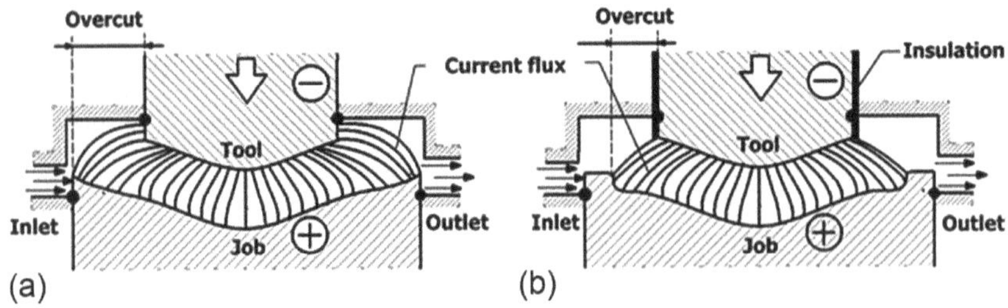

Fig. 5.2.18 Stray machining (a) and improvement of machining through insulation (b).

of insulation is important. The insulation material should be adhesive enough thus it sticks to the surface and resists high electrolyte pressure as well as temperature. Reinforced plastics, synthetic rubber coating, epoxy resin can be used as insulating materials. These types of insulations can be applied to the desired areas by spraying technique as well as chemical or physical vapor deposition techniques. High velocity electrolyte causes stripping of the insulation layer, which should take care of during insulation design.

5.2.6 Tool material and electrolytes

(i) Tool material

Choice of tool material plays an important role in ECM. Any conductive material can be used as tool. Several factors determine the selection of tool materials, e.g., high electrical and thermal conductivity, resistance to chemical corrosion, strength, stability, easy to machine and repair, availability, etc. Metals, e.g., copper, high grade stainless steel, brass, bronze, monel, gun metal can be used as a tool as well as fixture materials due to its resistance to chemical corrosion as well as these exhibit the above mentioned properties. For some special cases, e.g., titanium, copper-manganese, copper-tungsten, etc. are also recommended.

In practice copper or stainless steel are widely used due to their several advantages. However, copper is the most preferred tool materials due to its high conductivity which ensures distribution of machining current to all parts of the machining surface without overheating. Spark damage of the copper tool surface is limited which can be repaired easily by electroplating. With copper fabrication of complex profile tool is easier and its bright polished surface make it more acceptable in ECM. However, it is difficult to strongly bond the insulation to the copper tool surface.

(ii) Electrolytes

In ECM, electrolyte acts as a conductor, connects tool and the workpiece as well as carries current which is responsible for machining. Selection of electrolyte influences various major factors, e.g., metal removal rate, accuracy, surface finish and surface integrity during machining.

Electrolyte can be classified into three categories based on its pH value such as acidic, neutral, and alkaline. Neutral electrolytes such as NaCl,

$NaNO_3$ are commonly used in ECM. Electrolytes can also be classified based on the degree of dissociation and nature of passivation factor. The degree of dissociation is the fact that generates current carrying free ions, which are dissociated from the fraction of solute at a given concentration. Strong electrolytes are dissociated to a great extent for concentration ranging from very low to high. In weak electrolytes, the degree of dissociation tends to zero at high concentration. Strong electrolytes such as NaCl are widely utilized for ECM.

Whereas, the electrolytes based on passivation are classified into two categories: passivating electrolytes containing oxidizing anions, i.e., sodium nitrate, sodium chlorate, and nonpassivating electrolytes containing relatively aggressive anions such as sodium chloride.

Passivating electrolytes are known to give better machining precision, e.g., $NaNO_3$. In some applications alkaline electrolytes are preferable for ECM, e.g., NaOH, KOH, etc. However, sodium nitrate is more advantageous over other electrolytes due to its less throwing power, high and controlled metal removal which leads to high speed and accuracy in machining. NaCl is nontoxic, inexpensive and widely used electrolyte in ECM. However, it generates stray machining due to high throwing power. $NaNO_3$ generates better surface finish, improves accuracy and limits stray machining due to low throwing power and is also used extensively in ECM.

Mixed electrolytes which are the combination of more than one electrolyte such as NaCl and $NaNO_3$, NaCl and NaBr, NaCl and NaOH, etc., are also utilized in ECM for machining different metals. Mixed electrolytes provide combined advantageous effects of constituent electrolytes during machining. For some applications, additives are also mixed in the electrolyte. Dimensional accuracy can be improved by using additives such as NaHSO4 in the electrolyte. In ECM, normally machining gap is very small, hence to prevent formation of metal hydroxide precipitates, some complexing agents are also added in the electrolyte. Citric acid is a complexing agent which is added in NaCl electrolyte to restrict the formation of precipitates in the narrow gap. The use of complexing agent such as Ethylenediaminetetraacitic acid disodium salt (EDTA-Na2) has been reported, which prevents the formation of metal hydroxide in the IEG and provide encouraging results by restricting the occurrence of frequent short circuit during ECM operation [7]. Table 5.2.1 represents suitable electrolytes for different metals machining utilizing anodic dissolution [6].

Table 5.2.1 Suitable electrolytes for different metals machining [6]

Material	Electrolyte
Aluminum and aluminum alloys Copper and copper alloys Beryllium	Sodium chloride (NaCl), sodium nitrate ($NaNO_3$)
Tungsten	Sodium hydroxide (NaOH)
Molybdenum Titanium and titanium alloys	Sodium bromide (NaBr), sodium chloride (NaCl), mixed electrolyte, e.g., sodium chloride (NaCl) + sodium nitrate ($NaNO_3$)
Nickel and nickel alloys Cobalt and cobalt alloys	Sodium chloride (NaCl)
Brass Tungsten alloys Stainless steel Steel and iron based alloys	Sodium chloride (NaCl), sodium nitrate ($NaNO_3$), sodium perchlorate ($NaClO_4$), mixed electrolyte, e.g., sodium chloride (NaCl) + citric acid, mixed electrolyte, e.g., sodium chloride (NaCl) + hydrochloric acid (HCl) or sulfuric acid (H_2SO_4)
Magnesium and magnesium alloys	Sodium chloride (NaCl), sodium nitrate ($NaNO_3$), mixed electrolyte, e.g., phosphoric acid (H_3PO_4) + sulfuric acid (H_2SO_4)
Gold	Lithium chloride (LiCl), dimethyl sulfoxide $(CH_3)_2SO$
Tungsten carbide	Mixed electrolyte, e.g., sodium chloride (NaCl) + sodium hydroxide (NaOH) + triethanolamine ($C_6H_{15}NO_3$), mixed electrolyte, e.g., sodium chloride (NaCl) + sodium hydroxide (NaOH) + sodium tartarate ($Na_2C_4H_4O_6$)

Final selection of correct electrolyte from a group of acceptable electrolytes for a particular material can be made by verifying success in machining a component so that it fulfill all its design and quality requirements.

5.2.7 Influence of ECM parameters on machining performances

Metal removal and quality of machining in ECM is influenced by the nature of anodic dissolution where metals are removed from workpiece surface atom by atom. To achieve effective and high precision machining, the process parameters of the ECM system will have to be optimally controlled.

Various process parameters of the electrochemical machining such as applied voltage, current density, electrolyte type, concentration, flow rate, tool feed rate, etc., influence major machining criteria, i.e., metal removal rate, surface finish and quality, profile accuracy, etc. [8]. Some of the predominant factors, which have major influences on machining performances, are analyzed hereunder to control the ECM process most optimally. A fish bone diagram considering all these process characteristic of ECM is presented in Fig. 5.2.19.

Some of the predominant ECM parameters, which influence on ECM performances are listed as follows:

(i) Applied voltage
(ii) Current density
(iii) Type of electrolytes
(iv) Concentration, temperature, and flow rate of electrolyte
(v) Tool feed rate, etc.

Major machining performances which can be achieved by controlling above mentioned parameters are mentioned and discussed hereunder:

(i) Metal removal rate, i.e., machining rate
(ii) Overcut, i.e., machining accuracy
(iii) Surface finish, i.e., quality of machining, etc.

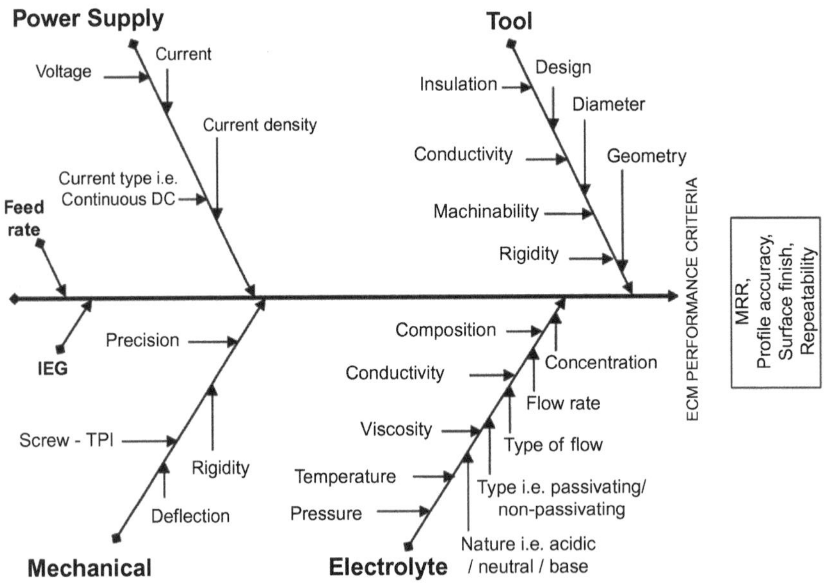

Fig. 5.2.19 Fish bone diagram considering the process characteristic of ECM.

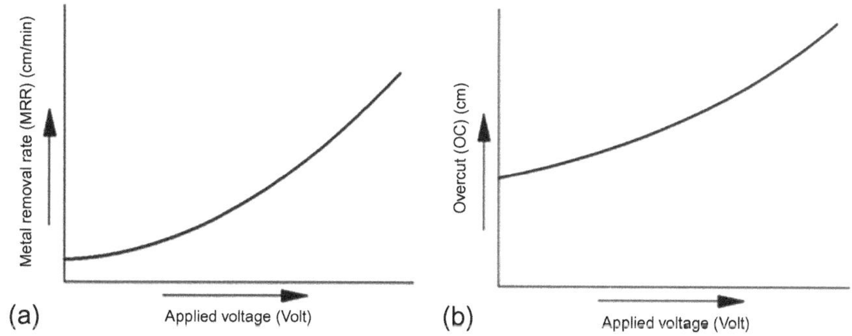

Fig. 5.2.20 Influence of applied voltage on (a) MRR (b) overcut.

In ECM, DC power supply with low voltage in the range of 0–40 V and high current is utilized for machining. Applied voltage directly influences the metal removal rate as well as quality of machining. Fig. 5.2.20(a) highlights the nature of change of metal removal which increases with applied voltage due to higher voltage causes higher machining current which leads to increment of metal removal. However, rate of increment of metal removal is reduced at the higher range of voltage due to changing in electrolyte conductivity for the formation of sludge, gas bubbles as well as electrolyte heating, etc.

Type of electrolytes, concentration and flow velocity control the current density which intern influence the machining rate, overcut as well as surface finish as shown in Fig. 5.2.20(b). For a particular electrolyte, current density increases with the increment of concentration as shown in Fig. 5.2.21(a). Current density also increases due to increment of its specific conductivity which depends on temperature. Metal removal and surface finish improve with the increment of current density as shown in Fig. 5.2.21(b). Fig. 5.2.22 indicates that with $NaNO_3$ electrolyte, current efficiency is maximum at the highest current density. Whereas, in NaCl electrolyte, changes of current efficiency is more or less same with current density [9]. As throwing power of $NaNO_3$ is less that the NaCl electrolyte, $NaNO_3$ is recommended for high accuracy machining due to less overcut. In the contrary, NaCl is recommended for higher metal removal however, accuracy is poor.

Electrolyte flow velocity also influences quality of machining. Increment in electrolyte flow velocity causes reduction of surface roughness due to the availability of fresh electrolyte which results in the increment of current density. However, at higher electrolyte concentration this tendency is quite

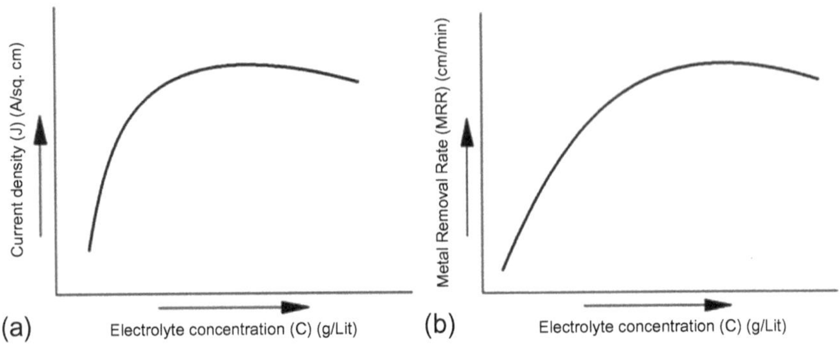

(a) (b)

Fig. 5.2.21 Influence of electrolyte concentration on (a) current density (b) MRR.

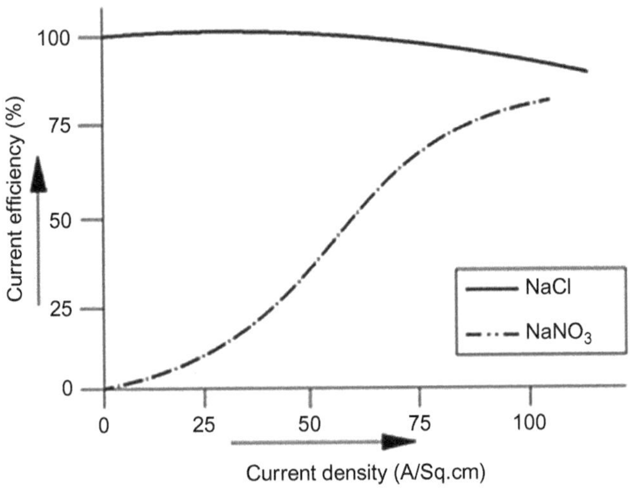

Fig. 5.2.22 Influence of current density on current efficiency.

smoothen out to yield better surface finish due to anodic smoothing at also low flow rate range as shown in Fig. 5.2.23.

As discussed earlier, tool feed rate greatly influences the ECM performance. Tool feed rate along with other parametric setting actually control the value of equilibrium gap. Accuracy in terms of overcut minimized at lower equilibrium gap which intern depends on higher tool feed rate. However, there is a limit for increment of the tool feed rate due to occurrence of sparking because of the inefficient sludge removal from the narrow gap.

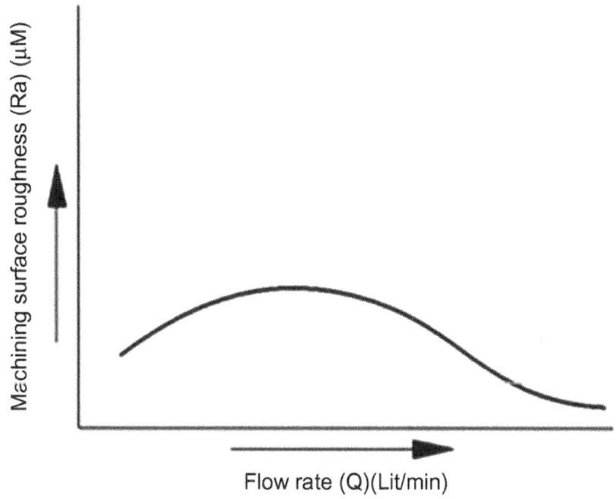

Fig. 5.2.23 Influence of flow rate on roughness.

Considering above mentioned discussions, there is a need of selection of optimal combination of parameter settings to achieve best responses in terms of higher machining rate, higher accuracy as well as best quality of the machined surface. Based on developed a mathematical model which presents relationships between various ECM parameters with different machining responses, designer can choose the optimal parameter settings to achieve design requirements to accomplish the desired machining objectives [10].

5.2.8 Surface integrity and accuracy

5.2.8.1 Surface integrity

Surface integrity signifies surface finish as well as the various machined surface conditions. It can be divided into surface texture and sub surface conditions of the machined component. As anodic dissolution is not produced any thermal damage on the machined surface hence, there is no heat affected zone (HAZ) on the top of the machined surface. However, in ECM, machined surface integrity depends on the various other issues, e.g., selective dissolution, formation of passive oxide film on the workpiece surface, uneven breakdown of passive film, evaluation of H2 gas and electrolyte flow related factors.

As discussed earlier, surface finish achieved by ECM depends on the current density as well as flow velocity. Surface finish in ECM depends also on the combination of anode material and electrolyte. The metallurgical characteristics of workpiece play an important role in ECM. Microstructure, i.e., size, structure and orientation of grains of the job material as well as level of finishing of the tool surface also influence the surface finish [11]. Smoothing rate is depended on the initial shape profile of the job surface and not the cumulative area machined. Electrolyte flow direction also greatly influences the smoothing affect in the case of machining asymmetrical job profiles. However, irrespective of initial shape profile of the job, the surface irregularities will smoothen out to a sinusoidal profile before attaining the final shape [12].

Variation of electrolyte flow velocity also causes generation of striations or ripples on the machined surface in the direction of flow. Oxide film on the work surface during machining becomes sensitive to small variation in flow velocity. Bright raised areas or spots on the work surface may also generate due to variation of flow velocity. Striations as well as bright spots can be eliminated by controlling the ratio of current density to flow rate for a particular material and electrolyte combination. It can be possible either by decreasing current density or increasing flow rate.

Selective dissolution of metal ions from work surface during ECM may occur due to different electrode potentials for dissimilar constituents of alloy. As differences in dissolution potential at grain boundaries are higher compare to inside grains which results in occurrence of faster dissolution of ions from the grain boundaries. Improper selection of the electrolyte as well as very low current density causes intergranular attack and pitting. In this situation, ion dissolves from grain boundaries before the actual metal grains and reduces the residual stress which causes reduction of fatigue resistance of the machined parts produced by the ECM. At higher current density, this effect reduces due to potential required for dissolution of the dissimilar constituents are more easily achieved.

5.2.8.2 Accuracy

Throwing power of the electrolyte influences the ability of dissolutions at remote area of anode workpiece from cathode tool. Hence, during polishing higher throwing power of electrolyte is desirable which generates uniform degree of smoothing from remote areas inspite of complex geometrical features. However, improvement of machining accuracy can be achieved utilizing low throwing power electrolyte. It is due to the fact that low throwing power will reduce the current falling in the remote areas of job thus less dissolution at the remote areas results in substantial profile conformity between anode workpiece and cathode tool.

Accumulation of hydrogen gas bubbles at the cathode tool increases along the downstream of the electrolyte flow channel within the machining gap. The effective conductivity of the electrolyte reduces due to presence of hydrogen gas bubbles which lessens the amount of controlled metal removal. Hence, this decrease of metal removal becomes more prominent in the exit side of the electrolyte flow path which influences the accuracy as well as quality of machined surface.

Flow separation occurs between the hills and valleys present on the irregular initial job surface. In ECM, metal removal is controlled by the diffusion-migration mechanism. However, nature of flow over an irregular job surface is quite different. Electrolyte flow adheres to the job surface around the hills. However different situation arises in the valleys which separate the nature of flow on the metal surface causing a rotating eddy. Due to this, local removal of the smallest irregularities cannot be achieved results in generation of mat or etched finished surface on the machined surface.

Occurrence of cavitation in the machining gap also deteriorates machining accuracy. Cavitation is the occurrence of bubbles of vapor in electrolyte solution. Low pressure pockets within the machining gap are responsible for the formation of vapor bubbles below boiling temperature which causes cavitation. As current does not passes through the cavitation zone, machin ing does not occurs and creates several inaccuracies. Cavitation phenomena can be reduced by increasing the tool feed rate as well as machining voltage. However, simplest strategy to overcome cavitation is use of back pressure in the electrolyte flow channel.

5.2.9 Tool design

To introduce ECM into the manufacturing industries, attention needs to be paid for designing cathode tool. In ECM, tool cathode design is a major challenge which deals with the determination of tool shape for producing a workpiece with desired dimension and accuracy. Tool design considering iterative procedure is time consuming results in long machine down time. Under equilibrium or steady state gap condition, the formulation of the tool design problem can be predicted by the steady state geometric relationship in between the job and the tool surface. The tool design problem can be subdivided into two groups as follows:

A. Prediction of the tool shape for a known anode profile, workpiece.

B. Determination of the tool shape for achieving a desired anode workpiece shape.

However, the second case is much more difficult than the first one. In ECM operation, gap condition becomes complicated due to the complex physico chemical phenomena and various hydrodynamic changes. So, for carrying out the tool design analysis in ECM following conditions need to be assumed:

(a) The tool and work surface is considered to be equipotential

(b) The electrolyte is homogeneous and isotropic

(c) The metal removal rate is governed by Faraday's laws of electrolysis

(d) Ohm's law is applied in the machining gap

(e) The distribution of potential and the current flux in the machining gap can be described by Laplace's equation

(f) The workpiece shape is invariant with time, i.e., an equilibrium state of machining is reached

The shape of the ECM tool is not just the inverse replica of the shaped to be machined in fact there is a complex relationship between tool shape and to the required shape on the work. Following are the methods by which design of tool in ECM can be done:

A. Prediction of the tool shape for a known anode workpiece profile

  (i) Tool design for known and mathematically defined machined surface by cosθ method [13]

B. Determination of the tool shape for achieving a desired anode workpiece shape

  (i) Finite element method

  (ii) Generalized approach to tool designs for complex job shapes

  (iii) Tool shape correction

5.2.9.1 Tool design for known and mathematically defined machined surface by cosθ method

Designing the ECM tool on basis of the operating conditions can be very much useful not only for the simple geometric flat tool but also for any type of complex tool profile design. The tool design problem can also be solved in a generalized way for the mathematically definable job surface considering the various boundary conditions and above mentioned assumptions.

Considering a two dimensional case where there is no variation in the Z direction. If the work surface geometry is described as:

$$Y = f(x)$$

Let's consider any point say, Q (x_w, y_w) on the anode and the corresponding point at the tool say P(x_t, y_t)

Then, the equilibrium gap in between two electrodes is PQ and this can be written as:

$$PQ = \frac{Ge}{cos\theta} \tag{5.2.54}$$

where,

$$Ge = \frac{kA(V - \Delta V)}{\rho_a ZFfcos\theta} \tag{5.2.55}$$

Here, $\boldsymbol{\theta}$ is the inclination of feed direction and V, ΔV and the f are applied voltage, over voltage and the feed rate respectively. As shown in Fig. 5.2.24 that the y-axis is parallel with the feed direction.

Now, it can be written as:

$$y_w - y_t = PQ \cos\boldsymbol{\theta} \tag{5.2.56}$$

and

$$x_t - x_w = PQ \sin\boldsymbol{\theta} \tag{5.2.57}$$

$$= Ge \tan\theta \tag{5.2.58}$$

$$= Ge\left(\frac{dy}{dx}\right) \tag{5.2.59}$$

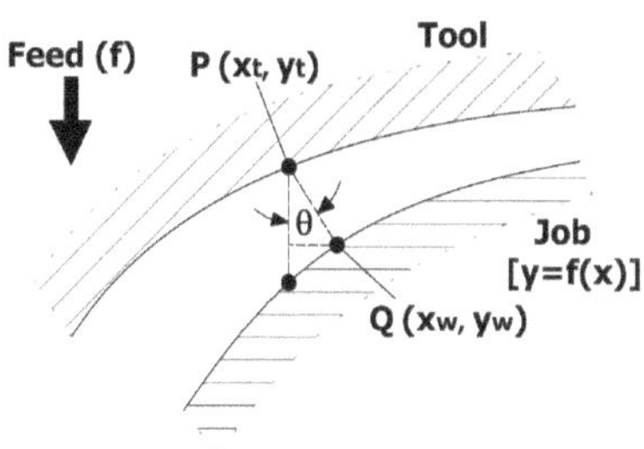

Fig. 5.2.24 Tool design of complex profile by cosθ method.

The required known workpiece shape may be definable by an equation and can be considered for defining section of the job. The tool profile can be determined for that section of the workpiece surface profile.

For example, if a section of the workpiece profile is defined by the following equation:

$$y_w = a + bx_w + cx_w^2 \tag{5.2.60}$$

Then, the above equation can be written by the following equation:

$$y_t = a + bx_t + cx_t^2 - b\frac{Ge}{f}\left(\frac{dy}{dx}\right) - 2cx_t\frac{Ge}{f}\left(\frac{dy}{dx}\right) + c\left(\frac{Ge}{f}\right)^2\left(\frac{dy}{dx}\right)^2 \tag{5.2.61}$$

From Eqs. (5.2.60), (5.2.61),

$$\left(\frac{dy}{dx}\right) = \frac{b + 2cx_t}{1 + 2c\frac{G_e}{f}} \tag{5.2.62}$$

Then, under steady state condition, the tool surface can be expressed after substitution of Eq. (5.2.62) in Eq. (5.2.61) and can be expressed as:

$$y_t = a + bx_t + cx_t^2 - \frac{Ge}{f} - \frac{Ge}{f}\left[\frac{(b + 2cx)^2}{1 + 2c.\frac{Ge}{f}}\right] + \frac{cGe^2}{f}\left[\frac{b + 2cx}{1 + 2c.\frac{Ge}{f}}\right]^2 \tag{5.2.63}$$

From the above equation, by changing the value of a, b and c, different workpiece profile can be expressed and corresponding tool profile can be determined. In the equation, last two terms are indicating the correction factor due to the formation of the workpiece surface profile and compare to first five terms the value of last terms is negligible thus the final equation of the required tool profile can be expressed as:

$$y_t = a + bx_t + cx_t^2 - \frac{Ge}{f} - \frac{Ge}{f}\left[\frac{(b + 2cx)^2}{1 + 2c.\frac{Ge}{f}}\right] \tag{5.2.64}$$

5.2.9.2 Determination of the tool shape by finite element method

Ideally, tool design can be easily implemented by performing a simple calculation at the outset of the design procedure, but in practice it can easily be implemented due to the complex gap configurations which affect by many process variables. Thus, tool shape is hardly a perfect replica of that of the workpiece. Prediction of the tool shape is a complex inverse boundary problem of Laplace's equation [14–16]. This problem requires adjustment of a free boundary (tool) to satisfy imposed boundary conditions at both electrodes. Difficulties in designing ECM tools stem from a lack of adequate understanding of ECM dissolution phenomena and mathematical complexities. In practice, subsequent adjustment of tool shape is still done empirically. The iterative procedure of tool design is a time consuming process, resulting in costly machine down and long lead-times [17].

Early investigations are limited to the simple methods, e.g., analytical solution, graphical, geometrical, and complex variable techniques, etc. These early attempts advanced knowledge of the field but did not yield the desired accuracy owing to over simplification of the current field in the inter electrode gap. Rapid advancements in computer technology and numerical techniques make it possible to develop more comprehensive solutions. Recent studies have primarily been directed to numerical solutions to the inverse boundary problem instead of simple geometry approximations [18, 19].

Determination of the tool shape for achieving a desired shape of work surface by Finite element method is based on Laplace's equation and other boundary conditions. It is used to determine the distribution of electric potential $\boldsymbol{\varphi}$ in the ECM inter electrode gap as shown in Fig. 5.2.25.

Where, **V** is machining voltage, $\boldsymbol{\delta E_a}$ is anode over potential, $\boldsymbol{\delta E_c}$ is cathode over potential, $\mathbf{V_f}$ is feed rate, **K** is electrolyte conductivity, $\mathbf{K_v}$ is

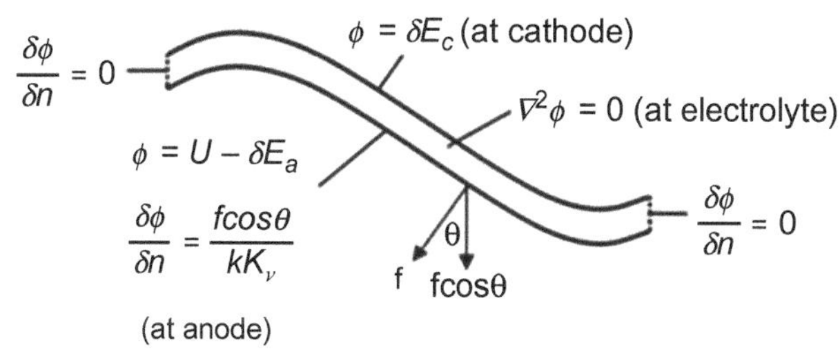

Fig. 5.2.25 Boundary conditions for tool shape design by FEM.

volume electrochemical machinability and **θ is** angle between feed rate and the normal to anode surface.

Based on this model, the problem of tool design is to find a cathode tool boundary which enables Laplace's equation and all boundary conditions to be satisfied for a desired anode workpiece shape. The problem, thus posed, differs from typical boundary value problems which basically deal with the solution of a potential function. The tool design problem seeks for boundary geometry instead of potential distribution, and is known as an inverse boundary value problem or free boundary problem of the Laplace's equation and it indicates how tool shape is affected by changing the boundary conditions. A change of electrolyte conductivity **K** with machining time and flow path caused by hydrogen bubbles and Joule heat generated in ECM variation of electrochemical machinability $\mathbf{K_v}$ with current density, and random change of some parameters such as machining voltage **U** and feed rate $\mathbf{V_f}$ alter boundary conditions and hence affect the potential distribution and consequently gap distribution, as illustrated in Fig. 5.2.26. These individual or interactive effects greatly complicate the tool design procedure.

Difficulties and complexities in modeling of ECM come with the variables such as **K** and $\mathbf{K_v}$ and lack of effective mathematical approach in addressing inverse boundary problems. Many attempts have been made to solve it by analytical methods and complex variable functions. Finite element method (FEM) and boundary element method (BEM) iteratively adjust the electrode boundary until no further appreciable change of boundary or optimal value of a target function is found. Of these methods, FEM and BEM show more flexibility in the treatment of the boundary.

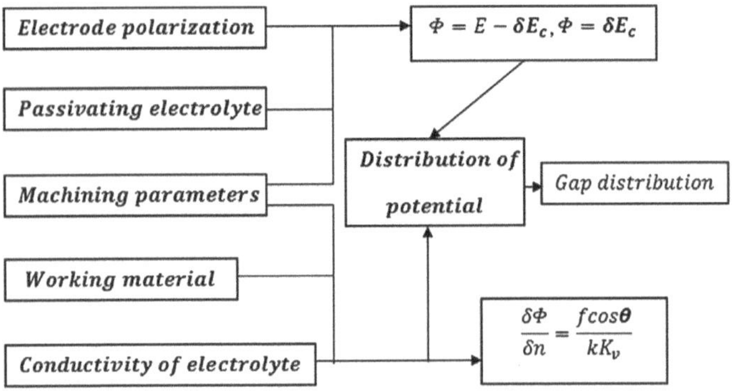

Fig. 5.2.26 Influence of ECM variables on gap distribution.

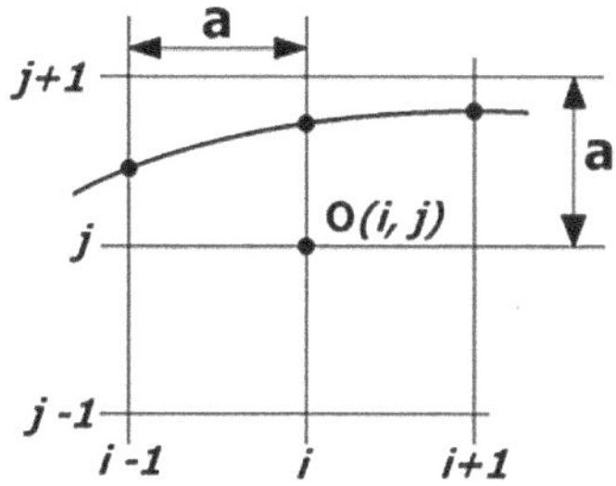

Fig. 5.2.27 Meshing for FEA analysis.

Determination of the tool shape for achieving a desired shape of work surface can be possible accurately by finite element method and widely used in practice. This method also includes the effects of electrolyte flow which render almost intractable general problem of electrochemical shaping. For solving the problem by this method requires specification of the coordinates of both cathode tool and the proposed anode workpiece shapes [20]. These coordinates can be identified as an equi-spaced rectangular mesh which contains a set of grid points of general coordinates (i, j).

Fig. 5.2.27 shows a square mesh of side a. Then, the finite difference equation corresponding to the Laplace's equation is

$$\frac{\delta^2 \Phi}{\delta x^2} + \frac{\delta^2 \Phi}{\delta x^2} = 0 \tag{5.2.65}$$

Considering, the general points $P(i,j)$ at the mesh side a,

$$\frac{\delta \Phi}{\delta x} = \frac{\Phi(i+1,j) - \Phi(i,j)}{a} \tag{5.2.66}$$

So,

$$\frac{\delta^2 \Phi}{\delta x^2} = \frac{\Phi(i+1,j) \ldots 2\Phi(i,j) + \Phi(i-1,j)}{a^2} \tag{5.2.67}$$

And

$$\frac{\delta^2 \Phi}{\delta y^2} = \frac{\Phi(i,j+1) \ldots 2\Phi(i,j) + \Phi(i,j-1)}{a^2} \tag{5.2.68}$$

From the Fig. 5.2.27, at any point (i,j), the potential is equal to the average of the potentials at the neighboring points. Thus, the Eq. (5.2.68) becomes

$$\Phi(i+1,j) - 4\Phi(i,j) + \Phi(i-1,j) + \Phi(i,j+1) + \Phi(i,j-1) = 0 \tag{5.2.69}$$

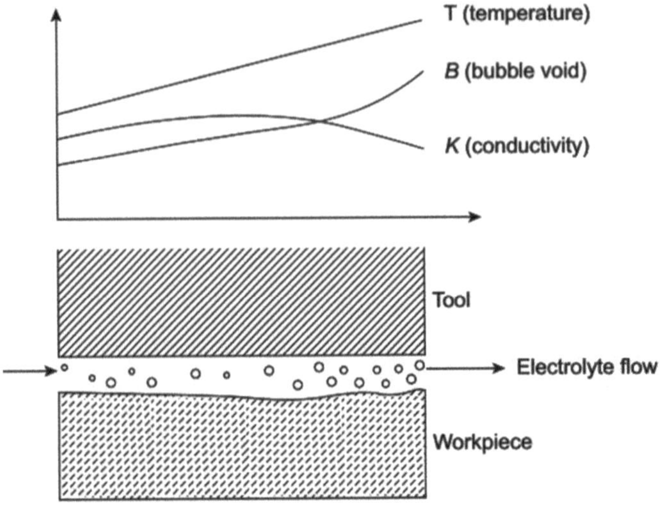

Fig. 5.2.28 Variation of electrolyte conductivity along the flow path.

The method of solution needs to adjust repeatedly the potential at each mesh point within the boundary. At each run, the deviation of the potential from its corrected value is found in the form of residuals, R. Thus, Eq. (5.2.69) can be written as

$$\Phi(i+1,j) - 4\Phi(i,j) + \Phi(i-1,j) + \Phi(i,j+1) + \Phi(i,j-1) = R \quad (5.2.70)$$

The value of R is zero only if the potential at any point (i,j) is the mean of the four neighboring points. The solution is achieved when the residual becomes lesser than the set value.

Methods utilized for tool design only allow determination of a first approximation to the final shape. The subsequent correction of the tool shape considering various factors is needed. A "correction factor" concept has been adopted to modify the tool shape generated by various tool design techniques in ECM.

Change in electrolyte conductivity due to generation of hydrogen gas bubbles, vapor bubbles as well as joule heating along the flow path with machining time causes variation of electrochemical machinability with current density. Fig. 5.2.28 shows variation of electrolyte conductivity along the flow path [21].

Random change of some major parameters, e.g., voltage, tool feed rate during actual machining may alters the initial conditions considered for tool design. Hence, it affects the potential distribution across the machining zone which results in consequent changes in gap distribution. The tool design procedures become more complicated due to the above mentioned individual as well as interactive effects. An effective tool design solution should give tool dimensions along with suitable electrolyte flow path as well as appropriate insulation to overcome stray machining in the undesired regions for reduction of overcut. These all factors should be taken into account for determination of the final tool shape through modification of the tool geometry as achieved by various tool design techniques.

5.2.10 Different variants of ECM

ECM is most flexible and versatile method of machining due to its basic mechanism of material removal. Anodic dissolution can be controlled precisely to generate different shape, size and finish of the products. It involves interaction of large number of parameters; hence it can be controlled by maneuvering those parameters to achieve different applications. ECM can be effectively utilized for various purposes, e.g., shaping, turning, milling, sawing, drilling, deburring, smoothing, finishing, etc. It can also be combined with other advanced machining processes to develop newer hybrid machining processes for achieving more advantages out of the constituent processes. Anodic dissolution can be effectively utilized also for micro machining applications. Fig. 5.2.29 represents basic mechanisms as well as versatile applications of ECM.

5.2.10.1 Electrochemical drilling

When ECM is used for fabricating holes only then the technique is known as Electrochemical Drilling. The electrolyte is sent through the hollow tool to the machining zone as shown in Fig. 5.2.30. The Electrochemical Drilling is very faster as compare to conventional drilling. In EC drilling tool feed rate is in the range of 2.5 to 6 mm/min which is comparatively very high with conventional drilling. The drilling accuracy, i.e., overcut, circularity, and straightness of the drilled hole depends on the various parameters of ECM. Higher electrolyte pressure improves the hole accuracy. The accuracy also can be improved by increasing tool feed rate, however, very high tool feed rate results in sparking. Hence, during EC drilling critical feed rate should be selected and controlled under a particular parametric combination. Selection of electrolyte also plays an important role. Recommended electrolyte for EC drilling is $NaNO_3$ as well as acidic electrolyte. This

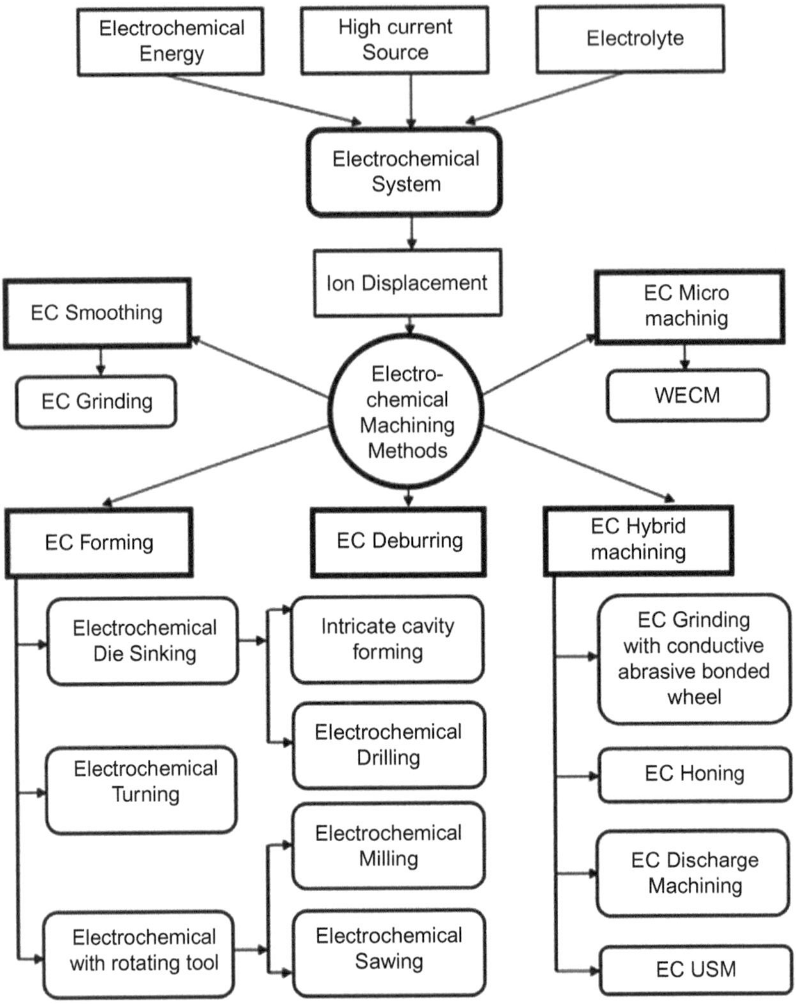

Fig. 5.2.29 Different types of ECM processes.

process is normally utilized for generation of multiple holes simultaneously thus improves productivity. Electrochemical Drilling requires precise tool feeding system controlled by servo mechanism to maintain the constant gap tool and the workpiece. As shown in Fig. 5.2.30, the circumference of the tool is coated with proper insulation to restrict stray machining along the wall of the hole. There is no formation of burrs in electrochemical drilling which produces straight hole length. However, it generates a bell mouth entrance and straight or tapered exit depending upon the dwell time, i.e., stay of the tool for a certain time at hole exit.

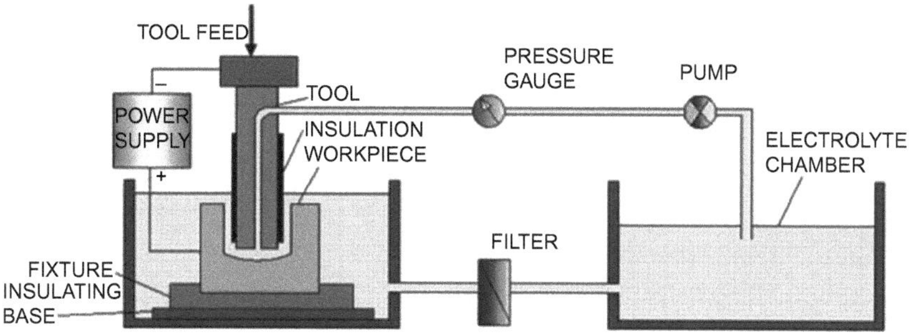

Fig. 5.2.30 Schematic diagram of EC Drilling.

However, holes with smaller diameters are difficult to fabricate by conventional ECM drilling techniques due to size restriction, i.e., diameter of the hollow metal tube used as cathode tool. For the drilling of smaller diameter holes, special types of electrochemical drilling methods are utilized which were first developed by General Electric Company, United States in the year 1960 for drilling of thousands of small cooling holes at an angle on super alloy used for aerospace engine. General Electric Company also invented techniques of ECM drilling which can generates high aspect ratio micro holes.

Microholes with precision on HSTR alloys can be fabricated by electrochemical drilling. Various microholes drilling methods are (i) Jet Electrolytic Drilling (JED), (ii) Capillary Drilling, (iii) Shaped Tube Electrolyte Machining (STEM).

(i) Jet Electrolytic Drilling (JED)

The tool size limitation of ECM drilling overcomes by utilizing a charged high velocity stream of electrolyte instead of hollow metal tube as shown in Fig. 5.2.31. This charged electrolyte stream of very small diameter is

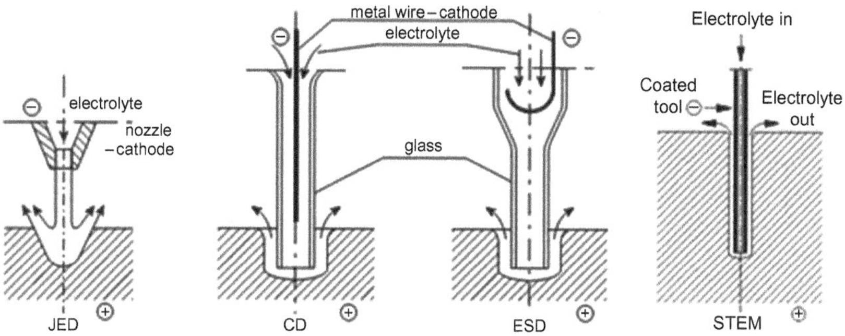

Fig. 5.2.31 Different micro hole drilling methods by ECM.

generated utilizing specially designed nozzle. This special type of drilling prefers sodium chloride, sodium nitrite, and dilutes acid depending upon the materials to be drilled. The performance of these types of drilling methods is also governed by heat generated in the electrolyte during drilling. Control of the electrochemical drilling process is difficult, which require suitable control strategy for correlating machining rate and different other parametric combinations considering smaller tool diameter as well as higher aspect ratio [22].

In this process, a jet of electrolyte with high pressure is directed to the workpiece surface to achieve electrochemical drilling. The nozzle through which the electrolyte jet emerges acts as a cathode and the workpiece is made anode during drilling. The gap between tool and workpiece is much larger compared to that in ECM. Electrolytes of high conductivity are used for achieving high current density. A high working DC voltage is applied between the jet and the workpiece which may be 10 times higher than ECM drilling. This technique is used for generating a small diameter hole of <0.8 mm. It is also used for generating narrow grooves and slots where the jet is held fixed and the sample is precisely moved by computer controlled X, Y, and Z stages [23, 24]. This technique is utilized also for generating microslots, maskless patterns in thin metal sheets for microelectronic applications. JED is used for machining microholes of low aspect ratio with maximum 5:1.

(ii) Capillary Drilling (CD)

Stream of electrolyte is passed through a fine glass capillary to the cutting zone as shown in Fig. 5.2.31. Acid electrolyte is used in capillary drilling because of better flushing due to dissolve of sludges. A dilute solution of nitric acid as electrolyte is preferred for the CD technique. The glass capillary contains very fine wire inserts such as platinum, which acts as a cathode. The diameter of the metal wire will be very small to suit the fine glass capillary tube bore without obstructing the electrolyte flow down to the drilling zone. The wire is positioned 2–4 mm above the tip of the capillary tube. The hole generated by the CD technique is in the order of <500 μm diameter. In this technique, the path of the electrolyte is shorter than in Jet Electrolytic Drilling, which reduces the electrical resistance across the gap and permits the application of a smaller machining voltage, i.e., 100–200 V to flow the current for metal removal. CD is widely used to produce smaller holes in turbine blades with the aspect ratio ranging around 40–80 [25]. A large number of microholes can be effectively drilled at a time in hard materials with multiple capillary tube drilling assembly. This technique can be effectively utilized for micro drilling brittle and HSTR alloys on curved surfaces at an angle.

When deep and accurate holes are to be generated by this technique, the capillary nozzle should be fed continuously into the workpiece. Hence, it requires nozzle feed system as well as gap sensing device to monitor and maintain the constant gap throughout the drilling cycle. This technique is used for generating a small diameter hole of <0.5 mm. CD is used for machining microholes of high aspect ratio with maximum 10:1. For much higher aspect ratio holes, CD technique can be utilized effectively with the aid of much smaller diameter capillary nozzles at the front end of the capillary tube. This modified CD technique is known as electrostream drilling.

(iii) Electrostream Drilling (ESD)

This method is similar to CD, which also uses a nozzle shaped glass capillary tube to direct an electrolyte jet as shown in Fig. 5.2.31. Capillary nozzle is made of drawn glass from a comparatively larger diameter glass tube. Platinum wires are inserted into the glass tube and placed in the larger diameter portion well above the fine capillary. The fine capillary on which the glass tube is fitted forms the cathode. The length of the reduced diameter of the capillary tube should be more than the depth of the hole to be machined. Electrolyte as dilute sulfuric acid is preferred for drilling stainless steel, cobalt alloys and hydrochloric acid is recommended for aluminum and titanium alloy. A much higher applied voltage in the range of 600 V which is 20 to 30 times compare to normal ECM drilling. Higher drilling voltage is needed to achieve a sufficient current flow at the machining zone due to a longer and thinner electrolyte flow path. The holes drilled by this technique are much smaller than in CD. With this technique small diameter hole of <0.2 mm. ESD is used for machining microholes of higher aspect ratio of maximum 40:1.

(iv) Shaped Tube Electrolyte Machining (STEM)

STEM is highly dedicated electrochemical drilling technique used for drilling of small holes with comparatively much higher aspect ratio upto 300:1. In this technique, a hollow shaped metal tube is used as a cathode tool through which electrolyte reaches the machining zone as shown in Fig. 31. The external surface of shaped tube is coated by an insulating organic coating except the front tip to reduced stray machining as well as corrosive action of the electrolyte. In STEM technique, only acid electrolyte is used.

Low voltage DC power supply ranging from 5 to 15 V is used in STEM due to the advantage of using metal tube as cathode tool. Generally, temperature controlled diluted sulfuric acid is used as electrolyte. STEM requires controlled feed servo system for maintaining a constant gap between workpiece and metal shaped tube tool. Titanium is used for fabricating shaped tube tool of much smaller diameter to overcome the corrosive action of acidic electrolyte. The tip of the shaped tube is dressed at an angle of 10 degrees to achieve better performance. Multiple electrodes of different shape and size can be used to fabricate several holes simultaneously. Due to use of shaped metal tube in STEM, there is a chance of electroplating on the tool which can be eliminated reversing the polarity in between workpiece and tool periodically. During drilling, overcut can be minimized by controlling the machining voltage and electrolyte pressure. Shaped tube electrodes are fed inside the job through guide arrangements to control depth penetration. STEM is widely used for drilling superalloys, e.g., fabrication of cooling holes in turbine blades.

As discussed above, various types of ECM drilling are in used as per the design requirements. Different ECM drilling techniques have utilized anodic dissolutions for removal of material, however they differs in methods implemented. The Table 5.2.2 exhibits some important characteristics of various ECM drilling techniques.

Table 5.2.2 Characteristics of various ECM drilling techniques

	STEM	CD	ESD	JED
Tool	Titanium tube	Glass capillary tube with titanium wire	Glass capillary tube with nozzle shaped end with titanium wire	Metal nozzle
Voltage (V)	5–15	100–200	200–500	300–600
Electrolyte	HNO_3, H_2SO_4	HCl, HNO_3, H_2SO_4	HCl, HNO_3, H_2SO_4	HNO_3, H_2SO_4
Electrolyte pressure (bar)	3–10	3–20	3–10	10–60
Feed (m/min)	1–3.5	1–4	1–3.5	0
Diameter of the hole (mm)	0.5–5	0.2–2	0.125–1	0.3–1
Max. aspect ratio, i.e. (L/D)	200–300	80–100	30–50	2–3
Precision (mm)	±0.05	±0.03	±0.03	±0.05

5.2.10.2 Electrochemical deburring

Burrs are thin ridges, usually triangular in shape that produced along the edge of a workpiece from various manufacturing operations, e.g., machining, shearing of sheet materials, trimming, forging, casting, etc. Removal of burrs in the drilling process occupies >40% of total machining time and reduces production efficiency, and increases cost [26]. Successful deburring process into manufacturing system with high efficiency and full automation is an extremely difficult problem. Generally, manual methods are often employed. But various internal burrs, which are complicated in shape, are difficult to be removed manually due to inaccessibility.

Removal of metal burrs produced in the finished items during conventional tool based machining can be achieved effectively utilizing same principle of metal removal mechanism of ECM and known as Electrochemical Deburring. However, in Electrochemical Deburring, electrolyte pressure and flow as well as current are relatively low compare to ECM. In Electrochemical Deburring, the cathode tool is generally kept stationary during operation. Fig. 5.2.32 shows schematic of Electrochemical Deburring system where, tool is totally insulated except the portion which is adjacent to the burrs. Here, electrolyte is supplied through the gap between burrs and the stationary tool. The position of the deburring tool and electrolyte flow path depends on the geometry of the burrs, e.g., height, width and

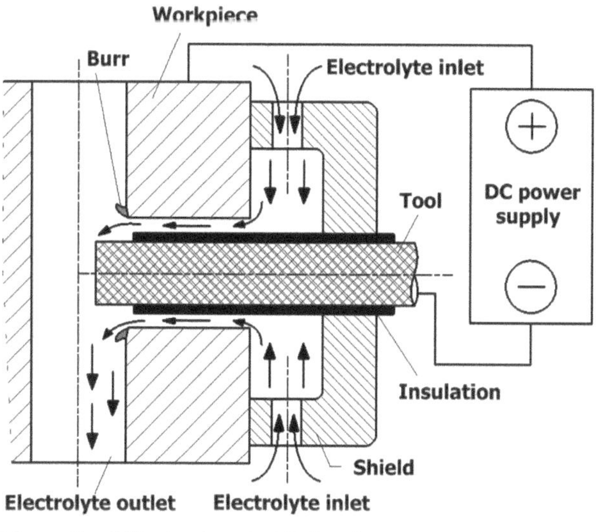

Fig. 5.2.32 Schematic of Electrochemical Deburring.

shape as well as location. The equipment is simple in construction due to absence of tool and workpiece movement; however electrolyte is to be supplied into the machining zone by pump.

In Electrochemical Deburring, the tool is kept in such a way that preferential anodic dissolution takes place on the burrs because the increment of localized current density. Here, special attention is taken to position the tool to concentrate the metal dissolution action at the root of the burrs. Hence, orientation of the burrs is one of the important factors during design of the electrochemical deburring tools. Design of electrolyte flow path is another aspect due to effective flushing of the free burrs after removal from the gap between tool and the workpiece surface other wise it may cause short circuit during electrochemical deburring operation [27]. Important parameters under consideration during Electrochemical Deburring are low applied voltage (5–20 V), low machining current (100–1000 A), low electrolyte pressure (60–300 kPa) at low flow rate of 3 to 10 L/min. Machining time plays an important role which can be calculated prior to deburring considering the characterization of the burrs, with the aid of mathematical formulation as discussed previously in dynamics of ECM. Depending upon the parameter settings during the deburring operation, removal of burrs is faster which takes around 5 to 30 s with a very good surface finish in the order of 0.08 to 1.2 μm.

5.2.10.3 Electrochemical milling

In the last few decades ECM has taken a keen industrial attention due to its various advantages over other non-traditional machining processes. However, design and fabrication of complex shaped tool takes lot of time and efforts which reflects to the cost of the final products. Larger projected area of the tool leads to high power consumption and at that high magnitude of current, pulsed DC is costly and sometimes not possible at all. Constant DC machining also leads to inaccuracy of the machined profile due to non-efficient sludge removal.

To overcome the above mentioned issues in ECM, electrochemical milling (EC milling) is introduced [28]. The principle of electrochemical milling process is alike to conventional ECM process and the strategy of machining is same as the conventional end milling process. Here, using a simple geometric tool, comparatively larger area can be milled by controlling the tool as well as the job movement thus; no extra effort is required to design a complex 3D shaped tool like conventional sinking ECM. In EC milling, same amount of current density can be achieved with relatively lesser amount of current thus utilization of pulsed DC power supply is much easier compare to die sinking ECM. This leads to high machining accuracy, low power consumption and reduction of cost of the final products. With the incorporation of tool rotation as well as flushing through tool, removal of sludge can be enhanced to a large extent which also improves the accuracy.

A simple shaped cylindrical tool tip can be utilized for realization of 3D features in the workpiece by controlling the scanning movement of the tool electrode along a predesigned tool path in an electrolytic cell as shown in Fig. 5.2.33. Generation of correct tool paths with the help of CAD and CAM considering the phenomena of stray current and microsparks in EC Milling ensures the accuracy of 3D shape generation. A simple shaped tool moves along the desired path and machining will be performed utilizing the bottom tip of the tool with a small downward tool feed at the end of each cycle [29, 30]. Machining will continue till the required shape and depth of the 3D feature is achieved. Complex 3D shape with high aspect ratio can be generated by this scanning type of strategy. Here, the machining gap is maintained within a smallest suitable range.

The machining efficiency in EC milling can be improved by dividing the machining into rough and finish operation. The major portion of the job material is removed by rough machining utilizing a large voltage and high current density. The final design requirements can be achieved by finishing operation utilizing pulsed DC power supply with lower voltage and short

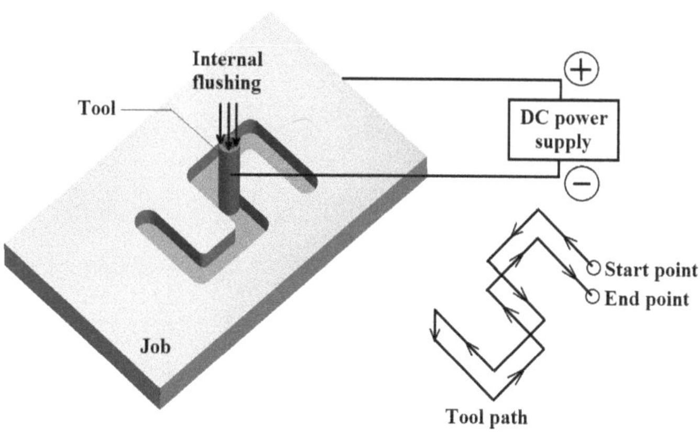

Fig. 5.2.33 Schematic of Electrochemical Milling with predefined tool path.

pulse width, lower duty ratio, and maintaining higher scanning movement of tool with a very smaller IEG. Based upon the tool movement strategy, EC milling can be classified considering milling layer thickness and feed rate during scanning layer-by-layer method as shown in Fig. 5.2.34 where point "S" denotes the start position of machining whereas instantaneous position of the tool indicates to the position "P."

In this technique, the to and fro movement of the tool is given while machining continues on a layer-by-layer basis between the front face of the tool and anodic workpiece as shown in Fig. 5.2.34. Thus, the simple shaped tool moves faster and repeatedly over the same area of the workpiece with a downward feed before each scan movement during generation of the required features. This technique has some limitations such as higher overcut, higher machining time, and deviations at both the ends. Moreover, due to the high scan speed during the machining operation, the chances of tool damage are always there [31].

In EC milling, apart from electrical parameters, two most significant process parameters are tool feed rate and milling layer depth. In EC milling, tool feed rate is defined by the velocity of the tool at which it moves over the workpiece and generally expressed in mm/min. For movements of the tool per each step is responsible for stepper motor which is actually imparting the motion to the tool. Milling layer depth is the depth which is given to the tool after completing each pass to achieve a desired depth, expressed in mm.

For making a simple groove with a cylindrical tool, curvature is formed at the junction area of two walls, i.e., side and bottom walls which is called radius of curvature. The sidewalls of the machined profile also become inclined due to stray current effects and side angle is the inclination of the side walls with vertical wall. Due to the unavoidable stray current effect, the length of the groove is obtained larger than the pre-set length.

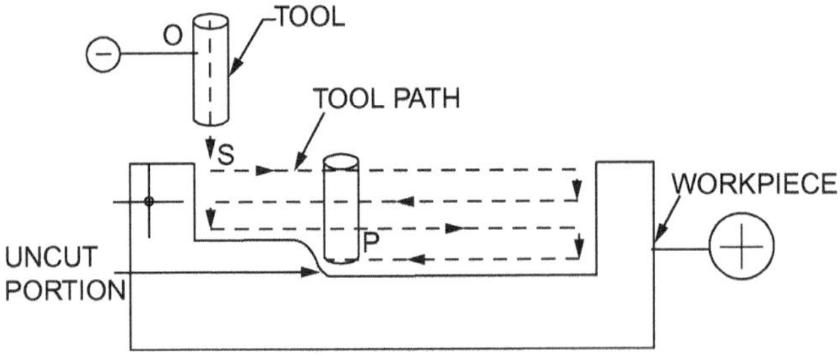

Fig. 5.2.34 Electrochemical milling using layer-by-layer method.

The formation of these characteristics like radius of curvature, side angle and obtained length are mostly dependent on those two most important process parameters of EC milling, i.e., milling layer depth and tool feed rate.

Let, α_c is the side angle along cross sectional direction then it can be written as:

$$\alpha_c = Sin^{-1}\left(\frac{R_c - D}{R_c}\right) \tag{5.2.71}$$

where, R_c is the radius of curvature along cross sectional direction and D is the average depth of the groove [32].

From Faraday's law, theoretical volumetric removal is

$$V_{th} = \frac{1}{\rho F} I.E_x.T \tag{5.2.72}$$

where, η is current efficiency, F is the Faraday's constant, i.e., 96,500 C/mol, I is the current, E_x is the electrochemical equivalence, T is the total machining time.

Let, t is time taken for one milling layer machining, n is the number of milling layer, l_t is the milling layer thickness, F_r is the tool feed rate along X-Y plane, L_g is the given length and d_g is the given depth during machining. So, total machining time during EC milling operation is $T = t_0 + t_1 + t_2 + \cdots + t_n$ where, t_0 is time taken to complete tool path at initial IEG and t_n is the time taken at nth milling layer to complete the tool path. When tool feed rate on X-Y plane is kept constant, time taken for machining each layer will be same. That implies, $t_0 = t_1 = t_2 = \ldots \ldots = t_n$, hence, $T = (n+1)t$.

Again, $t = \frac{L_g}{F_r}$ and $n = \frac{d_g}{l_t}$

So, theoretical volumetric removal can be written as:

$$V_{th} = \frac{1}{\rho F}.I.E_x.\left(\frac{d_g}{l_t} + 1\right)\left(\frac{L_g}{F_r}\right) \tag{5.2.73}$$

In order to ensure a good shape precision, the milling layer thickness it cannot be greater than the cylindrical electrode diameter d which is used in the machining process. In order to generate accurate 3D shaped features, the IEG must be as small as possible. However, some factors such as rigidity of the machine, electrolyte boiling and process inability, tool positioning errors limit the minimum gap size. The control of the side gap are especially important because they cannot be corrected by controlling the tool feed movement [32]. The side gap can be minimized, and better accuracy can be achieved by utilizing insulating side walls of the tools, passivating electrolyte, high frequency pulsed power, and lowering electrolyte concentration.

EC milling setup is necessary to alter parameters such as voltage, feed rate, rotational speed and electrolyte flow pressure and velocity during EC milling operation. Incorporation of internal flushing, current and rotation to the tool simultaneously make the design of electrolyte flow system more stringent and complex thus fabrication of this unit is one of the toughest challenges in the EC milling system. EC milling set up consists of several sub units, e.g., mechanical machine unit, power supply unit, electrolyte supply system and control unit. Electrolyte supply system and control unit are most important among all the other sub units. EC milling completely depends on the movement of the tool and the accuracy of tool motion which directly influences the accuracy of machined profile. This precise movement can be achieved with the help of X-Y-Z CNC controller. A rotary union is necessary in EC milling system to facilitate not only to supply the electrolyte through the tool but also to supply current at the rotating condition of the tool. DC power supply also plays an important role in EC milling system.

5.2.10.3.1 Advantages of EC milling

As discussed earlier, EC milling is advantageous process comparing other nonconventional as well as other electrochemical processes. For its various advantages, it is a versatile process. The major advantages of EC milling processes are as follows:

- **(i)** EC milling process does not require any complex tool for generating complex shape features. As stated earlier, with the help of a simple geometrical tool like square or cylindrical tool any surface profile can be generated.
- **(ii)** This process can machine material irrespective of their mechanical property so, any kind of conductive hard material can be machined by EC milling process.
- **(iii)** EC milling can machine larger area using smaller current range thus no requirement of special electrical design to control high current. As this process uses a smaller range of current thus, pulse type power supply can be applied, therefore better machining quality can be achieved.
- **(iv)** There is no requirement of designing special electrolyte flow system for various shape generation.

5.2.10.3.2 Challenges of EC milling

Though EC milling process is an easy way to generate complex profile, there various difficulties to perform the process. Find out the solution of difficulties and implementation into the process is necessary. The tool electrode is required to move on a complex path, it is difficult to control the motion of the tool along that predefined path. Electrolyte flow system is one of the most remarkable challenges of EC milling. It can be overcome by two ways. One is to supply the electrolyte through the tool and another way is to supply the electrolyte concentrically with the tool. In EC milling, tool is continuously moved along the required path, thus electrolyte jet should follow the tool to reach the electrolyte between tool and workpiece. Thus electrolyte flow system should be developed in such a way that it can reach in the intricate machining zone properly. In EC milling, relative movement in between tool and the workpiece is necessary, so it is difficult to achieve stable machining.

The global requirement at various field is rising day by day, thus requirement of manufacturing process is also increasing. With the variety and complexity of product manufacturing processes also to be developed to meet the challenges. In this scenario EC milling has various significant applications starting from aeronautical industry to medical industry.

5.2.10.4 Wire electrochemical machining (WECM)

With the advancement of technology and the requirement of high aspect ratio structures, a new kind of ECM technique has been introduced. This new technique of ECM is known as wire electrochemical machining (WECM). This technique has added advantages and eradicates the weakness of machining high aspect ratio features through combining with the idea of wire cutting. By introducing wire electrode of suitable diameter and by applying electrolyte of suitable concentration between the tool and workpiece, this process removes material by electrochemical reactions that occur during the machining. The wire electrode is taken as cathode and workpiece is taken as anode where both are separated by narrow IEG.

Machining is performed in WECM by wire as a tool, hence it can only be evaluated with ECM where the dimension of fabricated feature is within few mm. During ECM, within this smaller dimensional range, as the aspect ratio increases, the chances of arresting bubbles on the surface of the wire tool increases. It leads to increase in vibration which results in distortion of wire tool and deterioration of quality and consistency of machining. In WECM, wire is tensed and regulated accordingly which decreases the chance of wire distortion. Likewise, flowing electrolyte is used for flushing of machining products from the narrow IEG which may have detrimental effect on machining stability. However, use of stagnant electrolyte makes the removal of sludge and gas bubbles impossible after a certain depth of machining. Hence, with the increment of aspect ratio in WECM proper tension of wire and suitable flushing are compulsory. Moreover, in WECM the feed forward direction lies perpendicular to the direction of gravity for which simulation of machining gap is easier and improved flushing of machining products will be simple. In addition, WECM offers higher material removal rate with the presence of devoted flushing arrangements, because of machining is carried out along the length of wire using a single feed.

Basic features of WECM are exhibited in Fig. 5.2.35. With initiation of machining, tool wire of diameter in micron ranges is fed forward with the help of servo controlled feed unit in predefined path within the workpiece by maintaining proper IEG and metal starts dissolving anodically into metallic ions. The machined position on the workpiece varies with the position of wire tool and dissolution continues through the time which results in generation of desired shape. During the machining, the wire feed rate is kept balance with the rate of dissolution of workpiece in order to avoid sparking which leads to machining stability. In WECM, DC pulsed power is applied across the IEG to confine the anodic dissolution within the range of micron to sub-micron to improve the machining accuracy which leads to reduction of IEG below 10 μm.

Different process parameters of WECM such as feed rate, pulsed voltage, and pulse on time influence the size of the machining gap that affects the machining accuracy. Machining gap decreases with wire feed rate. Machining gap becomes larger as the pulse voltage and pulse on time increases. As

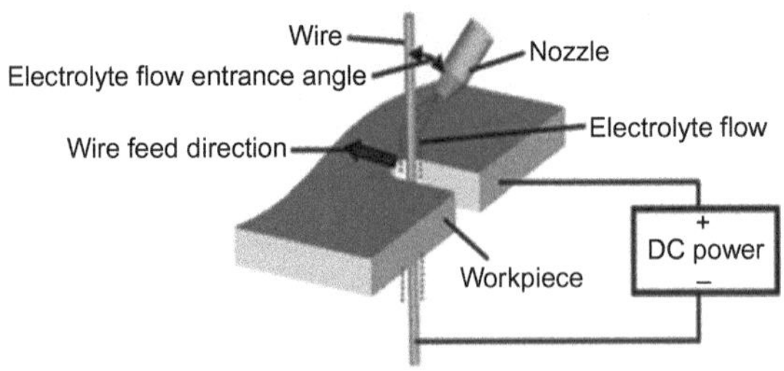

Fig. 5.2.35 Basic scheme of WECM [33].

this machining process does not make direct contact with workpiece, it can be applied for any conducting material regardless of their mechanical and chemical properties, which otherwise pose many challenges for conventional machining techniques. Moreover, this newer technique utilizes very thin wire by which smaller features with precise dimensions can be machined easily. The materials for smaller diameter wires are commercial platinum, tungsten, and copper. Among these materials, tungsten has the highest strength, which is about two times higher than that of platinum. Researchers have fabricated in situ tool electrode of 5 μm diameter from a 30 μm tungsten wire by the same anodic dissolution process [34].

Still, WECM is a transport-limited electrochemical dissolving process and as the tiny IEG of several micron or submicron dimensions become deeper and narrower with machining time. Removal of electrolysis products, namely the hydroxides and hydrogen gas, and renewal of fresh electrolytes become more difficult due to the narrow machining gap. Bubble characteristics also influence the machining quality by influencing the distribution of current density in IEG. It is a well known fact that, the potential distribution across the machining gap follows Laplace's equation when static electric field theory is applied. The electrical flux lines are distorted with bubble generation which influences the distribution of current density in IEG. Hence, the rate of dissolution is lower where bubbles are present in the machining gap which leads to inhomogeneous machining. Moreover, at the portions of the work surface, the current density distribution pattern becomes much more disturbed due to the accumulation of more bubbles thus, removal of gas bubble from the machining zone is essential. Also, the conductivity of electrolyte is adversely affected by the amount of bubbles and sludge formed in the IEG which can be expressed as:

$$k = 2k_1(1-\beta)/(2+\beta) \quad (5.2.74)$$

where, k_1 is the initial conductivity of electrolyte and β is the electrolysis product fraction [35].

Edge radius of fabricated micro features increases due to increase in current density and with the presence of gas bubbles at the vicinity of edges. Hence, gas bubbles must be removed from the machining gap rapidly in order to achieve smaller edge radius [36].

Moreover, the smaller diameter wire is deformed by the bubbles that are generated in the machining area which causes physical contact of the wire tool and the workpiece. Physical contact can be prevented by increasing the concentration of electrolyte, pulse on time, and pulse frequency of supply voltage. But this can also increase the side gap. The bubble formation depends on the pulse period. In a short pulse period, large bubbles are generated and deform the wire. In a long pulse period, small bubbles are generated and leave the machining area easily. Therefore, a long pulse period is suitable to prevent wire deformation. However, a too long pulse period causes a low machining rate and physical contact of the wire and the workpiece.

Joule heat is generated in the passage of machining current and heats up the electrolyte in the IEG [37]. During WECM operation, the generated bubbles move upward in the axial direction of wire and the rest of the machining products moves downwards [38]. Hence, bubbles at upper portion of the machining zone reduce the electrolyte conductivity. Joule heat from the middle of the machining zone disperses into the bottom due to diffusion. Hence, at the middle of the machining zone electrolyte conductivity is always higher than two end which results in inhomogeneous mass transfer leading to deterioration of process stability results in non-uniform micro structures.

Various types of electrolytes such as acidic, alkali, salt-based and mixed are used during WECM depending upon the type of work material. Acidic electrolytes like H_2SO_4 and HCl is extensively used for machining high aspect ratio features on stainless steel, nickel and cobalt based alloys, etc. [39, 40] as because acid electrolysis don't produce any insoluble reaction products during machining [41]. Acidic electrolyte also generates smoother surface. However, better surface quality and localization has been achieved by using H_2SO_4. HCl results in high material removal and non-uniform surface. The applications of salt based electrolytes like NaCl and $NaNO_3$ are limited for machining stainless steel substrate with tungsten, copper and molybdenum wire of diameter not $<20\,\mu m$ utilizing better flushing strategies like axial electrolyte flow.

Thickness of the workpiece greatly influences machining stability. The machining stability worsens as the workpiece thickness increases. To overcome this problem, means other than the application of pulsed power supply are highly required. Therefore, researchers have introduced different means to enhance the mass transport phenomenon during the machining that involves traveling of wire electrode in one direction, application of vibration of cathode wire, and renewal of electrolyte in the IEG by implementing axial flow [42]. The machining stability can be achieved with applying vibration frequency in the order of 10 Hz and small amplitude of vibration in the order of $8\,\mu m$ by enhancing the mass transport in machining gap. The machining stability increases as the electrolyte flow rate increases for the same wire feed rate. However due to radial swing of wire, high flow rate results in deterioration of machining quality. Feed rate can be increased by increasing the flow rate up to a certain limit. Hence, flow rate is effective only up to a critical value beyond which the flow rate in the IEG does not increase further and remains unchanged where maximum of electrolyte flows through the outside of machining area. It has been reported that optimal electrolyte flow rate and wire feed rate are 0.75 m/s and 0.5 mm/s, respectively, for most stable WECM operation. Apart from these, feed rate is also influenced by several process parameters, e.g., voltage, electrolyte pressure, frequency and machining conditions, nozzle diameter, etc. Higher feed rates can be achieved by employing higher voltages but this also causes increment of concentration of sludge and byproducts. Nozzle diameter also affects the feed rate where the critical electrolyte pressure at which maximum feed is achievable, increases with increasing feed rate and decreasing nozzle diameter. In WECM, feed rate is inversely proportional to the square root of workpiece thickness, however the experimental values falls shortly behind that [43].

In WECM, majority of research works focus to solve the experimental problems. Hence, mathematical modeling was formulated by the researchers to finding out the impact of process parameters especially frequency and duty ratio on responses to achieve a desired width of the fabricated slit. During modeling, double layer charging time and the resistances offered by double layer have been considered and the width of fabricated slit, W_s has been given by

$$W_S = d_w + 2\sqrt{2d_w \frac{\alpha K_e (Dr - \tau f) V}{f_W} + \left(\frac{\alpha K_e (Dr - \tau f) V}{f_W}\right)^2} \quad (5.2.75)$$

where, d_w is the diameter of the wire, α is the volumetric electrochemical equivalent, K_e is the conductivity of the electrolyte, D_r is duty ratio, τ is the double layer charging time, f is applied frequency, V is input voltage and f_W is feed of wire.

Feed rate of WECM should be maximize for a particular experimental setting for increasing the productivity and also to compete with other process like wire electrical discharge machining (WEDM) process. Hence, by the researchers another attempt was made to develop mathematical model to correlate maximum feed rate with other controllable factors, e.g., applied voltage and duty ratio. Then, the maximum feed rate, f_m can be written as:

$$f_m\lambda = \frac{\pi d_w L e_a K_e (Dr - \tau f) V}{2\rho_a \left[d_w L + \frac{\pi d_w^2}{8} \right]} \tag{5.2.76}$$

where, d_w is the diameter of the wire, L is the total travel length of wire, $e_a = A/(zF)$ is electrochemical equivalent, K_e is the conductivity of the electrolyte, D_r is duty ratio, τ is the double layer charging time, f is applied frequency, V is input voltage, ρ_a is density of anode material and λ is an unknown factor which has been determined by carrying out experimentations with selected duty ratio and voltage values and subsequent curve fitting technique.

In contrast to WEDM, the wire electrode of WECM is not worn out, and so very thinner wire can be used and wire feeding is not necessary. Moreover, this technique has several advantages, such as higher production rate, lower cost, and better surface integrity. No tool wear, better stability, and higher aspect ratio of microstructures are some of the added advantages of this process. On the other hand, tool fabrication has always been a major area where fabrication of different complex shaped micro tools, e.g., tapered, reverse tapered and disc shaped is problematic and sometimes impossible by conventional electrochemical etching process. Hence, research attempts were made to fabricate semi cylindrical tungsten tools by WECM with 20 μm diameter tungsten wire for micro machining applications [44].

With reciprocating traveling motion to the 100 μm diameter molybdenum wire tool, researchers were fabricated micro grooves, micro beam, micro key, key groove on 20 mm stainless steel plate during WECM operations. With the aid of vibrations on both cathode and anode, fabrication of micro structures like micro square helix with 10 μm diameter tungsten wire was fabricated as shown in Fig. 5.2.36 [33].

Also for increasing the productivity and reducing the machining time, an array of microfeatures can be produced simultaneously by introducing multiple wire electrodes [45]. Using multiple wires fabrication of arrays of microfeatures can be fabricated in a single pass which are useful for various micro machining applications as shown in Fig. 5.2.37 [46].

The wire electrochemical turning process by modifying WECM is another development [47]. WECM process can be extended to micro-turning in order to achieve a high degree of accuracy and surface finish in plain turning as well as in form turning of hard-to-machine materials with high slenderness ratio. One of the most promising developments in this area

Fig. 5.2.36 Fabricated micro square helix [33].

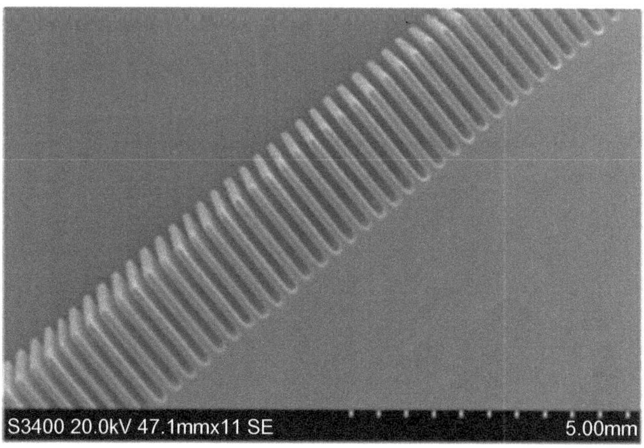

Fig. 5.2.37 Fabricated micro slits with 15 wire electrodes [46].

is abrasive electrochemical multiwire slicing which can be used for successful slicing of solar silicon ingots into thinner wafers [48]. This new hybrid technique has several advantages such as higher production rate, lower cost, and better surface integrity and can produce larger and thinner size silicon wafers.

5.2.10.5 Electrochemical sawing

Principle of ECM can also be used for sawing purpose. It is very difficult to cut very thin and fragile workpiece by the conventional cutting processes which mainly employed huge amount of mechanical force during cutting operation. Due to generation of cutting force, these type of workpiece breaks or deformed. As conventional sawing process is mechanical type which requires rigid fixture for work holding. However, in the case of sawing of very thin and fragile workpiece, it is not proper to hold the job with higher firmness.

In EC sawing, delicate workpiece is parting off utilizing anodic dissolution without producing any adverse effect of cutting on the job. In this process as shown in Fig. 5.2.38, a thin circular tool acts as cathode, rotated over the anode workpiece. Electrolyte is supplied through the nozzle in the cutting zone. Electrolyte is forced to the cutting area by the rotating tool; hence supply of electrolyte at high pressure is not essential in EC sawing. Removal of sludge and gas bubbles from the cutting area is effective due to the centrifugal effect of rotating tool. Due to generation of no cutting force, job holding fixture is of simple nature in EC sawing.

Here, the tool is generally disc shaped with minimum thickness and made of electrically conductive material, e.g., copper, brass and graphite, etc. The gap between rotating tool and the workpiece should be kept very low to achieve controlled metal removal from the workpiece. In EC sawing, low rotational speed is recommended for improvement of cutting efficiency. Rotating tool is feed forward across the job to achieve parting off the delicate

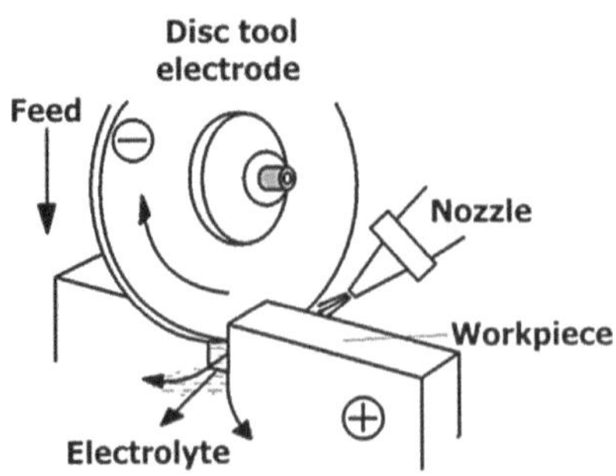

Fig. 5.2.38 Basic scheme of EC Sawing.

job through a single pass operation. Recommended electrolytes are NaCl, $NaNO_3$, etc. which also depends on nature of material to be parting off. Efficiency of cutting operation depends on the combination of parameter settings considering machining voltage, current, tool rotational speed and feed rate, etc.

EC sawing has several advantages over conventional sawing. It can be used for cutting off any conducting materials irrespective of its mechanical properties. It produces no thermal and mechanical damage to the workpiece. No burrs will produce during parting off operation hence; no post cutting finishing operation is needed. However, all parts of the machine should be protected from the corrosive environment which leads to higher maintenance cost.

5.2.10.6 Electrochemical grinding

This special type of grinding uses electrochemical material removal mechanism. This process is alike to mechanical grinding uses for generation of good surface finish. However, here metallic wheel is employed in place of abrasive bonded grinding wheel. In between metal wheel and workpiece, a small gap is maintained which is submerged in electrolyte solution. Unlike conventional grinding, there is no cutting force, heat generation, distortion or stress development on the workpiece because of no contact between tool and the workpiece.

A schematic diagram of EC Grinding is shown in Fig. 5.2.39. Here, metallic wheel acts as cathode and rotates over the anodic workpiece. Both are immersed in electrolytic chamber. This process is specially utilized for

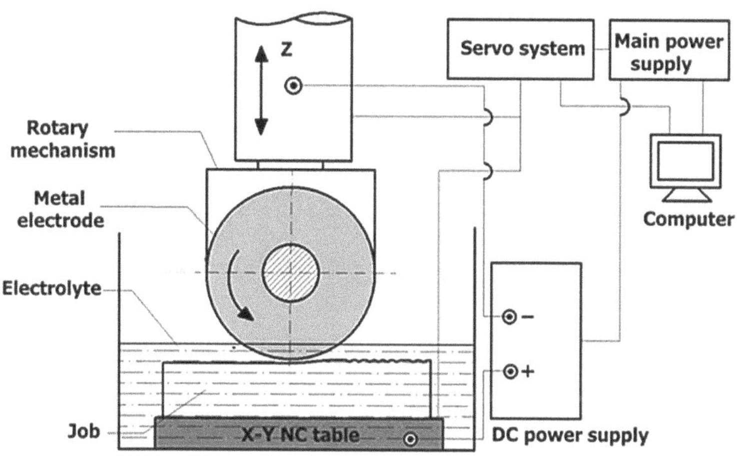

Fig. 5.2.39 Scheme of EC Grinding.

producing flat surface with very good surface finish hence it requires very small gap between job and the metallic wheel. Wheel is fed downwards by controlling Z drive of the setup. Servo feed system is utilized to maintain and control the small gap between metallic wheel and job. Job is given a horizontal motion under the metallic wheel by X-Y stage for which feed rate can be controlled utilizing servo system. Rotational speed of the metallic wheel force circulates the electrolyte solution into the machining gap.

This process is mainly utilized for good flat surface generation. However, abrasive bonded metallic wheel can be employed for faster metal removal and efficient machining. In this case, micro cutting by abrasive as well as electrochemical machining take place. This process is more popular for machining as well as grinding and discussed in details in Chapter 6 under hybrid machining.

5.2.10.7 Electrochemical honing

Electrochemical honing utilizes anodic dissolution and mechanical abrasion simultaneously for finishing of internal surface. This is similar to conventional honing, however here most of the metal is removed by anodic dissolution. Inner surface of bore of electrically conductive material of any hardness can be finished by EC honing.

Fig. 5.2.40 shows scheme of EC honing. Here, tool consists of a hollow stainless steel body on which non conductive spring loaded honing stones

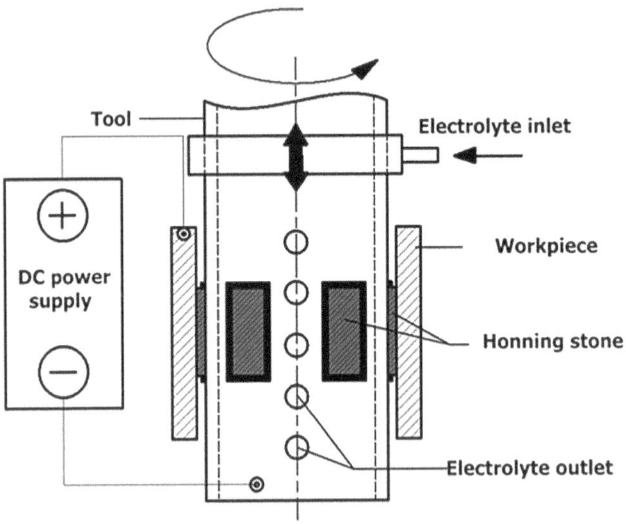

Fig. 5.2.40 Basic scheme of EC Honing.

are placed around the circumference. These stones are similar to conventional honing stone having property of corrosive resistance. Minimum number of stones is at least three and applies equal pressure against the workpiece due to spring action. Generally, length of the stone should be half of the length of the bore to be finished. The length of the main steel body acts as a cathode and should be long enough to cover full length of the bore during reciprocation. Electrolyte is supplied to the sleeve housed on the main steel body and spread out along the circumference of the inner wall of the hole through multiple outlets as shown in Fig. 5.2.40. The main steel body moves to and fro inside the bore at the same time it also rotates to achieve uniform metal removal from all over the inner surface of the bore.

During EC honing, metal oxide is formed due to electrochemical reaction which is removed by abrasives of the honing stones. Due to anodic dissolution, material is removed from the bore, the gap between the tools, i.e., steel body and the wall increases. However, the stones expand to maintain continuous contact with the inner wall of the bore due to spring loaded mechanism. NaCl and $NaNO_3$ solution are recommended as electrolytes during EC honing.

EC honing process is much faster than the conventional honing due to metal removal by anodic dissolution as well as mechanical erosion. Here surface finish is extremely good; no micro scratches remain on the finished bore. Dimensional tolerance is also better compare to the conventional honing. Unlike ECM, it produces compressive residual stress in the work surface due to mechanical action which improves the fatigue strength of the finished component by EC honing.

5.2.10.8 Electrochemical turning

Electrochemical turning is similar to traditional turning operation performed by single point cutting tool on a lathe. Here, material removal is achieved by anodic dissolution. It can be utilized to remove material from external surface, internal surface of rotating electrically conductive cylindrical work surface. It can also be employed for facing operation as performed in conventional turning. However, the shape of the tool is different from the conventional turning tool. Here the front shape of the tool should be same as the final shape to be formed on the job. Hence, it acts as a form tool during machining, and through single pass any contour shapes can be generated on the cylindrical workpiece during EC turning.

Schematic arrangement of EC turning is exhibited in Fig. 5.2.41 which represents an external turning. The setup is similar to conventional lathe

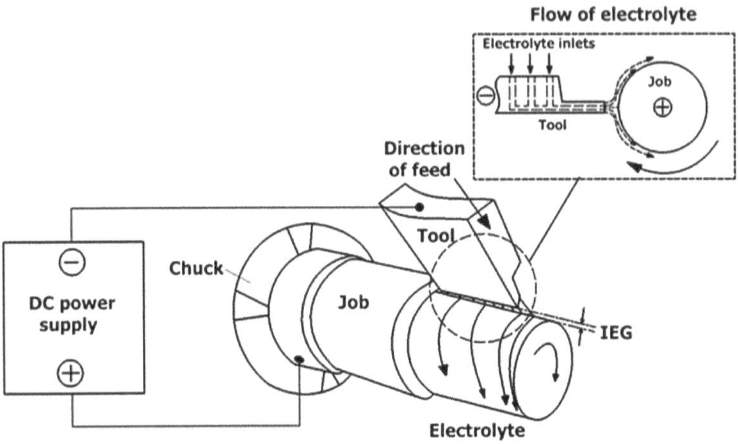

Fig. 5.2.41 Scheme of EC Turning.

except power supply and electrolyte supply system. During EC turning, a small gap should be maintained between tool and workpiece. The electrolyte is continuously supplied to the machining gap through the tool. As shown in the Fig. 5.2.41 electrolyte is delivered through several holes situated at the tip of the tool. Here, depending upon the requirements of shape machining, the tool shape as well as electrolyte flow path design should be arranged accordingly. The tuning zone should be enclosed in a chamber to restrict the electrolyte flushing out into the adjacent areas.

As the front surface of the tool in EC turning is large which, necessitates application of much high DC power supply in the order of 30,000 A. As there is no cutting force developed during turning, the process can be utilized for turning or facing of very smaller diameter workpiece without any deformation. The EC turning process is much economical due to absence of tool wear, hence same profile tool can be utilized for several turning operations.

5.2.11 Environmental impacts of ECM

Manufacturing engineers are trying to design machining processes which will generate minimum waste considering environmental friendly aspects. At present, sustainable manufacturing is expanding its popularity due to its positive impact on the society at large. ECM produces considerable amount of wastes, e.g., solid, liquid and gaseous bi products which are harmful for the human beings as well as ecological system.

ECM is gaining popularity due to its enormous advantages for machining difficult-to-cut materials in modern manufacturing applications. Hence, it is important to analyze the environmental impacts of ECM and attempts should be made to reduce the level of wastes by considering two approaches [21]. One is to minimize the amount of wastes generation during ECM operations, e.g., electrolytic sludge and harmful substances. Second is mainly focusing on reduction of wastes through sludge treatment and proper disposal. Normally second approach is usually considered during ECM. However, first approach is much more effective which mainly reduce the production of wastes during ECM operation. It can be accomplished by predicting minimum machining allowance, improvement of localized dissolution and online reduction of generated harmful by products.

Fig. 5.2.42 shows different hazardous wastes generated during ECM operation as well as its harmful impacts. Electrolyte splashing, contamination of eye and skin and a free expansion of toxic vapor needs to be evaded [49]. Critical impacts on the environment come from the electrolyte and ECM slurry. Presence of harmful byproducts in electrolyte is one of the major environmental issues. Spent electrolyte may contains acids, alkalines, heavy metals, chromate, nitrate, sulfate and oils, etc. Some HSTR materials machined by ECM contain 1–20% chromium [49], results in dissolved hexavalent chromium ions in electrolyte which are hazardous. Attempts have been made to remove hexavalent chromium ions from the waste electrolyte solution utilizing effective treatment. Through chemical reduction method, hexavalent chromium ions reduce to trivalent chromium which precipitates as hydroxide. Electrodeposition, solvent

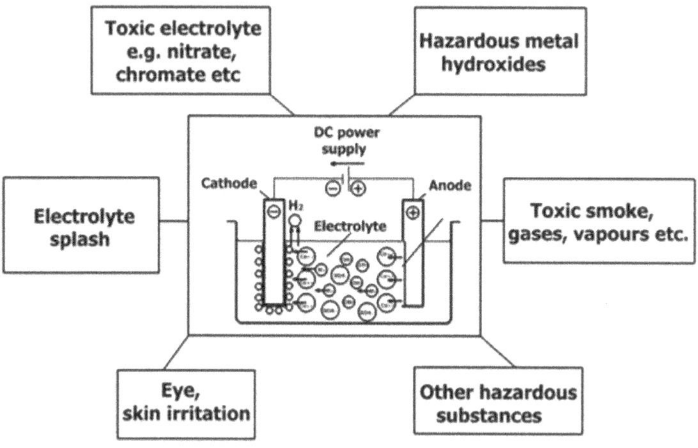

Fig. 5.2.42 Hazardous wastes generated during ECM operation.

extraction, and electrochemical reduction are common techniques through which waste electrolyte can be treated. These treatments are important, however, it finally influences the economic considerations of ECM.

Hence, for reduction of generation of pollutant, it is important to develop efficient pollution prevention techniques which can be utilized during ECM operation. A significant technique for reduction of waste generation is possible by reduction of effective material removal. Amount of sludge in the electrolyte can be decreased by reducing stray machining, improving localized dissolution, accurate prediction and minimizing removal of large amount of initial stock. Fresh aqueous solution of $NaNO_3$ is considered as physiologically safe and at the same time it increases the machining localization.

5.2.12 Advantages and limitations of ECM

Electrochemical machining is considered to be one of the most successful modern machining processes for several advantages. Out of which some of the major advantages are listed below:

(i) Complex 3D shaped can be easily generated
(ii) Machining is independent of mechanical and chemical properties of the job materials. Hence, any hard materials, e.g., super alloys can be easily machined
(iii) No tool wear, hence single tool can be utilized for number of times
(iv) Very good surface finish can be achieved, i.e., mirror finish
(v) No thermal effects during machining, hence absence of HAZ, recast layer, etc.
(vi) No mechanical effects on the job during machining
(vii) Highly flexible can be utilized for several applications

However, ECM process has some disadvantage as stated below:

(i) Job material must be electrically conductive
(ii) Material removal rate is comparatively low
(iii) Intergrannular attack on top of the machined surface
(iv) Reduction of fatigue strength of machined components due to low residual stress which needs post machining treatment
(v) Consumption of high specific energy
(vi) Tool design is complicated and time consuming
(vii) Initial investment is comparatively higher due to involvements of several other subunits

5.2.13 Applications of ECM

ECM is one of the most versatile and flexible machining processes which are fully utilized for different modern machining applications. Some of the major applications are listed below:

(i) Advantageously used in manufacturing and construction industries, e.g., different types of dies and molds by die sinking ECM, pistol barrel rifling, EC grinding of tungsten carbide cutting tool, fragile parts, e.g., thin wall tubes, hypodermic needles, removal of fatigue cracks as well as organic sea growth in offshore structures, etc.

(ii) Extensively utilized in aerospace and automobile industries for fabrication of different parts, e.g., rocket engine parts, jet engine parts: aero foils, platform, leading and trailing edges of turbine blades in a single operation, integral bladed rotors and turbine wheels, cooling holes in turbine blades, firing chamber of internal combustion gasoline engine, etc.

(iii) Widely used for fabrication of bio implants for biomedical applications, e.g., artificial hip joint, valve parts and other biomedical implants made of Titanium and Cobalt alloys, etc.

(iv) Applied for the fabrication of several parts for chemical industries for fabrication various parts of highly non corrosive metal in chemical process plants, e.g., super fine sieves, etc.

(v) Efficiently applied for micro fabrication of gears, actuators, sensors for MEMS applications, super thin metal strips; micro holes, slots and channels, metal masks, integrated circuits, etc.

(vi) Effectively applied for surface smoothing, improvement and texturing, etc., removal of recast layer, superfine surface generation with different integrity considering tribological aspects, micro surface texturing on uneven surface, etc.

5.2.14 Advancement in ECM

ECM is one of the most promising modern machining methods due to its versatility in applications. It has wide range of applications in modern manufacturing industries starting from shaping and finishing of large parts to fabrication of micro components. However, several issues, e.g., stray machining, tool design, electrolyte flow path design as well as control and monitor of machining gap, etc. need to be improved for making it more acceptable for commercial purpose in coming years. Although, researchers have attempted to solve some of the key issues through the advancement of the ECM process which are discussed hereunder:

(i) Control of end gap

The major problem in ECM is the online control of gap between tool and the workpiece. It needs development of proper sensors and strategies to control the end gap during machining [50]. For the involvement of several parameters during machining, it is difficult to achieve effective control of the gap. Thus, attempts have been made in this direction using different feedback signals from several sensors, e.g., eddy current sensors, current and voltage signal sensors, etc. In Pulsed ECM, it is possible to control the gap size by monitoring the instantaneous electrode position [51]. However, more research is needed for the development of online process monitoring and gap control strategies.

(ii) Tool design

In ECM, correct design of tool is still considered to be one of the key issues. Several methods for tool design are already proposed and utilized in real time practice [52]. However, all these methods usually give only the first approximation to the final tool shape. Newer methods with capable strategies and flexibility should be proposed considering not only steady state condition but also transition condition which involves insufficient machining allowance. An effective tool design method must also consider design of electrolyte flow path, fixture design as well as the pattern of insulation to reduce overcut [21].

(iii) Model development

It is difficult to model the ECM process due to occurrence of several physical and chemical phenomena simultaneously. The developments of various models and simulation strategies are urgently needed to control the process much more effectively. Interdependent phenomena those to be modeled are electrochemical reactions, electrolyte flow, thermal effects and anodic dissolution, etc. during ECM process [53]. Attempts have been made during development of the models considering mass balance as well as charge balance strategy [54].

(iv) ECM with pulsed power supply

Researchers have taken into consideration of application of pulsed DC power supply instead of continuous DC for the improvement the machining accuracy to a great extent. In pulsed DC power supply, several power parameters, e.g., type of pulse pattern, on time, off time, peak voltage, peak current, pulse frequency and duty factor play an important role during anodic dissolution. Pulse ECM (PECM) in which major units are power switching unit and pulse generator. The power switching unit is most important for the generation of pulsed power from smooth DC power supply. This power switching unit is rated by switching time, mode of control and finally the economical considerations. Power thyristors and transistors can be utilized for switching device in the circuit of pulsed DC power supply. However, for high speed switching, MOSFET is considered to be most appropriate. MOSFET is a solid state electronic device which is able to switch high power at much larger speed [21]. The limitation of all electronic devices applied for this switching purpose of PECM is that these are costly and not capable of withstand such high current as required in ECM operations. Hence, it can only be capable for small and medium scale pulse power supply in PECM. Another development for switching device is the insulated gate bipolar transistors (IGBT) which is cost effective for high power with medium frequency applications [51].

In PECM, machining is taking place during only on time in one cycle and during off time better flushing of produced byproducts is possible which leads to application of instant higher current density which results in improvement of machining accuracy. Effective process control of PECM is possible by adjusting various pulse parameters. The end gap can be monitored, controlled and reduced to a great extent in PECM which further improves the machining stability and accuracy.

(v) NC ECM

NC ECM incorporates the flexibility and automation of NC technology in ECM process [55]. Here, problems of tool design in case of conventional ECM can be eliminated. It can be effectively utilized for single and small scale production where frequent changes to the part shapes as well as shape of the tools are needed. This can be possible by utilizing simple shaped cylindrical cathode tool in NC ECM. The process is same as EC Milling method which has already been discussed in details. However, utilization of NC ECM is still in research stage which needs further development for make it more successful in modern manufacturing application.

(vi) Micro ECM

ECM process has already been successfully employed for fabrication of micro parts due to its ionic level metal removal mechanism [56]. In micro ECM, the gap between micro tool and the workpiece is very narrow in the order of $<50\,\mu m$ for improving the machining accuracy. It can be effectively utilized for fabrication of electronic components, e.g., print bands, meal masks, printed circuit board, etc. Micro ECM can also be employed for micro surface texturing. This technique has already been discussed in details in Chapter 7. However, successful utilization of micro ECM is still needed further research and development for commercialization in modern micro manufacturing applications.

(vii) WECM

WECM technique has already been discussed in details in this chapter. However, this technique of ECM is still in development stage. For establishment of WECM in modern manufacturing practice requires further improvement in the area of wire tool feeding, cutting speed, flushing of machined products from the narrow gap and in-depth parametric investigations, etc. [57]. Increment of the productivity of WECM, introduction of multiple wires is essential for which research is going on. In the near future, WECM may replace wire EDM in modern manufacturing practice.

(viii) Hybridization of ECM process

Several attempts have already been made to combine ECM with other modern machining processes. It has already been discussed in detail in Chapter 6. However, some special attention needs to be drawn on ECG for its several advantages and wide applications. ECM can be coupled with EDM, USM, LBM and also with many others existing process for adding its advantages and flexibility in the developed hybrid processes [58]. However, hybridization of ECM process needs further investigations to solve challenging problems for further opening up of newer possibilities in the field of modern machining.

Complex interactions between electrical, chemical as well as physical characteristics are not completely predictable during ECM operation till date. However, soon it will be most commercially successful machining process if various problems as discussed above can be resolved through extensive research efforts and continuing advancements.

5.3 Model questions

Chemical machining

Q1. What is "etch factor"? How it influences performance of CM?

Q2. Explain in brief the basic mechanisms involved in chemical machining.

Q3. Differentiate between chemical and electrochemical machining.

Q4. Illustrate about the design and fabrication of resist mask on job surface in CM.

Q5. Identify and describe different process parameters of CM.

Q6. Differentiate between chemical blanking and chemical milling.

Q7. Achieved surface finish by CM is poor. Why? Explain in brief.
Q8. Elucidate in brief the role of photochemical machining in CM.
Q9. How to improve the accuracy in CM?
Q10. Discuss about the different environmental impacts of CM.
Q11. Identify the advanced environment friendly etchant utilized in CM.

Electrochemical machining (ECM)

Q1. In ECM, which properties of the workpiece affect MRR?
Q2. What do you understand by over potential associated with ECM?
Q3. Which properties of an electrochemically machined surface contain.
Q4. Elaborate the "Self regulating" feature of ECM.
Q5. What are the different dynamics related to ECM? What is the practical application of ECM under "No feed" condition?
Q6. Using Faraday's law of electrolysis, derive and equation to compute maximum permissible feed rate. Modify this equation to take into consideration the effect of the temperature of the void fraction on the electrolyte conductivity.
Q7. What are advantages of pulsed DC power supply over constant DC power supply?
Q8. In ECM, what are the different types of flow path that you can design? Which one is best for the improvement of accuracy of the machined profile?
Q9. What do you understand by stray current effect? What actions can be taken to minimize it?
Q10. Elucidate in brief the influence of flow rate on the machined surface.
Q11. How the void fraction and the temperature of the electrolyte affect the slope of the machined surface?
Q12. Explain elaborately about the various tool design techniques involved in ECM.
Q13. Determine the equation of the required tool surface geometry for a given workpiece surface by different tool design techniques.
Q14. Compare the different types of electrochemical drilling techniques.
Q15. Explain the working principle of electro stream drilling (ESD). In what respect it is different from conventional ECM?
Q16. What is most interesting application of ESD?
Q17. What do you understand by STEM? What are the advantages of STEM over other Electrochemical drilling techniques?
Q18. In electrochemical deburring technique, how the deburring time can be estimated?

Q19. How can you protect the areas where dissolution of the work material during EC-deburring is not desirable?

Q20. Discuss about the advantages of EC milling technique over ECM process?

Q21. How to control different accuracy factors in EC milling?

Q22. Describe in brief the role of flushing in WECM.

Q23. What are the different types of flushing techniques in WECM?

Q24. What types of electrolyte is suitable for WECM? What are the reasons behind it?

Q25. What are the advantages of WECM over WEDM? Compare.

Q26. Explain the advantages of electrochemical sawing (ECS) technique.

Q27. Differentiate between ECG with abrasive assisted ECG?

Q28. Differentiate between ECH and mechanical honing technique.

Q29. Write down the different environmental impacts of ECM.

Q30. How to control the end gap in ECM?

Q31. What do you understand by NC-ECM and Hybridization of ECM process? With the help of an example explain your answer.

References

[1] J.F. Wilson, Practice and Theory of Electrochemical Machining, Wiley-Interscience, 1971.
[2] J.A. McGeough, Principles of Electrochemical Machining, Chapman and Hall, 1974.
[3] N. Taniguchi, Current status in, and future trends of, ultra-precision machining and ultrafine materials processing, Ann. CIRP 32 (2) (1983) 573–582.
[4] A.J. Bard, L.R. Faulkner, Electrochemical Methods—Fundamentals and Applications, second ed., John Wiley & Sons, 2001.
[5] J. Koryta, W. Dvorak, L. Kavan, Principles of Electrochemistry, second ed., John Wiley & Sons, 1993.
[6] B. Bhattacharyya, Electrochemical Micromachining for Nanofabrication MEMS and Nanotechnology, Elsevier, William Andrew Applied Science Publishers, Massachusetts, USA, 2015.
[7] R. Vidal, A.C. West, Aluminum and aluminum alloy dissolution in concentrated phosphoric acid, J. Electrochem. Soc. 145 (1998) 4067.
[8] B. Bhattacharyya, S.K. Sorkhel, Investigation for controlled electrochemical machining through response surface methodology-based approach, Int. J. Mater. Process. Technol. 86 (1999) 200–207.
[9] J.A. McGeough, Advanced Methods of Machining, Chapman and Hall, 1988.
[10] S.K. Sorkhel, B. Bhattacharyya, Parametric control for optimal quality of the workpiece surface in ECM, Int. J. Mater. Process. Technol. 40 (1994) 271–286.
[11] O.V.K. Chetty, R. Moorthy, V. Radhakrishnan, On some aspect of surface formation in ECM, J. Eng. Ind. ASME 103 (1981) 301–348.
[12] O.V.K. Chetty, V. Radhakrishnan, Surface studies in ECM using a relocating machining fixture, Int. J. MTDR 18 (1) (1978).
[13] H. Tipton, The determination of tool shape for ECM, Mach. Prod. Eng. 14 (1968) 325–328.
[14] H. Hardisty, A.R. Mileham, H. Shirvarni, A finite element simulation of the electrochemical machining process, Ann. CIRP 421 (1993) 201–204.

[15] R.H. Nilson, Y.G. Tsuei, Free boundary problem for the Laplace equation to ECM tool design, ASME J. Appl. Mech. 98 (1) (1976) 54–58.
[16] H. Nied, M.S. Lamphere, Electrochemical machining simulation using the boundary element method, in: Proceedings of the ISEM-11, 1995, p. 473490.
[17] V.K. Jain, K.P. Rajurkar, An integrated approach for tool design in ECM, Precis. Eng. 13 (2) (1991) 111–124.
[18] S. Das, A.K. Mitra, Use of boundary element method for the determination of tool shape in electrochemical machining, Int. J. Numer. Methods Eng. 35 (1992) 1045–1054.
[19] M.S. Reddy, V.K. Jain, G.K. Lal, Tool design for ECM: correction factor method, ASME J. Eng. Ind. 10 (1988) 111–118.
[20] P. Lawrence, Computer aided design for ECM electrodes, Int. J. Mach. Tool Des. Res. 21 (1981) 379–385.
[21] K.P. Rajurkar, D. Zhu, J.A. McGeough, J. Kozak, A. De Silva, New developments in electrochemical machining, CIRP Ann. 48 (2) (1999) 567–579.
[22] J. Kozak, K.P. Rajurkar, R. Balkrishna, Study of electrochemical jet machining process, Trans. ASME 118 (1996) 490–499.
[23] C. Clerc, M. Datta, L.T. Rominkiw, in: R. Gleason, G.J. Haffron, L.K. White (Eds.), High speed maskless patterning by electrolytic jet, patterning science and technology, Electrochemical Society, 1990, pp. 90–91 PV.
[24] M. Datta, L.T. Romankiw, Applications of chemical and electrochemical micro machining in the electronic industry, J. Electrochem. Soc. 136 (1989) 285c.
[25] G.E. Baker, Hole drilling processes: experiences, applications and selection, in: SME Non-traditional Machining Symposium, Orlando, Florida, 3rd–5th, Amchem Company, 1991, pp. 6–12.
[26] I.-H. Choi, J. Du Kim, Electrochemical deburring system using electroplated CBN wheels, Int. J. Mach. Tools Manuf. 38 (1998) 29–40.
[27] S. Sarkar, S. Mitra, B. Bhattacharyya, Mathematical modeling for controlled electrochemical deburring (ECD), J. Mater. Process. Technol. 147 (2004) 241–246.
[28] S. Hindujaa, J. Pattavanitch, Experimental and numerical investigations in electrochemical milling, CIRP J. Manuf. Sci. Technol. 12 (2016) 79–89.
[29] K. Mishra, D. Dey, B.R. Sarkar, B. Bhattacharyya, Modeling on volumetric material removal for fabrication of complex shapes by EC milling of Ti6Al4V, J. Electrochem. Soc. 165 (2018) 388–396.
[30] K. Mishra, D. Dey, B.R. Sarkar, B. Bhattacharyya, Experimental investigation into electrochemical milling of Ti6Al4V, J. Manuf. Process. 29 (2017) 113–123.
[31] B. Ghoshal, B. Bhattacharyya, Investigation on profile of micro channel generated by electrochemical micromachining, J. Mater. Process. Technol. 222 (2015) 410–421.
[32] De Silva, J.A. McGeough, Process monitoring of electrochemical micro machining, Int. J. Mater. Process. Technol. 76 (1998) 165–169.
[33] Y.B. Zeng, Q. Yu, S.H. Wang, D. Zhu, Enhancement of mass transport in micro wire electrochemical machining, CIRP Ann. Manuf. Technol. 61 (2012) 195–198.
[34] D. Zhu, K. Wang, N.S. Qu, Micro wire electrochemical cutting by using in situ fabricated wire electrode, Ann. CIRP 56 (1) (2007) 241–244.
[35] L. Meng, Y. Zeng, X. Fang, D. Zhu, Micropatterning of Ni-based metallic glass by pulsed wire electrochemical micro machining, Intermetallics 81 (2017) 16–25.
[36] K. Xu, Y. Zeng, P. Li, D. Zhu, Vibration assisted wire electrochemical micro machining of array micro tools, Precis. Eng. 47 (2017) 487–497.
[37] X.L. Fang, N.S. Qu, Y.D. Zhang, Z.Y. Xu, D. Zhu, Effects of pulsating electrolyte flow on electrochemical machining, J. Mater. Process. Technol. 214 (2014) 36–43.
[38] R.H. Nilsonz, S.K. Griffiths, Natural convection in trenches of high aspect ratio, J. Eletrochem. Soc. 150 (2003) 401–412.

[39] S. Wang, D. Zhu, Y. Zeng, Y. Liu, Micro wire electrode electrochemical cutting with low frequency and small amplitude tool vibration, Int. J. Adv. Manuf. Technol. 53 (2011) 535–544.
[40] K. Xu, Y. Zeng, P. Li, D. Zhu, Study of surface roughness in wire electrochemical micro machining, J. Mater. Process. Technol. 222 (2015) 103–109.
[41] S.H. Ahn, S.H. Ryu, D.K. Choi, C.N. Chu, Electrochemical micro drilling using ultra short pulses, Precis. Eng. 28 (2004) 129.
[42] S. Wang, Y. Zeng, Y. Liu, D. Zhu, Micro wire electrochemical machining with an axial electrolyte flow, Int. J. Adv. Manuf. Technol. 63 (1–4) (2012) 25–32.
[43] R. Maeda, K. Chikamori, H. Yamamoto, Feed rate of wire electrochemical machining using pulsed current, Int. J. Precis. Eng. 6 (1984) 193–199.
[44] Z. Liu, Y. Zeng, W. Zhang, Fabrication of metal microtool applying wire electrochemical machining, Adv. Mech. Eng. (2014).
[45] N. Qu, X. Fang, W. Li, Y. Zeng, D. Zhu, Wire electrochemical machining with axial electrolyte flushing for titanium alloy, Chin. J. Aeronaut. 26 (2013) 224–229.
[46] H. Ha, Y. Zeng, Y. Yao, N. Qu, Improving machining efficiency in wire electrochemical micromachining of array microstructures using axial vibration-assisted multi-wire electrodes, J. Manuf. Process. 25 (2017) 452–460.
[47] T.A. El-Taweel, S.A. Gouda, Performance analysis of wire electrochemical turning processed RSM approach, Int. J. Adv. Manuf. Technol. 53 (2011) 181–190.
[48] W. Wang, Z.X. Liu, W. Zhang, Y.H. Huang, D.M. Allen, Abrasive electrochemical multi-wire slicing of solar silicon ingots into wafers, CIRP Ann. Manuf. Technol. 60 (2011) 255–258.
[49] H. K. Tcinshoff, R. Egger, F. Klocke, Environmental and safety aspects of electrophysical and electrochemical processes, Ann. CIRP, Vol. 452 1–15.
[50] B. Wei, K.P. Rajurkar, S. Talpallikar, Identification of Interelectrode gap sizes in pulse electrochemical machining, J. Electrochem. Soc. 144 (11) (1997) 3913–3919.
[51] J.Y. Wang, et al., The investigation of the MOSFET power characteristics in PECM application, in: Proceedings of the ISEM-11, 1995, pp. 505–509.
[52] H. Tipton, The calculation of tool shapes for electrochemical machining, in: Fundamentals of Electrochemical Machining, The Electrochemical Society, Princeton, 1971.
[53] S. Hinduja, M. Kunieda, Modelling of ECM and EDM processes, CIRP Ann. Manuf. Technol. 62 (2013) 775–797.
[54] P.T. Pajak, et al., Virtual design of the shaving cap process by multi-physics ECM approach, in: Proc. 16th Int Symposium on Electrochemical Machining, 2010, pp. 693–697.
[55] N.Z. Yun, B. Wei, X.Y. Yang, J.W. Xu, Investigation on application of electrochemical contour evolution machining, in: Proceedings of the ISEM-9, 1989, pp. 143–145.
[56] B. Bhattacharyya, et al., Advancement of electrochemical micro-machining, Int. J. Mach. Tools Manuf. 44 (2004) 1577–1589.
[57] S. Debnath, J. Kundu, B. Bhattacharyya, Modeling and influence of voltage and duty ratio on wire feed in WECM: Possible alternative of WEDM, J. Electrochem. Soc. 165 (2) (2018) E35–E44.
[58] A. De Silva, J. McGeough, Hybrid electrodischarge-electrochemical process for roughing and finishing dies and moulds, in: Proceedings of the ISEM-1, vol. 2, 1998, pp. 397–406.

Further reading

[59] V.S. Bagotsky, Fundamentals of Electrochemistry, second ed., Wiley Interscience, 2006.

CHAPTER SIX

Hybrid machining technology

Contents

Modern Machining Technology
https://doi.org/10.1016/B978-0-12-812894-7.00006-2
© 2020 Elsevier Inc.
All rights reserved.

6 Introduction

With the advancement of the industries like aerospace, biomedical, electronics, environmental, communications and automotive, more and more advanced materials are being developed. Today's manufacturing industry is facing challenges from those advanced and difficult to-machine materials, i.e., Inconel-718, Ti6Al4V, PCD, γ-Ti aluminide, Al_2O_3, Si_3N_4, ZrO_2, etc. These materials are being machined neither by traditional machining processes nor by any nontraditional machining processes solely, effectively. Manufacturing scientists have developed some conventional or non-conventional machining processes, which can be utilized for machining these advanced materials as mentioned but those processes are not at all economical. This necessitates development of newer machining process by utilizing material removal mechanisms of two or more existing conventional or non-conventional machining processes. These attempts of the scientists gave rise to a new perception of machining technologies, which are currently well known as hybrid machining process (HMP).

The reasons for developing hybrid machining process (HMP) are not only to make use of the combined or mutually enhanced advantages but also to avoid or reduce some adverse effects on the constituent processes produce when they are individually applied.

CIRP collaborating working group on hybrid process in 2014 put forward following definition of hybrid machining:

> *"Hybrid manufacturing processes are based on the simultaneous and controlled interaction of process mechanisms and/or energy sources/tools having a significant effect on the process performance."*

Here, "simultaneous and controlled interaction" indicates that the process/ energy sources should interact more or less in the same processing zone and at the same time [1].

Hybrid machining processes (HMPs) are the machining techniques which can merge merits of two or more sorts of machining methods. Hybrid machining processes are based on the simultaneous and controlled interaction of machining mechanisms and/or energy sources/tools having a significant effect on the process performance. Hybrid machining processes are also effective for reducing adverse effects of the constituent processes produce when they are individually applied for machining purposes.

6.1 Need and basis of classification of HMP

Recently, researchers have characterized the machining capabilities, advantages, drawbacks, possible future efforts and developments of the state-of-the-art hybrid machining processes under major categories, i.e., assisted techniques and combined techniques. In general, hybrid machining processes can be classified into two main categories as follows:

(1) HMPs, in which all constituent processes are directly involved in the material removal.

(2) HMPs, in which only one of the participating processes directly removes the material while the others only assist in removal.

Fig. 6.1 shows the basic classification of hybrid machining [2]. In assisted hybrid machining, a main process or mechanism responsible for material removal is defined as primary process where secondary process only assists the primary process. In combined hybrid machining, several processing mechanisms are combined and have a synergetic effect in the machining zone for material removal.

In both the above catagories, plastic flow, mechanical abrasion, heating, melting, evaporation and dissolution interaction take place and changes physico-chemical conditions of the processes and workpiece material properties. By utilizing different interactions major hybrid machining processes are being developed across the world. The development of hybrid machining process by combining different interactions can be represented in Table 6.1. There are mainly three types of interactions, e.g., Mechanical Interaction, Thermal Interaction and Chemical and Electro-chemical

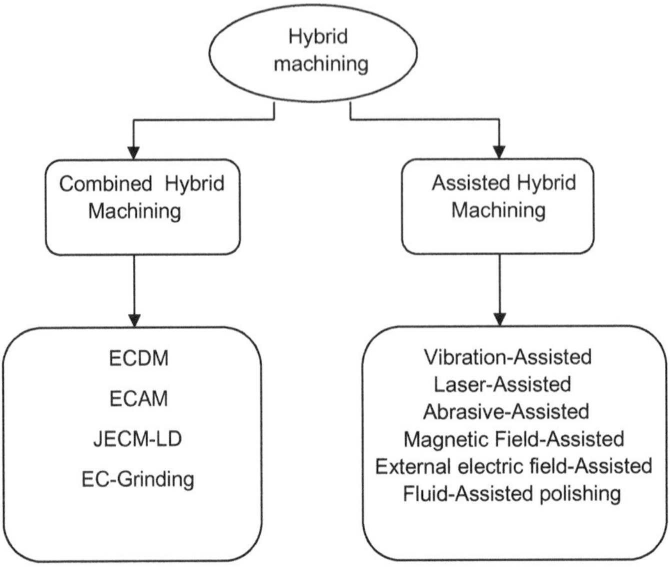

Fig. 6.1 Classification of hybrid micro-machining processes.

interaction [3]. These interactions with advantages and limitations are noted here under.

(i) Mechanical interaction

All cutting, grinding and abrasive finishing methods are based on mechanical interaction of cutting edge of tools and abrasives with workpiece material. Grinding and abrasive flow creates abrasion action on the workpiece. Turning, milling and drilling, etc. produce cutting action by plastic deformation of the workpiece. High pressure water jet, low pressure suspension jet, etc. generates flow fluid action on the workpiece. However, these mechanical interactions having some drawbacks, e.g., it produces higher cutting force leads to heat generation, tool wear, various mechanical and thermal effects on the surface. Rigidity of the machine tools kinematics structures should be higher to overcome the cutting forces as developed by mechanical interaction.

(ii) Thermal interaction

Laser beam machining (LBM), Electron beam machining (EBM), Plasma beam machining (PBM), Electrical discharge machining (EDM) and are thermal processes where material is removed through a phase change, either by melting or vaporization. Well guided and controlled electron beam, laser beam, plasma beam and electrical

Table 6.1 Hybrid machining process by combining different interactions

Interactions		Electrical discharge	Laser beam	Electron beam	Plasma beam	Chemical	Electro chemical	Abrasive	Turning/ cutting	Ultrasonic vibration	High pressure flow	Magnetic field
Thermal	Electrical discharge	EDM/ WEDM					ECDM/ WECDM	AEDG		UAEDM		MEDM
	Laser beam		LBM			LAE	LECM		LAT			
	Electron beam			EBM								
	Plasma beam				PBM				PAT			
Chemical	Chemical		LAE			CHM		EEM			CHFP	
	Electro chemical	ECAM	LECM				ECM	AECG		USECM	ECJM	MECM
Mechanical	Abrasive	AEDM				MCP	AECH/ ELIDG	GRINDING		USG	AWJM/ AJM	MAFF/ MRAFF
	Turning/ cutting		LAT		PAT				TURNING	UAT		
	Ultrasonic vibration	UAEDM	UALBM				UAECM	USG	UAT	USM	USFP	MUS
	High pressure flow					CHP	ECJM	AFM		USP	JM	
	Magnetic field	MEDM					MECM	MAFF/ MRAFF		MUS		MM

discharges produces the thermal effects on the work surface to remove material for achieving desired shape. However, secondary phenomena relating to surface quality which occur during machining, such as micro-cracking, formation of heat-affected zone and formation of striations due to the thermal effects are the major constraints.

(iii) Chemical and electro-chemical interaction
Chemical machining (CM), Electrochemical machining (ECM), Pulse electrochemical machining (PECM), Electropolishing (EP) are shaping and finishing processes based on application of dissolution of material utilizing chemical or electrochemical energy. Being a non-mechanical metal removal process, these processes are capable of machining any electrically conductive material regardless of their mechanical properties, such as strength, hardness, toughness and brittleness. However, these interactions unable to remove material from chemically inert work materials as well as passivation effects restrict the material removal process which leads to deterioration of the machining accuracy.

The main interactions show the significance of combining various existing conventional and non-conventional machining methods based on different kind of interactions for enhancing machining as well as finishing characteristics which are being depicted in Table 6.1.

Table 6.1 reveals that still now, in several areas combinations of different interactions have not been possible which needs further research. This table only represents thermal, chemical and mechanical interactions for developing existing hybrid machining processes. However, there may be other types of interactions which can also be utilized for developing hybrid machining processes, e.g., magnetic fields, electrical fields, etc.

Most important hybrid machining and finishing processes under development worldwide are as follows:

(i) Electrochemical Grinding—ECG
(ii) Abrasive Electrochemical Honing—AECH
(iii) Electrochemical Arc Machining—ECAM
(iv) Electrochemical Discharge Machining—ECDM
(v) Abrasive Electrical Discharge Grinding—AEDG
(vi) Abrasive Electrical Discharge Machining—AEDM
(vii) Magnetic Abrasive Finishing—MAF
(viii) Vibration assisted Electrochemical Machining—UAECM
(ix) Vibration Assistance Electrical Discharge Machining—UAEDM
(x) Thermal Assisted Turning such as Laser Assisted Turning—LAT, and Plasma Assisted Turning—PAT; etc.

(xi) Laser Assisted Electrochemical Machining—LECM
(xii) Laser Assisted Etching—LAE
(xiii) Mechano-Chemical Polishing—MCP
(xiv) Elastic Emission Machining—EEM
(xv) Electrolytic In-Process Dressing (EILD) grinding, etc.

Some of the hybrid machining processes, which are most popular and well developed at current stage, have been discussed in detail here under.

6.2 Electrochemical grinding

6.2.1 Introduction

Electrochemical grinding (ECG) is a hybrid machining process in which electrochemical machining (ECM) process and mechanical grinding process are combined to finish the surface of the work piece by removing materials by electrochemical anodic dissolution and mechanical abrasion. In ECG process a rotary metal bonded abrasive grinding wheel is used as cathode tool and workpiece is as anode as low voltage DC power supply is connected between grinding wheel and work piece. The electrolyte, usually a neutral salt solution with high velocity is supplied between the inter-electrode gap to complete the electrolytic cell and also to remove material particles from the machining gap. During ECM process, the insulating passive oxide layers (insoluble oxide layer) are formed in the gap and as a result machining rate decreases due to decrease in current flow. This difficulty of reduction of current flow due to passive layer is overcome by mechanical micro cutting actions of grinding as it removes the passive layers and as a result material removal rate increases. Electrochemical Grinding (ECG) process is also called as Mechanical Assisted Electrochemical Grinding Process. Electrochemical Grinding is an ideal machining process that provides a better, faster, and more cost effective metal cutting and grinding solution for today's toughest materials. Unlike conventional grinding techniques, Electrochemical Grinding offers the ability to grind difficult to machine materials such as stainless steel, high strength temperature resistance (HSTR) alloys, cutting tool of high speed steel, cemented carbides, WC, CBN, PCD, etc. irrespective of their hardness, toughness, ductility and microstructure, etc. This is because Electrochemical Grinding is an entirely different machining process in which electrical energy combines with chemical energy for metal removal. The problems of conventional grinding such as extreme wheel wear and rapid loading of grinding wheel for grinding super alloys can be overcome by the development of ECG process. The grinding wheel life in ECG is more compared to conventional mechanical grinding.

6.2.2 Working principle of ECG

In ECG there is a grinding wheel (cathode) which is connected with negative terminal of the DC power supply, as in conventional grinding wheel except that the bonding material is electrically conductive. Here electrolyte is supplied through inter electrode gap (IEG) between the wheel and workpiece. The height of abrasive particles protruding outside bonding material of the wheel helps in maintaining a constant IEG because the abrasive particles act as spacers. Life of the ECG wheel is about 10 times than the conventional type. There are mainly two reasons (i) only 10% distribution by abrasive action toward the total material removal and (ii) very short length of arc of contact.

In ECG, there are three fundamental modes of material removal involving mechanical abrasion and another mode which has no mechanical contribution such as [4]:

(i) Totally mechanical material removal during the initiation of the grinding process

(ii) Electrochemical dissolution combined with mechanical abrasion and a zero overcut

(iii) Electrochemical dissolution combined with mechanical abrasion and an overcut greater than zero

(iv) Totally electrochemical anodic dissolution while large inter electrode gap is created

The grinding area between the wheel and work piece can also be divided into three zones of interaction named as, zone I, zone II, and zone III as mentioned in the following discussion [5]. Figure 6.2 shows three zones of electrochemical grinding.

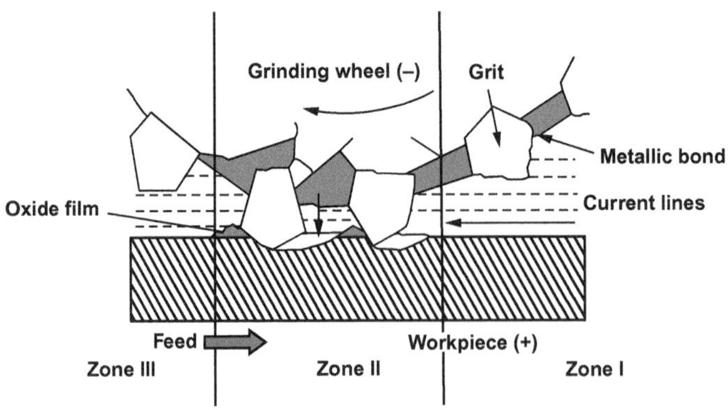

Fig. 6.2 Three zones of material removal modes in ECG.

Zone I: As grinding wheel rotates, the fresh electrolyte is drawn into initial larger inter electrode gap. Material is mainly removed due to electrochemical anodic dissolution. As the machining process is in progress, the electrolyte is contaminated with reaction products and gasses and oxide layer is formed which lowers the electrical conductivity of medium and as a result inter electrode gap deceases and also wheel is almost touching the workpiece. Material removal rate decreases due to less current flow. This zone ends while the wheel is in contact with workpiece.

Zone II: Material is removed from the surface of the workpiece by the abrasive particles of ECG grinding wheel as wheel is in contact with workpiece in this zone. The abrasive grains protruded over the metal bonded wheel maintains the interelectrode gap to avoid short circuit and remove materials as ground chips due to mechanical grinding action. After removing the oxide passive layer, the interelectrode gap increases. The fresh work pieces surface is exposed. The fresh electrolyte is further drawn by the cavitation effect generated due to rotation of wheel. The local electrolyte pressure increases and the formation of gas bubbles is suppressed. As a result electrochemical anodic dissolution starts at higher rate due to high current flow. This zone combines mechanical abrasion and partial electrochemical dissolution of metal oxides. Zone II ends where the wheel lifts of workpiece surface.

Zone III: Material removal takes place solely due to electrochemical anodic dissolution of workpiece. Inter electrode gap is larger and less material is removed. The electrolyte pressure is less. This zone is the finishing zone in which the scratches or micro burrs formed in Zone II is removed due to electrochemical dissolution.

The life of grinding wheel is about 8 to 10 times more than that of a conventional grinding wheel due to very less wheel wear and also very small contact length and time between wheel and workpiece [6, 7]. In ECG about 90–95% of the material is removed by electrochemical dissolution (ECD) and the remaining 5–10% material is removed by mechanical grinding (MG) action.

6.2.3 ECG machining set-up details

The ECG Machining set-up consists of metal bonded abrasive grinding wheel rotation and monitoring unit, workpiece holding and feeding machining table, DC power supply unit and electrolyte supply unit. The schematic diagram of ECG system is shown in Fig. 6.3. In ECG system,

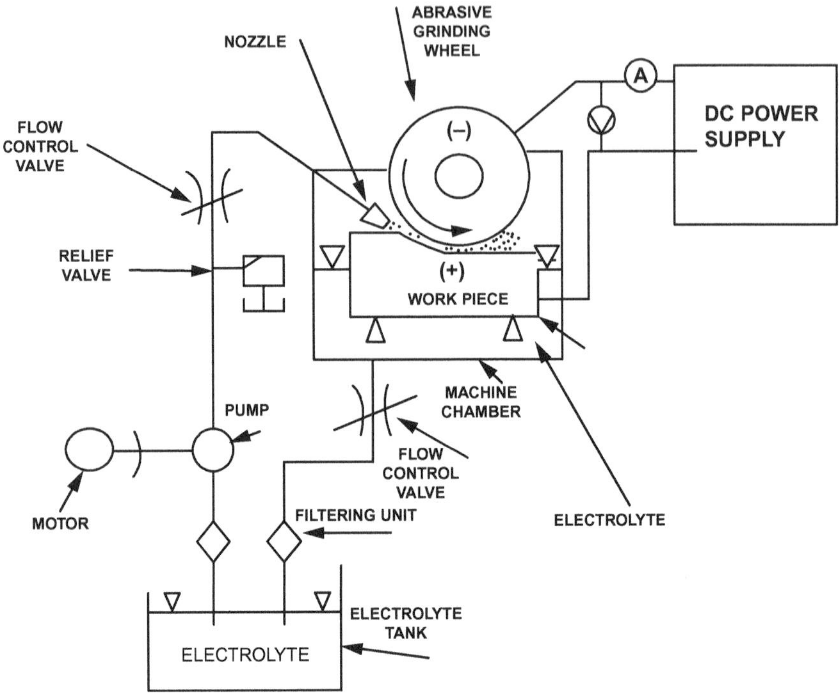

Fig. 6.3 The schematic diagram of ECG system.

rotating conductive metal bonded abrasive grinding wheel acts as cathode and workpiece acts as anode in electrolytic cell. The wheel rotates at a surface speed of 20–35 m/s. Bonding materials are copper, brass, bronze, nickel or copper impregnated resin. Alumina, Diamond, CBN grit of mesh size of 60–100 are used as abrasive for ECG wheel. Non-conducting abrasives are protruded over conducting metal bonded wheel surface which created inter-electrode gap between cathode (wheel) and anode (workpiece). There is less chances of short circuit condition due to the gap between metal bond and workpiece. Power is supplied to the wheel through spindle either with the help of brushes or mercury coupling. The current flow between wheel and workpiece in ECG ranges from 50 to 3000 A. The machining gap voltage in ECG generally varies from 2 to 35 V. The workpiece is firmly held on machine table which can be controlled by CNC Drive and computer software program. The workpiece feed rate usually varies from 10 to 30 mm/min. The depth of cut generally <2.5 mm. The pressurized electrolyte is supplied to the inter electrode gap through the nozzle. The electrolyte supply unit consists of pump, flow control valves, nozzle, electrolyte tank and filter and pipeline, etc. The electrolyte flow rate is usually varied from 10 to 15 l/min. The electrolyte concentration varies from 5% (5 g per 100 cc water) to 30%. Commonly used electrolyte is aqueous $NaNO_3$. Aqueous solutions of potassium carbonate, sodium nitrate or sodium nitrite have high conductivity, which is helpful in reducing corrosion. Other electrolytes such as KNO_3 and NaCl are rarely used because these have strong corrosive effects on machine elements. Associated equipment, control valves, fixtures, work holding tables and machining chambers are of suitable materials such as stainless steel, copper and corrosion resistant coated materials so as to capable of operating in the salt mist corrosive environment. Workpiece and machining system should be properly cleaned and checked against the adverse effect of electrolyte. All components connected with cathode tool remain electrically insulated from cathode components to avoid short circuit condition during ECG.

Various kinds of ECG operations such as surface grinding, face grinding, cylindrical grinding, internal grinding and form grinding can be done [8]. In ECG, generally the direction of wheel rotation is opposite to the direction of feed of workpiece. The contact area between wheel and workpiece is larger in EC face grinding process, hence it is a faster process. As the contact area is less, EC cylindrical grinding is a slower process.

Typical values of ECG operating parameters are given as follows:

DC Power Supply Voltage: 4 to 14 V; Current: 50 to 3000 A; Electrolyte Type and Concentration: NaNO3 (120 to 140 g/L), KNO3 (180 to 240 g/L), NaCl (60 to 180 g/L); Abrasive Wheel: Conductive; Abrasive: Al_2O_3 or diamond grit, Speed: 1000 to 7000 rpm (1200 to 2100 m/min).

6.2.4 Material removal rate in ECG

The metal removal by electrochemical grinding process comprises of the electrochemical dissolution (ECD) and the mechanical action of the abrasive grinding wheel. The mathematical modeling for the metal removal rate (MRR) evaluation through electrochemical grinding process has been developed considering the involvement of the electrochemical dissolution (ECD) component and mechanical abrasive grinding component in ECG.

Thus, the total material removal rate (W) in ECG can be evaluated as:

$$W = W_1 + W_2 \tag{6.1}$$

where, W_1, material removal rate due to electrochemical dissolution (ECD) in ECG; W_2, material removal rate due to mechanical action of the abrasive grinding wheel.

(a) *Mathematical modeling for evaluation of metal removal rate (MRR) in electrochemical dissolution (ECD) process*

The modeling for the metal removal rate (MRR) evaluation due to ECD is mainly based on the assumptions that the ohm's law and Faraday's laws of electrolysis validate in the machining gap.

Based on the Faraday's law of electrolysis, the mass material removal rate, W_1 (g/min) can be expressed as

$$W = \frac{M_x \eta_c}{nF} I \times 60 \tag{6.2}$$

where, M_x, atomic weight of the anodic material, g; n, valency of anodic material; F, Faraday's constant, 96,500 A s; I, current that passes through the machining gap, A; η_c, current efficiency (%).

Hence volumetric material removal rate, V_m (mm^3/min) can be represented as

$$V_m = \frac{M_x I 60 \eta_c}{nF\rho} \tag{6.3}$$

where, ρ, density of the workpiece material, g/mm^3;

From the Ohm's law, if the over potential that includes activation and concentration overpotential at the electrodes is ΔV and the applied voltage is V (volt), then the current passes through the gap between the electrodes is represented as:

$$I = \frac{V - \Delta V}{R} \tag{6.4}$$

where, R, the resistance in the working gap of the electrochemical grinding, ohms;

This resistance (R) consists of the two component as:

$$R = R_e + R_p \tag{6.5}$$

where, R_e, ohomic resistance of the layer of the electrolyte between the inter electrode gap, ohms; R_p, ohomic resistance of the oxide layer on the anode surface, (ohms) whose value is less compared to R_e.

Accordingly, the total resistance (R) can be expressed as:

$$R = K_p R_e \tag{6.6}$$

where Kp, the co-efficient that defines the degree in which the process occurs without the polarization phenomenon. This co-efficient can also be regarded as the loss of energy because the electrochemical dissolution rate decreases with the decrease in current density in the working gap due to the disturbance resulting from the polarization of the electrodes.

The electrolytic resistance (R_e) can be defined as:

$$R_e = \frac{rh}{A} = \frac{h}{K_e A} \quad (6.7)$$

where, r, specific resistance of the electrolytes, ohm mm; h, effective inter-electrode machining gap, mm; A, machining area, mm^2; K_e, conductivity of the electrolyte, ohm^{-1} mm^{-1}.

The current density, i.e., the current passing through the machining gap per unit area of machining, can be expressed as:

$$J = \frac{I}{A} = \frac{(V - \Delta V)K_e}{K_p h} \quad (6.8)$$

The velocity, u_t (mm/min) of the electrochemical penetration of the anode toward the cathode under steady state condition of grinding can be evaluated as:

$$u_t = \frac{60 M_x \eta_c}{nF\rho} \frac{I}{A} \quad (6.9)$$

$$= \frac{60 M_x \eta_c}{nF\rho} J \quad (6.10)$$

Substituting the value of J in Eq. (6.8), the velocity of electrochemical penetration (u_t) can be represented as:

$$u_t = \frac{60 M_x \eta_c (V - \Delta V) K_e \eta_c}{nF\rho K_p h} \quad (6.11)$$

From the grinding wheel and workpiece system configuration as shown in Fig. 6.3, the workpiece is fed toward the grinding wheel at a rate of $f\cos\theta$ in the opposite direction, then the rate of change of position of the gap is given by

$$\frac{dh}{dt} = \frac{60 M_x K_e \eta_c (V - \Delta V)}{nh\rho F K_p} - f\cos\theta \quad (6.12)$$

where, f, linear feed rate of the workpiece, mm/min.

In a peripheral grinding configuration, the gap at the leading edge (at $\theta = \theta_0$) will be the least for any set of operating condition, and therefore, the stable conditions will be determined by the minimum gap requirements corresponding to this position. Under steady conditions, dh/dt=0,

$$\text{i.e., } \frac{60(V-\Delta V)M_x K_e \eta_c}{n h_0 \rho F K_p} = f \cos\theta_0 \tag{6.13}$$

where, h_0, interelectrode steady state gap at the leading edge. If θ_0 is such that the gap formed at the leading edge, h_0 then, the following relationship [9] can be obtained.

$$\cos\theta_0 \approx \left(\frac{2ds}{R}\right)^{1/2} \tag{6.14}$$

where, ds is set depth of cut, mm; R, radius of the grinding wheel, mm.

Substituting the value of $\cos\theta_0$ in the Eq. (6.13), the set depth of cut can be expressed as

$$ds = \frac{R}{2}\left\{\frac{60(V-\Delta V)M_x K_e \eta_c}{n h_0 \rho F K_p}\right\}^2 f^{-2} \tag{6.15}$$

$$= \frac{R}{2}\{U_t\}^2 f^{-2} \tag{6.16}$$

The metal removal rate W_1 (mm^3/min) can be expressed as:

$$W_1 = U_t A = fA\sqrt{\frac{2ds}{R}} \tag{6.17}$$

Hence, for constant area of machining, A and for a particular size of grinding wheel, the metal removal rate depends on the linear feed rate of the workpiece and set depth of cut during ECG process.

The metal removal rate due to ECD, W_1 (g/min) can also be represented as:

$$W_1 = \frac{60(V-\Delta V)M_x K_e \eta_c A}{n h_0 \rho F K_p} \tag{6.18}$$

According to Eq. (6.18), the metal removal rate in ECG process also depends on supply voltage, the working gap at steady state condition, types of the workpiece material and electrolyte solution, etc.

(b) *Mathematical modeling for evaluation of material removal rate (MRR) due to mechanical action of the abrasive grinding wheel in ECG*

The metal removal rate by mechanical grinding is proportional to the machining area and velocity of penetration of the material. Metal removal rate, W_2 (g/min) can be expressed as:

$$W_2 = \rho dsBf \tag{6.19}$$

Where, ds, depth of cut, mm; B, width of cut, mm; ρ, density of the material, g/mm^3; f, velocity of penetration into the material, mm/min.

The penetration velocity, f due to mechanical action of grinding can be evaluated as:

$$f = Z_t t_z \tag{6.20}$$

Here, Z_t, number of grains participating in the mechanical grinding process during 1 min contact interval of the wheel with the material; t_z, mean thickness of the layer removed by one cutting point for a given pressure, mm.

The value of Z_t can be evaluated as:

$$Z_t = zn_s = \pi dcb' n_s \tag{6.21}$$

where, z, number of grains on the working circumference of the wheel; d, diameter of the grinding wheel, mm; c, surface density of the active grains, mm^{-2}; b′, average grain width of cut and n_s, rpm.

Hence, the mechanical metal removal rate due to grinding, W_2 (g/min) can be obtained as:

$$W_2 = \rho dsB\pi dcb' n_s t_z \tag{6.22}$$

If the electrochemical grinding is considered as a process compound of electrochemical dissolution and mechanical grinding, Eqs. (6.18), (6.22) may be combined to give the total metal removal rate (MRR) in electrochemical grinding as:

$$\begin{aligned} W &= W_1 + W_2 \\ &= \frac{60(V - \Delta V)M_x\eta_c K_e A}{nFh_0\rho k_p} + \pi\rho d_s BdCb' n_s t_z \end{aligned} \tag{6.23}$$

Therefore, the metal removal rate by electrochemical grinding is a function of the supply voltage, steady state gap, type of workpiece material, type of electrolyte, set depth of cut, wheel grain size, rotational speed of the wheel, etc.

At the initial stage of electrochemical grinding, the metal removal mainly takes place due to action of the grinding wheel. As the machining operation is in progress, the small machining gap is generated due to mechanical grinding. As soon as the steady state of interelectrode gap is attained, the metal removal takes place due to electrochemical dissolution only and W_2 component will become zero. As the passive layer is formed on the workpiece surface, the direct contact of the rotating grinding wheel with this layer results the mechanical removal of the material and passive oxide layer from the workpiece surface, then material removal occurs due to mechanical grinding followed by anodic dissolution.

6.2.5 Analysis of the process parametric influences on metal removal rate (MRR) in ECG

The effects of the various process parameters can be analyzed in the present research through the developed mathematical models and the experimental observations. For the purpose of analysis, the workpiece, tool and electrolyte combination have been chosen as C-20 steel, conducting diamond grinding wheel and $NaNO_3$ solution respectively.

It is found from the analysis that the metal removal rate in ECG increases with the machining voltage as exhibited through Fig. 6.4. This is in the

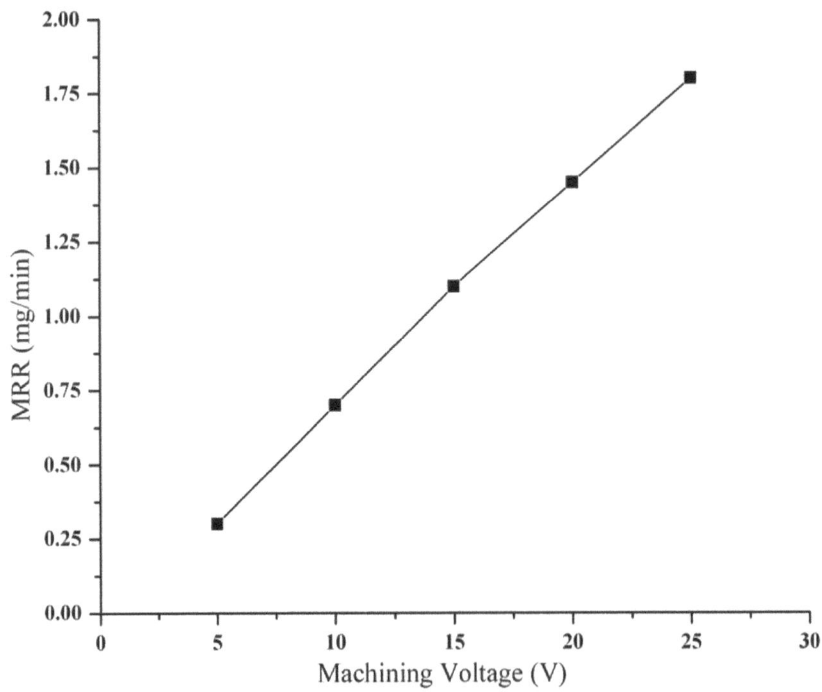

Fig. 6.4 Variation of MRR with machining voltage.

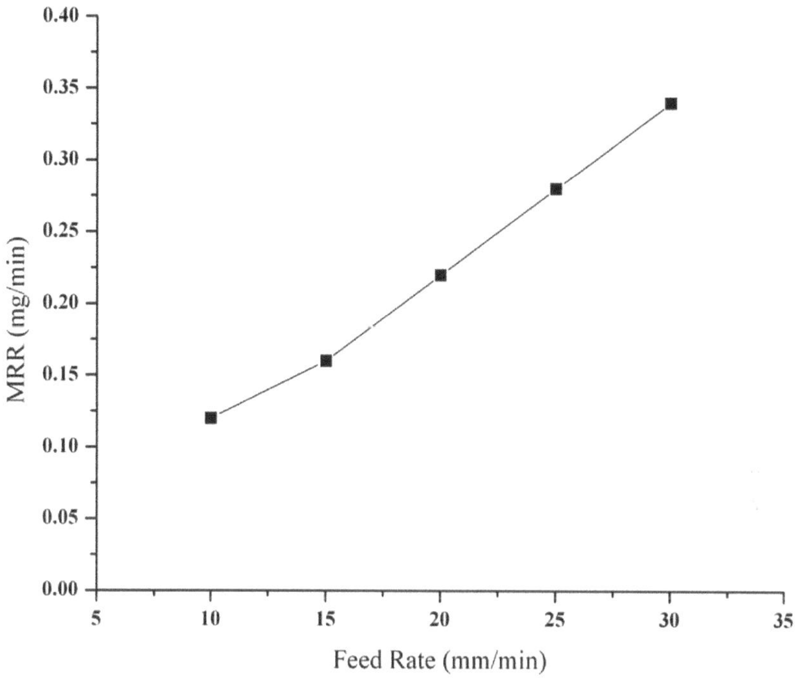

Fig. 6.5 Variation of MRR with feed rate.

conformity with the mathematical model relation for constant values of feed rate, set depth of cut, higher current will flow through the machining gap due to increase in machining voltage and hence the metal removal rate will be enhanced.

Fig. 6.5 shows that with the increase in feed rate of the workpiece, the metal removal rate increases because of the reasons that higher material removal takes place at a lower machining gap condition. The effects of feed rate on the machining gap has also been exhibited through Fig. 6.6.

Fig. 6.6 shows that the machining gap decrease in a parabolic fashion with the increment in feed rate and is obvious from the point of view of system designs for effective grinding operations. Fig. 6.7 exhibits that higher material rate can be obtained with setting higher depth of cut. At the initial stage, the mechanical grinding rate will also be enhanced with higher set depth of cut. The reduction in inter electrode machining gap with the increase in set depth of cut has been shown in Fig. 6.8. The metal removal due to electrochemical dissolution take place at a higher rate with reduced machining gap justifying the characteristic features of such systems and the correctness of the mathematical model developed for the purpose.

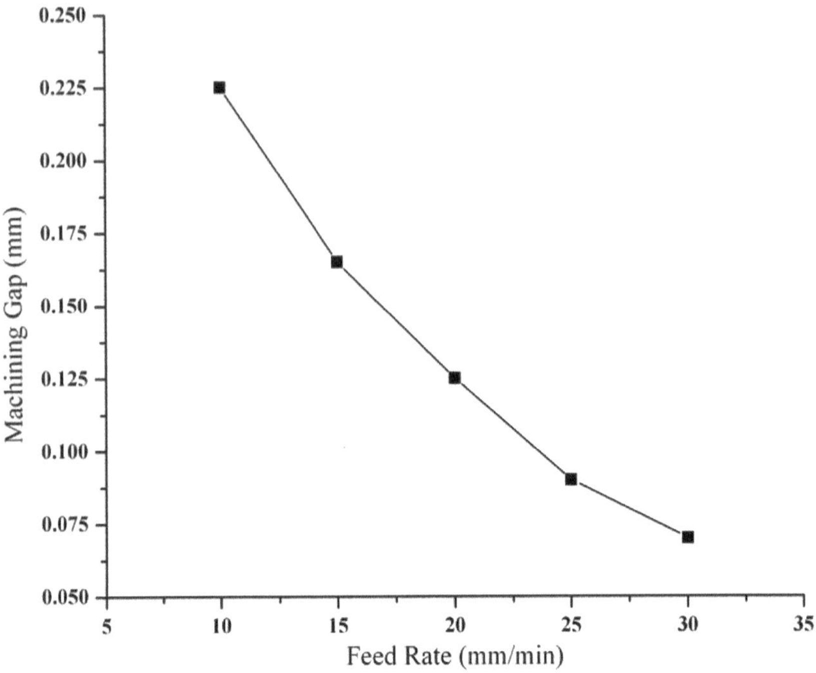

Fig. 6.6 Variation of machining gap with feed rate.

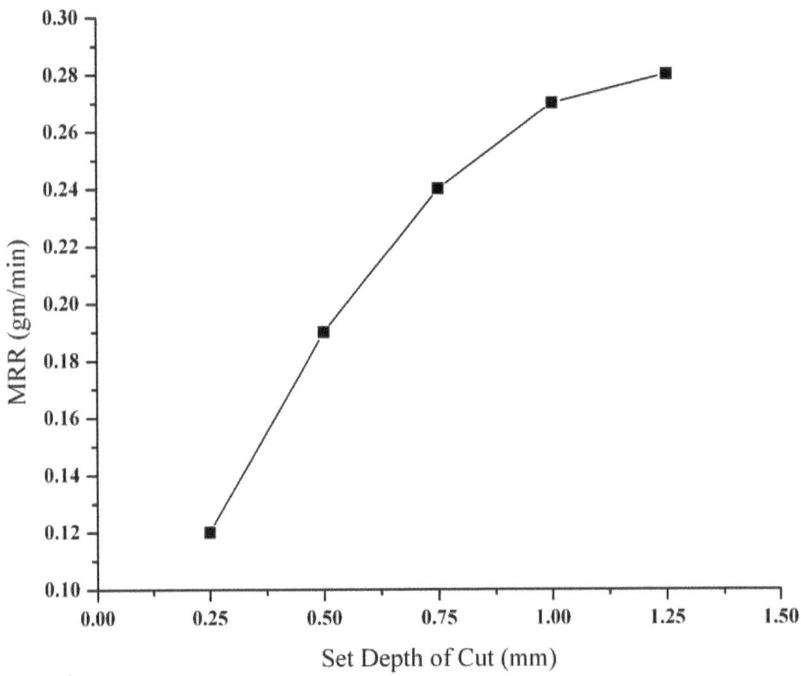

Fig. 6.7 Variation of MRR with set depth of cut.

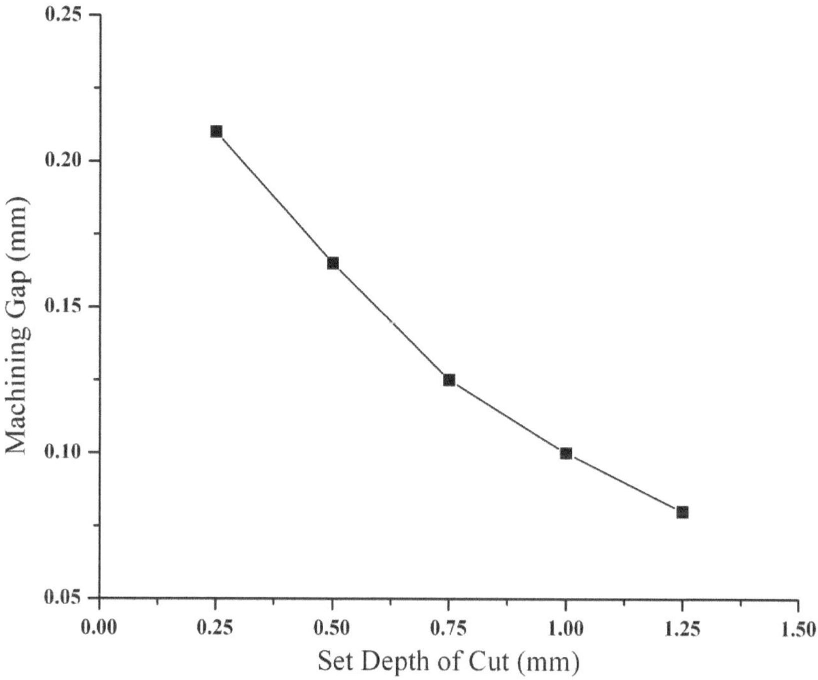

Fig. 6.8 Variation of machining gap with set depth of cut.

6.2.6 Process capabilities of ECG

Traditional grinding removes metal by abrasion, leaving tolerances of about ± 0.003 mm and creating heat and stresses that make grinding thin stock very difficult. In ECG however a production tolerance of ± 0.025 mm is easily obtainable [8]. Under special circumstances a tolerance of ± 0.008 mm can be achieved. The ability to hold closer tolerances depends upon the current, electrolyte flow, feed rate, and metallurgy of the workpiece itself. Accuracies achieved are usually about ± 0.125 mm. A final cut is usually done mostly by the grinding action to produce a good surface finish and closer dimensional tolerances. It is recommended that lower voltages be used for closer tolerances, reduced overcut, sharp edges, and a bright surface finish. ECG can grind thin material of 1.02 mm, which normally warp by the heat and pressure of the conventional grinding thus making closer tolerances difficult to achieve. In ECG there is little contact between the wheel and workpiece, which eliminates the tendency of the workpiece to warp as it might with orthodox grinding.

When a gap voltage of 4 to 40 V is applied between the cathodic grinding wheel and the anodic workpiece, a current density of about 120 to 240 A/cm^2 is generated. The current density depends on the material being machined, the gap width, and the applied voltage. Material is mainly removed by ECD, while the Mechanical Action (MA) of the abrasive grits accounts for an additional 5% to 10% of the total material removal.

The interactions of electrochemical process are mainly based on the influence on physical properties of workpiece surface layer. It changes dislocation density in surface layer of the workpiece material. Velocity of dislocation increases due to surface potential reduction during dissolution which facilitates plastic deformation during micro cutting in ECG. During machining of WC, differences in potentials between carbides and cobalt matrix increase which results in enhancement of dissolution rate in the order of 100–150%. Another promoting factor may be the generation of hydrogen during anodic dissolution in ECG. Hydrogen diffusion into materials changes residual stresses and increase material brittleness on the surface layer which also assists metal removal from the workpiece surface during ECG.

Removal rates by ECG are 4 times faster than by conventional grinding, and ECG always produces burr-free parts that are unstressed. The volumetric removal rate (VRR) is typically 1600 mm^3/min.

6.2.7 Advantages and limitations

The advantages of ECG process are stated hereunder:

(i) Absence of work hardening
(ii) Elimination of grinding burrs
(iii) Absence of distortion of thin fragile or thermo-sensitive parts
(iv) Good surface quality
(v) Production of narrow tolerances
(vi) Longer grinding wheel life
(vii) Low induced stress
(viii) Large depth of cut
(ix) High material removal rate, etc.

However, ECG process has some limitations as highlighted below:

(i) Higher capital cost than conventional machines
(ii) Process limited to electrically conductive materials
(iii) Corrosive nature of electrolyte
(iv) Requires disposal and filtering of electrolyte

(v) The process is not adapted to cavity sinking, and therefore it is unsuitable for the die-making industry
(vi) Higher maintenance cost, etc.

6.2.8 Applications of ECG

The ECG process is particularly effective for following applications:

(a) Machining parts made from difficult-to-cut materials, such as sintered carbides, creep-resisting (Inconel, Nimonic) alloys, titanium alloys, and metallic composites
(b) Applications similar to milling, grinding, cutting off, sawing, and tool and cutter sharpening
(c) Production of tungsten carbide cutting tools, fragile parts, and thin walled tubes
(d) Grinding of Inconel 718 using brazed diamond wheel
(e) Grinding of ceramic-metal tungsten plate by using diamond metal bonded wheel
(f) Machining of metal bonded diamond composite wheels
(g) Removal of fatigue cracks from steel structures under seawater. In such an application holes about 25 mm in diameter, in steel 12 to 25 mm thick, have been produced by ECG at the ends of fatigue cracks to stop further development of the cracks and to enable the removal of specimens for metallurgical inspection
(h) Producing specimens for metal fatigue and tensile tests
(i) Machining of carbides and a variety of high-strength alloys, etc.

6.2.9 Advancements in ECG

The dressing of the grinding wheel of ECG can be possible in the same system by reversing the connection of DC power supply. Work piece is connected with negative terminal and grinding wheel is connected with positive terminal and run for some time. The metal bond materials are removed due to electrochemical dissolution and the fresh abrasive grains are exposed. The material chips occupied is the inter abrasive gap can also be chemically dissolved by this process. Therefore the loading and glazing problems of conventional grinding can be solved in this electrochemical grinding process.

Electrochemical grinding (ECG) can be used to grind metal (bronze) boded diamond wheel by using nickel plated diamond composite form tool. This wheel was again used to grind ceramics. The ECG wheel wear is minimum and the wheel can be used for longer periods without redressing.

There is a risk of overcut with ECG but this can be controlled by optimizing the process parameters in ECG [10].

Electrochemical Grinding of composite materials faced major difficulties of achieving good surface integrity in grinding composite materials because of differential dissolution rates of different ingredients present in those materials. Analysis of the effect of voltage on surface texture in electrochemical grinding of cemented carbide of grade P-20 has been done using autocorrelation function to study the relative contribution of electrolytic dissolution of composite materials [11].

A hybrid process for grinding of precision small holes of stainless steel of 300 μm thickness has been developed by electrochemical material removal and mechanical grinding [12]. In the proposed process, nickel bonded coated diamond abrasive wheel as a cathode tool rotates at high speed and removes material electrochemically and mechanically from a pre-machined pilot hole. Aqueous Sodium Nitride is used as electrolyte. Machining voltage (3.5–7.5 V), tool rotation speed (10,000–3000 RPM), tool feed rate (4–6 μm/s), electrolyte concentration (10%) were taken for ECG process. Precision holes of diameters of 0.6 mm were obtained without burrs using .5 mm diameter tool. The finest surface roughness of R_a value was achieved as 0.21 μm. The schematic diagram of the hybrid process of ECG for micro-grinding operation is shown in shown in Fig. 6.9.

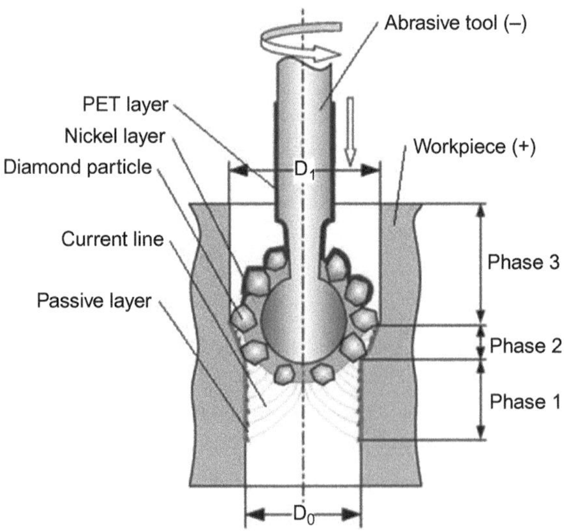

Fig. 6.9 Schematic view of hybrid process of ECG for micro-hole grinding [12].

Brazed diamond wheel in ECG has been used to grind Inconel 718 to prolong the wheel life instead of the electrodeposited diamond wheel [13]. Tool durability has been improved from 15 to 50 h when a brazed diamond wheel was employed. A electrode feed rate of 6.6 mm/min was obtained at applied voltage, 30 V and electrolyte ($NaNO_3$) temperature, 36 °C, respectively. Various mechanisms of abrasive tool wear in ECG such as molecular interactions, mechanical interactions, heat release, gases evolutions and electric discharge have been identified [14]. Different forms of cathode tool wear are abrasive wear, adhesion-diffusion wear, tribo-chemical wear, fatigue wear, abrasive-corrosion wear, hydrogenous and erosion wear in ECG. Wheel (cathode tool) wear in ECG is about 6 to 8 times lesser than that of mechanical grinding.

Diamond-ECM grinding of ceramic metal tungsten and sintered hard alloys of WC-Ni have been performed [15, 16]. Surface roughness R_a value <0.63 was achieved. No micro-cracks on the surface layer was observed at grinding speed of 12–18 m/s, operating voltage of 3–4 V, normal component of grinding force of 0.25–0.27 kN/cm^2. Aqueous solution of Na_2HPO_4 and Na_2CO_3 with additives was used as electrolyte for this purpose.

Electrochemical Grinding (ECG) is a mechanical assisted hybrid grinding process which has potential to remove the materials as well as finish the job surface and also to grind electrically conductive form tool and metal bonded grinding wheels, etc. The problems of ECM process such as oxide layer formation and slower removal of materials and demerits of mechanical grinding such as higher rate of wheel wear, loading and glazing problems are overcome by combining two processes in ECG. The grinding performances such as surface finish and rate of grinding in ECG can be improved by optimizing and controlling the process parameters of this hybrid process.

6.3 Electrochemical arc machining

6.3.1 Introduction

The rapid technological advancement in the field of Advanced Manufacturing Technology (AMT) has led to the development of newer and non-traditional techniques of machining in order to overcome the problems associated with the shaping of the harder, tougher and stronger workpiece materials such as high-strength-temperature resistant (HSTR) alloys, various types of carbides, fiber reinforced composite materials. Stellites, ceramic materials, various composite tool materials, etc. which are difficult to machine with the traditional methods of machining processes. However, tremendous attention is paid toward the non-traditional machining processes which are thermo-electric or physico-chemical characteristic oriented where the machinability has been more associated with the material properties such as electrical conductivity, melting temperature, electrical resistivity, atomic valence and other thermo-electric or physico-chemical properties of the energy transfer media rather than only on the mechanical properties of the workpieces.

The Electrochemical machining (ECM) and Electrodischarge machining (EDM) processes are widely used in modern manufacturing industry for the machining of the conductive materials irrespective of their hardness, toughness, ductility, shape features, etc. In Electrochemical machining (ECM), the dissolution of the anode takes place causing the shape of the machined part to confirm more or less to that of the cathodic tool. This process has many restrictions on the maximum attainable machining rate and surface finish, such as spark damage, cavitation, electrolyte boiling and difficulties in maintaining an equilibrium gap [17–19]. Again, the achievement of a high quality surface finish with reasonable machining rate is difficult in Electro discharge Machining (EDM) due to the limitations of the spark generators used in the process, the problems associated with the use of the liquid dielectric medium for the sparking action which, in turn, cause the metallurgical damage to the machined surface and also the other thermo-electrical and physical problems involved in the process.

A process that hold the potential to solve the above mentioned problems of ECM and EDM by substantially assisting the manufacturing systems for achieving improved product surface quality at an enhanced material removal rate condition has been called as Electro Chemical Arc Machining (ECAM) process. ECAM is developed through the optimal combination of the ECM and EDM process. In fact, in ECAM the ECM process is supplemented by a thermal erosion of workpiece material through electrical discharges in the machining gap filled with electrolyte and a very high metal removal rate as much as 5 to 40 times greater than ECM and EDM individually can be achieved [20] through optimal combinations of the process parametric conditions.

6.3.2 Working principle of ECAM

The Electrochemical Arc Machining (ECAM) is a machining process which combines the mechanism of Electrochemical Dissolution (ECD) and

Electrodischarge Erosion (EDE) by the application of a pulsed DC voltage between cathode (or tool) and anode (or workpiece) separated by a thin film of liquid electrolyte medium. On application of rectified DC pulsed voltage (20–45 V) between the cathodic tool and the anodic workpiece, the current passes across the machining gap between the electrodes through which an electrolyte solution, e.g., $NaNO_3$, NaCl, etc. is passed. The machining process is triggered by passing the current through the electrolytic cell and the anodic dissolution of the workpiece takes place through the electrochemical dissolution reactions. During electrochemical dissolution (ECD) process, the hydrogen, oxygen and chlorine gas bubbles and also the vapor bubbles are generated. However, the generation of electrolytic gas bubble, the growth of salt and gas bubble layers of low ionic conductivity in the comparatively small machining gap ($<100\,\mu m$) are some of the causes that lead to the induction of the discharges between the electrodes and the electrolyte solution, while a sufficiently high electric field is applied across the machining gap. When the discharge produced between the electrode and the electrolyte grows in size and bridges the gap between the electrodes, the arc discharges occurs instead of spark discharge. The workpiece material, thus, melted and vaporized due to the high temperature heating at the point of arc discharge. Though the ECAM process occurs at the constant machining gap between the tool and the workpiece, if the vibrational motion is given to the tool or the workpiece then the metal removal rate of the ECAM process will be increased.

So, in the ECAM process, the electrochemical dissolution (ECD) process is thermally assisted by the Electro-discharge Erosion (EDE) process and the idle ′off time′ intervals found in conventional EDM are replaced by the useful phases of the anodic dissolution. The electrochemical dissolution (ECD) and Electro discharge erosion (EDE) in ECAM process arise in discrete phases. Those can occur simultaneously in the course of ECAM in different regions of the machining gap or in succession in the same area. Hence, the material removal from the workpiece takes place due to both ECD and EDE actions so that the desired shape of the workpiece is generated with a predesigned tool used for the purpose of machining. Figure 6.10 shows the flow chart to represent ECAM as combined mechanism of ECM and EDM phenomena.

As already highlighted, in ECAM, the metal is removed from anodic work-piece with the help of a cathodic tool separated by a thin film of electrolyte by a succession of discrete phases of metal removal through Electrochemical Dissolution (ECD) and Electro-discharge Erosion (EDE).

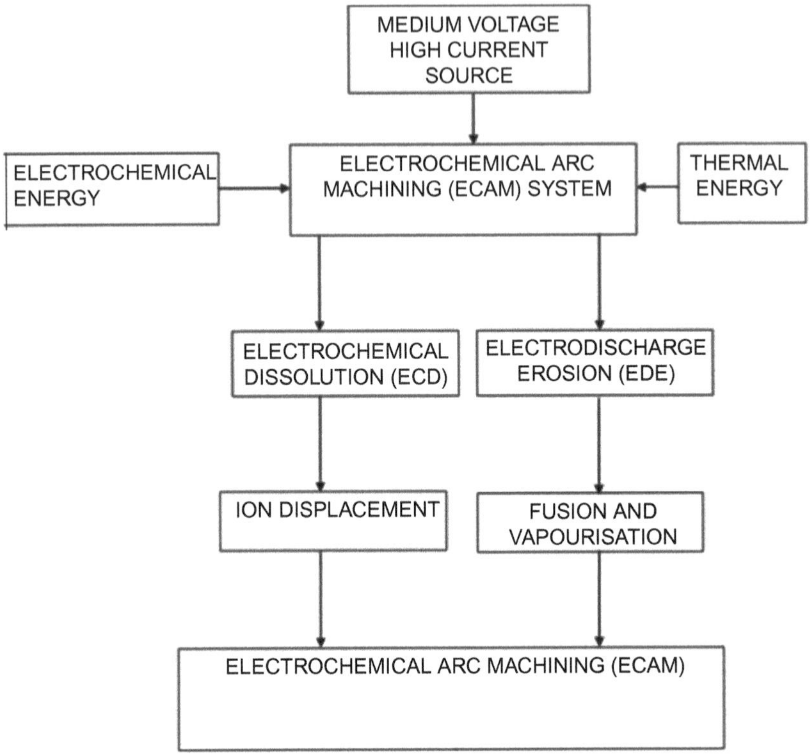

Fig. 6.10 Representation of ECAM as combined mechanism of ECM and EDM.

The latter phase occurs usually in the form of arc rather than spark discharge. The successive occurrence of the ECD and EDE process in ECAM are described in the subsequent stages. The scheme of material removal mechanism of ECD and EDE process in ECAM are shown in Fig. 6.11(a) and (b).

(a) Electrochemical dissolution (ECD) process in ECAM

The fundamental principle of the Electrochemical Machining process is based on the principle of Faraday's law of electrolysis. Since in ECM, the objective is to remove metal from the workpiece surface, hence, the workpiece and the tool are made the anode and the cathode of an electrolytic cell respectively and a DC voltage supply (10 to 20 V) is applied across them. The gap between the tool and the workpiece is filled with the suitable electrolyte solution (e.g., aqueous salt solution). When the current is passed and sufficient electrical energy is available, the metallic ion may be pulled out of the workpiece surface. The positive metallic ions will react with the negative hydroxide ions present in the electrolyte solution forming metallic hydroxides and other reaction compounds. Thus, the metal will be anodically

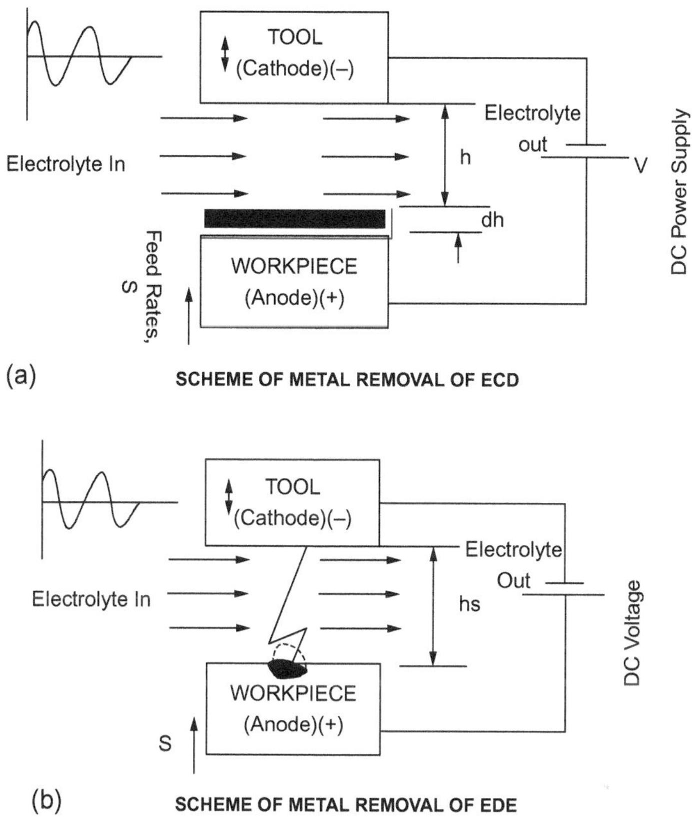

Fig. 6.11 The scheme of material removal mechanism of ECD (a) and EDE process (b) in ECAM.

dissoluted with the formation of sludge and precipitates. The dissolution rate is more when the gap is less and vice versa as the current density is inversely proportional to the gap.

The electrolyte is so chosen that the anode (workpiece) is dissoluted but no deposition takes place on the cathode (tool) as the cathode shape remains unchanged. The electrolyte is pumped at a high pressure and high flow velocity (e.g., 3 to 30 m/s.) through the small gap between the electrodes to remove the products of machining and to diminish the unwanted effects, such as those that arise with cathodic gas generation and electrical heating. As the machining proceeds with the simultaneous movement of the cathode (tool) toward the anode (or workpiece), the gap width gradually tends to reach a steady state value. Under these conditions, a shape, more or less complementary to that of the cathode (tool), will be reproduced on the anode.

A typical gap width should be maintained in the order of 0.1 to 0.5 mm and the average current density should be of the order of 50 to 600 A/cm^2.The tool and the workpiece are subjected to very large forces exerted by the high pressure of the electrolyte passing at a high flow rate through the small gap between the electrodes.

(b) The electro-discharge erosion (EDE) phenomena in ECAM

The ECAM process also involves controlled erosion of electrically conducting materials by the initiation of rapid and repetitive electrical spark discharge of relatively long duration or the arc discharge between the tool (or cathode) and the work piece (or anode) separated by the electrolyte solution as shown in Fig. 6.11(b).

When a suitable voltage is built up across the tool and the work piece, separated by a small gap filled with electrolyte the breakdown of the gas bubble, electrolyte vapor molecule, etc. formed due to electrochemical dissolution will take place owing to the development of a strong electrostatic field between the electrodes. The local ionization of the liquid molecules encourages the ignition of electrical discharges.

The cold emission of electrons will occur from the cathode surface toward the anode originating from the micro-irregularities on the electrode surface having the shortest gap in between. Thus, free electrons, liberated from the cathode surface, will be accelerated toward the anode by the electric field and will acquire a high velocity.

After gaining a sufficient velocity, the electrons collide with the molecules of the electrolyte medium breaking those into electrons and positive ions. The electrons so produced also accelerate and may ultimately dislodge the other electrons from the fluid molecules. Moreover, a small ionized fluid column will be formed owing to the establishment of an avalanche of electrons in the gap, with the larger and massive ions accelerating toward the cathode. The considerable flow of electrons along the ionized fluid column results a momentary current impulse or discharge.

Electrical discharge can take place in an electrolyte system of the ECAM process due to following reasons:

(i) Electrolytic gas generation at the electrode surface
(ii) Growth of layers with low ionic concentration near electrodes and formation of partially conducting oxide film on the anode surface due to ECD action
(iii) Steam generation and cavitation
(iv) Vapor blanketing of electrode surface
(v) Local stagnation of flow in the machining zone
(vi) Metal particles and sludge in the electrolyte, etc.

The liberation of energy accompanying the discharge leads to the generation of extremely high temperature, in the order of 8000 °C to 10,000 °C causing fusion or partial vaporization of the metal and the electrolyte at the point of discharge. The metal in the form of liquid drops is dispersed into the space surrounding the electrodes by the explosive pressure of the gaseous products in the discharge. This results in the formation of a tiny crater at the point of discharge in the workpiece surface. Since the spark occur at the spot where the tool and the workpiece surfaces are closest and the spot of sparking changes after each spark, the spark travels all over the surface. The spark discharge firstly occurs between the electrode and the electrolyte solution then it grows in size and bridges the gap between the electrodes. The series of sparking with long time duration are known as arc discharge. This results in a uniform material removal all over the surface due to arc discharge. So depending on the cathode (predesigned tool) shape, an identical impression is created on the anodic workpiece. To maintain the machining gap the feed motion is given to the workpiece (or anode). The electrolyte is passed through the small machining zone between the workpiece and the tool to remove the machining products and also the sludge.

Comparatively less material is eroded from the cathode (tool) as compared to the anode (workpiece) due to the following reasons:

(i) The momentum with which the positive ions strike the cathode surface is much less than the momentum with which the electron stream impinge on the workpiece surface.

(ii) A compressive force is generated on the tool surface by the arc which helps to reduce tool wear.

6.3.3 ECAM machining set-up details

The Electro-Chemical Arc Machining Set up consists of various subsystem in order to meet the objective of having maximized material removal rate in ECAM subject to the constraint of satisfactory surface finish and dimensional accuracy, etc. In order to control all the ECAM process parameters, ECAM machining set up consists of various systems such as:

(i) Mechanical hardware System, which includes the machining chamber, work piece and holding unit, work piece feeding unit and tool vibration unit

(ii) Electrolyte flow control system

(iii) Pulsed DC Power Supply Unit

The scheme of a typical electrochemical arc machining unit for workpiece drilling operation is shown in Fig. 6.12. A linear feed motion is given to the

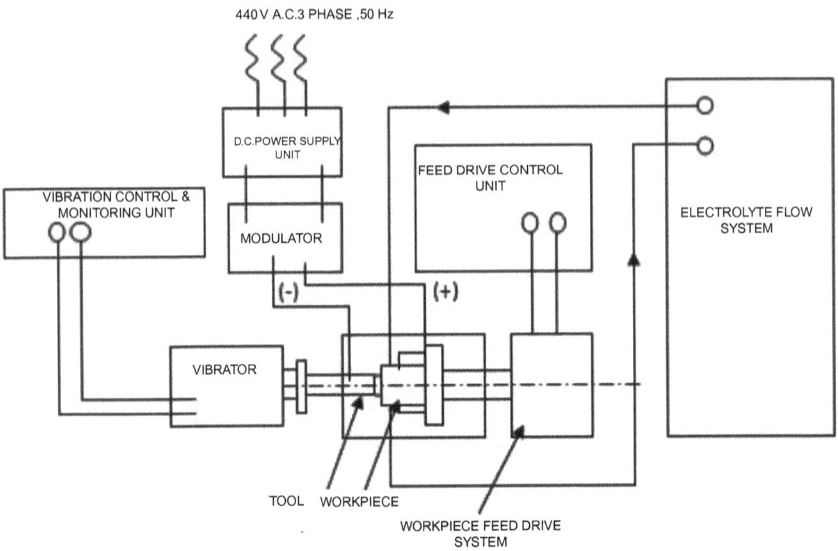

SCHEME OF ELECTROCHEMICAL ARC MACHINING UNIT

Fig. 6.12 The scheme of a typical Electrochemical Arc Machining unit.

anodic workpiece by the workpiece feed drive system so as to control the machining gap width between the tool and the workpiece that is increased due to the metal removal rate during machining. The feed drive control unit monitors the feed drive system to provide a wide range of feed rates that is required for controlled machining operation. The forced vibrational motion is given to the cathode tool by means of electro-mechanically driven vibrator. The vibrator control and monitoring unit monitors and controls the operation of vibrator so as to provide the different operating amplitude of vibration and frequency of the tool. The special types of power supply unit having the modulator coupled with a constant DC power source supply the pulsed DC voltage wave form (sinusoidal, rectangular or triangular) to the tool and the workpiece.

ECAM machining system controls over the following process parameters:

(a) The electrolyte (type, conductivity)
(b) The flow system (pressure, flow rate, flow pattern)
(c) The power supply
(d) Amplitude of tool oscillation
(e) Feed rate
(f) Inter electrode Gap dimension

The feature of controlling these process parameters are discussed as follows:

(a) Electrolyte

The Electrolyte used in ECAM performs many functions such as:

(i) Completing the electric circuit between the tool and the work piece
(ii) Allowing desirable machining reactions to occur
(iii) Carrying away products of reactions from the zone of machining
(iv) Allowing the formation of arc or spark discharge of electrolyte gas and electrolyte solution with smaller gap

The relative contribution of the Electrochemical Dissolution (ECD) and Electro-discharge Erosion (EDE) phases respectively decrease and increase with the increasing electrolyte conductivity. This effect is attributed to the corresponding increase in the rate of gas evolution thus promoting the discharge. The common electrolyte solutions used in ECAM process are aqueous $NaNO_3$ and NaCl solution, etc.

The electrolyte flow is necessary in ECD and in EDE phases for the removal of the machining products and evolved gases. The machined products generally consist of a hydroxide precipitate in ECD and small spherical metal particles of diameter upto 100 μm in EDE process.

(b) Electrolyte flow system

The evolved gas, while promoting discharges, is undesirable during ECM because it results in reduction of electrolyte conductivity. This, in turn, encourages pitting of the machined surface, especially in the downstream direction of flow. In addition, uniformity in the flow pattern is desirable in ECM, since it affects geometry and surface finish. The flow rates in ECAM will, therefore, depend on whether EC or ED action is to be promoted.

In the drilling operation, it was found that the electrolyte pressure could be present in the range 13 to 120 bar and flow rate of electrolyte was about 1 to 2.5 L/min [20].

(c) Power supply

Basically, the ECM uses a low voltage, high current DC power supply, where EDM uses a higher medium voltage, low current pulsed supply. Typically, an ECM power supply will be capable of providing several thousand amps. DC at 10 to 15 V and machining is carried out at a current density of 50 to 600 A/cm^2.

An EDM power supply might generate 90 V square wave pulses and have a current capability upto 200 A. The machining is carried out generally at a current density upto 10 amp/cm^2. There is considerable flexibility in choosing a power supply of ECAM and the choice will have a marked effect on

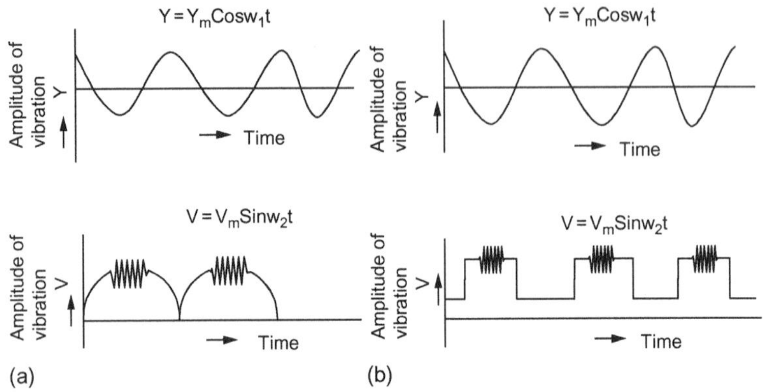

Fig. 6.13 Types of waveform of DC power supply and tool vibration wave form during ECD and EDE phases in ECAM process (a) sinusoidal (b) rectangular.

relative contribution of EC and ED action. Essentially, the power supply must be polarized and contain both DC and AC components.

Figure 6.13 shows the types of waveform of DC power supply during ECD and EDE phases in ECAM Process. Figure 6.13(a) shows the sinusoidal waveforms obtained full wave rectification of the mains voltage supply with tool vibration. Figure 6.13(b) shows the rectangular waveforms obtained full wave rectification of the mains voltage supply with tool vibration. These waveforms produce within each pulse, an EC phase followed by an ED phase, followed by another EC phase.

The earlier researchers have found that it is possible to remove large surface irregularities from plane parallel electrodes using either smooth and unsmooth DC in a "strong" (200 g/L $NaNO_3$) electrolyte solution [21].

(d) Amplitude of tool oscillation

The electrochemical arc machining process can occur with small gap between the electrodes without vibration of the tool. But, the vibrational motion is incorporated with the tool to improve the performance of the ECAM machining process. The amplitude of tool oscillation has a pronounced effect on the discharge process. Increasing the vibration amplitude decreases the time for which the tool is in the discharge zone during each cycle of the tool oscillation, which will limit the amount of metal removed.

The amount of ECM component present in ECAM is also limited by the amplitude of the tool oscillation since the average machining gap increases with a corresponding decrease in the metal removal rate with increasing amplitude. Figure 6.13(a) and (b) show the relative position between the voltage and vibration waveforms with sinusoidal waveform and rectangular waveform respectively. The suitable range of amplitude of vibration is generally used in the order of 0.1 to 0.5 mm. The frequency of tool vibration generally varies from 10 to 100 Hz. The forced vibration of tool electrode causes the increment of the machining efficiency, metal removal rate as well as decrement in the roughness of machined surface by controlling the stability of the process and also by synchronizing tool vibration waveform and electrical pulse wave form during electrical discharge phenomena [22, 23].

(e) Feed rate of electrode

As the machining due to the ECAM process is in progress, the gap between the electrodes increases ultimately causing the decrease in ECE and EDE actions. The linear feed motion is given to one of the electrodes (tool or workpiece) to maintain the effective gap thickness between the electrodes.

Analysis of the kinetics of the process and the review of the previous researches also reveal in the fact that as the federate increases, the rate of change of gap decreases and the ECE and EDE action will become more which will cause increase in the metal removal rate. At low feed rate, the material removal mainly takes place due to the ECD action and at high feed rate the EDE phase will be comparatively more. The further increase in feed rate may cause the contact between the electrodes and the tool may leads to greater tool wear.

(f) Inter electrode gap dimension

The gap thickness has a marked effect on Electrochemical (EC) and Electro discharge (ED) contribution. Duration of the EC and ED phases within a rectangular pulse respectively rise and fall with the increasing gap was observed [24]. The ECE and EDE can occur simultaneously in the course of ECAM, in different regions of the gap or in succession in the same area.

It was found from drilling experiment that the inter electrode gap in ECAM can be roughly divided into three regions as shown in Fig. 6.14 [25]. The first region occurs over the leading gap where the separation between the tool and the work piece varies in a sinusoidal manner. In this region, the metal is removed mainly by the thermal erosion. The electrical discharge arise because of the high contamination of the electrolyte in the gap by the machining products which are favorable to EDE. The formation of discharge is also promoted by the process conditions of high feed rates and small gaps which lead to the high current densities and the prospect of arcing. A slightly larger gap over the second region allows ECD to occur. The third region is similar to that found in the side gap of conventional ECM and the machining precedes by ECD.

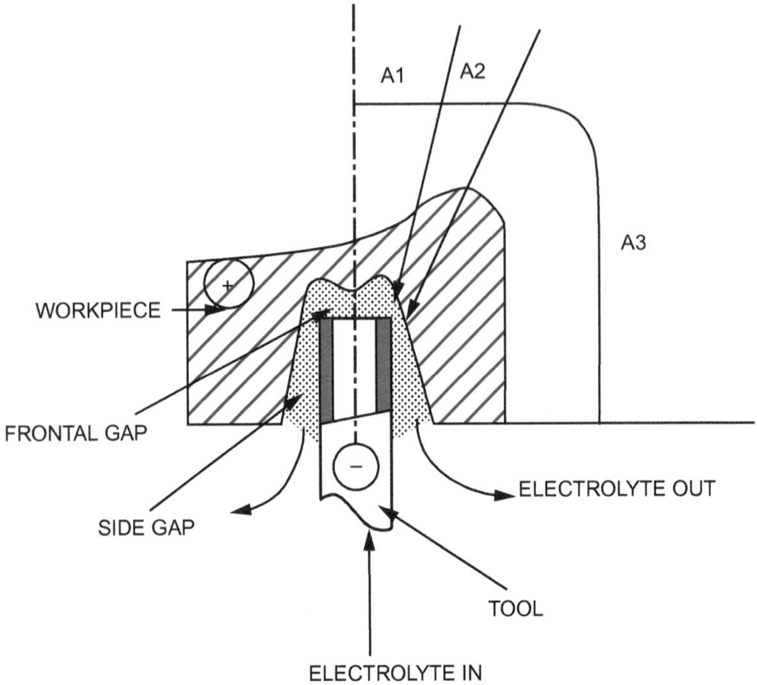

Fig. 6.14 Schematic representation of three regions of machining in inter electrode gap [25].

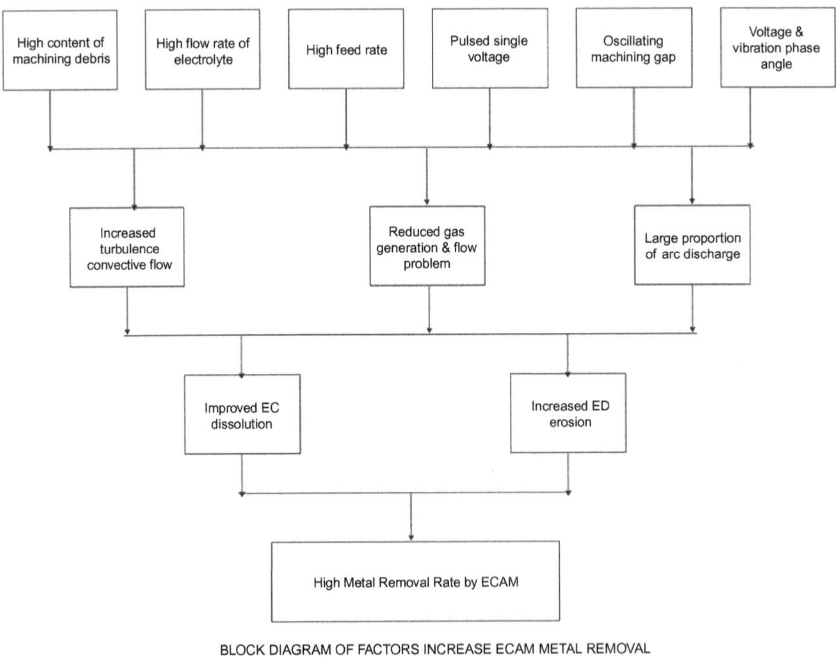

Fig. 6.15 Effect of level of various factors for enhanced MRR in ECAM.

Many other factors which contribute for the higher metal removal rate in ECAM are indicated in schematic diagram as shown in Fig. 6.15. The various factors such as high content of the machining debris, high flow rate of the electrolyte, oscillating machining gap, high feed rate, pulsed sinusoidal DC voltage supply and the phase difference are responsible for the improved Electrochemical (EC) action as well as increased Electro-discharge (ED) action so that the total metal removal rate of ECAM is increased.

6.3.4 Material removal rate in ECAM

Since, the ECAM process is very much complicated, it is not easy to predict theoretically the metal removal rate for the ECAM process. If the separate model for the ECD and the EDE components in ECAM are available and the relative duration of both the discrete ECD and EDE phases are known, then the total MRR of the ECAM can be evaluated by plugging the two separate MRR models of ECD and EDE components.

Hence, the total metal removal rate (MRR) in ECAM process is the sum total of the product of the partitioning factors and the metal removal rate of corresponding to both ECD and EDE phases.

Thus, the total MRR in ECAM can be represented as:

$$MRR_{ECAM} = K_1 x\, MRR_{ECD} + K_2 x\, MRR_{EDE} \tag{6.24}$$

where, K_1, partitioning factor for ECD phase in ECAM; K_2, partitioning factor for EDE phase in ECAM; MRR_{ECD}, metal removal rate through ECD in ECAM; MRR_{EDE}, metal removal rate due to EDE in ECAM.

(A) Mathematical modeling for metal removal rate (MRR) evaluation through ECD in ECAM

The mathematical modeling for the metal removal rate (MRR) evaluation through Electrochemical Dissolution (ECD) has been theoretically developed with a view to generating information based on the consideration of the involvement of the ECD components in ECAM. The analysis has been made to study the effects of the process parameters, such as feed rate, phase angle between the supply pulsed voltage waveform and vibration waveform, frequency of tool or work piece vibration, amplitude of vibration, machining voltage on the machining criteria, e.g., metal removal rate [26].

The modeling for the analysis is mainly based on the assumptions that ohm's law and faraday's laws of electrolysis validate in the machining gap.

While deducing the related mathematical modeling with Electro-Chemical Dissolution (ECD) in ECAM, the following assumptions are made in general [27].

(i) The configuration of the tool and the workpiece are plane and parallel
(ii) The electrolyte conductivity in the inter-electrode gap width is constant
(iii) The ohmic component of the applied voltage is really considered to be dominant
(iv) The edge effects due to the electric field at the sides of the electrodes are neglected, e.g., the current flux lines are parallel across the gap width

The variability of the following factors along the gap width is not taken into consideration, e.g., temperature, volumetric concentration of gas phase, electrolyte flow velocity, etc.

Based on the faraday's law of electrolysis, considering the current efficiency, η_c, the mass of ions dissolved (m in g) can be given by

$$\mathrm{m} = \frac{M_x.\eta c}{V_x.F}.\mathrm{I.t} \tag{6.25}$$

where, M_x, atomic weight of anodic material, g; V_x, valency of anode material; F, faraday's constant; I, the current passing through the machining gap, amps; t, machining time, s.

The geometrical features and conditions of the considered electrodes are shown in Fig. 6.11(a).

If the initial gap h increases by an amount dh in time dt, then the metal is removed by an amount dh in time dt. Thus, the metal removed (dm) during the period of time, dt can be written as:

$$\mathrm{dm} = \rho_a \mathrm{A.dh} \tag{6.26}$$

where, ρa, density of the workpiece (anode); g/cm^3; A, machined area, cm^2.

$$\text{Again, dm} = \frac{M_x.\eta c}{V_x.F}.\mathrm{I.dt}$$

$$\rho_a \mathrm{A.dh} = \frac{M_x.\eta c}{V_x.F}.\mathrm{I.dt}$$

$$\frac{dh}{dt} = \frac{M_x.\eta c}{V_x.F.\rho_a}.\frac{I}{A} \tag{6.27}$$

This equation shows the rate of increase in the gap thickness with respect to time, t.

When the workpiece is being fed with a constant velocity S toward the tool, then the rate at which the gap between the workpiece and the tool surfaces changes is given by

$$\frac{dh}{dt} = \frac{M_x.\eta c}{V_x.F.\rho_a}.\mathrm{J} - \mathrm{S} \quad (6.28)$$

The important consideration is that the relative motion between the surface is considered rather than the actual motion of an electrode. So even though the feed motion is normally given to the tool, the analysis remains valid.

The metal removal rate (MRR) due to ECD and the steady state gap thickness can be evaluated for possible combinations of pulsed DC supply voltage waveform with tool vibrational motion are as follows:

(a) Full wave rectified sinusoidal pulsed supply voltage waveform with tool vibration.

(b) Rectangular pulsed voltage supply waveform with vibrational motion of the tool.

Case (a): Sinusoidal pulsed supply voltage waveform with tool vibration When pulsed voltage (full wave rectified) is applied across the two electrodes, the supply voltage waveform is represented as

$$\mathrm{v} = \mathrm{v_m} \sin \mathrm{w_2 t} \quad (6.29)$$

where, v_m, the maximum voltage amplitude volt; fp, $\omega_2/2\pi$ is the frequency, Hz; ω_2, the angular velocity of the voltage waveform, rad/s; t, time, s.

If, vibrational motion is given to the cathode (tool), the form of vibrational motion can be considered to be represented as

$$\mathrm{y} = \mathrm{y_m} \cos \mathrm{w_1 t} \quad (6.30)$$

where, y_m, the amplitude of vibration, cm; ω_1, angular velocity of vibration, rad/s.

$$\mathrm{y_m} = 2\pi fv$$

where, fv, the frequency of vibration.

Now, over each full wave voltage pulse and oscillation of the cathode, the relative configuration of the voltage and oscillation waveforms are represented in Fig. 6.13(a). The preferable condition of ECAM arises when the inter electrode gap is at its minimum value even when the voltage has risen to its highest value. For these conditions, the vibration wave form of cathode has already been shown to be same as depicted through Eq. (6.30).

From the Ohm's law, the current density at any instant is represented as

$$J = \frac{I}{A} = \frac{Ke.V}{hi} = \frac{Ke.[V_m \sin w_2 t]}{[h + y_m \cos w_1 t]} \tag{6.31}$$

The rate of change of the gap can be represented from Eqs. (6.28), (6.31) as:

$$\frac{dh}{dt} = \frac{M_x.\eta c}{V_x.F.\rho_a} \frac{Ke.[V_m \sin w_2 t]}{[h + y_m \cos w_1 t]} - S \tag{6.32}$$

The steady state condition will arise after some times and gap becomes the steady state gap (hs) when the rate of change of gap dh/dt) will be equal to zero.

The effective feed rate at steady state condition, from Eq. (6.32) is represented as:

$$S = \frac{M_x.\eta c}{V_x.F.\rho_a} \frac{Ke.[V_m \sin w_2 t]}{[hs + y_m \cos w_1 t]} \tag{6.33}$$

The steady state gap (hs) at any time, t is given by

$$hs = \frac{M_x.\eta c.Ke}{V_x.F.\rho_a.S}[V_m \sin w_2 t] - y_m \cos w_1 t \tag{6.34}$$

The mass of metal removed per unit time (g/s) due to the Electrochemical dissolution can be expressed as:

$$\begin{aligned} MRR_{ECD} &= \frac{M_x.\eta c}{V_x.F}.I \\ &= \frac{M_x.\eta c}{V_x.F}.\frac{V.Ke.A}{hs} \\ &= \frac{\frac{M_x.\eta c}{V_x.F} A.Ke[V_m Sin w_2 t]}{\frac{M_x.\eta c.Ke}{\rho_a.S.V_x.F}[V_m Sin w_2 t] - y_m Cos w_1 t} \end{aligned} \tag{6.35}$$

If, Ø is the phase angle between the supply pulsed voltage waveform and vibration waveform, in radians, then the supply pulsed voltage waveform can be represented as

$$V = V_m \text{ Sin } (w_2 t + Ø) \tag{6.36}$$

Then, the steady state gap (hs) becomes:

$$\mathrm{hs} = \frac{M_x.\eta c.Ke}{V_x.F.\rho_a.S}.[\mathrm{V_m} \sin(\mathrm{w_2 t} + \text{Ø})] - \mathrm{y_m} \cos \mathrm{w_1}\,\mathrm{t} \tag{6.37}$$

The metal removal rate (MRR) due to Electrochemical dissolution becomes as

$$\mathrm{MRR_{ECD}} = \frac{\frac{M_x.\eta c}{V_x.F} A.Ke[V_m Sin\,(w_2 t + \text{Ø})]}{\frac{M_x.\eta c.Ke}{\rho_a.S.V_x.F}[V_m Sin\,(w_2 t + \text{Ø})] - y_m Cos\, w_1 t} \tag{6.38}$$

The Eq. (6.38) will help analyzing the effects of amplitude of vibration, voltage, feed rate and phase angle on the metal removal rate of ECD component in ECAM.

Case (b): A Rectangular pulsed supply voltage and vibration with the cathodic tool When a rectangular pulsed voltage is supplied between the electrodes and the vibrational motion is given to the cathode (tool).

So, the supply voltage, $\mathrm{V} = \mathrm{V_m}$ and the vibration waveform is as:

$$\mathrm{y} = \mathrm{y}_m \cos \mathrm{w_1}\,\mathrm{t}$$

The metal removal rate (MRR) due to ECD under such a case can be represented as

$$\mathrm{MRR_{ECD}} = \frac{\frac{M_x.\eta c}{V_x.F}.\mathrm{V_m.A.Ke}}{\frac{M_x.\eta c.Ke.V_m}{V_x.F.\rho_a.S} - \mathrm{y}_m \cos \mathrm{w_1 t}} \tag{6.39}$$

The steady state gap (hs) can be determined from Eq. (6.32) as:

$$\mathrm{hs} = \frac{M_x.\eta c.Ke.V_m}{V_x.F.\rho_a.S} - \mathrm{y}_m \cos \mathrm{w_1}\,\mathrm{t} \tag{6.40}$$

(a) Development of mathematical model for MRR evaluation due to EDE in ECAM

The metal is removed from the workpiece with the production of craters in the work surface due to electrodischarge erosion when the discharge takes place at the gap between the electrodes filled with the gas bubbles and vapor bubbles generated during ECD and also the electrolyte solution. The volume of crater depends on the pulse energy for each discharge. Moreover, the same set of assumptions as made for ECD modeling have been considered to validating while evaluating the EDE contribution to ECAM.

The scheme of the metal removal due to electro discharge erosion (EDE) is represented in Fig. 6.11(b).

Now, the energy delivered in a single pulse (wp) through the gap resistance due to Electro discharge erosion.

$$\text{wp} = \int_0^{tp} I^2 R dt \tag{6.41}$$

$$= \int_0^{tp} \frac{V^2}{R} dt$$

where, I, current passing through the electrodes, amp., tp, voltage pulse duration, seconds, R, gap resistance between electrodes, ohm, V, the applied voltage between the electrodes.

The pulse energy utilized in electro discharge erosion is equal to $\eta_p . w_p$ where, η_p, efficiency of pulse energy utilization. The volume of crater (vc) formed due to electro discharge action is proportional to the pulse energy of spark received by the work surface [28].

The mass material removal rate due to electro discharge erosion(MRR $_{EDE}$)can be represented as

$$\text{MRR}_{\text{EDE}} = \frac{1}{tp} . \rho_a . \text{C}. \eta p \int_0^{tp} \frac{V^2}{R} \, \text{dt g/sec} \tag{6.42}$$

where, ρ_a, density of anodic material.

C is a factor (cm^3/watts) whose value depends on material properties such as density, ultimate tensile strength, melting temperature, etc. and also the supply voltage [26].

Now, as the supply voltage can be either rectangular shaped or sinusoidal nature and the machining also can be done with vibration of the tool as well as without any tool vibration, hence, two different sets of condition as shown in Fig. 6.13 can be achieved under which the machining can be done, and in each case the form of the above derived Eq. (6.42) will be slightly different keeping in conformity with operating conditions.

The mathematical modeling for MRR $_{EDE}$ under the above mentioned conditions are derived as described here-in-under.

Case (a)I: Sinusoidal pulsed voltage and with vibrational motion to the tool

The pulsed voltage (full wave rectified) is applied between the electrodes whose waveform can be represented as V = Vm Sin (w_2t + Ø). The gap resistance, R can be represented as $R = \frac{hs}{Ke.A}$. Hence, the metal removal rate (MRR) in gm/s due to EDE can be written from Eq. (6.42) as:

$$MRR_{EDE} = \frac{1}{tp}.\rho_a.C.\eta_p \int_0^{tp} \frac{V^2 Ke.A}{hs}.dt \tag{6.43}$$

$$= \frac{1}{tp}\frac{\rho_{a.C.\eta_p.Ke.A}}{hs} \int_0^{tp} V_m{}^2 \sin^2(w_2 t + \varnothing)\, dt$$

$$= \frac{\rho_{a.C.\eta_p.Ke.A} V_{m^2}.}{2.tp.hs} \left[\frac{tp - \sin 2\,(w_2 tp + \varnothing)}{2w_2}\right] \tag{6.44}$$

Case (b): Rectangular pulsed supply voltage waveform between the electrodes and sinusoidal vibration motion with the cathodic tool

For rectangular pulsed voltage waveform supply voltage and sinusoidal vibration motion of tool, the metal removal rate (MRR) due to EDE in gm/s is represented as

$$MRR_{EDE} = \frac{1}{tp}.\rho_a.C.\eta p. \int_0^{tp} \frac{Vm^2.Ke.A}{hs}.dt.$$

$$= \frac{\rho_{a.C.\eta_p}.Vm^2 Ke.A}{hs} \tag{6.45}$$

Identification of the partitioning effects of ECD and EDE phases in ECAM The Electrochemical Arc Machining (ECAM) is a combined process of Electrochemical dissolution (ECD) and Electrodischarge erosion (EDE) action which occurs in discrete phases within a single pulse duration of pulsed DC supply voltage [29]. The relative duration of ECD and EDE phases within a single voltage pulse duration can be distinguished by experimental studies [30]. If the duration of ECD and EDE phases within a single pulse duration time (tp) are t_{ECD} and t_{EDE} respectively, then the partitioning factors which is the portion of time of occurrence of each phase with single pulse duration time are determined as analyzed in the subsequent stages. The partitioning factor (K_1) for the ECD phase can be represented as:

$$K_1 = \frac{t_{ECD}}{t_p} \tag{6.46}$$

Again, the partitioning factor (K_2) for the EDE phase can be determined as:

$$K_2 = \frac{t_{EDE}}{t_p} \tag{6.47}$$

Theoretically, $K_1 + K_2 = 1$, but experimentally

$$K_1 + K_2 + K_L = 1 \tag{6.48}$$

where, K_L, loss factor arising out of the initial electrical characteristic variations.

At the initial stage, when the voltage pulse is applied, the voltage and current oscillate at high frequency for a short period of time. At this condition, no material removal takes place. Thus, the loss factors can be minimized by power supply circuit modification.

After obtaining the partitioning factors for both ECD and EDE phases the combined mathematical model for evaluation of the metal removal rate (MRR) can be established.

The metal removal rate of the electrochemical dissolution (ECD) and electrodischarge erosion (EDE) components in ECAM for the purpose of analysis are evaluated on the basis machining input data. The values of the various input data based on some real time machining have been considered as follows:

$$A = 0.122\,\text{cm}^2, M_x = 56\,\text{g.}, V_x = 2, \rho_a = 7.87\,\text{g/cm}^3,$$
$$K = 0.1\,\text{ohm}^{-1}\,\text{cm}^{-1}, F = 96,500\,\text{amps}, fp = 50\,\text{Hz}, f = 100\,\text{Hz},$$
$$\eta_c = 1.0 \text{ and } \eta_p = 1.0$$

For the purpose of analysis, the workpiece materials and electrolyte combination have been chosen as C-20 steel and $NaNO_3$ salt solution respectively.

6.3.5 Influences of ECAM process parameters on machining characteristics

Influences of various process parameters on machining characteristics such as MRR, Surface Finish, etc. of ECAM process are to be studied to determine the range of process parameters for better machining performance. As material removal in ECAM occurs due to ECD and EDE phenomena, the influence of ECAM process parameters on ECD components of MRR and EDE components are discussed here in after.

(i) Effects of process parameters on ECD component of MRR in ECAM
The effects of the various process parameters on metal removal rate (MRR) due to ECD component in ECAM are quantitatively assessed through the

earlier developed mathematical model and experimental observations. The main process parameters which control the electrochemical dissolution (ECD) process are identified as:

(a) Feed rate of the workpiece
(b) Machining voltage
(c) Amplitude of tool vibration

The influences of the various process parameters on the ECD components of MRR for the considered operating condition have been analyzed with the consideration that the sinusoidal pulsed DC voltage is supplied between the electrodes and the vibrational motion is given to the tool.

Figure 6.16 shows that with the increase in feed rate, the metal removal rate due to ECD component increases because the machining gap reduces with high feed rate which increases the current flow and hence there is the increment of metal removal due to ECD with lower machining gap. The rate of increase in MRR due to ECD becomes less when the workpiece feed rate varies from 15 to 25 mm/min and the effect of the feed rate on MRR is less at higher rage of feed rate of workpiece.

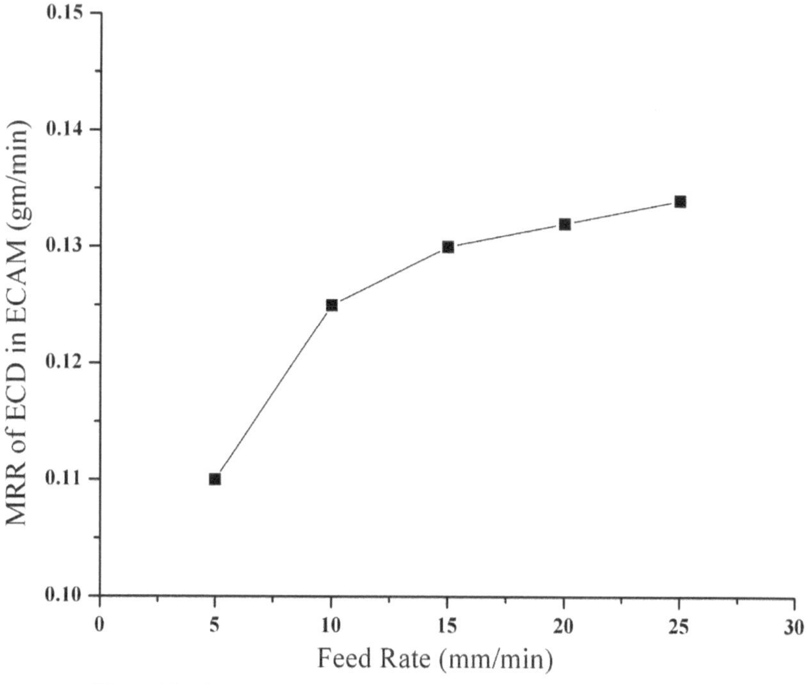

Fig. 6.16 Effect of feed rate on MRR of ECD phase.

The metal removal rate due to the electrochemical dissolution (ECD) in ECAM linearly increases with the machining gap voltages as has been exhibited through Fig. 6.17. For constant values of feed rate, phase angle, amplitude of vibration, more current will flow through the machining gap due to the increase in machining voltage. From the Faraday's laws of electrolysis. It can be explained that the amount of material dissoluted from the work surface (or anode) is more as high current passes through the interelectrode gap.

Again, the effect of the amplitude of the tool vibration on the metal removal rate (MRR) of electrochemical dissolution (ECD) phase present in ECAM is shown in Fig. 6.18. The amount of metal removed through ECD is reduced with the increase in the amplitude of tool oscillation because of the fact that the average machining gap increase with the increment of the amplitude of tool vibration which corresponds to the decrement in the metal removal rate. The curve of MRR is parabolic in nature and has high value of MRR at very low vibration tool amplitude.

(ii) The process parametric influences on EDE component on MRR in ECAM

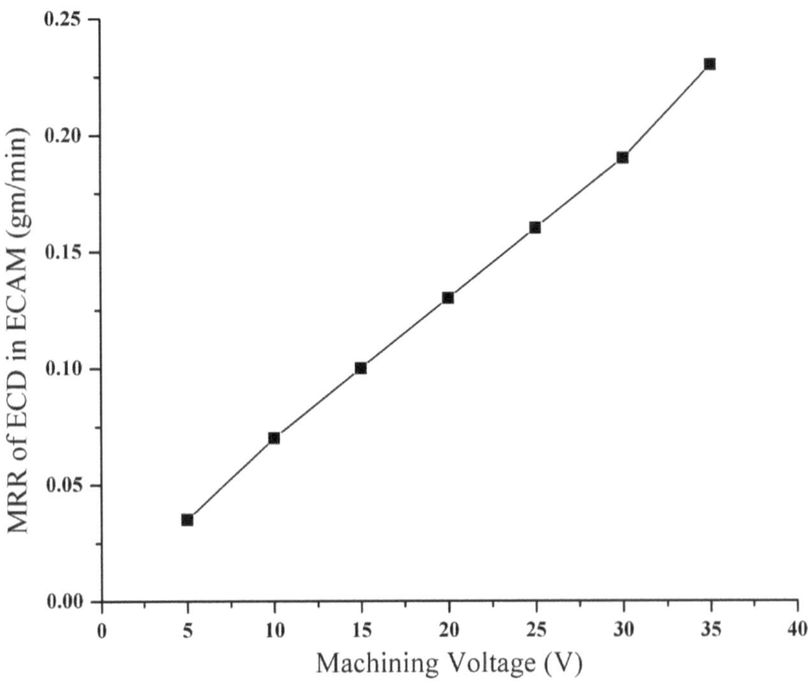

Fig. 6.17 Effect of machining voltage on MRR of ECD phase.

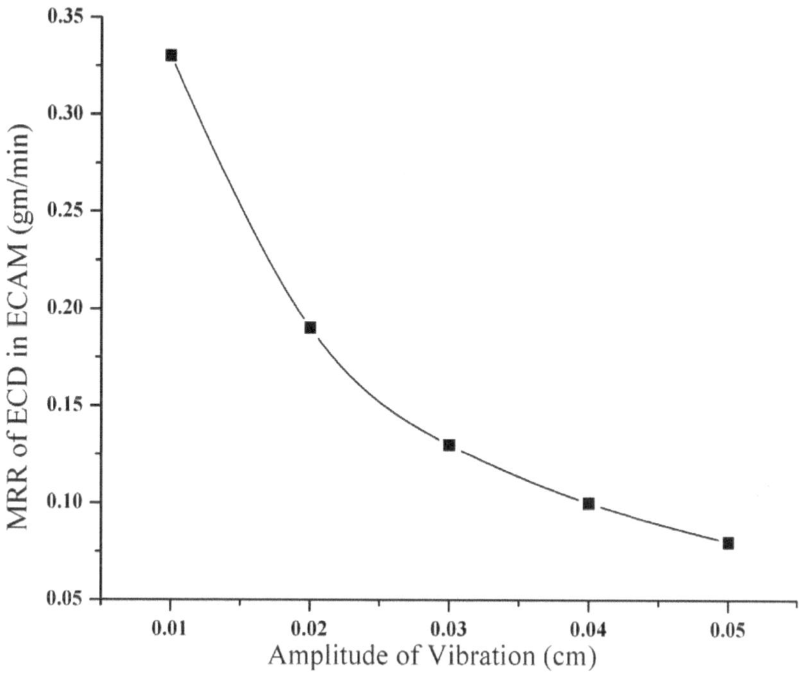

Fig. 6.18 Effect of amplitude of tool vibration on MRR of ECD phase.

The influence of the process parameters on metal removal rate (MRR) of electrodischarge erosion (EDE) components in ECAM are discussed for the sinusoidal pulsed voltage supply and with the tool vibration.

The results are obtained on the basis of the following machining conditions such as:

(i) Amplitude of vibration, $y = 0.04\,\text{cm}$.
(ii) Feed rate, $S = 5\,\text{mm/min}$.
(iii) Phase difference between the voltage waveform and vibration waveform $= 18°$.

The best fitted regression line is also obtained to establish the co-relation between the machining voltage and the value of C as follows [31]:

$$C = \left(197.98 - 72.5521\text{Vm} + 9.42772\,\text{Vm}^2 - 0.558815\,\text{Vm}^3 + 0.0154828\,\text{Vm}^4 - 0.000162485 \times \text{Vm}^5\right)/10{,}000 \qquad (6.49)$$

The MRR (EDE) is mostly affected by the main process parameters such as:

(i) Feed rate of workpiece
(ii) Amplitude of tool vibration
(iii) Machining voltage amplitude

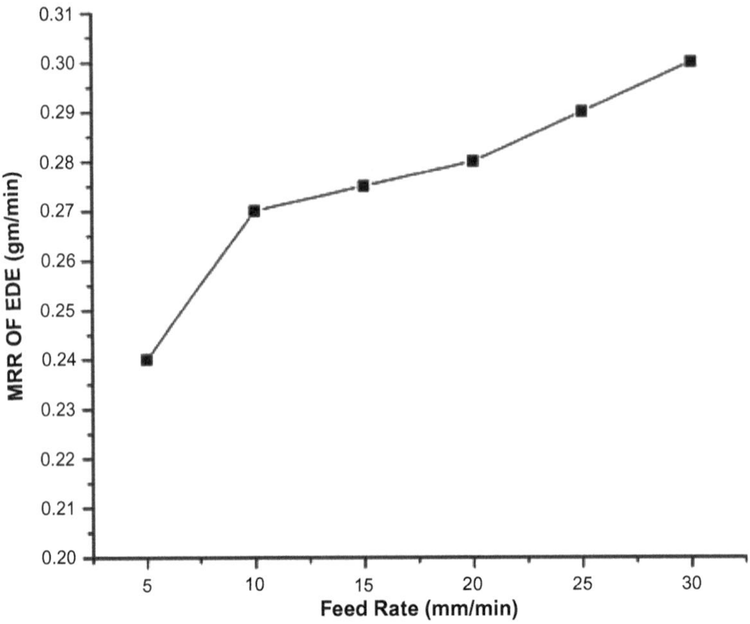

Fig. 6.19 Effect of feed rate on MRR of EDE phase.

From Fig. 6.19 it is observed that the metal removal rate of EDE component in ECAM increases with the increase in the feed rate and then reaches a maximum value. The increment of feed rate causes the decrement of the machining gap, which, in turn, causes an increase in the sparking component during each voltage pulse and thus, results in greater MRR. As the feed rate is further increased, the metal removal rate increases because of the fact that the amount of sparking action increases. The sparking action affects both the tool and workpiece and an increasing amount of metal is eroded from tool electrode. A rise in the cathode tool wear, which may increase the instantaneous inter-electrode gap width, with a reduction in the amount of sparking.

The amplitude of the tool oscillation has a pronounced effect on the electrodischarge erosion (EDE) phases. The increment in the vibration amplitude causes the time for which the tool is in the discharge zone during each cycle of the tool oscillation to decrease, which thus reduce the amount of metal removal. This argument is applicable because of the pattern as shown in Fig. 6.20 which exhibits that the metal removal rate drops off sharply as the amplitude of vibration is increased from 0.01 to 0.05 cm. An amplitude

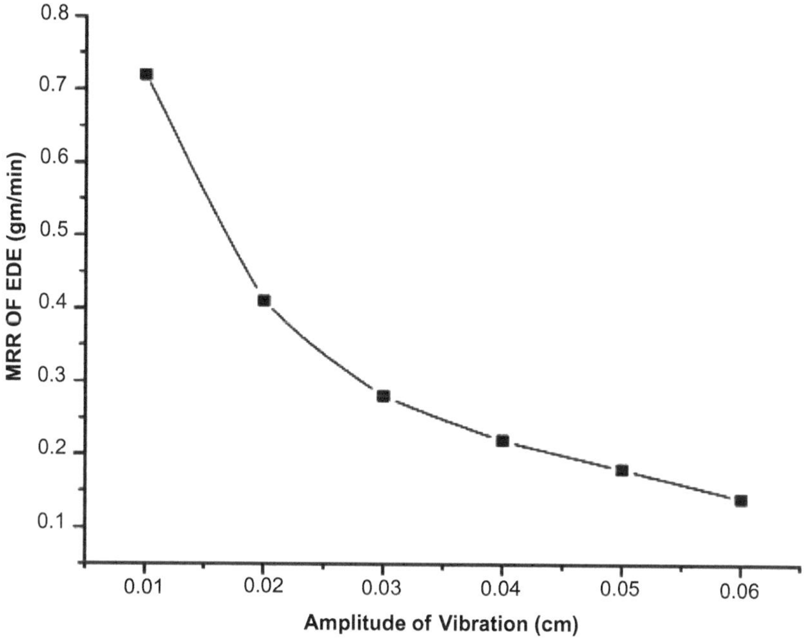

Fig. 6.20 Effect of amplitude of tool vibration on MRR of EDE phase.

of 0.03 cm has been chosen for the majority of experiments since it generates more stable machining [26].

Again, the effect of machining gap voltage on the amount of material removed per unit time due to electro discharge erosion (EDE) components as indicated in Fig. 6.21. At the low machining voltage ranging over 10 to 22 V, the material removal due to EDE component in ECAM is small. As the gap voltage increases, the more discharge action takes place between the electrodes, hence the MRR of EDE component increases.

The effects of the feed rate and the machining voltage on the partitioning factors (k_1 and k_2) for both the Electrochemical dissolution (ECD) and Electro Discharge Erosion (EDE) phases as well as loss factors (K_L) are shown in Fig. 6.22. Fig. 6.22 indicates that as the voltage increases, the partitioning factor for ECD phase (k_1) increases and the partitioning factor for EDE phase (k_2) decreases. This occurs due to the fact that at the constant feed rate, the machining gap increases with the increase in the machining voltage.

The Electrochemical dissolution phase predominates over the electro discharge erosion phases with the increase in the gap width between the tool and the workpiece. The loss of time during which no ECD and EDE action in ECAM takes place increases with the increase in machining voltage.

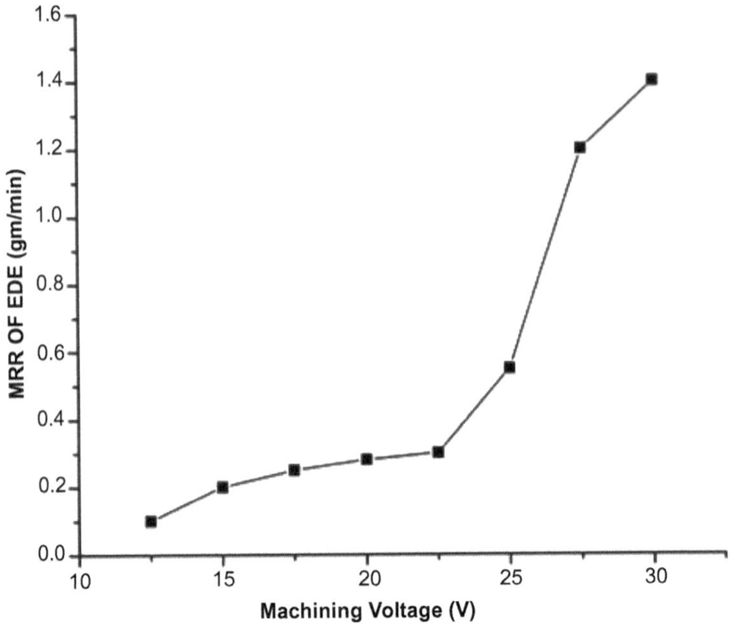

Fig. 6.21 Effect of machining voltage on MRR of EDE phase.

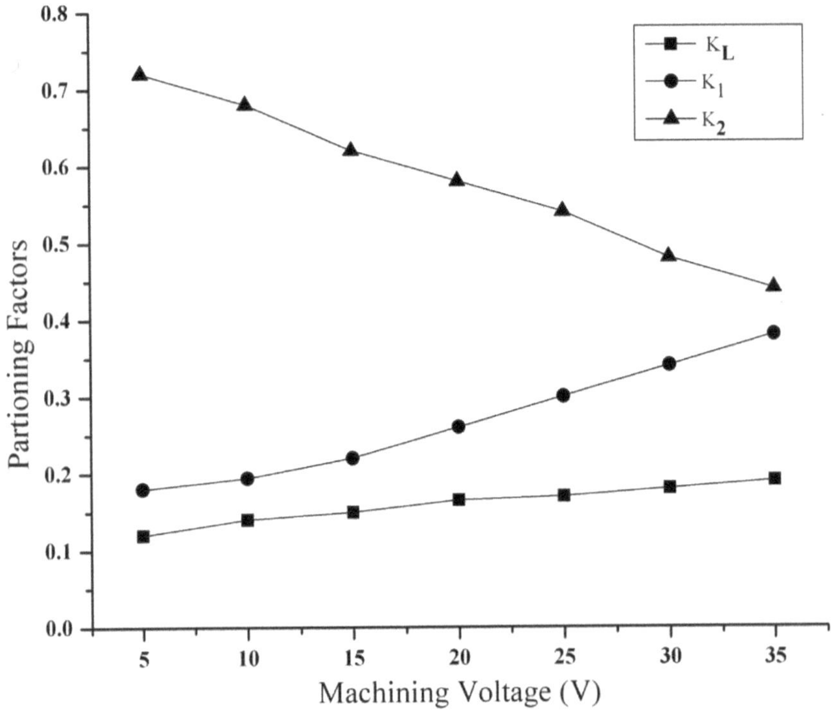

Fig. 6.22 Effect of machining voltage on partitioning factors in ECAM.

6.3.6 Process capabilities of ECAM

The ECAM process is capable of high material removal rate from electrically conducting and chemically active material irrespective of mechanical properties like hardness, toughness and brittleness, etc. The process is capable of micromachining operation using micro-tool as it has less tool wear characteristics. The ECAM process can be applied for cutting purpose using wire as electrode and this process can be termed as wire-ECAM. The rotary tool can also be used for enhancing the material removal rate in ECAM process.

6.3.7 Advantages and limitations

The advantages of ECAM process are as follows:

(i) MRR is higher compared to ECM as well as EDM
(ii) Tool wear is less
(iii) Surface finish is better
(iv) Less Heat Affected Zone

The limitations of ECAM process are as follows:

(i) The work piece material should be electrically conducting in nature
(ii) The work piece material should be chemically active
(iii) The vibration wave form and pulsed wave form of electrical supply should be synchronized properly
(iv) Accuracy is less

6.3.8 Applications

The Electrochemical Arc Machining (ECAM) process is comparatively a new development and expected to gain potential application at the shop floor level because of its obvious merits as characterized in earlier section and especially because of its relatively larger material removal rates, lower specific energy, good surface finish and dimension control. The use of ECAM process in the removal of surface irregularities from components such as dies and molds has been investigated by the earlier researcher [21]. Many researchers [20] have performed their experiments with ECAM apparatus for drilling holes and obtained the satisfactory indication for the possible industrial application of ECAM in drilling operations. The Arc discharge phenomena in aqueous NaNO3 electrolyte is applied for making grooves on silicon nitride ceramics [32]. Further research work is still required to develop the suitable ECAM machine for modern manufacturing industry, ECAM may be applied to any machining operations such as drilling, deburring, grinding, turning, cavity sinking, machining by wire, slitting, etc. which are

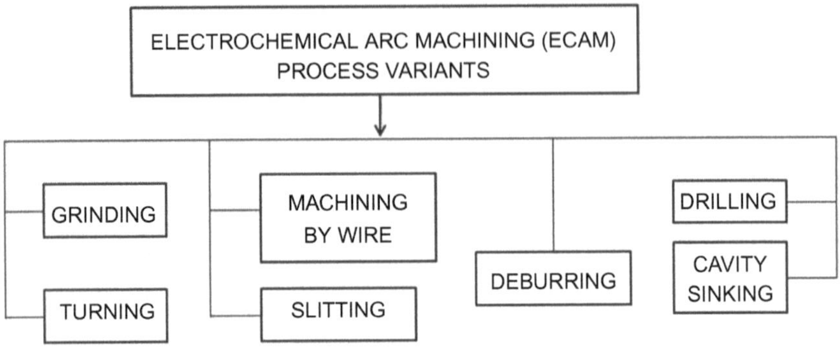

Fig. 6.23 Process variants of ECAM.

undertaken by EDM and ECM. The application modality of ECAM in multi-processing operations is exhibited through Fig. 6.23.

6.3.9 Advancements in ECAM

The high efficient hybrid machining on AISI 304 stainless steel has been performed by combining EDM milling and Arc Machining. Arc machining is a stable discharge process and continuous discharges are appeared in the same position which assists the debris expelling process effectively. MRR as well as surface finish are improved by this new hybrid machining process compounded by Arc Machining and EDM milling [33].

A novel compound machining of titanium alloy (Ti6Al4V) has been performed by high speed EDM and Arc Machining. A rotating tubular graphite electrode is used. A significantly higher MRR and lower electrode wear rate and lesser HAZ thickness were achieved by this hybrid process compared to individual EDM process. High current was supplied and efficient flushing improved the performance of this process [34].

A low cost high efficiency material removal process namely Blasting Erosion Arc Machining (BEAM) has been developed to remove bulk materials of difficult to cut materials such as Inconel 718. BEAM is a environment friendly machining process with high velocity flushing of water based working fluid medium breaks the arcing plasma column to blow of the molten material explosively. High MRR and low tool wear ratio (<1%) than that of EDM was achieved. BEAM is capable of machining complex shape 3D cavity by using bundled or laminated electrode [35].

Electrochemical Arc Machining (ECAM) is one of the hybrid machining process which has great potential to machine electrically conducting materials irrespective of mechanical properties such as hardness and ductility, etc. Because of the complex nature of the process, ECAM is not widely used in industry, but it has advantages over ECM and EDM. The developed model of MRR in ECAM is very much effective to combine the MRR due to ECD and EDE phenomena in ECAM. The parametric analysis on MRR during drilling operation by ECAM based on developed models and experimental observations discussed in the chapter provides knowledge base on this hybrid machining process for exploring the process applicability in various uses. The arcing phenomena in electrolytes in ECAM can enhance the MRR at very high rate which improves the performance of hybrid machining process.

6.4 Electrochemical discharge machining

6.4.1 Introduction

The last few decades have witnessed a rapid progress in the machining of harder and difficult to machine, high-strength-temperature-resistant (HSTR) alloys, various types of carbides, fiber reinforced composites, ceramic materials and various composite tool materials. The requirements of hybrid machining processes have been highly demanded in the present day manufacturing industry to take care the needs of work piece shape complexity, automation of data communications, surface integrity, precision and miniaturization of work piece. The advanced machining processes like ECM, EDM, Wire cut EDM processes are widely accepted in modern manufacturing industry but the major limitation of these processes is that these are applied to only electrically conducting materials.

Hard and brittle electrically non-conducting materials can be machined by advanced machining processes such as Abrasive Jet Machining (AJM), Abrasive Water Jet Machining (AWJM), Ultrasonic Machining (USM), Laser Beam Machining (LBM) and Electron Beam Machining (EBM), but these methods have their inherent demerits. The major limitations of AJM and AWJM are tapering of drilled hole, embedding of abrasives in workpiece and possibilities of stray abrasive action. Although in USM the surface finish is good but there is problem of tool wear and low material removal rate. In LBM, thermal damage is one of the major drawbacks besides the generation of smoke and fume and it has low machining efficiency and high operating cost. EBM needs vacuum medium and it is very expensive.

The tool-electrolyte sparking phenomena was extensively discussed in late 1950s. However, in 1968 Professor Kurafuji discussed a method of machining micro-holes in glass using an electrolyte discharge system [36]. Later this process of electrical discharge in electrolyte was known as Electrochemical Discharge Machining (ECDM). Since then a good number of research are being made to develop a thorough insight into various process features of ECM, EDM, and ECDM for successfully deploying those processes in the practical material machining field.

Presently various kinds of hybrid machining processes are coming up by combining mechanical machining and electro-machining processes, but most of these are applied for machining electrically conducting materials. For machining electrically non-conducting materials electro-machining processes like Electrochemical Grinding (ECG) is not applied. These is a need to generate complex shapes with higher geometrical accuracy and higher material removal rate during machining of electrically non-conducting engineering materials like ceramics, glass and composites.

Electrochemical discharge machining (ECDM) is an advanced hybrid machining process comprising the techniques of electrochemical machining (ECM) and electro discharge machining (EDM). The process is also referred as electrochemical spark machining (ECSM) process. The process is important since it can machine a variety of electrically non-conducting materials including ceramics, composites, alumina, glass, etc. Electrochemical discharge machining (ECDM) process has capability to machine the parts of *meso* or micron range dimensions. ECDM process can be applied for various machining operations such as drilling, cutting, milling, grinding, turning and marking, etc.

6.4.2 Working principle of ECDM

In ECDM, a general electrochemical cell consists of two electrodes, tool and auxiliary electrodes dipped in electrolyte. The workpiece is placed just below the tool electrode. The negative terminal of pulsed DC power supply unit is connected with tool and positive terminal with auxiliary electrode. No electrical connection is given to the workpiece. Generally the gravity feeding arrangement provides feed motion to the workpiece during machining. Application of external potential between the electrodes causes the flow of electric current through the cell, resulting in gas bubble generation due to electrochemical reactions and formation of gas bubble layer surrounding the tool electrode. Beyond a certain value of applied voltage, electric sparks appear across the gas bubble layer at electrode-electrolyte interface on the smaller electrode and the cell current drops. This is known as electrochemical discharge (ECD) phenomenon. Electrochemical Discharge Machining (ECDM) process, a hybrid technology, is a combination of two machining processes, i.e., Electro-chemical Machining (ECM) and Electro-Discharge machining (EDM). Electrochemical reactions help in the formation of gas bubbles, which are responsible for electro-spark discharge (ESD). As soon as an electrically non-conducting workpiece is placed in the close vicinity of discharges, the material of the workpiece is removed by melting, vaporization due to the energy released by sparks.

The material removal mechanism in ECDM are discussed here under:

6.4.2.1 Electrochemical reaction mechanism in ECDM

From the analysis of the electrochemistry of ECM it has been found that the following two types of reaction usually occur in the interelectrode gap as the appropriate power supply voltage is provided in the ECDM system.

(a) Electrochemical reactions at the electrode, e.g., gas evolution, electrode dissolution and oxidation, etc.

(b) Chemical reaction in the bulk of electrolyte, e.g., chemical combinations, complex formation or precipitation reaction to form precipitates and sludge, etc.

The electrochemical reaction occur at the material—electrolyte boundary layers while the transfer of ions in the electrolytic solution take place due to:

(i) Diffusion

(ii) Migration in an electrical field

(iii) Convection in the flow

Figure 6.24 exhibits a typical arrangement of the electrolytic cell of the ECDM process.

The different cathode and anode reaction takes place as soon as an appropriate potential is reached between the tool and the auxiliary electrode in the machining zone. The electrochemical reaction in ECDM is described here-in-under.

6.4.2.2 Reaction at cathode (or tool)

Usually the following reaction occurs at the cathode typically:

(i) Plating of metal ions

(ii) Evolution of hydrogen gas

The reaction of metal plating is

$$M^{+} + e^{-} = M$$

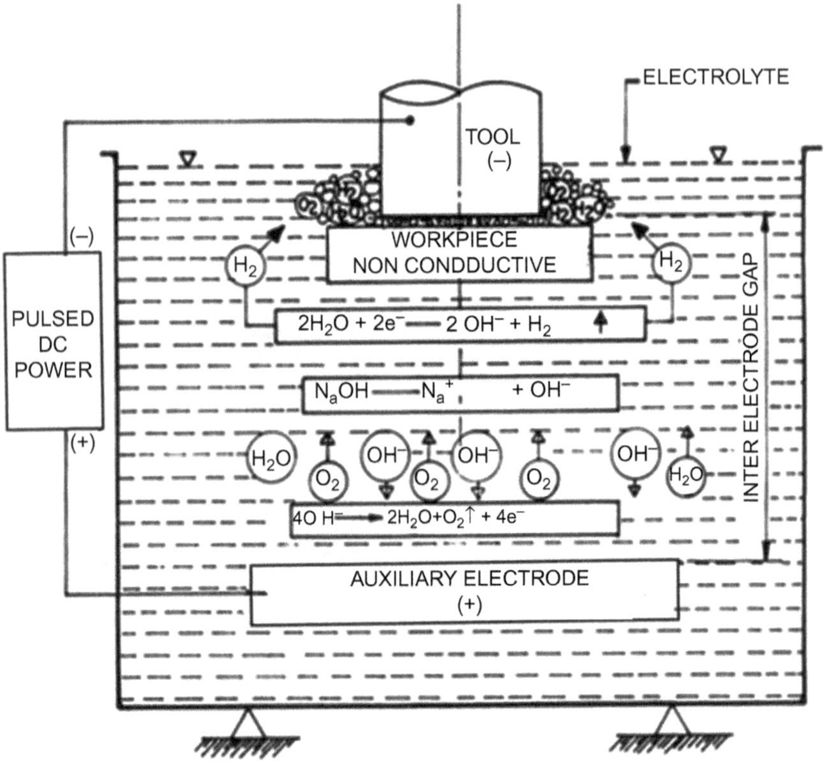

Fig. 6.24 Typical arrangement of the electrolytic cell of the ECDM process.

where M represents the anode material.

The reactions for hydrogen evolution are

$$2H^+ + 2e^- = H_2\uparrow \text{ (in acidic electrolyte solution)}$$

$$2H_2O + 2e^- = +\, 2(OH)^- + H_2\uparrow \text{ (in alkaline electrolyte solution)}$$

6.4.2.3 Reaction at anode (auxiliary electrode)

Two types of anodic reactions may be usually considered to take place at the anode (or auxiliary electrode):

(i) Dissolution of material ions in the electrolyte solution, and
(ii) Evolution of oxygen gas at the auxiliary electrode surface.

The anodic dissolution reaction can, therefore be

$$M = M^+ + e^- \text{ (in acidic electrolyte solution)}$$

The evolution reaction in the electrolyte is given by

$$2H_2O = O_2\uparrow + 4H^+ + 4e^- \text{(in acidic electrolyte solution)}$$
$$4(OH^-) = 2H_2O + O_2\uparrow + 4e^- \text{ (in alkaline electrolyte solution)}$$

In the ECDM, electrochemical reactions for gas bubbles generation are exhibited in Fig. 6.24.

The interelectrode gap is very large in Electrochemical Discharge Machining (ECDM) compared that of the ECM process. Hence, for the same applied voltage in ECDM, very low current passes through the interelectrode gap and thus the material removal due to electrochemical dissolution of auxiliary electrode is negligibly small. But a high DC power supply voltage, the positively charged metal ions are formed and move toward the cathode (tool) and there is a chance of plating of metal ions on the cathode. At high voltage power supply, electrochemical dissolution of auxiliary electrode may take place. But, in ECDM, anodic dissolution is not the mechanism of material removal from the workpiece because non-conducting workpieces does not act as anode in the ECDM process.

The electrochemical and chemical reaction mechanisms occurring in ECDM gap for machining Si_3N_4 ceramics using aqueous NaOH salt solution as electrolyte are discussed as follows:

The reaction that occur at the bulk of the electrolyte and at a high temperature are given by

$$NaOH = Na^+ + OH^-$$
$$2Na^+ + SiO_3^- = Na_2SiO_3$$
$$Si_3N_4 + 6H_2O = 3SiO_2 + 4NH_3$$
$$Si_3N_4 + 6H_2O = 3SiO_2 + 2N_2\uparrow + 6H_2\uparrow$$

The reaction at the cathode (tool) are as follows:

$$2H_2O + 2e^- = H_2\uparrow + 2(OH)^-$$
$$2H^+ + 2e^- = H_2\uparrow$$

The reaction at the anode (auxiliary electrode) are

$$2H_2O = O_2\uparrow + 2H^- + 4e^-$$
$$SiO_2 + OH = SiO_3^{--} + H^+\uparrow$$

With the existence of high temperature, SiO_2 and steam are formed on the work surface, and subsequently reacts with the high temperature electrolyte and dissolves as soluble ions SiO_3^{--} [37].

Therefore the electrochemical reactions in ECDM process generate various gas bubbles, e.g., hydrogen gas and vapor bubbles, etc., which are usually low ionic positively charged bubbles and these gas bubbles are accumulated in the vicinity of the tool. The rate of gas bubble generation increases with the applied voltage and conductivity of the electrolyte. The accumulated gas bubbles at tool, electrolyte and workpiece interfaces create a high potential gradient.

From the above set of electrochemical reactions which may occur at various regions in the electrolytic cell of ECDM during machining of non-conductive ceramics, It is revealed that the actual reaction mechanisms of material removal which is mainly depend on various complex chemical reactions occurred at the high temperature along the workpiece surface.

(a) Electrical discharge phenomena in ECDM

The electrical discharge phenomena is highly responsible for the removal of material from non-conducting engineering ceramics during electrochemical discharge machining (ECDM). Due to the passage of electric current in the electrolytic cell in ECDM, hydrogen gases are usually generated at the cathode and lead to bubble formation. The bubbles grow in size with passage of time and once the critical size is reached, bubbles detach from the electrode surface and move upward. When hydrogen bubbles become sufficiently large in number, the electrical resistance at the tool electrode interfaces increases substantially. This leads to higher ohmic heating of the electrolyte at the region, causing generation of vapor bubbles. As the current density increases, the number of nucleation sites of the bubble also increases. At the critical condition the numbers of nucleation sites are such that the fully developed gas bubbles cover the maximum possible surface area of the tool, leading to electrical isolation between the tool and electrolyte. Consequently, the current drops to zero in a very small time which is analogous to a switching off action of an electric circuit. Since change in current with respect to time is very large, an instantaneous spark discharge occurs between tool surface and electrolyte interface across the gap bubbles layer in the mean free path.

Once the bubble dislodges from the tool surface, contact between the electrolyte and the tool is again re-established and the cycle will be repeated. When the applied voltage is sufficiently larger than the critical voltage, the re-establishment of contact does not take place and a stable discharge is established.

However, electrical discharge may take place in an electrolyte system due to the following reasons:

(i) Electrolytic gas generation at the electrode surface
(ii) Growth of layers with low ionic concentration near electrodes and formation of partially conductive oxide film on the anode surface
(iii) Steam generation and cavitation
(iv) Vapor blanketing of electrode surface
(v) Local stagnation of flow

When the voltage gradient setup is sufficient enough to breakdown the gas bubble layer between the tool and the workpiece, a conducting path is developed for spark discharge owing to the ionization of the gas bubbles and, thereby, cause a current to flow. The initiation of spark generation in ECDM is, somewhat, similar to that of electro-discharge machining in ECDM, the insulating layer of gas bubbles on the electrodes provides the medium for the initiation of the breakdown. The value of the breakdown voltage at which the sparking starts is known as critical sparking voltage. The breakdown voltage depends upon the concentration and type of electrolyte, tool geometry, the type of gas in the bubble and the size of the bubble.

The possibility of the two consecutive sparks during ECDM taking place at the same location is low. The sparking takes place at random. As soon as sparking takes place, the bubbles present around the spark will be pushed away by the pressure wave initiated due to sparking and may result in bursting of some of the bubbles. Some of the bubbles merge with other bubbles and size of those increases to the extent that without sparking those die out. Obviously, bubbles of smaller size have high probability of sparking.

The sparking occurs between the tool and the electrolyte across the low ionic gas bubble layers. The non-conducting workpiece is placed in the sparking zone. During sparking the electric current flows from anode to cathode through the electrolyte. Due to the presence of electrically non-conducting workpiece and supporting platform the resistance to the flow of electric current increases. There are two main sources of heat generation during ECDM, i.e., heat generation due to sparking and also joule heating. Only the heat generated due to sparking is productive. The evolution of the heat due to sparking results in melting and evaporation of work material. Each electrical discharge causes a focused stream of electrons to move with a very high velocity and acceleration from the cathode (tool) toward the workpiece and ultimately creates compression shock wave on the workpiece surface. The kinetic energy of the electrons is converted into heat energy on the workpiece surface.

This phenomenon is accomplished within a few micro-seconds and the temperature of the spot heat by the stream of electrons may rise to a very high magnitude. At the high temperature above the melting point of the workpiece material. It melts and finally evaporates. The high pressure of compression shock wave creates a blast causing metallic vapor of wear products to form in the shape of metallic globules, leaving crater on the workpiece surface. Further, the mechanical erosion may also occur due to the cavitation effect of the gas bubbles rupturing on the workpiece because the discharge across the bubble will results in it violent rupture. Fig. 6.25 exhibits the mode of gas bubble formation and the occurrences of electrical sparking phenomena in ECDM process. Hence, the material removal from the workpiece in ECDM mainly takes place due to the electrical spark discharge followed by the mechanical erosion of the material from the workpiece.

(b) Thermal spalling phenomena in ECDM

The thermal spalling is considered to be another mechanism of material removal from the workpiece surface in ECDM process. The thermal spalling of material is usually defined as a mechanical failure of the material without melting, due to localized, thermally induced internal stresses which exceed the bond strength of the material. As the surface expands and contracts with the rapid temperature changes, the resulting tension and compression leads to tensile or shear failure respectively. In electrochemical discharge machining, the material goes through the thermal cycling under spark discharge condition when pulsed DC power supply voltage is applied and hence a complex temperature gradient is established resulting in internal thermal stresses which leads to thermal spalling. The thermal spalling phenomena is very much dependent on the material properties in terms of its thermal conductivity, thermal expansion co-efficient, tensile strength and Young's modulus. The material with high thermal expansion co-efficient and low thermal conductivity are subjected to higher internal stresses due to more severe temperature gradients. If a material has a low ultimate strength or Young's modulus as well, the tendency to spall is enhanced. In addition, under pulsed sparking at the same spot, there may be fatigue failure of the surface after a particular number of cycles. The thermal spalling is frequently observed on the surface of non-conducting engineering ceramics, during machining.

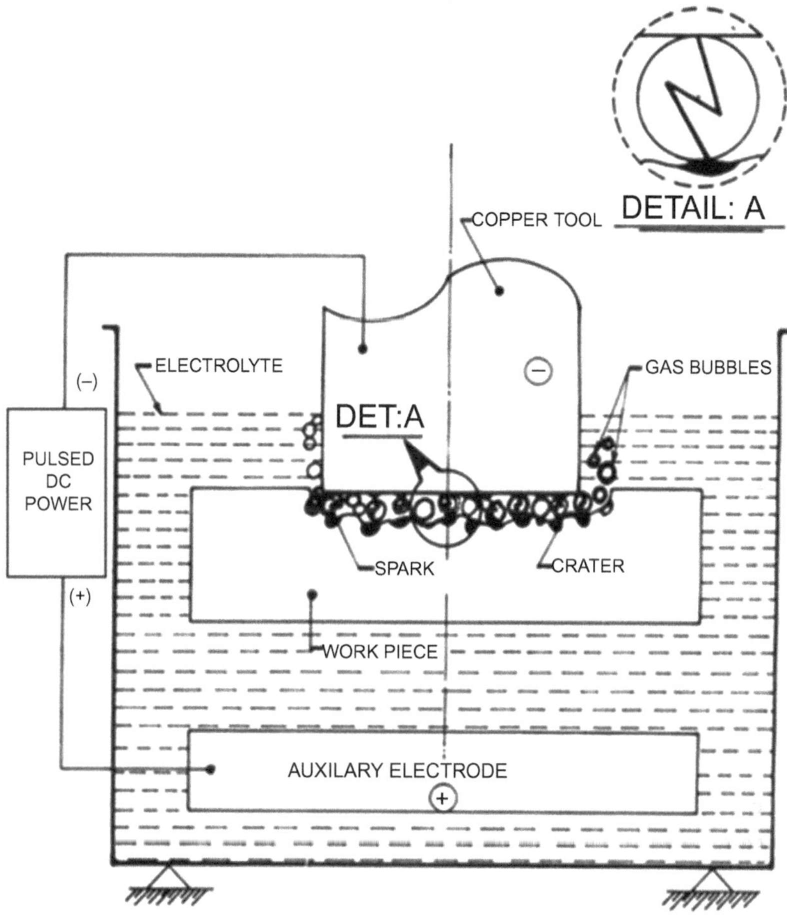

Fig. 6.25 The occurrence of electrical discharge phenomena in ECDM.

(c) Mechanism of tool wear in ECDM

In the ECDM process, the sparking occurs at the tool-electrolyte interface and the heat energy is released due to spark discharge. As a result the tool is being heated to red hot temperature. If the applied voltage is very high, the tool wear may takes place due to melting and vaporization of the tool material. The material removal from the tool surface may also occur due to the chemical reaction of the tool material with the electrolyte at high temperature.

Though there is no mechanical wear due to friction, but the compression shockwaves are developed during sparking which may cause tool wear. At high temperature, the workpiece material may be positively ionized. Due to electrochemical reactions, the positively charged gas bubbles are formed. With high voltage of DC power supply the material of the auxiliary electrode may be positively ionized. The positively ionized particles and gas bubbles strike the tool surface with a higher velocity and as a result the tool wear may take place. In ECDM, lesser tool wear takes place as low pressure compression shock waves are developed on the tool and the positive ions strike the tool surface with less momentum. If the tool acts as anode in ECDM process, the tool wear increases because higher portion of the heat energy goes to the anode during sparking. The tool wear rate increase with the increase in voltage and specific conductance of the electrolyte because the higher spark energy obviously results in higher tool wear. The input-output model of ECDM process is shown in Fig. 6.26.

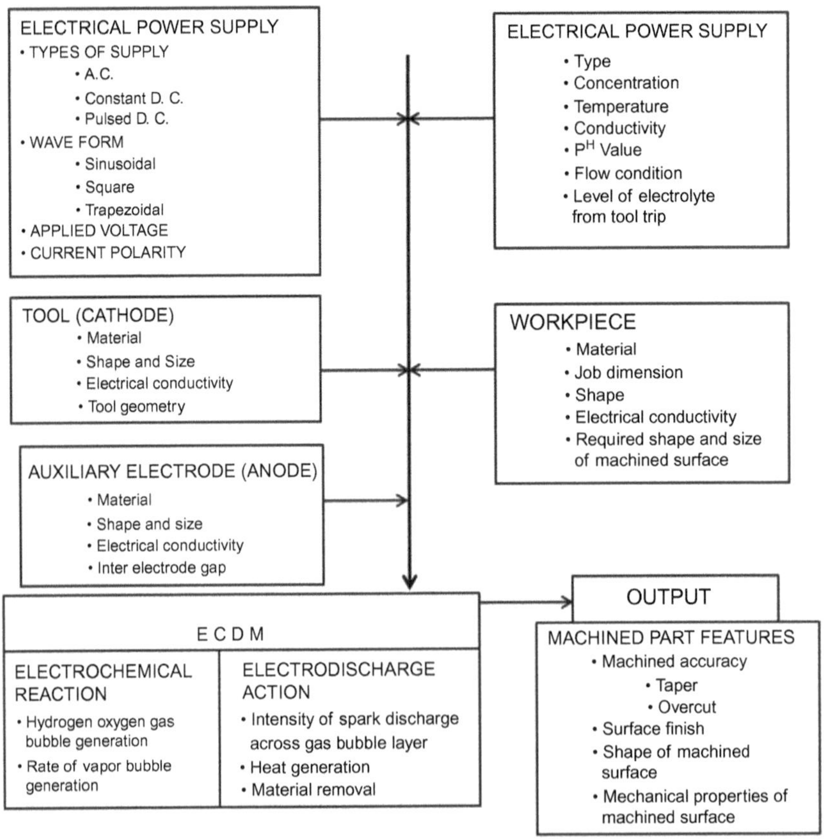

Fig. 6.26 Input output model of ECDM process.

6.4.3 ECDM machining set-up details

The Electrochemical Discharge Machining (ECDM) set-up consists of three major system as follows:

(a) Mechanical hardware system
(b) Electrolyte supply system
(c) Electrical power supply unit

The schematic diagram of ECDM set-up is shown in Fig. 6.27.

(a) Mechanical hardware system

The mechanical hardware of a typical ECDM set-up consists of various units are as follows:

(i) Main machine chamber
(ii) Tool holding and feeding unit
(iii) Job holding and feeding unit
(iv) Auxiliary electrode unit
(v) Inter-electrode gap control unit

The details of the above components are discussed as follows:

(i) Machining chamber

The job holding and job feeding devices are attached on one side wall of the machining chamber. The auxiliary electrode holding and the inter electrode gap control units are also fixed on to the same wall of the machining chamber. At the bottom of the chamber,

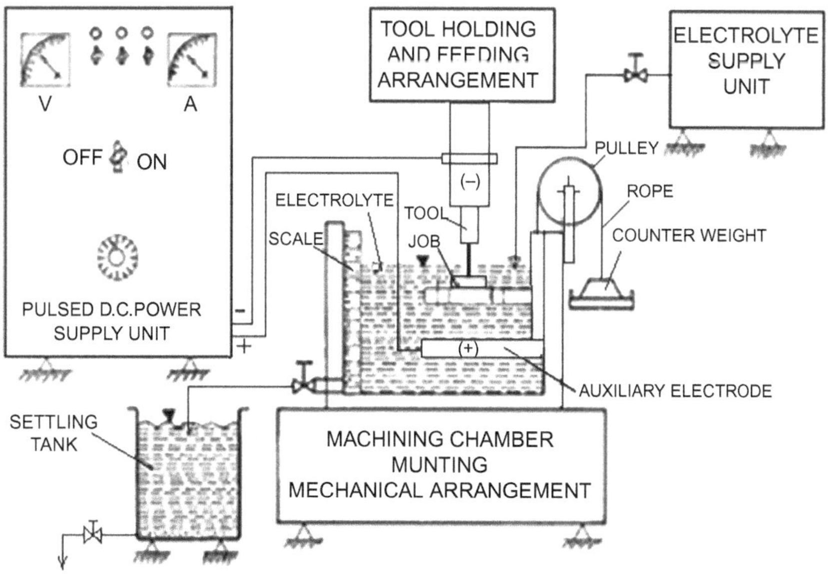

Fig. 6.27 Schematic diagram of an ECDM set-up [38].

there is an exit passage for electrolyte to flow out. One graduated scale is also attached at the side wall of the chamber to monitor the upper level of the electrolyte in the machining chamber. The machining chamber is generally made of stainless steel of the size 250 mm × 250 mm × 150 mm. and thickness of 2 mm. to avoid of the corrosive action caused by electrolyte. Besides, the stainless steel has enough strength, toughness and rigidity to withstand the desired volume and pressure of any type of electrolytes.

(ii) Tool holding and feeding unit

The tool holding unit consists of electrically non-conductive holder which is mounted on the main machine structure and the main tool holder which is having threaded portion at the end with the facility of holding electrical connector for the power supply. The tool (or cathode) is finally connected with the negative terminal of the pulsed DC power source through electrical lead wires which are locked firmly with the help of the slide clamp. The front side of the tool holder is internally threaded to facilitate for holding different sizes and shapes of the tools. As the ECDM process continues and the material removal takes place, the gap between the tool and the job increases which may decreases the machining rate.

For effective and continuous machining operation, the feeding of either the job or tool is required. In ECDM system, there is also a scope for tool feeding. The tool feeding device is attached with the help of the rotational movement of the stepper motor which, in turn, is connected with a reduction gear train so as to have a speed reduction. The stepper motor driven system is used to control the stepper motor speed to generate various controlled tool feed rate, which are required for different ECDM operations for machining different grades of ceramic materials.

(iii) Job holding and feeding unit

During the machining process, the jobholding unit plays a vital role for achieving the desired level of the machining accuracy. The dislocation of the job surface from the original job holding position may lead to the variation in the dimensional accuracy of the job. So, a rigid holding device of the job is needed for better machining accuracy. The job is mounted on a base plate which itself rest on a support structure. A cover plate made of acrylic glass is placed on the job and the rigid clamping is secured by tightening the nuts at four corners of the cover plate. There is a guide way on the cover plate for the easy positioning of the tool toward the job during machining. The guide ways on the cover plate should be designed according to the required profile which is to be generated in the ceramic work pieces. Four numbers of screw and nuts are used for the rigid fixation of the work piece in the experimental set up. The screw and nut mechanism can be altered according to the size and shape of the job or workpiece. The job holding unit should be able to withstand the high compressive shock waves of electrical discharge sparks which are generated during machining operation.

A gravity controlled job feeding arrangement is developed to achieve the vertical feed motion of the job holding device during the machining process. The gravity controlled job feeding arrangement is shown in Fig. 6.27. A string is made to pass through the pulley and one end of the string is connected with the guide bar of the job holding unit. Two rectangular blocks and the supporting plate guide the upward and downward movement of the guide bar. The other end of the string is attached to a pan and counter weight arrangement. An optimum counter weight is properly chosen and place on the pan to keep and maintain the tool and the work piece just in touching condition during the machining operation. The increased counter weight may result in a stress at the contact point of the tool and the work piece, which may also result the premature cracking of the workpiece during machining process. The excess counter weight on the pan may also prevent the electrolyte to flow into the micro gap between the tool and the work piece which may reduce the chance of occurrences of sparking. Hence the counter weight should be properly selected during machining in an iterative process of manipulation of the units.

(iv) Auxiliary electrode unit

As the non-conducting engineering ceramics are used as work piece samples, the auxiliary electrode unit is needed to complete the electrical circuitry in ECDM system. The auxiliary electrode is a flat rectangular plate made of highly electrically conductive and non-corrosive material, such as, stainless steel, copper and graphite, etc. The positive terminal of the pulsed DC power supply unit is connected with the auxiliary electrode through electrical lead wires locked firmly with the help of a clamp. In order to set the inter electrode gap, a screw-nut mechanism is provided for lifting and lowering the auxiliary electrode. The auxiliary electrode plate is mounted on the auxiliary electrode holding block which moves up and down through the grooved path provided in the guide blocks. The guide blocks are rigidly fixed with one side wall of the machining chamber.

(v) Inter electrode gap control unit

The separation length between the tool and the auxiliary electrode is called as the inter-electrode gap. The inter-electrode gap is a machining parameter in the ECDM process. To vary the inter-electrode gap, the position of the tool or auxiliary electrode should be changed. The position of the auxiliary electrode can be varied by a screw and nut mechanism. The upward and downward movement of the electrode can be adjusted according to the desired gap length. Two vertical long screws are fixed with the auxiliary electrode holding block and passed through the drilled hole of the two plates which are screwed on the upper portion of the guide blocks. Two nuts are engaged with the screw and rest on the plates. By rotating the nut, the movement of the auxiliary electrode in the upward and downward direction can be controlled to maintain the required interelectrode gap during the machining operation.

(b) Electrolyte supply system

In order to achieve better machining performance in terms of material removal rate (MRR) and accuracy, the gas bubbles generation due to electrochemical reactions in the machining gap is necessary at the bottom edge of the tool. Hence, in order to obtain high rate of gas bubble generation, stagnant electrolyte condition is preferred and the upper surface of the electrolyte level is maintained at a constant height from the base of the machining chamber so that the tool is just immersed at a constant depth from the upper level of the electrolyte. But due to heating, the electrolytes evaporates, as a result the electrolyte level is being reduced and, thus, causing the reduction of the sparking intensity during machining. In ECDM system, the electrolyte supply unit delivers the electrolyte to the overhead reservoir which is placed at a particular height. The electrolyte from the bottom storage tank is pumped through a strainer by a self-priming centrifugal pump. The flow rate of the electrolyte is controlled through a predesigned set of stainless steel throttle and relief valves fitted in the delivery pipe line. In order to maintain the electrolyte level in the machining chamber, the controlled flow of electrolyte to the machining chamber is required. A scale is used to measure and adjust the level of electrolyte. The pouring of the electrolyte into the machining chamber is to be controlled appropriately which is mainly govern by the evaporation rate of electrolyte at different conditions of machining. By adjusting the flow control valve, the flow rate of the electrolyte from the overhead reservoir can be controlled. The exit pipe is fitted with the control valve at the bottom level of the machining chamber for removing the used electrolyte containing sludge and precipitates. This can also be utilized for cleaning of the machining chamber. The used electrolyte is stored in the separate storage settling tank. The fresh electrolyte is generally used for machining.

(c) Electrical power supply system

In Electrochemical Discharge Machining (ECDM) constant polarity is to be maintained and hence a DC power supply is required. The applied voltage for machining operation generally varies between 30 to 160 V. The current rating is small and only few amperage serves the purpose.

Keeping in view the above requirements, a pulsed DC power supply unit is designed and developed for the voltage range of 0 to 200 V and current rating up to 10 amp. The main input power supply is a 3-phase, 440v AC but only a single phase is utilized. The reduction of applied voltage and subsequent conversion from AC to DC have been achieved by step down transformer and silicon-diode controlled rectifier unit respectively.

The voltage and current are recorded by voltmeter and ammeter respectively connected with the circuitry as shown in Fig. 6.28a. The voltage is

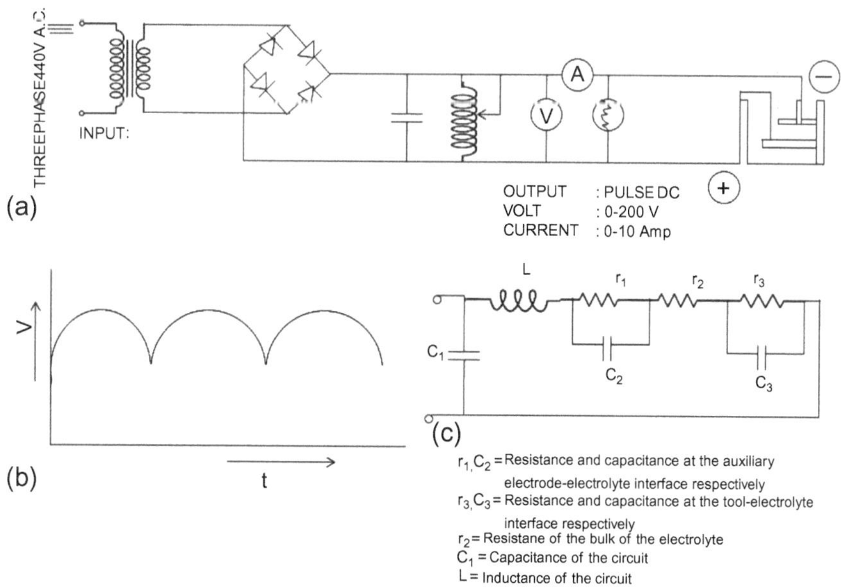

Fig. 6.28 (a) Electrical circuitry of pulsed DC power supply unit, (b) pulsed wave form, and (c) equivalent circuit of ECD cell.

varied and set to a particular value by variac connected with the power supply unit. Though the power supply is a DC, but it is not smooth DC rather pulsed DC which mainly activates the machining operation. The wave form is shown in Fig. 6.28(b). Figure 6.28(c) shows the Equivalent circuit of ECD cell. The arrangements for adequate protective circuit for the transformer, rectifier and the machine have been provided against short circuit and overload conditions.

6.4.4 Material removal rate in ECDM

Material is removed from workpiece due to melting and vaporization utilizing the energy of electric spark discharge in ECDM. Mathematical model of MRR has been developed based on material removal mechanism in ECDM.

The following assumptions are considered to simplify the development of the mathematical model:

(i) Material removal takes place at the bottom edge of the electrode only
(ii) Current through the other part of the electrode surface except the edge is negligible

In ECDM, the volume of workpiece material removed per spark (v) and the energy released per spark (Q) can be related approximately [39] as follows:

$$v \propto Q^{1.5}$$
$$\text{or}, v = C.Q^{1.5} \tag{6.50}$$

where, C, constant of proportionality.

The heat energy (Q) released per spark in joules can be obtained as:

$$Q = 1/2L.I_O{}^2 \tag{6.51}$$

where, L, the inductance of the circuit in Henry, I_C and I_O, the instantaneous current in amperes.

When the value of the critical current equal to I_O, the situation is just to initiate sparking, but hardly any material removal takes place. Hence the actual amount of energy available for material removal is given as:

$$Q = 1/2\,L\,(I_O - I_c)^2 \tag{6.52}$$

The resistance r_1 is negligible for a very sparse distribution of bubbles owing to the large size of the electrode.

The resistance r_2 is constant for any particular composition of electrolyte, electrolyte concentration and system configuration.

The resistance r_3 increased with the applied voltage, Vo. The maximum value of r_3 at the discharge is r_{3c}.

Hence the value of critical resistance r_c can be determined as:

$$r_c = r_2 + r_{3c}. \tag{6.53}$$

From the Eqs. (6.52), (6.53), the expression of Q can be obtained as:

$$Q = \frac{1}{2} L \left[\frac{V_o - V_c}{r_c} \right]^2 \tag{6.54}$$

The volume of workpiece material removed per spark can also be expressed as:

$$v = C. \left[\frac{1}{2} L \left(\frac{V_o - V_c}{r_c} \right)^2 \right]^{1.5}$$

$$\text{or, } v = C. \left(\frac{L}{2} \right)^{1.5} . \left(\frac{V_O - V_C}{r_c} \right)^{3.0} \tag{6.55}$$

The sparking frequency (f) can be expressed as [39]:

$$f = \frac{1}{t_1} = \frac{r_2 + 0.5 r_{3c}}{L.\ \ln \left(\frac{2r_c}{r_{3c}} \right)} \tag{6.56}$$

If the density of the work material is considered as ρ g/cm^3, the material removal rate (m), mg/min can be expressed as:

$$\dot{m} = \rho.v.f \tag{6.57}$$

substituting the value of v and f from Eqs. (6.53), (6.56), the Eq. (6.57) can be modified to be expressed as:

$$\dot{m} = \frac{2.121\, x 10^4 . \rho . C . L^{0.5} (V_o - V_c)^{3.0} (r_2 + 0.5 r_{3c})}{r_c^3 \ln \left(\frac{2_{rc}}{r_{3c}} \right)} \tag{6.58}$$

The value of C,L,V_c,r_2,r_c and r_3c can be evaluated for the given electrolyte concentration and system configuration after performing the pilot experiments.

The MRR model can be verified using the following input data for evaluating theoretical MRR. The variation of theoretical MRR from Eq. (6.58) and actual experimental MRR with different applied voltage has been

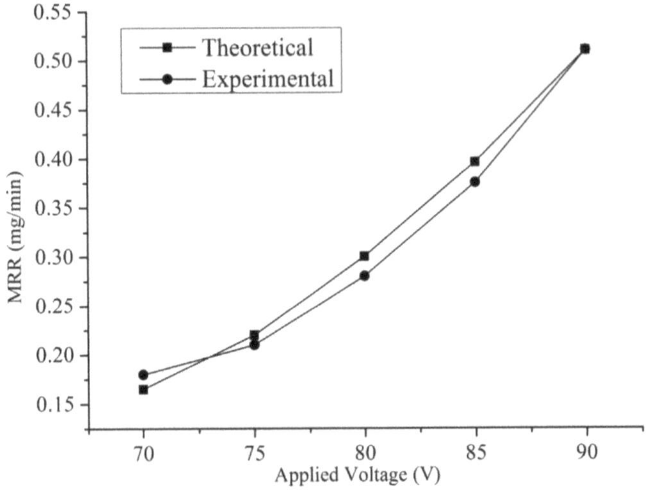

Fig. 6.29 Variation of theoretical MRR and actual experimental MRR with different applied voltage.

shown in Fig. 6.29. The input data for evaluation of theoretical MRR are listed below:

Input parameters	Values
(a) Inductance of the circuit, L	100 mH
(b) Critical voltage, V_c	27 V
(c) Total critical resistance, r_c	21.6 Ω
(d) Resistance of bulk electrolyte, r_2	10 Ω
(e) Critical resistance at tool electrolyte interface, r_{3c}	11.6 Ω

6.4.5 Influences of ECDM process parameters on machining characteristics

The influences of ECDM process parameters such as applied voltage, electrolyte concentration and inter-electrode gap on the machining criteria, i.e., material removal rate, overcut phenomena, heat affected zone and tool wear rate have been analyzed for finding better machining characteristics of electrically non-conducting materials. The taper wall and curvature front tip tools of different diameters, e.g., 0.5, 1.0, 1.5, and 2 mm are used for the present set of experimental studies. The interaction and level of ECDM process parameters for enhancing MRR has been listed in Fig. 6.30.

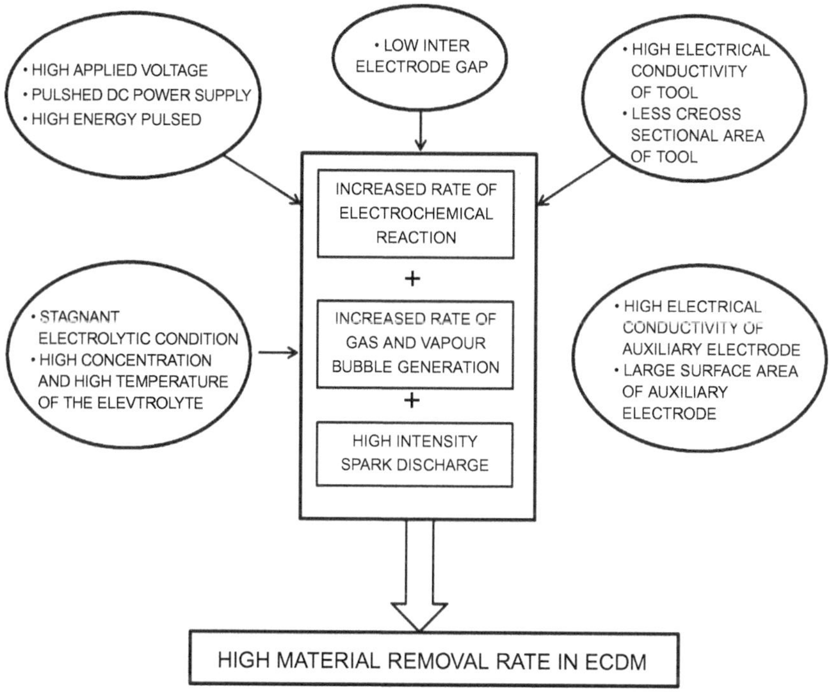

Fig. 6.30 The interaction and level of ECDM process parameter for increased MRR.

(a) Analysis of the influence of process parameters on MRR in ECDM

The Electrochemical Discharge Machining of Silicon Nitride ceramics has been carried out at various applied voltage levels, e.g., 70, 80, 90 V and using different electrolyte concentration, e.g., 20%, 22.5%, 25%, 27.5% and 30% NaOH solution. The effect of applied voltage and electrolyte concentration on MRR is presented in Fig. 6.31 which clearly indicates the MRR increases non linearly with the increase in the applied voltage. The increment in MRR at higher voltage is relatively larger than that at the lower voltage. At higher voltage and higher concentration of electrolyte, the MRR increases rapidly. This may be due to high rate of gas bubble and reaction product generation causing better sparking action. All the graphs approximately follow the same pattern. The limiting applied voltage and electrolyte concentration of maximum MRR are observed to be 90V and 30% concentration respectively because the increment of applied voltage beyond this maximum value has been observed to damage the workpiece.

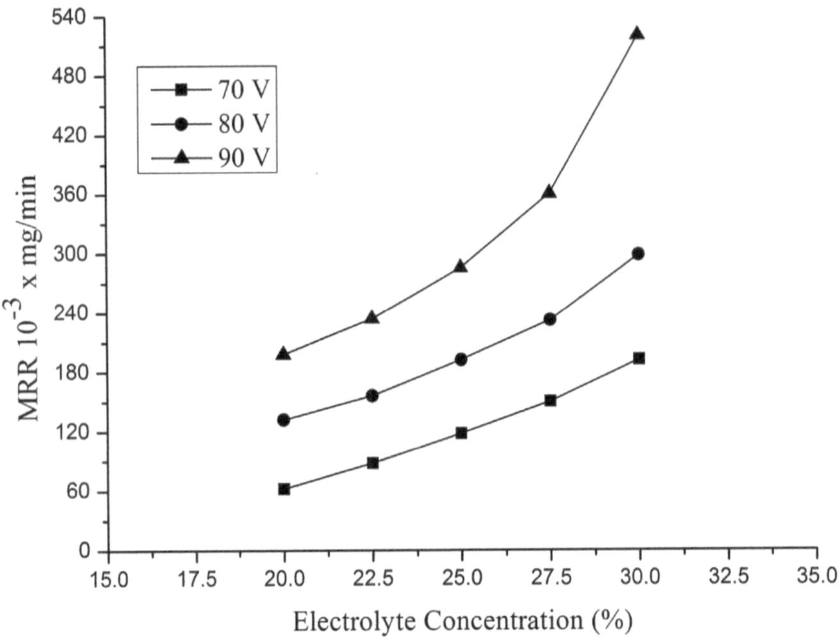

Fig. 6.31 Effect of electrolyte concentration on MRR at varying applied voltage

The type of electrolyte is an important factor that influences the material removal rate (MRR) in the machining of ceramics by ECDM process. Different electrolyte will cause different chemical reactions with the workpiece materials during ECDM process. Two different electrolytes, e.g., NaOH and KOH solution are used to investigate the effect of the type of electrolytes on the material removal rate. The test results show that aqueous NaOH solution is the most suitable electrolyte for machining Si_3N_4.Owing to its high conductance, the electrolyte allows more current to pass through the medium of electrolyte resulting in greater gas bubble generation, and, in turn, causing more sparking to occur in the machining zone.

To study the effect of inter-electrode gap on MRR the experiments have been planned at 70V using 30% NaOH solution. The different inter-electrode gaps are considered as 10, 20, 30, and 40 mm. The graph indicates that the material removal rate initially increases, then decreases with the increment of inter-electrode gap. Figure 6.32 exhibits the effects of inter-electrode gap on MRR. At small inter-electrode gap, the sparking is concentrated and relatively less side sparking occurs resulting low material removal. At 20 mm gap, the condition is more suitable for ECDM because concentrated sparking occurs at this gap during the machining. The material

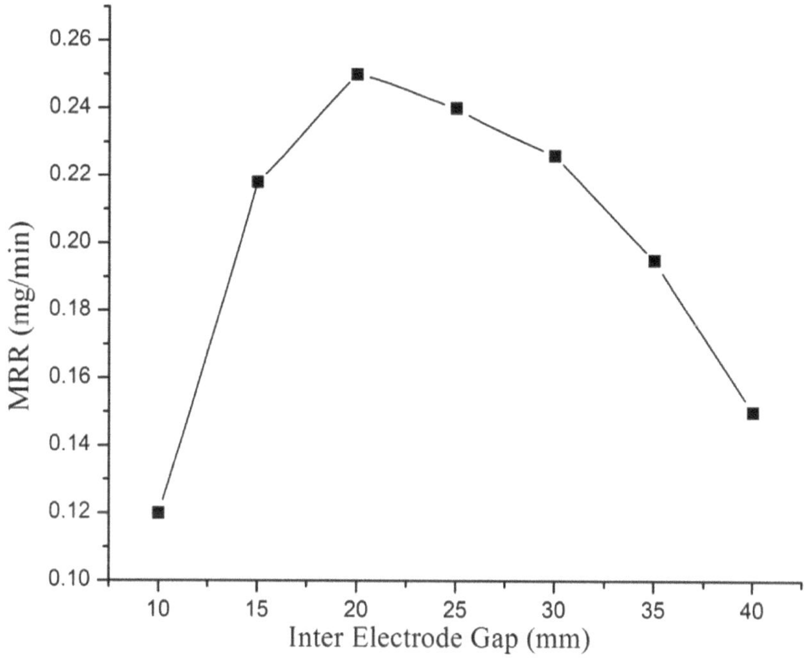

Fig. 6.32 Effect of inter electrode gap on MRR.

removal is also high, and the overcut is also minimum at this gap. At higher inter-electrode gap, the intensity of sparking are reduced and material removal rate also becomes low. As the tool diameter increases, the equivalent surface area of tool also increases which causes the generation of gas bubbles in large volume. As the material removal rate is proportional to rate of gas bubble generation, high MRR is obtained using large diameter of tools.

(b) Analysis of influences of process parameters on accuracy, i.e., overcut and machining depth criteria in ECDM

The influence of the dominant process variables, e.g., applied voltage, electrolyte concentration, type of electrolyte, interelectrode gap and tool diameter on the overcut criteria during machining of silicon nitride ceramics have been observed and analyzed for achieving controlled machining criteria conditions of ECDM process. The effects of the applied voltage, electrolyte concentration and interelectrode gap on machining depth in ECDM have also been reported.

Figure 6.33 exhibits the variation of overcut with electrolyte concentration at different applied voltage levels of 70, 80, and 90 V. Different

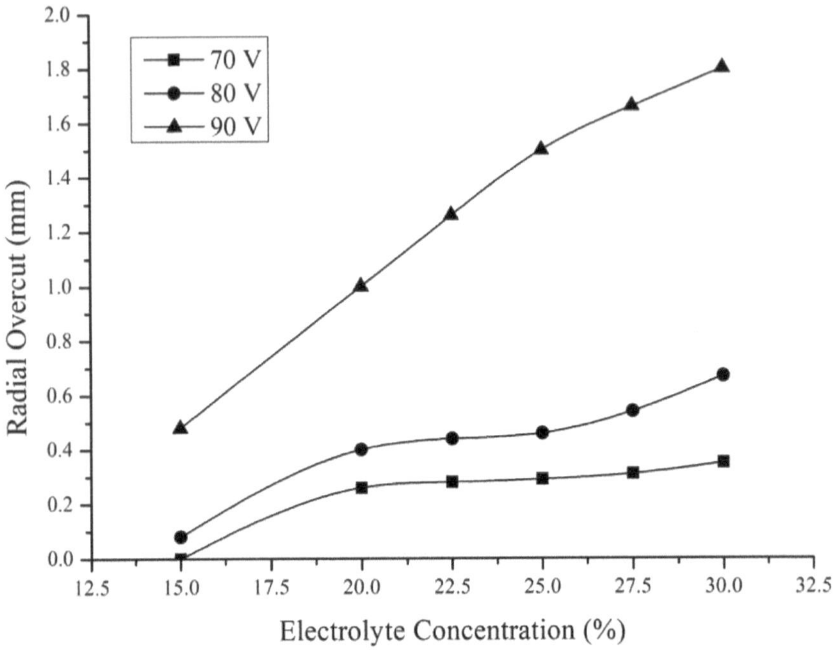

Fig. 6.33 Effects of electrolyte concentration on radial overcut at varying applied voltage.

electrolyte concentrations of 15%, 20%, 22.5%, 25%, 27.5%, and 30% NaOH solution are set during machining. It is clear that overcut increases with increase in electrolyte concentration and applied voltage. At higher voltage level, i.e., at 90 V, the overcut is more predominant in comparison to lower voltage level, i.e., at 70V.With the increase in electrolyte concentration and applied voltage, the rate of gas bubble generation around the tool tip increases causing the sparking rate as well as the energy per spark to increase which leads to a more overcut. From the graphs, it can be concluded that for precision machining with ECDM, lower applied voltage and electrolyte concentration are most suitable.

The conductivity of the electrolyte will also influence the ECDM process. KOH is suitable electrolyte as it produces smaller overcut due to lower amount of side sparking. The conductivity of KOH is lower and chemically less reactive than NaOH. The gas bubble generation rate is low in case of KOH electrolyte. Mostly gas bubbles are generated in KOH electrolyte and those are accumulated just below the tool tip. Due to this reason, the sparking is more concentrated in nature and helps producing lesser amount of overcut, leads to better accuracy.

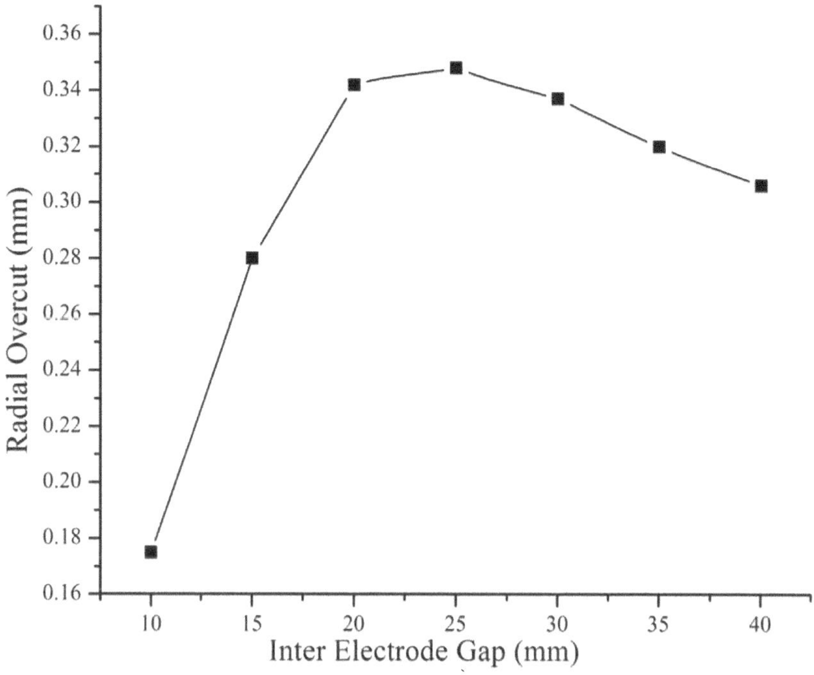

Fig. 6.34 Effects of inter electrode gap on radial overcut.

Figure 6.34 exhibits the influence of inter-electrode gap on overcut at 80 V applied voltage and 25% NaOH solution. At smaller inter-electrode gap the sparking is concentrated and relatively lesser amount of side sparking occurs causing lower overcut. The overcut is minimum at 10 mm inter-electrode gap and relatively large at 20 mm inter-electrode gap. At 20 mm inter-electrode gap, violent side sparking is observed with cracking sound, which may be due to bursting of gas bubbles. At large inter-electrode gap 40 mm, the intensity of side sparking is reduced resulting in less overcut to occur.

The tool diameter has effect on overcut during ECDM. Radial overcut decreases with the increase in tool diameter. The current distribution at the tool electrode is not uniform because of the sharp edge at the end face, as a result of which the sparking process is mostly confined to this region. In case of small diameter tool, the side sparking is more which is responsible for the larger overcut.

Figure 6.35 exhibits the variation of the machining depth with the applied voltage. The set of graph indicates that the machined depth first increases, and then decreases considerably in non-linear fashion with increase in applied voltage. At an applied voltage of 70 V, the spark generation and

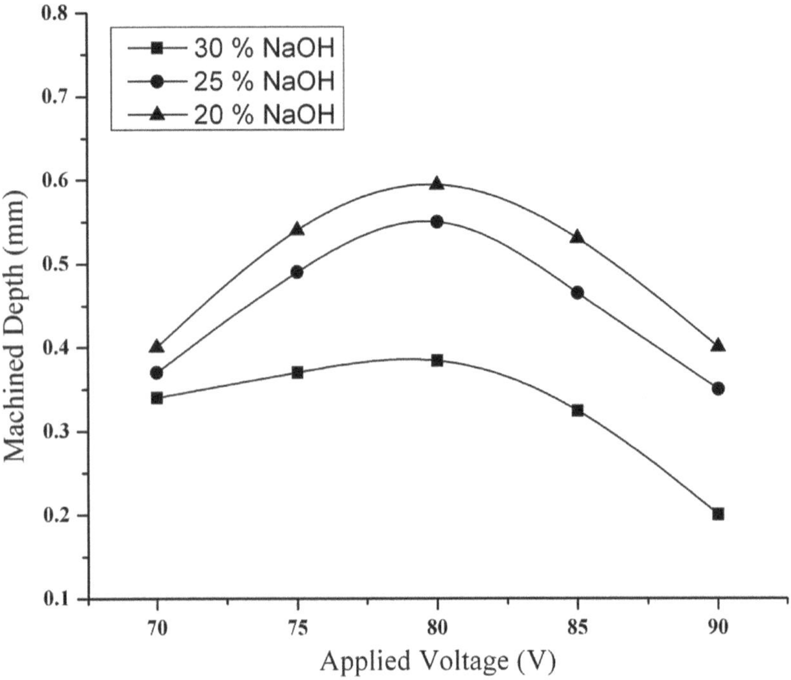

Fig. 6.35 Effects of applied voltage on machined depth at varying electrolyte concentration.

chemical reaction rate is less, and hence, the machining rate is less. At 80 V, the concentrated sparking is noticed and better machining depth is obtained. At 90 V, violent sparking is observed and greater side sparking is observed due to large volume of gas bubble generation. Though the material removal is high but the machining depth is less because the overcut is high at this condition. In general, the machining depth in ECDM is less due to the fact that the majority of gas bubbles are accumulated on the side face of the tool because tool and workpiece are physically in contact. When the voltage is increased sufficiently high, the machined depth does not increase proportionately, and because of this, the overcut on the upper face of the machined hole increases appreciably. This phenomenon occurs due to the fact that the upper face of the workpiece receives heat earlier than material at lower face of the workpiece. As a result, an increase in thermal conductivity at top surface is more due to temperature rise. This produces larger overcut at the top face as more material is removed from the top surface.

Figure 6.35 also exhibits the effect of electrolyte concentration on the machining depth. The machined depth decreases with the increase in the electrolyte concentration. The chemical and electrochemical reaction rates or gas bubble generation rate increases with the electrolyte concentration causing violent sparking and more side sparking. Greater machining depth is obtained only when the concentrated sparking occurred. Hence, the machining depth reduces with the increase in the electrolyte concentration because of the fact that the sparking is not concentrated at the bottom face of the tool rather than at the side edge of the tool.

The variations of machining depth with inter-electrode gap have been plotted in Fig. 6.36. It is clear from the graph that the machining depth increases with the inter-electrode gap and becomes maximum at 40 mm, which is the maximum level of inter-electrode gap. At the lower inter-electrode gap the machining depth is less as the electrolyte surrounding the tool gets contaminated with the reaction products including the gas bubbles. These factors contribute to achieve lesser amount of machining depth. The machining depth is limited due to the fact that as the depth of penetration of the tool inside the workpiece increases, the amount of fresh electrolyte available between the tool and the workpiece decreases. This results in reduced current density at the bottom part of the tool as compared

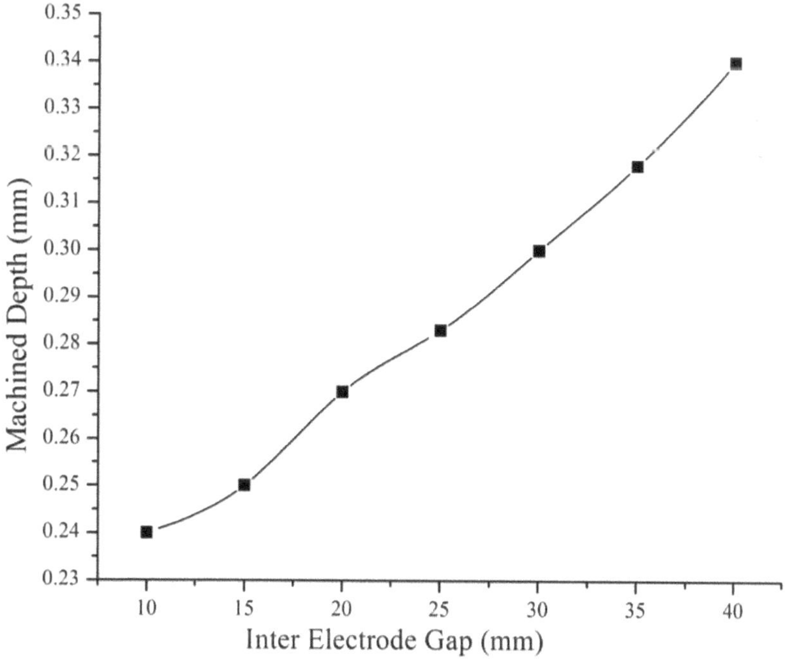

Fig. 6.36 Effects of inter electrode gap on machined depth.

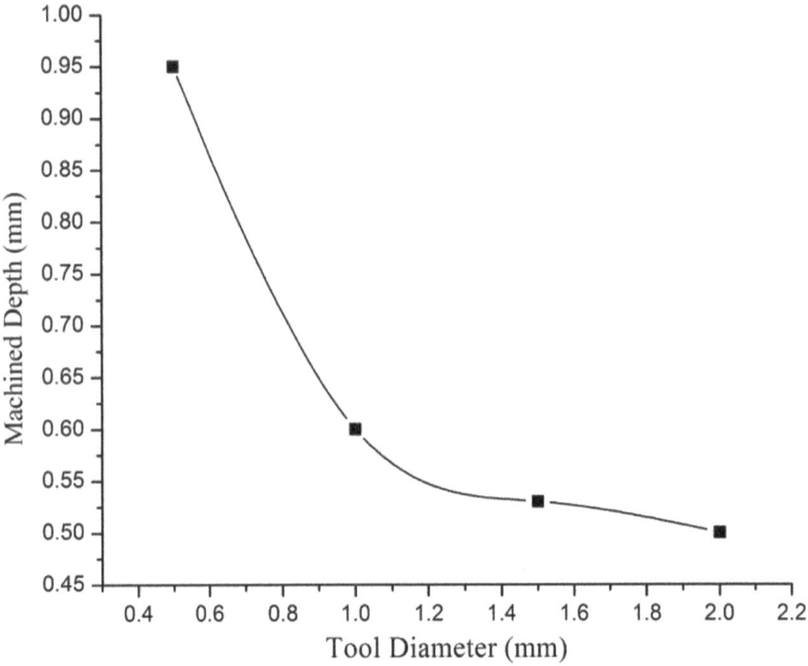

Fig. 6.37 Effects of tool diameter on machined depth.

with the upper part. This is also one of the reasons, why the machining rate decreases with the penetration depth.

The influence of the tool diameter on the machining depth has been presented in Fig. 6.37. The machining depth decreases with the increase in tool diameter. The larger machining depth is observed using small diameter of the tool. Probably the high current density is responsible for high machined depth. As the same amount of spark energy is released for a particular combination of voltage, electrolyte concentration and inter-electrode gap, the heating effect is confined within a small area in case of small diameter tool which causes faster material removal from the workpiece materials just below the bottom face of the tool.

6.4.6 Process capabilities of ECDM

The Electrochemical Discharge Machining (ECDM) process has capabilities for its wider applications in various machining operations such as drilling, cutting, turning, milling, grinding, trueing and dressing of hard and brittle materials. ECDM can be applied for micro-machining operations such as micro-drilling, micro-grooving, micro-milling for producing

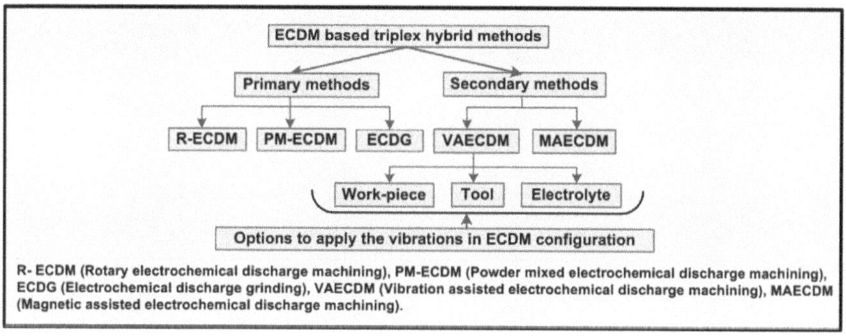

Fig. 6.38 ECDM based triplex hybrid methods [43].

various micro-features (<100 μm dimension) like micro-hole, micro-wall, micro-pillar, micro-groove and micro-pyramids on glass, ceramics, quartz and metal matrix composite materials.

For generating small and shallow die cavity, die sinking ECDM process has been applied using hollow bronze tool at 3–18 mm/min feed rate with pulsed DC voltage range of 20–30 V in 120 g/L $NaClO_3$ electrolyte. The dimensional tolerance of dies machined by ECDM is close to EDM and better than ECM [40].

ECDM process has been applied for drilling 2 mm diameter hole on non-conducting zirconium oxide ceramics using aqueous KOH electrolyte by setting the applied voltage 50 V, electrolyte concentration of 15% and inter-electrode gap of 30 mm to achieve minimum overcut of 0.01 mm [11].

Machining of non-conducting materials in MEMS applications is a challenging task. The spark assisted chemical engraving (SACE) utilizing electrochemical discharge phenomena has been proposed by R. Wurthrich has potential to machine glass for various applications in the field of MEMS technology [42].

The ECDM process has capability of possible hybrid machining at triplex level as shown in Fig. 6.38.

6.4.7 Advantages and limitations

Advantages of ECDM process are as follows:

(i) Electrically non-conducting material, e.g., glass, ceramics, fiber reinforced composites, quartz, etc. can be machined
(ii) Low heat affected zone as machining zone is immersed in electrolyte
(iii) Operating cost of machining is low
(iv) Chemically inert material can be machined

(v) Complex shape (non-circular) cavity can be generated, etc.

Limitations of ECDM process are as follows:

(i) Limited depth of machining
(ii) High overcut due to side sparking
(iii) Poor surface finish
(iv) Tool wear, etc.

6.4.8 Applications

The electrochemical discharge machining (ECDM) process can be utilized effectively for the machining of non-conductive materials such as glass, ceramics and advanced composites considering the capability of machining complex profiles. The ECDM process has the potential for macro- and micro-hole drilling operations on ceramic components. The ECDM technology has the possibilities and feasibilities to slice non-conductive materials. The ECDM technology can be used for engraving on electrically non-conducting materials. The ECDM process variants are exhibited in Fig. 6.39 [43].

6.4.9 Advancements in ECDM

The penetration depth during electrochemical discharge drilling is increased by using an orbital rotational tool [44]. Using abrasive cutting tools, cylindrical and conical in ECSM, machining performance in terms of MRR and machined depth have been improved during machining alumina and borosilicate glass [45]. The proper selection of tool tip shape is

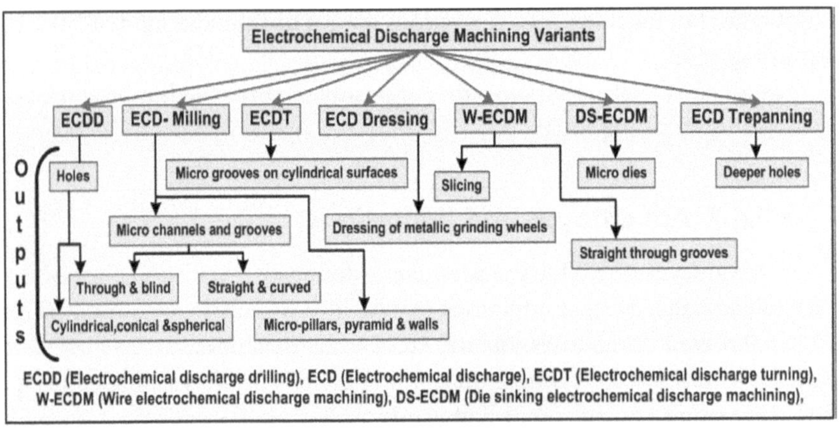

Fig. 6.39 Application variants of ECDM process [43].

important for controlling the availability of electrolyte for gas bubble generation and also for spark generation in ECDM. Taper side wall curvature front tool tip causes maximum amount of electrolyte availability in the machining zone which generates maximum number of sparks and thus results higher MRR. Taper side wall flat front tip increases machining rate as well as reduces overcut due to concentrated sparking [46]. The machining performance of ECDM can be improved by using spherical tip tool electrode. Due to the curve shape of spherical tool electrode, contact surface area between tool and workpiece is reduced and flow and availability of electrolyte is increased for gas bubble generation and it results in efficient machining in terms of larger machining depth and lower machining time during ECDM [47]. Electrochemical discharge turning (ECDT) is one of the variants of Electrochemical Discharge Machining to machine cylindrical parts [48]. Electrochemical discharge machining principle has been applied for trueing and dressing of metal bonded diamond grinding tools [49].

Electrochemical discharge dressing of metal bond micro-grinding tools can be done for reducing the normal grinding force and surface roughness of workpiece. In this process, fresh abrasive grains are protruded due to erosion of metal bonds and workpiece debris, as a result the micro-grinding tool is sharpened [50]. Magnetic field assisted electrochemical discharge machining method has been applied using a special magnetic tool chuck to enhance the circulation of electrolyte for generating deep holes with less overcut [51].

Electrochemical discharge grinding (ECDG) is a triplex hybrid machining method in which three different processes namely, mechanical grinding, ECM and EDM are combined [52]. In ECDG, metal bonded rotary grinding wheel embedded with abrasive particles is used as tool. The materials are removed from the workpiece due to combined action of mechanical abrasion and spark discharge action. For profile cutting and grooving operation, Traveling wire Electrochemical Discharge Machining (TWECDM) is one of the variants of ECDM. The traveling wire is used as tool and the movement of wire helps in removal of debris during machining. In order to increase the machining rate, SiC abrasive powders are mixed with electrolyte during micro-grooving using TWECDM [53]. The performance of ECDM process has been enhanced using rotary tool electrode of different shapes such as needle, cylindrical, flat side wall and helical during drilling and machining of micro-channels. Deep hole micro-drilling with high aspect ratio on thicker glass plates are successfully machined by ECDM process [54].

Electrochemical Discharge Machining (ECDM) is a special form of hybrid machining technology which can be applied for especially electrically non-conducting materials. The research on ECDM in various directions has improved the machining performance and enhanced the scope of its wider applications in various fields of industry where electrically non-conducting materials such as glass, ceramics and composites are used. ECDM process is a very complex hybrid machining process which needs thorough understanding for its applications in milling, turning and grinding, etc. The ECDM has also been widely used for various micro-machining operations.

6.5 Abrasive assisted advanced machining

6.5.1 Introduction

Abrasive powder can assist various advanced machining process when they are supplied into the machining zone. Abrasive particles, which may have sharp edges, electrically conductive or non conductive interact with the primary machining operations and enhance the capability of hybrid machining processes. It has been noticed that addition of other different additives like metal powders, different chemical agents, nanoparticles, etc. can also be used to improve various process characteristics of different major advanced machining methods. Here, loose abrasives particles or different additives are not directly involved in the primary machining mechanisms but assist it in other different ways. Abrasive grains mixed in the machining media generate mechanical actions in removing material by erosion, abrasion, etc. On the other hand, additives like metal powder, chemical agents, and nanoparticles added in the working media of primary machining processes modify the properties of the media by changing its characteristics, which helps in achieving more stable machining with enhanced capabilities. Manufacturing scientists are carrying out investigations to explore different possibilities in this context. Some of the major research outcomes in this direction, which shows tremendous potential and better commercial applicability, are reported here under to provide detail insight of different adaptive assisted hybrid machining process.

6.5.2 Abrasive water jet machining

Abrasive water jet machining (AWJM) is a hybrid machining process which combines both the principles of abrasive jet machining (AJM) and water jet machining (WJM). The problems of AJM are low MRR, generating airborne dust, large taper and more overcut due to stray abrasive action. WJM has also problems of cutting very hard materials and the applications of WJM are generally limited to softer materials. AWJM process has been developed to overcome the problems of AJM and WJM and also to get the merits of both the individual processes. Abrasive water jet machining (AWJM) is a well-established non-traditional machining process used for cutting softer as well as harder difficult-to-machine materials. In abrasive water jet machining, a stream of fine grained abrasive particles is mixed with water in suitable proportion, which is focused on a work piece surface through a nozzle. Material removal occurs due to erosion caused by the impact of abrasive particles on the work surface. In 1979 Dr. Mohamed Hashish worked on the process of adding abrasives to a standard waterjet. After testing a variety of options, garnet of mesh size 80 is commonly used as abrasive particle in AWJM. The abrasive water jet machining, which is based on the principle of rapid erosion by high-speed abrasive water jet combined with rapid cooling by the water jet, is a powerful jet in processing various materials. AWJ can machine a wide range of materials including titanium, stainless steel, aerospace alloys, glass, plastics, ceramics, etc. on thin sheet as well as thicker blocks. AWJ can cut net-shape parts. As there is no heating and chemical interaction of the material being cut, thermal and chemical damage free machined surface is produced. High-quality cut surface without burr which does not require any additional machining. AWJM can cut materials of thickness ranges from foils to very thick and layered products with very complex characteristics of individual layers and composites.

6.5.2.1 Mechanism of material removal in AWJM

The mechanisms of material removal in AWJM are cutting, fatigue, melting and brittle fracture under solid particle erosion process [55]. In AWJM of ductile behaving materials, material is mainly removed by low angle impact by abrasive particles leading to plowing and cutting. Further at higher angle of impact, the material removal involves plastic failure of the material at the sight of impact, which was studied initially by Bitter [56, 57]. Bitter considered two modes of material erosion from ductile behaving materials such as cutting wear that occur at low impact angle and deformation wear that happen at high impact angle. There are also two modes of material removal under micro-cutting such as cutting deformation for sharp edged angular abrasive particles and plowing deformation for spherical particles. Hashish [58] unified such models as applicable under AWJM at a later stage. Kovacevic [59] identified wear tracks plowed by abrasive grains due to scratching in scooping during AWJ cutting of stainless steel. Chip formation and plowing are the material removal mechanism in AWJ. In case of AWJM of brittle behaving materials, other than the above models, material would be removed due to lateral crack initiation and propagation because of brittle failure of the material. Zeng and Kim [60] proposed erosion mechanism of material removal from polycrystalline ceramics in the context of AWJM. The material removal mechanisms of AWJM for different materials [61] are listed in the following Table 6.2.

Two distinct mechanisms of material removal such as ductile erosion and brittle erosion are common in AWJM depending upon workpiece materials whether it is ductile or brittle. In ductile erosion the sharp edged abrasive particles progressively cut ductile material causing material removal. In brittle erosion, brittle materials are removed by the propagation and intersection of cracks ahead of and around the abrasive particle with high velocity.

Wang [62] proposed a model on material removal for predicting the depth of cut in abrasive waterjet contouring of alumina ceramics of 25.4 mm thickness. The depth of cut in AWJM depends on process parameters such as stand of distance, abrasive flow rate, water pressure and traverse speed and contour radius during contour cutting.

Table 6.2 Material removal mechanisms of AWJM for different materials

Materials	Mechanism of material removal	Proposed by
Stainless steel	Plowing and micro-cutting due to scratching and scooping	Kovacevic
Inconel	Plowing and micro-cutting	Webb and Rajurkar
Titanium	Chip formation and plowing	Arola and Ramulu
Aluminum	Micro-cutting at low impact angle	Ramulu
Aluminum alloy, titanium alloy, ferrite steel	Micro-cutting	Guo
Metal matrix composite	Scooping and plowing	Hamatani and Ramalu
Fiber reinforced composite	Brittle fracture of fibers and abrasive wear of matrix	Ramulu and Arola
Polycrystalline ceramics (aluminum oxide)	Brittle fracture and plastic deformation	Zeng and Kim
Refractory ceramics	Trans-granular fracture at the top and intergranular fracture at the bottom	Momder, Eusch, and Kovacevic
Glass	Lateral cracking	Momder and Kovacevic

6.5.2.2 Details of the AWJM setup

The abrasive waterjet machining system consists of high pressure water generation system, cutting head, abrasive delivering system, workpiece holding table and motion control unit, abrasive water jet catcher, and supporting accessories such as hoses and control valves, etc. The schematic diagram of abrasive water jet machining system is shown in Fig. 6.40.

(i) High pressure water generation system

Water is purified in reverse osmosis (RO) plant and supplied to the intensifier pump. Intensifier amplifies low pressure water to high pressure water of about, 600 MPa (6000 bar) using hydraulic cylinders and plungers of hydraulic cylinders of different cross-sections at different stages and also using control valves and hoses, etc. The attenuator reduces the pressure fluctuation of high pressure water from intensifier and supplies a steady flow of high pressure water to cutting head. Water and oil flow in two fluid circuits of a typical intensifier pump. Water RO filter, booster pump, intensifier and shock attenuator are the components of the water circuit. The oil circuit consists of electric motor, oil pump, oil reservoir, piston plunger and oil manifold. The high pressure oil is supplied by the pump to the hydraulic cylinders through manifold and moves the piston back and forth for delivering high-pressure water to the attenuator.

(ii) Cutting head

The cutting head consists of an orifice and nozzle arrangement for high velocity water jet, abrasive feeding nozzles/tube, abrasive water

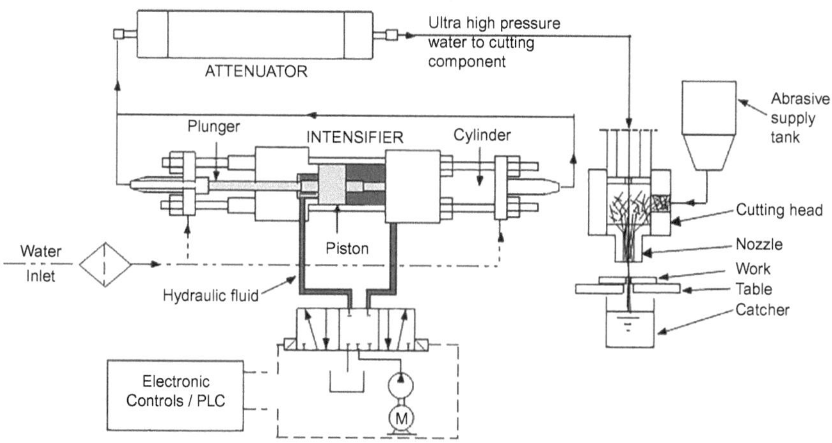

Fig. 6.40 The schematic diagram of abrasive water jet machining system [63].

mixing chamber and focusing or mixing tube. When water at such pressure is issued through a suitable diamond or sapphire orifice (generally of 0.2–0.4 mm dia), at the jet exit a nozzle made up of Tungsten carbide (0.76 mm tip diameter) is used to direct and increase the velocity of water. The potential energy of water is converted into kinetic energy, yielding a high velocity jet (1000 m/s). Abrasives are supplied to the cutting head from abrasive delivering tank for mixing with water in the nozzle. The abrasive flow rate is controlled by changing the entrance diameter of control orifice of abrasive feeding unit. Abrasive particles are dragged from abrasive feeder at very low velocity to the mixing chamber. Cutting head serves the purpose of mixing of abrasives with water and also forming high velocity abrasive water jet. Mixing chamber is a metallic compartment lies in between orifice and focusing tube. Focusing tube of cutting head is also the place where abrasive particles are mixed up with high velocity water jet effectively and also focused to an efficient and coherent abrasive water jet for machining. The length and diameter of focusing tube depends on wall friction of the jet and size of abrasive particles respectively. A typical focusing tube made up of tungsten carbide, inner diameter of 0.76 mm and 100 mm length. The size of focusing tube controls the final jet velocity, jet diameter as well as kerf width during cutting operation.

(iii) The workpiece holding table and motion control unit

Depending upon the types of AWJ machining operations such as drilling, profile cutting and milling, etc. either cutting head or workpiece holding table moves as per required movement controlled by CNC hardware and software program. For AWJM turning operation the workpiece is rotated as well as moved along axial direction. The workpiece rotating and holding arrangement is mounted on the machining table while abrasive water jet is perpendicular to the workpiece. The cutting head can be tilted for changing the angle of incident of jet to machine the inclined surface of the job. The cutting head is required to move up and down to control the nozzle tip distance from the workpiece and also to accommodate the workpiece of different thickness.

(iv) Catcher

A catcher captures the used abrasive water jet under the cutting point which has as much as 75% of initial energy of jet. By catching the abrasive water jet efficiently, the noise level can be reduced to the great extent. The catcher should be robust and can absorb high energy of jet. While the cutting head is stationary and workpiece is moving, then a long narrow tube can be used as catcher. While cutting head moves and workpiece is stationary, a settling tank filled with water and energy absorbent materials is used as catcher. The size of catcher is at least equal to the working area of machining. When both workpiece and cutting head move in opposite direction, slot catcher with energy absorbent materials is used.

6.5.2.3 Process parametric studies on machining performance criteria of AWJM

In order to achieve the desired machining performance such as material removal rate (MRR), surface finish, kerf width and taper angle, geometrical and dimensional accuracy of AWJM, the following process parameters are to be controlled:

(i) Hydraulic parameters: Water pressure, water flow rate and water jet diameter.
The water jet diameter again depends on water pressure and stand off distance.

(ii) Cutting parameters: Traverse speed, stand off distance or nozzle tip distance, number of passes and impact angle, mixing tube length and diameter, mixing ratio, number of passes, hardness and thickness of workpiece materials.
The most influential cutting parameter is traverse speed. Stand off distance affects the cutting performances and kerf width. Exit jet divergence angle increases with increase in stand off distance, which affects the top and bottom kerf width. To achieve the better surface finish the numbers of passes are to be controlled.

(iii) Abrasive parameters: Abrasive mass flow rate, abrasive particle size, abrasive particle size distribution, abrasive particle shape, abrasive particle hardness and abrasive recycling capacity.

Type and size of abrasive particle is one of the most influential parameter in AWJM cutting. Size should not be greater than the nozzle tip diameter. The shape of the particle should be irregular and sharp edged. These abrasive particles act like multipoint cutting tool. Popularly AWJM is used for cutting of hard and brittle material. The choice of abrasive material depends on hardness. Commercially Garnet ($Fe_2Al_2(SiO_4)_3$) is used for less cost and abundance in nature. Silicon carbide (SiC), Boron Carbide (B_4C), Silicon nitride (Si_3N_4) can be used for better finish. Using of diamond abrasives are more effective on finishing surface, which is a costlier. However, using of these harder materials decrease the costly tungsten carbide nozzle life and increase the production cost. Recycling of the abrasives makes the technology more economical, effective, and environmentally friendly. Abrasive materials of size 106 μm can be used for rock cutting.

The influences of process parameters such as water pressure, abrasive flow rate, stand off distance and traverse speed on material removal rate, kerf width and surface roughness during AWJM cutting of silicon nitride has been discussed here in after.

The influences of water pressure on material removal rate, kerf width and surface roughness are shown in Fig. 6.41. Material removal rate increases with the increase in water pressure. Kerf width initially increases and then becomes almost constant with increase in water pressure. Surface roughness, R_a value decreases continuously with the increase in water pressure. As the jet pressure increases, surface becomes smooth. With increase in jet pressure, brittle abrasives break down into smaller ones which again corrode the inner surface of the material. As a result of reduction of size of the abrasives the surface finish becomes better. Though, due to increase in jet pressure, the kinetic energy of the particles increases which ought to results in smoother machined surface and MRR increases. It is found that the machined surface is smoother near the jet entrance and gradually becomes rougher toward the jet exit. This is due to the fact that as the particles moves down they lose their kinetic energy and their cutting ability deteriorates.

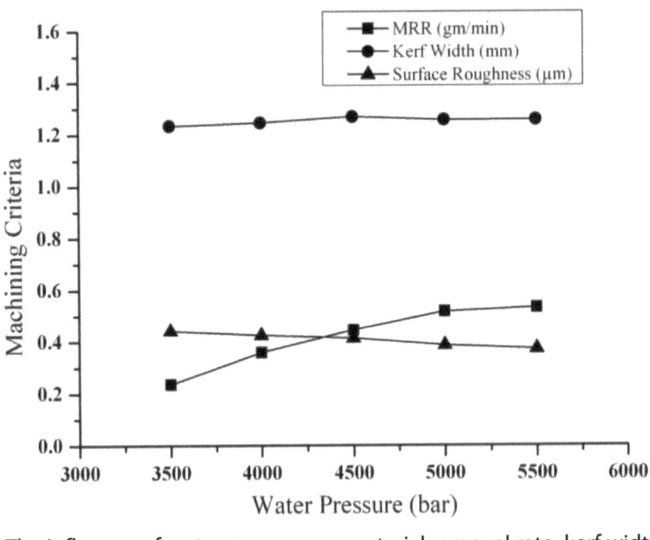

Fig. 6.41 The influence of water pressure on material removal rate, kerf width and surface roughness.

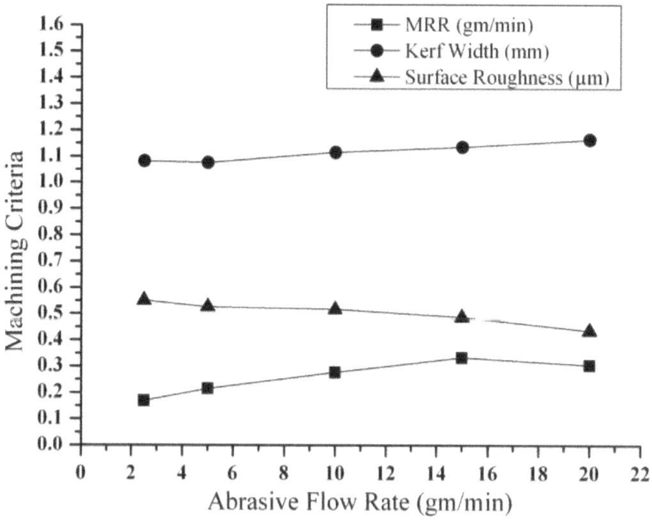

Fig. 6.42 The influences of abrasive flow rate on material removal rate, kerf width and surface roughness.

Influences of abrasive flow rate on material removal rate, kerf width and surface roughness are shown in Fig. 6.42. From this graph, it has been observed that MRR increases with the increase of abrasive flow rate upto a certain value and then MRR decreases. Kerf width first decreases then it increases. More abrasive makes the kerf width larger. For narrow kerf width, lower value of abrasive flow rate is recommended. Surface finish increases with the increase in abrasive flow rate. More abrasive particles helps in good surface finish. It needs a large number of impacts per unit area under a certain pressure to overcome the bonding strength of any material. With the increase in abrasive flow rate, surface roughness decreases. This is because of more number of impacts and cutting edges available per unit area with a higher abrasive flow rate. Abrasive flow rate determines the number of impacting abrasive particles as well as total kinetic energy available. Therefore, higher abrasive flow rate, higher should be the cutting ability of the jet. But for very high abrasive flow rate, abrasives may collide among themselves and lose their kinetic energy.

The influences of traverse speed on material removal rate, kerf width and surface roughness are shown in Fig. 6.43. From the above graph it is observed that MRR increases with the increase in traverse feed rate. Kerf width increases continuously with the increase in traverse speed. R_a value increases with the increase in traverse speed. After a certain value it

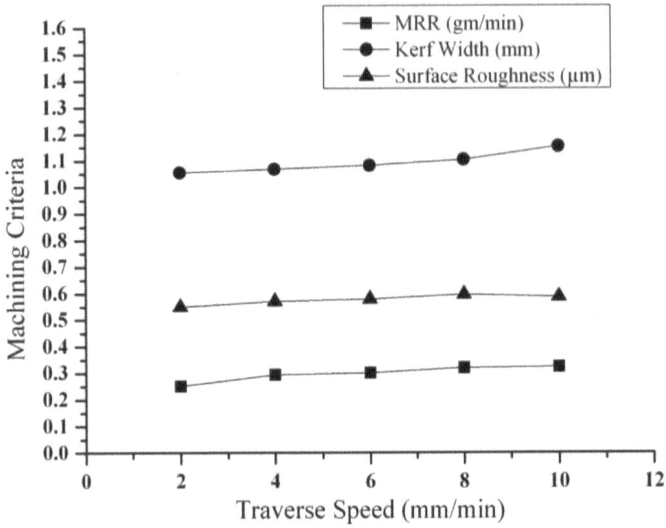

Fig. 6.43 The influences of traverse speed on material removal rate, kerf width and surface roughness.

decreases. For decreasing of the machining costs every user try to choose the feed rate of the cutting head as high as possible, but increasing the traverse speed always causes increasing of inaccuracy and surface roughness. This is due to the fact that as the work moves faster, less number of particles are available that pass through a unit area. Therefore, less number of impacts and cutting edges are available per unit area, which results a rougher surface.

The influences of stand off distance (SOD) on material removal rate, kerf width and surface roughness are shown in Fig. 6.44. MRR initially increases with increase in stand off distance from 2 to 4 mm. After that material removal rate starts to decrease. Maximum MRR is 0.29 g/min, when the SOD is 4 mm while other input parameters are kept constant. Kerf width increases continuously with the increase in the SOD because stray cutting happens with the increase in nozzle tip distance. R_a value increases with the increase of SOD. More distance from nozzle tip will cause stray cutting. So, surface roughness increases. SOD is one of the important factors for surface roughness. Generally, higher stand off distance allows the jet to expand before impingement. Therefore, increase in the standoff distance results an increased jet diameter as cutting is initiated and in turn, reduces the kinetic energy of the jet at impingement. So surface roughness increases with increase in stand off distance. It is desirable to have a lower stand off distance which may produce a smoother surface due to increased kinetic energy.

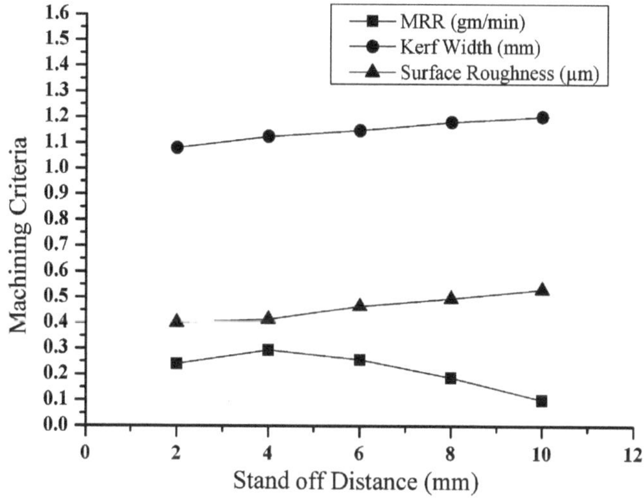

Fig. 6.44 The influences of stand off distance (SOD) on material removal rate, kerf width and surface roughness.

The machined surface is smoother near the top of the surface and becomes rougher at greater depths from the top surface.

Water Pressure of 5500 bar, abrasive flow rate of 15 g/min, traverse speed of 2 mm/min and the standoff distance of 10 mm is optimum setting for the best performance of AWJM during cutting dense Si_3N_4 using −80 BS SiC grits and 0.75 mm dia nozzles. At this optimum machining condition during cutting of Si_3N_4 ceramic maximum material removal rate of 0.3510 g/min, minimum R_a of 0.2359 μm, and minimum kerfs width of 1.01 mm have been achieved.

Abrasive Water Jet can also cut dense Silicon Carbide [SiC], very hard ceramic materials, used for both room and high temperature structural applications, like strategic, automobile, electrical and electronics industries. Silicon carbide grits (80 BS) can be used in replacement of commercially used garnet particle and process parameters such as water pressure, abrasive flow rate, traverse speed and stand off distance are controlled to cut SiC workpiece by AWJM process. Water pressure is varied from 2500 bar to 4500 bar. Abrasive flow rate is varied from 5 g/min to 25 g/min. Traverse speed is varied from 2 mm/min to10mm/min. Standoff distance is varied from 2 to 10 mm.

Glass of thickness 4 mm were drilled by AWJM process using abrasive particle as silicon carbide of 180 grit size with working pressure of 60–80 N/mm^2.The stand of distance varies from 2 to 5 mm. The chemicals such as acetone, phosphoric acid, polymer (polyacrylamide) with different chemical concentration were mixed with abrasive and water mixture to improve machining performance. The taper of the hole was very less and MRR was higher while polymer was mixed [64].

The general domain of parameters during typical AWJ machining process can be considers as orifice of Diamond or Sapphires of diameter ranges from 0.2 to 0.4 mm; focusing tube of Tungsten carbide (WC) of diameter ranging from 0.6 to 2.4 mm; water pressure upto 6200 bar; abrasive materials of garnet and silicon carbide of 80 grit size; abrasive flow rate ranging from 2 g/min to 1 kg/min; stand-off distance from 1 to 350 mm; machine Impact Angle of generally 90^0; traverse Speed of max 5 m/min and depth of cut of 1 mm to 250 mm.

6.5.2.4 Advantages of AWJM

Abrasive Water Jet Machining Process is a cold machining process, so there is absence of thermal damage and heat affected zone and thermal cracks on machined surface. AWJ machining can be used as supplementary process for wire EDM, LBM and Plasma cutting where thermal effects on work piece materials are detrimental. It can cut any conducting as well as non-conducting materials. It can cut softer as well as harder materials. Modern AWJ machines enable the cutting of materials which are difficult for conventional machining such as Inconel alloys, titanium, and composites with a tolerance of ± 0.05 mm [65]. AWJ can cut materials ranging from 1.6 to 300 mm in thickness with an accuracy of ± 0.13 mm [66]. The ultra high pressure (UHP) abrasive water jet can cut virtually any hard material like ceramics and composites. This technique is also suitable for very soft, brittle and fibrous materials. AWJ cutting is particularly environmentally friendly as it does not generate any cutting dust or chemical air pollutants like fume and gases. As no tool is used, so there is no problems of tool wear, tool design and fabrication like other hybrid machining processes. It is a faster machining process compared to EDM. It has various distinct advantages over the other advanced cutting technologies, such as no thermal distortion, high machining versatility, minimum stresses on the work piece, high flexibility and small cutting forces. AWJM can perform profile or contour 2D/3D cutting of thicker materials by controlling the motion of abrasive water jet or workpiece by CNC control. It can cut non-conductive and reflective materials such as glass and quartz those are especially difficult to cut by thermal process because they have a tendency to fracture. Deep hole drilling with tight tolerance can also be achieved by AWJM process. AWJM is very much effective for cutting honeycomb structures as well as layered composite materials.

6.5.2.5 Limitations of AWJM

Abrasive Water Jet Machining (AWJM) process has many advantages for potential applications but it has some limitations such as stream lag, cone, and taper effect [67]. The abrasive water jet deflects as it enters into the workpiece. The exit point of the stream (jet) lags behind the entrance point during cutting of thicker materials. For cutting contour, a cone shape is generated by the effect of stream lag. The cut surface also becomes taper. The cone shape and taper cutting edge are the errors of AWJM process. However these can be minimized by controlling the process parameters such as traverse speed to cut complex shapes with varieties of corners and curves. The abrasive water jet contains air, water and abrasive particles. Because of air entrainment in the jet, jet diverges with increase in standoff distance and kerf width and surface waviness is more. Therefore, it is difficult to apply AWJM for micro-machining applications. Although some polymer powder is mixed with water to increase the cohesiveness of three phases (air, water and abrasives) of jet to reduce the divergence angle of the jet for better machining actions. High pressure abrasive slurry jet machining (HASJM) can be used for reducing the surface waviness and also kerf width of micro-channel in micro-machining applications [68]. Nozzle wear is present in AWJM, but this can be reduced using suitable design and nozzle materials. There are some environmental issues of AWJM such as spent water disposal, chip and abrasive recovery and reuse and water recycling, those lead to environmental pollution. High pressure cryogenic jet machining can reduce the environmental degradation where liquid nitrogen replaces the water phase and dry ice crystals replace the abrasive phase leading to no need of disposal or waste generation.

6.5.2.6 Application of AWJM

Abrasive Water Jet Machine (AWJM) can be used for machining wide range of softer as well as harder materials of large thickness for different applications. Some of the typical applications [67] are as follows:

(a) AWJM is used for cutting metallic sheet of titanium, aluminum, stainless steel, high tensile strength steel, super alloy, etc. for aircraft industry, rolling stock industry, automobile industry, ship building industry, mechanical engineering industry, steel frame products, bridge manufacturing, ferrous industry, non-ferrous industry and manufacture of metallic products, etc.

(b) AWJM is used for cutting of varieties of glass such as wire glass, stained glass, laminated glass for glass industry, housing industry, interior decoration, advertising and medical appliances manufacture, etc.

(c) AWJM can cut advanced materials such as composite materials (FRP, FRM, FRC), ceramics, magnetic materials, etc. for aircraft industry, rolling stock industry, automobile industry, sporting goods industry, ceramic industry, electronic parts industry and optical fiber industry etc.

(d) AWJM can cut building material such as board, light weight concrete, granite and tiles, etc. for construction industry, housing industry and tile industry, etc.

(e) AWJM can be used to cut spent nuclear fuel pipe and various kinds of graphite for atomic power industry and also for processing of graphite.

(f) AWJ cutting of reinforced concrete, cement, mortar, rocks, stones is highly demanded for building industry, gardening and mining applications.

(g) AWJ cutting of scraps such as tank, tire, FRP, etc. for scrap handling and robotized AWJM is used to dismantle nuclear plants.

(h) AWJM can be used for burr removal of casting in foundry and also for cleaning purposes such as paint removal.

(i) Besides cutting, AWJM is used for pocket milling, drilling, peening operations, etc.

Research and development on abrasive water jet machining has been made to apply AWJM as hybrid machining method for polishing and cleaning with high pressure abrasive water jet for better surface finish [69]. Abrasive water jet turning (AWJT) method also becomes very popular for hybrid machining of metals as well as non-metals [70].

6.5.3 Abrasive assisted EDM

Electrical discharge machining (EDM) is an advanced machining method which is based on the thermal energy produced by series of repetitive sparks generates between the tool and workpiece emerged in dielectric fluid. EDM has some outstanding capabilities which include capability to machine conductive HSTR materials for generating complex geometrical features. Details of this process have already been discussed in chapter 4.

In spite of these advantages, applications of EDM are constrained due to tool wear, relatively low material removal and its comparatively poor surface quality. Since the machining characteristics are highly dependent on the nature and properties of dielectrics, significant attention has been made by the manufacturing scientists to modify the properties of dielectrics by introducing additives, e.g., abrasives, metal powders and different chemicals to achieve higher machining performance and productivity.

Utilizing powder mixed dielectric in the EDM process is called powder mixed EDM (PMEDM). Suitable host dielectric for enhancing machining characteristics by adding powder for PMEDM applications is one of the main criteria. PMEDM facilitates producing parts with predominantly high surface quality. Some investigations also indicate that proper powder selection increases machining efficiency in terms of material removal rate. Hence, the role of powder addition in the discharge characteristics and its influence on machining output are prime important. In addition, the influence of the thermo-physical properties and concentration of powder particles, the performance of various powder materials and abrasives are investigated extensively.

Fig. 6.45 illustrates a basic scheme of PMEDM circulation system which contains magnetic filter that enables constant use of powder mixed dielectric [71]. Maintaining constant powder concentration in a dielectric is a

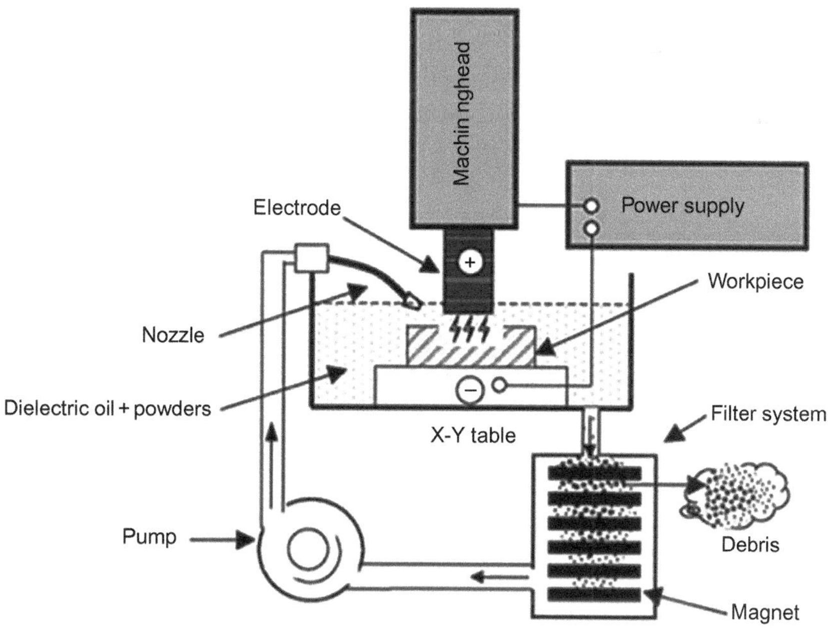

Fig. 6.45 Basic scheme of PMEDM circulation system [71].

problematic area in PMEDM. Stirring, vibration, addition of surfactant to powder mixed dielectric are some of the proposed techniques which are utilized for increasing powder homogeneity in the dielectrics [72].

The suitable dielectric which hosts the powder can be in the form of liquid, gas or a mixture of liquid and gas, e.g., hydrocarbon oil, deionized water, air, helium, argon, etc. In EDM, normal pulse discharges cause arcing due to contaminated dielectric which contains too much local debris [73]. Adding adequate metal powder or abrasive to the dielectric decreases the electrical resistivity and expands the spark gap which consequently stabilizing the process through better flushing and servo-hunting [74]. A wider discharge gap diminishes the heat flux [75], which leads to reduction of material removal and enhances the surface quality. Spark gap expansion depends on powder parameters. Upon addition of conductive powder, the reduced electrical resistivity of the dielectric enhances the ionization and spark frequency between the tool and workpiece [76]. This phenomenon overcomes the lower MRR resulting from gap expansion. Moreover, discharge distribution reduces the generation of surface ridges because of the creation of shallower craters with lower borders [77]. Hence, smaller craters and debris particles facilitate the gap flushing which intern enhances the material removal [78].

Powder parameters which influence PMEDM performance are types of powder material, density, electrical resistivity and thermal conductivity along with particle size and concentration. Various powder material, e.g., graphite, aluminum, silicon, silicon carbide, titanium, copper, chromium, tungsten, boron carbide and titanium carbide, etc. are mixed as additives in the dielectrics to improve discharge characteristics of PMEDM like energy intensity, spark frequency, discharge gap and spark size. Graphite and aluminum powders are widely used material which improves MRR and surface quality. Some metal powders, e.g., aluminum, titanium, graphite, tungsten and copper results in surface modification during machining. Generally powder concentrations are kept below 20 g/L. Micro size powders ranging from 1 to 55 μm have been used PMEDM.

Abrasive powder, e.g., SiC and B_4C which are semiconductive ceramics thus generates smaller gap expansion compare to metallic powders. Abrasive partials reduces the gap size and increases the tool wear and material removal depth. The effects of erosion of abrasive particles during PMEDM enhance the surface finish. Addition of abrasives also helps to remove molten material from the craters thus also reduces the recast layer thickness.

Nano size powders ranging from 20 to 150 nm are also shown better performances in PMEDM. During micro EDM, size of powder additives are comparatively in lower range. Carbon nano fibers can also be used in micro EDM. Due to the excellent electrical conductivity of carbon nano fibers, during micro EDM, it reduces insulating strength of the dielectric and increase the spark gap. Fig. 6.46 compares schematically conventional EDM and carbon nano fiber assisted micro EDM where carbon nano fiber arranges themselves in the form of micro chain which generates breezing

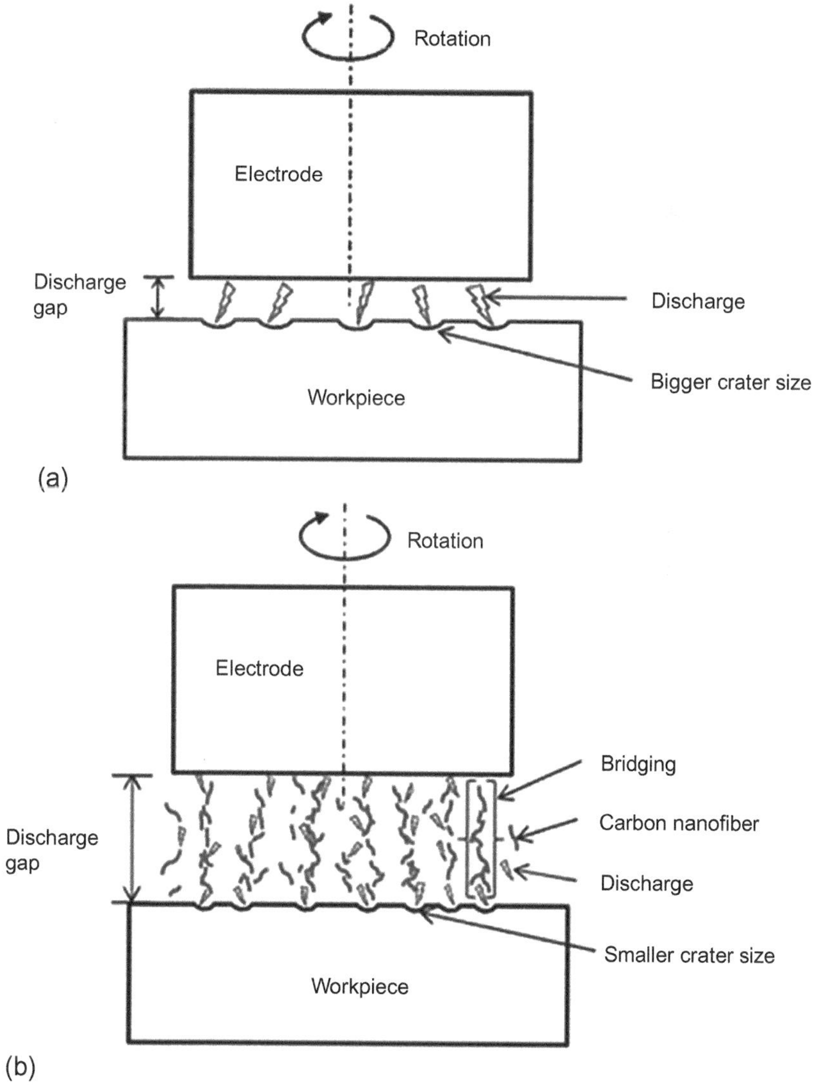

Fig. 6.46 Schematic model for (a) conventional EDM and (b) carbon nanofiber-assisted micro-EDM [79].

networks between tool and the workpiece. Addition of carbon nano fibers in dielectrics during micro EDM improves surface finish as well as reduced tool wear due to generation of multiple fine discharges [79].

6.5.4 Abrasive assisted ECM

Electrochemical machining is one of the most promising non conventional machining processes where material removes by means of anodic dissolution. This process is already explained in details in chapter 5. It has various advantages however some limitations of this process hinder its wider applications in the manufacturing industries. Main drawbacks of ECM are the electrochemical passivation of the workpiece surface during machining which restricts the smooth machining process for a longer period of time. This phenomenon is most prominent during machining of some HSTR alloys, e.g., steel and supers alloys and other metallic materials, titanium super alloys, aluminide, metal matrix composite, etc.

Electrochemical passivation generates a thin oxide layer on the workpiece surface which is electrically non conductive as well as adhere to the workpiece surface. Electrolyte flow is not capable to remove this thin oxide layer as well as potential difference utilized during normal ECM is also not sufficient to convert this oxide layer from passive to transpassive condition. However, this thin passive oxide layer in some cases are sticky in nature.

Addition of abrasive grains in the electrolyte is one of the possible solutions to overcome these above mentioned drawbacks of ECM. Abrasive grains which are mixed into the high velocity flowing electrolyte impinge on the oxide layer of the workpiece and expected to remove that layer by abrasion. The abrasive grains are mainly acts only on the oxide layer but not on the material of the workpiece. For the removal of passive oxide layer, the anodic dissolution process of ECM is accelerated. The machining can be continued for a longer time due to depassivation of the material surface by the abrasive grains. Fig. 6.47 shows the schematic of abrasive assisted ECM where anodic dissolution as well as abrasion of the abrasive grains is responsible for machining.

For machining metal matrix composites (MMC), abrasive assisted ECM can be utilized. Especially in the case of Aluminum-MMC reinforced with silicon carbide (SiC) and boron carbide (B_4C) nano particles which are one of the most advanced MMC. This MMC cannot be machined well utilizing different advanced machining processes due to presence of harder semi

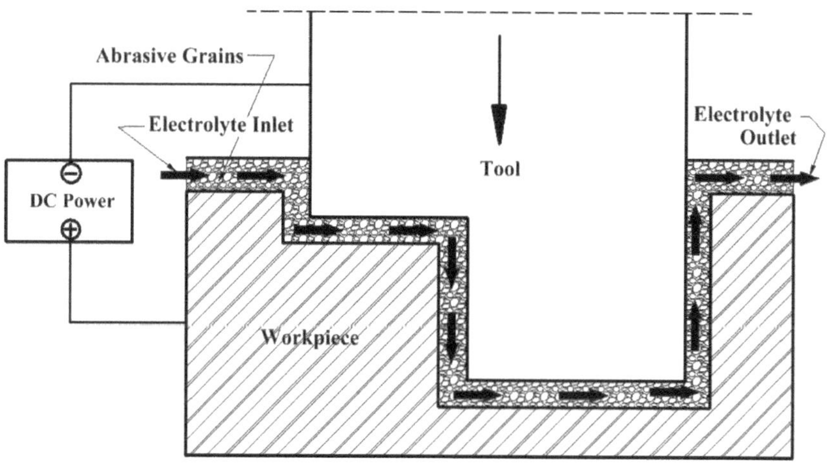

Fig. 6.47 Scheme of abrasive assisted ECM.

conductive SiC or B_4C abrasive grains. In case of conventional ECM, aluminum matrix dissolved in a slower rate and addition of abrasives, e.g., SiC in the flowing electrolyte solution during machining operation enhances the MRR effectively due to abrasion.

In case of finishing operation with abrasive assisted ECM, tool rotates and job reciprocates. A lump of abrasive grains are accumulated around the tool like formation of a brush. This accumulation is formed due to the electrophoresis phenomenon which is caused due to the abrasive particles being positively charged and attracted to the negative tool while suspended in a low conductivity electrolyte. Rotation of the accumulated abrasives with the rotating tool of ECM brings about effective removal of the metal hydroxides as well as minor asperities and burrs from the machined surfaces. The work piece is finished by anodic dissolution and the abrasion action of the accelerating abrasive particles. The abrasives also remove the undesirable passive oxide layers which formed on the surfaces and thus enhance the efficiency of the ECM operation.

However, type of abrasive, its grain size and concentration are some of the important factors which influence the process of abrasive assisted ECM. Normally non conductive abrasives, e.g., Al_2O_3, WC, SiC, etc. are widely used in abrasive assisted ECM. Finer abrasive grains size ranging from 25 to 70 μm and concentration in the order of 5–8 g/l are recommended for abrasive assisted ECM. However, due to the abrasion abrasive grains may cause tool wear as well as wear of machine parts. Abrasive grains change the properties of the electrolyte. These effects have to be overcome by proper selection of the types of abrasive, grain size and concentration as well as other parameters setting of ECM.

6.6 Vibration assisted advanced machining

6.6.1 Introduction

Vibration assisted machining utilized vibration ranging from low to high frequency with small amplitude. Generally, ultrasonic vibration is used for machining especially brittle materials with the help of abrasives slurry in which material removal is achieved by brittle fractures. The details of ultra sonic machining (USM) is already discussed in chapter 3. However, ultrasonic vibration is also directly employed in many of the advanced machining applications for assisting material removal mechanism. Trend has been shifted to use low to high frequency vibrations as assisting media to enrich process performance of different machining processes. Normally piezoelectric, magnetostrictive or piezomagnetic transducer is used as a vibrating unit to vibrate the tool or workpiece or machining media. Vibration assisted machining also combines advance machining processes with small amplitude to tool, workpiece or work media vibration. This approach has been applied to number of machining processes from conventional machining, e.g., turning, drilling, milling and grinding, etc. to non conventional advanced machining process, e.g., EDM, ECM and LBM, etc. The advantages of vibration assisted machining are low tool wear and chipping, reduction of the cutting force, improves the surface quality of the product, better geometrical shape and improves process productivity. Ultrasonic assisted machining system consists of a high-frequency signal generation system, transducer to convert the signal to high-frequency mechanical vibrations and acoustic horn to amplify the ultrasonic vibrations and transfer the vibration to attached tool or workpiece or media which facilitating the machining operation of different advanced processes. Different vibration assisted mechanical machining as well as non conventional machining are discussed here under.

6.6.2 Vibration assisted mechanical machining

Vibration assisted cutting is employed for turning, drilling, milling as well as grinding. In turning, the vibration can be applied in uni-direction or bi-directions which generates an elliptical movement. Various mode of vibration assisted turning are exhibited in Fig. 6.48 [80]. In vibration-assisted

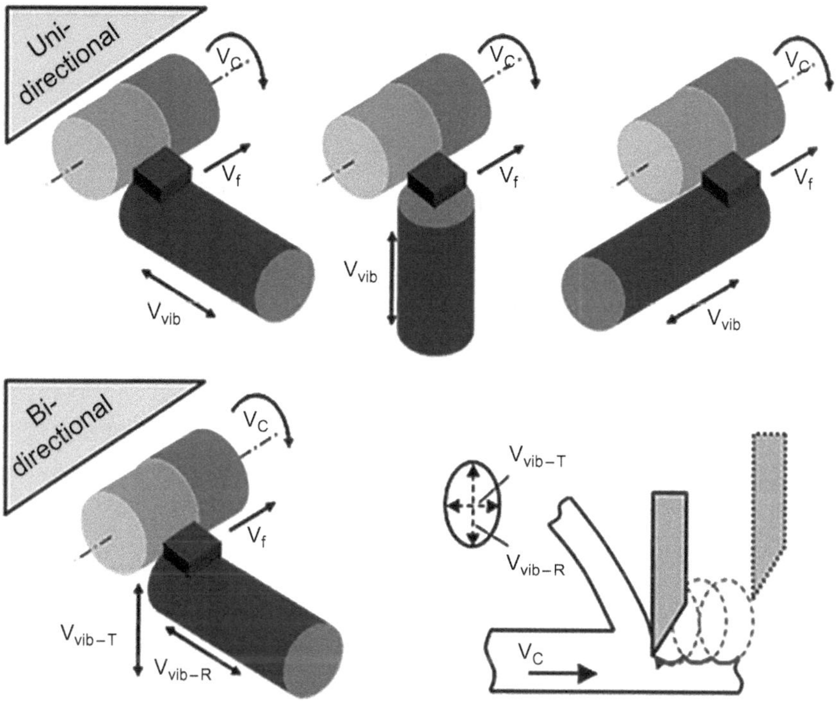

Fig. 6.48 Various mode of vibration assisted turning [80].

turning, the cutting tool loses contact with the job and chips on specified amplitude, resulting in decreasing of machining forces and improving tool life and surface finish [81]. In consequence, surface cracks may develop due to the hammering of the tool on workpiece surface [82]. Ultra-precision machining of hardened steel glass and even other ceramic materials can be achieved by the assistance of cutting tool vibration where periodic disengagement of single crystal diamond tools during vibration reduces cutting forces and increases surface qualities [83].

Generally better surface qualities can be achieved with ultrasonic vibration frequency. Short or even discontinuous chips can be produced by chip breakage due to vibration assistance cutting. Additionally the advantages of vibration assisted cutting are the reduction of built-up edges and burrs during cutting operation. Vibration assisted turning also capable of generating predefined surface microstructures without using any additional textured tool [84] where, the vibration should be applied in the radial or in the feed direction.

The vibration assisted drilling is also possible by either vibrating tool or workpiece. Due to the vibration kinematics, lower cutting forces and intern

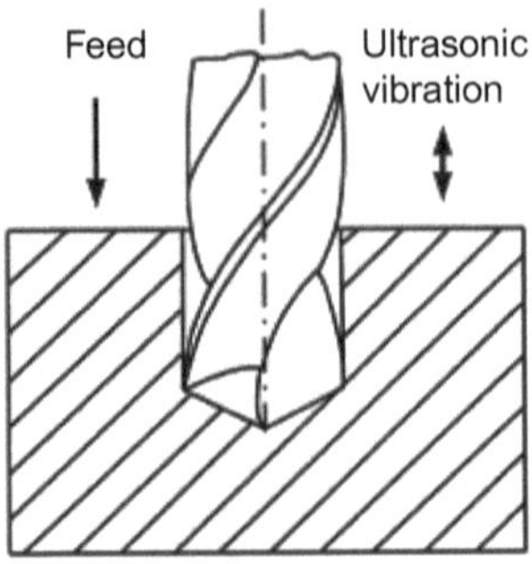

Fig. 6.49 Scheme of vibration assisted drilling [86].

drilling torques, better chip breakage and chip removal are generally achieved [85]. Fig. 6.49 shows scheme of vibration assisted drilling.

The feed force decreases up to 20% for high amplitudes of vibration at higher temperatures as compared to conventional drilling of Ti6Al4V. Vibration assisted deep hole drilling of Inconel 738LC produces lower process forces, better surface roughness due to chip breakage [87]. The resultant forces and torques are also reduced in vibration assisted drilling. Same type of results are also observed during vibration assisted drilling of metal matrix composites [88] and carbon fiber-reinforced polymer (CFRP) composites [89]. Vibration assisted milling and grinding is less applicable because of the complication in machine structure kinematics which are needed for complex shaped geometry of the workpiece. The amplitude of the vibration is an important parameter which should be optimized to achieve good surface roughness [90].

6.6.3 Vibration assisted EDM

Electro-discharge machining (EDM) is one of the earliest non-traditional machining processes. EDM process is based on thermoelectric energy employed between the workpiece and electrode and is one of the major manufacturing processes widely used in die and mold making industry to generate deep and 3D complex cavities in many different classes of materials under roughing and finishing operations [91]. The material is removed with the erosive effect of the electrical discharges from the tool and workpiece. The material removed in the form of debris is accumulated in the machining gap and leading to reduction of resistance in the gap, hence boosts the formation of abnormal discharges through short circuits, which eventually results in significant tool wear and slows down the material removal rate

(MRR) [92]. Hence, in order to maintain efficiency and consistency of the process, effective evacuation of debris from the sparking gap plays a commanding role in EDM process. Development of effective debris evacuation or flushing methods to remove debris from sparking gap has been the focus area of research in EDM since last decade. Application of vibration is one of the methods for debris evacuation in EDM.

Vibration-assisted electro-discharge machining (VAEDM) is hybrid machining process, wherein small-amplitude vibrations of specific frequency are employed to either of the electrodes (tool or wire and workpiece) or dielectric fluid to improve flushing and hence machining efficiency [82]. The basic principle of this process clarifies that the ultrasonically vibrating electrode or workpiece changes the sparking gap rapidly and generates a high frequency alternating pressure assists in better ejection of the molten materials. In addition, the ultrasonic vibration improves dielectric circulation, which in turn prevents the conglomeration of the debris particles in the working gap. It facilitates effective debris removal and hence improves the discharge efficiency and results in high MRR significantly [93]. The investigation on VAEDM have been carried out with various modes of applying the ultrasonic vibrations in EDM system like vibrations applied to workpiece, tool/wire and dielectric fluid.

The common mode of employing the ultrasonic vibration in EDM system is to vibrate the workpiece with the help of vibrating system. Generally piezoelectric/magnetostrictive transducer is used as a vibrating system for die sinking EDM, wire cut EDM and micro EDM. The schematic diagrams of an ultrasonic-assisted micro-EDM system where the workpiece is directly attached to a piezoelectric transducer to secure the vibration and a wire electro-discharge machining (WEDM) where piezoelectric actuator was used to vibrate the workpiece are shown in Fig. 6.50(a) and (b), respectively.

The influence of ultrasonic vibration on machining rate with drilling or cavity sinking is highly remarkable on EDM performance due to improved stability of sparking because of ultrasonic field forces and micro streaming. When vibration is applied to workpiece, the cutting rate is 1.5 times higher than when vibration is applied to the wire and 2.5 times higher than conventional WEDM. Larger area of the workpiece is involved to circulate dielectric and flushing away the melted material in case of workpiece vibration, resulting in higher machining efficiency. In the micro EDM assisted by workpiece vibration an increase in discharge frequency is directly proportional to the vibration frequency. In addition, an increase in effective discharge ratio results in more spark explosive force to accelerate dielectric

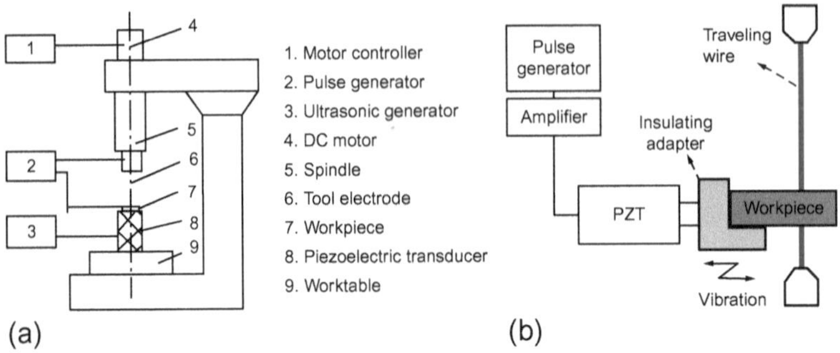

Fig. 6.50 (a) Ultrasonic-assisted micro-EDM [94], (b) ultrasonic-assisted WEDM [95].

liquid circulation and debris removal so that the tool electrode wear is reduced.

The other mode of applying the vibration in EDM system is to employ the ultrasonic vibration to tool electrode or wire. One end of the vibrating system is directly connected to the Z axis of EDM and the other end to the electrode. In WEDM the ultrasonic vibration is superimposed in the wire by the use of a transducer, wire holder and an ultrasonic generator mounted on WEDM as shown in Fig. 6.51. The vibrating tool electrode results in improved process performance compared to non-vibrating tool EDM. Even

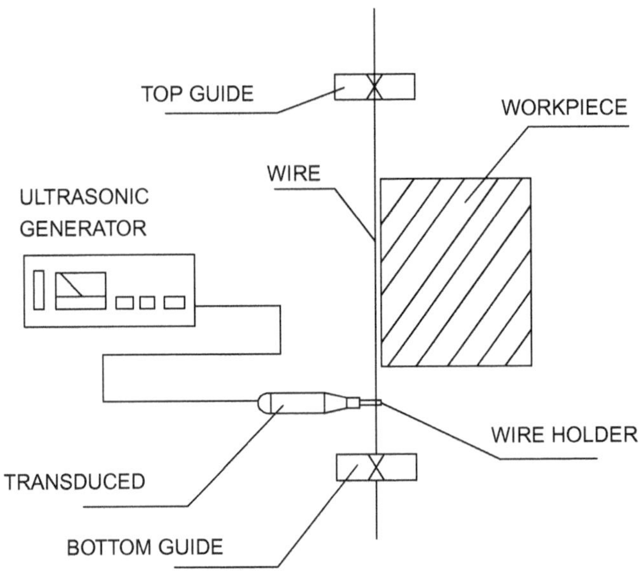

Fig. 6.51 Ultrasonic-assisted wire vibrated WEDM [96].

ultrasonic vibration enhances the cutting efficiency of WEDM by 30% and reduces the surface roughness of the machined surface remarkably. Some researchers introduced simultaneous vibration of the tool and the workpiece as a new idea in ultrasonic vibration assisted EDM. It has been reported that simultaneous vibration of the tool and the workpiece produces higher MRR in comparison with the conventional EDM and VAEDM with ultrasonic vibration of the tool. Further combining ultrasonic vibration with planetary movement of an electrode is the new method to drill micro-holes with high aspect ratios by micro-EDM for improvement of MRR [97].

The concept of applying ultrasonic vibration to dielectric fluid is the third mode of introducing the concept of vibration in EDM system. Here ultrasonic vibration enhances MRR significantly and improves surface quality due to increase in the kinetic energy of discharge debris in ultrasonically activated bath. The vibrating sonotrode is immersed in the dielectric in such a way that high intensity near the field of ultrasonic is aligned to machining zone and eventually enriched the process speed.

Till now most of the researches have been carried out with the application of ultrasonic vibration to the tool electrode or wire in comparison with other two modes of vibrations, i.e., vibration applied to workpiece and dielectric fluid. Ultrasonic vibration of the workpiece effects the larger area and trends to increase dielectric circulation and eventually flushing efficiency, resulting higher machining efficiency than the vibration of the wire [94]. On the other hand, the vibration of tool or wire requires smaller piezoelectric ultrasonic device with small size, low cost and lower energy consumption compared to the vibration of the workpiece. While the vibration to the dielectric fluid creates pressure variation in spark gap for better debris evacuation without change in gap distance. All methods have potential to crater in one or more aspects and have unique advantages compared to other modes and hence have been tried at different research levels.

Ultrasonic vibration assisted EDM not only improves surface finish and sparking efficiency but also helps molten globule to get detached easily, form smaller size and spherical shape. It results in thinner heat affected zone and hence smaller surface micro cracks. Ultrasonic assisted EDM has proved to be an effective and successful means to improve fatigue behavior, tribological performance and wear characteristics of advanced and difficult-to-cut materials.

Manufacturing of 3D cavities with electro-discharge milling is possible for simple geometric features only due to inefficiency of debris flushing from sharp corners and intricate features. Ultrasonic-assisted electro-discharge

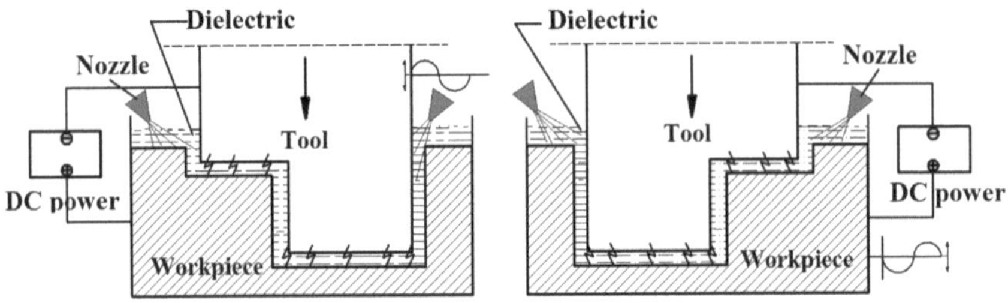

Fig. 6.52 Scheme of vibration-assisted micro-EDM process.

milling is expected to provide the better flushing, uniform dispersion of debris in dielectric by resisting formation of debris bridge and conglomeration. Computational fluid dynamics modeling in plasma channels formation and bursting for ultrasonic EDM may provide more interesting findings. Application of ultrasonic vibrations in powder and additive mixed dielectric may ensure uniform suspension and prevent conglomeration to ensure stable and favorable sparking cycles. Ultrasonic vibration assistance in dry and near-dry EDM processes is anticipated to address the debris reattachment, poor debris ejection and non-uniform erosion rates, etc. Fig. 6.52 shows details scheme of vibration-assisted micro-EDM process.

In vibration-assisted micro-EDM process, the perpendicular vibration has reduced machining time than the parallel vibration [92]. Employing vibration to the wire electrode directly in wire electrical discharge grinding (WEDG), improves surface integrity of tungsten carbide which is better to that by vibrating dielectric fluid or without ultrasonic assistance [98].

There is a noteworthy improvement in dimensional accuracy, surface quality and MRR with substantial reduction of EDM gap, machining time and micro tool wear rate during drilling deep micro-holes by means of workpiece vibration-assisted micro-EDM. The machining results also reveal that the low frequency vibration is more effective at the low discharge energy level, thus making it more suitable for micro-EDM. Ultrasonic transducer can also be attached to the work tank to vibrate dielectric fluid. Work fluid vibration is an effective way to circulate debris and increase material removal rate and geometrical accuracy in micro-EDM [99, 100].

6.6.4 Vibration assisted ECM

During ECM operation, gas bubbles and sludge generates in the gap between tool and the workpiece which changes the characteristics of the electrolyte. Due to this change in properties of the electrolyte in the machining zone, controlled machining cannot be continued for a longer period of time. By vibrating tool or workpiece, the electrolyte in the machining zone can be stimulated. This vibration creates cavitational effect in the zone of narrow inter electrode gap. This phenomenon generates pressure waves across the gap and facilitates flushing out of the gas bubbles, sludge from the machining zone. Hence, effects of vibration create better flushing of machining zone which intern helps controlled machining.

In ECM, electrochemical passivation is also one of the main drawbacks when utilized for machining specially titanium based alloy. Electrochemical passivation generates thin oxide layer on the substrate surface, which possesses lower mechanical strength than the core material. It can be removed easily creating vibration in the electrolyte media in inter electrode gap. The lower or higher frequency vibration of tool during ECM operation can also helps to remove passive oxide films from the workpiece surface during machining of alloy steels and super alloys, titanium alloys and other composite materials. Hence, the depassivation of the workpiece surface assists the machining process to continue further. A schematic details of the vibration assisted ECM is presented in Fig. 6.53. It consists of DC power supply, tool electrode and its feed mechanism; components of ultrasonic system viz. ultrasonic transducer coupled with ultrasonic generator, horn with a tool holder and an electrolyte or abrasive mixed electrolyte. DC voltage ranging from 10 to 30 V is applied across the anodic workpiece and cathodic tool.

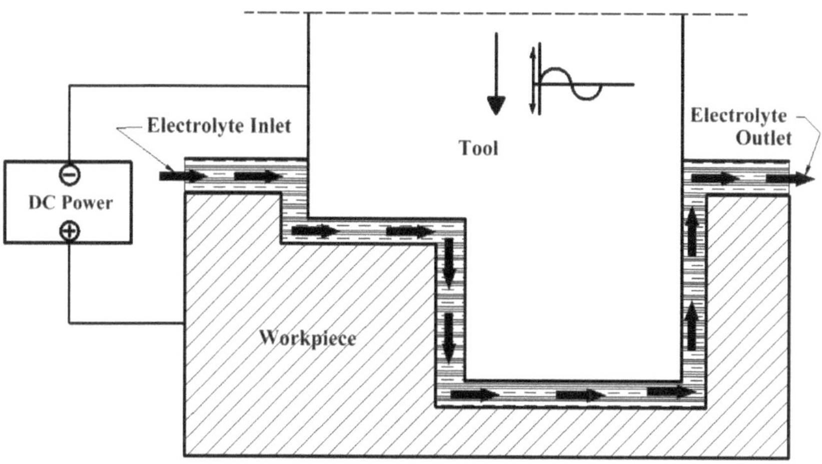

Fig. 6.53 Scheme of the vibration assisted ECM.

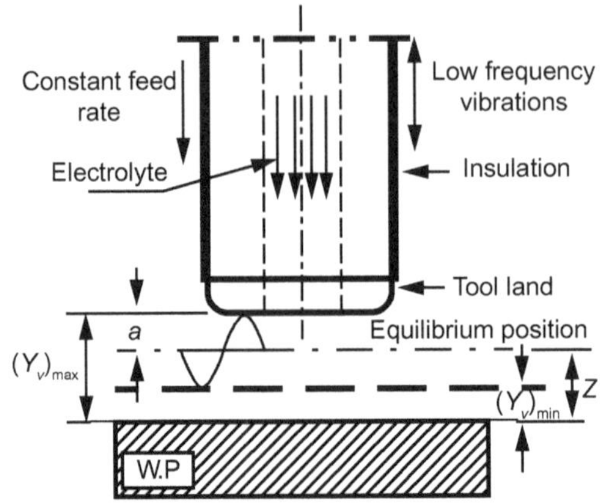

Fig. 6.54 Tool-workpiece configuration during ECM with tool vibration [101].

Vibrations, with frequency and amplitude ranging from 5 to 20 kHz and 10–60 μm, respectively, are applied to the tool. The abrasive particles are accelerated by the motion of vibrating tool electrode as well as electrolyte flow velocity. The passivating layer is removed by these accelerated abrasives that impinge on the work-piece surface.

ECM with low frequency tool vibrations provides a positive effect by changing the physical conditions in the inter-electrode gap as shown in Fig. 6.54 [101]. Gap pressure change leads to the removal of the sludge and allows restoration of the electrolyte in the machining gap. The reciprocal motion between the tool and the workpiece surface as shown in Fig. 6.54 improves the circulation of the electrolyte through the inter electrode gap which allows the use of higher current in order to improve the quality of the machined surface.

As well as, uniform flow of electrolyte is also a challenging task when ECM is utilized for generating complicated shaped profiles on titanium or nickel based super alloys. In the intricate corners of complex profiles, the availability of electrolyte is very less during machining. Thus, fabrication of complex profile is not easily possible by ECM. However, this problem can also be overcome with the assistance of vibrations which may help to create the pressure wave in the inter electrode gap and force the electrolyte media to reach each corner of machining zone thus, uniform machining from every corners of the complicated profile can be possible. Hence, vibration of tool or work piece during machining may help ECM process to produce complex job profile.

Influences of the ultrasonic vibrations on electrochemical machining are as follows [102]:

(i) Improvement of the heat and reactions products removal out of machining area
(ii) It helps diffusion which intern decrease the rate of electrochemical passivation
(iii) Increment of the electrochemical machining efficiency
(iv) It generates most favorable hydrodynamic conditions from surface layer point of view
(v) It also has influence on composition and proprieties of the electrolyte due to sono-chemical reactions

In case of micro ECM, the size of the tool is very small, diameter may be in the order of microns. Hence, as in the case of conventional ECM, electrolyte flow with high velocity may vibrate the micro tool and deteriorates the micro machining accuracies in the case of micro ECM. To overcome this difficulty, stagnant electrolyte is recommended in case of micro ECM. In micro ECM, the removal of gas bubbles and sludge from the very narrow inter electrode gap is a challenging task. It can also be solved by vibrating the micro tool with lower frequency and amplitude which creates pressure waves in the electrolyte at the machining zone and helps to clean the inter electrode gap without utilizing flowing electrolyte [103].

For electrochemical micro-hole machining in EMM, the aspect ratio and the machining rate are still small. To overcome these restrictions, vibration assistance can be utilized as an alternative method during deep micro hole drilling. When ultrasonic vibration is applied on the micro-tool, it vibrates up and down which helps the electrolyte to be sprayed out and reaching into the narrow inter electrode gap. This facilitates flushing out the reaction products which enhances the electrolyte refreshment. This phenomenon prevents the chances of occurrence of micro sparks due to tool-workpiece contact. The machining speed, stability and accuracy are also improved. Ultrasonic vibration is also significant for the generation of good machined surface quality. High aspect ratio micro-holes >10 can be successfully fabricated by disk micro-tool with ultrasonic vibration assistance [104].

6.6.5 Vibration assisted LBM

Laser Beam Machining (LBM) is one of the most popular non conventional machining processes used in the area of advanced machining. LBM process has already been discussed in details in previous chapter 4. Laser beam machining is a thermal process having different advantages such as it can machine any material irrespective of its electrical conductivity as well as mechanical properties. However, due to it uses heat energy for melting, vaporization and some time ablation of workpiece material, it produces some adverse effects on the machined products. Some of the main drawbacks of LBM are thermal damage, heat affected zone (HAZ), recast layer, cracking and melt or re-deposition of the molten metal, formation of tapering effect as well as formation of dross and spatter. LBM cannot be effectively machined highly reflective material as well as transparent material due to low ablation rate. To improve the performance of LBM process and to overcome the above mentioned limitations laser beam machining can be assisted by vibration.

Use of LBM in the micromachining range is becoming more acceptable due to its various advantages. Researchers are investigating to transform LBM in such a way that the limitations can be reduced to a great extent. In this direction, vibration of the workpiece can prevent surface oxidation and formation of recast layer by deactivate the agglomeration of nanoparticles. Moreover, nanoparticals of the material are more sensitive to heat energy than bulk materials due to the high distribution of surface atom in the nanomaterial [105]. Ultrasonic vibration is an effective way to release the local energy concentration because a high frequency repetitive motion has an consequence to normalize the spatially concentrated energy evenly. Fig. 6.55 represents the concept of this process. The surface characteristics of a machined surface can be improved in the ultrasonic vibration-assisted laser micro-machining compare to without assistance of vibration [106, 107].

Femtosecond lasers are increasingly used for micron and sub-micron machining applications, due to its improved machining quality as well as machining capacity compared with other lasers [108]. However, the major limitations of femtosecond laser machining is that the side wall of the machined surface and the adjacent machined areas are usually covered with resolidified and redeposited nanoparticles material as a result of high energy laser beam interactions with the workpiece material.

Attempt has been made to improve the LBM process by introducing objective lens vibration with a low frequency in the order of 500 Hz with amplitude in the range 0–16.5 μm. The effectiveness of vibration of an objective lens in vibration-assisted femtosecond laser micro-drilling of copper substrate has been improved considerably [109, 110]. Vibration is superimposed on the normal objective lens movement to increase the blowing

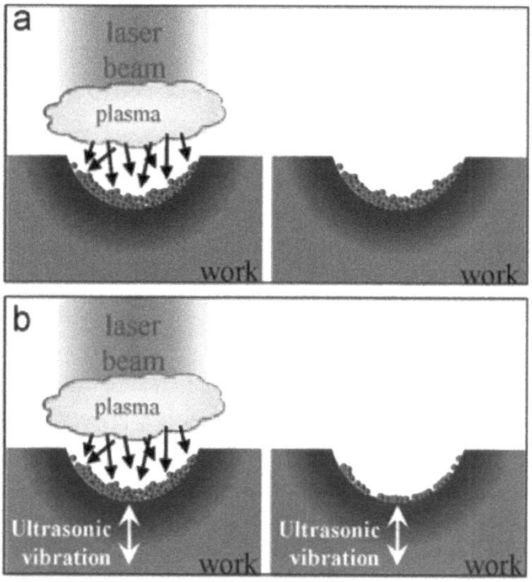

Fig. 6.55 Concept of laser ablation processes by (a) conventional method and (b) ultrasonic-assisted method [106].

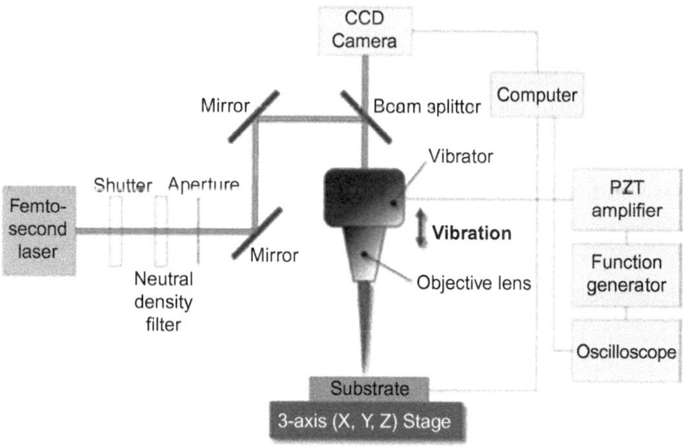

Fig. 6.56 Schematic diagram of vibration-assisted femtosecond laser micro-drilling system [111].

effect during the femtosecond laser machining process. The scheme of experimental system setup is shown in Fig. 6.56 [111].

Surface finish of the machined structures and the aspect ratio have been improved. This is due to a stirring effect generating from vibration of the objective lens which changes the focused beam point. Focusing and defocusing state at the machining zone occurring simultaneously. Due to local force convection, this effect increases the heat transfer of the ablated particles. Furthermore, this increased cooling assist to reduce the chances of particles being joined to the wall surfaces and simultaneously particle agglomeration. The wall surface becomes much cleaner than conventional laser micro-drilling. The aspect ratio is also increased by 154% with the assistance of objective lens vibration compared with no vibration during micro drilling utilizing LBM.

6.7 Laser assisted advanced machining

6.7.1 Introduction

Laser beam when focused into machining zones of various machining methods can produce significant outcomes which results in the possibility of achieving more efficient machining. Laser beam having different advantages such as it can be focused into a small area and can also be focused or defocused to generate diverse level of heating as well as it can penetrate any transparent medium to reach the machining zone. Laser assisted machining (LAM) is one of the important and most widely used classes of the heat-assisted hybrid machining processes, which can be classified into two categories as follows:

(i) Laser-assisted mechanical machining (LAMM) process, a defocused laser beam is utilized to soften the work-piece during conventional machining, i.e., turning and milling, etc.

(ii) Laser-assisted advanced machining (LAAM) process; here focused laser beam assistance also enhances the machining performance in advanced machining processes such as in electrochemical dissolution and electro-discharge based machining.

Laser beam machining can also be coupled with the other non-traditional machining processes such as ultrasonic machining, water jet [112] and liquid film etc. The common motives to these assisted hybrid machining approaches are to minimize the drawbacks of various conventional and advanced machining processes such as unable to machine various advanced materials economically, high tool wear, poor surface integrity, reduced machining accuracy, low productivity, etc. Pulsed wave as well as continuous wave (CW) laser, e.g., Nd: YAG laser, CO_2 lasers, Fiber laser and Excimer laser are commonly employed in laser assisted machining process.

6.7.2 Laser assisted mechanical machining (LAMM)

6.7.2.1 Laser assisted turning

Because of physical and mechanical properties such as high strength and low thermal conductivity, high strength and heat resistant materials are often difficult to machine. During mechanical machining of these super alloy, the cutting forces and cutting temperature becomes very high which lead to short tool life. Due to the flow stress and strain hardening rate of materials normally decrease with the increase of temperature owing to thermal softening, thermal-assisted machining becomes a possibility when machining the hard-to-machine materials [1]. Fig. 6.57 shows how mechanical strength reduces in the case of hard material as temperature increases.

The LAMM mechanism is very complex in nature due to the complexity of influencing parameters and their mutual interactions [113]. The local heating and softening of the work-piece during LAMM by an external heat source, changes the work-piece materials deformation behavior from brittle to ductile. This enables the difficult-to-machine materials such as nickel super alloys, titanium alloys, as well as ceramics to be machined more effectively. As the laser beam heat source has controllable spot size and high power density over other heat sources like gas torch, induction heating, furnace preheating and electrical current heating, it is more suitable during the thermally enhanced machining. The laser heat source also facilitates heating of the work-piece surface locally in front of the cutting tool which further creates a large thermal gradient through the thickness. The volume of material is removed only from soften zone which results in narrow heat affected zone and minimum thermal distortion without affecting the integrity of the machined subsurface and increase energy efficiency of the process.

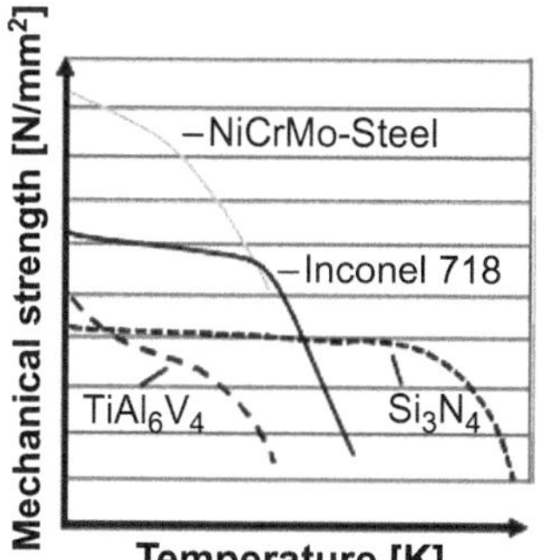

Fig. 6.57 Mechanical strength as a function of temperature [1].

Laser assisted mechanical machining, where the laser beam is directly focused in front of the cutting tool, resulting in easier machining and higher process performance. In this process, the main material removal mechanism is plastic deformation which occurring in conventional cutting, but the laser action softens the workpiece material, so machining of high HSTR alloys or some ceramics becomes easier. Due to the laser softening effect the cutting force decrease significantly and superior surface quality can be achieved.

Laser assisted turning and laser assisted milling are widely recognized by the researchers as the LAMM process for machining of difficult-to-machine materials. In laser-assisted turning, a single cutter and a laser beam are stationary in relation to the work-piece. Fig. 6.58 shows the basic principle of laser assisted turning [114].

6.7.2.2 Laser assisted milling

Whereas in the laser assisted milling or grinding process, the cutter follows a complex predetermined tool path to ensure that the laser beam also follows this tool path and the material is heated directly ahead of the cutter. Laser produces intense localized heating to the workpiece ahead of the cutting region. Strength of the workpiece is reduced and its machinability is improved. Schematic view of this process is shown in Fig. 6.59. Laser-assisted micro-milling or grinding can result in decreased machining forces and improved machining accuracy and surface finish. Experimental researches on laser-assisted end milling of tool steel with the aid of fiber laser, sintered steel material by CO_2 laser and aluminum and steel respectively using Nd: YAG laser consistently showed a significant reduction of the cutting forces by applying laser assistance.

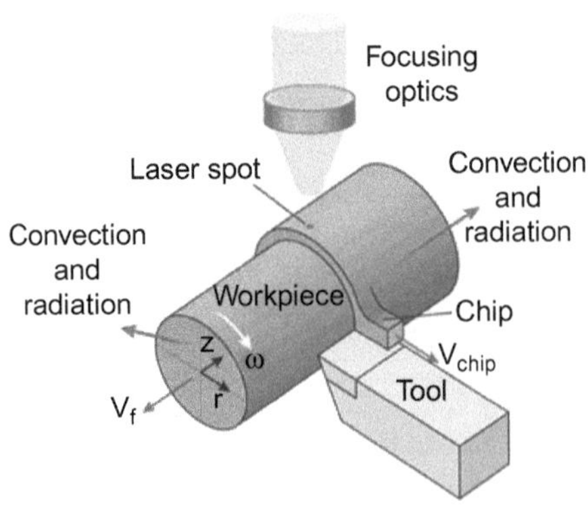

Fig. 6.58 Basic principle of laser assisted turning [114].

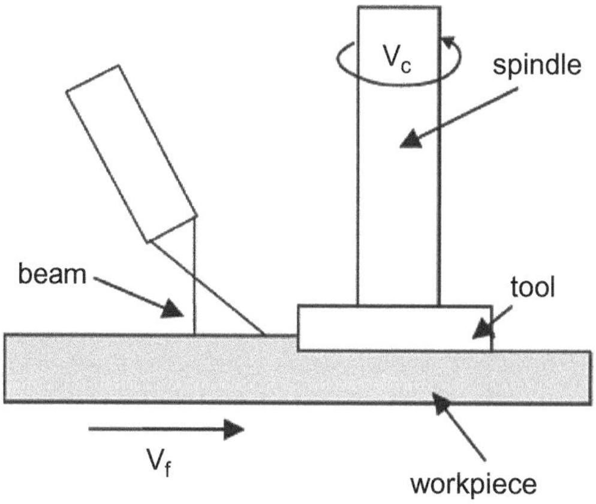

Fig. 6.59 Schematic view of laser-assisted milling [115].

Position and orientation of the incident beam on the work-piece are also critical as the minor axis of laser spot along the work-piece rotation direction is effective during laser-assisted turning of Ti-6Al-4V alloy with a carbide tool, whereas the major axis of laser spot along the work-piece rotation direction is effective during laser-assisted turning of Inconel 718 alloy with ceramic tool. Laser heating produces two types of segmented chips: one at low and the second at high cutting speeds. The segmented chip produced by LAMM at low cutting speed is sharper, the tooth depth and spacing are larger compared to conventional machining while the segmented chip produced by LAMM at high cutting speed is similar to that produced by conventional machining.

Laser assisted turning of hardened steels can be used as an alternative to grinding and conventional hard turning [116]. In addition of larger material removal rate, laser assisted turning produces good surface finish. The thermal expansion consequently produces more compressive surface axial residual stresses than hard turning. LAMM of difficult-to-cut Ni- and Ti-based alloys has also been investigated which also demonstrated to be effective. 40% to 60% reduction in cutting forces and improved tool life allowing the use of higher cutting speeds during LAMM of Inconel alloys [117].

Due to the importance of temperature distribution at the cutting zone in the LAMM process, a lot of efforts have been made to simulate the temperature distribution due to laser radiation with consideration of material removal processes in turning operation on various work-piece materials, such as semi-transparent ceramic, mullet, Inconel 718 alloy, steel, compacted graphite iron, and Ti-6Al-4V alloy. Thermal modeling has also been performed with laser-assisted milling of Inconel 718, steel, and Ti-6Al-4V alloy [116]. The yield strength of work-piece is one of the main components of the cutting force, and can be effectively reduced by increasing the temperature. Therefore, the cutting force is reduced in LAMM, which leads to a reduction in the specific cutting energy. Not only the cutting forces, but also the ratio of the feed force to the main cutting force decreases with increasing laser power during LAMM of Magnesia-Partially-Stabilized Zirconia (PSZ) and mullite. Cutting forces or specific cutting energy are also found to be reduced in LAMM of titanium alloys, Inconel 718, Cobalt-based Stellite 6 alloy, 6061-T6 aluminum alloy, Iron, and steels in both turning and milling operations.

6.7.2.3 Laser assisted ceramic machining

Ceramic machining utilizing mechanical machining processes is another challenge. However, researchers have also successfully applied LAMM for machining of different types of ceramics. Laser assisted machining finds its applications for the machining of ceramic materials. Different ceramic materials such as Si_3N_4, SiC, ZrO_2, Al_2O_3, etc. and carbides which contain some amorphous glass phases, can efficiently be machined by LAMM. Heating up the glass phase surrounding the crystals at elevated temperature over 1000 °C results in reduction of the deformation resistance and local softening of the material in the shear zone, facilitate machining of the ceramic with a geometrically defined cutting edge. An improvement of the surface roughness is also obtained during machining of Al_2O_3 ceramics [118], compared to conventional machining. Turned surfaces demonstrate a smooth texture with surface roughness, R_a <1 μm without fracture, indicating that plastic deformation is the dominant mode in material removal. However, it is observed that the adjacent front area of the cutting zone is mostly affected. This zone is characterized by cracks where crack density increases with temperature.

Work-piece crack may occur when LAMM of ceramics. The sudden impact and stress release between cutting tool and work-piece when the cutting tool enters and leaves the work-piece during milling of ceramics (Si_3N_4) results in work-piece edge chipping at both the entry and exit edges. Local cracks present in the heat affected zone (HAZ) without cutting indicate that the laser heating cycle introduces high thermal stresses in Zirconia, PSZ, etc.

Significant improvement of surface roughness can also be achieved with laser-assisted mechanical micromachining (LAMM) of ceramics. Surface quality can be improved with the aid of high power laser. The cutting zone stress is found to decrease with increasing material removal temperature and feed, but is not significantly dependent on the cutting speed during LAMM of Si_3N_4. The cutting forces and specific cutting energy are reported to decrease with increasing laser power or surface temperature when cutting ceramics such as mullite, PSZ and Si_3N_4 because of the increasing material removal temperature with laser power, but are not significantly affected by the laser-tool lead distance. LAMM sample only showed very small heat affected zone and no sub surface cracks.

6.7.3 Laser assisted advanced machining (LAAM)

Although, laser assisted mechanical machining process has advantageous than the conventional machining process, but some of the drawbacks such as heat effected zone, presence of sub surface cracks, tool wear, low dimensional accuracy, cost and complexity of tooling and difficult to generate complicated shapes, etc. have made the researchers to develop new laser assisted hybrid process associated with advanced machining processes. The energy sources of LAAM can be combined with advanced machining processes, e.g., chemical machining (CM), EDM, ECM and even water jet, etc. These LAAM processes are not affected by chatter or the force of the cutting tool but can be controlled by various parameters of different advanced machining processes as well as the parameters of laser sources such as the input energy, wavelength, spot diameter, processing speed and focusing characteristics, etc.

6.7.3.1 Laser assisted WJM

Focused laser beam can also be assisted WJM process. Laser beam can be utilized in focused or defocused condition during water jet cutting. In WJM, chip formation and plowing are the basic mechanism of material removal. In this process of laser assisted WJM, the main material removal mechanism is plastic deformation but the laser action softens the workpiece material, so machining of HSTR alloys or some ceramics becomes easier. In laser assisted WJM, cutting force decreases significantly due to the laser softening effect of the material in the machined surface which intern produces superior surface quality. Laser beam can assists WJM process during machining in three different ways as represented in Fig. 6.60 remote laser cutting with WJM is exhibited in Fig. 6.60(a). Unique advantage of remote laser assisted WJM cutting is the high cutting speed. Some of the others advantage of this technique of cutting is reduced heat-affected-zone, reduced dross height, improved surface quality with an increase in the cutting efficiency [119, 120]. Fig. 6.60(b) exhibits laser assisted WJM where, laser beam is focused on the workpiece along with the water jet [121]. Fig. 6.60(c) represents laser

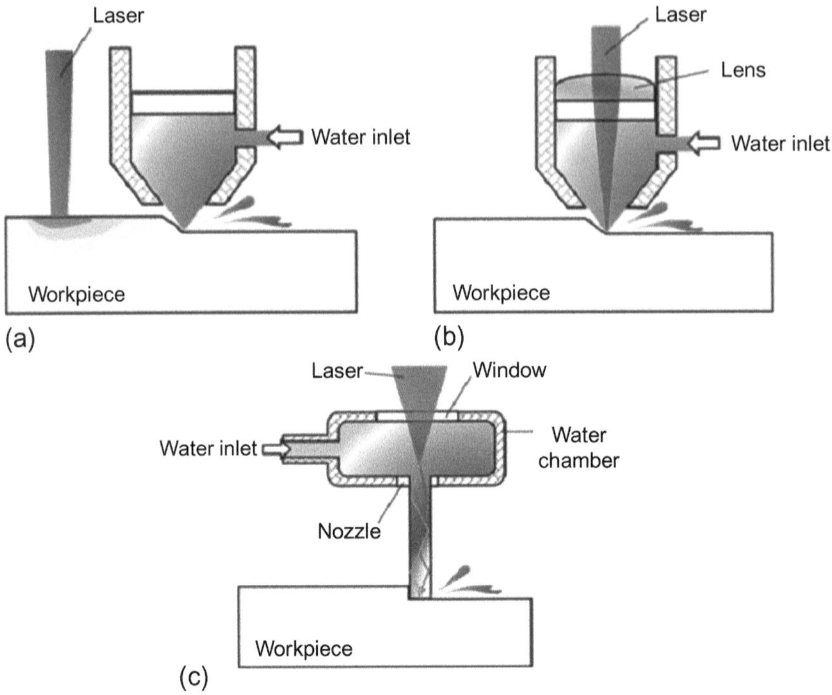

Fig. 6.60 Concept of laser-assisted water jet machining (a), (b) and (c) [119–121].

assisted WJM where, laser beam is guided by the total internal reflection at the water-air interface [122]. The surface roughness can be achieved in the range of 150 nm.

6.7.3.2 Laser assisted EDM

Laser-assisted electric discharge machining is a hybrid process of combining both laser beam machining (LBM) and electro-discharge machining (EDM) in order to address the problems associated with the both EDM and laser. Main limitations of EDM are longer machining times, high tool wear. Major drawbacks of laser are formation of recast layer and heat-affected zones (HAZ) and low surface quality of the machined work-piece. In laser assisted EDM, a short-pulsed laser beam ranging from nanoseconds to microseconds with high ablation rates are utilized to pre-machine the basic part features, i.e., groove, hole, etc. A schematic view of laser assisted EDM is shown in Fig. 6.61.

This process is subsequently followed by EDM within a suitable dielectric to remove the surface defects caused by the thermal effects of laser

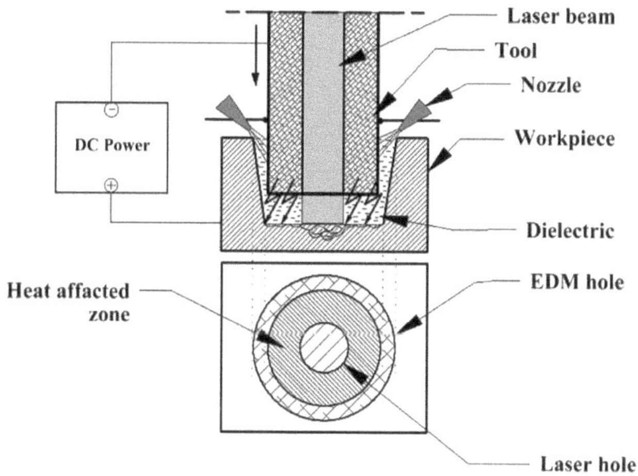

Fig. 6.61 Scheme of laser assisted EDM.

ablation and to finish the feature in order to attain the required shape and quality. Nd-YAG lasers are most widely used. In EDM voltage, electrode type, material and feed rate, type of dielectric and in LBM, laser power and pulse frequency are some of the important process parameters during laser assisted EDM. Researchers have studied a hybrid process which combines a laser and the micro-EDM process for micro-drilling [123]. Optimum machining conditions when combining micro-EDM and laser machining were identified. As a result, the machining time which does not affect the quality of the micro-holes was decreased by approximately 50–65%. Complicated 3D micro-structures, micro-grooves and diesel fuel injection nozzle holes of 140 μm diameter have been precisely machined by Laser-assisted electric discharge machining [124–126]. In laser assisted EDM, the surface finish can be improved to a higher level, i.e., in the range of 450 nm.

6.7.3.3 Laser assisted chemical and electrochemical machining

Laser assisted chemical and electro-chemical machining have been developed in order to increase the local material removal rate by increasing local chemical activity or electrical conductivity to promote the chemical reactions on the work-piece. Laser assisted electrochemical machining or laser assisted jet electrochemical machining is a process in which laser is coupled with the electrochemical machining (ECM) so as to improve the localization of the dissolution process of ECM. Laser assisted ECM consists of three main systems: (a) electrolyte supply system; (b) power supply system; and (c) laser assistance system. The electrolyte and the focused laser beam are coaxially aligned and pass through a thin tube-shaped nozzle which acts as the cathodic tool and creates a jet of electrolyte. This combined coaxial laser beam and electrolyte jet impinge the anodic work-piece surface at the same location. The laser beam causes a localized increase of temperature in the work-piece and electrolyte. The laser supports the kinetics of dissolution and has no melting effect. Upon a suitably applied voltage between the tool and the work-piece, the ECM process is initiated more readily and the diffusion of ions and the transportation of the removed work-piece material are enhanced. This improves process productivity and machining quality.

Jet electrochemical machining assisted with laser combines two different sources of energy concurrently, i.e., energy of photons of laser drilling and energy of ions of ECM. The electrolyte jet and the focused laser beam coaxially aligned with it creating a contactless tool-electrode and strike against the same spot on the workpiece surface. During the process, the work-piece material is removed by melting and vaporization by laser beam during pulse width, and subsequently during the inter-pulse of laser recast and spatter are removed by anodic dissolution due to electrochemical reaction. Fig. 6.62 shows the principles of Jet electrochemical machining assisted with laser. It is found that the spatter is reduced >95%, and recast is reduced >90% utilizing this hybrid process during machining of 321 stainless steel compared with laser drilling in ambient atmospheric conditions [127–129].

The important process control parameters of laser assisted ECM are the inter-electrode gap, voltage, electrolyte concentration, electrolyte jet speed and laser power. The main applications of LAECM include manufacturing of small holes, cavities and precise micro-shapes especially in the difficult-to-machine metals and alloys, composites and ceramics. Significant improvements in material removal rate up to 54% were observed using LJECM for Hastelloy, titanium alloy, stainless steel and aluminum alloy [130]. Average roughness up to 20 nm in 25 μm-deep cavities produced in titanium and Hastelloy when using LAECM [131]. Overall quality of laser drilled hole can be improved by removing recast layers and spatters by LAECM [132]. The laser radiation enables to accelerate the electrochemical processes by 10 times and to localize the machining into the area of several microns.

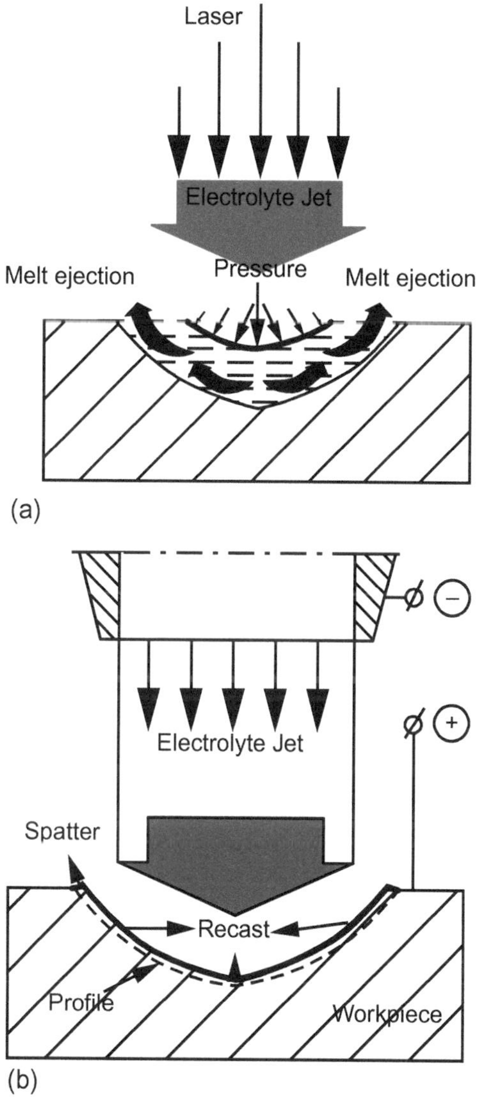

Fig. 6.62 Principle scheme of JECM-LD (a) in pulse width and (b) in the inter-pulse [127].

6.7.3.4 Laser-assisted etching (LAE)

An especially attractive feature of laser assisted etching process is the ability to obtain maskless patterns in the form of material deposition or dissolution, dictated by the path of the focused laser beam on the sample. Laser-assisted Etching (LAE) utilizes a maskless photoreactive or thermal reactive process.

The etching of ceramics, semiconducting materials, metals and high polymers in an etchant is enhanced using a laser beam.

Various phenomenon involves in laser assisted etching includes laser energy absorption, chemical reaction, heat diffusion, hydrodynamics and mass transfer. Reduction oxidation reaction is the fundamental phenomenon of LAE. Laser energy just supports the kinematic of dissolution and has no melting effects. Although, in some applications direct laser ablation mainly removes material as vapor but laser chemical allows surface heating without surface melting or vaporizing. Therefore, many drawbacks of LBM can be avoided.

Advanced precise lithography without applying photoresist mask can be possible by utilizing focused laser beam. This technique of lithography is known as Oxide Film Laser Lithography (OFLL). Using a well focused laser beam (10 μm) to form a pattern on an oxide film (200–300 nm) on Titanium, the photoresist could be entirely eliminated [133]. Chemical etching or electrochemical dissolution can be localized at the irradiated areas, the remainder of the oxide film acting as a mask. After etching or anodic dissolution, the freestanding oxide film at the edge of machined cavities is removed by ultrasonic cleaning.

6.8 Advantages and limitations of hybrid machining processes

Based on above mentioned discussions on different hybrid machining processes, it is clear that the various process combinations provide considerable enhanced advantages as well as minimize potential disadvantages of individual processes. Some of the advantages of combined hybrid machining processes have already been elaborated in previous discussion. To specify some more advantages, following discussion has been made hereunder.

Major advantage of ECDM is electrically non-conducting material also can be machined. By Electrochemical Arc Machining (ECAM) process, MRR can be increases to 5–50 times than individual ECM or EDM and higher surface finish also can be achieved by this hybrid process. With the application of ECG, productivity can be increased remarkably. Surface layer properties are also improved, where as tool wear and energy consumption are reduced by this process. In this process oxide film can be removed by mechanical depassivation which leads to increase the MRR by 100–150%

when grinding WC. In EDG, electric discharges cause considerable decrease in grinding forces, lower grinding wheel wear and provides effective method to dressing grinding wheel during machining process.

In case of Abrasive assisted EDM, mixing silicon powder into dielectric reduces electrical capacitance across the discharge gap by increasing the gap size. As a result, sparking intensity is more and this leads to improvement in the discharge characteristic, especially in machining of large workpiece area. Application of powder-mixed working media is allowed AEDM to obtain mirror finishing of complex shape and more uniform and crackless affected layer of the machined surface. In vibration assisted EDM, it is recognized that the role of the acoustic wave and cavitation phenomena improves the flushing and material removal from the surface craters. These process conditions are significant for micro drilling and production of slots and grooves. Improve dielectric circulation results in a better ejection of the molten metal from craters. This increases the removal rate, but also lets less liquid material recast on the surface. Vibration assisted ECM is employed in the steel and super alloys, titanium alloys and other metallic materials. Anodic dissolution of electrically conducting materials is accompanied by the formation of a brittle oxide layer. The vibration along with abrasive grains is expected to act not on the material of the workpiece but mainly on the brittle oxide layer. The oxide film is removed by abrasion; the dissolution process is accelerated due to depassivation of material surface.

Hybrid machining processes facilitate to improve performance capabilities in terms of the surface roughness, material removal rate, tool life and shape accuracy, etc. However, based on suitable process-material interaction, process parameters should be selected to overcome adverse consequences such as poor surface finish, accuracy and HAZ with micro cracks, etc.

Thermal assisted hybrid mechanical tool based machining process leads to increment of density of the cracks as well as it's high temperature may results in accelerated dissolution-diffusion and adhesion wear which shorter tool life. The disadvantage also includes increment of burr heights due to thermal softening. Thermal assisted advanced machining processes also have different thermal effects on the machined components. Laser assisted machining (LAM) involves several parameters, i.e., laser parameters, machining parameters, along with the properties of the work-piece. Therefore, the optimization of the output criteria has been a challenging task. In addition to this, correlation of the different process parameters needs to be searched out for better optimized values. However, most of the LAM studies till date have reported either on turning or surface milling operations. Application of LAM to the other machining processes, such as drilling, is still challenging due to the complexity of the integration of the laser beam and the machine. Further research work needs to be undertaken in this area to make the process more attractive from an industrial perspective.

Vibration assisted hybrid machining processes may deteriorate the machining accuracy due to the influence of vibrating machine elements. The tool, workpiece as well as various machined elements may wear out due to abrasion of abrasive grains in abrasive assisted hybrid machining processes.

6.9 Advancements in hybrid machining processes

Hybrid Machining Processes can be utilized for wide applications especially for machining advanced materials and making various kinds machine and tool parts, electronic devices and micro machine parts due to their high accuracy and quality of surface, productivity and automatization. The complex interactions of the electrical, chemical and mechanical characteristics of Hybrid Machining Processes are still not completely understood and its potential manufacturing capabilities are not completely recognized.

In hybrid machining processes different forms of energy are used at the same time at the same zone of impact which make these process more complex. Mechanisms involved in these processes need to be investigated in depth in order to fully understand the processes which is mostly influenced by simultaneous amalgamation of processes with various energy sources. Hybrid machining processes are based on the simultaneous and controlled interaction of process mechanisms which necessitate multi-axis ultra precision machine tools. Some of the major requirements in this direction are high thermal stability, high static and dynamic stiffness, precise feed drives and control systems with high accuracy and short response, etc.

Process monitoring systems are also to be developed to characterize, control, and improve hybrid machining processes. Monitoring should be applied to various parameters and their interactions. Numerical modeling of hybrid machining processes approaches are also required in order to better understanding the processes and also optimizing these processes.

In situ tool fabrication methods can be developed and integrated with hybrid machining processes. On-machine metrology system is essential to achieve and maintain high quality. Most of the hybrid machining processes are still at the research stage, and only a few of them successfully commercialized for industrial applications. More research on industrial implementation of hybrid machining processes is urgently required to make these processes more cost effective.

The growth of hybrid processes is constantly increasing, from basic development toward industrial implementation and finally to its level of technology maturity. Further developments are urgently needed to cater the requirements of industries to produce highly complex product features made of advanced materials and on the other hand to fabricate parts in a more productive, cost effective and energy efficient way. In the near future hybrid machining processes are expected to enhance manufacturing capabilities further for wider industrial applications.

6.10 Model questions

Q1. What are the basic differences between mechanical grinding and electrochemical grinding?

Q2. Why is wheel life more in ECG than Mechanical Grinding?

Q3. What are the applications of ECG?

Q4. What are the precautions to be taken for ECG operations?

Q5. What are the bonding materials and abrasives used in ECG wheel?

Q6. What are the different zones of material removal in ECG?

Q7. What are the effects of process parameters on grinding performance of ECG?

Q8. What are the importance of Electrolyte Supply in ECG?

Q9. What are the wheel (cathode) wear mechanisms and different forms of it in ECG?

Q10. Derive the mathematical models of MRR in ECG.

Q11. Discuss the working principle of ECAM process.

Q12. How is ECAM different from ECM and EDM?

Q13. What are the system requirements of ECAM?

Q14. What are the advantages of ECAM over ECM and EDM?

Q15. Distinguish between spark discharge and arc discharge.

Q16. Derive the expressions of MRR in ECD phase and EDE phase of ECAM.

Q17. What are the applications of ECAM process?

Q18. Why is tool wear rate less that material removal rate from workpiece in ECAM?
Q19. How is ECDM different from ECM and EDM?
Q20. Explain the basic principles of ECDM with sketch.
Q21. What are the process parameters of ECDM? What are the effects of process parameters on the machining performance of ECDM?
Q22. What are the ECDM process variants?
Q23. What are the advantages and limitations of ECDM?
Q24. Derive the expression of MRR in ECDM?
Q25. What are the applications of ECDM?
Q26. What are the requirements of ECDM for micro-machining applications?
Q27. State the advantages of AWJM over AJM and WJM.
Q28. State the material removal mechanism of AWJM for ductile and brittle materials.
Q29. Discuss the effects of process parameters on machining performance of AWJM.
Q30. What are the advantages of AWJM over other conventional cutting processes?
Q31. What are the functions and types of catchers used in AWJM?
Q32. What are the functions of different components of cutting head of AWJM?
Q33. What are the criteria for selecting focusing tube of AWJM system?
Q34. What are the various industrial applications of AWJM?
Q35. How abrasive assisted EDM improves quality of the machined parts? Describe some of the application cases of abrasive assisted EDM employing different abrasives/metal powders.
Q36. How abrasive assisted ECM improves quality of the machined parts? Describe some of the application cases of abrasive assisted ECM employing different abrasives.
Q37. Explain different types of vibration assisted mechanical machining and its advantages.
Q38. How vibration at tool or workpiece improves EDM performances?
Q39. How application of vibration improves ECM performances?
Q40. Describe applications of vibration for improving laser machining performances.
Q41. How mechanical machining performances can be improved with the help of assisted laser?
Q42. Describe application of laser assistance for the improvement of advanced machining performances.

References

[1] B. Lauwers, F. Klocke, A. Klink, A. Erman Tekkaya, R. Neugebauer, D. Mcintosh, Hybrid processes in manufacturing, CIRP Ann. Manuf. Technol. 63 (2014) 561–583.

[2] S.Z. Chavoshi, X. Luo, Hybrid micro-machining processes: a review, Precis. Eng. 41 (2015) 1–23.

[3] J. Kozak, K.P. Rajurkar, Hybrid machining process evaluation and development, in: Proceedings of 2nd International Conference on Machining and Measurements of Sculptured Surfaces, Keynote Paper, Krakow, 2000, pp. 501–536.

[4] J. Atkinson, C.F. Noble, Residual stresses in workpieces after peripheral electrochemical grinding, in: Proc. of 19th IMTDR Conference Manchester, September, 1978, pp. 525–532.

[5] V.K. Jain, Advanced Machining Processes, Allied Publishers, New Delhi, 2007.

[6] R.E. Phillips, What is electrochemical grinding and how does it work, in: Proc. of Non-traditional Machining Conference, American Society for Metals, Ohio, 1985, pp. 65–70.

[7] R.E. Iihan, G. Sathyanaryanan, R.H. Storer, R.E. Phillips, A study of wheel wear in electrochemical surface grinding, Trans. ASME J. Eng. Industry 114 (1) (1992) 82–93.

[8] G.F. Benedict, Non-Traditional Manufacturing Processes, Marcel Dekker, London, 1987.

[9] A. Geddam, C.F. Noble, Peripheral and electrochemical grinding with a formed wheel, in: Proc. of 13th IMTDR Conf., Birmingham, September, 1972, pp. 315–321.

[10] T.M.A. Maksoud, A.J. Brooks, Electrochemical grinding of ceramic form tooling, J. Mater. Process. Technol. 55 (1995) 70–75.

[11] S. Roy, A. Bhattacharyya, S. Banerjee, Analysis of effect of voltage on surface texture in electrochemical grinding by autocorrelation function, Tribol. Int. 40 (2007) 1387–1393.

[12] D. Zhu, Y.B. Zeng, Z.Y. Xu, X.Y. Zhang, Precision machining of small holes by the hybrid process of electrochemical removal and grinding, CIRP Ann. Manuf. Technol. 60 (2011) 247–250.

[13] N.S. Qu, Q.L. Zhang, X.L. Fang, E.K. Ye, D. Zhu, Experimental investigation on electrochemical grinding of Inconel 718, Procedia CIRP 35 (2015) 16–19.

[14] S. Zaborski, M. Lupak, D. Poros, Wear of cathode in abrasive electrochemical grinding of hardly machined materials, J. Mater. Process. Technol. 149 (2004) 414–418.

[15] A.V. Mogilnikov, M. Ya, Y. Chmir, S. Timofeev, V.S. Poluyanov, Diamond-ECM grinding of ceramic metal tungsten, Procedia CIRP 6 (2013) 407–409.

[16] A.V. Mogilnikov, M.Y. Chmir, Y.S. Timofeev, V.S. Poluyanov, Diamond-ECM grinding of sintered hard alloy of WC-Ni, Procedia CIRP 42 (2016) 143–148.

[17] J.A. Mc Geough, Principles of Electrochemical Machining, Chapman and Hall, London, 1974.

[18] C.N. Larsson, E.M. Baxter, Tool damage by sparking in ECM, in: 18th IMTDR Conf., 14 to 16th September, 1977, pp. 499–606.

[19] S.I. Ebeid, E.M. Baxter, C.N. Larson, Further effects of process parameters on the incidence of sparking in electrochemical machining, in: 19th IMTDR Conf., 13-15th Septemebr, Manchester, 1978, pp. 511–516.

[20] T.K. Drake, J.A. Mc Geough, Aspects of drilling by electrochemical arc machining, in: Int. Proc. of 22nd IMTDR Conference, London, 1981, pp. 361–369.

[21] J.A. McGeough, N.M. Levy, W. Munro, Investigations of surface finishing of dies and moulds by electrochemical arc machining, in: Proc. of ISEM, April, 1983.

[22] O.C. Enache, G. Stoica, E. Strajescu, The study of EDM with forced vibration of tool electrode, CIRP Ann. 39 (1) (1990) 167–170.

[23] D. Kremer, C. Lhiaubet, A. Moisan, A study of the effect of synchronizing ultrasonic vibrations with pulses in EDM, CIRP Ann. 40 (1) (1991) 211–214.

[24] I.M. Crichton, J.A. McGeough, Studies of the discharge mechanism in electrochemical arc machining, J. Appl. Electrochem. 15 (1985) 113–119.
[25] A.B.M. Khayry, J.A. Mc Geough, Analysis of electrochemical arc machining by stochastic and experimental methods, Proc. R. Soc. Lond. A412 (1987) 403–429.
[26] J.A. Mc Geough, A.B.M. Khayry, W. Munro, Theoretical and experimental investigations of the relative effects of spark erosion and electrochemical dissolution in electrochemical arc machining process, CIRP Ann. 33 (2) (1983).
[27] J. Kozak, K. Lubkowski, A computation of physical conditions and geometry of the gap in the steady state of electrochemical machining, in: Proc. of 27th IMTDR Conf., April 20–21, 1988, pp. 269–274.
[28] J.A. Mc Geough, H. Rasmussen, A macroscopic model of electrodischarge machining, Int. J. MTDR. 22 (4) (1982) 333–339.
[29] A.B.M. Khayry, J.A. Mc Geough, Modeling of electrochemical arc machining by use of dynamic data systems, in: Proc. of IMTDR Conference, vol. 25, 1985, pp. 321–328.
[30] I.M. Crichton, J.A. Mc Geough, Theoretical experimental and computational aspects of the electrochemical arc machining process, CIRP Ann. 33 (2) (1984) 429–432.
[31] B. Doloi, Analusis and Design of the Electrochemical Arc Machining (ECAM) System, (Dissertation thesis), Jadavpur University, India, 1993.
[32] K. Chikamori, Grooving on silicon nitride ceramics arc discharge in electrolyte, Int. J. Jpn Soc. Precis. Eng. 25 (2) (1991) 109–110.
[33] Y. Sen, Y. liu, W. Sun, High efficient dry hybrid machining of EDM and arc machining, Procedia CIRP 42 (2016) 149–154.
[34] F. Wang, Y. Liu, Y. Zhang, Z. Tang, R. Ji, C. Zheng, Compound machining of titanium alloy by super high speed EDM milling and arc machining, J. Mater. Process. Technol. 214 (2014) 531–538.
[35] W. Zhao, L. Gu, H. Xu, L. Li, X. Xiang, A novel high efficiency electrical erosion process-blasting erosion arc machining, Procedia CIRP 6 (2013) 525–621.
[36] H. Kurafugi, H. Suda, Electrical discharge drilling of glass, Ann. CIRP 16 (1) (1968) 415.
[37] K. Chikamori, Grooving on silicon nitride ceramics with arc discharge in electrolyte, J. Mech. Eng. Jpn 47 (4) (1991) 130–137.
[38] B. Doloi, Analysis of Multi-Parametric Mixes for Controlled Electrochemical Discharge Machining (ECDM) of Non-Conducting Engineering Ceramics, (PhD thesis), Jadavpur University, India, 2001.
[39] I. Basak, A. Ghosh, Mechanism of material removal in electrochemical discharge machining: a theoretical model and experimental verification, J. Mater. Process. Technol. 71 (1997) 350–357.
[40] A.B.E. Khairy, J.A. Mcgeough, Die-sinking by electro-erosion dissolution machining, CIRP Ann. Manuf. Technol. 39 (1990) 191–195.
[41] B. Doloi, B. Bhattacharyya, S.K. Sorhkel, Electrochemical discharge machining of non-conducting ceramics, Def. Sci. J. 49 (4) (1999) 331–338.
[42] R. Wuthrich, V. Fascio, Machining of non-conducting materials using electrochemical discharge phenomena-an overview, Int. J. Mach. Tool Manuf. 45 (2005) 1095–1108.
[43] T. Singh, A. Dvivedi, Developments in electrochemical discharge machining: a review on electrochemical discharge machining, process variants and their hybrid methods, Int. J. Mach. Tool Manuf. 105 (2016) 1–13.
[44] N. Gautam, V.K. Jain, Experimental investigations into ECSD process using various tool kinematics, Int. J. Mach. Tool Manuf. 38 (1) (1998) 15–27.
[45] V.K. Jain, S.K. Choudhury, K.M. Ramesh, On the machining of alumina and glass, Int. J. Mach. Tool Manuf. 42 (2002) 1269–1276.

[46] B. Bhattacharyya, B.N. Doloi, S.K. Sorkhel, Experimental investigations into electrochemical discharge machining (ECDM) of non-conducting ceramic materials, J. Mater. Process. Technol. 95 (1999) 145–154.
[47] C.K. Yang, K.L. Wu, J.C. Hung, S.M. Lee, J.C. Lin, B.H. Yan, Enhancement of ECDM efficient and accuracy by spherical tool electrode, Int. J. Mach. Tool Manuf. 51 (2011) 528–535.
[48] K. Furutani, H. Maeda, Machining a glass rod with a lathe type electrochemical discharge machine, J. Micromech. Microeng. 18 (2008) 8.
[49] M. Schopf, I. Beltram, M. Boccadoro, D. Kramer, ECDM (electrochemical discharge machining), an new method for trueing and dressing of metal bonded diamond grinding tools, CIRP Ann. Manuf. Technol. 50 (2001) 125–128.
[50] C. Wei, D. Hu, K. Xu, J. Ni, Electrochemical discharge dressing of metal bond micro-grinding tools, Int. J. Mach. Tool Manuf. 51 (2011) 165–168.
[51] C.P. Cheng, K.L. Wu, C.C. Mai, Y.S. Hsu, B.H. Yan, Magnetic field assisted electrochemical discharge machining, J. Micromech. Microeng. 20 (2010) 7.
[52] H. El-Hofy, Advanced Machining Processes, McGraw-Hill, New York, (2005).
[53] C.T. Yang, S.L. Song, B.H. Yan, F.Y. Huang, Improving machining performance of wire electrochemical discharge machining by adding SiC abrasive to electrolyte, Int. J. Mach. Tool Manuf. 46 (2006) 2044–2050.
[54] M. Goud, A.K. Sharma, C. Jawalkar, A review on material removal mechanism in electrochemical discharge machining (ECDM) and possibilities to enhance the material removal rate, Precis. Eng. 45 (2016) 1–17.
[55] H.C. Meng, K.C. Ludema, Wear models and prediction equations, Wear 181–183 (1995) 443–457.
[56] J.G.A. Bitter, A study of erosion phenomena part I, Wear 6 (1963) 5–21.
[57] J.G.A. Bitter, A study of erosion phenomena part II, Wear 6 (1963) 5169–5190.
[58] M. Hashish, A model for abrasive water jet machining, J. Eng. Mater. Technol. 111 (1989) 154–162.
[59] R. Kovacevic, Surface texture in abrasive water jet cutting, J. Manuf. Syst. 10 (1991) 32–40.
[60] J. Zeng, T.J. Kim, An erosion model of polycrystalline ceramic in abrasive water jet cutting, Wear 199 (2) (1996) 275–282.
[61] A.W. Momber, R. Kovacevic, Principles of Abrasive Water Jet Machining, Springer, 1998. ISBN 3-540-76239-6.
[62] J. Wang, A new model for predicting the depth of cut in abrasive water jet contouring of alumina ceramics, J. Mater. Process. Technol. 209 (2009) 2314–2320.
[63] A. Akkurut, M.K. Kulekci, U. Seker, F. Ercan, Effect of feed rate on surface roughness in abrasive waterjet cutting applications, J. Mater. Process. Technol. 147 (2004) 389–396.
[64] M. Palleda, Astudy of taper angles and material removal rates of drilled holes in the abrasive water jet machining process, J. Mater. Process. Technol. 189 (2007) 292–295.
[65] M. Hashish, Characteristics of surface machined with abrasive water jet, J. Eng. Mater. Technol. 113 (1991) 354–362.
[66] P.J. Singh, W.L. Chen, J. Munoz, Comprehensive evaluation of abrasive waterjet cut surface quality, in: 6th American Waterjet Conference, USA, 1991, pp. 139–156.
[67] M.K. Kulekci, Processes and apparatus developments in industrial waterjet applications, Int. J. Mach. Tool Manuf. 42 (2002) 1297–1306.
[68] N. Hagbin, F. Ahmadzadeh, J.K. Spelt, M. Papini, Effect of entrained air in abrasive water jet micro-machining: reduction of channel width and waviness using slurry entrainment, Wear 344–345 (2015) 99–109.
[69] R. Kovacevic, M. Hashish, R. Mohan, M. Ramalu, T.J. Kim, E.S. Geskin, State of the art of research and development in abrasive waterjet machining, Trans. ASME 119 (1997) 776–785.

[70] F. Kartal, A review of the current state of abrasive water-jet turning machining method, Int. J. Adv. Manuf. Technol. 88 (2017) 495–505.
[71] T. Yih-Fong, C. Fu-Chen, Investigation into some surface characteristics of lectrical discharge machined SKD-11 using powder-suspension dielectric oil, J. Mater. Process. Technol. 170 (2005) 385–391.
[72] K.L. Wu, B.H. Yan, F.Y. Huang, S.C. Chen, Improvement of surface finish on SKD steel using electro-discharge machining with aluminum and surfactant added dielectric, Int. J. Mach. Tool Manuf. 45 (2005) 1195–1201.
[73] Y.F. Luo, The dependence of interspace discharge transitivity upon the gap debris in precision electrodischarge machining, J. Mater. Process. Technol. 68 (1997) 121–131.
[74] Y.S. Wong, L.C. Lim, I. Rahuman, W.M. Tee, Near-mirror-finish phenomenon in EDM using powder-mixed dielectric, J. Mater. Process. Technol. 79 (1998) 30–40.
[75] A. Kojima, W. Natsu, M. Kunieda, Spectroscopic measurement of arc plasma diameter in EDM, CIRP Ann. Manuf. Technol. 57 (2008) 203–207.
[76] H.-M. Chow, B.-H. Yan, F.-Y. Huang, J.-C. Hung, Study of added powder in kerosene for the micro-slit machining of titanium alloy using electro-discharge machining, J. Mater. Process. Technol. 101 (2000) 95–103.
[77] B. Ekmekci, White layer composition, heat treatment, and crack formation in electric discharge machining process, Metall. Mater. Trans. B Process Metall. Mater. Process. Sci. 40 (2009) 70–81.
[78] H.-M. Chow, L.-D. Yang, C.-T. Lin, Y.-F. Chen, The use of SiC powder in water as dielectric for micro-slit EDM machining, J. Mater. Process. Technol. 195 (2008) 160–170.
[79] P.J. Liew, J. Yan, T. Kuriyagawa, Carbon nanofiber assisted micro electro dis-charge machining of reaction-bonded silicon carbide, J. Mater. Process. Technol. 213 (2013) 1076–1087.
[80] A. Schubert, A. Nestler, S. Pinternagel, H. Zeidler, Influence of ultrasonic vibration assistance on the surface integrity in turning of the aluminium alloy A2017, Mater. Sci. Eng.Technol. 42 (7) (2011) 658–665.
[81] D.E. Brehl, T.A. Dow, Review of vibration-assisted machining, Precision Eng. 32 (2008) 153–172.
[82] B. Lauwersa, Surface integrity in hybrid machining processes, Procedia Eng. 19 (2011) 241–251.
[83] B. Bulla, F. Klocke, O. Dambon, M. Hünten, Influence of different steel alloys on the machining results in ultrasonic assisted diamond turning, Key Eng. Mater. 523–524 (2012) 203–208.
[84] D.E. Brehl, T.A. Dow, 3-D micro-structure creation using elliptical vibration assisted machining, in: ASPE Proceedings, Vibration Assisted Machining Technology, The Carolina Inn, Chapel Hill, NC, 2007.
[85] R. Neugebauer, A. Stoll, Ultrasonic application in drilling, J. Mater. Process. Technol. 149 (1–3) (2003) 633–639.
[86] B. Lauwers, F. Klocke, A. Klink, CIRP Questionnaire "Hybrid Processes" (2009). CIRP Internal Document.
[87] V. Baghlani, P. Mehbudi, J. Akbari, M. Sohrabi, Ultrasonic assisted deep drilling of Inconel 738LC super alloy, in: 17th CIRP Conference on Electro Physical and Chemical Machining, vol. 6, 2013, pp. 571–576.
[88] V.A. Phadnis, F. Makhdum, A. Roy, V.V. Silberschmidt, Ultrasonically assisted drilling: machining towards improved structural integrity in carbon/epoxy composites, Key Eng. Mater. 569–570 (2013) 49–55.
[89] M.A. Kadivar, J. Akbari, R. Yousefi, Investigation of ultrasonic-assisted drilling of Al/SiC metal matrix composite with Taguchi method, Key Eng. Mater. 523–524 (2012) 215–219.

[90] J.H. Ko, K.C. Shaw, H.K. Chua, R.M. Lin, Cusp error reduction under high speedmicro/meso-scale milling with ultrasonic vibration assistance, Int. J. Precision Eng. Manuf. 12 (1) (2011) 15–20.
[91] F.L. Amorim, W.L. Weingaertner, Die-sinking electrical discharge machining of a high-strength copper-based alloy for injection molds, J. Braz. Soc. Mech. Sci. Eng. 26 (2) (2004) 137–144.
[92] T. Endo, T. Tsujimoto, K. Mitsui, Study of vibration-assisted micro-EDM—the effect of vibration on machining time and stability of discharge, Precis. Eng. 32 (2008) 269–277.
[93] J. Zhixin, Z. Jianhua, A. Xing, Ultrasonic vibration pulse electro-discharge machining of holes in engineering ceramics, J. Mater. Process. Technol. 53 (1995) 811–816.
[94] C. Gao, Z. Liu, A study of ultrasonically aided micro-electrical-discharge machining by the application of workpiece vibration, J. Mater. Process. Technol. 139 (2003) 226–228.
[95] K.T. Hoang, S.H. Yang, A study on the effect of different vibration-assisted methods in micro-WEDM, J. Mater. Process. Technol. 213 (2013) 1616–1622.
[96] Z.N. Guo, T.C. Lee, T.M. Yue, A study of ultrasonic-aided wire electrical discharge machining, J. Mater. Process. Technol. 63 (1997) 823–828.
[97] Z.Y. Yu, Y. Zhang, J. Li, High aspect ratio micro-hole drilling aided with ultrasonic vibration and planetary movement of electrode by micro-EDM, CIRP Ann. Manuf. Technol. 58 (2009) 213–216.
[98] A.W.J. Hsue, J.J. Wang, C.H. Chang, Milling tool of micro-EDM by ultrasonic-assisted multi-axial wire electrical discharge grinding processes, in: Proceedings of the ASME 2012 International Manufacturing Science and Engineering Conference MSEC2012, 2012.
[99] S. Je, D.W. Kim, H.S. Lee, C.N. Chu, Micro EDM with ultrasonic work fluid vibration for deep hole machining, J. Korean Soc. Precision Eng. 22 (7) (2005) 47–53.
[100] D.J. Kim, S.M. Yi, Y.S. Lee, C.N. Chu, Straight hole micro EDM with a cylindrical tool using a variable capacitance method accompanied by ultrasonic vibration, J. Micromech. Microeng. 16 (2006) 1092–1097.
[101] M.S. Hewidya, S.J. Ebeidb, T.A. El-Taweela, A.H. Youssefa, Modelling the performance of ECM assisted by low frequency vibrations, J. Mater. Process. Technol. 189 (2007) 466–472.
[102] S. Skoczypiec, Research on ultrasonically assisted electrochemical machining process, Int. J. Adv. Manuf. Technol. 52 (2011) 565–574.
[103] B. Ghoshal, B. Bhattacharyya, Shape control in micro borehole generation by EMM with the assistance of vibration of tool, Precis. Eng. 38 (2014) 127–137. 2013.
[104] M. Wang, Y. Zhang, Z. He, W. Peng, Deep micro-hole fabrication in EMM on stainless steel using disk micro-tool assisted by ultrasonic vibration, J. Mater. Process. Technol. 229 (2016) 475–483.
[105] P.A. Buffat, J.P. Borel, Size effect on the melting temperature of gold particles, Phys. Rev. A 13 (1976) 2287–2298.
[106] B. Kang, G.W. Kim, M. Yang, S.H. Cho, J.K. Park, A study on the effect of ultrasonic vibration in nanosecond laser machining, Opt. Lasers Eng. 50 (2012) 1817–1822.
[107] H.Y. Zheng, H. Huang, Ultrasonic vibration-assisted femtosecond laser machin-ing of microholes, J. Micromech. Microeng. 17 (2007) N58–N61.
[108] J. Koch, F. Korte, T. Bauer, C. Fallnich, A. Ostendorf, B.B. Chichkov, Nanotexturing of gold films by femtosecond laser-induced melt dynamics, Appl. Phys. A Mater. Sci. Process. 81 (2) (2005) 325–328.
[109] J.K. Park, J.W. Yoon, M.C. Kang, S.H. Cho, Surface effects of hybrid vibration-assisted femtosecond laser system for micro-hole drilling of copper substrate, Trans. Nonferrous Met. Soc. Chin. 22 (2012) 801–807.

[110] J.K. Park, J.W. Yoon, S.H. Cho, Vibration assisted femtosecond laser machining on metal, Opt. Lasers Eng. 50 (2012) 833–837.
[111] W.-S. Chu, C.-S. Kim, H.-T. Lee, J.-O. Choi, J.-I. Park, J.-H. Song, et al., Hybrid manufacturing in micro/nano scale: a review, Int. J. Precis. Eng. Manuf. Green Technol. 1 (1) (2014) 75–92.
[112] J. Lu, R.Q. Xu, X. Chen, Z.H. Shen, X.W. Ni, S.Y. Zhang, C.M. Gao, Mechanisms of laser drilling of metal plates underwater, J. Appl. Phys. 95 (8) (2004) 3890–3894.
[113] F. Tagliaferri, G. Leopardi, U. Semmler, M. Kuhl, B. Palumbo, Study of the influences of laser parameters on laser assisted machining processes, Procedia CIRP 8 (2013) 170–175.
[114] Y. Shin, Laser assisted machining, 26 (1) (2011). www.industrial-lasers.com/Articles.
[115] S. Sun, M. Brandt, M.S. Dargusch, Thermally enhanced machining of hard-to-machine materials—a review, Int. J. Mach. Tool Manuf. 50 (2010) 663–680.
[116] H. Ding, Y.C. Shin, Laser-assisted machining of hardened steel parts with surface integrity analysis, Int. J. Mach. Tool Manuf. 50 (1) (2010) 106–114.
[117] C. Brecher, M. Emonts, C.J. Rosen, J.P. Hermani, Laserunterstützte Fräs-bearbeitung, WT Werkstattstechnik Online 102 (6) (2012) 353–356.
[118] C. Chang, C. Kuo, An investigation of laser-assisted machining of Al2O3 ceramics planning, Int. J. Mach. Tool Manuf. 47 (2006) 452–461.
[119] K. Hock, B. Adelmann, R. Hellmann, Comparative study of remote fiber laser and water-jet guided laser cutting of thin metal sheets, Phys. Procedia 39 (2012) 225–231.
[120] A. Melaibari, P. Molian, P. Shrotriya, Two-dimensional contour cutting of polycrystalline cubic boron nitride using a novel laser/water jet hybrid process, Int. J. Adv. Manuf. Technol. 63 (5–8) (2012) 641–649.
[121] Y.K. Madhukar, S. Mullick, S.S. Chakraborty, A.K. Nath, Effect of water-jet on laser paint removal behaviour, Procedia Eng. 64 (2013) 467–472.
[122] C.A.A. Rashed, L. Romoli, F. Tantussi, F. Fuso, M. Burgener, et al., Water jet guided laser as an alternative to EDM for micro-drilling of fuel injector nozzles: a comparison of machined surfaces, J. Manuf. Process. 15 (4) (2013) 524–532.
[123] A. Al-Ahmari, M.S. Rasheed, M.K. Mohammed, T. Saleh, A hybrid machining process combining micro-EDM and laser beam machining of nickel-titanium based shape memory alloy, Mater. Manuf. Process. 31 (4) (2015) 447–455.
[124] C.L. Kuo, J.D. Huang, H.Y. Liang, Fabrication of 3D metal microstructures using a hybrid process of micro-EDM and laser assembly, Int. J. Adv. Manuf. Technol. 21 (2003) 796–800.
[125] S. Kim, B.H. Kim, D.K. Chung, H.S. Shin, C.N. Chu, Hybrid micromachining using a nanosecond pulsed laser and micro EDM, J. Micromech. Microeng. 20 (2010) 015037–015038.
[126] L. Li, C. Diver, J. Atkinson, R.G. Wagner, H.J. Helml, Sequential laser and EDM micro-drilling for next generation fuel injection nozzle manufacture, Ann. CIRP 55 (2006) 179–182.
[127] H. Zhang, J.W. Xu, Laser drilling assisted with jet electrochemical machining for the minimization of recast and spatter, Int. J. Adv. Manuf. Technol. 62 (9–12) (2012) 1055–1062.
[128] H. Zhang, J.W. Xu, J.M. Wang, Investigation of a novel hybrid process of laser drilling assisted with jet electrochemical machining, Opt. Lasers Eng. 47 (11) (2009) 1242–1249.
[129] Z. Hua, X. Jiawen, Modeling and experimental investigation of laser drilling with jet electrochemical machining, Chin. J. Aeronaut. 23 (4) (2010) 454–460.
[130] P.T. Pajak, A.K.M. Desilva, D.K. Harrison, J.A. McGeough, Precision and efficiency of laser assisted jet electrochemical machining, Precis. Eng. 30 (2006) 288–298.

[131] A.K.M. Desilva, P.T. Pajak, J.A. Mcgeough, D.K. Harrison, Thermal effects in laser assisted jet electrochemical machining, Ann. CIRP 60 (2011) 243–246.
[132] H. Zhang, J. Xu, Modelling and experimental investigation of laser drilling with jetelectrochemical machining, Chin. J. Aeronaut. 23 (2010) 454–460.
[133] P.F. Chauvy, P. Hoffmann, D. Landolt, Application of laser lithography on oxide film to titanium micro machining, Appl. Surf. Sci. 208–209 (2003) 165–170.

CHAPTER SEVEN

Micromachining processes

Contents

Modern Machining Technology
https://doi.org/10.1016/B978-0-12-812894-7.00007-4
© 2020 Elsevier Inc.
All rights reserved.

7 Introduction

Fabrication involves producing structures, equipments and components by casting, machining, forming, welding, assembling, etc. Among them and like other fabrication techniques, machining which is defined as the removal of material from work substrate may be categorized in two types on the basis of the dimensions of the feature produced, i.e., macro machining and micromachining. The former one denotes machining of features that are measurable clearly in naked eye, i.e., up to 1 mm, whereas, the later one is correlated with machining of features possessing an end dimension below 1 mm and up to 1 μm. However, in recent years the range of obtainable feature dimensions by micromachining techniques has been modified by CIRP committee where, dimensions between 1 and 500 μm has been denoted as acceptable range. In this connection, it is worthy to mention that, the present day notation which defines micromachining as machining of very small (micro) parts or miniature components only, is not correct. Because micromachining literally means removal of material in micro/nano level with high accuracy and no constraint on the size of the workpiece. Hence, micromachining can be implemented on micro, macro and nano work substrates without considering their dimensions.

The implementations of macro machining technologies are rather old and traditional. On the other hand, with rapid advancement of technologies, the demand of miniaturized parts, components as well as micro structures is increasing. Hence, fabricating and downscaling these components as per design criteria is an early requirement and needs adoption of suitable micromachining processes.

As, removing the surplus material and attaining desired shape, size as well as to create intricate feature is the sole aim of today's micro manufacturing technologies, therefore, micromachining technologies utilizes various material removal mechanisms to fabricate three dimensional micro features as well as formed surfaces in micron/sub micron dimensions. However, the fabrication processes of such precise and intricate micro features needs more stringent control and scaling down of magnitudes of material removal with controlled fashion by suitable selection of operating parameters for achieving successful results. Besides, the need of producing structures with functional requirements such as tighter tolerances and implementation of various new engineering materials are also playing a great role and pushing the implementation of micromachining techniques.

7.1 Need and basic of classification

During past few years, the capabilities and applicability of different micromachining techniques has been addressed as well as justified [1], however, the implementation of such methods are limited in batch or industrial scale due to various hindrances caused by lack of proper skill as well as experience and weaker knowledge base of micro machinability of materials [2]. On the other hand, perceived with newer ever flourishing key technologies of the future, the emerging micromachining technologies is opting toward sea change of implementer by revealing new ways how people and machines interact with the physical world. Existing micro parts produced by micro machining processes include micro-electromechanical systems (MEMS), micro-actuators, micro-sensors, micro fuel cells, micro grooves, micro holes, micro channels, micro valves, micro pump, medical implants and devices, and different other microfeatures. The involvement of micromachining methods for producing integrated circuits (ICs) is playing a major role in electronics industries that houses utmost importance starting from computer components like CD players, relay switches to lighter aircrafts, satellites, etc. The painless microsurgery and crafted devices for better health inspection also requires the presence of huge amount of micro components that urge toward rapid and efficient implementation of micromachining techniques. Beside the urgent need, the additional benefits of implementing micromachining techniques involves less requirement of power, energy, material and floor space.

The faster growing micromachining technology is demanding much more with components of durable, harder or nearly unmachinable materials being crucial, recent twentieth century tools employ extensive uses of various harder materials such as carbide, diamond, metal matrix composites (MMC), ceramics, etc. The increasing use of numerous tooling also invoked the use of advanced and adequate power sources which gradually enhances the overall effectiveness and capability of manufacturers. Apart from the use of conventional source like energies from electric motors, non-traditional micromachining processes utilizes energies in its direct forms to remove surplus material and generate complex features by controlling operating parameters without depending on mechanical properties of substrate. Hence, in addition to the accompanying research trends in miniaturization of equipments as well as metrology systems and as per requirements of manufacturing miniaturized products, emphasis should be given to various modern micromachining processes.

Based on inherent mechanism and basic requirements, type of energy required and medium of energy transfer, modern micromachining processes can be broadly classified as shown in Fig. 7.1. Mechanical micromachining techniques use mechanical forces to drive abrasive particles, water or accelerated ions toward the workpiece for subsequent striking and material removal. For example, AJM uses kinetic energy of loose abrasive particles whereas, WJM uses kinetic energy of water jet for removing material. USM uses ultrasonic frequency of transducer to drive the abrasive slurry

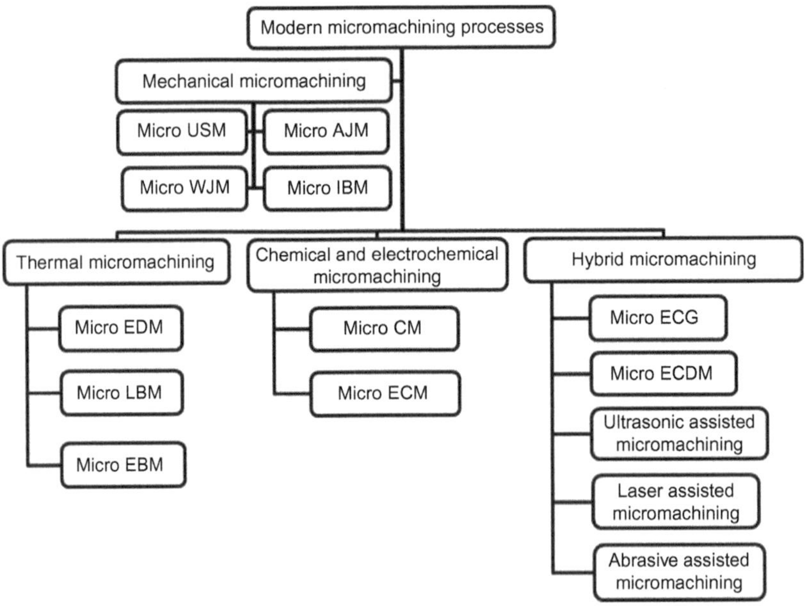

Fig. 7.1 Classification of non-traditional micromachining processes.

and remove material from workpiece. IBM utilizes ion sputtering and material removal occurs atom by atom. On the other hand, producing intense heat followed by melting and vaporization is the sole technique of material removal in thermal micromachining processes. For example, EDM uses electrical discharge for producing heat, LBM uses converging laser rays for generating high temperature, and EBM uses the kinetic energy of incident electrons to convert into heat energy for removal of material. Chemical and Electrochemical micromachining involves removal of parent material by subsequent chemical reactions and by electrochemical reactions respectively.

7.1.1 Mechanical micromachining

Mechanical Micromachining process involves ultrasonic machining (USM), abrasive jet machining (AJM), water jet machining (WJM), etc. These micromachining processes mainly use mechanical forces to remove materials. Out of these processes, micro USM is most acceptable machining process which is commercially available and used in modern industries for various micromachining applications. Detail discussion on macro domain of USM, AJM, WJM and FIB has already been included in previous Chapter 3. However, micro USM is discussed in detail in this chapter considering its wide range of successful implementation in the area of micro manufacturing. Basic concepts of USM, AJM, WJM as well as FIB for micromachining applications have also been briefed here under.

USM utilizes ultrasonic vibration on primarily produced, tough and ductile coaxial microtool for machining microfeatures on various conductive as well as non-conductive and semiconductive hard or brittle materials like glass, ceramics, etc. During machining the small cutting gap between the microtool and workpiece is filled up with continuously flowing abrasive slurry where the tiny abrasive particles moves and strike the worksurface during downward movement of each highly frequent repetitive linear reciprocating motion of microtool. The simultaneous striking of thousands of abrasive particles leads to formation of craters and removal of millions of chips from the worksurface during each downward motion. Few applications of USM include fabrication of inclined and square holes, micro devices for MEMS, micro structures on quartz and synthetic ruby, trenches [3], etc. for use in the ever-growing fields of aerospace, medical and electronics.

The mechanism of AJM includes use of kinetic energy of loose abrasive particles for removing material. A wear resistant mask is used prior operation to coat the work surface except the potion left for machining to be carried out. Hence, the geometry and dimension of mask that is used during machining also determines the geometry and dimension of final shape to be produced. During machining, whole work surface is being made exposed under jet of gas or air carrying loose abrasive particles. The loose abrasive particles, having grit size of approximately 50 μm strike with a velocity of 150–400 m/s and impinge the bare area of the work surface leading to removal of material by causing brittle fracture and formation of desired shapes. Unlike normal sand blasting, the material removal rate that can be controlled more precisely during AJM involves size and type of abrasive particles, flow rate of abrasives, pressure of accompanying gas, work material type, etc. The abrasive particles generally utilized during machining include silicon carbide, aluminum oxide, boron carbide, etc. The application of AJM includes cutting, etching, deburring and polishing or cleaning of different brittle materials like ceramics, glass, etc. for fabricating fine intricate features [4], whereas, machining of ceramic materials with other process like wet chemical etching is comparatively expensive. Other applications of AJM include generation of hole and channels on polymethyl methacrylate (PMMA) [5, 6], fabrication of components used in micro-electro-mechanical-systems (MEMS) [7] as well as in opto-electronics, marking on electronic equipments and rubber stencils, frosting of glass surfaces, engraving numbers, codes on car windows, fabrication of free standing and complex 3D micro features on glass [8], etc.

WJM utilizes kinetic energy of water jet to remove material from workpiece. The Jet has a high velocity of up to 900 m/s which strikes the work surface where the kinetic energy instantaneously changes into pressure energy up to a scale of three times higher than normal stagnant pressure. Material removal occurs when the generated pressure is high enough to overcome the binding force of constituting elements and molecules forming up the workpiece. To create and maintain sufficient amount of pressure for carrying out machining operations, a typical WJM system comprises of proper pumping unit, high pressure hydraulic accumulator for storing pressure energy during idle periods and introducing smoother flow for next operation, different types of nozzles, valves, tube fittings, etc. WJM, as a highly eco friendly process, is extensively utilized for cutting porous materials like woods, corrugated boards, different softer materials like lead, asbestos, rubber, etc. with superior surface finish and straightness. Other applications of WJM include rock excavation [9], cutting fibers, machining features on sandstones, cutting polyethylene to be used in automotive industries, drilling of coals and hydraulic mining, descaling and cleaning in oil refineries as well as chemical plants, etc. Multipass operations in WJM is utilized for generating features having higher thickness whereas for machining little harder materials like brass, copper, aluminum, etc. tremendously pressure of water jet is required which is detrimental for commercial uses.

FIB or IBM is another machining process for removing material atom by atom utilizing ion beam of specific diameter and high intensity. The whole process is required to be carried out within a vacuum chamber to avoid interaction between ions and atmospheric gas molecules. Nowadays mainly gallium is used as liquid metal ion source where the gallium ions, with high potential, strikes the worksurface and machining occurs when these sputtered as well as accelerated gallium ions dislodge the workpiece atoms, positive ions as well as secondary electrons. The displaced ions, atoms or electrons transfers their momentum into their surroundings and material removal continues. The rate of material removal in FIB machining is highly dependent on sputtering yield which can be controlled by controlling different influencing parameters like energy of accelerated ions, types of atom and ions of substrate, their impingement direction and density [10]. The main advantage of this ion sputtering process comes as its negligible thermal effect on workpiece during operation; however, the process is far from being economical. Various advantages of FIB involves machining of micro/nano parts like micro tools and dies for micro EDM, fabrication of micro end mill cutters, formed micro parts, fabrication of surgical micro tool, diamond micro indenters, tissue stabilizer [11] as well as different nano components used for bio-technology, etc.

7.1.2 Thermal micromachining

Thermal micromachining processes remove material by generating intense and high localized heating followed by subsequent melting and vaporization. Thermal micromachining processes mainly include electro discharge machining (EDM), laser beam machining (LBM) as well as electron beam machining (EBM). All these processes in macro scale have already been included in previous Chapter 4. However, out of these processes, micro EDM and micro LBM are most acceptable for micro manufacturing applications especially in the area of micromachining. Hence, micro EDM and micro LBM has been discussed elaborately in this chapter. The operating principles of various thermal micromachining processes such as EDM, LBM, EBM, etc. have been briefly described as below.

EDM involves removal of material by discrete electrical discharge or microsparks where during machining the tool and the workpiece is separated by a small gap of few microns and is dipped inside suitable dielectric like deionized water, kerosene, air, etc. The discrete electrical sparking is due to the application of DC pulsed power across the tool and the workpiece which results in formation of plasma channel owing to very high electric field gradient. The sparking increases the temperature of both tool tip and workpiece resulting in melting and vaporization where some molten debris is flushed away from the machining zone by continuous flow of dielectric and remaining molten material resolidifies as recast layer over the workpiece. Machining continues due to thermal erosion when the tool is gradually fed toward the workpiece maintaining a certain discharge gap. Various applications of micro EDM involve drilling of high aspect ratio micro holes [12], fabrication of micro nozzles, fabrication of high precision metal mask and various other micro parts useful in different fields like electronics, automobile, etc. In comparison to macro counterpart, micro EDM can easily machine various intricate micro features of different geometries with enhanced accuracy, precision and surface quality by increasing the sparking efficiency, decreasing the spark size, etc., however, alike EDM, micro EDM suffers from detrimental thermal effects such as deterioration of surface quality and work material quality, formation of recast layer requiring additional techniques for removal, etc. Moreover, as the tool is also affected during EDM, it leaves a small crater on the tool which can also degrade the final product quality.

LBM is another thermal micromachining process of removing work material and generation of precise intricate shapes. During LBM, the target material residing on the focal point of the irradiated and pulsating laser absorbs the laser energy and heats up to its melting point followed by instantaneous vaporization and material removal occurs via ablation technique. This melting and vaporization of the target material is highly dependent on laser intensity of work material property and duration of laser pulses. As LBM is a thermally induced material removal process, hence, implementation of laser has some adverse effects like generation of thermal stress and formation of heat affected zone (HAZ) surrounding the machining area that leads to degradation of surface quality and material property by inducing surface as well as sub surface cracking. On the part of different types of laser pulses, long pulse laser is mainly used for commercial purposes, whereas, ultra fast or very short pulse laser is ideal for fabricating printer heads and hard disk drive heads, grating on optical fiber [13], various types of micro channels, holes as well various types of micro parts or components applicable in the field like microfluidics [14], for engraving works, marking, etc.

EBM involves removal of material by implementing kinetic energy of incident electrons that are emitted due to the application of high potential gradient on the cathode and accelerated toward workpiece due to the potential of anode. Upon striking the work surface with high velocity, the high kinetic energy of the electron pool converts to very high heat energy at the beam area that is responsible for subsequent localized melting and vaporization leading to removal of workpiece material. The rate of material removal in EBM is highly dependent on energy and width of incident beam, thermal property of work material, thickness of specimen to be machined, etc. A vacuum chamber is needed to carry out EBM operations to avoid collisions with surrounding gas molecules. However, as the beam area is considerably small, span of HAZ as a result of adverse thermal effects is also minimal. Moreover, EBM is commercially implemented in areas like nuclear, thermal, automotive, aerospace, etc. for faster production of micro holes, micro slots of various shapes as well as for micro welding. However, fabrication of contoured internal micro features, several other intricate shapes are difficult to achieve by EBM.

7.1.3 Chemical and electrochemical micromachining

Chemical and electrochemical micromachining involves removal of material by chemical and electrochemical reactions. It includes mainly chemical micromachining (CMM) and electrochemical micromachining (EMM), etc. Both these machining processes on macro scale have already been discussed in detail in previous Chapter 5. Out of them, EMM is most acceptable, flexible and widely used for micromachining applications in modern industries. Hence, EMM is elaborated in detail in this chapter. However, these two processes are discussed in brief here under for introducing micromachining utilizing chemical and electrochemical principle.

In CMM, work material is removed by controlled chemical reaction of strong alkaline or acidic solution. Before starting operation, the workpiece is covered with chemically inert mask and except the area onto which machining needs to be carried out. Machinimng occurs when the etchant react with workpiece that forms metallic salt which dissolves in the medium. CMM does not need any kind of external power source to machine workpiece, however, the process includes various sequential steps and strict control of parameters like temperature, composition and type of etchant for enhancing removal of material and obtaining the desired feature. Some applications of CMM include fabrication of IC, panels for different instruments, printed circuit boards, enclosure screens, etc. However, CMM also suffers from few drawbacks like lower MRR and accuracy, incapable to machine materials that are chemically resistive, use of etchants that are highly corrosive and hazardous toward environment, etc., that requires further in-depth research to be dealt with.

EMM is an anodic dissolution process that removes material by corresponding electrochemical reactions from the workpiece when both cathode and anode is dipped inside suitable electrolyte solution separated by an inter-electrode gap and pulsed DC is applied. As a non-thermal material removal process, EMM offers several advantages like, absence of thermal stress, deterioration of surface quality and material property, formation of HAZ, etc. [15]. Moreover, in comparison with CMM, EMM gives better accuracy and surface finish, whereas, CMM is literally uncontrolled in micromachining domain. Moreover, by controlling process parameters as well as decreasing the IEG to $<50\,\mu m$, dissolution becomes highly confined and localized that results in generation of precise shapes with desired geometries. Besides being a eco friendly process, several other benefits of EMM include better precision, good MRR, better flexibility, control and reliability, ability of machining chemically resistive copper alloys, stainless steel, titanium, titanium alloys, super alloys, etc. [16]. Moreover, EMM is a cost effective process for manufacturing micro parts required in MEMS as well as highly sophisticated fields like electronics, biomedical, optics, etc. and can even enhance the feature quality and geometry as per desired requirements that has been previously fabricated with CMM.

7.2 Conditions for micromachining

The top down approach which is followed generally for micromachining requires involvement of both conventional and non-conventional techniques for achieving desired criteria [17]. However, more specific and systematic approach is required in accordance with following two guidelines:

(i) Applying the concept of unit material removal that defines the tiniest amount of material removal as well as smallest dimension that can be adjusted per unit time or per pulse during fabrication of micro feature.

(ii) Implementation of precise machineries for decreasing the dimensional disparity proportionate to the size of the micro feature.

Unit removal can also be expressed in one dimensional, 2D or 3D value like the units of length, area and volume. Also, as unit removal limits the tiniest adjustable dimension of a feature, its value should be considerably smaller compared to the bulk size of the feature. For this, when removing a material from a component with dimensions in millimeter ranges, unit removal must be in sub micron range or even lower to ensure required precision and accuracy. On the other hand, the dimensional disparity should be down scaled with same proportion when the size of the component is down scaled. For example, if a product is downscaled to one fifth of the original, its dimensional disparity should be decreased to one fifth accordingly. For this, more precise equipments comprise of precision machine elements are required when machining is carried out in micron ranges. However, maintaining such accuracy in very low dimensional ranges is often problematic and sometimes impossible.

7.3 Micro ultrasonic machining (micro-USM)

7.3.1 Introduction

Micro ultrasonic machining (MUSM) is a well known micromachining technique that is gaining popularity in fabricating different intricate and complex micro features on various harder ceramic materials like, quartz, glass, composites, etc. as well as different stronger and tougher materials like nickel, titanium and their alloys [17]. The highly frequent vibration driven abrasive slurry involved in MUSM has even been responsible for machining micro holes of diameter as low as 5 μm [18] on silicon to be readily used in PCBs with ICs required in electronics industries. The common use of MUSM process for fabricating micro holes has also led the implementation of this process in producing diverse components like fluidic filters, biomedical filters, ink-jet printer nozzles, fuel injection nozzles, grids, optical apertures, high pressure orifices, pneumatic sensors, micropipettes, manipulators, etc.

7.3.2 Basic mechanism

Micro-USM constitutes the successive process of generating a micro feature right from techniques required for fabricating microtool. In comparison with usual USM process, MUSM differs in size of tool used and attaching it with horn or tool holder as well as the implementation of process in generating precise and intricate micro features. The tool in regular USM process is firstly prepared by soldering, brazing, screwing and then is attached to the horn, whereas, in case of MUSM the same approach does not guarantee proper tooling as the system can lose stability by risking the concentricity as well as accuracy with which the tool approaches the workpiece. For this reason, the micro tools of spiral, polygonal or cylindrical shapes required for MUSM process are primarily produced on the same setup by wire electrode discharge machining (WEDM)/wire electrode discharge grinding (WEDG) after which machining is carried out. The combination of both WEDG and EDM can also fabricate 3D micro tools whereas; the lowest dia of tool has been 3 μm [19]. Moreover, other processes like chemical micromachining or electrochemical micromachining are also suitable for fabrication of micro tools for MUSM.

During MUSM, low frequency signal is converted into high frequency which again converts into high frequency linear vibration by magnetostrictive or piezoelectric transducer. This high frequency vibration of ultrasonic range (30–50 KHz) is being transmitted to tool holder followed by tool or sonotrode. The tool for machining in MUSM is such that, it represents the mirror of the shape to be generated at the end product. Tool is also produced with all the required design criteria in mind to achieve better machining efficiency and to deliver constant resonance of high amplitude. Prior to machining, a narrow gap is created and maintained between tool and workpiece which is afterwards flooded with abrasive slurry. The abrasive slurry consists of loose abrasives starting from harder diamond dust, boron carbide, silicon carbide to softest alumina particles, etc. These abrasives remain suspended mainly in water due to its capability of transferring ultrasonic wave efficiently. During each vibrating cycle, the downward stroke of sonotrode drives the abrasive particles toward work surface. The abrasive particles accelerate and simultaneously strike resulting in removal of material in the form of small chips due to crater formation. The constant cyclic downward thrust carries out the machining operation and the tool is feed into workpiece and in any relative direction for fabrication of required shape.

The basic material removal mechanism during micro-USM can be categorized as:

(i) Microchipping where free and moving abrasive particles removes material by impact.

(ii) Removal of material by mechanical abrasion between the abrasive particles and the work material.

(iii) Removal of material due to the effect of cavitation of abrasive slurry generated from the application of ultrasonic vibration.

(iv) Removal of material by chemical action of liquid.

The effect of cavitation erosion and chemical reaction have secondary influences owing to proper circulation of slurry where most of the removed work piece particles again assist in machining. On the other hand, material is primarily removed by direct hammering of abrasives during mechanical abrasion and microchipping. Although all these mechanisms simultaneously aids in removing material but erosion by cavitation is better found during machining porous materials like graphite in comparison to ceramic material or different grades of hardened steels. The end result is removal of material by shearing, fracturing of harder materials as well as by plastic deformation resulting in relative dislodgment of work material from the surface in transitory region.

7.3.3 Setup details

The MUSM setup consists of microtool rotating mechanism with spindle, ultrasonic frequency generator, acoustic head, CNC axis, micro tool, slurry circulation unit and machining chamber for holding the workpiece as well as slurry. Fig. 7.2 shows a schematic diagram of MUSM system. The tool rotating unit consists of motor, mandrel and bearing to provide required rotations per minute. The ultrasonic frequency generator consists of a resonance circuit. The resonance circuit converts the low frequency electrical power to high frequency electrical power which is then fed to the transducer for converting into mechanical motion. Modern advanced frequency generators

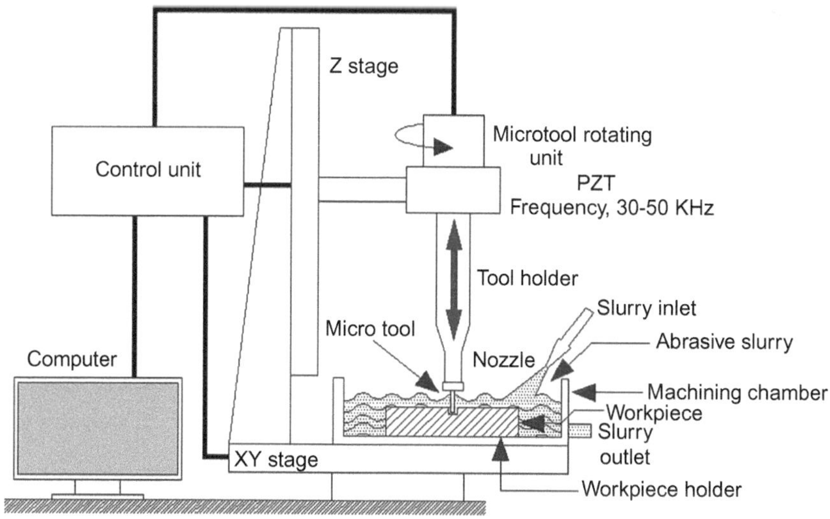

Fig. 7.2 Schematic diagram of MUSM system.

come with means for adjusting the high frequency electrical power to the exact output mechanical motion and other means for different error correction occurring due to tool wear and small heat generation. Also, the generator comes with necessary adjustments for minimizing the loss of power. The acoustic head consists of transducer and horn. The transducer converts the high frequency electrical impulse to high frequency mechanical vibration whereas; the horn or tool holder amplifies and focuses the vibration to drive the tool end with necessary and required vibration amplitude by minimizing the loss in energy. The CNC axis controller XY stage moves the workpiece in desired direction and required position where Z stage moves the micro tool. The tool design in MUSM should be such that, it can produce maximum vibration amplitude at given frequency. Generally, the tool materials are relatively ductile as because tool having brittle materials will wear out rapidly. Also, fabrication as well as fixing of micro tool in MUSM is quiet problematic for which micro tool is being fabricated on machine by WEDG/EDM/ECM/ChM and is used later for machining. Slurry circulation unit drives the abrasive slurry to the front end of the tool by pumping it and throwing like a jet or by sucking the slurry or by combining the both.

Apart from vibrating the tool, MUSM can also be carried out with vibrating the workpiece which is a new technique and recently been introduced. When machining with vibrating the workpiece, vibration is not given to the tool that eventually results in need of no horn for amplifying vibrations making the machining setup more stable, compact and simple. Furthermore, the workpiece vibration method eradicates tool wear that is severe in more common tool vibration method. Different parameters which influence the final machining quality and provides better dimensional accuracy, surface finish and MRR includes vibration frequency, vibration amplitude, RPM of rotary motion in case of its implementation on workpiece or tool head, load, abrasive particle size and type, type of liquid component used, slurry flow rate and concentration, geometry of work material, its thickness, type, etc.

7.3.4 Comparisons of USM and micro-USM

The range of process parameters and conditions involved in MUSM is substantially different from its macro counterpart and a brief comparison is necessary for better understanding associated with the feasibility and adaptability of MUSM process to achieve the specific target or end use. Table 7.1 shows a

Table 7.1 Comparison between USM and MUSM

Process characteristics	USM	Micro-USM
Power of machine	200–1000 watt	200–600 watt
Fabrication of tool	Can be fabricated on different setup	Uses WEDG/EDM, ECM/ChM on same setup
Tool diameter (typical)	>1 mm	As low as 4 μm
Static force	In tens of N	0.1 mN to 0.5 N
Gap	30–400 μm	<25 μm
Slurry flow rate (typical)	>30 L/min	<30 L/min
Vibration frequency	≥ 20 kHz	30–50 kHz
Vibration amplitude	15–50 μm	0.1–15 μm
Tool rotation	Not mandatory	Required, 300–5000 rpm
Abrasive particle size	15–300 μm	0.1–20 μm
Vibrating part	Tool	Tool or workpiece
Penetration rate	2.0–100 μm/min	0.025–25 μm/min
Machined hole diameter (typical)	>1 mm	As low as 5 μm
Aspect ratio (typical)	200	7
Best tolerance	± 7 μm	± 0.5 μm
Roughness (R_a)	0.25–0.75 μm	0.2–0.4 μm, 7 nm in polishing

brief comparison between the values of prime operating parameters as well as other process characteristics of both USM and MUSM process.

7.3.5 Influence of process parameters

In micro-USM, the parametric relationship is complicated due to the involvement of several factors and related parameters which influence the process outputs. The major influencing parameters in micro-USM can be grouped into mainly four categories such as acoustic system, abrasive and slurry, microtool, workpiece, etc. Various operating parameters, machining conditions and their influence can be represented through a fishbone diagram as shown in Fig. 7.3.

Machining rate in micro-USM is significantly influenced by type and size of abrasive particle. The machining rate increases with the size of abrasives. The machining speed is also improved with an increase in the average static load. Tool rotation improves the machining speed noticeably. The tool rotation facilitates faster removal of debris that results in increase in machining speed.

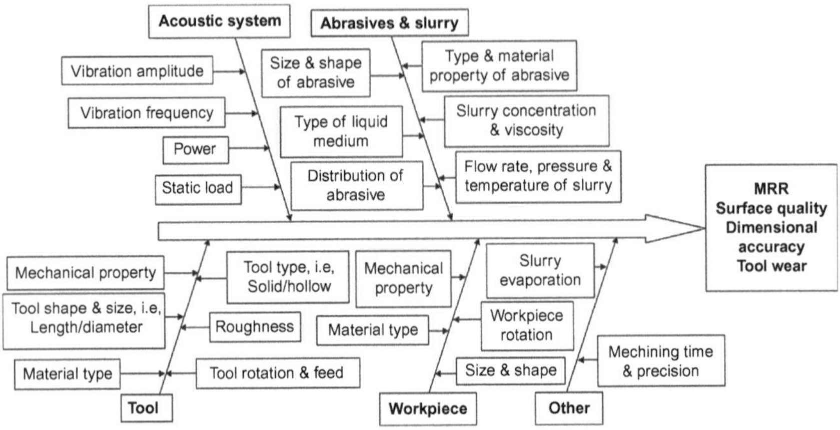

Fig. 7.3 Fish bone diagram showing various operating parameters, machining conditions and their influence.

Tool wear increases with the use of harder and coarser abrasive [3]. Due to larger tool wear, stainless steel and thoriated tungsten tool which are used in conventional USM are not suitable in micro-USM. In micro-USM, tool wear increases extremely with a decrease in tool diameter [20, 21]. Machining speed and tool wear increases with increase in oscillation amplitude and machining load.

The type and size of abrasive particle affect the surface finish and it increases as the size of abrasive increases. It have been reported that tungsten carbide causes the worst surface as compared to titanium carbide and boron carbide [3]. Abrasive particles of smaller size are used in micro-USM. Smaller size abrasive particles get embedded in the workpiece or in the tool during machining. This improper abrasive particle distribution in the gap restricts further movement of abrasive and leads to increase in surface roughness.

Shape accuracy factors of micro-USM are limited as observed during machining of high aspect ratio microfeatures. Ultrasonic machining has limitations in its application to micromachining because there are difficulties in fixing microtools to the machine due to the tool's size restriction.

In micro-USM, interactions between abrasive particles and workpiece are very complicated. The contributions of probable material removal mechanisms like microchipping, abrasion, cavitation, and chemical reaction are still not clear. Hence, further research for understanding of the material removal mechanism for micro-USM is needed.

7.3.6 Advancements and challenges

MUSM, as a micromachining process of generating features on both electrically conductive as well as on non-conductive material, poses numerous applications in the ever changing and developing fields like electronics, medical, aerospace, etc. The superior possibility to fabricate micro tool and its use in the same setup enhances the applicability of MUSM in fabricating micro holes, 3D micro cavities and micro slots with required geometries. By using MUSM, micro holes of 20 μm diameter and 50 μm depth have been fabricated on quartz and silicon [3] previously. On the other hand, the use of rotary tool during MUSM enables tool rotation accompanied with tool vibrations which have resulted in rolling and simultaneous sliding of abrasives. This material removal mechanism comprised of both material removal mechanism of both diamond grinding and MUSM has led to the increase in impact as well as indentation and rate of material removal [22]. Use of rotary MUSM also enhances machining accuracy and tool life. However, in accordance with the requirement of sophisticated setup for ensuring non-eccentricity as well as the need of micron sized tool, very small load is required to be applied on the microtool itself to obtain controlled, effective and stable rotation with the necessary feedback mechanism attached. Moreover, the addition of MUSM with turning has resulted in strain hardening, removal of generated residual stress as well as reduction of machining time, enhancement of surface quality and tool life, etc. [23]. Research work involved fabrication of multi micro tool which have been used for machining 20 μm diameter micro holes on soda lime glass utilizing MUSM. The whole process constitutes fabrication of micro hole array by tungsten micro tool on copper foil implementing electro discharge machining. The tungsten tool has been primarily produced by WEDG and the generated micro hole arrays has been used for fabricating multi micro tool of MUSM by reverse EDM [24]. Additionally, MUSM, in the field of microfluidics are gaining an upper hand for developing various types of micro channels, micro cavities, etc. on both electrically conductive and non-conductive material required for storing or delivering fluid in micro sensors and micro heat exchangers. Also, the applicability of MUSM on other various aspects like fabricating intricate 3D micro features or channels on newer polymer and ceramics materials for micro-electromechanical system (MEMS) is quite recognizable.

Apart from the above mentioned recent advancements as well as the increasing applicability, MUSM is not devoid of limitations that pose few challenges lying in process feasibility and capability for modern manufacturers. The required micro tool for MUSM needs stringent control during handling and fixing, which if not carried out properly can degrade the performance of machining as well as process stability. There are also few challenges in aligning the microtool properly for achieving desired accuracy and geometry of the end product. Here the traditional approach such as micro milling for tool fabrication and soldering or brazing for tool fixing don't pay off due to very small size of micro tool and requires in situ tool fabrication by WEDM or WEDG. Other challenges lie in the application of tool itself, the tip of which gets shortened over time due to gradual wear. This wearing out of tool tip hampers the consistency of tool vibration, messes up with tool position and decrease overall accuracy of the end product. The solution lies in the implementation of piezo vibration in the workpiece that has been introduced recently which also influences machining efficiency by improving the flushing condition of slurry as well as inducing better stirring. However, for this to achieve, some compromise is required to be made in the size of the workpiece, otherwise the amplitude of workpiece vibration and resulting accuracy may be badly affected. Other various issues lie in accuracy of measurement, quality of surface, implementation of online monitoring and controlling system, etc. that need to be addressed as early as possible to extend the capability of MUSM. Besides, better understanding of abrasive particle and workpiece interaction as well as technical knowhow of mechanism of material in light of various phenomena like micro chipping, cavitation and concerned chemical reaction is urgently needed for broadening the horizon and exploring further capabilities of ultrasonic material removal process in micromaching domain. Issue like temperature rise during micro cutting can add a surplus problem whereas; further research is also needed in light of high frequency low amplitude vibration and noise reduction.

7.3.7 Applications

MUSM has been widely applied for fabricating 3D micro-air turbines, tiny holes required in power transmission gears and shafts of choppers and helicopters, machining of micro holes of high aspect ratio on glass and silicon wafers required in flow and pressure sensors, fabrication of different micro components of MEMS, etc. [21]. The generated micro feature in the tip of the micro tool can also help in machining complex shapes where researchers have used gear shaped micro tool to fabricate an array of patterned microfeature obtained by feeding the tool inwards and replicating the patterned surface of tool [20]. Other applications of MUSM involve fabrication of

Fig. 7.4 Micrograph of (a) pin type micro-channel machined with MUSM on silicon with a pitch of 300 μm and a pin height of 100 μm, (b) detailed view [25].

pin type micro-channel on silicon [25], center-pin bearing airturbine, micro force-balance airturbine, spiral groove on low-melting glass by contouring mode [20] and micro pillars having high aspect ratio on Lead Zirconate Titanate [26], complex and contoured 3D micro cavity where tool path has been designed previously by CAD/CAM [27], spiral groove, wafers on borosilicate glass, micro feature in alumina that has been used later for fabricating complex shaped micro tool [28], micro channels required in bio-MEMS, machining and polishing of steel surface of tool [29], processing of numerous brittle materials as well as slicing of different semiconductor components required in electronics industries, etc. The application field of MUSM can further be enlarged by using it for machining relatively ductile metal-based materials with proper adjustments and modifications for implementation in the field of heat transfer as well as microfluidics. Fig. 7.4 shows micrograph of pin type micro channel machined by MUSM on silicon.

7.4 Micro electro discharge machining (micro-EDM)

7.4.1 Introduction

Micro electro discharge machining (EDM) is one of the most popular non-traditional machining processes utilized for fabricating features within micro ranges. This process is applicable for softer as well harder and stronger materials but is limited to electrically conducting only. This thermal induced process is being widely used and removes material with discrete electrical discharge. Before operation, the cathode tool and anode workpiece is dipped inside suitable dielectric and are connected to a Resistance Capacitance (RC) circuit maintaining a suitable small discharge gap. Upon feeding the tool toward the workpiece, discharge occurs and a plasma column is formed. The flux of electron produces huge amount of heat that melts the work material. As the feeding of tool continues, the work material melts accordingly generating a replica of tool. Micro-EDM is an efficient process of machining microfeature even on enormously hard carbide materials as well as on HSTR alloy work surface due to the absence of any cutting force. Whereas, the process slightly differs from general EDM in terms of energy range, size of tool and resolution of XYZ axis that is required to produce desired micro features effectively. However, micro-EDM suffers from thermal effects like deterioration of surface quality as well as material property and tool wear due to melting of tool material itself that restricts further application.

7.4.2 Basic mechanism

As mentioned previously, micro-EDM, being a thermoelectric material removal technique, melts and vaporize workpiece material by controlled spark or discharge between tool and workpiece that are dipped inside suitable dielectric medium and separated by small discharge gap [30]. The dielectric medium primarily involves deionized water, kerosene, air, etc. Upon application of suitable potential difference between cathode tool and anode workpiece from proper pulse power supply, high electric field generates at the positions where the distance between tool and workpiece is smaller. This high electric field propels the movement of electron from cathode to anode. During movement, the electrons collide with neutral dielectric molecules which lead to formation of more free electrons. This increased ionization results in building up of plasma channel followed by constant flow of ions toward tool and electrons toward workpiece and generates current impulses to form distinct spark or discharge. The high heat associated with the spark or discharge heats up the workpiece resulting in removing of material from desired sites by melting and vaporization. Also, during every discharge few amount of tool material melts and vaporizes that is one of the basic adverse thermal effects of this material removal process. During operation, the high pressure existing in plasma channel actually helps in removal of molten metal. Material removal from workpiece as well as tool material, on the other hand, depends upon the kinetic energies of both ions and electrons where, electrons being the lighter posses relatively higher velocities than what the ions have. As a result, when both tool and work material is same, material removes faster in workpiece when it is selected as anode. After the discharge and subsequent removal of material, the gap between the tool and workpiece increases. Hence, discharge occurs in the next closest positions between tool and workpiece and continues. Thus, the spark travels all the way across the tool surface resulting in uniform removal of work material. Upon feeding the tool into workpiece, same phenomenon repeats leading to formation of desired shape on the work surface.

Micro-EDM can be carried out with both straight and reverse polarities depending on the requirements to be achieved during machining. In straight polarity workpiece is made positive and in reverse polarity tool is positive. The small spark radius, high frequency and associated high temperature removes tiny chunk of material in each discharge making this machining suitable for micromachining applications. However, during machining, a portion of evaporating metals escapes through the dielectric whereas, remaining portion resolidifies and forms recast layer over the work surface. Moreover, as because, both micro tool and workpiece are exposed to severe temperature and material removal occurs from both micro tool and workpiece, thermal stress generates and crater forms which degrades the integrity of both anode and cathode material. This also limits further use of micro tool and adds with the disadvantages [31]. In contrast to regular EDM, micro-EDM differs in implementation of servo controlled feed mechanism for accurate positioning with increased sensitivity to maintain a spark gap of 1–5 μm [32, 33], whereas, other differences include type of power circuitry, process of erosion and proper control of operating parameters, as because, in regular EDM the motto is to maximize the material removal and in micro-EDM, high erosion efficiency is needed that is obtained from reduction of spark energy by limiting it to a single pulse electrical discharge for achieving material removal of desired micro components.

7.4.3 Setup details

The setup of micro-EDM is basically a modified version of regular EDM capable of fulfilling micromachining criteria as per requirements. Hence, the basic requirements of micro EDM involves precise control over spark gap with improved positioning and accuracy as well as discharge energy, means for tool wear compensation, better flushing of dielectric as well as in process measurement. Fig. 7.5 shows schematic diagram of micro electro-discharge machining setup.

The main components of micro-EDM setup consists of a micro tool rotating unit, piezoelectric actuator, XYZ stage and power supply unit.

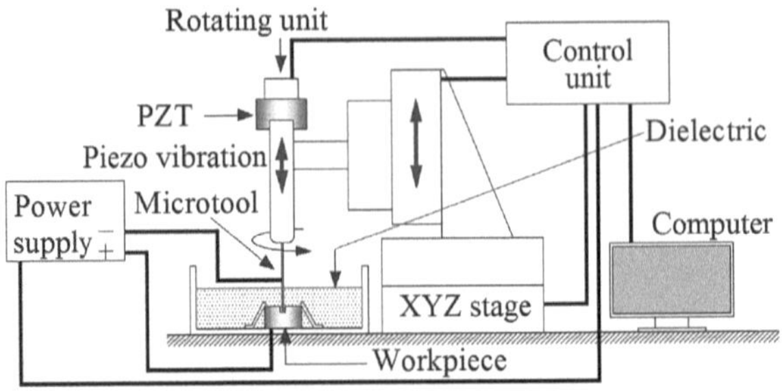

Fig. 7.5 Schematic diagram of micro electro-discharge machining setup.

The stepper motor controlled XYZ stage is a low response drive and results in coarser movements during positioning the workpiece, whereas, piezo actuator is a high response drive, e.g., response frequency, 1 kHz and travel, 15 μm, helps in finer movement of micro tool. Moreover, electrode rotating unit comprises of motor that rotates the electrode during machining and is primarily required in high aspect ratio micro hole drilling. The RC single pulse power supply in micro-EDM ensures minimal erosion in single discharge allowing the machining of micro features. The difference in power circuitry between regular and micro-EDM originates from their specific objectives where in regular EDM, the sole aim is to increase the spark energy as well as maximize the erosion and in micro-EDM the aim is to minimize the spark energy as well as erosion and enhance the erosion efficiency.

7.4.4 Comparisons of EDM and micro-EDM

Due to the different application criteria and sole objective of EDM and micro-EDM process, the parameter settings and machining condition varies in both macro as well as micromachining techniques. Table 7.2 reveals a brief comparison between different values of process parameters as well as process characteristics for EDM and micro-EDM.

7.4.4.1 Types of micro-EDM

Micro-EDM can be classified into five categories:

(i) Die-sinking micro-EDM: Here, the micro feature on the tool tip is generated on the work surface.

(ii) Micro wire EDM: In this case, a wire of diameter as low as 20 μm is used as a tool instead of rod for cutting different shapes on conductive workpiece.

Table 7.2 Comparison between EDM and micro-EDM process

Process characteristics	EDM	Micro-EDM
Power supply	Pulse generator	RC single pulse discharge/ advanced pulse generator
Power condition	Constant gap voltage and current	Single capacitive discharge, voltage and current is not constant
Dielectric	Transformer oil, paraffin oil, kerosene, lubricating oil, deionized water	Deionized water preferable for avoiding carbon layer formation
Discharge energy	0.1–4J	10^{-6}–10^{-7}J
Discharge gap	10–125 μm	1–5 μm
Current	Above 10 amp	Up to 5 amp
Voltage	40–400 V	20–180 V
Pulse on time	Constant which is user defined	Not constant
Pulse frequency	2–200 kHz	100–500 kHz
Recast layer thickness	1–8 μm	Below 1 μm
Volumetric MRR	0.3–30 mm^3/min	0.000125–0.02 mm^3/min
Feature accuracy	±25–127 μm	±1 μm
Feature width/dia	200 μm (typical)	20–40 μm
Aspect ratio	Up to 100	Up to 50
Roughness (R_a)	0.8 3.1 μm	0.015 0.8 μm

(iii) Micro-EDM drilling: Here, micro holes are drilled with tool of diameter as low as 5–10 μm.

(iv) Micro-EDM milling: Here 3D cavities are generated on the workpiece by implementing micro tool of diameter as low as 5–10 μm and tool movement strategies like conventional milling.

(v) Wire Electro Discharge Grinding (WEDG): In WEDG, wire of diameter as low as 70 μm grinds and reduces the diameter of cylindrical micro tool to lower micron ranges by implementing electrical discharge.

All the above mentioned micro-EDM processes come with own specific applications as well as different capabilities for microfabrication in terms of complexity in feature geometry, minimum size of feature, aspect ratio and surface roughness. These capabilities and applications have been identified in Table 7.3.

Table 7.3 Application capabilities of micro-EDM process

Types of micro-EDM	Complexity of feature geometry	Minimum size of feature	Aspect ratio	Surface roughness	Applications
Die-sinking micro-EDM	3D	20 μm (approx.)	15–20	0.05–0.3 μm	Embossing of tools, replication of molds, etc.
Micro wire EDM	2½D	30 μm (approx.)	100–125	0.07–0.2 μm	Fabrication of stamping tools and opto-electronic components
Micro-EDM drilling	2D	5 μm	25–50	0.05–0.3 μm	Fabrication of injection nozzles
Micro-EDM milling	3D	20 μm (approx.)	10 (approx.)	0.2–1 μm	Embossing or coining, generation of complex micro features
Wire electro discharge grinding	Axial or rotational symmetrical	3 μm	30	0.015–0.8 μm	Fabrication of steeped electrodes

In situ fabrication of micro-electrode from an electrode thicker than the required one during micro-EDM, micro-EDG process is employed. For the microelectrode fabrication process, three different technique of EDG are generally utilized where sacrificial electrodes are used as the cathode and workpiece as the anode, which will be eroded to form a microtool [34, 35]. Three techniques of EDG for fabrication of microtools during micro-EDM are as follows:

(i) Stationary block
(ii) Rotating disk
(iii) Guided running wire

Three different types of EDG with sacrificial electrodes are shown in Fig. 7.6. Fig. 7.6(a) shows a stationary block, which is the simplest EDG method to machine microtool electrode. Microsparking takes place between the sacrificial block and the cylindrical workpiece. The sacrificial block is kept stationary and the cylindrical workpiece rotates about its axis and moves upward and downward. It maintains a small gap between the two electrodes in a dielectric medium. Due to sparking, the material erodes from the cylindrical workpiece. The diameter of the cylindrical workpiece electrode is reduced uniformly. The finishing of the microelectrode is comparatively good. Fabricated microtool may form a taper as the lower portion of the cylindrical electrode undergoes greater wear due to longer interaction with the sacrificial block during EDG.

EDG utilizing rotating disk electrode is shown in Fig. 7.6(b). The disk electrode rotates about its axis, and cylindrical workpiece electrode moves up and down as well as rotates maintaining a small gap in between them. During EDG, disk electrode acts as sacrificial electrode, that is, the cathode and cylindrical electrode as the anode. This technique can fabricate microelectrodes with good surface finish as well as uniform diameter.

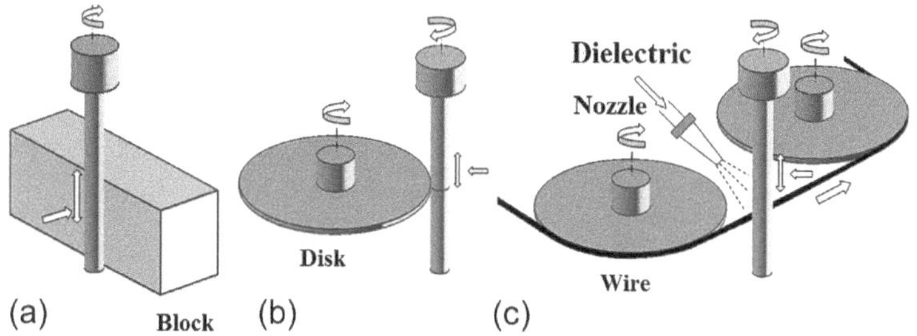

Fig. 7.6 Schematic diagram showing fabrication of micro tool by (a) Stationary sacrificial block, (b) Rotating sacrificial disk and (c) Guided running wire.

EDG with a guided running wire as a sacrificial electrode is shown in Fig. 7.6(c) and is also known as wire electro-discharge grinding (WEDG). During the microtool fabrication process, the cylindrical electrode rotates to move the wire and acts a wire guide and provides fresh wire for interaction during sparking. Dielectric fluid is injected into the machining zone through nozzles in between the running wire and cylindrical electrode. The spark gap is maintained by the movement of the spindle, which holds the cylindrical electrode. Once the cylindrical workpiece reaches one end of its stroke movement, it moves further toward the electrode to a given depth of cut, and the process continues. The surface finish of the fabricated microelectrode is not so good enough which requires effective selection of process parameter settings.

7.4.5 Influences of process parameters

Process parameters of EDM such as voltage, current, duty ratio, types of dielectric, flushing pressure, spark gap, etc. have to be controlled to achieve different machining requirements such as MRR, surface quality and profile accuracy, etc. However, some other related factors as well as few parameters are to be considered most influencing during micro-EDM. Influencing machining parameters and other related factors are exhibited in the fish bone diagram as shown in Fig. 7.7.

Among machining parameters, circuit type, voltage, peak current, and duty ratio are most predominant for micro-EDM. In general straight polarity

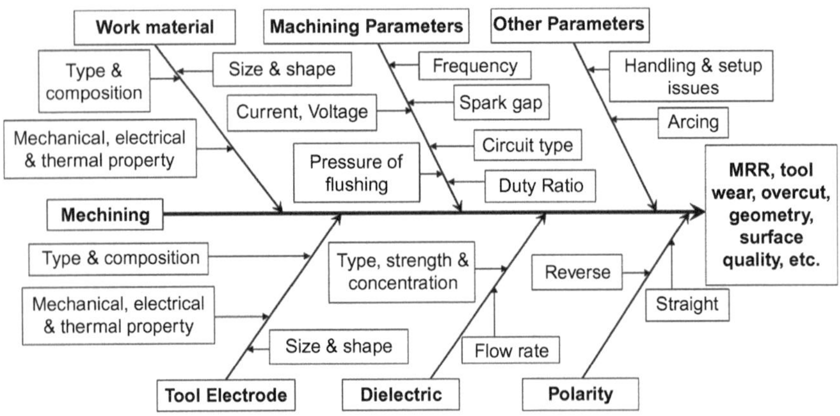

Fig. 7.7 Fishbone diagram showing influence of process parameters and machining condition.

is used in micro-EDM. Type of dielectric and flushing pressure also play important role during micromachining. Shape, size and material of micro tools influence the machining performance.

Material Removal Rate (MRR) increases with increase in voltage. At higher voltage, spark energy increases that result in higher temperatures between the microtool and workpiece leading to higher MRR. In case of RC circuit, the capacitance is more influencing factor and it mainly affects the MRR. MRR decreases with the increase in spark gap as because, when the distance between electrodes increases, the discharge energy concentration nearer to workpiece becomes less.

Peak current is most important parameters which controls the machining rate and accuracy. However, higher peak currents have adverse impacts on the surface finish and tool wear. On time influences the machining rate. During on time, material removal occurs due to sparking and during off time flushing of debris takes place. Hence, for machining rate and accuracy, both on and off time play important role which can be controlled by setting appropriate duty ratio.

Different types of pulse waveform such as rectangles, square and trapezoidal, etc. can be introduced utilizing controlled pulse generator circuit during micro-EDM. By properly selecting pulse shape, micromachining can be controlled which results in very low value of relative tool wear.

To prevent arcing between the microtool and the workpiece, new strategy like periodic reversal of the polarities can be an effective measure [36]. Tool Wear and MRR are affected by the type of dielectric and the method of its flushing [37]. However, deionized water as dielectric is preferred during micro-EDM. Dielectric mixed with fine abrasive powder, metal powder, etc. significantly affects the performance of the micro-EDM process [38].

7.4.6 Advancement and challenges

As an efficient process of fabricating microfeatures, micro-EDM consists of simple setup with added flexibility which provides means for further modifications as per end requirement. This underlying flexibility has lead researchers to modify setup with two drivers, one for stepper motor controlled coarser movement of workpiece with lower response and other with piezo electric capabilities for finer movement as well as higher response for tool electrode to introduce precise single electrical discharge [39]. In another development, two different local actuator modules have been introduced, one for holding and positioning of electrode as well as better control of thrust and consists of two electromagnetic actuators whereas, other module comprises of eight electromagnetic actuators for minimizing tool wear as well as accurate, smoother and stable machining in radial direction [40]. The flexibility and feasibility of micro-EDM to be combined with other material removal techniques like micro-USM and punching have resulted in generation of different shapes and micro holes in borosilicate glass required for MEMS [41, 42] where, other research work involve implementation of variable capacitance to reduce the tool wear. Moreover, the sequential micro-EDM and electrochemical polishing has been proved to be advantageous for achieving better surface quality and precise geometry [43]. The potential to machine microfeatures in micro-EDM by rotating tool has also resulted in achieving better process stability, enhanced process efficiency, higher material removal rate as well as increment of flushing efficiency and improvement of surface quality. Additionally, micro-EDM can aid present researchers as well as fabricators in enlarging the implementation field by understanding the fundamental science and applicability when machining various stronger carbide materials.

On the other hand, with the increase in need for miniaturization, several newer approaches in process design are required. In this context, the use of regular EDM process and machining setup that have been modified accordingly to meet micro manufacturing requirements, in absence of any specifically designed micromachining system and setup have given rise to several problems. These problems needs to be addressed as early as possible and requires further research. For example, more advanced and newer versions of pulse generator is needed to develop that generates much lower discharge energy for enhancing the process capability in machining sub micro features. The automated micro EDM milling employed in fabricating micro dies and molds needs more controlled and optimized process parameters for ensuring generation of required features as per end use criteria. Other research challenges involve higher machining time associated with production of tungsten carbide micro structures which is detrimental for mass scale production, improper circulation of dielectric medium when the spark gap is in the order of few microns followed by inappropriate removal of machining products that is detrimental for high aspect ratio machining, etc. Tool wear is another drawback in micro-EDM.

Also, during fabricating very high aspect ratio and complex 3D micro features as well as micro molds in die sinking micro-EDM, tool cannot be rotated, that results in bad circulation as well as coagulation and sticking of machining products in the spark gap. The generation of poor surface is another problem that increases with pulse on time current and machining current [44]. The thermal effect associated with micro-EDM results in deterioration of surface integrity by forming thermal stress and surface as well as sub surface cracking. Due to the formation of resolidified layer, the microstructure as well as mechanical property changes whereas hydrogen as well as carbon diffusion increases in the periphery of the machining zone. Also, residues like crater and recast layer on the work surface needs additional operations for removal that points toward stringent and precise control of operating parameters as well as further investigation [45] shows different problematic areas in micro-EDM process Figure 7.8.

7.4.7 Applications

As one the prime machining methods, micro-EDM is being extensively used for fabricating micro slots, micro holes of high aspect ratio, inject printer nozzle, precision blanking dies, complex shaped micro gear wheels, minuscule equipments related to neurosurgery such as micro forceps of nickel and titanium, several components for aerospace applications, etc. Moreover, different research works include fabrication of microelectrode by block micro EDM process [47], combination of LIGA and micro EDM to fabricate gear patterned WC-Co micro structures of high aspect ratio [48], fabrication of micro turbine impeller and micro compressor [49], generation of array of micro features of high aspect ratio with reverse micro-EDM [50], fabrication of micro ball joint on bio compatible magnesium alloy [51], generation of micro electrode by using single electrical discharge [52] and fabrication of 2.8 μm diameter micro electrode of cemented tungsten carbide made of super fine particles (SWC) using straight polarity [53], etc. All these applications corresponds to ready and diverse use of micro-EDM for generating precise, intricate and desired micro features on different harder and conducting materials in modern high precision manufacturing industries show SWC optical micrograph of 2.8 μm diameter SWC microtool and tungsten microtool achieved with single electrical discharge Figures 7.9 and 7.10.

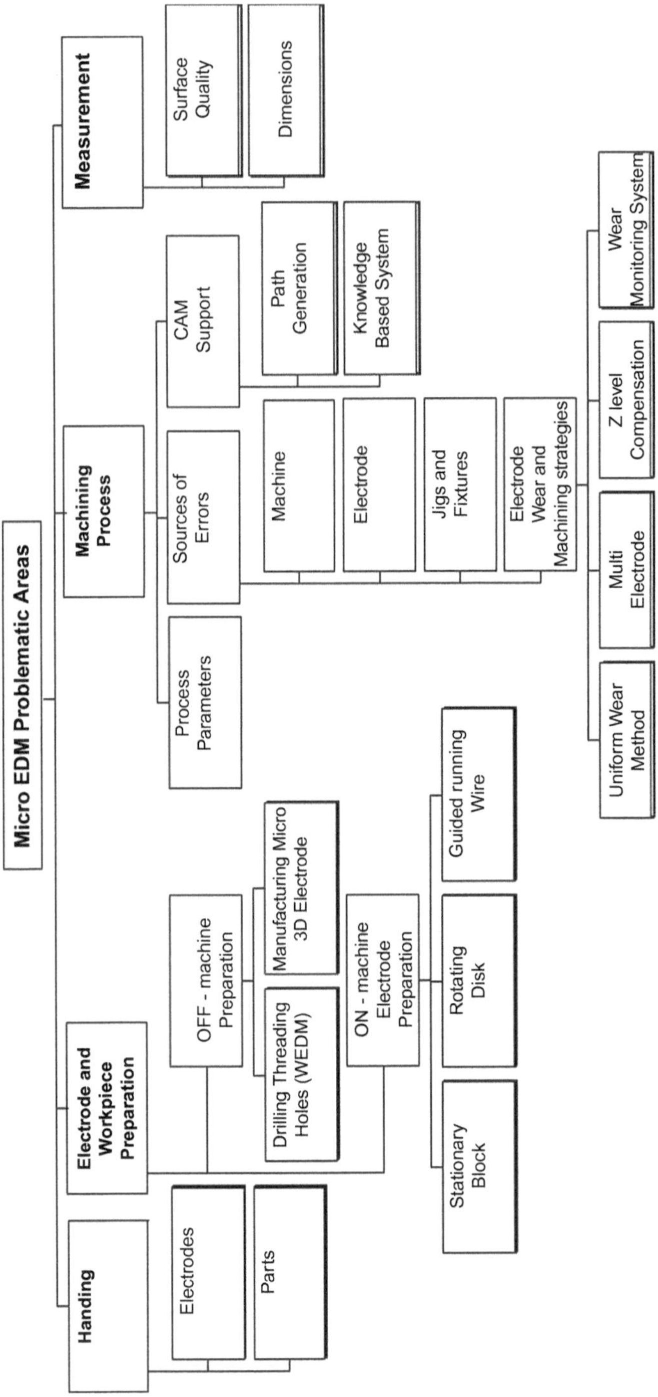

Fig. 7.8 Different problematic areas in micro-EDM [46].

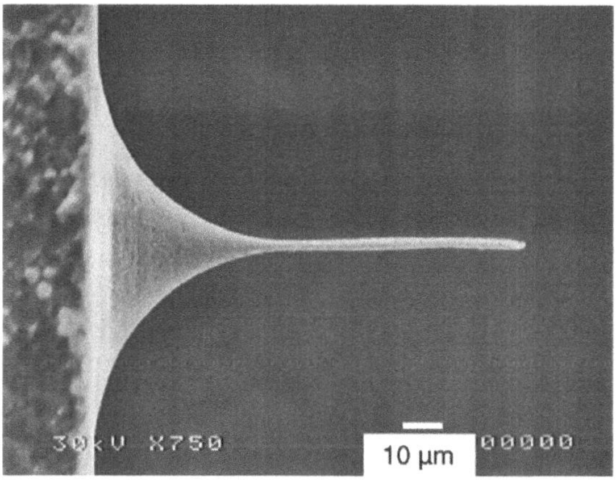

Fig. 7.9 SWC microtool of 2.8 µm diameter achieved with straight polarity [53].

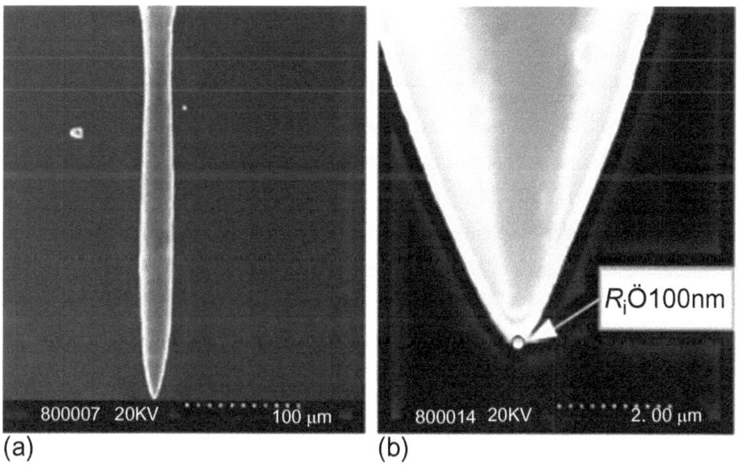

Fig. 7.10 Optical micrographs of 30 µm dia tungsten microelectrode and tip obtained with single electrical discharge [52]. (a) microelectrode and (b) tip of microelectrode.

7.5 Micro laser beam machining (micro-LBM)

7.5.1 Introduction

Micro-laser beam machining (LBM) is one of the most popular, versatile, contact as well as wear less methods of removing material. Machining with laser irradiation offers superior flexibility, possibility as well as automation of high degree that makes this process readily applicable on both electrically conducting and non-conducting materials like different ceramics, metal as well as polymer. Moreover, today's semiconductor industries also rely on laser micromachining for fabricating solar cells. The flexibility to change pulse period from micro to femto second, change in wavelength as well as frequency have enabled laser micromachining to fabricate micro parts with higher aspect ratio. Also, other than implementation in machining techniques like drilling, cutting, surface texturing, laser is also used for welding purposes. However, quiet obviously, the detrimental thermal effects like change in microstructure of parent material and deterioration of surface quality can be reduced to some extent during micromachining by careful control of different associating parameter settings as well as implementation of shorter pulses.

7.5.2 Basic mechanism

Upon focusing the laser on the work surface, the workpiece absorbs that energy and heats up. The light pairs with the free electrons depending upon the transfer of photonic energy to electron vibrational energy. This results in excitation of electrons and their subsequent collision with surrounding atoms. The atoms knock off from the vicinity and the process is repeated for a number of times that leads to high heat generation. This high heat ultimately melts, vaporize and thermally ablate the workpiece material if it is a strong absorber. The penetration depth of laser rays is highly dependent on absorptivity of workpiece material. Higher absorptivity results in higher penetration depth where, the absorptivity again depends upon the optical property of the workpiece material. In case of material with low absorptivity, the pairing of incident laser rays and electron of workpiece becomes very low and the temperature cannot rise to that extent when machining can occur. On the other hand, with increase in power density, the vaporization rate of workpiece material increases. This high vaporization rate causes a shockwave and ejects the molten material after each pulse.

The mechanism of material removal during laser micromachining can be achieved by three ways:

(i) Direct writing technique
(ii) Mask projection technique
(iii) Interference technique

Direct writing technique—In direct writing, the laser is focused on the substrate and desired micro features is generated by moving the workpiece with respect to fixed laser beam or by scanning the laser beam by keeping the workpiece fixed. The additional scanning mirror is controlled by galvanometer and helps in continuous deflection of beam over the required pattern. Moreover, the use of flat field lens offers suitable focal conditions at different deflection conditions. The process of direct writing in laser micromachining is very simple, though the process needs scanning of entire work surface for micromachining. The main parameters that influence the direct writing technique involves size of focal point, working distance and the depth of focus. The added automation in movement by incorporating CNC controlled micro positioner can help further. The process is mainly used for drilling of micro holes, cutting micro features as well as scribing materials.

Mask projection technique—In mask projection, a mask is patterned with the desired shape of the workpiece and is illuminated by laser which is shaped and homogenized for uniform intensity distribution across the uncovered region of mask. The pattern is then projected on the worksurface using projection lens. Generally the masks are made of quartz covered by a thin chrome layer. The mask projection technique offers fabrication of widely spread microfeatures by using single or multiple shots of laser beam. The feature resolution is highly dependent on the projection system and the size of mask used. During fabrication of micro holes via mask projection technique, the mask remains stationary whereas, fabrication of micro channels or micro grooves can be achieved by using multiple laser shots, stationary mask but moving substrate in desired direction. This method of laser micromachining is ideal for excimer laser.

]Interference technique—In interference technique, laser beam is split and super-positioned with beam splitter for producing interference patterns. These interference patterns have unique intensity variation that enables periodic micromachining over work surface. The periodicity of beam interference is dependent on period, wavelength and angle between beams. In this technique, the size of generated micro feature ranges from 50 to 300 nm.

Material removal by thermal ablation is very useful method for machining metal as well as ceramics where, smaller thermal diffusivity and higher value of absorption coefficient is required to maintain for achieving better ablation efficiency. Laser micromachining can be carried out in two ways; one with ultra fast or ultra short pulse and other with long pulse.

During machining with long pulse laser, the time needed for diffusing the heat is smaller than pulse period. As a result, heat can be easily diffused in the surrounding that reduces the machining efficiency. Moreover, during operation and due to boiling, the molten material splashes in the vicinity of machining zone forming droplets. These droplets gradually cool down in the form of machining debris and require addition techniques for removal. Additionally, high heat associated with machining results in the formation of heat affected zone and due to subsequent heating and cooling, thermal stress generates that leads to surface as well as subsurface cracking.

In contrast, implementation of short pulses in nano, pico, and femto seconds results in enhanced machining efficiency due to better ablation. Also, use of such shorter pulse period entraps the most of laser energy which results in faster transition from melting to evaporation. As a result, laser with ultra short pulses can effectively and efficiently melt even various refractory materials like molybdenum, rhenium, etc. Additionally, no splashing of molten material occurs, there is no machining debris and heat effected zone and other adverse thermal effects are minimal. Other associating parameters like reflectivity of work material, ablation threshold, wavelength and pulse frequency affects the ablation during machining.

7.5.3 Setup details

The laser micromachining system consists of a laser generator, CNC stage, video monitoring system, etc. The laser generator produces ultra short pulse laser where, a programmable galvanometer scanner is used to guide the laser beam with high velocity and appropriately on the desired area of the workpiece. A galvo shutter is used to switch the laser and an expander is used to increase the laser beam diameter. The aperture shapes the laser beam where, quarter wave plate ensures circular polarization to poses similar absorption properties throughout the machining area. Attenuation of pulse energy can be done by rotating half wave plate and then a linear polarizer. The laser is focused by a focusing lens where an aperture is attached before it to eliminate low intensity laser beam. The computer controlled stages position the workpiece under focal point of laser beam and controls the necessary vertical position as well as alignment to maintain focus. The controller is also responsible for carrying out operations like pulsating and halting of laser beam as per requirement. It also controls the pulse frequency appropriate to the movement speed of workpiece and maintains similar spaced pulses when the velocity of workpiece changes. The video monitoring system helps in monitoring the machining area during operations whereas, the power monitor helps in monitoring as well as adjusting the power parameters of the conditioned laser beam. A schematic diagram of laser micromachining system has been shown in Fig. 7.11.

7.5.4 Comparisons of LBM and micro-LBM

Though LBM itself can be used for generating features that have dimensions nearer to that in micro ranges; still some differences in process parameter setting arises when the main objective is the applicability of this process

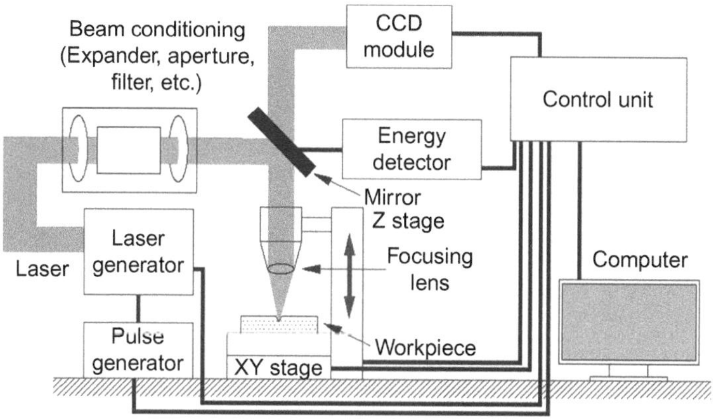

Fig. 7.11 Schematic diagram of laser micromachining system.

in micro/nano domain only. Table 7.4 shows such comparisons between process parameters when LBM is used for micromachining rather than its traditional uses.

7.5.5 Influence of process parameters

Process parameters of LBM such as Laser type, pulse frequency, wavelength, pulse duration, fluence, spot diameter, cutting speed and factors related to assisted gas, etc. have to be controlled to achieve different machining requirements such as machining rate, surface quality and profile accuracy, etc. However, for micromachining, some factors as well as few parameters are to be considered most influencing. The most dominant process parameters for Laser micromachining are types of Laser, its pulse frequency, wavelength, pulse duration, spot diameter and fluence, etc. Figure 7.12 exhibits influencing machining parameters and other related factors through a fish bone diagram.

Various types of lasers are used for micromachining purposes. These include microsecond CO_2 lasers having wavelengths ranging from 9 to 11 μm, nanosecond and femtosecond solid-state lasers such as Nd: YAG and Ti: Sapphire with wavelengths ranging from 1030 nm to 1064 nm having possibility of higher harmonic generation in visible (515–535 nm) as well as ultraviolet (UV) spectrum (342–355 and 257–266 nm) [54].

However, ultrashort pulse Laser is most efficient for micromachining that generates very high peak intensity, i.e., $>10^{15}\,Wcm^{-2}$ can deliver energy before thermal diffusion occurs in the parent material during machining. Hence, without significant thermal degradation, i.e., melting,

Table 7.4 Comparison between process characteristics of regular LBM and LBM used for micromachining applications

Process characteristics	LBM	Micro-LBM
Type of laser	CO_2, Nd: YAG, fiber, diode laser, etc.	CO_2, Nd: YAG, fiber, excimer, Ti-sapphire, copper vapor laser, etc.
Beam quality number	Lower, 10–50	Higher, 1
Influence during machining	Higher thermal effect forming recast layer of thickness 0.002 to 0.1 mm	Very lower thermal effect and more apparent optical distortion
Wavelength	Broader, in μm	Shorter, in nm
Power density	Lower, in the range of 10^{11} W/cm^2	Higher, maybe up to 10^{15} W/cm^2
Pulse duration	Higher	Lower, in tens of picosecond
Pulse frequency	In Hz	In kHz and MHz
Spot size	Bigger, in μm and above	Smaller, in nm and μm
Ablation rate	Higher	Lower ablation rates such as 0.05 μg per pulse
Feature size	0.127 to 1.27 mm (typical)	In nm, 30 (typical)
Aspect ratio	Lower	Comparatively high
HAZ	0.002 to 0.1 mm	< 0.01 mm
Surface finish	Poor	Better, in nm

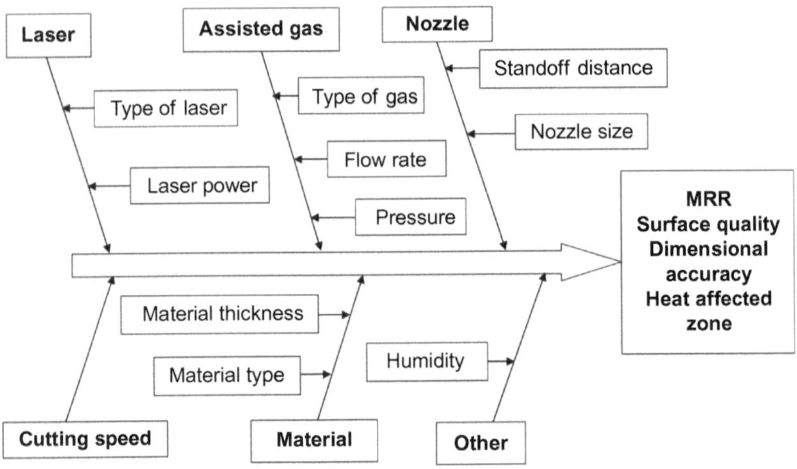

Fig. 7.12 Fishbone diagram showing influence of process parameters and machining condition on quality of machining.

spatter, recrystallization, etc. occurring to the machining zone, these Lasers can achieve higher efficiency and precision. As a consequence, the ablation leaves no droplets condensing onto the surrounding material and results in removal of material with extremely precise edges [55]. Lasers which are also most effective for micromachining include copper vapor lasers, diode lasers, and excimer lasers emitting at UV spectrum (157–353 nm) [54].

The influence of laser fluence on the machining depth and feature dimension during micromachining is predominant. Due to plasma absorption of laser energy, depth increases rapidly when the surface temperature increases slightly. The peak surface temperature does not occur at the center of laser spot during micromachining. The advantages of using nanosecond laser pulse having Gaussian beam profile during fabrication of microgrooves on biomedical materials such as Ti-6Al-4V has also been demonstrated [56].

Number of shots and total energy also affect the microhole diameter. Multiple pulses with low peak fluence is preferable to generate small hole compared to single pulse possessing same total energy [57]. Kerf width increases with the increase of laser power, pulse width and pulse frequency [58]. However, kerf width decreases with the increase in cutting speed. Moreover, small craters with uniform line width can be obtained at high frequency and lower pulse energy. Hence, the pulse frequency plays major role in the formation of HAZ during micromachining.

For high peak power, femto second laser can be used for various other microfabrication applications like surface micro-structuring, material modification, etc. Transparent materials can also be machined effectively by femto second laser for its extremely short pulse width that generates high density of electron which can induce high absorption. Moreover, the femto second laser is suitable for fabricating microholes with special geometries as well as superior quality and high reproducibility [59].

In order to enhance the machining speed, the use of double pulse technique is introduced [60]. However, the surface roughness decreases drastically with the increase of number of pulses. Focus position also plays an important role during micromachining. When the focus position is varied from positive to negative, the hole diameter at the entrance gradually decreases to minimum value and then it begins to increases. On the other hand, as opposed to this trend, the hole diameter at the exit decreases gradually [61].

Air and oxygen can be used as assist gas during laser micromachining. It has been is observed that for the same laser fluence the use of assist gas jet generally reduces the ablation rates but increases the feature dimensions compared to machining without assisted gas. Minimum recast layer formation has been observed using air as assist gas due to high pressure gradient at the edge of the microprofile. The assist of oxygen during machining decreases ablation rate due to plasma shielding. The supersonic gas jet assisted laser micromachining results in enhanced feature quality by removing the recast layer; however, it reduces the ablation rate [62].

With decrease in laser pulse duration from nano to femto second, thermal effects become less apparent while optical distortion effects become more apparent. On the one hand, the melting and formation of recast layer will decrease with pulse duration but cannot be eliminated even when the pulse duration is reduced to femto second regime. Recent research works recommend the use of 1 to10 ps pulse duration for achieving high micromachining precision [63].

7.5.6 Advantages and challenges

Laser micromachining is a versatile process and is used widely for machining plastic, glass, metal as well as preparing thin foils. The process comprises different mechanisms like cutting, drilling, marking, turning, threading, etc. The recent advances in laser micromachining includes its utilization for controlled fracture in fine items, enhancement in drilling high aspect ratio micro holes on the basis of helical drilling where, material is removed by helical passes by dividing the total machining length into several single passes, trepanning where laser spot moving in circular fashion and polarization control by changing linear polarization to circular is required [55, 64]. Also, polarization control can aid the helical drilling to fabricate high quality, precise and high aspect ratio micro holes. Laser cutting also uses trepanning by traveling the laser in predefined path during machining big sized holes on different ceramic materials. CNC is applied for fabricating non-linear holes whereas; laser marking is utilized to remove materials with small quantities having tiny machining depths to form patterns.

Nonetheless, laser micromachining utilizing ultra short pulses poses various possibilities for carrying out further research activities. Besides, the use of ultra short pulses in excimer, copper vapor and diode pumped solid state laser, scopes are also there in investigation of change in surface morphology as well as proper fabrication of precise 3D micro features as per requirements, surface structuring to be used in micromechanics and the field of microfluidics, implementation in different other emerging field like biomedical and fabrication of photovoltaic cells for semiconductor industries, etc. [65].

As laser micromachining is thermally induced process, it comes with different problems and research challenges. Formation of heat affected zone, alteration of material properties and generation of thermal stress are the foremost problems. Others include formation of machining debris from solidified molten material, generation of recast layer that requires involvement of additional methodologies for removal, etc. Though ultra short pulse laser can be utilized to prevent pile up of machining debris and generation of heat effected zone, it cannot completely protect against the formation of recast layer. Additionally, implementation of ultra fast laser is not at all economical and leads to optical problems. Moreover, the high intensity associated with ultra fast laser results in strong deformation of laser in focal plane that can deteriorate machining accuracy [66]. All these, various drawbacks and problems need extensive research works for minimization or eradication. Research activities are also required in different areas like underwater laser machining for exploring the process feasibility and uncovering its beneficial outcomes like absence of any detrimental thermal effects, etc. However, problems in machining different reflective materials like aluminum and copper still exists. Other research issues involve engraving larger areas, reducing the higher machining time associated with short pulsed laser, minimization of machine-machine variation as well as reducing the amount the fumes and odors produced, investigation of ablation characterization of different composite materials, etc.

7.5.7 Applications

Micro-LBM is extensively used in micro texturing of hard disk drives, video heads and diamond films, preparation of silicon wafers, machining of very thin glass sheet, microfeature generation on different optical materials such as gallium arsenide, lithium niobate, lithium tantalite, etc., micromachining of fluoropolymers for miniature lab-on-chip technologies, drilling of micro holes on polymers for circuit boards and chip packages, patterning, scribing, dicing of produced optical wafers and substrate, polishing of CVD diamond to be used for various thermal applications as well as infrared detectors and sensors, fabrication of ink jet nozzles, etc. [65, 67, 68]. Other biomedical applications involve manufacturing of stents and micro holes for patient monitoring probes, fabrication of different micro features in MEMS, rapid prototyping of micro fluidic channels combined with optical wave guides, different textured surfaces to adhere cells, etc. [65, 69]. Other various applications consist of micro manufacturing of different polymeric structures to

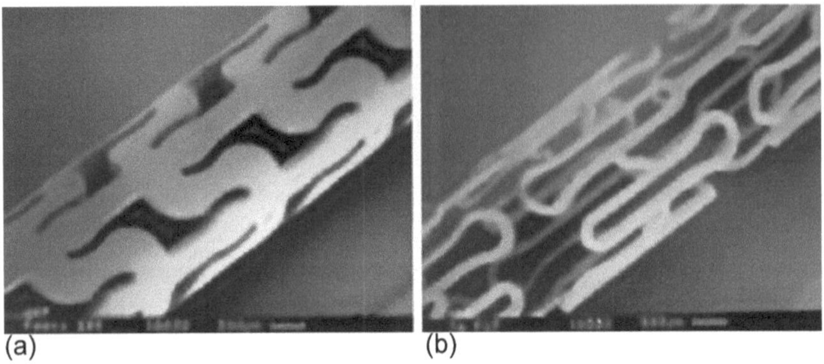

Fig. 7.13 Micrograph of prototypes of stents made on (A) bio-resorbable polymer and (B) tantalum [65].

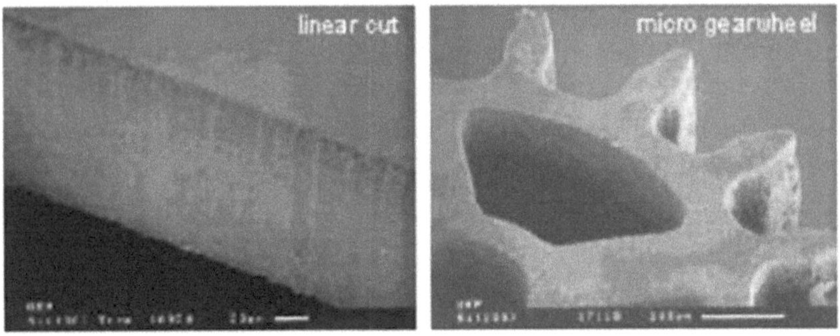

Fig. 7.14 Micrograph of machined silicon samples [65].

transfer patterns on metallic substrate, repairing of circuit boards, patterning on circuit boards, organic, inorganic and biological materials, fabrication of bio sensors, micro manufacturing of OLED display and thin film transistors, mesh drilling in metallic foils, trimming of various components, underwater ablation on materials like silicon, percussion drilling, micro manufacturing of intravascular rotor micro turbine on nickel, fabrications of mold inserts required in photo molding, etc. [65, 70–74]. Figures 7.13 and 7.14 show prototypes of stents on bio-resorbable polymer and tantalum as well as machined silicon samples.

7.6 Electrochemical micromachining (EMM)

7.6.1 Introduction

Electrochemical micromachining is another popular process of fabricating micro features which is fulfilling the growing needs of modern fabrication industries. The potential to machine different chemically resistive and stronger materials like titanium, nickel and their alloys, other super alloys, stainless steel has led electrochemical micromachining highly applicable in the various emerging fields like aerospace, automotive, medical, optics, electronics, etc. The high accuracy in the order of $\pm 1\,\mu m$ on $50\,\mu m$ associated with EMM allows it to fabricate precise products with desired geometries without undergoing any thermal effects, tool wear and burr formation [75]. EMM, as a downscaled version of regular ECM, removes surplus material by anodic dissolution where unit metal removal not larger than the size of ions is quiet achievable that results in generation of delicate, precise and intricate micro components as per requirements.

7.6.2 Basic mechanism

The basic mechanism of EMM is similar to that of ECM with few changes in some aspects. During EMM, the cathode microtool as well as anode workpiece remain submerged in the suitable electrolyte solution with a narrow inter-electrode gap (IEG) smaller than that of ECM and falls in the range of 5–20 μm. IEG governs the resolution of feature to be produced as well as governs the localization of anodic dissolution that enhances with the decrease in mentioned gap. Mainly two types of electrolytes are used during EMM; one is passivating electrolyte like as sodium chlorate, sodium nitrate [76], etc. that has sufficient oxidizing anions and results in enhanced accuracy whereas; the other electrolyte is of non passivating type having aggressive anions like sodium chloride. Additionally, the employment of different acid electrolytes such as hydrochloric acid, sulfuric acid, etc. is advantageous as no insoluble reaction debris is generated during machining. Apart from regular ECM, EMM introduces the use of stagnant electrolyte as because any kind of movement or any kind of flow can vibrate the tiny micro tool and deteriorate the stability as well as machining accuracy of the end product. During operation, the instantaneous current density reaches as high as $75–100\,A/cm^2$ resulting in high mass transport when the dissolution is confined and localized within sub micron ranges, IEG is shortened in the mentioned range and ultra short pulse DC instead of regular DC with a voltage range of 1–10 V is applied. Implementation of pulse DC primarily helps with improved flushing of electrolyte as well as efficient removal of electrolysis products that is crucial and must for EMM process. Tiny hydrogen bubbles as well as reaction debris leaves the machining gap during pulse off time where, fresh electrolytes fill up the machining gap for ensuring steady and stable machining [77]. With the primarily set IEG, the micro tool starts feeding toward or forward within the workpiece leaving a mirror image of the tool itself by dissolving the workpiece material electrochemically. The continuation of this anodic dissolution lastly results in formation of desired micro shape and geometries as per requirements.

During machining, to maintain uniform current flow between tool and workpiece and to achieve uniform material removal, the feed of the micro-tool remains constant with the rate of material removal where higher feed values than that can decrease the machining gap and hinder the total process by disturbing the flushing efficiency, forming frequent short circuits and sparks as well as deteriorating machining stability. However, still there are chances of coagulation of reaction products in the tiny machining gap even when the process parameters are adequately chosen and maintained. To encounter that problem, rotary micro tool instead of stationary tool can be used. Implementation of piezoelectric transducer (PZT) [78] either in micro tool or workpiece also helps in achieving better flushing by generating small pressure waves that can clean the machining gap and renew fresh electrolytes. There are other different parameters that seek the attention of researchers such as, pulse period, applied frequency, electrolyte concentration, electrolyte temperature, PZT vibration amplitude and frequency, etc. [79].

Different anodic reactions take place when adequate pulse power is applied depending on the operating conditions and metal-electrolyte combinations. Rate of anodic reactions is also influenced by the supply of fresh electrolyte, which facilitates the removal of reaction products as soon as they generates into the narrow machining zone. The electrolyte flow velocity is negligible in case of EMM. So there is not adequate transfer of mass from one electrode to the other. This gives rise to the formation of diffusion layer at the electrode-electrolyte interface at anode workpiece.

Hence, in EMM, all types of resistance factors within the narrow machining gap must be considered. An electrical equivalent circuit model of EMM can be proposed which has been already developed incorporating all factors [80]. The applied potential has to overcome various types of potential drops in addition to electrolyte resistance such as double-layer capacitance, Warburg impedance, and charge transfer resistance, etc. At the electrode-electrolyte interface the metal goes to the electrolyte as ions, but cannot cross the energy barrier till the applied voltage is greater than the activation potential. Charges remain polarized as double-layer capacitor. There is no electrolyte flow in EMM operation and hence there is not sufficient transfer of mass from one electrode to the other electrode. This gives rise to the diffusion component called Warburg impedance. Hence, unlike conventional ECM used for macro range of machining, for finding out the MRR in case of EMM, the following resistances and impedances factors have to be considered:

(i) Double-layer capacitance
(ii) Warburg impedance
(iii) Charge transfer resistance
(iv) Electrolyte resistance

7.6.3 Setup details

A schematic diagram of EMM setup has been shown in Fig. 7.15. The EMM setup consists of various sub components such as, mechanical machining unit, power supply unit, machining chamber, power monitoring system, etc. The mechanical machining unit provides a rigid support to its different subsystems. Mechanical machining unit consists of precise XYZ movement stage with controller which helps in positioning the microtool and moves in the required direction as per desired criteria and microtooling system that is responsible for holding the microtool in desired position. Depending upon the arrangement of XYZ axis, stage for EMM can be of two types; one is column and knee type and another is gantry type [80]. In gantry type structure, the workpiece and machining chamber are mounted on X axis only. The rest two axes that are only responsible for movement remain free and only hold the microtool. For this, X axis of a gantry stage should have high

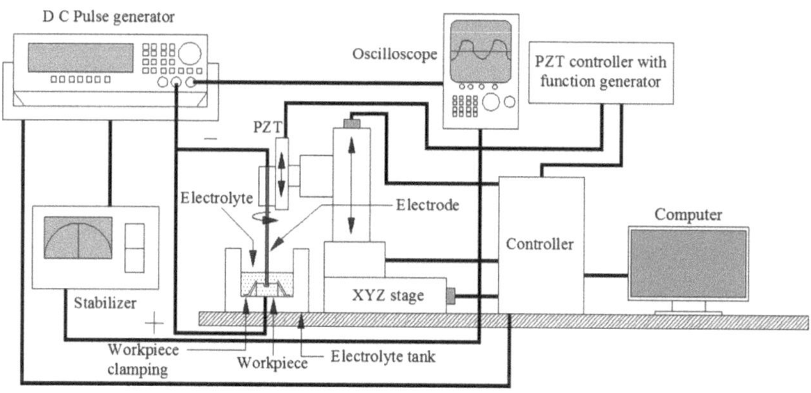

Fig. 7.15 Schematic diagram of EMM process.

load carrying capacity where other axis takes fewer loads and reduces the cost of the stage. In column and knee type, the workpiece and machining chamber are mounted outside the stage where, X axis carries the microtool and weight of other two axes. Hence, all the axes are made with high load bearing capacity. Column and knee type stage provide added advantages and more flexibility in mounting complex microtool holding assembly as this type of stage provides free space in either side of column.

Micro tool rotary unit can be mounted on the Z axis of the stage to improve flushing during operations. Also, the microtooling system has means for attaching PZT that can generate small pressure waves in the machining gap and ensures renewal of fresh electrolyte. The power supply unit consists of DC pulse power supply and function generator to generate and control necessary voltage pulses with required period and pulse on time. The power monitoring system helps in monitoring the pulse power parameters during machining and consists of oscilloscope, multimeter, etc. The machining chamber holds the suitable electrolyte and workpiece is at the desired position. This chamber is generally made of transparent perspex so that the machining operation as well as the movement of tool can be monitored properly. Moreover, accessory such as stereo zoom microscope can be used to visualize the machining operations online.

Types of EMM: EMM can be classified into three preliminary groups based on the material removal mechanism, degree of localization and the implementation of solid tool or metallic wire. These are through mask EMM, maskless EMM and wire ECM. Different types of EMMs are represented in Fig. 7.16.

Through mask EMM—In through mask EMM, a photoresist patterned mask is placed on the work surface excluding the portions where machining is needed to be carried out. The removal of material or dissolution occurs at the bottom of the cavity formed by the application of photoresist mask. Through mask EMM can be of carried out in two ways, one by dissolution of metal from single side and other by removal of material simultaneously from both sides. The steps to be implemented during machining with through mask EMM includes generation of master artwork, preparation of surface, selection of suitable photoresist and imaging. As the removal of work material is isotropic, hence it results in undercut of work material below the photoresist mask. The ratio of workpiece thickness or the amount of through straight cut to the amount of undercut is known etch factor. Etch factor depends upon various factors like type of substrate material, type of electrolyte, choice of adequate current density, aspect ratio, film thickness

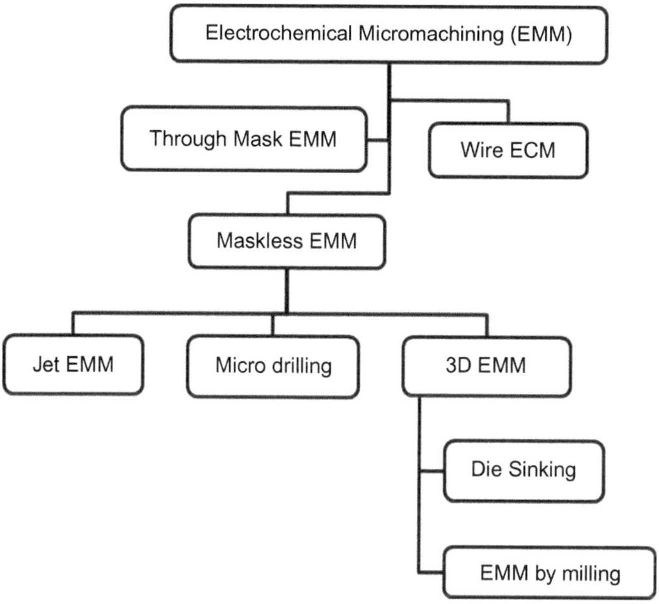

Fig. 7.16 Classification of EMM process.

ratio, spacing to opening ratio, etc. The through mask EMM offers faster material removal, though the process suffers from limitation like generation of micro feature of low aspect ratio resulting from isotropic material removal and formation of undercut. Fig. 7.17 shows schematic diagram of one sided and two sided through mask EMM process.

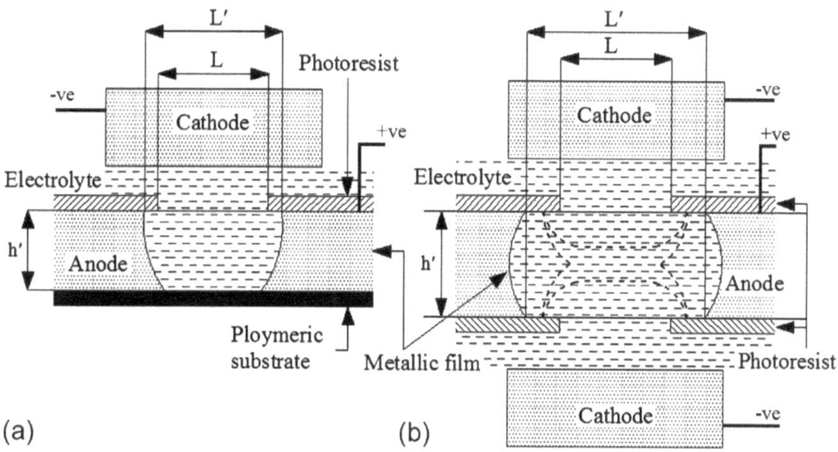

Fig. 7.17 Schematic diagram of through mask EMM (a) one sided and (b) two sided [80].

Maskless EMM—In maskless EMM, no mask is used during machining and dissolution is carried out in very localized and controlled manner on the selected portions of the workpiece. As a result complex and intricate 2D and 3D micro features can be generated by very precise movement of micro tool in XYZ direction. To minimize the effect of stray current and to confine the dissolution in a very small area, IEG remains very small. Here, current density is highest at the closest points between micro tool and workpiece and decreases with the increment of the tool-workpiece distance. This leads to selective and highly localized material removal. Moreover, the use of passivating electrolyte helps in formation of transpassive oxide films and limit the extent of stray machining. In contrast to through mask EMM, maskless EMM can machine microfeatures with higher aspect ratio and does not need additional operations for masking. Maskless EMM can further be divided into three categories depending upon the geometric complexity and depth of machining. These are, Jet EMM, Micro Drilling and 3D EMM.

Jet EMM—In Jet EMM, instead of solid tool electrolyte jet is used as a tool. Here, jet of electrolyte strikes the workpiece through a nozzle of 50–200 μm diameter which is considered as cathode. The distance between the nozzle and workpiece is selected and maintained accordingly during the operation. By moving the workpiece in desired direction, different complex profiles can be generated; however, the process has limitations in fabricating very high aspect ratio micro features. Depending upon the arrangement and direction of electrolyte jet, jet EMM can be of two types; one is horizontal jet EMM and another one is vertical jet EMM. In contrast to horizontal jet EMM, vertical jet EMM can be controlled easily and can fabricate microfeatures with greater accuracy and less overcut.

Microdrilling—Electrochemical microdrilling involves machining of high aspect ratio microholes with desired quality at a faster rate and with lower cost. Electrochemical microdrilling can further be categorized as jet electrolyte drilling, capillary drilling and electrostream drilling. In jet electrolyte drilling, jet of electrolyte from cathode nozzle strikes the anode workpiece and produces high aspect ratio microholes of diameter <1 mm due to the application of very high voltage value up to 500 V and electrolyte with very high conductivity for achieving high current density. Capillary drilling where, the electrolyte which is preferably dilute nitric acid is made to pass through a small glass capillary consisting of fine platinum wire insert at 2–4 mm above the tip to machine micro holes of diameter <500 μm. Electrostream drilling, where glass capillary consisting of platinum wire well above the workpiece is utilized to direct the electrolyte jet and machine micro holes having diameter less than that can be achieved by capillary drilling.

3D EMM—It involves controlled and localized anodic dissolution to form complex 3D microstructures by appropriately eliminating the effects of stray current and enhancement of flushing. Various factors which influence the machining process involves coating of microtool, control of IEG as well as nature of DC power supply used, etc. 3D EMM can be carried out by die sinking and milling. Die sinking EMM involves machining of complex profiles with microtools having complex patterns on the tip. The micro tool with patterned tip can be fabricated by LIGA or FIB process and is directly applied for single run die-sinking operation. The end result is the formation of complex shapes that reflects a mirror image of the microtool. As there is no horizontal movement associated with the microtool, the setup is simple; however, producing patterned microtools requires implementation of additional operations, increases complexity of the process and overall manufacturing time. Moreover, the process is not suitable for machining high aspect ratio microfeatures. EMM by milling involves micromachining of complex shapes by scanning of simple shaped microtool via predesigned tool path which if generated appropriately ensures minimization of stray current effect as well as sparking and enhanced accuracy. EMM by milling can be carried out in two ways, one by simple scanning and other by sinking and milling where the former one offers larger overcut, machining time and end deviation. Fig. 7.18 shows generation of tool path in 3D EMM process.

Wire ECM—Here, very small diameter metallic wires instead of solid microtool is employed in machining microfeatures. During machining, the wire is tensed in appropriate fixture for ensuring its straightness and minimizing the wire vibration. The wire is fed forward into workpiece maintaining a suitable inter-electrode gap for machining the required micro feature. Fig. 7.19 shows a schematic diagram of wire ECM process. Mainly, tungsten is used as wire material due to its superior tensile strength and

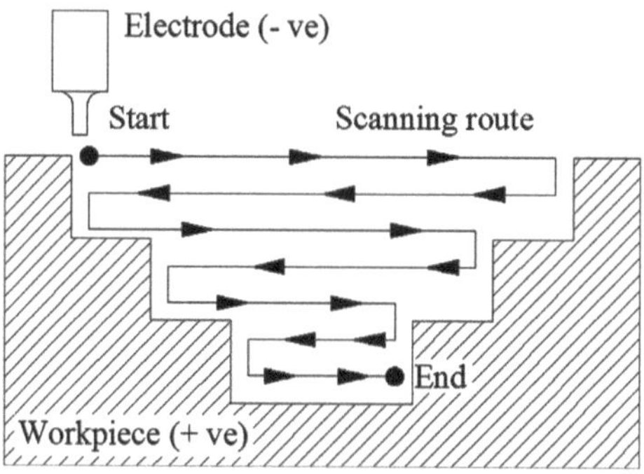

Fig. 7.18 Generation of tool path in 3D EMM [80].

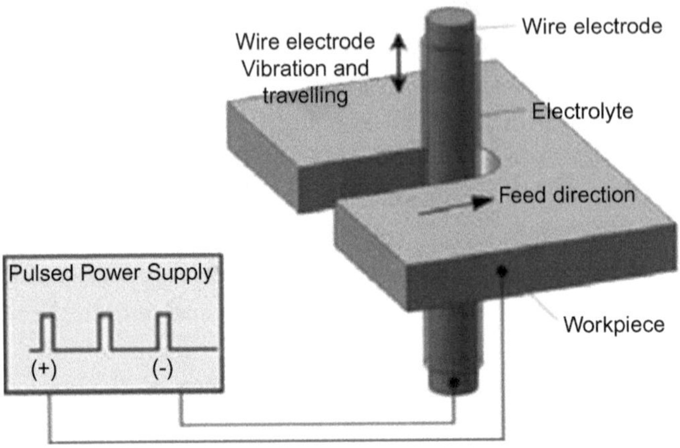

Fig. 7.19 Schematic diagram of wire electrochemical machining [81].

enhanced chemical resistance. As the wire is tensed, hence, very thin wire can not be utilized during machining that also results in generation of thinner slits. Moreover, as the wire traveling direction remains perpendicular to the direction of gravity, hence, it results in improved flushing and removal of sludge, i.e., metal hydroxides. Moreover, this type of EMM specializes in machining high aspect ratio micro features, however, suitable electrolyte flushing strategy is required to implement otherwise the reaction debris may clog the IEG as machining is carried out across the thickness of workpiece in a single run. The methods used for enhancement of mass transport in the machining zone involves axial electrolyte flow, monodirectional and reciprocating traveling of wire with change in wire shape, using wire of low hydrophilic material, implementing vibration on tool wire as well as employing intermittent feed directional vibrations.

Micro tools—In 3D EMM, the shape of the micro tool shank and tip greatly influences the shape of the end product. Type of microtools that are used now includes cylindrical microtool, conical microtool, reverse conical microtool, cut edge microtool, spherical end microtool and disk end microtool which have been shown in Fig. 7.20. Flat end cylindrical microtool can machine microgrooves, micro holes, micro cavities by controlling its movement. On the other hand, flat end and cut edge microtool can machine completely cylindrical microholes when rotary arrangement is attached with it. Implementation of rotational movement results in better flushing and improvement accuracy when compared with the cylindrical one. Conical microtools can be used for machining conical holes, micro nozzles utilized for ink jet printers, etc. Reverse tapered microtool, spherical as well as disk end microtool helps

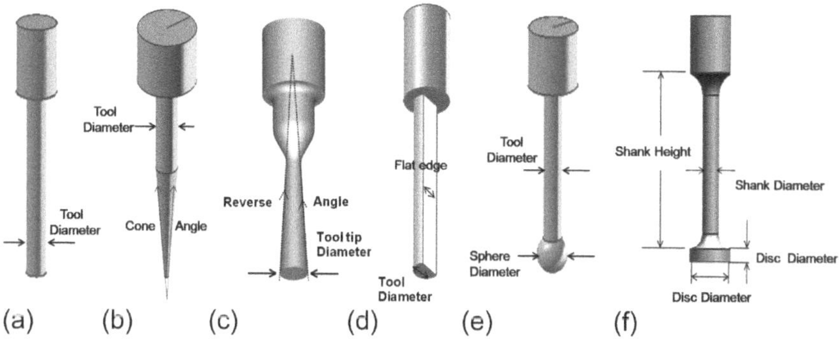

Fig. 7.20 Different microtools used in EMM; (a) cylindrical microtool, (b) conical microtool, (c) reverse conical microtool, (d) cut edge microtool, (e) spherical end microtool and (f) disk end microtool [80].

in reduction of taper during machining cylindrical microholes and microgrooves with straight vertical walls or edges.

7.6.4 Comparisons of ECM and EMM

Process characteristics of EMM are differs notably to that of regular ECM process. The differences between ECM and EMM are listed in Table 7.5 to acquire a deeper understanding of EMM process.

7.6.5 Influence of process parameters

Different process parameters of ECM such as voltage, current, types of electrolyte, concentration, flow rate, IEG, etc. have influences on different machining requirements such as MRR, surface finish and profile accuracy, etc. However, following parameters are to be considered most influencing during EMM and are presented at the fish bone diagram as shown in Fig. 7.21.

Applied voltage is generally kept at below 10 V. MRR increases with the applied voltage. However, at higher voltage Joule heating occurs in the IEG that causes change in electrolyte conductivity. This results in non-uniform current flow in the narrow gap which results in generation of inaccurate profile and deteriorates the surface quality of the machined job. Moreover, at higher voltage large amount of sludge and gas bubbles generate in the narrow machining zone resulting in the occurrence of micro spark which further deteriorates the machining accuracy.

Duty ratio of pulsed power supply influences the machining performance. During pulse on time machining occurs and during pulse off time,

Table 7.5 Comparison process characteristics between ECM and EMM

Process characteristics	ECM	EMM
Operation type	Maskless	Mask/maskless
Applied voltage	10–30 V	1–12 V
Power supply	Continuous/ pulsed	Pulsed
Current density	20–200 A/cm^2	75–100 A/cm^2
Electrolyte type	Natural salt solution	Salt/dilute acid/alkali
Electrolyte concentration	>20 g/L	<20 g/L
Electrolyte flow	10–60 ms	<3 m/s
Electrolyte temperature	24–65 °C	37–50 °C
Inter-electrode gap	100–600 μm	5–50 μm
Tool size	Large or medium	Micro
Machining rate	0.2–10 mm/min	5 μm/min
Accuracy	±0.02–0.1 mm	±0.01 mm
Side gap	>20 μm	<10 μm
Surface finish	Good, 0.1–1.5 μm	Excellent, 0.05–0.4 μm
Problems occurred from waste disposal/toxicity	Low	Low to moderate

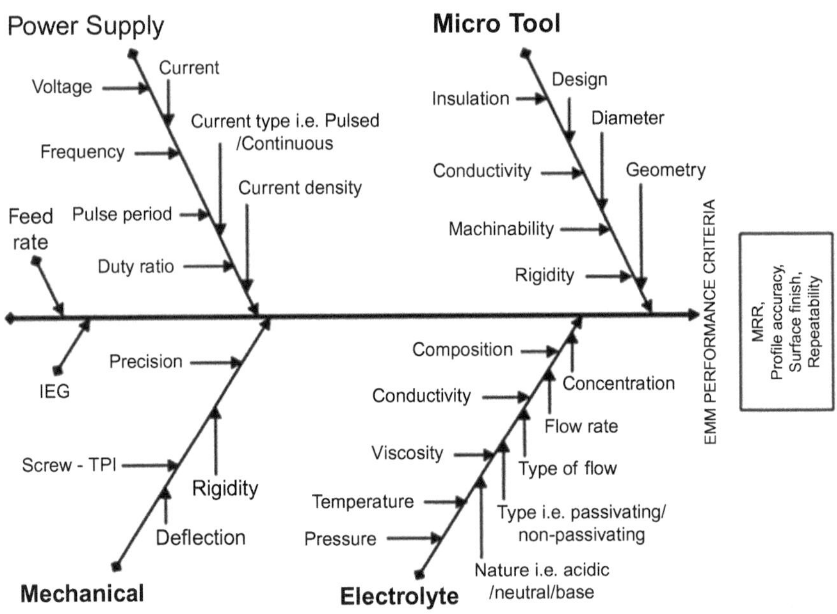

Fig. 7.21 Fish bone diagram showing influencing process parameters [80].

no dissolution take place and only reaction products are flushed away from the machining zone. It has been observed that, 60% duty ratio is most effective for higher unit removal. However, to improve accuracy, lower duty ratio is preferred which maintain adequate flushing.

At higher frequency, the pulse period is very small. During pulse on time of each pulse period, the dissolution of metal starts after only the charging of double layer. Hence, at very small IEG, ultra short charging time at high frequency pulse generates localized and controlled machining.

Commonly used electrolyte of EMM involves NaCl, $NaNO_3$, H_2SO_4 solutions. Mixed electrolytes such as NaCl and $NaNO_3$; NaCl and NaBr or NaCl and NaOH are also utilized in EMM during machining of different metals. However, acidic electrolyte which dissolves sludge is mostly preferable for EMM. Higher current density can be achieved at higher concentration of electrolyte for a particular applied voltage. However, for higher accuracy and better surface finish, dilute electrolyte with higher machining voltage and smaller IEG is preferable instead of low voltage and higher concentration of electrolyte.

Microtool feed rate also influence the machining performance in EMM. Implementing higher feed rate can improve machining accuracy due to lesser interaction time between tool and job. However, it may result in generation of micro spark and short circuiting if implemented in very high ranges.

Selection of optimum combination of parameter settings is very important for fabricating desired microfeatures utilizing EMM. Hence, better understanding of the role of various influencing parameters of EMM is vital to achieve best machining performance considering the accuracy as well as productivity.

7.6.6 Advantages and challenges

As an eco friendly and economical process for fabricating micro shapes effectively and efficiently, EMM is very advantageous in offering better control, high precision, flexible, reliable as well as burr free machining of different harder and chemically resistive materials without showing any adverse thermal effects that is prominent in different thermal induced material removal processes. The flexibility and versatility of EMM has led its wide implementation where different types of EMM include capillary EMM, jet EMM, through mask EMM, 3D EMM, etc. Recently, reverse EMM has been implemented for fabricating high aspect ratio micro tools with enhanced surface quality and various tip shapes such as conical, cylindrical, disk shape [82], spherical, wedge-shape [83], etc. These in situ fabricated electrodes have been used to machine desired micro shapes with required dimensions on-machine leading to minimizing the problems in micro tool handling and holding. Other advancement that recently adds up include fabrication of multi micro tool arrays that can machine arrays of micro feature in a single run and is very effective in reducing the total machining time. However, further research is needed admitting proper selection criteria of electrolyte, regarding enhancement of existing micro tool feeding system with incorporation of feedback mechanism for encountering and correcting the occurrence of short circuiting and sparking, online monitoring and control of IEG as well as enhancement of profile accuracy, etc. In contrast, research challenges that are still existing and requires urgent attention include further enhancement of machining performance and increment of feed rate, generating highly intricate and more complex 3D micro as well as nano structures, adequate optimization of controlling parameters and betterment of machining stability, etc. However, inability to machine non-conductive materials counts toward the process's disadvantage.

7.6.7 Applications

EMM is becoming widely employable process where researchers have fabricated different components for MEMS and different composite materials, semiconductors [84] and have been applied for fabricating plate for ink-jet printer head, micro nozzles as shown in Fig 7.22, finishing of print bands required in high-speed impact printers [16, 85], deburring different micro parts, mask less cutting of metal by electrochemical saw, manufacture of diaphragm for electro optical systems, fabrication of contoured 3D micro components, different types of high aspect ratio micro structure, shapes and profiles, different high aspect ratio micro structures on silicon and other various materials [78,86], etc. Researchers have also fabricated single and arrays of micro grooves, complex micro holes and slots having different internal features [87–89], micro milled structures [90], micro turned components [91], micro gear [92], micro reactor, micro patterns in flat and on cylindrical surfaces [91], square helix, cavity with sub micron dimensions of varying roughness [93], nano structured electrode surface by employing STM tip [94], various 3D nickel nano structures [95] as well as numerous micro/nano

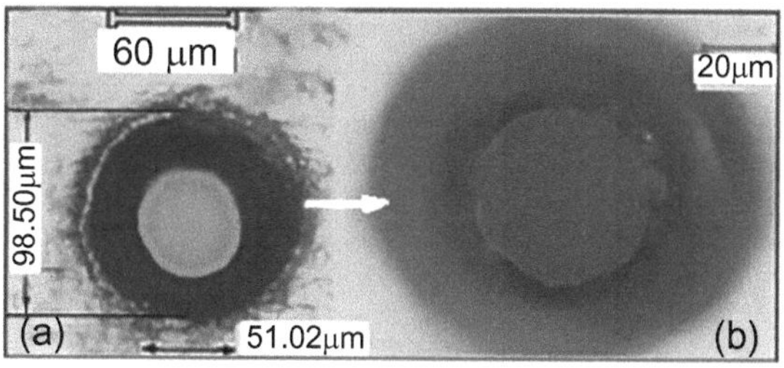

Fig. 7.22 (a) Micronozzle fabricated by EMM (b) Surface quality of internal wall [16].

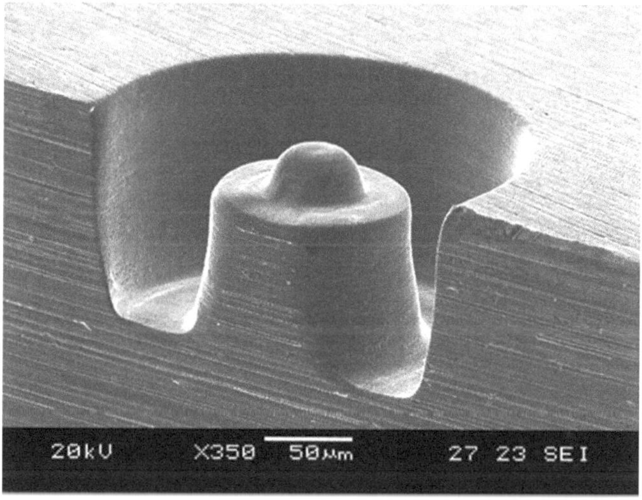

Fig. 7.23 SEM micrograph of micro hemisphere fabricated by EMM milling [96].

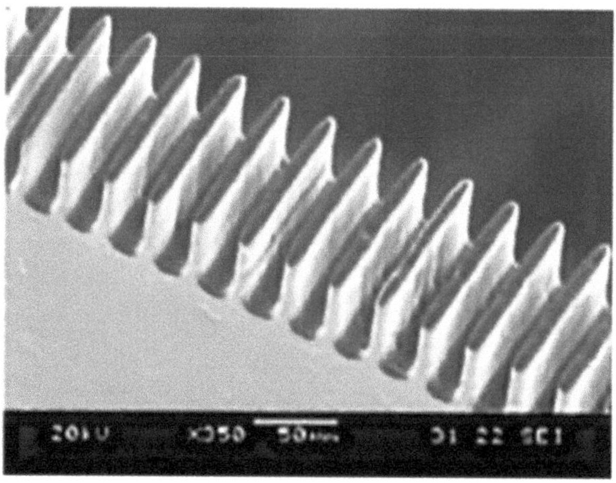

Fig. 7.24 Micrographs of microgrooves by Wire ECM [96].

features. Moreover, researchers have fabricated 3D intricate features like micro hemisphere and array of microgrooves by employing EMM milling and Wire ECM, high-aspect-ratio microcolumn by disk type electrode, microholes on Ti6Al4V surface by jet-EMM, etc. [96,97] which have been shown in Figs. 7.23–7.26 respectively.

Fig. 7.25 High-aspect-ratio microcolumn by disk type electrode [96].

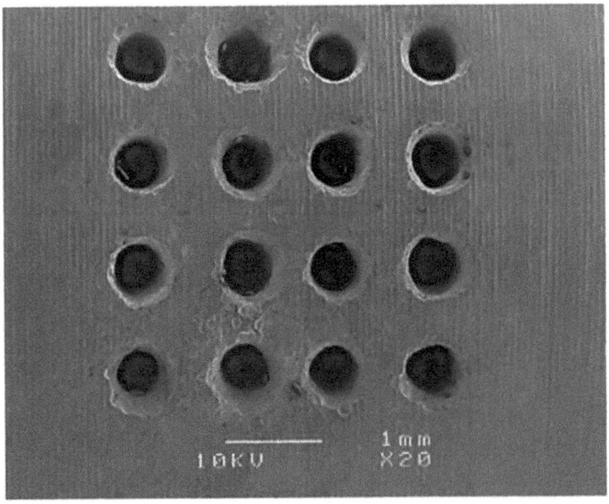

Fig. 7.26 SEM micrographs of microholes produced on the Ti6Al4V cylinder surface using jet-EMM [97].

7.7 Hybrid micromachining

7.7.1 Introduction

Hybrid micromachining involves combination of two separate micromachining processes for utilizing the advantages of both processes simultaneously in achieving better surface quality, accuracy as well as higher MRR. Fundamental mechanism of hybrid machining has already been discussed in details in Chapter 6. Same concept of machining can also be utilized for the purpose of micromachining to achieve the benefit of two or three micromachining techniques. For example, traditional micro grinding process gives better dimensional accuracy with lower tolerance values and higher MRR, but suffers from detrimental thermal effects like formation of HAZ that deteriorates the surface quality and material property, generation of thermal stress, formation of burrs, etc. On the other hand, electrochemical micromachining process is devoid of any adverse thermal effects and provides smoother surface finish. Hence, their combined machining process, i.e., micro electrochemical grinding offers the benefits of both constituent machining processes by surpassing their intrinsic limitations. Likewise, electro chemical micro discharge machining process uses the advantages of both electrochemical machining and electrodischarge machining for generating different complex microfeatures on both electrically conductive and non-conducting material like glass, quartz, etc. Some of the hybrid micromachining combining anodic dissolution with mechanical grinding, ultrasonic vibrations, magnetic field, EDM, LBM, etc. have been demonstrated to be very successful [98–100]. Few of the attempts undertaken by the researchers have been mentioned hereunder.

7.7.2 Micro-EDM and micro ECM hybridization

EMM has some inherent process limitations such as uncontrolled machining for poor localization of current flux, improper flushing of IEG, the formation of passive oxide film on metal workpiece, deposition on microtool, higher process time, etc. To overcome these limitations and to make EMM more effective by achieving controlled machining within shorter time, hybridization of anodic dissolution with various other metal removal processes is needed. Attempts have been made by the researchers to combine anodic dissolution with other metal removal processes. Hybrid EMM is developed to enhance its dissolution process with higher precision by minimizing above-mentioned limitations by combining the advantageous effects of another process.

By combining anodic dissolution with spark erosion mechanism of metal removal, a new hybrid micromachining can be established through the combination of micro electrical discharge machining (EDM) with EMM. Both these processes utilize different mechanisms for material removal. The micro-EDM removes the material by melting and vaporization from the workpiece by utilizing the thermal effects of small energy electric discharges. The EMM employs the electrochemical dissolution of material from the workpiece. However, the rate of material removal is slower in EMM which can be augmented by employing spark energy for faster material removal. By combining micro-EDM with EMM, the machining rate and accuracy of EMM can be improved. It also improves the surface quality with higher surface finish by removing thermal defects and discharged craters from the machined surface [101]. The following advantages can be achieved as follows:

(i) Faster material removal due to micro-EDM
(ii) Lesser heat affected zone due to EMM
(iii) Improved surface finish due to EMM
(iv) Improved accuracy due to micro-EDM

The basic scheme of hybridization of EMM with micro-EDM is shown in Fig. 7.27. Pulsed power supply with low-resistivity deionized water as electrolyte is utilized for the combined process [102]. Initially microelectrode is fed down to generate smaller critical gap such that spark can be generated by breaking down the deionized water. Higher amount of material is removed from the workpiece by melting and vaporization due to generated sparks which also produce rough surface. Gap between electrodes increase due to metal removal, hence no further spark takes place. Thereafter, anodic dissolution occurs in the gap owing to the slight conductivity of deionized water which improves the surface roughness. Thereafter, the microtool is again lowered down to repeat the process.

7.7.3 Micro-ECM and LBM hybridization

The combination of laser beam with EMM also contributes to improve the machining rate and accuracy [103]. In EMM, it is essential to circulate the electrolyte in narrow IEG for the continuation of process, but is

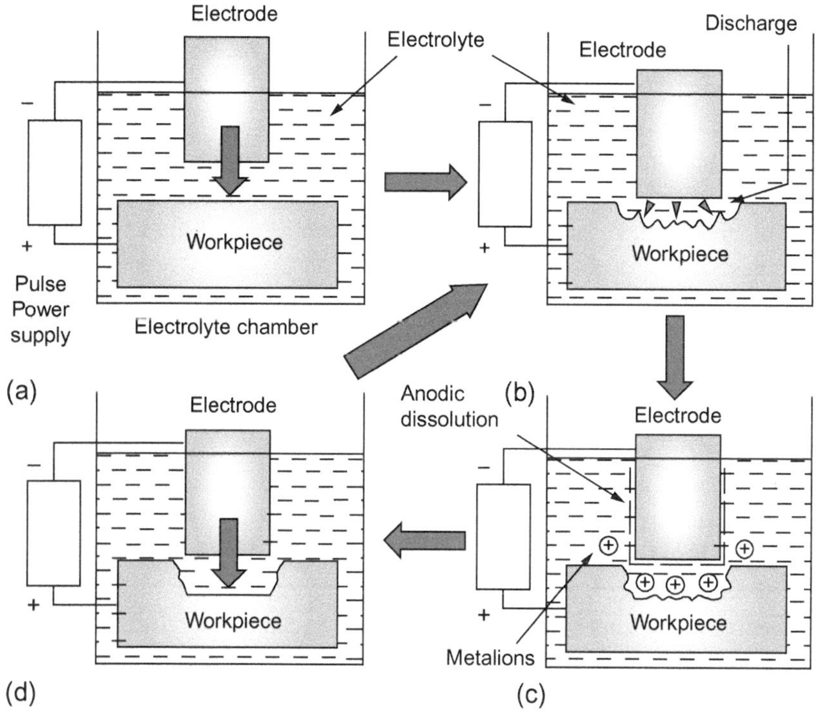

Fig. 7.27 The basic scheme of hybridization of EMM with micro-EDM: (a) Process begins, (b) discharge, (c) dissolution, and (d) feeding [80].

very difficult due to the micro size of the tool. It can be improved by increasing the temperature of the electrolyte at IEG, so that due to temperature difference electrolyte flow takes place which can be achieved by applying laser energy in the narrow machining zone and in turn increases the current density. The rate of anodic dissolution will increase and due to temperature difference in the machining zone and surrounding results better electrolyte circulation in the narrow gap. Low energy defocused laser beam is preferred for this purpose which only increased the temperature of the electrolyte without melting and vaporization of the workpiece. During EMM, oxide layer is formed on the workpiece due to anodic reaction, which is passive in nature and minimizes the rate of current flow and thus lower the machining rate. Laser beam can also minimize the formation of oxide layer by increasing the temperature of the upper surface of the workpiece which restricts the chemical reaction responsible for the formation of passive metal oxide layer. Hence, the combination of laser beam with EMM will prove to be beneficial for higher material removal.

It will also improve the localization effect in the small area where laser beam focused during the process. Laser beam helps in focusing the applied current into the machining area thereby increasing the effective machining rate and precision. By applying laser beam in EMM following advantage can be achieved:

(i) Better circulation of electrolyte in the IEG due to Laser interaction
(ii) Removal of passive oxide layer from the workpiece due to Laser interaction
(iii) Enhancement of electrochemical reaction rate for higher material removal
(iv) Improved accuracy due to more localized anodic dissolution due to laser interaction

Fig. 7.28 shows schematic diagram containing different stages of EMM hybridized with laser. Fig. 7.28(a) shows the beginning of the anodic dissolution process where ring electrode acts as cathode. Fig. 7.28(b) exhibits the introduction of laser beam in the machining zone which improves the electrolyte circulation and removes the passive oxide layer, and accelerates the anodic dissolution by enhancing the ionic movement. Fig. 7.28(c) shows the accelerated and controlled metal removal by electrochemical dissolution. Fig. 7.28(d) exhibits only anodic dissolution in the absence of laser beam for achieving finish machining and reoccurrence of oxide layer. The process repeats until the final shape is generated. Fig. 7.28(e) represents the final shape of the workpiece. Thus, controlled and faster material removal is achieved which will improve the accuracy and productivity of EMM.

Other hybrid micromachining process involves laser, ultrasonic as well as abrasive assisted micromachining. ECDM micromilling has been used to machine intricate 3D microfeatures on pyrex glass. ECM micro grinding which has been used to attain a surface roughness of approximately 0.21 μm on SS321 work material. EDM-electro rheological fluid assisted hybrid micromachining has been used to achieve surface roughness of 0.06–0.08 μm on SUS 304 and aluminum alloy. The feasibility of micromachining utilizing laser beam in electrochemical jet machining has already been demonstrated [104]. Combination of Laser drilling and Jet-ECM has also been utilized for reduction of recast layer formation on SS321 by >90%, etc.

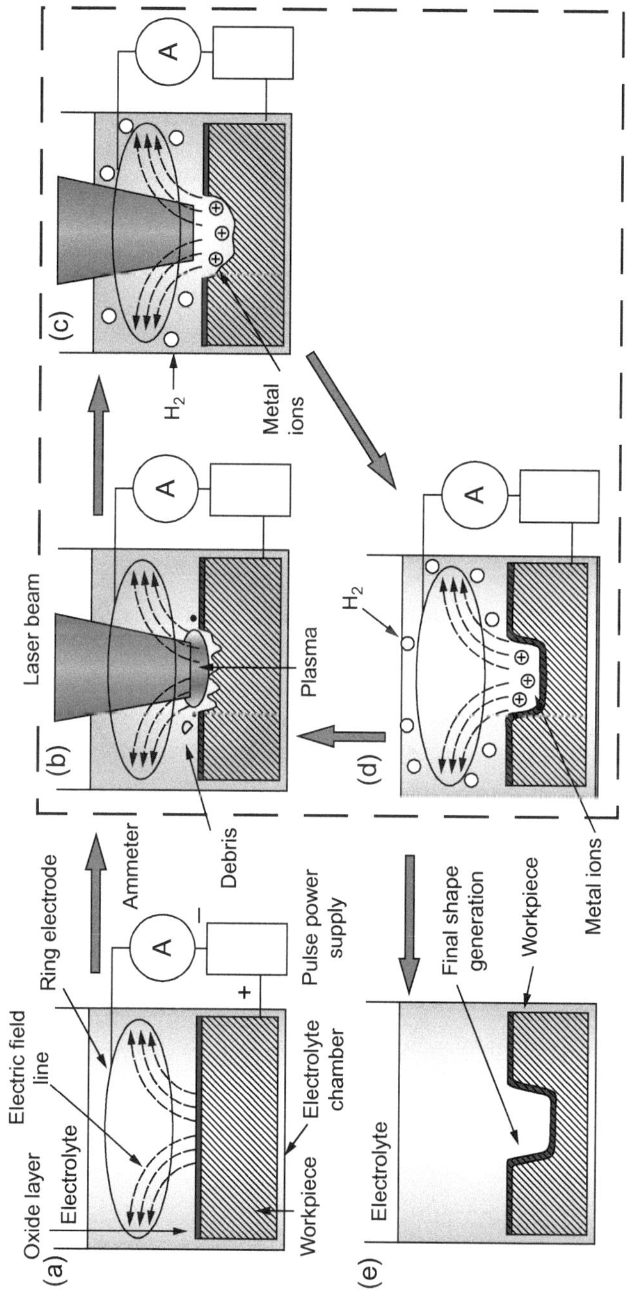

Fig. 7.28 Schematic diagram containing different stages of EMM hybridized with Laser [80].

7.8 Sequential micromachining (SMM) processes

7.8.1 Introduction

In the modern age of micromachining, there is an urgent requirement of highly sophisticated machines, tools and gadgets which may fulfill the growing expectations of the technical and industrial community. To achieve the desired goal and improve the accuracy and efficiency of manufacturing processes by overcoming the stringent requirements of tighter tolerances provided on micro parts, manufacturing industries are searching for the different innovative methods, processes and techniques for the production of micro components. Sequential micromachining technology is considered to be an important link in the chain of innovative and advanced micro part manufacturing methods. It is relatively a new technology which provides the added advantage of implementation of two or more machining techniques at single workstation without changing the tool, workpiece and set up.

Once the product design part is over, in order to inculcate the different features such as cutting of cylindrical, rectangular shapes, machining channels, grooves, walls, holes, etc. to the micro part, manufacturing strategy is decided. In non-conventional micromachining, while deciding the machining strategy, the sequence of machining operations to be carried out on the micro-part is decided on the basis of various characteristics. Some of the aspects of sequential machining such as achievable highest material removal rate, type of surface finish required on the final product, etc. are quite important. It is quite vivid that implementation of different machining techniques in sequential manner provides many added advantages such as completion of shaping as well as finishing operations in single setup without altering the position of tool or workpiece, reduction in machining time, reduction in surface roughness, machining of difficult to cut material with great ease, increase in dimensional accuracy, improvement in surface quality, etc.

7.8.2 Basic concept and mechanism

Sequential micromachining (SMM) is basically a machining strategy whereby two or more micromachining techniques are implemented in a sequence on the same machine tools. SMM system acts as a multifunctional machine tool as it tries to encompass more and more machining processes on a single workstation. So, right from initial shaping, sizing to final finishing each and every operation is performed sequentially without changing tool and workpiece positions. This type of system promotes lot many benefits. Sequencing of different machining processes drastically reduces the cost associated with independent machine tools and other ancillary mechanisms related to the system. Hence along with technical ingredients cost effectiveness is one of the most important aspects which have to be taken care of while designing and developing the multifunctional sequential machining units. Sequencing of machining reduces the workpiece handling time, as unloading and reclamping/reloading of the workpiece to carry out various machining operations are lessened. This phenomenon indirectly assist the workpiece to maintain its desired dimensional accuracy, which in turn reduces the production of scrap, which gives impetus to the effective utilization of workpiece material and enhancement of machining efficiency. The main intention behind applying various techniques in a sequence is to maximize the advantages and to minimize the disadvantages of each individual machining system.

Let's take an example of two stand alone machining processes such as electro-discharge micromachining (micro-EDM) and electrochemical micromachining (EMM), which are commonly applied in a sequence. The mechanism of material removal in both the techniques is based on different physical phenomenon. Micro-EDM needs insulating dielectric medium to occur whereas EMM process cannot take place without the presence of conductive electrolyte medium. These two individual processes have their own advantages and disadvantages. The micro-EDM process is considered as more precise from the machining point of view than EMM process, but it has some shortcomings. The micro-EDM is a thermal process, which is extensively used to machine difficult to cut materials, in which material from the workpiece as well as from the tool surface is removed due to the repetitive action of electric spark discharges. Therefore the surface machined by micro-EDM process consists of numerous overlapping discharge craters in the heat affected zone (HAZ). In this region of thermally damaged white layers, micro cracks are formed; residual stresses also develop which reduces the fatigue strength of the material. The predominant tool wear rate also influences the machining accuracy. In EMM process, the material removal from the surface of the workpiece takes place due to ionic dissolution of anodic material, where the process occurs in less temperature. The micro parts generated by this process have high surface finish. There is no recast layer, no burrs, no micro cracks, no heat affected zone occur, but during electrochemical machining, dissolution of anodic material occurs in an area larger than the facing zone of the micro tool electrode, therefore maintaining accuracy is a bit difficult task in EMM process. Thus both the machining processes have their own pros and cons. Implementation of both the processes in sequential manner minimizes their demerits and increase advantages collectively.

Most of the time, terminology of sequential micromachining is misapprehended with hybrid micromachining. Actually hybrid micromachining and sequential micromachining processes are recognizably different in nature. In hybrid micromachining, two or more than two micromachining techniques are applied simultaneously on a single machine tool that means the processes are interacting in the same processing zone and at the same time.

The trend of applying different micromachining techniques in a sequence on a single machine tool which acts as a multifunctional micromachining center is growing day by day. Application of micro turning before micro-EDM for reshaping the tool electrode, block micro-EDM and then micro-EDM, for in situ fabrication and application of tool, Laser micro ablation and then micro-EDM for machining micro part sequentially, manufacturing of micro tool by micro-EDM process and then applying the same tool for producing precise and high aspect ratios micro holes on difficult to cut material by using micro ultrasonic machining (micro-USM) process, are some of the examples of sequential machining. Classification of sequential machining processes into five major groups on the basis of purpose of operations [105] is exhibited in Fig. 7.29.

(i) *Micro tool making oriented processes*

In non-lithography based micromachining processes, micro tool is the most important element without which micromachining is not possible. Since the micro tool diameter starts from 0.999 mm. up to few microns, the manufacturing and handling of such tools is a difficult task. The micro tools are generally made of high strength materials such as titanium, tungsten, tungsten carbide, etc. Preparation of micro tools up to a diameter of 500 μm can be conveniently done by applying conventional micro turning processes, but to prepare micro tools less than this diameter requires specialized manufacturing processes, that's why the cost of the micro tool surges to a higher level. Sequential micromachining proves to be the best instrument for micro tool manufacturing. Turning followed by micro-EDM, shaping the tool by micro-EDM and then reducing it to the required size by ultra sonic machining (USM) or machining initially with micro-EDM and then

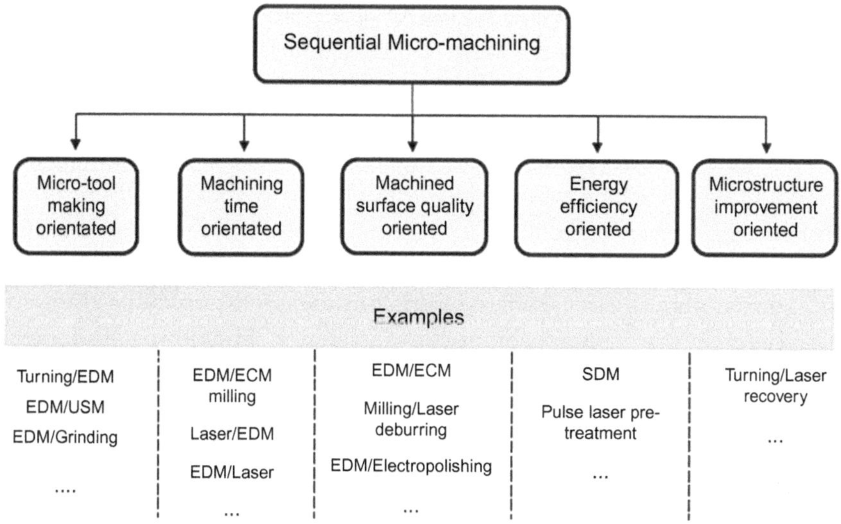

Fig. 7.29 Classification of sequential machining processes [105].

finishing the tool to required size either by EMM or grinding, are some of the methods used in sequential micromachining for micro tool manufacturing.

(ii) *Machining time oriented*

The sequential machining techniques are effectively utilized to reduce the machining time. The combination of the processes such as EMM and micro EDM milling, Laser beam machining followed by micro-EDM in a sequence improves the metal removal rate. In these processes during machining, the maximum amount of material from the work piece surface is removed by the EMM process, the second processes is applied just to bring the profile of the cut portion to the required shape and size. In some cases of EMM and micro-EDM milling, about 80% of the material is removed by electrochemical dissolution process whereas remaining 20% material from workpiece surface is removed precisely by applying electro-discharge machining [106]. This sequence makes it possible to reduce the overall machining time by achieving the maximum material removal rate with higher part accuracy.

(iii) *Machined surface quality oriented*

Sometimes the sequence of machining is devised based on the type of surface quality required on the workpiece surface. In micro part manufacturing, higher surface finish is always desired. Processes such as micro-EDM followed by EMM or electro-polishing, milling and then laser deburring are some of the examples of the processes falling in this category. From the point of view of achieving high surface finish on micro part, application of micro-EDM and EMM processes in a sequence are already discussed. The EMM imparts high surface quality on the part which was machined previously by micro-EDM.

(iv) *Energy efficiency oriented*

In these types of processes, the energy required for overall machining of the part is reduced considerably by applying the two different machining techniques in a sequence. The approach adopted is surface defect machining (SDM) followed by usual machining; SDM is a process of generating surface defects on the workpiece surface at depths less than the uncut chip thickness. The surface defects may be initially formed by applying various methods such as manual drilling, mechanical forming, threading or laser ablation on the workpiece to be machined [107]. SDM changes the nature of chips, in general, continuous chips formed during machining may rub against either with the tool surface or with the machined surface and may result in generation of higher amount of heat due to the friction occurred at the cutting tool workpiece interface, this phenomenon distort the machining surface. SDM facilitates the formation of fine surface defects which results in the reduction of the shear plane angle that increases the magnitude of tangential cutting forces over the thrust forces. This phenomenon contributes in the streamlined deformation of metal along the cutting direction that causes less side flow of the material and helps formation of segmented chips which results in improved surface finish. Thus the machining efficiency of difficult to cut materials increases significantly.

(v) *Microstructure improvement oriented*

In these types of processes, quality and integrity of machined surface can be improved by employing sequential machining. A sequential machining approach was adopted to prepare defects free surface of single-crystal silicon, where diamond grinding process was followed by laser ablation [108]. In the process of preparing silicon components, diamond grinding generates an amorphous sub surface layer which contains micro cracks which are few sub microns in depth. This sub surface defect was ameliorated by applying pulsed laser irradiation technique.

7.8.3 Development of different sequential micromachining strategies

In conventional as well as non-conventional machining processes, there is a general trend of applying the shaping process first before the finishing process. In sequential machining, one particular process which is used as a finishing process in one combination of sequencing can be used as a shaping process in other combination of sequencing. In sequential combination of micro-EDM and EMM, the micro-EDM is used as a shaping process, whereas, the same μEDM process is used as a finishing process in sequential combination of laser and micro-EDM. In Sequential micromachining more than one process are taking part, based on the first method applied in a sequence, some SMM processes are described here under.

(i) *Micro turning and micro-EDM*

In conventional micro turning, the tool and workpiece are in contact with each other and the cutting forces are exerted by the tool on the workpiece, hence by considering the stiffness of the rod material to be machined, the diameter of rod is reduced to certain extent by applying micro turning process. Further reduction of diameter by micro turning may impart the flaws due to wobbling effect; hence it is difficult to fabricate micro tools below 300 μm diameter by conventional machining processes. Similarly high aspect ratio tools cannot be manufactured by micro-turning. The maximum tool length achieved by micro turning is not more than 1 mm. Since micro-EDM has the capability to machine extremely fine featured micro-structures, it is used to reduce the tool diameter to a micro level. Fig. 7.30 represents the sequential machining where, micro turning is followed by micro EDM. The micro-tool is initially fabricated by using micro turning process. The same tool is then used as tool electrode in micro EDM process. Thus Fig. 7.30 represents unique combination of conventional and non-conventional micromachining processes taking place in a sequence. The micro tool is fabricated by conventional micro turning process and thereafter micromachining is carried out by micro-EDM process.

(ii) *Micro-EDM plus micro grinding*

In this method, the polycrystalline diamond (PCD) tool is fabricated on machine by applying block micro-EDM process [109]. In this process, the PCD tool is connected to positive terminal and the precisely shaped sacrificial conductive block of high wear resistance tungsten is connected to negative terminal of the power supply. The machining is carried out

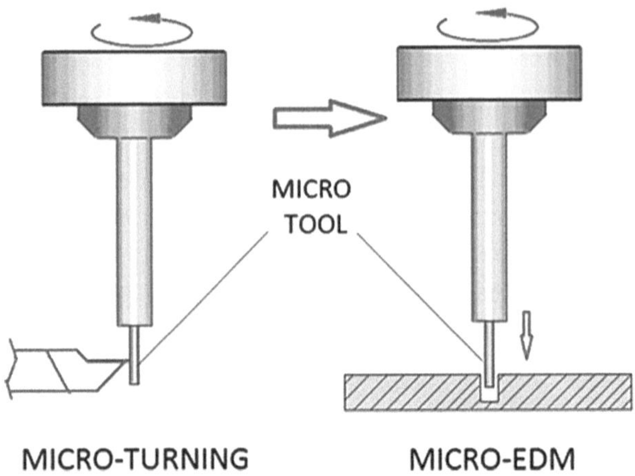

Fig. 7.30 Micro turning followed by micro-EDM.

due to spark discharges occurring between the tool and the block which removes the material from both the electrodes immersed in dielectric oil. The reaction products between the inter electrode gap is removed due to side flushing of dielectric. The surface roughness of PCD tool is controlled by varying capacitance and voltage. Once the PCD tool is brought to the desired shape and size, the same tool is then used for micro grinding difficult-to-machine glass materials. The fabrication of micro grinding PCD tool as well as the micro grinding operation is carried out sequentially on the same machine. Fig. 7.31 represents schematic setup arrangement implemented for glass grinding where, grinding tool is attached to rotary spindle.

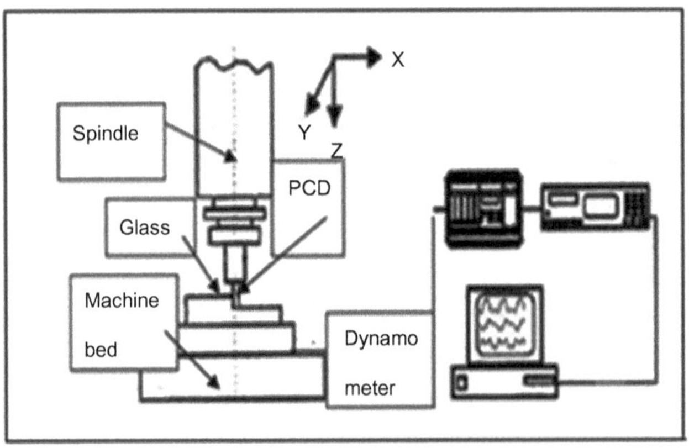

Fig. 7.31 Schematic diagram for glass grinding [109].

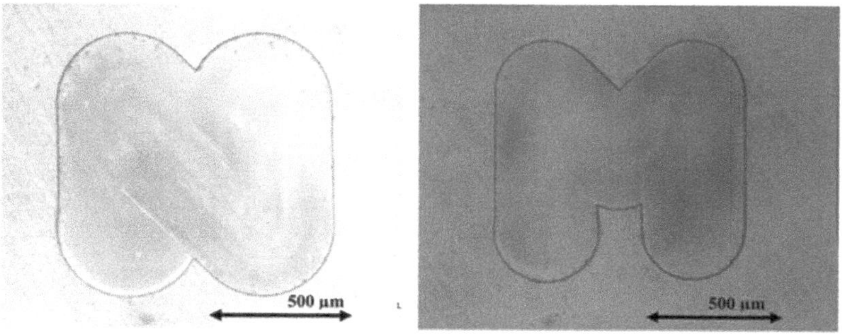

Fig. 7.32 Micro-EDM plus micro-grinding applied in a sequence on BK7 glass to produce "N" and "M" shapes [109].

The micro tool manufactured by micro-EDM process has rough surface due to the formation of micro craters. Such tools can improve MRR but enhances the surface roughness of grinded glass. The surface roughness of the glass during grinding reduced to a greater extent by imparting rotary motion to grinding tool. In Fig. 7.32, the 'N' and 'M' shaped smooth micro-cavities engraved on glass material by rotating PCD tool are shown.

(iii) *Laser micro ablation plus micro-EDM*

The combination of nanosecond pulsed laser and micro-EDM can be effectively used sequentially for micromachining [110]. The methodology applied for machining is rough cutting by pulsed laser beam and then finishing precisely by micro-EDM process. Initially the micro feature to be produced is machined by using high ablation power of nano second pulsed laser. The high ablation power of laser makes it possible to cut the micro profile in a very short duration of time, and reduces the machining time to a great extent. From the machining point of view, most of the material from the part to be generated is removed by laser ablation process. As it is well known that laser being a thermal process introduces certain flaws such as generation of highly rough surface, piling of recast layer in the machining zone, etc. Though micro-EDM has comparatively low metal removal rate, but due to its high precision it can be effectively implemented in removing the surface defects formed due to laser ablation. Thus micro-EDM process is used sequentially for final finishing. Figure 7.33 represents the methodology of sequential application of laser ablation and micro-EDM, where maximum material removal takes place due to pre machining by laser ablation and finishing by precise post machining using micro-EDM.

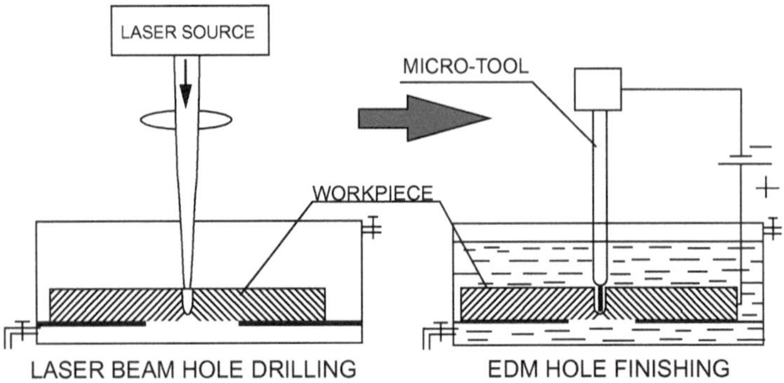

Fig. 7.33 Sequential applications of laser ablation and micro-EDM.

Fig. 7.34 Micro feature generated on SS 304 by sequential Laser ablation and micro-EDM [110].

Figure 7.34 exhibits the workpiece samples where, sequential machining is applied. The laser beam machining generates surface with high speed, but being a thermal process, it forms recast layer which is piled up surrounding the cut area, as shown in Fig. 7.34(a). The laser structured micro feature is then finished as shown in Fig. 7.34(b). The application of micro-EDM finishing process removes all the surface irregularities generated due to previous machining.

(iv) *Micro-EDM plus EMM sequential machining process*

This sequential machining process is used for milling cavities in SS-304 grade stainless steel plate of 0.4 mm thickness. Initially, the cavities are generated by micro-EDM process and then the same cavities are finished by EMM process. In this sequential machining process, the diameter of the micro tool which is to be used for combined machining was 0.5 mm was then reduced to the 0.1 mm by anti-copying block method [111]. The shaping is done by micro-EDM process in which the material is removed layer by layer to form cavity in the dielectric medium of EDM oil. The tool follows programmed path and moves in downward direction after completion of each layer. During micro-EDM machining the parameters chosen are of higher values than the normal, which drastically reduces the time required for machining.

As EDM is a thermal process, it generates recast layer, craters, micro pores, etc. in the machined part zone. The finishing of cavity is then carried by EMM process. In EMM, the path of tool electrode was maintained same as it was maintained during micro-EDM process. Fig. 7.35 depicts the sequencing operation procedure. Three operations are carried out at one station with retaining tool and workpiece at same location. In Fig. 7.35(a) the online fabrication of micro-tool is done by reverse copying block method. The tool located at the same position then used for micro-EDM milling, as shown in Fig. 7.35(b). The final finishing of cavity was achieved by ECM milling as shown in Fig. 7.35(c).

(v) *EMM and micro-EDM milling in sequence*

There is a general trend of implementation of EMM after micro-EDM process for improved surface characteristic of micro part. The sequential combination of EMM and micro-EDM for milling the cavity by die sinking and formation of groove is also implemented successfully. This combination is useful in milling where maximum material is removed by EMM process initially. The final finishing and geometric accuracy of micropart is then obtained by subsequent application of controlled micro-EDM process. The machine tool is equipped with tool electrode preparation as well as renewal unit. A microscope unit

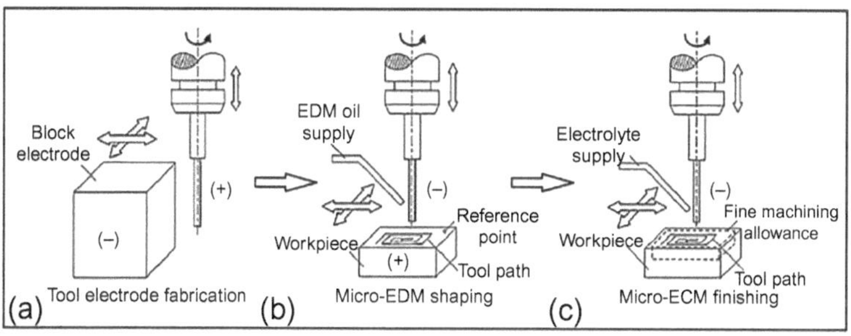

Fig. 7.35 Application of micro-EDM and micro-ECM or EMM processes in sequence [111].

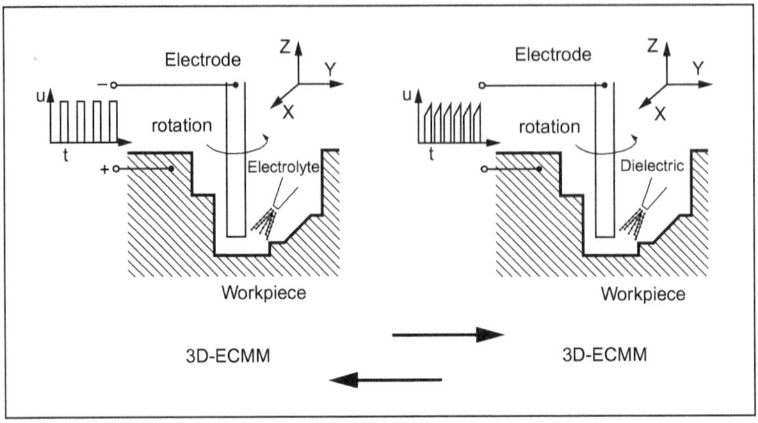

Fig. 7.36 Sequential machining of EMM and micro-EDM [106].

is mounted for precise positioning of tool electrode with respect to workpiece. The strategy of implementing electro-chemical and electro-discharge milling processes in sequence is designed and optimized with the aid of developed simulation software. In EMM and micro-EDM sequential combination, the same tool can be used for further machining as tool wear do not take place in initial EMM process. The micro milling operations performed by sequencing of EMM and micro-EDM processes on developed prototype with the primary aim of reducing the machining time. Fig. 7.36 depicts 3D milling operation, in which the order of machining processes can be changed from ECM to EDM and vice a versa. The 3D micro-EDM milling which was used as a finishing process in first operation can be used as a shaping process in next operation.

It has been observed that application of electrochemical and electric discharge micro-milling (EC/EDMM) processes in sequence reduces the machining time nearly by 50% in comparison with the same task performed simply by electric discharge micromachining. Figure 7.37 shows the comparison of machining time taken when the EMM and micro-EDM processes applied independently and sequentially [106].

The Fig. 7.37 reveals that the machining time required by sequential machining approach is almost 50% less than the time taken by conventional EDM process. Thus carrying out EMM and micro-EDM processes sequentially permits an essential reduction of the disadvantages of individual methods and enhancement of the advantages of both methods.

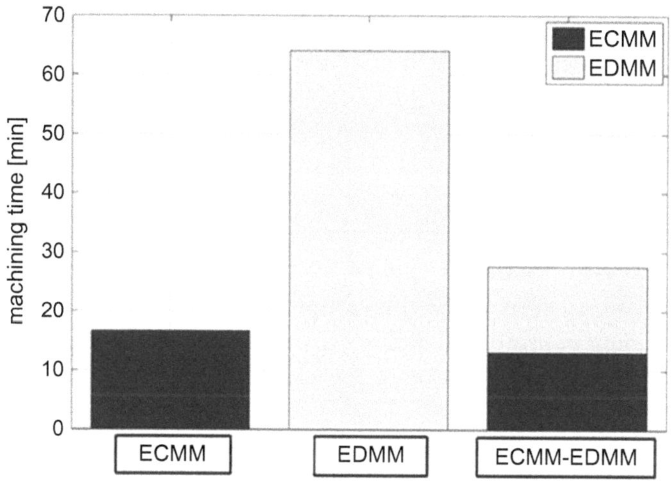

Fig. 7.37 Comparison of machining time [106].

7.8.4 Process capabilities of SMM

Machining of highly sophisticated, newer, harder and difficult to cut materials with improved surface characteristics have become possible due to the newly introduced machining technology known as sequential micromachining. It offers lower part handling time and machining time as well, leading to the reduction of processing cost which provides an impetus to the productivity. It offers an opportunity to manufacture micro parts with intricate shapes, high aspect ratio with reasonable level of dimensional accuracy and remarkably excellent surface finish by controlling the process parameters precisely. In most of the past research of sequential micromachining technique, the three stages mentioned hereunder are mostly followed; such as

(i) In situ manufacturing of micro tool

It is an essential requirement in sequential machining to maintain accuracy of the parts produced. The various methods such as block electric discharge grinding (BEDG), turning, reverse ECM, etc. can be utilized as a technique for in situ fabrication of micro-tool during sequential machining.

(ii) Shaping an object

In sequential machining, generally shaping or cutting operation is carried out initially for removal of material from the workpiece. These result in high material removal rate in a short span of time, which drastically reduces the machining time and enhances the machining efficiency.

(iii) Finishing the previously machined components

It is an important operation which is applied post shaping. The objectives of implementation are to remove burrs, cracks, recast layer from the heat affected zone, in case of EDM and laser, etc. and to smoothen the machined surface area. In sequential combination of ECM and EDM, it is applied for truing the profile and to impart dimensional accuracy to the part produced.

Micro grinding of brittle and difficult to cut glasses such as BK7, lithosil and N-SF 14 becomes possible due to the application of sequential machining approach where, PCD, i.e., polycrystalline diamond tool is initially manufactured in the same set up and then micro-grinding operations are performed on different types of glasses successfully [109].

7.8.5 Advantages and challenges

In micropart manufacturing, different micromachining operations are required for the formation of micro feature or complete structure, most of the time it may not be possible to perform all operations on a single machine tool or by a single machining process. In such cases, for further processing there is a need to shift the workpiece from one machine to the other suitable machine. Thus, transferring of workpiece from one place to the other work station may introduce some errors in the previously machined region of a component. Hence, repositioning the workpiece under the micro tool exactly at the same location in another set up is a challenging task and may introduce machining error due to realignment. The activities such as unloading the workpiece from one workstation, transferring it and then loading or repositioning it accurately to the next station are not advisable as it increases the lead time. As it is known that improper machining leads to loss of man machining hours, loss of power, material, etc. Thus the idea of integrating the two or more than two standalone machining technologies at single workstation and applying them sequentially is being developed. This new methodology of machining has many benefits; sequencing of different machining techniques reduces the disadvantages of each individual manufacturing method and adds the essential complimentary advantages. This reflects in the improved quality of microfeatures produced in terms of its dimensional accuracy and surface finish. The overall reduction of machining time drastically shortens the production chain and improves the machining efficiency. The elimination of repositioning of workpiece at different workstations minimizes the possibility of scrap produced due to realignment error by saving lot of material. The sequential machining has a capability of fabricating different complex shaped micro features on difficult to cut materials.

Nonetheless the sequential machining is still in rudimentary form because till now very few micromachining processes are brought under the roof of sequential machining, the three machining processes such as EDM, ECM, and Laser are used predominantly, in sequential machining, there are several other micromachining processes using electrical, mechanical, chemical and electro-chemical energies, the integration of which may form novel sequential machining combination. There are various difficulties which one has to come across while sequencing of different machining methods, such as combination of hardware of separate machining techniques on one stage, design and development of switching mechanism required for actuating two or more than two power sources for uninterrupted sequential machining. The past research shows that till now only two different machining methods are utilized in sequence, putting more and more machining methods in a sequence for producing better quality product is required. The machining methods which are put in a sequence are generally applied to machine either conducting or non-conducting materials only, integration of sequential machining methods which can machine both conducting as well as non-conducting materials at single workstation is a challenging work which is required to be done. In order to produce precise, high quality micro components, there is a need of designing and developing integral machining systems like multi-axis machine tool which may perform multiple tasks with the aid of optimized tool path algorithm. In order to improve the machining efficiency, development of fully automatic sequential micromachining systems will be the next step in the direction of achieving excellence.

7.8.6 Applications

The sequential machining can be used for machining difficult to cut high strength and temperature resistant materials and microstructures with better surface finish, which will be very useful in the field of micro-electro mechanical systems (MEMS), aerospace, bio-medical, electronics and optics, etc. Manufacturing of hot embossing dies used for bio-fluidic applications is a challenging task as die material has to withstand high processesing forces along with very high temperature, therefore materials such as high heat resisting alloy steel or ceramics are generally used to manufacture dies, and machining such objects at micro level with conventional machining to get a better geometric accuracy is a difficult task. Micro-fluidic dies are fabricated by applying sequential machining combination of laser machining for micro structuring and micro milling for finishing. Similarly micro-features as shown in Fig. 7.38 such as star and square shaped cavities are machined by sequential application of electro discharge milling followed by electrochemical milling.

The sequential machining strategies as discussed are important from the point of view of micro-part manufacturing. Most of the sequential machining combinations experimented are in rudimentary form, and still lot of research is required to be done in this area. Some multifunctional machining units and laboratory prototype setups are developed by the researchers, but commercial production of such sophisticated systems integrated unit is required urgently. This emerging technology which recommends initial micropart shaping and final finishing in one setting has tremendous future scope in the area of micro manufacturing. Thus, by considering the potential

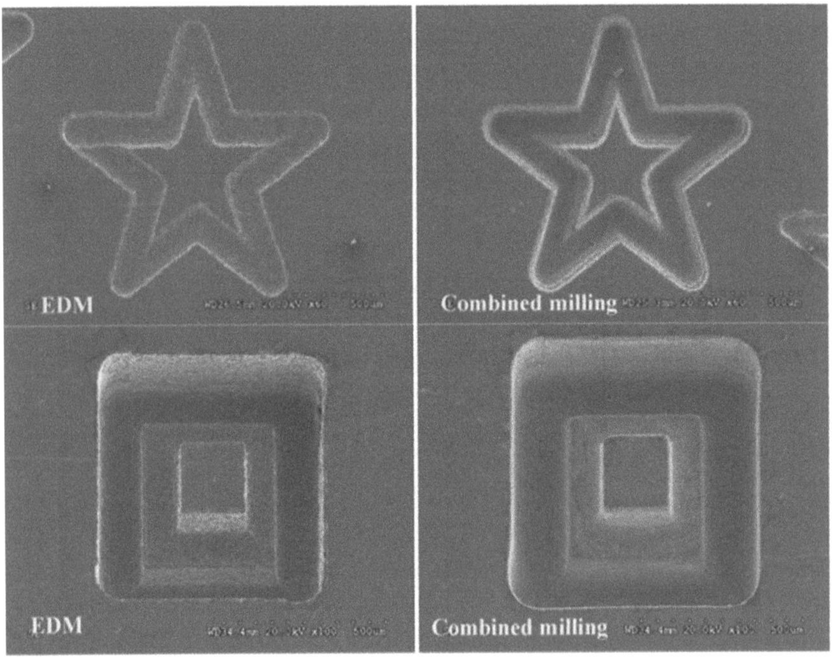

Fig. 7.38 Different features generated by sequential electro-chemical and electro discharge milling [111].

capabilities of this technique, researchers are putting an effort in the direction of developing and making the technology which will be commercially viable.

7.9 Model questions

Q1. What is 'micromachining'? Explain your answer with an example.

Q2. What do you mean by masked based and tool based micromachining processes? Which process should be suggested for the fabrication of a 3D micro component and why?

Q3. What the differences between 'recast layer' and 'Heat affected zone'?

Q4. How many types of electrolytes are used in EMM? What are those? Describe.

Q5. In AJM, for applying metal mask, what should be done or avoided to make sure an excellent transfer of the pattern?

Q6. The depth of the structure is controlled by the uniformity of the blasting profile. For this reason, nozzle is the key part. Compare the purpose of cylindrical as well as Laval nozzle for achieving best AJMM results.

Q7. Can it be considered abrasive waterjet cutting as a cold cutting process? Explain your answer by the events occur in the cutting zone.

Q8. Why abrasive waterjet machining is not so suitable to machine super hard materials? Explain your answer in the light of tube materials.

Q9. Explain about the different types of ultrasonic micromachining systems. Elaborate it with the help of diagrams.

Q10. With the help of diagram discuss about the mechanisms of material removal in ultrasonic micromachining process.

Q11. With the help of Ishikawa diagram, discuss the classification of the process parameters in micro-USM.

Q12. Identify the suitable power circuit for micro-EDM and describe the reasons for the selection.

Q13. What type of dielectric is suitable for micro-EDM?

Q14. Write down the input process parameters that can influence the micro EDM process performance.

Q15. Discuss about the different techniques and procedures of micro tool fabrication in micro-EDM.

Q16. What type of materials cannot be machined by micro-EDM process and why?

Q17. In micro-EDM, to control the feed rate which device should be used?

Q18. Which micro EDM characteristics can be improved by the efficient removal of debris and dielectric circulation?

Q19. Focus on the different problematic areas of micro-EDM and their preventive measures.

Q20. What do you mean by anti reflecting coating? Why it is used in micro-LBM?

Q21. In the light of interaction with material, differentiate between femto second and longer pulses or continuous-wave laser pulse.

Q22. What are the differences in between ECM and EMM?

Q23. What are the advantages of EMM over other types of micromachining?

Q24. In EMM, why pulsed DC power supply is preferred instead of constant DC power supply?

Q25. Describe about the different types of resistance and impedance factors associated with EMM during dissolution of metal?

Q26. In EMM, how overcut can be reduced?

Q27. How deposition on tool can be prevented in EMM?

Q28. What is Hybrid micro machining? Why it is advantageous for micro machining?

Q29. Discuss about the other different types of hybridization applicable for micro machining.

Q30. What is Sequential micro machining (SMM)? Differentiate between Hybrid and SMM.

Q31. Discuss about the different machining strategies applicable in SMM?

Q32. Elucidate in brief general classification of SMM.

Q33. Discuss about the other different types of machining strategies applicable for SMM.

Q34. Elaborate about the three adopted steps which are mostly followed in SMM.

Q35. Identify various challenges encountered during development of different SMM processes.

References

[1] J. Chae, S.S. Park, T. Freiheit, Investigation of micro-cutting operations, Int. J. Mach. Tools Manuf. 46 (2006) 313–332.

[2] S. Filiz, et al., An experimental investigation of micromachinability of copper 101 using tungsten carbide micro-endmills, Int. J. Mach. Tools Manuf. 47 (2007) 1088–1100.

[3] M.S. Puranik, S.S. Joshi, Analysis of accuracy of high-speed aspect ratio holes generated using micro-EDM drilling, J. Eng. Manuf. 222 (2008) 1453–1464.
[4] K. Egashira, et al., Application of USM to micromachining by on-the-machine tool fabrication, Int. J. Electr. Mach. 2 (1997) 31–36.
[5] G.F. Benedict, Non-Traditional Manufacturing Processes, Marcel Dekker, New York, 1987.
[6] H. Getu, et al., Abrasive jet micromachining of polymethyl-methacrylate, Wear 263 (2007) 1008–1015.
[7] Getu, H., et al. Abrasive jet micromachining of acrylic and polycarbonate polymers at oblique angles of attack. Wear 2008. https://doi.org/10.1016/j.wear.2008.01.013.
[8] E. Belloy, et al., The introduction of powder blasting for sensor and microsystems applications, Sens. Actuators, A 84 (2000) 330–337.
[9] A. Pawlowski, et al., Powder blasting patterning technology for microfabrication of complex suspended structures in glass, Microelectron. Eng. 67 (2003) 557–565.
[10] W.E. Cooley, L.L. Clipp, High pressure water jets for undersea rock excavation, s.l.: Trans. ASME, Series B, J. Eng. Ind. 92 (1970) 281.
[11] A.A. Tseng, Topical review: recent developments in micro milling using focused ion beam technology, J. Micromech. Microeng. 14 (2004) R15–R34.
[12] M.J. Vasile, et al., Micro fabrication techniques using focused ion beam and emergent applications, Micron 30 (1999) 235–244.
[13] H. Alemohammad, E. Toyserkani, A.J. Pinkerton, Femtosecond laser micromachining of fiber Bragg gratings for simultaneous measurement of temperature and concentration of liquids, J. Phys. D. Appl. Phys. 41 (2008) 1–9.
[14] K. Sugioka, Y. Cheng, K. Midorikawa, Three-dimensional micromachining of glass using femtosecond laser for lab-on-a-chip device manufacture, Appl. Phys. A 81 (2005) 1–10.
[15] V. Rathod, B. Doloi, B. Bhattacharyya, Influence of electrochemical micromachining parameters during generation of microgrooves, Int. J. Adv. Manuf. Technol. 76 (2015) 51–60.
[16] B. Ghoshal, B. Bhattacharyya, Shape control in micro borehole generation by EMM with the assistance of vibration of tool, Precis. Eng. 38 (2014) 127–137.
[17] T. Masuzawa, State of the art of micromachining, Ann. CIRP 49 (2000) 473–488.
[18] K. Egashira, T. Masuzawa, Microultrasonic machining by the application of workpiece vibration, CIRP Ann. Manuf. Technol. 48 (1999) 131–134.
[19] K. Egashira, K. Mizutani, Micro-drilling of monocrystalline silicon using a cutting tool, Precis. Eng. 26 (2002) 263–268.
[20] X.Q. Sun, T. Masuzawa, M. Fujino, Micro ultrasonic machining and self-aligned multilayer machining/assembly technologies for 3D micromachines, in: Proceedings of the IEEE Micro Electro Mechanical Systems (MEMS'96), 1996, pp. 312–317.
[21] X.Q. Sun, T. Masuzawa, M. Fujino, Micro ultrasonic machining and its applications in MEMS, Sensors Actuators A 57 (1996) 159–164.
[22] M. Komaraiah, P.N. Reddy, A study on the influence of work piece properties in ultrasonic machining, Int. J. Mach. Tools Manuf. 33 (1993) 495–505.
[23] R. Singh, J.S. Khamba, Ultrasonic machining of titanium and its alloys: a review, J. Mater. Process. Technol. 173 (2006) 125–135.
[24] K. Egashira, et al., Microultrasonic machining using multitools, in: Proceedings of the 7th International Conference on Progress Machining Technology (ICPMT '04), December, 2004, pp. 297–301.
[25] P. Heung-Kil, et al., Development of micro-diamond tools through electroless composite plating and investigation into micro-machining characteristics, Precis. Eng. 34 (2010) 376–386.

[26] J.J. Boy, et al., Developments in microultrasonic machining (MUSM) at FEMTO-ST, Int. J. Adv. Manuf. Technol. 47 (2010) 37–45.
[27] Z.Y. Yu, K.P. Rajurkar, A. Tandon, Study of 3D microultrasonic machining, s.l.: Transactions of the ASME, J. Manuf. Sci. Eng. 126 (2004) 727–732.
[28] C. Zhang, R. Rentsch, E. Brinksmeier, Advances in micro ultrasonic assisted lapping of microstructures in hard-brittle materials: a brief review and outlook, Int. J. Mach. Tools Manuf. 45 (2005) 881–890.
[29] A. Curodeau, et al., Ultrasonic abrasive μ-machining with thermoplastic tooling, Int. J. Mach. Tools Manuf. 48 (2008) 1553–1561.
[30] E.C. Jameson, Description and development of electrical discharge machining (EDM), in: Electrical Discharge Machining, Society of Manufacturing Engineers, Dearbern, 2001, p. 12.
[31] D.M. Allen, A. Lecheheb, Micro electro-discharge machining of ink jet nozzles: optimum selection of material and machining parameters, J. Mater. Process. Technol. 58 (1996) 53–66.
[32] S.T. Newman, K.H. Ho, The state of art—electrical discharge machining, Int. J. Mach. Tools Manuf. 43 (2003) 1287–1300.
[33] O. Koch, et al., Recent progress in micro-electro discharge machining technology—part 1, in: Proceedings of the 13th International Symposium for Electromachining ISEM XIII, Bilbao, Spain, 2001.
[34] T. Masuzawa, M. Fujino, K. Kobayashi, Wire electro-discharge grinding for micromachining, CIRP Ann. 34 (1985) 431–434.
[35] N. Ravi, H. Huang, Fabrication of symmetrical section microfeatures using the electro-discharge machining block electrode method, J. Micromech. Microeng. 12 (2002) 905–910.
[36] B.B. Pradhan, B. Bhattacharyya, Improvement in micro-hole machining accuracy by polarity changing technique for micro-EDM on Ti-6Al-4V, Proc. Inst. Mech. Eng. B J. Eng. Manuf. 222 (2008) 163–173.
[37] Y.S. Wong, L.C. Lim, L.C. Lee, Effect of flushing on electro-discharge machine surfaces, J. Mater. Process. Technol. 48 (1995) 299–305.
[38] A.P. Tiwary, B.B. Pradhan, B. Bhattacharyya, Investigation on the effect of dielectrics during micro-electro-discharge machining of Ti-6Al-4V, Int. J. Adv. Manuf. Technol. 95 (2017) 1–14.
[39] H. Takezawa, et al., Development of micro-EDM with rapidly sharpened electrode, J. Mater. Process. Technol. 149 (2004) 112–116.
[40] Y. Imai, et al., Local actuator module for highly accurate micro-EDM, J. Mater. Process. Technol. 149 (2004) 328–333.
[41] B.H. Yan, et al., Study of precision micro-holes in borosilicate glass using micro-EDM combined with micro ultrasonic vibration machining, Int. J. Mach. Tools Manuf. 42 (2002) 1105–1112.
[42] G. Chern, Y.E. Wu, S. Liu, Development of micro-punching machine and study on the influence of vibration machining in micro-EDM, J. Mater. Process. Technol. 180 (2006) 102–109.
[43] J.C. Hung, et al., Using a helical micro-tool in micro-EDM combined with ultrasonic vibration for micro-hole machining, J. Micromech. Microeng. 16 (2006) 2705–2713.
[44] D.K. Aspiwall, et al., Workpiece surface roughness and integrity after WEDM of Ti-6Al-4V and Inconel 718 using minimum damage generator technology, CIRP Ann. Manuf. Technol. 57 (2008) 187–190.
[45] K.M. Patel, P.M. Pandey, P.V. Rao, Surface integrity and material removal mechanisms associated with the EDM of Al2O3 ceramic composite, Int. J. Refract. Met. Hard Mater. 27 (2009) 892–899.

[46] D.T. Pham, et al., Micro-EDM—recent developments and research issues, J. Mater. Process. Technol. 149 (2004) 50–57.

[47] M.P. Jahan, et al., On-machine fabrication of high-aspect-ratio microelectrodes and application in vibration-assisted micro-EDM drilling of tungsten carbide, J. Eng. Manuf. 224 (2010) 795–814.

[48] K. Takahata, N. Shibaike, H. Guckel, High-aspect-ratio WC–Co microstructure produced by the combination of LIGA and micro-EDM, Microsyst. Technol. 6 (2000) 175–178.

[49] K. Liu, B. Lauwers, D. Reynaerts, Process capabilities of Micro-EDM and its applications, Int. J. Adv. Manuf. Technol. 47 (2010) 11–19.

[50] Y.S. Liao, et al., Fabrication of high aspect ratio microstructure arrays by micro reverse wire-EDM, J. Micromech. Microeng. 15 (2005) 1547.

[51] C.S. Lin, et al., Fabrication of micro ball joint by using micro-EDM and electroforming, Microelectron. Eng. 87 (2010) 1475–1478.

[52] G. Gao, et al., Instantaneous fabrication of tungsten microelectrode based on single electrical discharge, J. Mater. Process. Technol. 168 (2005) 83–88.

[53] F. Han, et al., Experimental attempts of sub-micrometer order size machining using micro-EDM, Precis. Eng. 30 (2006) 123–131.

[54] D. Karnakis, Ultrafast Laser Nanomachining: Doing More With Less, Commercial MicroManufacturing, Oxford, 2008.

[55] F. Dausinger, H. Hugel, V.I. Konov, Micromachining with ultra short laser pulses: from basic understanding to technical applications, Proc. SPIE 5147 (2003) 106–115.

[56] A.Y. Fasasi, et al., Nano-second UV laser processed micro-grooves on Ti6Al4V for biomedical applications, Mater. Sci. Eng. C 29 (2009) 5–13.

[57] C. Mullan, et al., Estimating spot size and relating hole diameters with fluence and number of shots for nanosecond and femtosecond laser ablation of polyethylene terephthalate, J. Laser Appl. 19 (2007) 158–164.

[58] H. Meng, et al., Laser micro-processing of cardiovascular stent with fiber laser cutting system, Opt. Laser Technol. 41 (2009) 300–302.

[59] G. Kamlage, et al., Deep drilling of metals by femtosecond laser pulses, Appl. Phys. A 77 (2003) 307–310.

[60] A.C. Forsman, et al., Double pulse machining as a technique for the enhancement of material removal rates in lase rmachining of metals, J. Appl. Phys. 98 (2005) 1–5.

[61] W.C. Wang, et al., Femtosecond laser drilling of alumina ceramic substrates, Appl. Phys. A Mater. Sci. Process. 101 (2010) 271–278.

[62] A.H. Khan, et al., Influence of micro supersonic gas jets on nanosecond laser percussion drilling, Opt. Lasers Eng. 45 (2007) 709–718.

[63] D. Breitling, A. Ruf, F. Dausinger, Fundamental aspects in machining of metals with short and ultrashort laser pulses, Proc. SPIE 5339 (2004) 49–63.

[64] C. Föehl, F. Dausinger, High precision deep drilling with ultra short pulses, Proc. SPIE 5063 (2003) 346–351.

[65] J. Meijer, Laser beam machining (LBM) state of the art and new opportunities, J. Mater. Process. Technol. 149 (2004) 2–17.

[66] D. Breitling, A. Ruf, F. Dausinger, Fundamental aspects in machining of metals with short and ultra short laser pulses, Proc. SPIE 5339 (2004) 49–63.

[67] N.H. Rizvi, Femtosecond laser micromachining: current status and applications, Riken Rev. 50 (2003) 107–112.

[68] U. Loeschner, et al., Micromachining of glass with short ns-pulses and highly repetitive fs-laser pulses, in: Proceedings of the 27th International Congress on Applications of Lasers and Electro-Optics (ICALEO '08), 2008, pp. 193–201.

[69] M.C. Gower, Industrial application of laser micromachining, Opt. Express 7 (2000) 56–67.

[70] S. Mishra, V. Yadava, Laser beam micro machining (LBMM)—a review, Opt. Lasers Eng. 73 (2015) 89–122.
[71] J. Ren, M. Kelly, L. Hesselink, Laser ablation of silicon in water with nanosecond and femtosecond pulses, Opt. Lett. 30 (2005) 1740–1742.
[72] A. Kruusing, Underwater and water-assisted laser processing: part 2—etching, cutting and rarely used methods, Opt. Lasers Eng. 41 (2004) 329–352.
[73] Y.L. Tao, H. Chen, W. Zhang, Time scale effects in laser material removal: a review, Int. J. Adv. Manuf. Technol. 26 (2005) 598–608.
[74] T. Hanemann, et al., Laser micromachining and light induced reaction injection molding as suitable process sequence for the rapid fabrication of micro component, Microsyst. Technol. 7 (2002) 209–214.
[75] B. Bhattacharyya, J. Munda, M. Malapati, Advancement in electrochemical micromachining, Int. J. Mach. Tool Manu. 44 (2004) 1577–1589.
[76] M.M. Lohrengel, et al., Microscopic investigations of electrochemical machining of Fe in NaNO3, Electrochim. Acta 48 (2003) 3203–3211.
[77] J. Kozak, K.P. Rajurkar, B. Wei, Modeling and analysis of pulse electrochemical machining, J. Eng. Ind. 116 (1994) 316–323.
[78] J. Munda, M. Malapati, B. Bhattacharyya, Control of micro-spark and stray—current effect during EMM process, J. Mater. Process. Technol. 194 (2007) 151–158.
[79] M. Kock, V. Kirchner, R. Schuster, Electrochemical micro machining with ultra short voltage pulses—a versatile method with lithographic precision, Electrochim. Acta 48 (2003) 3213–3219.
[80] B. Bhattacharyya, Electrochemical Micromachining for Nanofabrication, MEMS and Nanotechnology, s.l.: William Andrew, an imprint of Elsevier, 2015. 978-0-323-32737-4.
[81] J. Lei, et al., Fabrication of 3D microelectrodes by combining wire electrochemical micromachining and micro-electric resistance slip welding, s.l.: Procedia CIRP, in: 18th CIRP Conference on Electro Physical and Chemical Machining (ISEM XVIII), vol. 42, 2016, pp. 825–830.
[82] V. Rathod, B. Doloi, B. Bhattacharyya, Parametric investigation into the fabrication of disk microelectrodes by electrochemical micromachining, J. Micro Nano-Manuf. 1 (2013) 041005-1–041005-11.
[83] J.J. Wang, et al., Fabrication of wedge-shape tool via electrochemical micromachining with diamond-like carbon coating, J. Mater. Process. Technol. 187 (2007) 264–269.
[84] D. Landolt, P.F. Chauvy, O. Zinger, Electrochemical micro machining, polishing and surface structuring of metals: fundamental aspects and new developments, Electrochim. Acta 48 (2003) 3185–3201.
[85] M. Datta, D. Landolt, Fundamental aspects and applications of electrochemical micro fabrication, Electrochim. Acta 45 (2000) 2535–2558.
[86] M. Bassu, L.M. Strambini, G. Barillaro, Advances in electrochemical micromachining of silicon: towards MEMS fabrication, Proc. Eng. 25 (2011) 1653–1656.
[87] M. Wang, et al., Electrochemical machining of the spiral internal turbulator, Int. J. Adv. Manuf. Technol. 49 (2010) 969–973.
[88] C.H. Jo, B.H. Kim, C.N. Chu, Micro electrochemical machining for complex internal micro features, CIRP Ann. Manuf. Technol. 58 (2009) 181–184.
[89] K.H. Bo, P.J. Byung, C.N. Chong, Fabrication of multiple electrodes by reverse EDM and their application in micro ECM, J. Micromech. Microeng. 16 (2006) 843.
[90] M.D. Nguyen, M. Rahman, Y.S. Wong, Enhanced surface integrity and dimensional accuracy by simultaneous micro-ED/EC milling, CIRP Ann. Manuf. Technol. 61 (2012) 191–194.
[91] M. Kunieda, et al., Electrochemical micromachining using flat electrolyte jet, CIRP Ann. Manuf. Technol. 60 (2011) 251–254.

[92] S.S. Hong, K.B. Hyun, C.N. Chong, Analysis of the side gap resulting from micro electrochemical machining with a tungsten wire and ultrashort voltage pulses, J. Micromech. Microeng. 18 (2008) 075009.
[93] O. Zinger, P.F. Chauvy, D. Landolt, Scale-resolved electrochemical surface structuring of titanium for biological applications, J. Electrochem. Soc. 150 (2003) B495.
[94] M. Kock, V. Kirchner, R. Schuster, Electrochemical micromachining with ultrashort voltage pulses—a versatile method with lithographical precision, Electrochim. Acta 48 (2003) 3213–3219.
[95] A.L. Trimmer, et al., Single-step electrochemical machining of complex nanostructures with ultrashort voltage pulses, Appl. Phys. Lett. 82 (2003) 3327–3329.
[96] B.H. Kim, et al., Micro electrochemical machining of 3D micro structure using dilute sulfuric acid, CIRP Ann. 54 (2005) 191–194.
[97] X. Lu, Y. Leng, Electrochemical micromachining of titanium surfaces for biomedical applications, J. Mater. Process. Technol. 169 (2005) 173–178.
[98] K.P. Rajurkar, et al., New developments in electrochemical machining, CIRP Ann. 48 (1999) 567–579.
[99] P.S. Pa, Super finishing with ultrasonic and magnetic assistance in electrochemical micro-machining, Electrochim. Acta 54 (2009) 6022–6027.
[100] D. Zhu, et al., Precision machining of small holes by the hybrid process of electrochemical removal and grinding, CIRP Ann. Manuf. Technol. 60 (2011) 247–250.
[101] T. Masuzawa, S. Sakai, Quick finishing of WEDM products by ECM using a mate-electrode, Ann. CIRP 36 (1987) 123–126.
[102] M.D. Nguyenn, M. Rahman, Y.S. Wong, Simultaneous micro-EDM and micro-ECM in low-resistivity deionized water, Int. J. Mach. Tools Manuf. 54 (2012) 55–65.
[103] A.D. Davydov, Laser electrochemical machining of metals, Russ. J. Electrochem. 30 (1994) 871–881.
[104] M. Datta, et al., Jet and laser-jet electrochemical micromachining of nickel and steel, Electrochem. Soc. 136 (1989) 2251–2256.
[105] S.Z. Chavoshi, S. Goel, P. Morantz, Current trends and future of sequential micro-machining processes on a single machine tool, Mater. Des. 127 (2017) 37–53.
[106] S. Skoczypiec, A. Ruszaj, A sequential electrochemical-electro-discharge process for micropart manufacturing, Precis. Eng. 38 (2014) 680–690.
[107] W.B. Rashid, et al., The development of a surface defect machining method for hard turning processes, Wear 302 (2013) 1124–1135.
[108] J. Yan, F. Kobayashi, Laser recovery of machining damage under curved silicon surface, CIRP Ann. 62 (2013) 199–202.
[109] A. Perveen, et al., A study on microgrinding of brittle and difficult-to-cut glasses using on-machine fabricated poly crystalline diamond (PCD) tool, J. Mater. Process. Technol. 212 (2012) 580–593.
[110] S. Kim, et al., Hybrid micromachining using a nanosecond pulsed laser and micro EDM, J. Micromech. Microeng. 20 (2009) 015037.
[111] Z. Zeng, et al., A study of micro-EDM and micro-ECM combined milling for 3D metallic micro-structures, Precis. Eng. 36 (2012) 500–509.

CHAPTER EIGHT

Advanced finishing processes

Contents

Modern Machining Technology
https://doi.org/10.1016/B978-0-12-812894-7.00008-6

© 2020 Elsevier Inc.
All rights reserved.

8 Introduction

Traditional super finishing processes such as grinding, lapping, polishing and honing are commonly used for industry, but there are some limitations such as sub-surface damage, residual stress and difficult to finish complex and free form surfaces. But for achieving damage free, nano-level or angstrom level surface finish on difficult to finish materials such as ceramics, glass, silicon wafer, diamond wafer, sapphire wafer etc., and also for finishing complex and free form surfaces with high accuracy, the advanced ultra precision finishing processes are very much needed. In this chapter various advanced ultra precision finishing processes with respect to their working principles, setup details, process parametric effects, advantages and their applications are discussed. This chapter includes the discussion on Abrasive Flow Finishing (AFF), Chemical Mechanical Polishing (CMP), Elastic Emission Machining (EEM), Magnetic Abrasive Finishing (MAF), Magnetorheological Finishing (MRF) as well as Plasma Assisted Polishing (PAP) process. During abrasive based advanced finishing processes, normal force is responsible for penetration of abrasive particle into the workpiece and the depth of penetration is nanometer or angstrom range. The cutting or axial force is responsible and very much important for removal of the materials in the form of micro or nano chips or atoms. During abrasive flow finishing (AFF) and chemical mechanical polishing (CMP), there is no external control of forces during finishing operation, but nano level surface finish can be generated by controlling finishing process parameters. During magnetic abrasive finishing (MAF) and magnetorheological finishing (MRF), there is proper control of forces by externally applied magnetic field and therefore the nano-level finishing of internal and external flat as well as curved surface takes place comparatively at faster rate. For generating atomically flat and damage free surface on optics and semiconductors materials, elastic emission machining (EEM) and plasma assisted polishing (PAP) processes are effectively utilized.

SUBCHAPTER 8.1

Abrasive Flow Finishing (AFF)

8.1.1 Introduction

Abrasive flow finishing (AFF) is a non-conventional micro and nano finishing process, which is characterized by flowing an abrasive-laden abrasive medium through a workpiece. It is also well-known as abrasive flow deburring or extrude honing. Abrasive flow machining was first patented by the Extrude Hone Corporation in 1960. The abrasive flow finishing processes are widely used as finishing solutions in high-tech industries like defense, aerospace, automobile, tool and die, and prosthetics. The used fluid in AFF is characteristically high viscous having the constancy of putty [1, 2]. In this process, the flowing semi-solid abrasive-laden putty removes a small quantity of material from the workpiece surface to be finished. During AFF process the rubbing and plowing action of abrasive grain on the workpiece materials is the mechanism of material removal process [3]. AFF process facilitates to finish difficult to reach areas, to follow complex contours, and to concurrently work on many edges and surfaces, making it more flexible than other finishing processes. Abrasion occurs only where the media flow is constrained, other areas remain unaffected. AFF process provides a high level of surface finish and close geometric tolerances for a wide range of industrial components. This process can be effectively micro machined from soft aluminum to tough nickel alloys, ceramics, and carbides [4]. AFF process has been applied to smooth extrusion dies, welding nozzle, aerofoil surfaces of impellers, spring collects and micro channels [5]. AFF smoothes and finishes rough surfaces and is specifically used to polish surfaces, form radii, and even remove burrs [6]. The finishing process of AFF is ideal for slots, cavities, interior surfaces, holes, and other areas that may be difficult to reach with other polishing or grinding processes [5]. AFF is not typically used for large stock-removal operations due to its low material removal rate [5].

8.1.2 Basic working principle

In abrasive flow machining, the abrasive-laden fluid and abrasive particles pass through the workpiece and successfully performs erosion. The mixed fluid with abrasive particles generates different features on the surface of the workpiece. A hydraulic ram is used to force out the fluid through the workpiece where it acts as a flexible file, or slug, molding itself accurately to the form of the workpiece [1]. The highest amount of material removal takes place in the constrained areas where the flow of the fluid is controlled. According to Bernoulli's Principle, the flow pressure and speed of the fluid increase the material removal rate (MRR) [4]. The exerted pressure by the fluid on all contacting surfaces results in a very good surface finish. AFF performs the super finishing on the workpiece by the flow of abrasive laden polymer media as a one-way flow process or repeatedly as a two-way flow process. In one way flow process, the abrasive media is passed in one direction and it returns from the other end. In the two-way flow process, two vertically opposite hydraulic cylinders drive the abrasive laden polymer media up and down. The process is utilized for super-finishing of intricate cavities, edges, etc. At the same time, this process is able to fine finish a number of parts or different areas of the same workpiece. This process has capability for handling multiple parts of various workpieces per day for fine finishing operation reducing the production costs and tedious physical work. This process can also be used for different super-finishing operations acquiring very homogeneous and accurate results through proper control of the process parameters of AFF [7].

AFF process is classified into three categories viz. one way AFF, two way AFF and orbital AFF. One-way AFF process supplies the abrasive media through the workpiece in only one direction and allowing the abrasive media to exit freely from the part for easy cleaning, fine finishing, fast processing, and simple quick-exchange tooling. It is also able to process larger parts for super finishing. Fig. 8.1.1 shows the schematic diagram of one way AFF system [8].

Two ways AFF process exploits two vertically opposed cylinders which extrude an abrasive media to and fro through passages shaped by the workpiece and tooling. Abrasive action performs suitably where the media passes through the most restricted passages. The extrusion pressure is controlled between 7 and 200 bar. This process has faster setup, fully automated system capabilities,

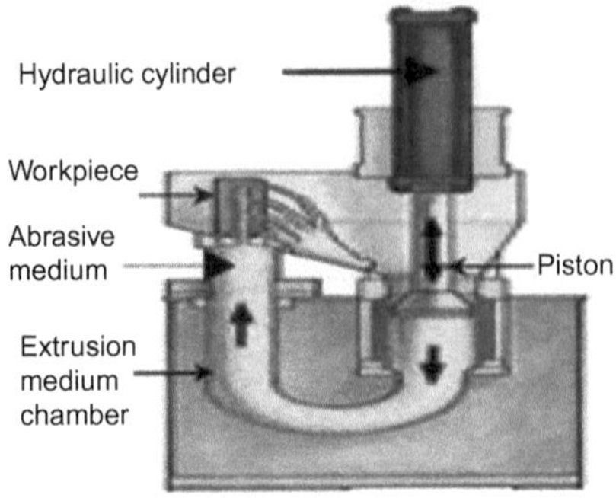

Fig. 8.1.1 Schematic diagram of one way AFF system [8].

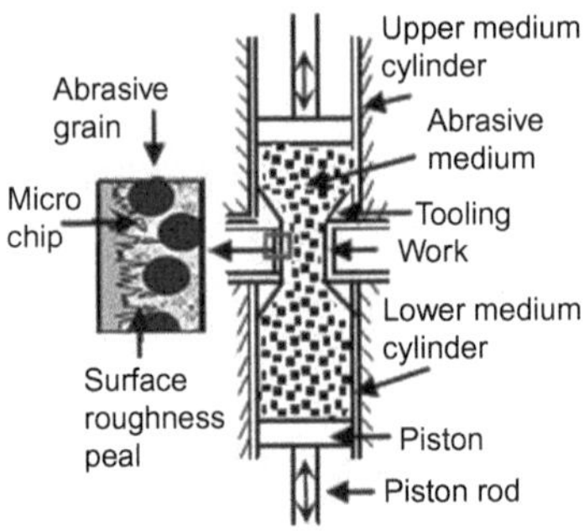

Fig. 8.1.2 Schematic diagram of two way AFF system [8].

excellent process control, quick-change tooling capacity, good control of radius generation. It can also finish both inner and outer diameter of the components. Fig. 8.1.2 shows the schematic diagram of two way AFF system [8].

In orbital AFF process, surface and edge finishing are performed by rapid, low amplitude and oscillations of the workpiece with respect to a self-forming elastic and plastic abrasive polishing tool. The tool is a layer of abrasive- laden elastic and plastic medium, but typically higher in viscosity and more in elastic. Fig. 8.1.3 shows the schematic diagram of orbital AFF system [8].

8.1.3 Setup details

Abrasive Flow Finishing (AFF) set-up consists of three elements, i.e. AFF machine, tooling, and abrasive media. Most machines allow for the

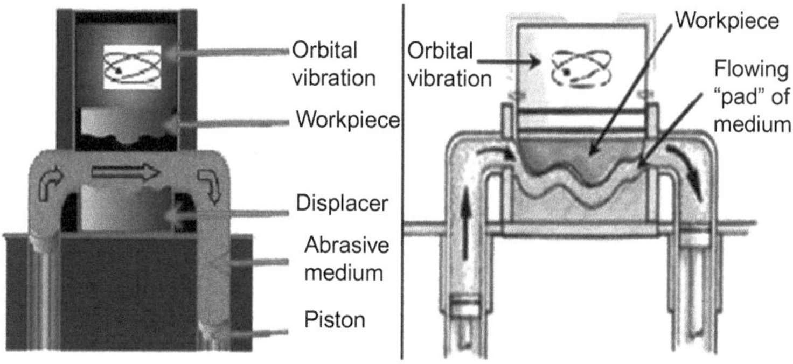

Fig. 8.1.3 Schematic diagram of orbital AFF system [8].

loading of various abrasive mediums and comprise the capacity to adjust the pressure utilized in extruding the medium through the workpiece. These machines may be manually operated or automated using CNC. For high production volumes, some important accessories such as unloading and reloading stations, part-cleaning stations, media heat exchangers and media re-feed devices may be included [1].

AFF system controls various process parameters such as number of cycles, extrusion pressure, abrasive media flow rate, viscosity, and concentration, etc. and component geometry related parameters such as aspect ratio, length of flow path, passage shape, etc. which affect the process performance and product quality such as surface integrity, surface finish, and material removal rate [8].

(i) AFF Machine

AFF set-up has various components such as hydraulic cylinder, medium cylinder, piston, working fixture, frame and housing. The medium cylinder supplies sufficient quantity of abrasive medium and guides to pass the abrasive medium through the passage of workpiece by the action of medium cylinder piston. The aluminum alloy is used as piston materials due to its light weight and good strength., Teflon piston ring is mounted on the piston head. A connecting rod connects the hydraulic cylinder rod with the medium cylinder piston. AFF machine is constructed with two opposite medium chambers which are hydraulically or manually clamped together. The pressure for extruding medium ranges from 7 bar to 220 bar. Lower pressure is preferred for fragile part or loose component. Little heat is generated due to friction hence medium coolants are recommended. Fig. 8.1.4 shows the schematic diagram of AFF set-up [9].

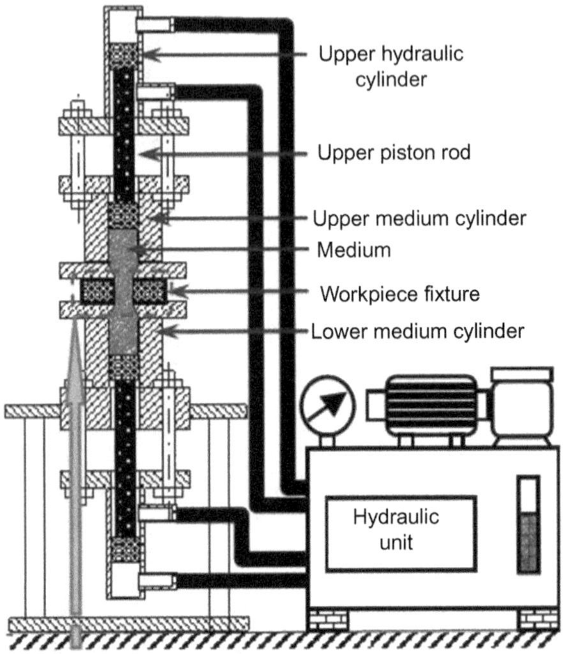

Fig. 8.1.4 Schematic diagram of AFF set-up [9].

The R-AFF set-up consists of specially designed rotary tooling, variable frequency drive along with high torque motor, speed reduction gear box, external geared tooling workpiece fixture, hydraulic drive and supporting frame. The input to reduction gear box is given from the variable frequency drive (VFD) via a motor. The workpiece can be rotated at different rpm in rotary AFF system for improving the performance of AFF process [10].

(ii) Fixtures

Fixtures are used only to hold the parts in position and also to contain or direct the flow of abrasive media during finishing operation. For guiding the abrasive medium through the workpiece properly without any leakage, fixtures provide a restriction passage. Fixture controls the medium flow uniform across the surface to be finished. Any numbers of parallel restrictions can be processed with uniform result. Generally replaceable inserts made of nylon, Teflon or hardened steel is used, can last for up to thousands of parts.

(iii) Media

Abrasive media in AFF consists of base polymers, abrasive particles and additives. The base viscoelastic polymer is made up of organic polymer and hydrocarbon gels. The base polymer is reinforced with the abrasive particles. The base polymer acts as a carrier medium and determines the stiffness of AFF media. Silicon Rubber (P- Silicone), Silicon Rubber with Additives (A- Silicone), Natural Rubber, Butyl Rubber, Ethylene Propylene Diene Monomer, Styrene Butadiene Rubber are used as base polymer in AFF medium [11]. The commonly used polymer media are polyborosiloxane and silicone rubber. The abrasive particle acts as a cutting tool which removes the material from the workpiece. The commonly used abrasives are silicon carbide, aluminum oxide, boron carbide, cubic boron nitride and polycrystalline diamond powders. The abrasive grit size varies from 8 to 1000 mesh size. The hardeners, plasticizers and lubricants are used as additives in AFF medium. The most commonly used AFF medium is polyborosiloxane, also known as "silly putty" used in AFF [11]. The stiffness of AFF media plays an important role in AFF process. AFF media acts as a deformable grinding stone. High stiffness AFF media results in uniform finishing whereas low stiffness AFF medium produces large radius at the passage opening [12]. Fig. 8.1.5 shows the photographic view of AFF machine.

8.1.4 Influence of process parameters

The controllable process parameters of AFF process are extrusion pressure, flow rate and viscosity of abrasive media, number of cycles and abrasive type, size and concentration. There are other process parameters such as hardness, composition of workpiece materials and flow path geometry of workpiece. Fixture design, size of cylinder and clamping pressure are AFF machine and tooling related process parameters. The influential AFF process parameters and their effect on the machining characteristics are important to control the process capability and productivity. The influences of significant parameters on the performance of AFF process are analyzed through the subsequent discussion.

(i) Extrusion pressure

Extrusion pressure is one of the important parameters, which has a major influence on the AFF machined characteristics. The range of extrusion pressures is varying from 0.7 to 22 MPa and material removal is quicker at an increased extrusion pressure with other constant parameters. The exerted force of extrusion pressure on abrasive particles during AFF is partly lost

Fig. 8.1.5 Photographic view of AFF Machine. *(Courtesy: Extrude Hone Inc.)*

within the media due to its internal resistance and the rest is imparted to the abrasive particles contacting the workpiece surface [13]. With increasing extrusion pressure, MRR increases [14].

(ii) Abrasive media flow rate

The abrasive media flow rate is more responsible for uniform material removal and the formation of an edge radius in AFF process [13]. The flow rate is the ratio between the flow volume and the processing time. Slow speed of media flow rate is more suitable for uniform material removal and higher speed of media flow rate creates a bigger edge radius [13]. In case of brass, there is a reasonably smaller effect of media flow speed on material removal [15]. The abrasive media flow volume has the most significant impact for controlling the surface finish and the amount of abrasion in the AFF process. Higher amount of media flow causes more abrasion on the workpiece keeping other process parameters constant. The rate of material removal by abrasion is directly proportional to slug length of media flow. The slug length of media flow is decided by the formula of the media flow volume divided by the restricted area of passage.

(iii) Abrasive media viscosity

The media viscosity has the significant effect on the MRR and surface roughness in AFF process. Higher viscosity enhances surface finish and material removal rate keeping all other parameters constant. One thumb rule recommends that if the passage length is considerably smaller than two times the original passage length, the higher media viscosity is used [16]. Conversely, if the passage length is significantly longer than two times the passage width, the lower viscosity is suggested. The media viscosity is also influenced by abrasive particle size and concentration. Higher temperature decreases the media viscosity, which may influence the overall abrasion process and flow properties. If the media viscosity lies below 30 Pa-s the abrasive particles remain in suspension within the media. There is a probability for abrasive particles to sink under gravity at the bottom of the media cylinder and this may cause the failure of media.

(iv) Abrasive particle size and concentration

The used abrasive particles in the AFF process are ranging from 8 mesh (grit) size (for roughing and stock removal applications) to 500 mesh size (for small hole applications) [15]. For better surface finish, higher mesh size or smaller abrasive grain size is suggested into complex and narrow passages. For fine surface finish of the work surface, finer abrasives are 1000 mesh size [14]. MRR as well as surface roughness level decreases with increasing the mesh size. MRR decreases because the depth of penetration and width of penetration decreases with increasing mesh size. However, this effect was suitable up to a certain percentage of abrasive concentration beyond which it becomes unimportant.

(v) Number of cycles

In the AFF process, this parameter is significant to choose the exact number of cycles that are required to attain the final surface finish. The total number of process cycles is ranging from one to several hundred. The abrasion is more comprehensive for a number of initial cycles, after which the material removal rate (MRR) and enhancement in surface finish are steady [14]. The material removal increases non-linearly with an increase in the number of cycles. Conversely, the material removal diminishes with an increase in the number of cycles [15].

8.1.5 Advantages, disadvantages and challenges

Advantages of AFF are as follows:

(i) AFF is an automated finishing process.
(ii) AFF can deburr or polish in one operation.
(iii) It can reduce surface roughness up to 0.05 μm.
(iv) AFF can produce true round radii even on complex edges.
(v) AFF can finish multiple number of holes or passages of parts simultaneously with uniform results. This process has simplified process capability for single component arrangement to multi-component arrangement.
(vi) It has a more uniform flow of abrasive media over complex shapes, multi-part fixtures and long passages.
(vii) This process decreases the risk of machine and tooling failure at high flow rates and high temperatures, and enhances media life due to controlled stresses.
(viii) It stabilizes media viscosity throughout long process periods.

Disadvantages of AFF are as follows:

(i) Fixture may be expensive.
(ii) Cannot process blind hole.
(iii) Large surface irregularities like deep scratches or large bumps cannot be removed.
(iv) Out of roundness and taper cannot be corrected.

Challenges of the AFF process are highlighted as follows:

(i) Utilization of AFF for finishing of different components having intricate contoured geometry such as different gears (i.e., spiral, hypoid, non-circular gears, worm, etc.), 3D contoured surfaces, impellers, micro-sized components, blind cavities, valves of an aero-engine, fuel injection nozzles, biomedical implants, etc. still requires to be explored. Finishing of these complicated components by the traditional finishing processes is a very difficult task and provides some unwanted side effects and results. AFF is one of the alternative super-finishing processes which can afford better material removal, surface integrity and surface quality economically.
(ii) Finishing of other hard materials such as high speed steel, functionally graded materials, shape memory materials, glasses and other ceramics, titanium alloys, tool steels, and composites by AFF desires concentration of the researchers.

(iii) More research is needed to develop efficient, economic, and non-toxic finishing medium and control of the properties of the finishing medium during the process.

(iv) Online monitoring of the surface quality produced and adjustment of the AFF process parameters and properties of the finishing medium is still unexplored area for future research.

(v) More concentration is still required for successful commercialization by adopting hybrid processes of AFF.

(vi) Development of various flexible workpiece fixtures is still needed for super- finishing of a variety of components.

(vii) Development of modeling and simulation of AFF process is still required to predict most favorable values of process parameters for achieving the desired improvement in workpiece as well as to reduce time and exhaustive experiments.

8.1.6 Applications

AFF process produces high-quality machined components with improved surface quality, enhanced cycle fatigue strength, reduced production costs, increased production throughput, improved functionality, enhanced uniformity and cleanliness of surfaces. This process is specially used in aerospace industry, dies and mold industry, automotive industry and medical industry. This makes AFF ideal for the following surface machining applications:

(i) Radiusing process produces continuous true-edge or rounded-edge radii.

(ii) Surface stress relief in AFF process removes stress that may lead to crack propagation.

(iii) AFF Polishing improves the surface finish uniformly while preserving the component geometry.

(iv) Geometry Optimization in AFF process improves entry and exit holes and fine-tuning surfaces to improve the flow characteristics of the finished component.

(v) AFF Deburring removes the unwanted material attached to a component as a result of boring, drilling or cutting operations.

AFF process has been applied for finishing various components used in industrial and biomedical applications to the required level of accuracy. The various applications of AFF process are presented in this section.

(a) Finishing of micro-channels

Micro channels are basic elements of micro-fluidic technology in biomedical applications. A self-modulating abrasive medium has been used in AFM to remove the recast layer from the micro channels produced by wire-EDM [17]. The quality of the channels has been improved by removing burrs, straightness, recast layer etc., with low cost and high efficiency.

(b) Finishing of micro-bores

AFF process has been applied to achieve the high quality of micro bore. Micro bores of diameter smaller than 500 μm in various sensors of miniaturized parts such as fuel injectors, micro filters, ink-jet printer nozzles, micro pumps. AFM technology is applied to polish micro bores of 400 and 500 μm bores in steel, 500 μm bores in stainless steel and 304 and 260 μm bores in zirconia ceramic material [18].

(c) Finishing of gears

Finishing of gears is challenging task in many industrial applications. Ultrasonically Assisted Abrasive Flow Machining (UAAFM) technique is applied to finish bevel gears to maximum level of accuracy. Ultrasonic frequency, extrusion pressure, time and media flows were considered as the input variables to improve the surface finish of bevel gears [19]. AFF process finished the burrs present at the interaction between the tooth surface and the end surface of the helical gears to improve the quality of helical gears [20]. AFF process is also applied for polishing plastic gear [21].

(d) Finishing of nozzles

AFF process is applied to finish Direct Injection (DI) diesel engine fuel injector nozzles and improved quality of the nozzle characteristics are found in AFM processed injectors resulting in enhancement of engine performance and improved emissions [22]. A prototype abrasive flow machine was used for finishing of convergent divergent nozzle [23].

Abrasive flow finishing is an advanced finishing process which has wide scope of application in different fields such as automobile, aerospace, die and mold finishing, biomedical, MEMS etc. AFF process is applied for finishing, deburring, edge rounding and improving the general surface quality of complex geometrical workpiece. As AFF is an automated flexible finishing process, it is applied for finishing tooth of helical and bevel gears of soft, ductile as well as hard and brittle materials by proper selection of abrasive medium and working conditions.

References

[1] L.R.K. Gillespie, Deburring and Edge Finishing Handbook, SME, 1999. ISBN: 978-0872635012, Chapter 12.

[2] G.F. Schrader, A.K. Elshennawy, L.E. Doyle, Manufacturing Processes and Materials, fourth ed., SME, 2000, ISBN: 978-0-87263-517-3, p. 626.
[3] V.K. Gorana, V.K. Jain, V.K. Lal, Forces prediction during material deformation in abrasive flow machining, Wear 260 (2006) 128–139.
[4] S. Kalpakjian, S.R. Schmid, Manufacturing Engineering and Technology, Pearson Education, Asia, 2000.
[5] G.F. Benedict, Non-traditional Manufacturing Processes, Marcel Dekker, New York, 1987.
[6] J.D. Kim, K.D. Kim, De-burring of burrs in spring collets by abrasive flow machining, Int. J. Adv. Manuf. Technol. 24 (2004) 469–473.
[7] G. Kibria, B. Bhattacharyya, J.P. Davim (Eds.), Non-Traditional Micromachining Processes, Springer International Publishing AG, 2017, ISBN: 978-3-319-52008-7, pp. 46–47.
[8] S. Santhosh Kumar, S.S. Hiremath, A review on abrasive flow machining (AFM), RAEREST 2016, Procedia Technol. 25 (2016) 1297–1304.
[9] M.R. Sankar, V.K. Jain, K.P. Rajurkar, "Nano-finishing studies using elastically dominant polymers blend abrasive flow finishing medium" 19th CIRP conference on electro physical and chemical machining, Procedia CIRP 68 (2018) 529–534.
[10] M.R. Sankar, V.K. Jain, J. Ramkumar, Experimental investigations into rotating workpiece abrasive flow finishing, Wear 267 (2009) 43–51.
[11] S. Singh, M. Ravi Sankar, "Design and performance evaluation of abrasive flow finishing process during finishing of stainless steel tubes", 4th international conference on materials processing and characterization, Mater. Today Proc. 2 (2015) 3161–3169.
[12] V.K. Jain, Advanced Machining Processes, Allied Publishers, New Delhi, India, 2002.
[13] L.J. Rhoades, Abrasive flow machining: a case study, J. Mater. Process. Technol. 28 (1991) 107–116.
[14] R.E. Williams, K.P. Rajurkar, Stochastic modelling and analysis of abrasive flow machining, Trans. ASME J. Eng. Ind. 114 (1992) 74–81.
[15] V.K. Jain, S.G. Adsul, Experimental investigations into abrasive flow machining (AFM), Int. J. Mach. Tools Manuf. 40 (2000) 1003–1021.
[16] T. Kohut, Surface finishing with abrasive flow machining, in: Proceedings of the Fourth International Seminar on Aluminium Extrusion Technology, Washington, DC, SME Technical Paper, 1998, pp. 35–42.
[17] H.J. Tzeng, B.H. Yan, R.T. Hsu, Y.C. Lin, Self-modulating abrasive medium and its application to abrasive flow machining for finishing micro channel surfaces, Int. J. Adv. Manuf. Technol. 32 (11 – 12) (2006) 1163–1169.
[18] L. Yin, K. Ramesh, S. Wan, X.D. Liu, H. Huang, Y.C. Liu, Abrasive flow polishing of micro bores, Mater. Manuf. Process. 19 (2) (2004) 187–207.
[19] G. Venkatesh, A.K. Sharma, P. Kumar, On ultrasonic assisted abrasive flow finishing of bevel gears, Int. J. Mach. Tools Manuf. 89 (2015) 29–38.
[20] Y.C. Xu, K.H. Zhang, S. Lu, Z.Q. Liu, Experimental investigations into abrasive flow machining of helical gear, Key Eng. Mater. 546 (2013) 65–69.
[21] J. Kenda, J. Duhovnik, J. Tav, J. Kopa, Abrasive flow machining applied to plastic gear matrix polishing, Int. J. Adv. Manuf. Technol. 71 (2014) 141–151.
[22] D. Jung, W.L. Wang, A. Knafl, T.J. Jacobs, S.J. Hu, D.N. Assanis, Experimental investigation of abrasive flow machining effects on injector nozzle geometries, engine performance, and emissions in a diesel engine, Int. J. Automot. Technol. 9 (1) (2008) 9–15.
[23] S. Santhosh Kumar, S.S. Hiremath, Finishing of convergent and divergent nozzles using the developed abrasive flow machine, in: Proceeding of International Conference on Precision, Meso, Micro and Nano Engineering (COPEN-9) IIT-Bombay, 10–12 December, 2015.

SUBCHAPTER 8.2

Chemical Mechanical Polishing (CMP)

8.2.1 Introduction

There is a great demand of component parts made of glass/dielectric, quartz, ceramics, metal, macro-molecular materials, silicon with highly finished surfaces in advanced electronics, mechanical and optical industries. Since some of the materials such as semiconductors possess unique functional properties, they need to be machined to the required shape, size and geometry in a highly precise manner so that they can be utilized in high performance devices and equipments used in optics, microelectronics industries. Superior quality nano-scale surface finish on such important materials is required for incorporating them in the various important micro-parts such as integrated circuits, microprocessors etc. Chemical Mechanical Polishing (CMP) was first introduced by IBM in 1980s by Klaus D. Beyer to generate a highly planar surface so that the lithographic imaging after planarization can be done without much distortion. The first commercial chemical mechanical polisher was designed by Cybeq in 1988 in Japan. The planarization using CMP technique has been extended to different materials, including tungsten, glass, titanium, dielectrics and semiconductors. CMP is a highly sophisticated planarization technique and it is said that there is no other practical processing method exists that can produce better surfaces than chemical mechanical polishing (CMP). By using the ultra-precision polishing, CMP can impart almost perfectly planar surface to each interconnection layer of an Ultra-Large Scale Integration (ULSI) device, that's why it is also termed as chemical mechanical planarization (CMP) technique. The term CMP is used interchangeably to indicate either of the two processes. Initially the CMP process was used for filling the trenches with silicon dioxide (dielectric) and then the removal of excess material was accomplished by applying polishing.

CMP is broadly divided in two categories such as dielectric and polysilicon CMP and metal CMP [1]. Chemical mechanical polishing (CMP) is a hybrid finishing process; this process uses a combination of chemical reaction and mechanical polishing which is used simultaneously for the removal of unwanted material from the workpiece surface [2]. From this point of view, CMP process is different from purely mechanical polishing process which includes lapping; grinding, etc. similarly it is also different from purely chemical etching process such as electro polishing and electro-etching. In the advanced CMP process, removal of material directly from a wafer surface due to effect of mechanical abrading or by chemical etching is reduced to the greatest extent. Otherwise the wafer surface would be damaged severely due to scratching and corrosion. In the early days, processing of silicon wafers using chemo-mechanical polishing in semiconductor industry was a tedious and filthy job. This difficulty associated with CMP was overcome by continuous improvement in its manufacturing strategy.

The main purpose of CMP in the semiconductor industry is to planarize the uneven wafer surface with its global planarization ability. The conventional polishing methods are used for thinning and smoothing the surfaces of variety of materials such as dielectric, ceramic, polymers, glass, semiconductors, metals, composites, etc. Planarization is defined as the process of removing surface topologies by flattening, and smoothing the rough surface. The smoothness or flatness of the surface represents the degree of planarization. Along with local and global planarization methods, thermal flow and etch back methods such as sputtering, photoresist, spin-on glass (SOG) are emerging as the premier methods for achieving ultra-flat surfaces.

CMP is an important semiconductor processing technique which has a capability to achieve the performance goals of a modern microprocessor industry. CMP produces excellent local and global planarization at low cost, in comparison with the other planarization techniques. The phenomenon of the actual mechanism of material removal in CMP is too complex; therefore understanding its chemo-mechanical polishing behavior is a strenuous activity for the engineers and researchers.

8.2.2 Basic principle of CMP process

The phenomenon of material removal from wafer surface during CMP is generally carried out in various steps; lapping and polishing, is a preprocessing step, then ultra-precision polishing and finally the post cleaning or finishing step. The particular chemicals mixed in slurry plays a decisive role of selective machining, where it decides which one material has to be removed and which other material should not have to be removed. For example, copper is initially deposited on the silicon dioxide wafer, but the copper is not supposed to remain present in the entire area of silicon dioxide. Hence, to retain copper on selective areas, etching of unwanted copper carried out with the help of chemical slurry containing nano size abrasive particles. During planarization, the copper present on a silicon dioxide surface will be polished whereas the silicon dioxide material will not be removed. When the copper is deposited on the wafer, it may not be deposited uniformly. The abrasive particles present in slurry generate planar surface by removing higher levels (peaks) at a faster rate and lower levels (valleys) at a slower rate. The slurry chemicals react with the wafer material and the oxidation or hydration mechanism between the slurry's chemical and wafer material generates a chemically-reacted layer on the wafer surface. Generation of passivation layer at the interface of pad and wafer minimizes mechanical abrasion and chemical etching, this is the favorable condition for achieving better polishing. The oxidizers such as potassium ferricynate, hydrogen peroxide, $FeCl_3$, ferric iodate etc. are used for the formation of passivation layer during copper CMP. The nonuniformity within wafer can be minimized by maintaining thickness of passivation layer thinner than the height difference between high and low regions on wafer surface [3].

Fig. 8.2.1(a) indicates the patterned dielectric (silicon dioxide) wafer deposited with copper on its top surface. Whereas Fig. 8.2.1(b) shows schematic presentation of a planarization of a silicon dioxide wafer which is held upside down with the portion deposited with copper film facing towards the polishing pad. Chemical slurry mixed with abrasives is present between the wafer and pad. The CMP is a highly reliable process to maintain stringent dimensional tolerances, as material removal from the wafer surface in the range of micro and nano scale up to the atomic level is possible [4].

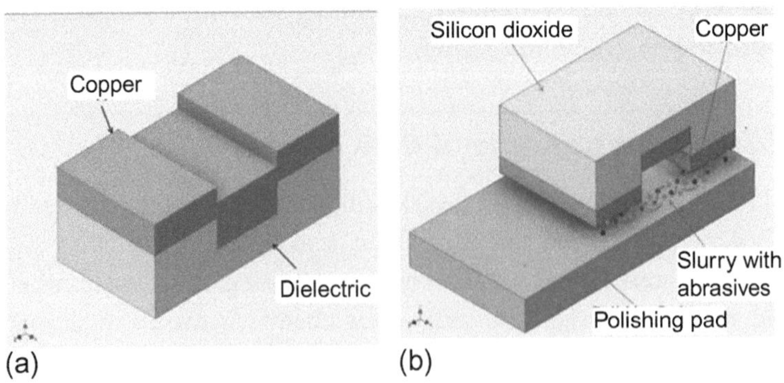

Fig. 8.2.1 Schematic of (a) a wafer and (b) wafer pad arrangement at the beginning of CMP.

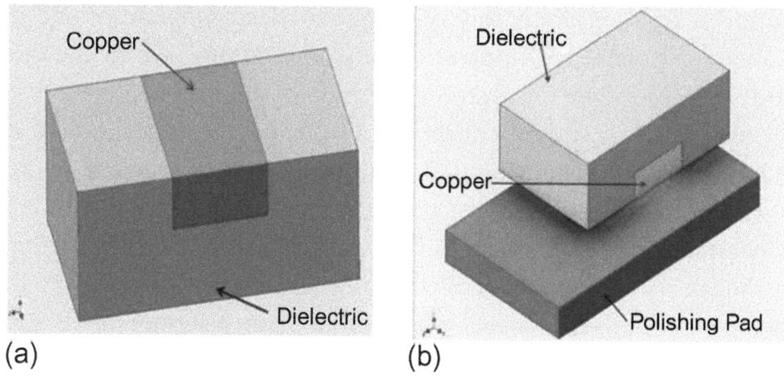

Fig. 8.2.2 Schematic of (a) a polished wafer (b) wafer and pad after CMP.

Fig. 8.2.2 shows the schematic of (a) a polished wafer (b) wafer and pad after CMP process.

The CMP is mainly used in the microchip manufacturing industry, where chips are formed by layering of thin films of different types on substrate (Silicon wafer). The layers of electrically conducting materials are separated by the layers of electrically insulating materials. The method of addition of thin layers of various materials to the substrate is known as film deposition. Two technologies named oxidation and epitaxy are used for layer formation. Chemical vapor deposition (CVD) is a low temperature film deposition technique used for depositing thin layers on the wafer surface. The physical vapor deposition (PVD) process is used for depositing metal on the wafer surface. The conducting wires which are used for metal interconnects are formed by using PVD technique. Electrochemical deposition (ECD) or electroplating is the commonly used method for copper deposition. The dielectric material is used to insulate the layers from one another. In order to connect these functional elements, metal interconnects are required. Therefore, to make the electrical connections between these layers, holes are cut in the dielectric film through different layers and filled with Metal. In Metallization, an electrical connection of the three terminals of transistors viz. Source, drain and gate are made with the other elements of the chip circuit.

When the various circuit elements are connected via local or global wiring through on-chip interconnections it is termed as interconnect. Originally the CMP was developed by considering the two metals such as aluminum (Al) and tungsten (W) [12]. In the early stages, the aluminum interconnects were used in the semiconductor manufacturing industry. The rate of digital signal transmission slowed down due to the high electrical resistivity of aluminum at sub micron level device dimensions. The sub micron sized aluminum interconnects have relatively low resistance to electromigration, due to this they cannot withstand high power density applications. When the currents carrying electrons hit the metal atoms, diffusion of the atoms takes place which causes metal thinning, this phenomenon is called as electromigration. In the mid-1990s, the underperforming aluminum metal was replaced by copper metal. Copper has low resistivity and high immunity to electro-migration. The processing speed of the electronic devices increased to a great extent with the introduction of copper material. Since the removal of copper during machining was not so easy as that of aluminum, hence copper CMP technology was developed. The increased performance of IC's, microprocessor units due to the introduction of copper as the interconnect material increased the popularity of CMP technique. Now the copper wiring is used in the manufacture of highly advanced microprocessors/ICs.

The copper diffuses quickly as soon as it comes in contact with oxide or silicon, hence, during patterning it cannot be etched using conventional methods. In Damascene process instead of etching the copper, it is planarized using chemical mechanical planarization technique. In this process, initially the dielectric (silicon dioxide) is deposited. The dielectric then etched based on the photoresist pattern. After that a barrier layer is deposited on the dielectric to prevent the diffusion of copper into the dielectric material.

Though copper is suitable for fabricating interconnects, but during chemical mechanical planarization (CMP), it poses many problems. Copper has a tendency to diffuse quickly within silicon and oxides. The conventional etching techniques cannot be used to pattern copper, as, the copper spikes penetrate through the junction resulting in junction spiking For forming copper IC interconnects, the IBM (International Business Machines) invented an additive processing technique in 1990 known as Damascene Process.

(a) Damascene process

Damascene is an advanced multilevel metallization process. There are two methods where CMP is applied in the fabrication of copper interconnects; the Damascene process and Dual Damascene process. The name damascene originates from Damascus. Damascus is one of the oldest city in the middle east and the capital of modern Syria. The name Damascene was given because the metal inlay technique which was used in the Middle East resembles the planarizing process.

The materials such as Tantalum, Titanium nitride etc. are generally used for making barrier layers. On this barrier layer a thin seed layer of copper is deposited because without this seed layer, deposition of copper electrochemically on the wafer is not possible. The electroplating process is used for copper

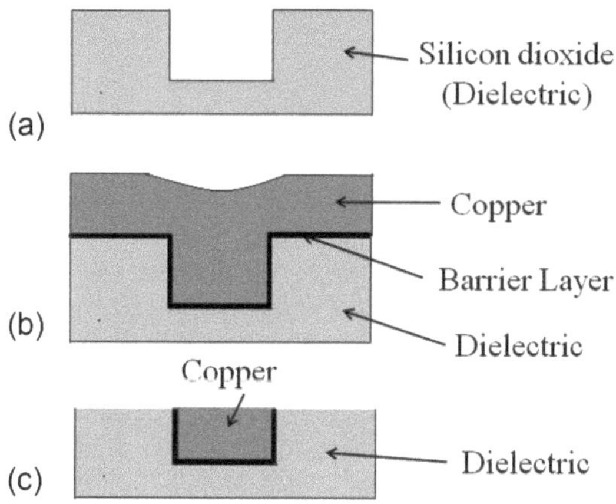

Fig. 8.2.3 Schematic view of steps in Damascene process (a) initial step; (b) intermediate step; (c) final step.

deposition. Finally, the chemical mechanical polishing is applied to remove the excess metal. Fig. 8.2.3(a) indicates the initial step of implementation of damascene process where the etching is done to obtain the shape of the photoresist pattern of the dielectric material (silicon dioxide). In intermediate step as shown in Fig. 8.2.3(b), to avoid the dissolution of copper with dielectric material a barrier of silicon nitride (SiN) or silicon carbide (SiC) is added by CVD [13] and after that copper is deposited. Fig. 8.2.3(c) shows the final step of formation of copper interconnects, where the highly planarized surface is obtained by using CMP process.

(b) Dual Damascene process

The dual Damascene process is similar to single Damascene process and used in microprocessor manufacturing for making metal interconnects. The dual Damascene process can be performed by using two etching steps; trench first, then via or via first, then trench, but there are difficulty in patterning using the trench first method hence via first then trench method is preferred. The dual damascene process is performed in two steps, in the first step the copper is deposited electrochemically by using electroplating technique. In the next step of processing, the excess amount of deposited copper removed by applying CMP technique. The processes such as mechanical abrasion and electrochemical dissolution occurring simultaneously are responsible for the copper wafer planarization material [14].

In via first method etching high aspect ratio contact holes through inter metal dielectric is a challenging task. The dual damascene is performed in two steps; initially via i.e. (vertical interconnect access) or holes are made

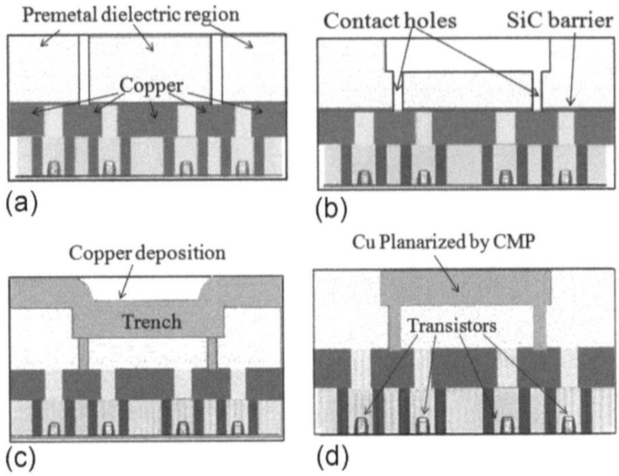

Fig. 8.2.4 Schematic view of steps (a), (b), (c), and (d) in dual damascene process [15].

to connect the underlying metal layers with copper wires. The holes as shown in Fig. 8.2.4(a) are etched throughout the pre-metal dielectric (PMD) and the silicon carbide barrier. In the second step, trench is made which connects the holes/via as shown in Fig. 8.2.4(b). The trench made is then filled with copper to make connections with underlying metal layers. A thin film of SiC covering the copper at the bottom of the holes is also removed during the process as shown in Fig. 8.2.4(c), as this barrier will lead to poor or no connections between conducting layers. Earlier silicon nitride was used as a barrier layer, but silicon carbide (SiC) is preferred for the formation of barrier layer, as SiC can be deposited at low temperature by applying copper vapor deposition (CVD) process. SiC posses lower k which helps in enhancing the performance of the device. Similarly, SiC sticks well to most of the inter-metal dielectric (IMD) and to copper as well. Finally the CMP process applied to obtain the planarized surface as shown in Fig. 8.2.4(d).

The two planarization methods used predominantly in CMP are local planarization and global planarization as shown in Fig. 8.2.5.

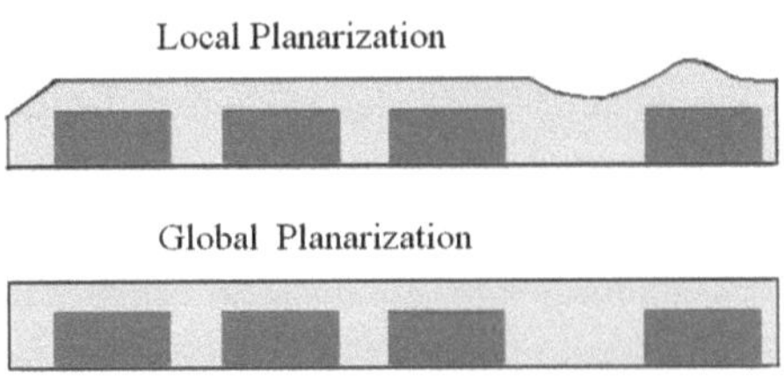

Fig. 8.2.5 Schematic of local and global planarization methods.

For planarization of feature size greater than 0.35 μm, local planarization and other methods are used. The feature size smaller than 0.35 μm can be achieved by applying the global planarization. The term global planarization is used when the uniform planarity is required for multiple chips and across the wafer surfaces. In local or partial planarization, some area planarize faster than other this phenomenon leads to dishing and erosion.

8.2.3 CMP setup details

The CMP system consists of a round table (platen) which can be rotated. A polishing pad is fastened to the top surface of the table. In usual arrangement, the polishing pad and the wafer carrier rotates in the same direction. Before initiating the CMP process, there are a number of factors needs to be taken care of, such as; material to be machined, pattern density, slurries used, platen speed, film stacks, pads, sweep profile, slurry flow rate, slurry distribution, consumables, etc. The complexity of CMP increases due to the adaptation of planarization of different materials in the same system. Based on the type of material to be processed, there is a prerequisite such as to select optimum process parameter settings, modification of the work holding pad, content and combination in abrasive slurry for both planarizing and post-CMP cleaning. Hence all these factors required to be individually tailored to the application for accomplishing desired results. Figure 8.2.6 represents the schematic diagram of Chemical Mechanical Polishing (CMP) setup.

The polishing pad is made of polyurethane polymer material; the pad hardness can be altered by adding filler material. The pads are of two types;

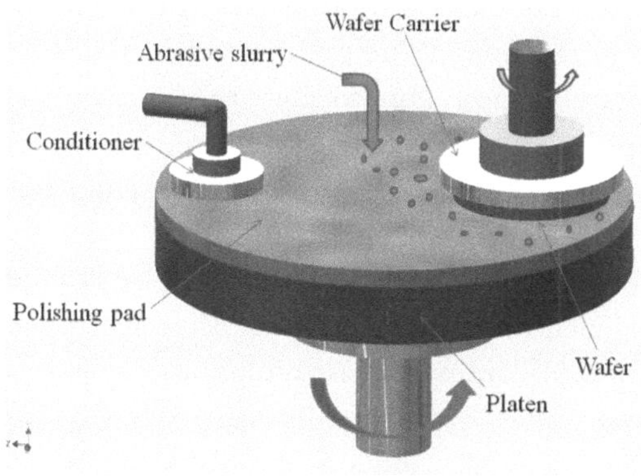

Fig. 8.2.6 Schematic diagram of Chemical Mechanical Polishing (CMP) setup.

hard pad and soft pad. Hard pads are used for polishing materials such as silicon dioxide, whereas soft pads are used for removing materials like copper. A soft pad resembles to a "puff," which is generally used for applying powder, whereas, hard pad may have a feel of a leather bag. The repetitive polishing tends to form glazing layer due to removal of asperities and filling of micro-cavities by polished material [5]. During CMP the temperature at processing zone raises in the range of 20 to 30 °C [6]. If the polishing pad is exposed to the increased temperature for a long time, its hardness reduces; it becomes soft which affects the material removal rate [7]. Using the polishing pad without conditioning deteriorates pad and shortens its life. For effective polishing, the surface of a pad should have to be maintained rough. The inbuilt pad conditioner resurfaces the pad and also removes the used slurry. The pad conditioners are conventional and diabond type. The conventional conditioners have a stainless steel plate coated with nickel-plated diamond grits. Whereas stainless steel plate of diabond conditioner is faced with a CVD diamond film plated diamond grits. The frequent conditioning improves the surface properties, but reduces the thickness and life of the pad. The wafers of semiconducting materials such as silicon, germanium, gallium arsenide, etc. are initially prepared. The wafer or a workpiece material which is to be polished is held upside down in the wafer holding unit via back pressure or surface tension. The restraining ring surrounding the wafer holds it exactly and pressure in a downward direction on the wafer is applied through the carrier membrane. The wafer holding unit is provided with rotary motion and at the same time it can be pressed down against the rotating polishing pad which is supported on rotating platen. The wafer held in a carrier slide on the pad surface with a relative velocity generated by the rotation of the carrier and the platen. Abrasive slurry consisting of a mixture of chemicals and abrasive particles is introduced in between the wafer and the pad surface. The platen holding polishing pad and wafer holding unit are provided with rotary motion. Due to the rotary motion imparted to both the units, a complex interaction between chemical and mechanical forces takes place. It has been observed that the material removal in CMP occurs due to the effect of a combination of chemical reaction of the slurry chemicals with the wafer surface materials and also due to the repeated sliding, rolling, or indentation of the abrasive particles against the wafer surface.

Slurries used in CMP are water based solutions which consist of abrasive nano particles, pure water, organic or inorganic acids or bases, surfactants, corrosion inhibitors, polymers, buffers, oxidizers, anti coagulating agents, complexing agents, bactericides and fungicides etc. An abrasive present in the slurry imparts mechanical energy to the wafer surface. Cerium dioxide (CeO_2) is a metal oxide, which is used extensively for polishing glass, similarly the abrasives such as aluminum oxide (Al_2O_3) and silicon dioxide (SiO_2) are used frequently. The composition and content in the slurry engineered and formulated depending upon the type of material and the polishing process chosen. The slurry used for copper polishing consists of chemicals such as strong acids, alkalis etc. [8]. Chemicals in the slurry react with workpiece material and form soft chemical compound that can be removed easily by abrasive particles. Slurry influences the uniformity, planarity and metal removal rate. Slurries are important components in the manufacture of semiconductor chips and other electronics devices. The abrasive nano-particles generally made of alumina, silica, polymer, zirconia, ceria abrasives.

CMP polishing pads made up of hard polyurethane foam and laminated on to a sub pad. Pads are generally circular in nature. The rotary motion is provided to the pad and slurry is poured on the rotating surface of the pad. They spin at a speed of about 100 rpm. It causes the rolling, sliding and bouncing of the nano particles which slowly wears away the wafer surface.

After the polishing is over, post cleaning is required to remove chemical contamination and particles from the wafer surface. There are about four methods by which the post cleaning is accomplished, they are; (i) Buffing, (ii) Brush cleaning, (iii) Megasonic cleaning and (iv) Spin rinse drying.

The most widely used and commercially available CMP machines are grouped under four major heads based on the type of motions present between the wafer carrier and the polishing pad such as; (a) Rotary pad with reciprocating wafer carrier, (b). Rotary pad with oscillating carrier, (c). Pad with orbital motion with rotary carrier and (d). Pad with linear motion with rotary carrier [9]. Figure 8.2.7 shows the photographic view of CMP Machine Tool.

8.2.4 Influence of process parameters of CMP

The type of surface finish required and the degree of planarization to be achieved depend on the number of input process parameters such as velocity and pressure on the wafer, speed of rotation of polishing pad with respect to the rotation of wafer pad, material of the wafer, hardness of the pad, composition of slurry, abrasive size etc. The applied pressure on the wafer and movement of wafer with respect to the polishing pad has a great impact on material removal rate.

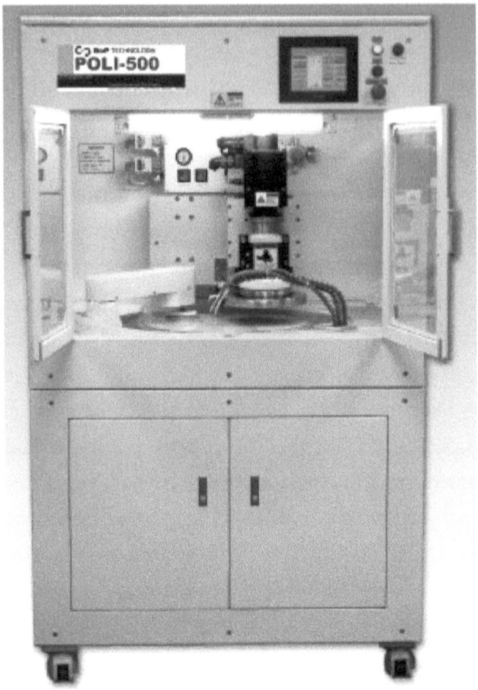

Fig. 8.2.7 CMP machine tool (model: POLI-500). *(Courtesy: G&P Technology.)*

CMP is a very complex process, it involves so many other influencing process parameters, such as geometry of pattern, temperature, abrasives, slurry composition; pH value, flow rate, concentration of chemicals, pad stiffness, wafer material type, size and curvature, thickness and properties of coatings, etc.

The material removal rate (MRR) in CMP is calculated by using Preston's equation according to it, $MRR = K_p P_o V$ [10], MRR is directly proportional to applied pressure, P_o and relative velocity of wafer and pad, V. K_p is the Preston's constant which represents the effect of other process parameters. Though Preston's equation is used extensively in CMP industry with some modifications it has some drawbacks such as it is suitable for mechanical polishing only, as it doesn't consider dominating process parameters such as slurry type, concentration, pH, abrasive shape, size, proportion, properties of pad and wafer materials etc. [11].

As CMP is considered as a highly complex process and the final outcome of the process is dependendent on the number of input parameters. Figure 8.2.8 shows the fishbone diagram representing the various input process parameters and output of CMP process.

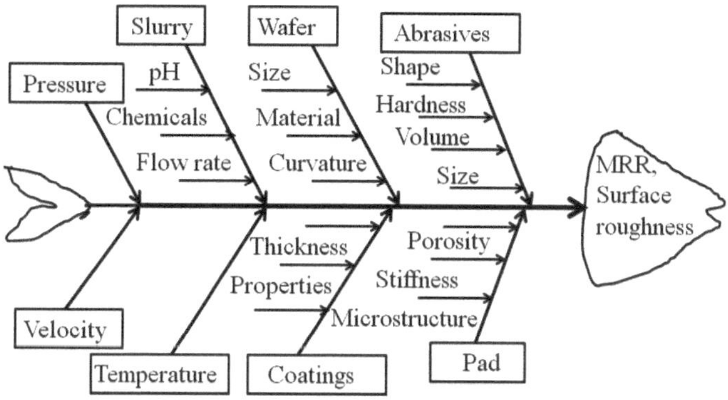

Fig. 8.2.8 Fishbone diagram of input process parameters and output (responses) of the CMP system.

8.2.5 Advantages and limitations

CMP is an important and extensively used planarization and polishing technique, which has following advantages.

(a) CMP is widely used for processing silicon wafers which are used in the production of integrated circuits in the semiconductor industry. CMP is a subtractive process which has increased performance, speed and reliability of the electronic circuits. It is predominantly used for multi-level metal interconnect structures for chips, inter-layer dielectric (ILD) planarization, shallow trench isolation (STI) and the formation of contacts and interconnects. Manufacture of highly integrated devices during semiconductor processing is possible because of the local and the global planarization technique.

(b) Any integrated circuit accommodating with more than one million transistors on a microchip is considered as an Ultra-large-scale integrated (ULSI) circuits. With the CMP technique, manufacture of ULSI with 10^8 or more devices on silicon semiconductor wafer substrates resulted in increased performance of electronic products with the reduction in cost.

(c) Along with glasses and semiconductor material, CMP can be used effectively for polishing variety of materials such as plastics, crystals, special metals and computer discs.

(d) CMP has an edge over the other existing surface finishing methods. This technique offers minimum surface and subsurface damage, high metal removal rate, enhanced machining efficiency, removal of surface defects to a great extent, low cost finishing etc. CMP is becoming popular because of the production of high quality surfaces of the component parts.

(e) The CMP has good selectivity as it smoothes the layers of selected material only, whereas non-selected layers won't be polished.

(f) The CMP promotes precise deposition and planarization of metal interconnects such as copper, aluminum, titanium, etc. in subsequent layers, which improves the step coverage. This helps in faster and reliable functioning of the devices due to the increased interconnectivity among the various layers in the multilayer chips.

Most of the limitations associated with CMP process are due to the mechanical stress and aggressive chemical action during polishing. The following major defects of CMP such as dishing, scratching, erosion and corrosion are discussed.

(a) Dishing

Figure 8.2.9(a) shows the silicon wafer with the layer of copper deposited before planarization. The CMP process is applied to the copper laden silicon wafer and excess Cu material is removed and the surface is planarized as shown in Fig. 8.2.9(b).

The polished surface is uniformly planarized in ideal case. Figure 8.2.10 shows the formation of dishing effect on the planarized surface. Thus the formation of the dish type cavity is known as dishing. The excessive dishing takes place due to the elastic deformation of the polishing pad and the applied pressure. As the pad is not a perfectly straight and rigid, therefore due to the slight bending of pad during polishing excess copper material get removed. This phenomenon generates non-planer surface resulting in dishing. A harder polishing pad can planarize the surface at faster rate and causes less dishing of the copper interconnect [16]. The higher aerial density of a deposited material is also a reason for dishing.

In Fig. 8.2.11, the two copper interconnects with different aerial densities are shown. The percentage of area on which the copper wire is spread in

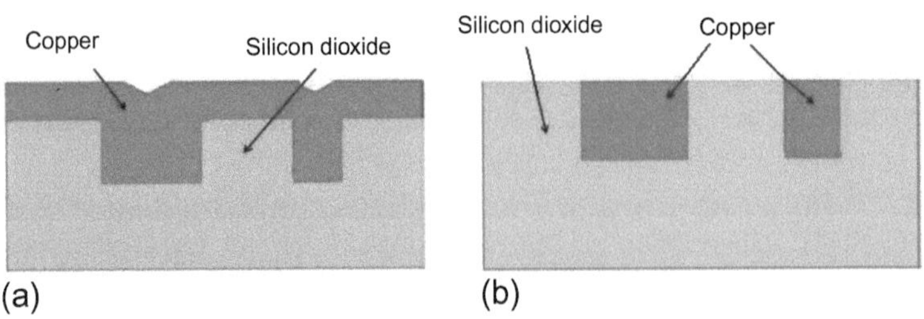

Fig. 8.2.9 Schematic of a wafer before (a) and after (b) CMP in ideal case.

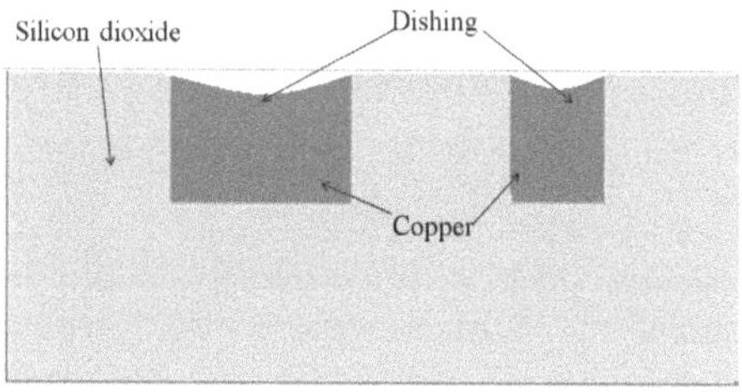

Fig. 8.2.10 Schematic of a dielectric wafer with dishing effect.

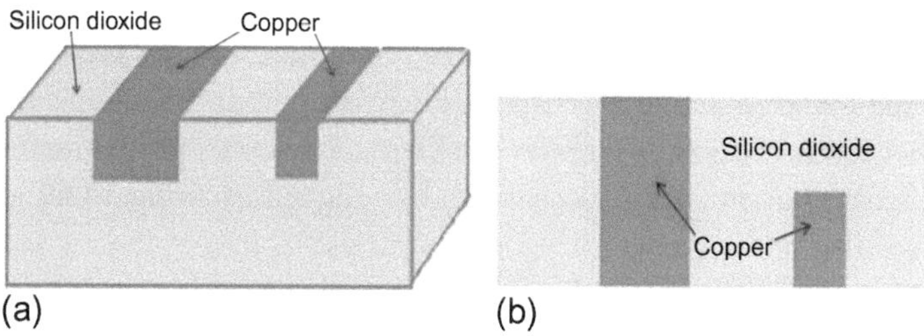

Fig. 8.2.11 Schematic of copper interconnects with (a) 3-D view and (b) top view.

the top view is termed as its aerial density. The aerial density is usually calculated by considering the size of $10\,\mu m^2$. The methods such as "slotting" and "dummy fill" are used for maintaining uniform aerial density of copper. If the line width of interconnect in top view is decreased, it increases the electrical resistance. Therefore, in the wide copper line a narrow line of oxide is introduced which is called as slotting. The slotting reduces the aerial density and stops the dishing during CMP. Figure 8.2.12(a) shows the slotting performed on dielectric material. In dummy fill process, along with the main wiring, few more copper lines are added in the top portion of wafer. Since these lines are not connected to any device, it won't serve any purpose from the functioning point of view. The dummy lines are introduced so that

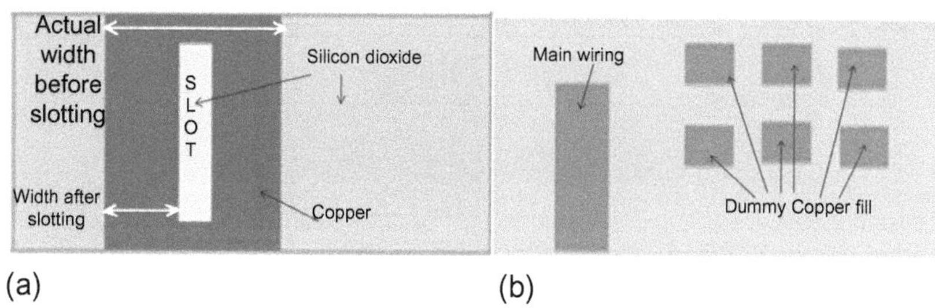

Fig. 8.2.12 (a) Slotting; (b) Dummy fill process.

the CMP process can planarize the main wiring properly without causing the dishing effect. Figure 8.2.12(b) shows the top view of the wafer where dummy copper fills are located in the vicinity of main wire.

(b) Scratching

During the CMP process, the abrasives under high down force creates extremely high localized mechanical stress, which forms scratches and other mechanical damages. Scratching takes place due to the abrasives present in the slurry, the application of the strong mechanical forces on abrasive particles create scratches. The size of the scratches can range from several angstroms to macro scratches [17]. Thus size, shape, material and the concentration of abrasives in slurry has an effect on surface scratching. Polishing pads are also responsible for scratching due to inadequate diamond dressing and break-ins. The other factors which contribute scratching are surfactants, contaminants, slurry distribution etc. the scratches can be reduced by using additives.

(c) Erosion

The unwanted dissolution of dielectric and copper leads to erosion. In high metal pattern density, many copper wires are placed in close vicinity. The dielectric present in these copper lines also experiences high pressure during CMP process. Therefore the erosion of oxide between the copper wires takes place which results in thinning of oxide and metal in a patterned area. Sometimes for achieving the high metal removal rate, the chelating agents and oxidizers are used which are also responsible for erosion. Figure 8.2.13 (a) and (b) shows the planarized silicon dioxide wafers with high metal pattern

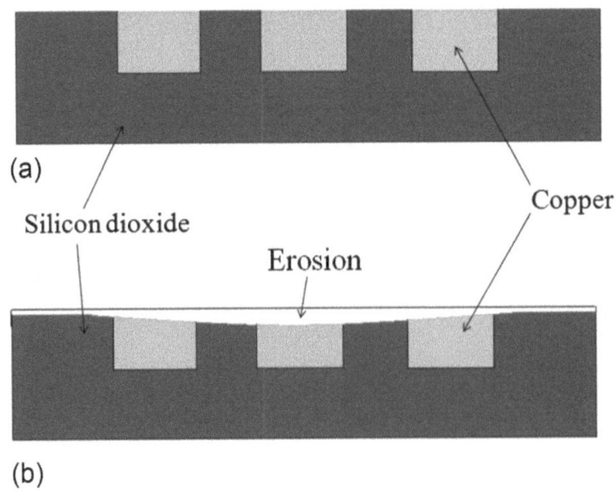

Fig. 8.2.13 Schematic of a wafer (a) Ideal CMP, (b) Erosion effect.

density. Generally the rate of erosion of dielectric with high metal pattern density is faster than low metal pattern density. With the increased pattern density, the dielectric density decreases due to which erosion of dielectric takes place rapidly.

(d) Corrosion

The abrasive slurry used for etching consists of a combination of various chemicals, the composition and the concentrations of the chemicals vary depending on the type of material to be polished. Though the chemicals assist in the removal of material by dissolution they are sometimes also responsible for corrosion of the wafer surface. Due to the excessive etching the unwanted dissolution of material takes place which also adds the corrosion as surface defect [18]. The corrosion inhibitor such as benzotriazole (BTA) is used to intercept corrosion, during copper polishing [19].

CMP is a time consuming and the most expensive process as it requires highly precise and costly equipments for its functioning. The consumables such as abrasives, chemical slurries needs frequent replacement for better performance, which escalates the maintenance cost [20].

8.2.6 Applications

CMP process in applied for finishing interconnects of different materials such as aluminum, copper, titanium etc. Some of the important applications are mentioned as follows.

(i) CMP used for finishing aluminum interconnects

In manufacturing ULSI, CMP has two major applications; to smooth the topography of inter-level dielectrics (ILD) usually silicon dioxide, or to remove excess material to produce an inlaid metal structure or isolation trenches. During conventional aluminum metallization using CMP, aluminum is deposited on a silicon dioxide layer (ILD) then patterned and finally etched to form interconnects. The aluminum interconnects thus formed is insulated by depositing another layer of ILD to build 3-D electrical wiring. To provide a smooth surface for depositing aluminum, CMP is employed for some top few layers on each ILD. The final position is determined based on the thickness of the ILD layer required for electrical isolation of aluminum line and the surface planarity obtained.

(ii) CMP used for finishing copper interconnects

The copper CMP has made it possible to use copper interconnects in IC manufacturing. It is because the copper wires that used in ICs, the small electronic devices such as laptops, mobiles etc. can run on a battery for a long period of time.

(iii) CMP used in hybrid finishing process (CMMRF)

An integrated circuit (IC), a micro-scaled electronic device integrates millions of different electronic components such as transistors, capacitors, resistors, diodes, rectifiers etc. in a functioning order. The CMP finishing plays an important role in the fabrication of IC interconnects which requires extremely flat and smooth surfaces. Jain et al. [21] developed a new hybrid finishing process for polishing silicon blanks as shown in Fig. 8.2.14. The process includes a combination of chemo-mechanical and magneto-rheological finishing. The new process invented has produced a superior quality nano-metric finish on silicon without any defects.

(iv) Other applications of CMP

(a) In optical micro electromechanical systems (MEMS), the thin films with a highly smooth surface finish are of great importance because it enhances the device performance. The implementation of CMP in planarizing the thin films in optical (MEMS) systems imparts a mirror like

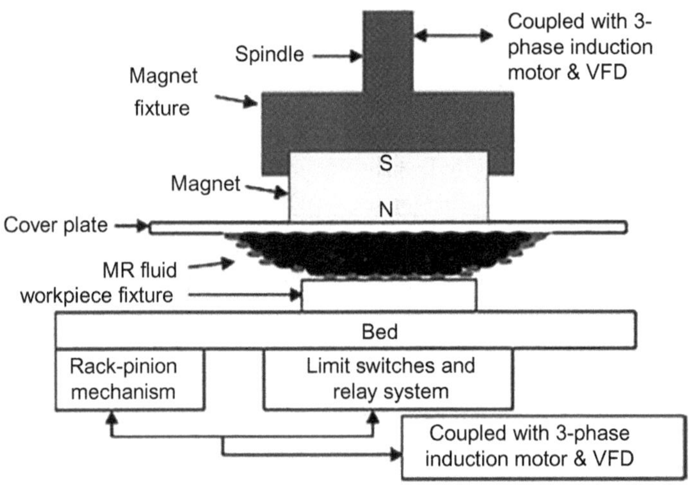

Fig. 8.2.14 Schematic of the CMMRF set up [21].

surface finish on the layers. It is also used in manufacturing flat monitor screens, photo diodes, optical fiber, telecommunication laser, etc.

(b) CMP can also be used for manufacturing actuators, sensors and micromachining of hard disc drives (HDD).The light emitting diodes (LED lights) are used in traffic signals.

(c) The CMP has become the most significant process due to its capability of imparting excellent global planarization to the wide range of materials.

(d) The application of CMP process has widened the scope of its application to micro-electro-mechanical systems (MEMS), through-silicon via (TSVs), printed circuit boards (PCBs), and more.

(e) Along with polishing the wafers in initial stage, the CMP technique is applied for processing semiconductor devices at various stages. By applying CMP, the different electronic component parts such as transistors, capacitors, diodes, resistors, rectifiers etc. are made by arranging p-type and n-type regions next to each other in various fashion.

CMP is an established method of planarization and it is widely used for removing surplus material from the earlier processing step. The uniform surface is essential for performing high resolution photolithography processes so that the multilayer wiring can be done. CMP is mainly dependent on pattern density. As small area can be polished in a faster way to reduce dishing in dense pattern, polishing takes more time. The CMP is a very complex technology in which the planarization is achieved by mechanical abrasion and electrochemical dissolution of the wafer material. The electrical insulator materials having relatively low dielectric constants are low-k dielectric materials. The technology associated with low-k dielectric is considered as the futuristic technology. The low-k and high-k is special dielectric materials used for replacing oxide inter level dielectric (ILD) layer to reduce the loss due to capacitance and increases the rate of signal transmission of the circuit. The application of low-k materials in ULSI manufacturing encourages scaling down the device dimensions which increase the packing density. The importance of CMP process increased tremendously when it was applied successfully for planarization of the copper interconnects. Two important methods of fabricating copper interconnects such as damascene and dual damascene are discussed elaborately. Since the CMP process is basically applied in semiconductor industry, the role of CMP in semiconductor manufacturing is discussed.

References

[1] A.K. Sikder, Mechanical and tribological properties of interlayer films for the damascene-Cu chemical-mechanical planarization process, J. Electron. Mater. 31 (10) (2002) 1016–1021.

[2] H. Lee, D.L. Lee, H. Jeong, Mechanical aspects of the chemical mechanical polishing process: a review, Int. J. Precis. Eng. Manuf. 17 (4) (2016) 525–536.

[3] R.K. Singh, R. Bajaj, Advances in chemical-mechanical planarization, MRS Bull. 27 (2002) 743–751.

[4] A.B. Kahng, K. Samadi, CMP fill synthesis: a survey of recent studies, IEEE Trans. Comput. Aided Design 27 (1) (2008) 3–19.

[5] G. Banerjee, R.L. Rhoades, Chemical mechanical planarization historical review and future direction, ECS Trans. 13 (4) (2008) 1–19.

[6] Y. Liu, K. Zhang, W. Di, Fang, investigation on the final polishing slurry and technique of silicon substrate in ULSI, Microelectron. Eng. 66 (2003) 438–444.

[7] H. Lu, Y. Obeng, K.A. Richardson, Applicability of dynamic mechanical analysis for CMP polyurethane pad studies, Mater. Charact. 49 (2002) 177–186.

[8] Z. Zhanga, J. Cuia, J. Zhangb, D. Liua, et al., Environment friendly chemical mechanical polishing of copper, Appl. Surf. Sci. 5 (11) (2019) 467–468.

[9] D. Zhao, L.U. Xinchun, Chemical mechanical polishing: Theory and experiment, Friction 1 (4) (2013) 306–326.

[10] F. Preston, The theory and design of plate glass polishing machines, J. Soc. Glas. Technol. 11 (1927) 214–256.

[11] R. Xu, G. Smart, M. Zhang, Particle characteristics and removal rate in CMP process, in: Proceedings of Fourth International Chemical-Mechanical Planarization for ULSI Multilevel Interconnection Conference, Santa Clara, CA, February 11–12, 1999, pp. 253–255.

[12] F.B. Kaufman, D.B. Thompson, R.E. Broadie, et al., J. Electrochem. Soc. 138 (1991) 3460.

[13] www.erc.arizona.edu/.../CMP%20Module/CMP%20Tutorial.ppt.

[14] A.R. Sethuraman, J.-F. Wang, L.M. Kook, Review of planarization and reliability aspects of future interconnect materials, J. Electron. Mater. 25 (10) (1996) 1617–1622.

[15] R. Yanda, M. Heynes, A. Miller, Demystifying Chipmaking, Elsevier, 2005.

[16] J. Steigerwald, S.P. Murarka, R.J. Gutmann, Chemical Mechanical Planarization of Microelectronic Materials, John Wiley & Sons, Inc, New York, 1997.

[17] N.K. Penta, Abrasive–free and ultra-low abrasive chemical mechanical polishing (CMP) processes, Adv. Chem. Mech. Planariz. (2016) 213–227.

[18] T.H. Sai, Y.F. Wu, S.C. Yen, Study of copper chemical mechanical polishing in urea-hydrogen peroxide slurry by electrochemical impedencespectroscopy, Appl. Surf. Sci. 412 (2003) 120–135.

[19] A. Misra, Stabilized slurry compositions, (2003)US Patent 6, 530,967.

[20] P.B. Zantyea, A. Kumar, A.K. Sikder, Chemical mechanical planarization for micro-electronics applications, Mater. Sci. Eng. R 45 (2004) 89–220.

[21] V.K. Jain, P. Ranja, V.K. Suri, R. Komanduri, Chemo-mechanical magneto-rheological finishing (CMMRF) of silicon for microelectronics applications, CIRP Ann. Manuf. Technol. 59 (2010) 323–328.

SUBCHAPTER 8.3

Elastic Emission Machining (EEM)

8.3.1 Introduction

There is requirement of fabrication of extremely precise surface on optics materials as advanced optical industries are constantly growing because they utilize mirrors for emitting light with extremely short wavelengths. Optical components have been used in various equipment such as advanced optics fields such as extreme ultraviolet (EUV) lithography, X-ray microscopy, telescopes, cameras, astronomical mirrors, metrology, navigation, vision systems, modulation of energy beams (laser, electron, etc.) and many more. Applications of optics in these components require control on profile and surface roughness. Manufacture of stress free surface of these components is urgently needed. Only some finishing process is offered to manufacture optical components with extremely precise surface. Elastic emission finishing process is one of them. Elastic Emission Machining (EEM) is a non-contact ultra-precision machining process having performances to create atomic level finished surfaces. In EEM, surface chemical activities of water-suspended ultra-fine powders are utilized. Fine oxide powders such as SiO_2 or Al_2O_3 or ZrO_2 are usually supplied. Ultra-fine powders are transported and chemically contact to the work surface. And when they are separated, atomic removal occurs from the work surface. EEM process produces mirror finish surface with high accuracy and angstrom level surface finish on optical components.

8.3.2 Basic working principle

Elastic Emission Machining (EEM) is extremely precise and non-contact finishing process where ultrafine abrasive powder particles are supplied to the workpiece surface by a flow of pure water. The chemical reactions occur between work and abrasive to remove the surface atoms from the workpiece [1]. In EEM, the quantity of surface atoms removed depends on the number of active powder particles supplied to workpiece surface. A spherical ball made of polyurethane is rotated with different speed with the help of motor. The different speed is set for keeping uniform working gap and these gaps adjust the thin film lubrication. The use of polyurethane sphere is due to its better elastic properties, stability in water and good surface roughness (less than 0.1 μm). Polyurethane sphere is manufactured by diamond turning, high precision grinding and equivalent high precision technique. Figure 8.3.1 shows the working principle where abrasive particles are drag over the workpiece surface by rotation of the polyurethane sphere with little normal load and without physically pressing abrasive particles on the work-piece surface.

The material removal in elastic emission finishing process occurs only when, atoms from both surfaces create bonding. In addition, the binding energy of uppermost layer must be smaller than the adjacent layer of uppermost layer and so on. Both the surfaces are composed different kind of atoms having different properties. Thus it may be possible that their interface electrically polarized and local charge induced due to various conditions of electrons. Hence, it helps to release atoms present in uppermost layer of workpiece during finishing [3, 4].

The possible steps of material removal in elastic emission finishing, are shown in Fig. 8.3.2 [5]. When both the surfaces try to contact with each other they are finished by hydroxide (OH) species as shown in Fig. 8.3.2(a). When both abrasive and workpiece come to contact to form a hydrogen bonding through the hydroxide species as shown in Fig. 8.3.2(b), then H_2O is detached by dehydration as shown in Fig. 8.3.2(c). The flow of water lifts up the abrasive and breaks the bond between both surfaces at the contact as shown in Fig. 8.3.2(d). The cycle is repeated continuously and material removes from workpiece atom-by-atom.

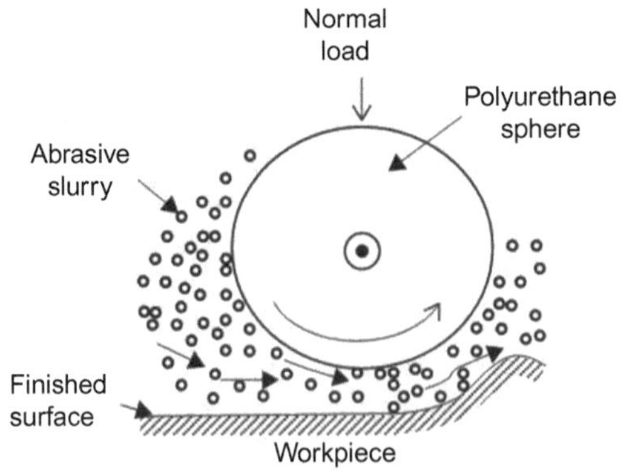

Fig. 8.3.1 Schematic diagram of the elastic emission finishing process [2].

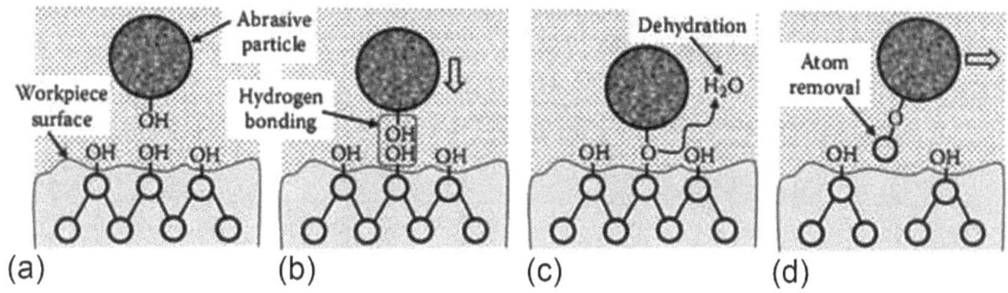

Fig. 8.3.2 Schematic illustration of steps in material removal during elastic emission finishing (a) before contact, (b) Hydrogen bonding, (c) formation of H_2O by dehydration (d) separation of material atom-by-atom [5].

8.3.3 Setup details

Depending upon the method of slurry flow, there are two types of EEM process. One is nozzle type EEM in which a nozzle delivers the jet of compressive slurry over the workpiece surface. Nozzle type EMM is suitable for NC machining with control of dwelling time because this process can finish the surface at stable material removal rate. The other type is rotating sphere EEM in which a rotating sphere is loaded over the thin film of fine abrasive slurry on the workpiece surface [6].

Elastic Emission Machining experimental setup consists of rotating arrangement for rotating sphere, spherical ball loading arrangement, abrasive slurry tank and supply arrangement and workpiece holding arrangement. The rotating sphere is not in direct contact on workpiece. The gap between the sphere and workpiece is filled up by thin elasto-hydrodynamic lubrication film of fine abrasive slurry. The fluid film thickness is maintained larger than the diameter of powder particle so that the work surface is not damaged. The part of spherical surface of polyurethane sphere reaches a sub-micrometer distance from the work surface, where the EEM process takes place [7, 8]. A schematic view of the elastic emission finishing setup is shown in Fig. 8.3.3 in which rotating spherical disc is used for finishing vertical surface of the workpiece. EEM experimental setup consists of a frame, a spindle suspended from the frame and a work tank filled with slurry. The sphere is fixed on the spindle tip and immersed in the slurry with the work. The sphere is a plastic disk with suitable elasticity. All of the sphere except for the equator is removed because it does not come close to the work surface. The sphere surface is finished to mirror surface with sub micrometer circularity. In addition to these basic elements, the spindle has a vertical feed

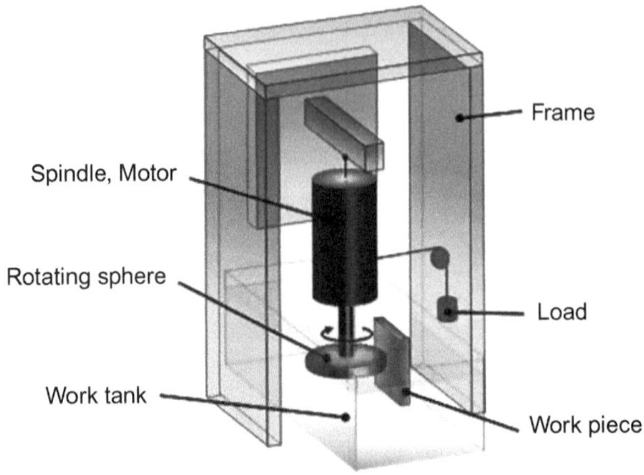

Fig. 8.3.3 Schematic view of the elastic emission finishing setup for finishing vertical surface [6].

mechanism and the work holder has a horizontal feed mechanism to enable scan processing. The sphere moves up and down and it is fed left to right on the workpiece for polishing the entire workpiece surface. This system consists of a three-axis control system, which makes it possible to process any desired area on a workpiece surface. A work spindle with a rotating-sphere head is suspended by a cross spring, similar to a pendulum, and the head is immersed in a tank of fluid mixture. The load acting on the rotating sphere is generated by applying a load to the work spindle. Figure 8.3.4 shows the schematic diagram of elastic emission machining experimental setup in which the spherical ball is mounted on inclined link for loading and for finishing horizontal surface of the workpiece mounted on NC table.

8.3.4 Influence of process parameters

The material removal rate in EEM process depends on the concentration of powder particle, spherical head rotational speed during finishing operation. As the concentration of particle increases the material removal rate initially increases and then decreases. Initially with the increase in concentration of abrasive particles in the slurry, the number of active particles supplied to the workpiece surface increases, as a result material removal rate increases. With higher concentration of powder particles, the resistance of the lubrication film increases, as a result MRR decreases. As the spherical head rotational speed increases, the material removal rate decreases due to

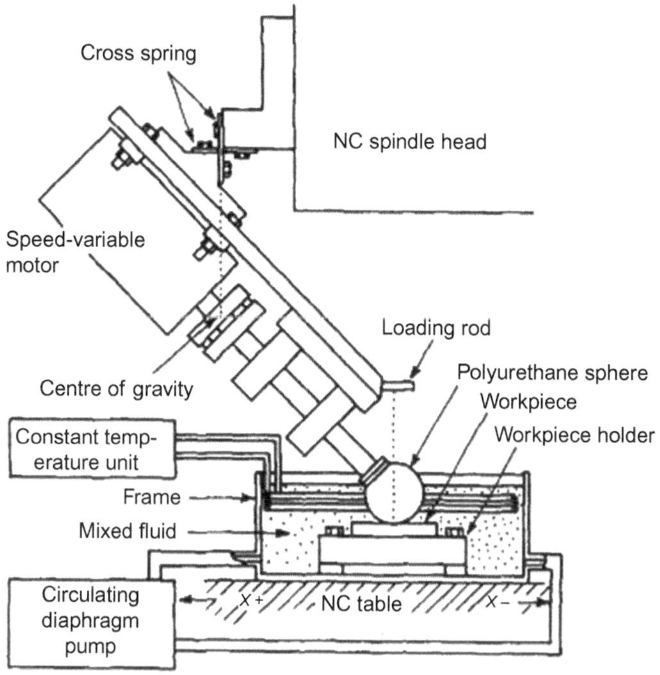

Fig. 8.3.4 Schematic view of the elastic emission finishing setup for finishing horizontal surface [2].

increase of near surface flow velocity. The number of powder particles supplied to the workpiece surface decreases as the near surface flow velocity increases. The material removal rate depends on type of abrasive powders. The material removal increases with the increase in machining time, but the machining rate is different for different workpiece materials. The thickness of abrasive slurry film is an important parameter to control the performance of the EEM process. This fluid film thickness is maintained by controlling the applied load and the rotation of spherical head. The surface roughness and geometrical accuracy during EEM process are influenced by the applied load, size and concentration of abrasive powders and rotational speed of spherical head. By controlling the process parameters of EEM, atomic level surface finish can be achieved during finishing of flat as well as curved optical surface. The surface roughness (root mean square value) decreases with increase in depth of machining. The removal depth of more than 15 nm is required to reduce the surface roughness to 0.1 nm rms value during EEM finishing of flat surface of Zerodur optics materials [8].

8.3.5 Advantages

The advantages of Elastic Emission Machining (EEM) process are given as follows:

(i) EEM is a non-contact finishing process;
(ii) EEM process can produce mirror finish with atomic level surface;
(iii) EEM can produce stress free surface;
(iv) EEM process can generate flat as well as curve surface;
(v) EEM is most efficient finishing process for optical materials.

8.3.6 Applications

The Elastic Emission Machining (EEM) process has wide scope of applications for generating atomic level surfaces as follows:

(i) Elastic Emission Machining (EEM) is applied for producing atomically smooth and stress free surfaces of low thermal expansion optical materials for Extreme Ultraviolet (EUV) lithography.
(ii) EEM process is applied for finishing ellipsoidal focusing mirrors used in X-ray microscopy.
(iii) EEM process is applied for producing atomically flat highly reflective mirrors used in SR beam lines.
(iv) The rotating sphere EEM system is used for smoothing flat surface as well as steeply curved surfaces on glass cylinder for fabricating mirrors [6].

Elastic Emission Machining (EEM) is a super finishing process in which the materials are removed atom by atom. During EEM process, the phenomenon of semi-spontaneous removal of atomic size materials makes it possible to produce geometrically perfect finished surface on optics materials. The major application of EEM process is to generate mirror finish on optical components. The rotating sphere EEM process has capability of polishing steeply curved surfaces of mirrors used for X-rays microscopy. EEM is one of the important finishing techniques for finishing surfaces of low thermal expansion materials used for extreme ultraviolet (EUV) lithography. The stability and accuracy of computer numerically control (CNC) EEM system is very much important during this super finishing process.

References

[1] Y. Mori, N. Ikawa, T. Okuda, K. Yamagata, Numerically Controlled Elastic Emission Machining, Vol. 26, Technology Reports of the Osaka University, 1976, pp. 283–294.
[2] Y. Mori, K. Yamauchi, Elastic emission machining, Precis. Eng. 9 (3) (1987) 123–128.
[3] Y. Mori, N. Ikawa, K. Sugiyama, Elastic Emission Machining – Stress Field and Fracture Mechanism, Vol. 28, Technology Reports of the Osaka University, 1978, pp. 525–534.
[4] K. Yamauchi, H. Kikuji, G. Hidekazu, S. Kazuhisa, I. Koji, Y. Kazuya, S. Yasuhisa, M. Yuzo, First-principles simulations of removal process in EEM (elastic emission machining), Comput. Mater. Sci. 14 (1) (1999) 232–235.
[5] K. Inagaki, K. Yamauchi, H. Mimura, K. Sugiyama, K. Hirose, Y. Mori, First-principles evaluations of machinability dependency on powder material in elastic emission machining, Mater. Trans. 42 (11) (1999) 2290–2294.
[6] T. Hirata, Y. Takei, H. Mimura, Machining property in smoothing of steeply curved surfaces by elastic emission machining, Procedia CIRP 13 (2014) (2014) 198–202.
[7] K. Yamauchi, T. Kataoka, K. Endo, K. Inagaki, K. Sugiyama, S. Ma kino, Y. Mori, A study on EEM (elastic emission Ma chining)-influence of dissolved oxygen to Si wafer surface, J. Jpn. Soc. Precis. Eng. 64 (1998) 907–912.
[8] M. Kanaoka, C. Liu, K. Nomura, M. Ando, H. Takino, Y. Fukuda, H. Mimura, K. Yamauchi, Y. Mori, Figuring and smoothing capabilities of elastic emission machining for low-thermal-expansion glass optics, J. Vac. Sci. Technol. B: Microelectron. Nanometer Struct. 25 (2007) 2100–2113.

SUBCHAPTER 8.4

Magnetic Abrasive Finishing (MAF)

8.4.1 Introduction

Magnetic abrasive finishing (MAF) is one of the advanced finishing processes, which produces a high level of surface quality and is primarily controlled by a magnetic field. MAF was first mentioned by Henry P. Coats in 1938 in a patent [1]. In MAF process, ferromagnetic abrasive particles act for finishing the workpiece surface as the action of these materials are controlled by magnets. The magnetic field is applied on the ferromagnetic abrasives across the gap between workpiece and magnet. A flexible magnetic abrasive brass (FMAB) is formed as magnetic abrasive particles are joined magnetically due to dipole-dipole interaction between the magnetic poles along the lines of magnetic force. FMAB acts like a multipoint cutting tool to remove materials from the workpiece surface. By controlling the magnetic field and finishing time, MAF process can be applied for micro and nano level finishing of varieties of workpiece materials. MAF process is also used for deburring and polishing of flat as well as cylindrical surface.

8.4.2 Basic working principle

In MAF the work piece is kept between the two poles of a magnet. The working gap between the work piece and the magnet is filled with magnetic abrasive particles. The abrasive particles such as silicon carbide or alumina and ferromagnetic particles such as iron grits of steel grits are mixed with a small amount of metal working fluids such as mineral oil or kerosene etc. to form magnetic abrasive particle. These magnetic abrasive particles are used either in the bonded form in which the abrasive particles are held in ferromagnetic matrix or in the unbonded form of homogeneous mixture of ferromagnetic particles, abrasives and additives. As the magnet is energized, the ferromagnetic particles are aligned along the magnetic lines of force and a chain like structure is formed surrounding the abrasive particles. A magnetic abrasive flexible brush (MAFB) is formed, acting as a multipoint cutting tool, due to the effect of the magnetic field in the working gap.

When inserting a cylindrical work piece in such a processing field giving revolution, feed and vibration in axial direction, surface and edge finishing are carried out by magnetic brush. In the application of ferromagnetic substance of work, for instance, work piece is also magnetized and the magnetic force acts on the top of the brush between the work piece and the abrasive grains resulting in pressing the abrasive grains to work surface. The MAF process removes a very small amount of material by indentation and rotation of magnetic abrasive particles in the circular tracks. Figure 8.4.1 shows the fundamental principle of MAF process for finishing external surface. For driving the abrasive particles for finishing the workpiece surface, magnetic force component, Fy is responsible. The magnetic force component Fx is responsible for penetrating action of abrasives into the workpiece along the line of magnetic force. Figure 8.4.2 shows the fundamental principle of MAF process for finishing internal surface. For machining internal cylindrical surface, the magnetic abrasive particles inside the workpiece surface are in action of magnetic field and workpiece rotates to finish the entire cylindrical surface. Figure 8.4.3 shows the fundamental principle of MAF process for finishing flat surface. For finishing flat surface the workpiece is held on the work holding table or fixture which is magnetized as south pole and a magnetic north pole rotates. The magnetic abrasive particles are

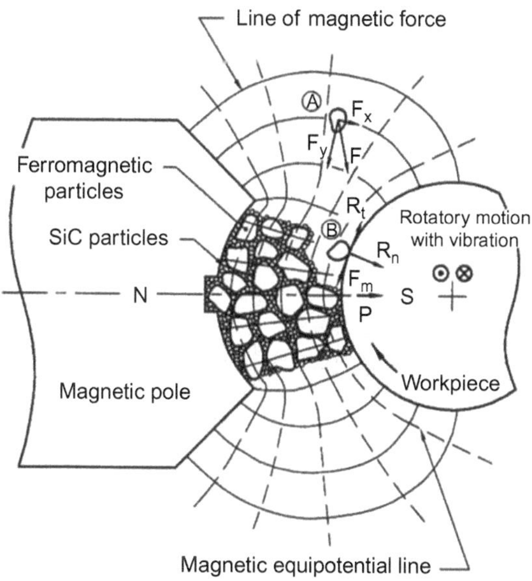

Fig. 8.4.1 Schematic diagram of MAF process for finishing external surface [2].

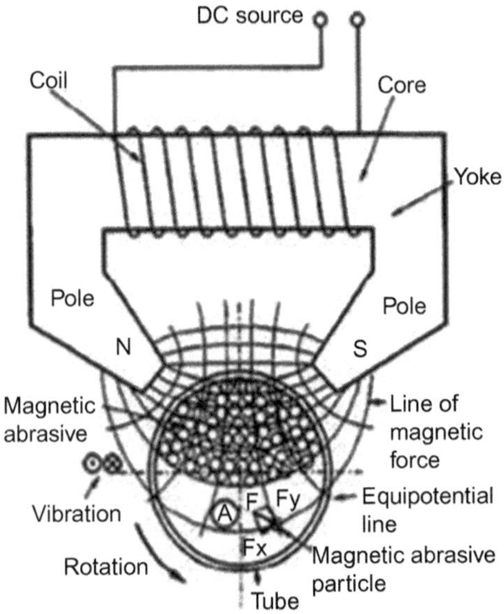

Fig. 8.4.2 Schematic diagram of MAF process for finishing internal surface [3].

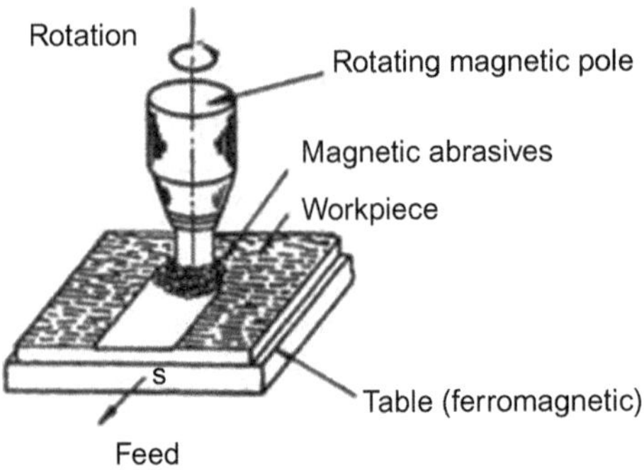

Fig. 8.4.3 Schematic diagram of MAF process for finishing flat surface [4].

supplied in the gap between workpiece and rotating magnetic pole. The energized abrasive particles surrounding the rotating magnetic poles acts like flexible magnetic brush to finish the flat surface of the workpiece.

8.4.3 Setup details

For finishing flat surface, experimental setup of MAF is shown in Fig. 8.4.4. Singh et al. [5] designed plane surface MAF process setup in

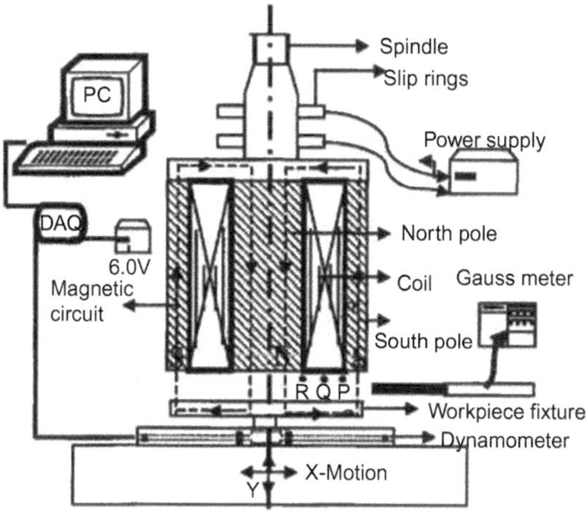

Fig. 8.4.4 Schematic diagram of experimental setup of MAF for finishing flat surface [5].

which flat faced electromagnet was fabricated in such a way that external shell was acted as south pole and middle portion was acted as north pole. The magnetic field was generated by DC power supply. The magnetic flux density in the gap between magnet and workpiece is varied by changing the current in the magnetic coil. The magnetic flux density is measured by using a digital gauss meter by placing the probe perpendicular to the direction of magnetic field. The magnetic abrasive particles are supplied in the gap between magnet and workpiece. The magnet assembly rotates during finishing process. The workpiece is held by workpiece fixture and it faces towards the magnet. The workpiece with fixture is placed on force dynamometer to measure the force applied on the workpiece during finishing process. The force data obtained from dynamometer can also be stored in PC and analyzed. The workpiece holding arrangement along with dynamometer is mounted on x-y table for moving the workpiece with respect to magnetic brass for finishing the selected area of workpiece.

During cylindrical magnetic abrasive finishing (MAF) process, the cylindrical workpiece is rotated by DC motor and vibrated by eccentric cam mechanism which is driven by induction motor combined with a frequency converter. The electromagnet is placed around the cylindrical work-piece. It consists of two poles that are surrounded by magnetic exciting coils arranged in such a manner as to provide the maximum magnetic field near the entire surface of the work-piece. Figure 8.4.5 shows 3D view of experimental set-up of MAF process [2]. For better quality finishing, the abrasive

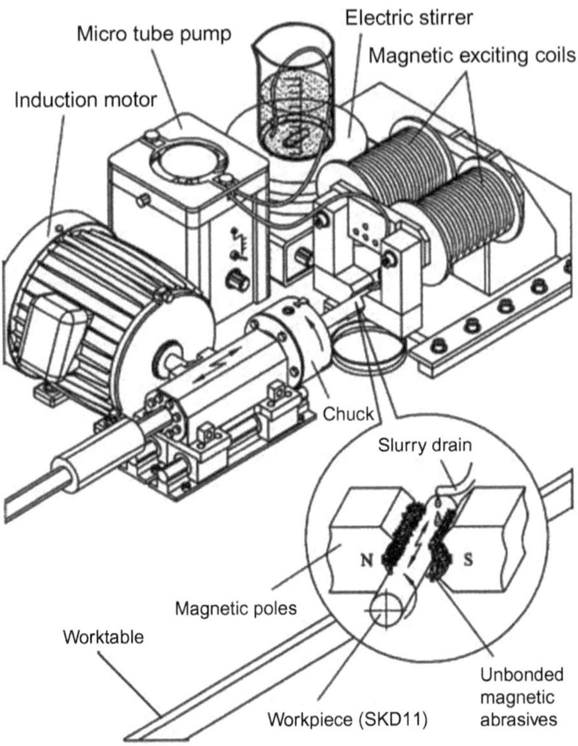

Fig. 8.4.5 3D view of experimental set-up of MAF process [2].

slurry which consists of magnetic abrasive particles with distilled water is supplied through an electric stirrer and a micro-tube pump as shown in Fig. 8.4.5.

8.4.4 Influence of process parameters

The material removal rate and surface roughness of work piece during MAF process are influenced by process parameters such as magnetic flux density, working gap, workpiece circumferential speed and size, type and volume fraction of abrasive particles. In magnetic abrasive finishing process, magnetic force is influenced by the material, shape and size of work, and shape and size of magnetic pole, work-pole gap distance, and composition of magnetic abrasives.

The performance criteria of MAF process are material removal rate, machined depth, surface roughness, out of roundness and surface integrity. Using larger sized magnetic abrasive particles, higher material removal takes place, but surface becomes poor. With the increase in finishing time, surface roughness decreases rapidly initially and then become steady. Surface finish is improved by using smaller size abrasive particle mixed with irregular shaped ferromagnetic particles.

The magnetic abrasives used in MAF process are of two types such as unbonded and bonded. The unbonded abrasives are the simple mixture of abrasives and ferromagnetic particles whereas the bonded abrasives are formed by sintering of abrasives in a ferromagnetic matrix. The bonded magnetic abrasives produce better surface finish with lower material removal rate, whereas unbonded magnetic abrasive particles produces comparatively lower surface finish with higher material removal rate. To improve the material removal rate, the addition of machining fluid in MAF process is recommended, but for achieving better surface finish, the machining fluid is not recommended [6]. With the increase in magnetic flux density and ferromagnetic particle size, the machined depth increases. With the increase in working gap between workpiece and magnet, the machined depth decreases. With the increase in magnetic flux density, surface roughness also decreases for flat surface finishing process [7].

8.4.5 Advantages

The advantages of MAF process are mentioned as follows:

(i) MAF process can be successfully used to eliminate the out of roundness error for cylindrical parts during finishing process.

(ii) MAF process is applied for finishing flat as well as internal and external cylindrical surfaces.

(iii) In MAF process, the finishing tool does not require dressing and it is a self-sharpening process.

(iv) In MAF process, the forces acting on the workpiece surface by the abrasive particles are controlled externally by generated magnetic field.

(v) MAF process produces mirror finish surface.

(vi) MAF process can finish workpiece surface of magnetic as well as non-magnetic materials.

8.4.6 Applications

The applications of MAF process has been discussed as follows:

(i) Magnetic abrasive finishing process can be used for surface finishing as well as surface modification of hard to finish surfaces of materials such as brass, stainless steel, etc.

(ii) Magnetic abrasive finishing can be successfully used for finishing of internal as well as external surfaces of complicated design.
(iii) MAF process is applied for finishing of additively manufactured components of 316L Stainless Steel made by selective laser melting process [8].
(iv) Magnetic Abrasive machining can be applied for reproducing the defined radii of the cutting edges and corner edges of twist drill of high speed steel with improved tool surface quality [9].
(v) MAF process can be applied for deburring and chamfering operation [10].
(vi) MAF process is applied for uniform internal finishing of SUS 304 stainless steel bent tube by controlling the size of the ferrous particles and magnetic field [11].
(vii) Ultrasonic assisted MAF can be applied for polishing of copper alloy.
(viii) Chemical assisted MAF can be applied for finishing Tungsten by the combined effect of chemical oxidation and magnetic force.

The MAF process is a relatively new and advanced finishing process in which the cutting force is controlled by the magnetic field. MAF process is applied to produce optical, electronic and mechanical components with good form accuracy and nanometer level surface finish with no defects. MAF process can finish shafts, automotive components, bearing parts and artificial hip joints etc. with surface finish in the order of few nanometer. The hybrid MAF processes such as ultrasonic vibration assisted-MAF which combines USM and MAF, Electrolytic MAF which combines ECM and MAF, Ultrasonic and Magnetic field assisted ECM etc. have been conceptualized for overcoming the limitation of individual processes and utilizing the advantages of the constituent processes to enhance the overall performance of the hybrid process.

References

[1] G.Z. Kremen, E.A. Elasayed, V.I. Rafalovich, Mechanism of material removal in magnetic abrasive process and the accuracy of machining, Int. J. Prod. Res. 34 (1996) 2629–2638.
[2] G.W. Chang, B.H. Yan, R.T. Hsu, Study on cylindrical magnetic abrasive finishing using unbonded magnetic abrasives, Int. J. Mach. Tools Manuf. 42 (2002) 575–583.
[3] R.K. Singh, D.K. Singh, S. Gangwar, Advances in magnetic abrasive finishing for fruturistic requirements – a review, Mater. Today Proc. 5 (2018) 20455–20463.
[4] N.K. Jain, V.K. Jain, Modeling of material removal in mechanical type advanced machining processes: a state of art review, Int. J. Mach. Tools Manuf. 41 (2001) 1573–1635.
[5] D.K. Singh, V.K. Jain, V. Raghuram, Parametric study of magnetic abrasive finishing process, J. Mater. Process. Technol. 149 (2004) 22–29.

[6] T. Shinmura, Study on magnetic abrasive finishing, Bull. Jpn. Soc. Precis. Eng. 20 (1) (1986) 52–54.
[7] T. Shinmura, K. Takazawa, E. Hatano, T. Aizawa, Study on magnetic abrasive process-process principle and finishing possibility, Bull. Jpn. Soc. Precis. Eng. 19 (1) (1985) 54–55.
[8] Y.W. Pei, H. Yamaguchi, Material removal mechanism of additively manufactured components finished by magnetic abrasive finishing, NAMRC 46, Procedia Manuf. 26 (2018) 394–402.
[9] B. Karpuschewsk, O. Byelyayev, V.S. Maiboroda, Magneto-abrasive machining for the mechanical preparation of high-speed steel twist drills, CIRP Ann. Manuf. Technol. 58 (2009) 295–298.
[10] T. Shinmura, K. Takazawa, E. Hatano, Study on magnetic abrasive finishing, Ann. CIRP 39 (1990) 325–328.
[11] H. Yamaguchi, T. Shinmura, M. Sekine, Uniform internal finishing of SUS 304 stainless steel bent tube using a magnetic abrasive finishing process, ASME J. Manuf. Sci. Eng. 127 (2005) 605–611.

SUBCHAPTER 8.5

Magnetorheological Finishing (MRF)

8.5.1 Introduction

Magnetorheological finishing (MRF) is one of the magnetic field assisted finishing process in which magnetorheological (MR) fluid is used. The MRF process was primarily conceived at the Luikov Institute of Heat and Mass Transfer in Minsk, Belarus in 1988 and further developed at the Centre for Optics Manufacturing at University of Rochester. Later on the commercial machine was developed by QED Technologies in 1996. MRF process provides better control over rheological properties of abrasive laden magnetorheological fluid medium. Magnetorheological (MR) fluid comprises of carbonyl iron powder and silicon carbide abrasives dispersed in the visco-plastic base of grease and mineral oil. MR fluid exhibits change in rheological behavior in presence of external magnetic field. This smart behavior of MR fluid is effective to precisely control the finishing process. Magnetically controlled magnetorheological fluid cuts surface peaks due to generation of high shear stress over the portion of the workpiece surface. MRF process has potential for non-finishing of freeform surfaces of complex shaped parts such as bio-implants. MRF is a flexible finishing process for internal as well as external surfaces. The variants of MRF process such as magneto-rheological abrasive flow finishing (MRAFF) and rotary magneto-rheological abrasive flow finishing (R-MRAFF) processes have been developed to finish internal and external cylindrical surfaces.

8.5.2 Basic principle of MRF process

The MRF process is a finishing process in which MR fluids are continuously supplied on the rotating carrier wheel to form a MR fluid ribbon. An MR fluid circulatory system consists of one pump for delivery of MR fluid and another pump for suction of MR fluid from carrier wheel. A MR fluid storage system collects the used MR fluid via suction pump and maintains the MR fluid temperature to avoid the sedimentation of solid particles and to mix the suspended particles for uniform delivery of MR fluid

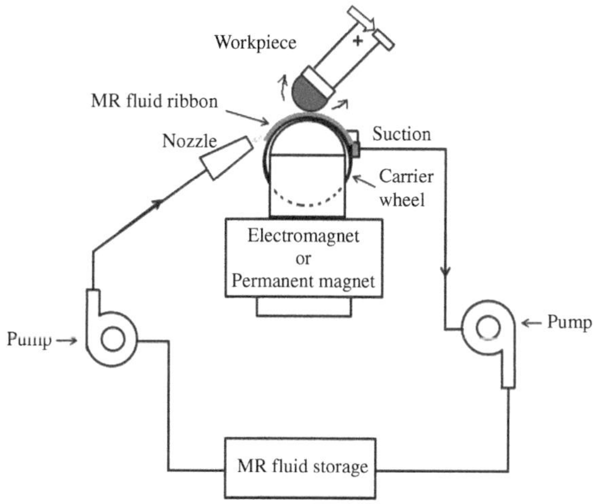

Fig. 8.5.1 Schematic diagram of MRF process.

via deliver pump. The surface of the workpiece to be finished in kept in contact with the MR fluid ribbon. The swinging and rotating action of workpiece and MR fluid ribbon results in material removal due to normal and tangential cutting forces applied by the abrasive particles present in MR fluids. Electromagnet or permanent magnet is placed below the carrier wheel and controls the cutting forces due to magnetic action of magnetic particles present in MR fluid. Figure 8.5.1 shows the schematic diagram of MRF process.

In MRF process, as magnetic field is applied, magnetic particles form chain like stricture to grip abrasive particles and MR fluid becomes stiffened. Under the action of magnetic field the abrasive particles interacts with the workpiece and materials are removed from the roughness peaks by cutting action of abrasive particles those are driven by magnetic particles. Figure 8.5.2 shows finishing actions of abrasive particles present in MR fluid during MRF process.

The MRF process depends on smart and magnetically controllable fluid called as Magnetorheological (MR) fluid. MR fluid was first developed by Rabinow in the late 1940s. MR fluid consists of micro sized magnetic particles, non-magnetic carrier medium, abrasive particles and surfactants or additives. Carbonyl iron particles (CIPs), electrolyte iron powder, iron cobalt alloy powder are generally used as magnetic particles. The non-magnetic carrier medium such as water, silicone oil, mineral oil etc. are used in MR fluids. The abrasive particles like silicon carbide, aluminum oxide,

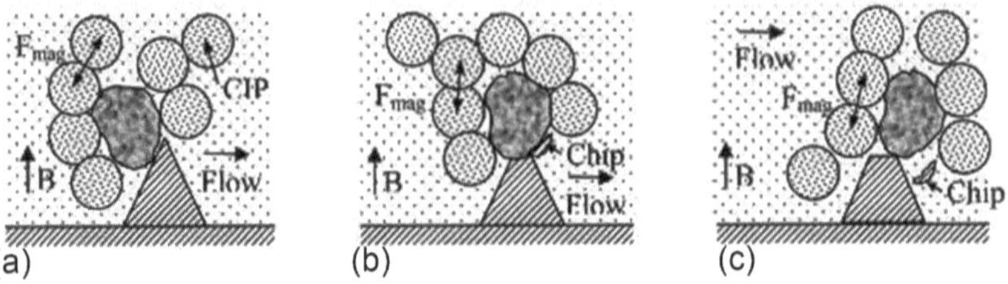

(a) (b) (c)

Fig. 8.5.2 Finishing actions of abrasive particles present in MR fluid. (a) abrasive grain along with CIP magnetic particles approaching towards surface peak; (b) abrasive grain cutting the peaks; (c) material is removed in the form of micro chips [1].

cerium oxide, zirconium oxide, diamond powder, boron carbide etc. are used. The surfactants (additives) such as grease, glycerol, sodium carbonate, oleic acid, xanthum gum are used in MR fluid [2]. On application of external magnetic field, the iron particles in carrier fluid aggregate into columnar chains aligned in the magnetic field direction and MR fluid exhibits its magnetorheological effect and behaves like Bingham plastic fluid. Ideal MR fluid behaves like Newtonian fluid when there is no action of magnetic field. The magnitude of magnetic force acting between the iron particles in MR fluid depends on iron particle concentrations, applied magnetic field intensity, size and magnetic permeability of the particles. Under the action of magnetic field the iron particles encompass abrasive grains and moves towards the workpiece surface. The essential features of a magnetorheological fluid are high concentration of magnetic and abrasive particles; high yield stress increment in magnetic field; resistance to corrosion for water based fluid and high finishing ability.

A calibrated measuring glass tube is used to measure the volume percentage of CI particles, abrasives and engine oil. MR fluid was prepared by mixing the compositions as per the predetermined percentage volume. The fine carbonyl iron (CI) powder is collected from the saw dust of pure iron by sieving.

The magneto-rheological abrasive flow finishing (MRAFF) process is a hybrid process based on principle of magnetorheological abrasive finishing in which MR fluid reciprocates to finish the workpiece held in workpiece fixtures. The magnets are placed surrounding the workpiece and these can be rotated in rotary MRAFF process. For rotary MRAFF process the MR fluid also rotates under the action of rotating magnets. Figure 8.5.3(a) shows the mechanism of MRAFF process and Fig. 8.5.3(b) shows the changes of magnetorheological behavior of MR fluid under the magnetic field.

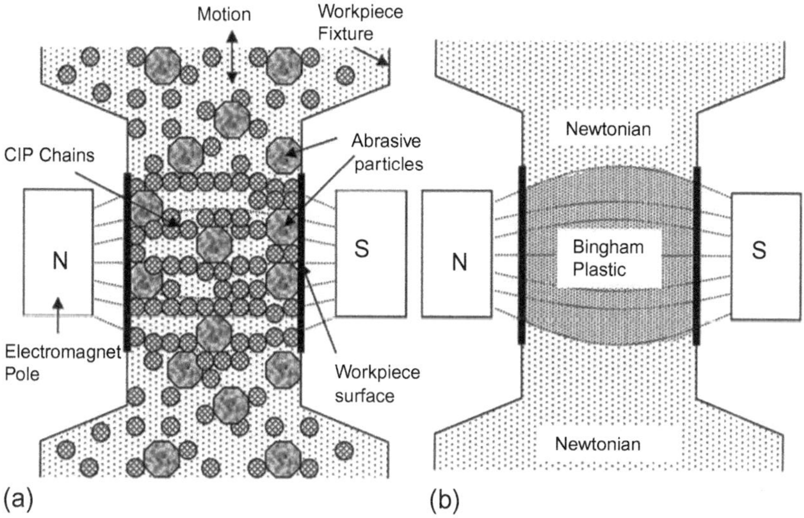

Fig. 8.5.3 (a) The mechanism of MRAFF process and (b) the changes of magnetorheological behavior of MR fluid under the magnetic field [1].

8.5.3 Setup details

The magneto-rheological abrasive flow finishing (MRAFF) is one of the hybrid finishing process based on the principle of magneto rheological finishing in which MR fluid is used. The MRAFF process is one of the versatile variant of MRF process for nano-level finishing of workpiece with complex internal as well as external geometries. The MRAFF setup consists of MRF cylinders with pistons, electro-magnets or permanent magnets, hydraulic drive and controls, workpiece fixture and supporting frame. The hydraulic drive and control unit consists of pump, motor and pressure gauges, piping etc. which imparts the reciprocating motion of the pistons and maintains the desired pressure during finishing process. The schematic diagram of MRAFF experimental setup is shown in Fig. 8.5.4 The MR fluid is extruded from cylinders by the pistons over the workpiece surface. The MR fluid reciprocates between the cylinders across the workpiece. The workpiece was held by fixture in the flow path of MR fluid. The pole faces of electromagnet are fixed to produce the desired magnetic flux density across the workpiece and MR fluid. The rotary MRAFF system is another variant of MRF process for nano-finishing of free form surface. In rotary

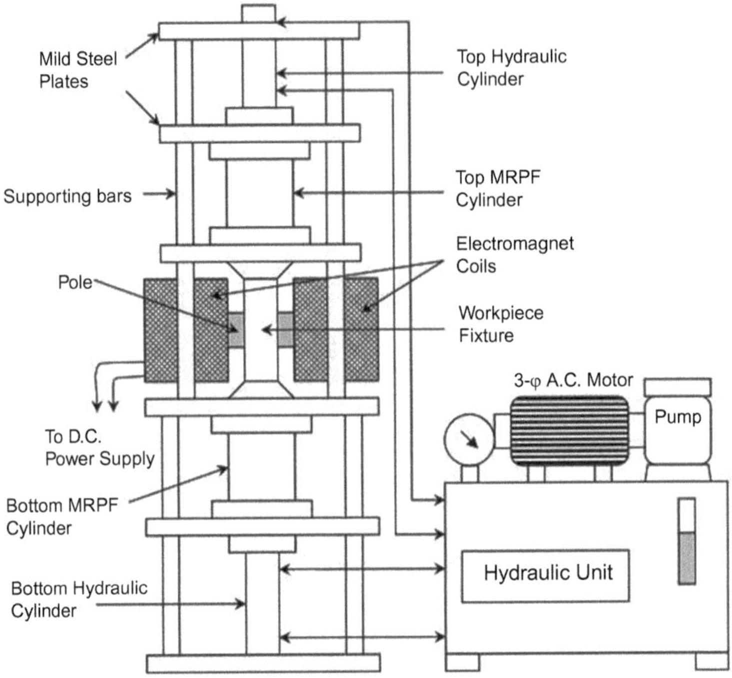

Fig. 8.5.4 Schematic diagram of rotary MRAFF system [1].

MRAFF experimental setup MR fluid reciprocates due to reciprocating motion imparted by pistons as well as rotates due to rotation motion imparted by the rotating magnets surrounding workpiece fixture. The rotational motion of magnets is controlled by a variable frequency drive of AC synchronous motor. The length of magnet is longer than that of the workpiece to avoid the concentration of magnetic field at the ends of workpiece. Under the action of the rotating magnetic fields, the stiffened MR fluid finishes the workpiece surface uniformly and effectively.

8.5.4 Influence of process parameters

The performance of MRF process is influenced by various process parameters such as type, size, concentration, hardness of abrasive particles; type, size, concentration, hardness of magnetic particles; type and concentration of carrier fluid; pH, viscosity and yield stress of MR fluid; working gap; magnetic flux density and wheel speed etc. The performance of MRF process is generally expressed in terms of Material removal rate (MRR) and surface roughness.

(a) The effect of magnetic flux density on MRR and surface roughness

The magnetic flux density is one of the important process parameters of MRF process. The stiffness of MR fluid is controlled by magnetic flux density. The magnetic flux density can be varied by changing the electric current in an electromagnet. The magnetic flux density can be varied by changing the working gap between the workpiece and permanent magnet. As the magnetic flux density increases, the yield stress as well as viscosity of MR fluid increases. The stiffness of MR fluid increases with the increase in magnetic flux density as magnetic particles are firmly aligned along the magnetic lines of force. Highly stiffened MR fluid interacts with workpiece surface with high finishing forces. As a result, MRR increases steadily with the increase in magnetic flux density [2]. MR fluid removes materials from workpiece surface without yielding as yield stress and viscosity becomes high with increase in magnetic flux density. Finally surface roughness R_a value sharply decreases with increase in magnetic flux density. Figure 8.5.5 exhibits the variation of MRR and surface roughness R_a value with magnetic flux density.

(b) The effect of magnetic particle concentration on MRR and surface roughness

The magnetic particle in MR fluid produces chain like structures to grip abrasive particles so that abrasive particles interact with the workpiece surface during MRF process. With the increase in concentration of magnetic particles, the chain structure becomes thicker and denser. As result the yield strength and viscosity of MR fluid increases with the increase in concentration of magnetic particles [4]. As yield strength increases, the abrasive particles are firmly gripped by chain structure of magnetic particles. In this

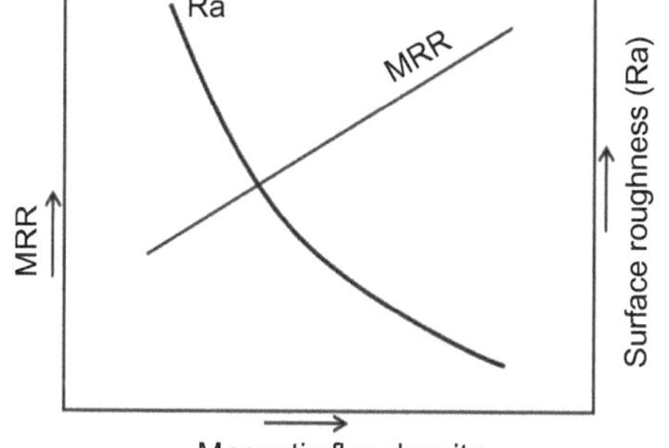

Fig. 8.5.5 Variation of MRR and surface roughness with magnetic flux density.

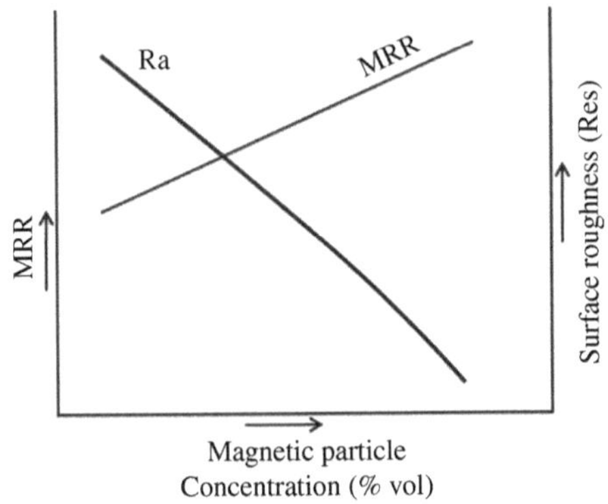

Fig. 8.5.6 Variation of MRR and surface roughness with magnetic particle concentration.

condition the abrasive particles interacts with the workpiece with high finishing force without sliding over the workpiece surface. As a result the MRR steadily increases and surface roughness R_a value also steadily decreases with increase in concentration of magnetic particles [2]. Figure 8.5.6 shows the variation of MRR and surface roughness R_a value with magnetic particle concentration.

(c) The effect of abrasive particle concentration on MRR and surface roughness

Initially the addition of abrasive particles in MR fluid for low concentration of abrasive particles increases MRR. But after reaching the optimum concentration of abrasive particles in MR fluid MRR decreases. With the increase in concentration of non-magnetic abrasive particles, the viscosity and yield strength of MR fluid decreases. With high concentration of abrasive particles, the abrasive particles come in between two magnetic particles. The magnetic force between magnetic particles decreases as a result yield stress and viscosity decreases [2]. The chain structure of magnetic particles to grip abrasive particle becomes week when the concentration of abrasive particles is high. Therefore, the finishing performance of abrasive particles in terms of MRR and surface finish becomes poor with high concentration of abrasive particles. The surface roughness R_a value initially decreases, the increases with the increase in concentration of abrasive particles in MR fluid during MRF process. Figure 8.5.7 shows the effects of abrasive particle concentration on MRR and surface roughness R_a value.

(d) The effect of carrier wheel speed on MRR and surface roughness

The speed at which MR fluid interacts with the surface of workpiece is termed as carrier wheel speed. At no-slip condition between MR fluid ribbon

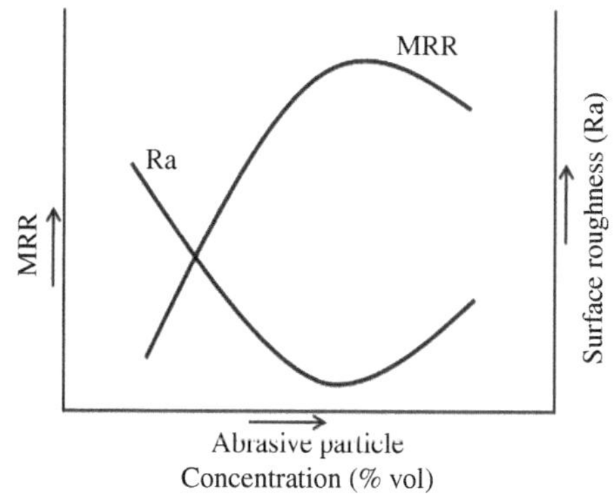

Fig. 8.5.7 Effects of abrasive particle concentration on MRR and surface roughness.

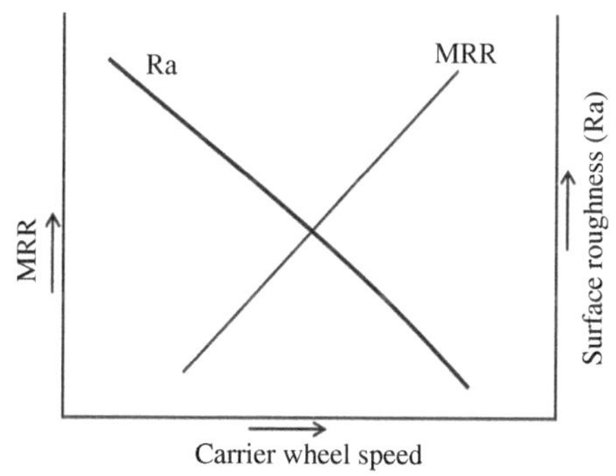

Fig. 8.5.8 Effects of carrier wheel speed on MRR and surface roughness.

and carrier wheel, speed of carrier wheel is same as MR fluid ribbon. The speed of MR fluid increases with the increase in carrier wheel speed. The cutting force applied by the abrasive particles on the workpiece increases and finishing action improves with increase in the carrier wheel speed. As a result MRR increases and surface roughness, R_a value decreases with the increase in carrier wheel speed [2]. Figure 8.5.8 shows the effects of carrier wheel speed on MRR and surface roughness R_a value.

8.5.5 Advantages and challenges

The advantages of MRF are given as follows:

(i) MRF process uses MR fluid for nano finishing applications.

(ii) MRF process uses MR fluid for finishing flat, curve and free form surfaces.

(iii) MRF process can finish soft as well as hard materials.
(iv) There is no tool wear related problems in MRF process.
(v) MRF process uses flexible MR fluid for finishing internal as well as external surfaces.
(vi) As MR fluid is used, the temperature rise during finishing is low to avoid thermal effects on the surface of workpiece such as optics.
(vii) MRF process finishes non-magnetic materials efficiently.

The limitation of MRF process is that a very less amount of materials are removed during the process as a result the finishing or processing time is more. The hybrid finishing processes such as chemical mechanical MRAFF process has been developed where chemical mechanical and magneto-rheological action improves the surface finish, MRR and also reduces the finishing or processing time.

8.5.6 Applications

The applications of MRF are mentioned as follows:

(i) MRF process has industrial applications in polishing optical lenses and mirrors.
(ii) MRF process is applied for fabrication and finishing of curved silicon based micro-structures.
(iii) MRF process is applied for honing applications where the abrasive mixed magnetorheological fluid pushed up and down against the curved surface of rotating workpiece. Magnetoreological Abrasive Honing (MRAH) has been performed for finishing aluminum and austenite stainless steel workpiece [5].
(iv) MRF process is applied for nano-finishing of almost all types of glasses and ceramic materials [6].
(v) MRF process can be applied for finishing hard and soft single crystals, hard and soft polycrystalline materials, metals and plastics. MRF process is applied to correct residual error of single point diamond turned electroless nickel part. MRF is applied for finishing for spherical and aspheric optics [7].
(vi) MRF process is applied for finishing Silicon nitride molds for precision glass molding [8].
(vii) MRAFF, variant of MRF process is applied for finishing bio-implants like artificial knee joints and hip joints [3, 9].
(viii) MRF process can be applied for finishing composites by proper selection of composition of MR fluid and related process parameters.

(ix) Magnetic field assisted finishing (MFAF) process is applied for nano-finishing of bio-titanium alloy using magnetic finishing tool and varieties of magneto-rheological polishing fluid [10].

Magneto-rheological finishing (MRF) process is very much useful for nano-finishing of internal and external, flat, curved and freeform surfaces of varieties of materials such as glass, ceramics, metals etc. for varieties of applications in the engineering and bio-medical fields. The different variants of MRF process such as MRAFF, rotary MRAFF, Magnetoreological Abrasive Honing (MRAH), Ball end MRAFF etc. have been developed using the principle of magneto-rheological finishing for specific applications such as finishing bio-implants and optical mirrors etc. Research activities are going on to overcome the limitation of the MRF process particularly to high finishing time of components with complex geometrical shapes.

References

[1] S. Jha, V.K. Jain, Design and development of the magnetorheological abrasive flow finishing (MRAFF) process, Int. J. Mach. Tools Manuf. 44 (2004) 1019–1029.

[2] A. Sidpara, V.K. Jain, Chapter 9: Magnetorheological finishing, in: Nanofinishing Science and Technology: Basic and Advanced Finishing and Polishing Processes, CRC Press, Taylor and Francis Group, 2017. ISBN: 978-1-4987-4594-9.

[3] S. Kumar, V.K. Jain, A. Sidpara, Nanofinishing of freeform surfaces (knee joint implant) by rotational magnetorheological abrasive flow finishing (R-MRAFF) process, Precis. Eng. 42 (2015) 165–178.

[4] A. Sidpara, V.K. Jain, Rheological properties and their correlation with surface finish quality in MR fluid-based finishing process, Mach. Sci. Technol. 18 (3) (2014) 367–385, https://doi.org/10.1080/10910344.2014.925372.

[5] A. Sadiq, M.S. Shunmugam, Investigation into magnetorheological abrasive honing (MRAH), Int. J. Mach. Tools Manuf. 49 (2009) 554–560.

[6] A. Sidpara, Magnetorheological finishing: a perfect solution to nanofinishing requirements, Opt. Eng. 53 (9) (2014) 092002, 1–7.

[7] D. Golini, W.I. Kordonski, P. Dumas, S. Hogan, Magnetorheological finishing (MRF) in commercial precision optics manufacturing, Proc. SPIE 3782 (1999) 80–91.

[8] A. Geiss, R. Rascher, J. Slabeycius, M. Schinhaerl, P. Sperber, F. Patham, Material removal study at silicon nitride molds for the precision glass molding using MRF process, Proc. SPIE 7060 (2008) 706007–706011.

[9] A. Sidpara, V.K. Jain, Analysis of forces on the freeform surface in magnetorheological fluid based finishing process, Int. J. Mach. Tools Manuf. 69 (2013) 1–10.

[10] A. Barman, M. Das, Nanofinishing of bio-titanium alloy to generate different surface morphologies by changing magnetorheological polishing fluid compositions, Precis. Eng. 51 (2018) 145–152.

SUBCHAPTER 8.6

Plasma Assisted Polishing (PAP)

8.6.1 Introduction

The conventional method of mechanical polishing using diamond as abrasive can be used for flattening of semiconductor materials with high material removal rate, but this process introduces many scratches, pits and considerable subsurface damages etc. The Chemical Mechanical Polishing (CMP) has ability for nano-finishing of semiconductor materials such as silicon, single crystal silicon carbide, single crystal diamond wafer and single crystal sapphire etc., but there is introduction of little subsurface damages with pits and scratches etc. CMP is a slow finishing process with low material removal rate. An atomically flat and damage free substrate surface is required to be generated to enhance the performance of electronic and optical devices. On the other hand, the Plasma Assisted Polishing (PAP) is a novel chemical dry polishing method which combines the surface modification by plasma irradiation and the removal of modified surface layer by soft abrasive polishing. The plasma assisted polishing process has potential to produce atomically flat surfaces of difficult to polish semiconductor materials. PAP process is applied for polishing of very hard semiconductor materials such as 4H-SiC, GaN, alumina, single crystal diamond wafer and single crystal sapphire etc.

8.6.2 Basic working principle

In Plasma Assisted Polishing (PAP), soft abrasive polishing is assisted by plasma irradiation. Plasma irradiation softens the hard surface of base substrate materials. Due to plasma irradiation on the hard material, the hardness of the surface of substrate material decreases and it is easier to remove the modified surface layer of material by soft abrasive polishing.

During plasma assisted polishing of 4H-SiC substrates, water vapor plasma containing a high concentration of hydroxyl (OH) radicals is used for surface modification. SiC substrate surface is irradiated by water vapor plasma and oxidized by OH radicals. The oxidation product SiO_2, and a thin transition layer (silicon oxycarbide) was found to be located between SiO_2 and SiC. The hardness of the substrate surface decreases after plasma irradiation. A soft oxide layer generated by plasma irradiation is polished using a soft abrasive. The selection of the abrasive material in the PAP process depends on base materials. The hardness of the abrasive material must be lower than that of the base material (SiC) so that there will be no scratches or subsurface damage. Generally cerium (CeO_2) is used as an abrasive to remove the oxidized surface layer because of its excellent polishing characteristics for silica glasses [1]. In the PAP process plasma irradiation and abrasive polishing are simultaneously conducted as a result scratch-free and damage-free base substrate material surface is produced. Figure 8.6.1(a) shows the removal of pit and scratches from SiC substrate by plasma irradiation and HF dipping process. In this process silicon oxycarbide layer generated by plasma irradiation is not completely removed although the flat surface is achieved. Figure 8.6.1(b) shows the PAP process in which both SiO_2 layer and silicon oxycarbide layer generated by plasma irradiation are completely removed by abrasive polishing and atomically flat surface is produced [2].

During plasma assisted polishing of gallium nitride (GaN) substrate materials, a mixture of helium gas with CF_4 was used for plasma generation. The aluminum alloy is used as powered electrode for plasma generation. Atmospheric pressure helium based CF_4 plasma at high radio frequency irradiates on gallium nitride (GaN) substrate materials. The plasma modified substrate surface is polished by using rotary resin bonded cerium (CeO_2) grindstone under the application of constant load [3].

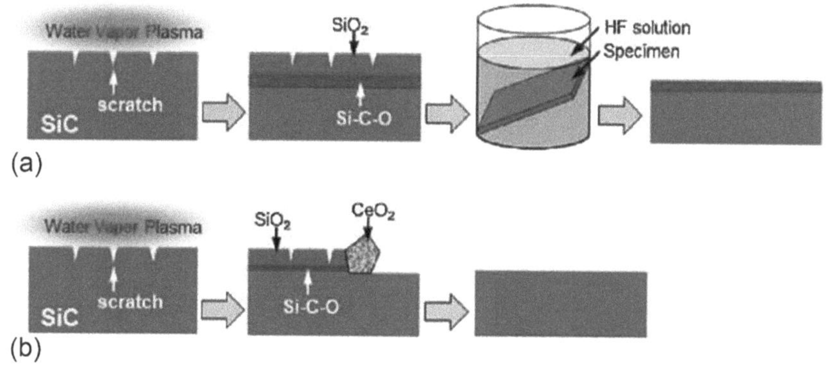

Fig. 8.6.1 Mechanism of pit and scratch removal from SiC substrate (a) by plasma irradiation and HF dipping; (b) by PAP process [2].

During plasma assisted polishing of single crystal sapphire, an atmospheric pressure low temperature helium gas plasma at radio frequency irradiates the surface of single crystal sapphire. Water vapor is introduced in the plasma, as a result hydroxyl species are generated which leads to the hydration of surface of sapphire wafer. The aluminum hydroxide soft layer is formed on the top surface and softened layer is then removed by buffing with hard fibrous polyurethane buffing wheel [4].

During plasma assisted polishing of single crystal diamond (SCD) wafer, argon plasma containing water vapor is used to modify the surface of polishing plate and single crystal diamond and SCD wafer is polished under high polishing pressure. Due to plasma irradiation and generation of hydroxyl radicals (OH), both the surface of SCD wafer and quartz glass polishing plate are modified. The surfaces of both the SCD and the polishing plate are terminated by OH through the direct irradiation of plasma. A dehydration condensation reaction between the OH terminated SCD and the polishing plate results in C—O—M (M is the metal atom of polishing plate material) bond formation in the region of contact. The relative motion between the SCD and the polishing plate removes carbon atoms from the C—O—M bonds on the SCD surface without graphitization. A damage free highly efficient polishing of single crystal diamond wafer is achieved by plasma assisted polishing without using loose abrasives. Figure 8.6.2

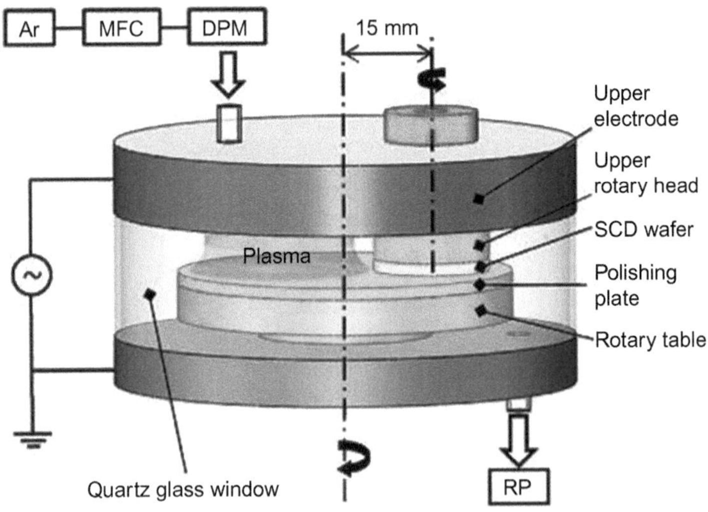

Fig. 8.6.2 Schematic diagram of working principle of plasma assisted polishing of SCD wafer [5].

shows the schematic diagram of working principle of plasma assisted polishing of SCD wafer. In this system the mass flow rate of Ar gas is controlled by mass flow controller (MFC) and water vapor concentration is measured by dew point meter (DPM). The rotating plate (RP) rotates the rotary table in which the quartz glass polishing plate is fixed. Typically the polishing pressure, the rotation speed of the polishing plate and radio frequency of the power supply and power are 52.6 kPa, 310 rpm, 13.56 MHz and 30 W, respectively [5].

8.6.3 Setup details

The plasma assisted polishing setup consists of plasma irradiation system and abrasive polishing system. Figure 8.6.3(a) shows the schematic diagram of plasma irradiation system. The plasma gas is supplied surrounding the aluminum electrode which has alumina ceramic casing. The radio frequency power supply is given to electrode for generation of plasma. The impedance matcher is used in the power circuit to maximize the power from power source to the electrode. Generally atmospheric pressure helium based CF_4 plasma is generated between the electrode and workpiece. The workpiece is mounted on X-Y stage of CNC system. Figure 8.6.3(b) shows the schematic diagram of abrasive polishing system. A resin bonded CeO_2 abrasive grindstone is mounted with rotary spindle. The workpiece is polished by abrasive grindstone under the application of the constant load and also the workpiece is scanned and polished by the movement of X-Y table under action of soft abrasives.

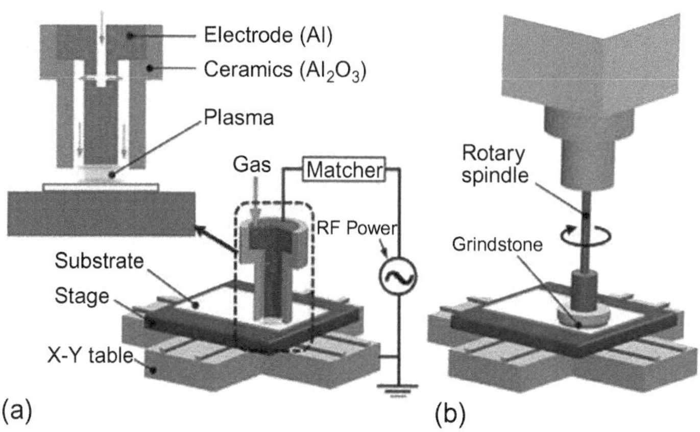

Fig. 8.6.3 Schematic Diagram of PAP Setup (a) plasma irradiation system and (b) abrasive polishing system [3].

8.6.4 Influence of process parameters

The performances of Plasma Assisted Polishing are controlled by plasma generation related process parameters and abrasive polishing related process parameters. The plasma related process parameters are type of plasma gas, plasma gas composition and mass concentration, plasma generation power and frequency and plasma exposure time. The abrasive polishing related process parameters are type and mass concentration of abrasives, load, rotational speed of polishing plate and polishing time. Helium or Argon with water vapor is generally used as reactive gas for plasma generation.

For plasma assisted polishing of SiC wafer, the gas flow rate varies from 1.5 to 3.0 l/min. The applied RF electric power varies from 18 W to 30 W. The rotational speed of substrate is 120 rpm and the rotational speed of polishing head is 500 rpm. The plasma exposure time is 1 h. The ceria (CeO_2) abrasive of average diameter 1 μm is generally used for polishing. The surface roughness of 0.1 nm rms of flat surface is obtained during PAP process. The MRR of PAP process is about 80 nm/h, which is quite low. The load during PAP process typically varies from 50 to 80 g [2, 6].

During PAP process of GaN, resin bonded ceria (CeO_2) abrasive is generally used. The rotational speed of polishing pad is 2000 rpm and polishing time is 3 h. As the polishing time increases, the surface roughness rms value decreases. For 3 h polishing the surface roughness less than 0.1 nm rms can be achieved. The helium carrier gas with O_2 or CF_4 reactive gas is used for plasma irradiation on GaN substrate [3].

During PAP process of sapphire wafer, a gas mixture of He and water vapor is supplied at the flow rate of 20 l/min. A constant electrode gap of 1.4 mm and the applied power of 150 W is maintained for generation of plasma arc. The plasma exposure time varies from 5 to 120 s. As the plasma exposure time increases, the generated soft layer thickness increases from 10 to 50 nm. The hardness decreases with the plasma radiation. During PAP process the MRR of 600 nm/h is possible to achieve. The surface roughness of 0.55 nm rms can be achieved [4].

8.6.5 Advantages

The plasma assisted polishing has the following advantages:

(i) Damage free surface;
(ii) Dry process;

(iii) No chemical is used;
(iv) Generation of atomically flat surface;
(v) Precision finishing;
(vi) For some applications of PAP, no loose abrasive is used.

8.6.6 Applications

The applications of plasma assisted polishing (PAP) are as follows:

(i) PAP process is applied for polishing single crystal diamond (SCD) wafer with polishing rate of 2.1 μm/h using quartz glass polishing plate with the irradiation of an Ar based plasma containing water vapor. Using sapphire polishing plate, a surface roughness of 0.13 nm root mean square is obtained during PAP of SCD wafer. Highly efficient and damage free surface is required to be generated on SCD for its wide applications in fabrication of semiconductor power devices [5].

(ii) The Chemical Mechanical Plasma polishing (CMPP) of single crystal sapphire has been performed by removing thin hydration layer formed on sapphire due to atmospheric plasma produced in a stream of helium gas and water vapor. For fabrication of microelectronic devices and precision high energy optics, polishing of sapphire or single crystal alpha alumina is highly needed [4, 7].

(iii) PAP process is applied for polishing gallium nitride (GaN) and producing a scratch free and pit free atomically flat surface by removing the plasma modified layer using a cerium grindstone. The surface roughness of less than 0.1 nm rms is obtained by plasma assisted polishing on GaN material surface which is used as materials for light emitting diodes, field effect transisters and high power, high frequency electronic devices [3, 8].

(iv) The atomic scale flattening of single crystal silicon carbide is needed as it is used for high power, high temperature electronics devices. The PAP process is applied for polishing very hard and chemically inert 4H-SiC material surface. The atomically flat surface with surface roughness of 0.2 nm rms and a well ordered terrace structure is obtained by PAP process [9, 10].

The Plasma Assisted Polishing (PAP) process is a hybrid finishing process in which plasma irradiation assists soft abrasive polishing by softening the difficult to polish materials. For producing atomically flat surface on semiconductor wafer materials, the plasma assisted polishing process is highly efficient with or without loose abrasive powders. PAP process overcomes the limitation such as subsurface damage of mechanical as well as chemical mechanical polishing process. PAP process can be combined with CMP process for enhancing the material removal rate and generating damage free and scratch free surface on single crystal semiconductor materials used for power electronics and optics devices.

8.7 Model questions

Abrasive flow finishing (AFF)

Q1. What are the components of AFF system?
Q2. What are different constituents of abrasive media used in AFF process?
Q3. Explain the working principle of two way AFF process with sketch.
Q4. What are process parameters involved in AFF process?
Q5. What are advantages of AFF process?
Q6. What are the various area of application of AFF process?

Chemical mechanical polishing (CMP)

Q1. Why is CMP called as hybrid finishing process?
Q2. What are the various elements of CMP setup?
Q3. Represent various process parameters and responses of CMP process with the help of fish bone diagram.
Q4. State the advantages and limitations of CMP process.
Q5. State the major areas of application of CMP process.

Elastic emission machining (EEM)

Q1. What is basic principle of EEM process?
Q2. What are the fine oxide powders used in EEM?
Q3. What are the components of EEM system?
Q4. What are advantages of EEM process?
Q5. What are the applications of EEM process in Optics?

Magnetic abrasive finishing (MAF)

Q1. How is a flexible magnetic abrasive brass (FMAB) formed during MAF process?
Q2. How is MAF process applied for finishing internal and external curve surfaces?
Q3. What are the advantages of MAF process over AFF process?
Q4. What are the applications of MAF process for nanofinishing?

Magnetorheological finishing (MRF)

Q1. What are the various constituents of magnetorheological fluid used in MRF process?

Q2. What is the material removal mechanism in MRF process?

Q3. How is magnetic rheological abrasive flow finishing (MRAFF) different from MRF?

Q4. What are the effects of process parameters on MRR and surface roughness in MRF process?

Q5. State the advantages and applications of MRF process.

Plasma assisted polishing (PAP)

Q1. Why is plasma assisted polishing (PAP) applied for atomic level finishing?

Q2. What are applications of PAP in semiconductor industry?

Q3. What are the types of plasma beam used in PAP process?

Q4. What is the working principle of plasma assisted polishing of single crystal diamond (SCD) wafer?

Q5. Why is PAP called as hybrid finishing process?

Q6. What are the advantages and applications of PAP process?

Advanced finishing processes

Q1. What are the components of AFF system?

Q2. What are different constituents of abrasive media used in AFF process?

Q3. Explain the working principle of two way AFF process with sketch.

Q4. What are process parameters involved in AFF process?

Q5. What are advantages of AFF process?

Q6. What are the various area of application of AFF process?

Q7. Why is CMP called as hybrid finishing process?

Q8. What are the various elements of CMP setup?

Q9. Represent various process parameters and responses of CMP process with the help of fish bone diagram.

Q10. State the advantages and limitations of CMP process.

Q11. State the major areas of application of CMP process.

Q12. What is basic principle of EEM process?

Q13. What are the fine oxide powders used in EEM?

Q14. What are the components of EEM system?

Q15. What are advantages of EEM process?

Q16. What are the applications of EEM process in Optics?

Q17. How is a flexible magnetic abrasive brass (FMAB) formed during MAF process?
Q18. How is MAF process applied for finishing internal and external curve surfaces?
Q19. What are the advantages of MAF process over AFF process?
Q20. What are the applications of MAF process for nanofinishing?
Q21. What are the various constituents of magnetorheological fluid used in MRF process?
Q22. What is the material removal mechanism in MRF process?
Q23. How is magnetic rheological abrasive flow finishing (MRAFF) different from MRF?
Q24. What are the effects of process parameters on MRR and surface roughness in MRF process?
Q25. State the advantages and applications of MRF process.
Q26. Why is plasma assisted polishing (PAP) applied for atomic level finishing?
Q27. What are applications of PAP in semiconductor industry?
Q28. What are the types of plasma beam used in PAP process?
Q29. What is the working principle of plasma assisted polishing of single crystal diamond (SCD) wafer?
Q30. Why is PAP called as hybrid finishing process?
Q31. What are the advantages and applications of PAP process?

References

[1] T. Hoshino, Y. Kurata, Y. Tarasaki, K. Susa, Mechanism of polishing of SiO_2 films by CeO_2 particles, J. Non-Cryst. Solids 283 (2001) 129–136.
[2] H. Deng, K. Yamamura, Atomic-scale flattening mechanism of 4H-SiC (0001) in plasma assisted polishing, CIRP Ann. Manuf. Technol. 62 (2013) 575–578.
[3] H. Deng, K. Endo, K. Yamamura, Plasma-assisted polishing of gallium nitride to obtain a pit-free and atomically flat surface, CIRP Ann. Manuf. Technol. 64 (2015) 531–534.
[4] A.F. Bastawros, A. Chandra, P.A. Poosarla, Atmospheric pressure plasma enabled polishing of single crystal sapphire, CIRP Ann. Manuf. Technol. 64 (2015) 515–518.
[5] K. Yamamura, K. Emori, R. Sun, Y. Ohkubo, K. Endo, H. Yamada, A. Chayahara, Y. Mokuno, Damage-free highly efficient polishing of single-crystal diamond wafer by plasma-assisted polishing, CIRP Ann. Manuf. Technol. 67 (2018) 353–356.
[6] H. Deng, K. Monna, T. Tabata, K. Endo, K. Yamamura, Optimization of the plasma oxidation and abrasive polishing processes in plasma-assisted polishing for highly effective planarization of 4H-SiC, CIRP Ann. Manuf. Technol. 63 (2014) 529–532.
[7] K. Monna, H. Deng, K. Endo, K. Ya Mamura, Investigation of removal mechanism of sapphire in plasma assisted polishing, Key Eng. Mater. 625 (2015) 458–462.
[8] C. Kageyama, K. Monna, H. Deng, K. Endo, K. Ya Mamura, Study on removal mechanism of sapphire in plasma assisted polishing, Adv. Mater. Res. 1136 (2016) 317–320.

[9] K. Ya mamura, T. Takiguchi, M. Ueda, H. Deng, A.N. Hattori, N. Zettsu, Plasma assisted polishing of single crystal sic for obtaining atomically flat strain-free surface, Ann. CIRP 60 (1) (2011) 571–574.
[10] H. Deng, T. Takiguchi, M. Ueda, A.N. Hattori, N. Zettsu, K. Yamamura, Damage-free dry polishing of 4H-SiC combined with atmospheric-pressure water vapor plasma oxidation, Jpn. J. Appl. Phys. 50 (2011) 08JG05.1-08JG05.4.

Index

Note: Page numbers followed by *f* indicate figures and *t* indicate tables.

B

C

G

H

I

M

R

T

FSC
www.fsc.org
MIX
Papier | Fördert gute Waldnutzung
FSC® C083411

Kolibri 360 GmbH
Ferdinand-Jühlke-Straße 7
99095 Erfurt, Deutschland
produktsicherheit@kolibri360.de